PALEOZOIC PALEOGEOGRAPHY
OF THE
WESTERN UNITED STATES - II

Volume Two

Editors

JOHN D. COOPER
Dept. of Geological Sciences
Cal-State University, Fullerton
Fullerton, CA 92634

CALVIN H. STEVENS
Department of Geology
San Jose State University
San Jose, CA 95192

March 6, 1991

SEPM
1926
PACIFIC SECTION

Published by
The Pacific Section Society Economic Paleontologists and Mineralogists
Los Angeles, California
U.S.A.

ISBN 1-878861-02-6

For copies of this publication write to:
Treasurer, Pacific Section S.E.P.M.
P.O. Box 10359
Bakersfield, CA 93389

Officers - 1991

PRESIDENT
HUGH McLEAN
USGS, Menlo Park

PAST PRESIDENT
STEVEN BACHMAN
BCI Geonetics, Inc.

NEWSLETTER EDITOR
JON KUESPERT
Chevron USA, Inc.

MANAGING EDITOR
REINHARD SUCHSLAND
DEKALB Energy Co.

Printed by
Comet Reproduction Service
Santa Fe Springs, CA 90670

CONTENTS

VOLUME 1

Regional Paleogeography

Mojave - Southern Great Basin Region

Central Great Basin

Eastern Great Basin

VOLUME 2

Washington, Idaho, Montana

Klamath Region

Sierra Nevada

Antler and Golconda Allochthons

Interregional Perspectives

PREFACE

Fourteen years ago in 1977, the milestone Pacific Section SEPM volume, *PALEOZOIC PALEOGEOGRAPHY OF THE WESTERN UNITED STATES*, was published. Since that time, major advances have been made in all aspects of geology that relate to paleogeographic interpretations of of this vast region. New ideas, new approaches, and new models have been employed in a great variety of investigations on Paleozoic rocks. Thus, the Pacific Section SEPM thought it timely to revisit Paleozoic Paleogeography by producing an updated publication that builds on the previous volume. This *PALEOZOIC PALEOGEOGRAPHY OF THE WESTERN UNITED STATES - II* is intended to be a companion, not a replacement, for the 1977 volume. We anticipate that it will enjoy the same measure of success and will rest alongside the '77 volume on your reference shelf.

This project did not get off the ground until April 1990, but even with this short time frame, there was little difficulty in generating the interest for 34 presentations at a Paleozoic Paleogeography Symposium for the March 4-8, 1991 Pacific Section annual meeting in Bakersfield, California, and the 55 papers that appear in this two-volume set. This is testimony to the enthusiasm and interest in publishing new data relating to Paleozoic paleogeography. The response has been tremendous in all aspects.

The organization of this publication is generally similar to that of the '77 volume, but is more tightly structured into regional categories. In Part 1, the lead article gives examples of the paleogeographic maps for each of the Paleozoic periods on a new palinspastic base. The seven articles that follow present regional overviews of each of the Paleozoic periods, generally west of 111° longitude. Authors of these articles are mostly those who wrote the lead articles for the '77 volume. The remaining 47 papers present new data on a diverse spectrum of topics and approaches, including isotope geology, igneous and metamorphic geology, terrane analysis, sedimentology, cyclostratigraphy, sequence stratigraphy, regional geology, paleontology, structure and tectonics. Unlike the '77 volume, these topical papers have been organized into eight separate parts. Volume 1 includes, in addition to Regional Paleogeography of the Paleozoic Periods, the Mojave-Southern Great Basin Region; Central Great Basin; and Eastern Great Basin. Volume 2 includes: Washington, Idaho, and Montana; Klamath Region; the Sierra Nevada; Antler and Golconda Allochthons; and Interregional Perspectives, wherein papers relate various aspects of the Paleozoic record of the Western U. S. to other regions such as the Appalachians, Mexico, Alaska, and central Asia.

As attractive as the idea is to achieve perfect uniformity throughout, this proved to be virtually impossible because of the late date at which the project was initiated, the difficulty in coordinating two geographically separated editors and authors of 55 papers, and the specific desires or requirements of authors or the data they wished to present. In spite of these difficulties and the diverse nature of the subjects, we feel the presentation of papers is remarkably uniform and consistent.

As editors we are grateful to the contributors for submitting so many papers of high quality. We benefitted greatly from the contacts and communication with them and from the education in reading the papers. We greatly appreciate the high degree of cooperation in adhering to deadlines and guidelines, and for the tireless efforts that helped ensure a timely publication. Without their time, effort, and interest, the completion of this ambitious project would not have been possible. It has been a rewarding experience for us and we thank all participants for this opportunity to work with them. We also gratefully acknowledge the editorial assistance of Ken Aalto, Jerry Brem, Gary Girty, Mitch Harris, and Don Zenger.

The Editors

John D. Cooper   Calvin H. Stevens

March 1991

COVER: illustration from J. B. Mahoney and others, this volume; for caption see Fig. 9, p. 573.

CAMBRIAN PALEOGEOGRAPHIC FRAMEWORK OF NORTHEASTERN WASHINGTON, NORTHERN IDAHO, AND WESTERN MONTANA

John W. Bush
Department of Geology and Geological Engineering
University of Idaho
Moscow, Idaho 83843

ABSTRACT

Several paleogeographic features relate to the evolution of the Cambrian passive margin in northern Idaho and western Montana. These features include the Lemhi Arch, Montania, Central Montana Trough, algal-shoal complexes, and the shelf margin. The Lemhi Arch and Montania aligned with the Kicking Horse Rim in the southern Canadian Rockies, and collectively were part of a narrow belt of intermittent basement highs at the margin's edge. Algal-shoal complexes developed on a ramp with their central axes over Montania during Middle and Late Cambrian transgressions. These complexes backstepped from the outer edge of the margin in north-eastern Washington towards the craton in Montana. Backstepping may have been related to the effects of long-term sea-level rise and increased flexure of the cratonic edge. Middle and Late Cambrian intrashelf basins developed in the Central Montana Trough behind barriers of shoal complexes and the Lemhi Arch.

Syndepositional faulting in northeastern Washington caused oversteepening that occurred along portions of the outer ramp. This faulting produced a patchwork of below-wave base basinal and subtidal environments and associated debris flow breccias. The breccias were part of a transition from a ramp-like edge in northeastern Washington to a steeper and more sharply defined platform edge in Canada.

The Cambrian passive margin in the region was, in many aspects, similar to other portions of the margin of the North American plate. Rapid subsidence in the Early Cambrian was followed in the Middle Cambrian by a decrease in the rate of subsidence coupled with a decrease in both siliciclastics and increased rate of transgression onto the craton over basement highs. The rate of subsidence continued to decrease in the Late Cambrian, although distal steepening did not occur along portions of the margin.

INTRODUCTION

Passive margin Middle and Late Cambrian carbonates crop out in northeastern Washington, northern Idaho, and western Montana (Fig. 1). These carbonates overlie Early Cambrian marine-shelf siliciclastics in northeastern Washington and Belt Supergroup rocks in northern Idaho and western Montana. The stratigraphic and sedimentologic framework is similar to other early Paleozoic carbonate platforms and ramps on passive continental margins that formed throughout the world after an episode of continental breakup in latest Proterozoic time (Heckel, 1974; Wilson, 1975; Read, 1982; Bond and others, 1983; Bond and Kominz, 1984; Read, 1985). The stratigraphic relations for several of these carbonate platforms in North America have been well documented. Recent papers detail their evolution in terms of thermal and mechanical processes, utilizing goephysical models, sequence stratigraphy, computer modeling, and eustatic cycles (Read, 1989; Bond and others, 1989; James and others, 1989). The lack of exposure and the structural complexity prohibit similar detailed modeling of the Cambrian in northeastern Washington, northern Idaho, and western Montana. However, platform evolution has been outlined by Bush and Hayden (1987) and several Cambrian paleogeographic features on the developing passive margin have been recognized.

Three principal Cambrian shelf environments of the Cordilleran miogeocline have been established by Palmer (1960, 1971), Robison (1960), and Oriel and Armstrong (1971) as the inner detrital, middle carbonate, and outer detrital belts. These belts existed over much of what is now western Montana and northern Idaho and have been used for paleogeographic reconstructions in those areas by Lochman-Balk (1972), Aitken (1978), and Bush (1989). The inner detrital belt contained terrigenous

Figure 1. Distribution of Cambrian rock outcrops in portions of Washington, Idaho, and Montana.

In Cooper, J.D., and Stevens, C.H., eds., 1991, **Paleozoic Paleogeography of the Western United States-II:** Pacific Section SEPM, Vol. 67, p. 463-473.

sediments deposited along the shore where clastics derived from the craton accumulated in shoreface and bay environments. The middle carbonate belt consisted of limestone or dolostone deposited in shallow water offshore from the inner detrital belt. Typical lithologies of the middle carbonate belt include algal bindstone, mottled lime mudstone, bioclastic, oolitic, and oncolitic wackestone and grainstone, and intraformational flat-pebble conglomerate. Offshore of the carbonate belt, the outer detrital belt consisted of deposits of the outer shelf, continental slope, and basinal environments. Siltstone, shale, and dark limestone are characteristic lithologies of this outer belt. Detritus of the outer detrital belt was derived from the inner detrital belt by sediment by-passing and from material that slumped along the seaward edges of the middle carbonate belt.

Early Cambrian deposition in northeastern Washington consisted primarily of a thick sequence (3000-4000 m) of predominantly marine siliciclastics that accumulated during the early post-rift stages of basin development. These siliciclastics were bordered by Montania, which was centered over western Montana and extended into southern British Columbia and Alberta (Diess, 1941). Early Cambrian isopach maps, compiled by Stewart (1972) for the western Cordillera, illustrate a possible cratonward deflection of the shelf's inner edge north of Montania near what is now the U. S.-Canadian border in southern British Columbia (Fig. 2). Transition to overlying platform carbonates occurs at approximately the Early-Middle Cambrian boundary, although the biostratigraphic control for precise timing is poor. Middle Cambrian transgression over Montania onto the stable craton produced upward-shallowing carbonate sequences and algal-shoal complexes that extended from central Montana to northeastern Washington.

These algal-shoal complexes contracted and expanded through three major depositional cycles during the Middle and Late Cambrian (Bush, 1989). At times they formed a barrier between intrashelf basins in the Central Montana Trough (Fig. 3) and seaward basins over northeastern Washington.

In east-central Idaho, the Lemhi Arch (Ruppel, 1986) (Fig. 3) significantly influenced deposition in adjacent intrashelf basins, while Montania's influence was minor. The adjoining platform edge in the southern Canadian Rockies was well defined by the development of an accretionary rim named the Kicking Horse Rim by Aitken (1971). Shallow-water carbonates extended seaward to northeastern Washington beyond and between the Lemhi Arch and the Kicking Horse Rim. Although a persistent steep, sharp platform edge in northeastern Washington did not form, slope deposits did develop at the shelf margin (Fischer, 1980, 1981; Morton, 1989). In this environment, abrupt facies changes into dark carbonates, fossil content, and laminated mudrocks indicate proximity to basinal environments. These basinal deposits are interbedded with, and overlie platform

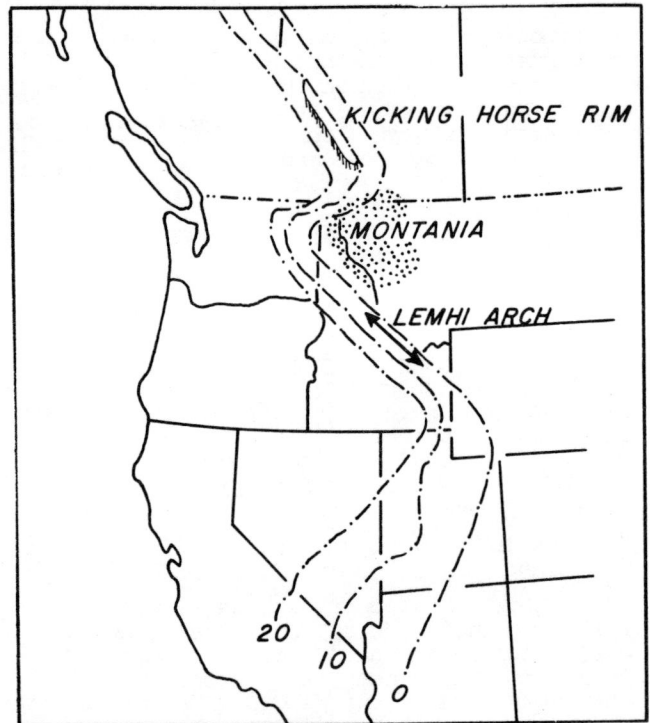

Figure 2. Isopach map of Early Cambrian Sedimentary rocks in thousands of feet and location of Kicking Horse Rim, Montania, and Lemhi Arch (modified from Stewart, 1972).

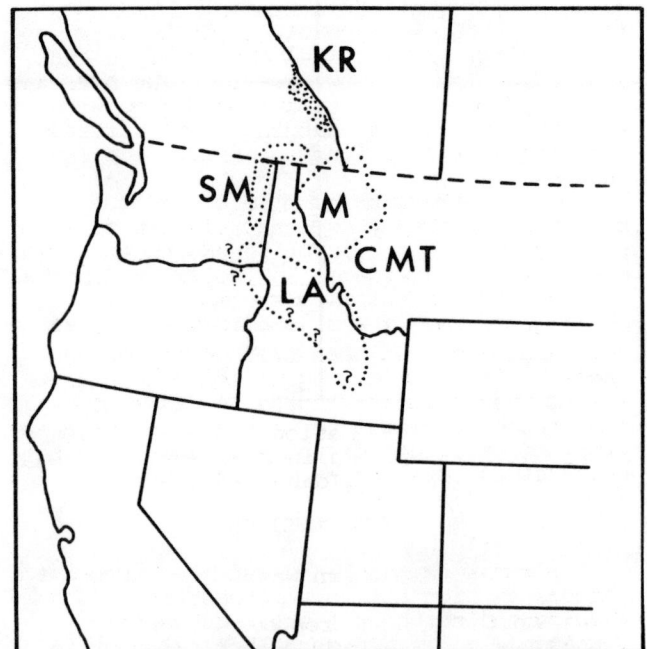

Figure 3. Map illustrating approximate location of Cambrian paleogeographic features: KR=Kicking Horse Rim; M=Montania; SM=shelf margin; LA=Lemhi Arch; CMT=Central Montana Trough.

carbonates and, where dominant, indicate the onset of drowning during the Early Ordovician.

This paper describes several Cambrian paleogeographic features as they relate to the development of the Cambrian cratonic margin in northeastern Washington, northern Idaho, and western Montana. These features include the Lemhi Arch, Montania, Central Montana Trough, algal-shoal complexes, and the shelf margin (Fig. 3).

GENERAL STRATIGRAPHY

Introduction

Cambrian rocks are exposed at several locations in northeastern Washington, northern Idaho, and western Montana. These outcrops represent portions of the Sauk Sequence of Sloss (1963). Most of these outcrops have been noted and described in earlier works and were reexamined during the past decade by students from the University of Idaho and Washington State University. Previously undescribed outcrops in northwestern Montana were detailed by Aadland (1979), Bush (1989), and Seward (1990). Additional work has been done by Morton (1989) and Carter (1989a, 1989b) in northeastern Washington.

Figure 4 is a correlation and nomenclature chart for the regions discussed. These sequences were measured and compared using standard straigraphic and paleontologic techniques. In addition, correlations have been assisted by utilizing Grand Cycles (Bush, 1989) and examination of isolated small outcrops (Hayden and Bush, 1988).

	NORTH EASTERN WASHINGTON	LAKEVIEW IDAHO	LIBBY TROUGH MONTANA	LEWIS CLARK RANGE MONTANA	SOUTH WESTERN MONTANA
LOWER ORDOVICIAN	LEDBETTER		UNNAMED SEQUENCE		
UPPER CAMBRIAN	UPPER (METALINE FORMATION)		LITHOFACIES 6		RED LION
			LITHOFACIES 5	DEVILS GLEN	PILGRIM
	MIDDLE	LITHOFACIES 6	LITHOFACIES 4	SWITCH BACK	PARK
MIDDLE CAMBRIAN		LITHOFACIES 5	LITHOFACIES 3	STEAMBOAT POGODA DEARBORN DAMNATION	MEAGHER
		LITHOFACIES 4	LITHOFACIES 2		
	LOWER	LITHOFACIES 3	LITHOFACIES 1		
		LITHOFACIES1-2			
		RENNIE GOLD CREEK	GORDON FLATHEAD	GORDON FLATHEAD	WOLSEY FLATHEAD
LOWER CAMBRIAN	MAITLEN GYPSY				

Figure 4. Correlation and nomenclature chart for the Cambrian in northwestern Washington, northern Idaho, and western Montana.

Northeastern Washington

Cambrian strata crop out at numerous localities in Pend Oreille and Stevens Counties, northeastern Washington (Joseph, 1990). In Pend Orielle County, near the town of Metaline Falls (Fig. 5), the Cambrian has been subdivided, from base upward, into the Gypsy Quartzite (1740-2790 m thick), Maitlen Phyllite (1250-1640 m thick, and the Metaline Formation (985-2000 m thick). Lindsey (1987) has detailed the basal sandstone units, dividing them into several lithofacies.

The Metaline Formation traditionally has been subdivided into informal lower, middle and upper members. Morton (1989), noting the original caution of Dings and Whitebread (1965), indicated that these designations are misleading because the lithologic subdivisions do not have time-stratigraphic boundaries. The "lower" limestone-shale unit (290-390 m thick) consists of dark, subtidal lime mudstone and shale with trilobites of late Middle Cambrian age (Lochman-Balk, 1972). The "middle" bedded dolostone unit (360-1200 m thick) consists of subtidal dolomudstone with numerous peritidal features. The "upper" massive limestone unit (196-450 m thick) consists of subtidal bioturbated lime mudstone and intraclastic packstone. Graptolite work by Carter (1989a, 1989b) indicates that, in places, the upper Metaline is as young as Middle Ordovician, but is predominantly Late Cambrian and Early Ordovician in age.

In additon to these lithologic units originally subdivided and described by Dings and Whitebread (1965), other units of the Metaline have been described and mapped. Yates (1964) recognized the "intraformational" breccia which was later named the Fish Creek Member by Fischer (1980, 1981). Fischer (1981) detailed its lithology and concluded that it was a lateral equivalent of portions of the "middle" and "upper" Metaline. The Fish Creek Member (450 m thick) consists of black, thinly bedded dolomudstone interbedded with black and gray dolofloatstone. The dolomudstone is commonly laminated and contains black chert nodules. Dolofloatstone occurs in 2-3 m thick sedimentation units composed of poorly sorted, tabular clasts in a black dolomicrite matrix.

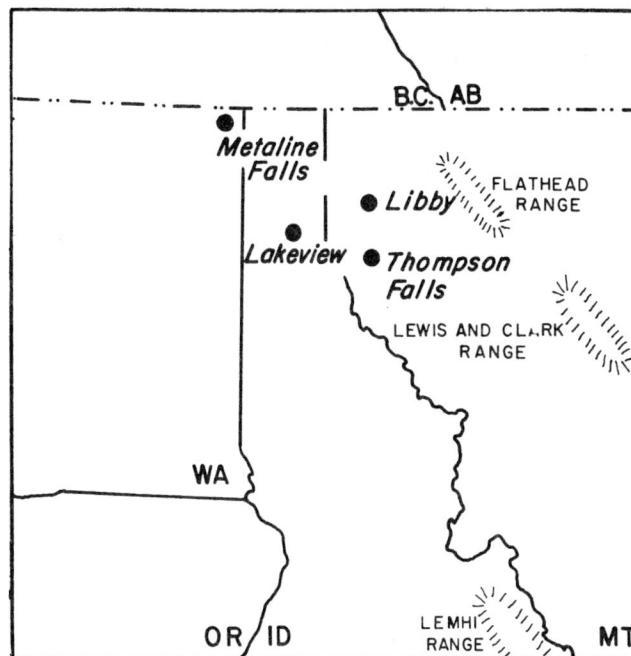

Figure 5. Location map illustrating geographic features referred to in the text.

Another important carbonate sequence that contains an abundance of matrix-supported breccias occurs in the northeastern portion of northeastern Washington and is referred to as the Josephine unit (McConnell and Anderson, 1968). The unit occupies an irregular stratigraphic position at the top of the Metaline, directly in contact with the overlying Ledbetter Formation. Review of outcrops, mine exposures, and drill core indicates the Josephine is most persistent in the immediate vicinity of the Pend Oreille, Grandview and Metaline mines of the Metaline mining district (Morton, 1989).

The Josephine comprises a wide variety of lithologies in an irregular interbedding of carbonate breccia and non-brecciated dolostone, lime mudstone and argillite. Many of the lithologies are common to other Cambrian and Ordovician units in the area. As a result, the Josephine cannot be defined by the presence of a single lithology. It is best described as a heterogeneous assemblage of black or dark gray lithologies dominated by dolostone, with locally abundant chert, limestone, and argillite. Abrupt lateral variation in lithology and character of bedding is common, and undulating, discontinuous bedding surface predominate.

Overlying the Metaline in northeastern Washington is the Ledbetter Formation, which consists mostly of dark, thinly laminated siliciclastic basinal mudstones. The Ledbetter ranges in age from Early Ordovician to Silurian (Carter, 1989a, 1989b). The contact between the Ledbetter and the underlying Metaline Formation has been interpreted as varying from unconformable to conformable at different localities (Hurley, 1980). Carter (1989a) noted that the age of the Metaline-Ledbetter contact is not the same everywhere. Numerous core holes and mine exposures in the Metaline mining district indicate that the Ledbetter and Metaline Formations are in conformable contact (Morton, 1989).

Northern Idaho

Middle Cambrian outcrops in northern Idaho are limited primarily to small areas along the southern shores of Lake Pend Oreille where they are complexely faulted and locally metamorphosed. From the base upward, the sequence consists of the Gold Creek Quartzite (130 m thick), the Rennie Shale (30 m thick), and the Lakeview Limestone (600 m thick). Younger Paleozoic rocks have not been identified in northern Idaho.

The Gold Creek Quartzite rests unconformably on the lower part of the middle member of the Wallace Formation of the Proterozoic Belt Supergroup and consists primarily of massive cross-bedded quartz arenite with minor interbedded shale and a basal pebble conglomerate. The Rennie Shale consists of fossiliferous, olive brown, terrigenous mudrock with interbedded nodules of fossiliferous lime mudstone. Collection of trilobites, one from near the base and one from the upper part, were examined and both collections correspond with the *Albertella* Biozone (Harrison and Jobin, 1965).

The Lakeview Formation is divided into a lower Limestone and an upper dolostone. The lower Lakeview (335 m) consists primarily of dark gray, laminated, argillaceous, pyritic, fossiliferous lime mudstone with interbedded calcareous shale and non-laminated lime mudstone. The upper Lakeview (252 m) consists primarily of light gray to tan, non-pyritic, unfossiliferous dolomudstone with interbedded oolitic dolopackstone and cryptalgal dolobindstone.

The Lakeview Formation can be subdivided into six major lithofacies (Fig. 4). From the base upward, these are: 1) parallel-laminated lime mudstone; 2) nodular shale; 3) mottled lime mudstone; 4) peloidal-oolitic dolopackstone; 5) cryptalgal dolobindstone; and 6) crystalline cryptalgal dolomudstone (Bush, 1989). The lower Lakeview consists of lithofacies 1-3 and the 4-6.

The lower Lakeview contains faunas of the *Glossopleura* and *Bathyuriscus-Elrathina* biozones (Lochman-Balk, 1971). Robison (1964) noted that a portion of the faunas of the lower Lakeview Formation belong to the *Bolaspidella* Biozone. Motzer (1980), using Robison's (1976) revision of biostratigraphic zones for the Great Basin, placed the boundary between lithofacies 1 and 2 to approximate the boundary between the underlying *Ocryctocephalus* and overlying *Bolaspidella* biozones for open shelf polymeroid faunas.

Northwestern Montana

In northwestern Montana, between the towns of Thompson Falls and Libby, Cambrian rocks are preserved in a northwest-southeast trending syncline referred to as the Libby Trough. The basal unit is the Flathead Sandstone (8 m thick), which is overlain by the Wolsey Formation (75 m thick). These units are lithologically equivalent to the Gold Creek Quartzite and the Rennie Shale. The Wolsey is overlain by a thick, predominantly carbonate sequence referred to as the Fishtrap Dolomite (Aadland, 1985). In the Missoula area, southeast of the Libby Trough, the Fishtrap equivalent is the Hasmark Formation.

Aadland (1979) subdivided the Fishtrap into a basal portion that consists of a bioturbated mudstone unit (244 m thick), a dolomitized oolitic and peloidal grainstone unit (61 m thick), and an algal dolomudstone unit (92 m thick). This succession is overlain by a light-colored shale unit (92 m thick) and an upper algal dolobindstone-dolomudstone unit (407 m thick). Aadland (1979, 1985) subdivided the Fishtrap into several lithofacies that Bush (1989) reorganized into six lithofacies (Fig. 4). The Fishtrap is overlain conformably by approximately 65 m of quartz arenite and by 37 m of dolostone. This sequence has been dated as Early Ordovician (Bush and others, 1985).

West-Central and Southwestern Montana

The Cambrian System of southwestern Montana contains a variety of formational names. The generally accepted nomenclature for the stratigraphic sequence begins with the basal Flathead Sandstone, overlain by the Wolsey, Meagher, Park, Pilgrim and the Red Lion Formations. Complete sections occur at several localities: Melrose, Ruby Range, Three Forks, Helena, and Missoula. Thicknesses exceed 500 m in places. The sequence consists of packages of alternating units of terrigenous clastics and upward-shallowing carbonates (Bush, 1989).

In west-central Montana, the Cambrian System is best exposed in the Flathead and Lewis and Clark Ranges. Most of these sections have been described by Deiss (1939) and more recently by Enterline (1978) and Hilty (1989). The 500 m thick sequence along the Dearborn River in the southern Lewis and Clark Range, from the base upward, consists of the Flathead, Gordon, Damnation, Dearborn, Pagoda, Steamboat, Switchback and Devil's Glen Formations.

GRAND CYCLES

Aitken (1966, 1978) defined "Grand Cycles" as depositional cycles, each of which is represented by 90 to 600 m of strata, one or more formations, and two or more biostratigraphic zones. Each cycle has an abrupt basal contact and consists of a lower, shaly half-cycle that is gradationally overlain by a carbonate half-cycle. Grand Cycles, as defined by Aitken (1966, 1978), are easily recognized in southwestern (Healy, 1986) and west-central (Healy, 1986; Hilty, 1989) Montana. In southwestern Montana a basal cycle (Fig. 6) is represented by the Wolsey to Meagher sequence (early Middle to mid-Middle Cambrian), a second cycle by the Park to Pilgrim sequence (late Middle to early Late Cambrian), and a third cycle by the Red Lion Formation (middle Late to late Late Cambrian). In the Dearborn River area of west-central Montana, a basal cycle is represented by the Flathead to Damnation sequence, a second cycle by the Dearborn to Steamboat sequence, and a third by the Switchback to Devil's Glen sequence.

In northwestern Montana and northern Idaho, the tops of Grand Cycles are difficult to recognize because deposition was dominated by middle carbonate belt environments that were isolated from the influence of the inner detrital belt. However, detailed lithologic and paleontologic data have been used to determine approximate cycle boundaries at several localities (Bush, 1989).

Although the cycles are well defined in southwestern Montana, the length of time represented by each is difficult to determine accurately because paleontologic data are absent in places and most of the Upper Cambrian is eroded. However, comparison of the available data suggests 9-15 million years for each. Even with this imprecise estimate, comparisons of thicknesses for each cycle suggest decreasing rates of subsidence of the margin over time. This decrease is similar to that reported for Middle and Late Cambrian sequences in the northern Appalachians (James and others, 1989).

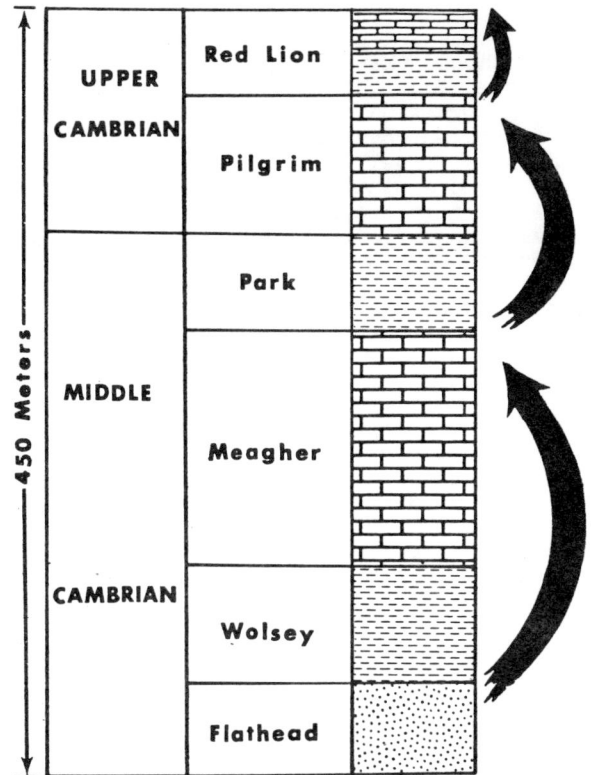

Figure 6. Middle and Upper Cambrian stratigraphic column for southwestern Montana. Grand Cycles are indicated by the arrows.

PALEOGEOGRAPHIC FEATURES

Algal-Shoal Complex

A large algal-shoal complex existed on the outer edge of the shelf from the Middle to Late Cambrian (Aitken, 1978; Bush and Fischer, 1981). The complex was several hundred kilometers in width and separated an outer, deeper basin over northeastern Washington from an inner, shallow basin over central Montana. The complex expanded and contracted during three Grand Cycles (Bush, 1989). The size and shape of the complex influenced sediment distribution, salinity, dolomitization, current flow, faunal distribution, evaporation, and tidal range. Sedimentary sequences deposited in the outer basin include the Ledbetter, the "lower" and "upper" Metaline, portions of the "middle" Metaline, and the Fish Creek Member and Josephine unit of the Metaline Formation. The shoals centered over western Montana progressively migrated landward and were primarily responsible for deposition of the peritidal portions of "middle" Metaline, upper Lakeview, Fishtrap, Hasmark, Meagher, Pilgrim, and Red Lion Formations. Sequences deposited in the inner basin include the Park Shale and subtidal portions of the Hasmark, Red Lion, Meagher, and Pilgrim Formations.

The algal-shoal complex was part of a much larger outer shelf geographic feature that, according to Aitken (1978), extended from northern Alberta southward to at least the southwestern corner of Wyoming. Aitken (1971) showed that portions of the shoal, refererred to as the "Kicking Horse Rim" in southwestern Alberta, controlled the distribution of sedimentary facies. At maximum shoal size, tidal exchange with the inner basin was prevented. At minimum shoal size, tidal exchange occurred through many small passes (Aitken, 1978). In Utah, Robison (1976) stated that during the Middle Cambrian, faunal exchange between nearshore environments and outer shelf areas was limited by shallow-water carbonate barriers. Lack of exchange with the outer basin also increased salinity in the inner basin.

In northeastern Washington, northern Idaho, and western Montana, dolomitization can also be related to shoal development. A sharp boundary presently exists between north-south striking Middle Cambrian limestone and dolostone lithofacies in western Montana. This boundary is visible on lithofacies maps for the *Bathyuriscus-Elrathina* and *Bolaspidella* biochrons presented by Lochman-Balk (1972), and corresponds to the inner edge of the shoal complex. When the shoal was extensive, maximum subaerial exposure and minimum exchange with the open sea resulted. At that time, algae flourished, evaporation rates were high, and early penecontemporaneous dolomitization was extensive. Simultaneously, development of normal marine fauna and subtidal sedimentation was reduced. At its minimum development, the shoal would have had a "leaky" rim that allowed more exchange with basinal waters which, in turn, would have reduced evaporation rates and algal growth, allowing wider distribution of normal marine fauna and subtidal sediments.

Reflux and mixing were common dolomitization mechanisms as the supratidal and intertidal portions of the shoals migrated over deeper intertidal and subtidal deposits. The shoals' influence on dolomitization can be documented for the "middle" Metaline, upper Lakeview, Fishtrap, Hasmark and portions of the Meagher, Pilgrim and Devil's Glen Formations.

The algal-shoal complex in central Montana reached its maximum development in the Middle Cambrian following transgression onto the craton. The central axis for the complex during maximum development was over the buried Montania in northwestern Montana while the seaward edge was at times in northeastern Washington. The complex exhibited major expansion and contraction at least two more times in the Cambrian and possibly once again in the Early Ordovician (Bush, 1989), with the central axis and its seaward edge moving cratonward each time. Aitken (1978) and Bush (1989) related changes in shoal configuration to Grand Cycles.

In the southern Canadian Rockies, the shelf edge was a carbonate platform with an accretionary rim. This feature was named the Kicking Horse Rim by Aitken (1971), who noted that peritidal deposition persisted at the shelf edge even during minimum shoal develop-ment. Peritidal and algal-shoal environments did migrate to the shelf edge in northeastern Washington, depositing portions of the "middle" Metaline. However, a persistent rim was not maintained as the shoals' seaward edge migrated cratonward in the late Middle and Late Cambrian. This migration of the seaward edge suggests that the shelf over northern Idaho and western Montana was more ramp-like in contrast to a rimmed platform in Alberta. Such differences in shelf evolution indicate slightly different tectonic development within a passive margin framework for the two regions.

Montania

Walcott (1915) first described a Cambrian positive area in northwestern Montana and southern British Columbia that he called the "Montana island". Diess (1941) renamed the uplift "Montania" and believed it extended over 130,000 square kilometers at times, with most relief present in northwestern Montana. In fact Diess (1941) believed northwestern Montana to be emergent throughout the Cambrian and until at least the Middle Devonian. Lochman-Balk (1971) provides several paleogeographic reconstructions for the Middle and Late Cambrian with Montania shown as an emergent island.

Aadland (1979), Bush (1989) and Seward (1990) describe sequences that indicate northwestern Montana was not emergent from early Middle Cambrian to Early Ordovician. Collectively these rocks document continued deposition with minor siliciclastic influx from local source areas. Earlier work by Fritz and Norris (1966) noted the absence of the basal Middle Cambrian Flathead Sandstone in the southwestern corner of British Columbia. This absence, along with thinning of the Flathead to less than 3 m in the Libby Trough area (Aadland, 1979; Apgar, 1986), suggests the existence of a slight high during early Middle Cambrian. However, Middle and Upper Cambrian units in southern Canada do not show any direct evidence of the presence of Montania (Aitken, 1966), nor do Middle and Upper cambrian units above the Flathead in northwestern Montana. Therefore, Montania was above water only in the Early Cambrian and possibly during the early Middle Cambrian as a marginal uplift supplying siliciclastic sediments to a rapidly subsiding basin over northeastern Washington.

Although Montania was not an active area during the Middle and Late Cambrian, it still may have indirectly influenced carbonate deposition. Though buried, Montania is believed to have controlled the location of the central axes of the algal-shoal complexes during their development in the Middle Cambrian.

Lemhi Arch

The Lemhi Arch was a relatively large landmass in central Idaho that influenced sedimentation patterns from late Proterozoic to late Paleozoic (Sloss, 1954; Ruppel, 1986). During the Cambrian, the Lemhi Arch, in conjunction with the algal-shoals, separated the outer shelf from cratonic shallow-water platform environments.

Trace fossils from the Wilbert Formation in the southern Lemhi Range indicate non-emergence of the arch during the Early Cambrian (Derstler and McCandless, 1981). However, on the east flank near the Idaho-Montana border, all Middle and upper Cambrian units are either absent or depositionally thinned (Ruppel, 1986), indicating the active nature of the arch during that interval. Evans and Zartman (1988) note Cambrian and Ordovician alkalic plutonism in east-central Idaho and suggest that this plutonism and the coeval topographic high, the Lemhi Arch, had a common tectonic control.

Aadland (1979) and Hayden (1990) present stratigraphic details indicating that the arch affected deposition in northwestern Montana and in intrashelf basins over western and central Montana. Aadland (1979) describes a Middle and Late Cambrian carbonate-clastic sequence in the Libby Trough of northwestern Montana in which the siliciclastic content increases southward, primarily in response to tectonic activity of the Lemhi Arch. Hayden (1990) compares regional facies changes of the Late Cambrian Red Lion Formation of west-central and south-central Montana and notes an increase in siliciclastics southwesterly towards the arch. These siliciclastics curtailed carbonate production during deposition of portions of the Red Lion (Hayden and Bush, 1988). The Lemhi Arch and the shoal complex discussed earlier collectively produced a partial barrier which, in turn, created an intrashelf basin over central and southwestern Montana where units such as the Park Shale and Red Lion were deposited.

The Lemhi Arch, Montania, and the Kicking Horse Rim delineate an intermittent, narrow, outer shelf "positive area" along the western Cordillera. In Alberta the Kicking Horse Rim existed as a stationary area of shallow-water deposition at the very edge of the shelf throughout Middle and Late Cambrian time. Localization of the rim was caused by a subtle Early Cambrian anticlinal feature along the shelf edge (Aitken, 1981). In western Montana and northern Idaho, Middle and Late Cambrian deposition occurred over the position once occupied by Montania.

Shelf Margin

The Cambrian shelf margin existed in the vicinity of northeastern Washington (Fischer, 1980, 1981). The nature of that margin has been difficult to reconstruct because of structural complexities, lack of exposures, poor biostratigraphic control, sparse detailed mapping, and the presence of abrupt facies changes. In the Early Cambrian, the margin was a rapidly subsiding basin that accumulated a thick sequence of primarily siliciclastic sediments (Gypsy Quartzite, Maitlen Phyllite) similar to numerous other Early Cambrian passive margin sequences noted by Bond and others (1989), Read (1989), and James and others (1989).

Rapid subsidence of the shelf margin slowed in the Middle and Late Cambrian, but exceeded subsidence rates of depocenters on the craton in western and central Montana. Subtidal ramp and basinal sediments such as the dark, pyritic, thinly laminated and non-bioturbated sediments of the Metaline and Ledbetter Formations accumulated in a patchwork of predominantly below-wave base environments. Within this setting debris flows formed the Fish Creek Member and Josephine unit (Morton, 1989).

Bush and Fischer (1981), Fischer (1982), Bush and Hayden (1987), and Bush (1989) considered the shelf margin to be a ramp with a major shoal buildup that reached the shelf edge during the Middle Cambrian. The shoal was considered to be primarily responsible for forming most of the middle Metaline dolostone. However, Morton (1989) noted that the abundance of the breccias in a region considered to have been a ramp is somewhat incongruous. He also noted that although there are some aspects of the "middle" Metaline which could be interpreted as "shallow", most of the others suggest deeper environments. Reevaluation of the data leads to the conclusion that major shoal development rarely reached the edge of the shelf, and that below-wave base, subtidal and basinal environments, in concert with associated slope deposits, dominated the shelf edge in northeastern Washington from the Middle Cambrian to at least the Middle Ordovician.

Comparison of the distribution of the Josephine unit to the previously noted deflected position of the shelf edge may explain the ramp-like setting in association with debris flows. It is believed that ramp-like conditions existed but there was a gradation into a distally steepened, tectonically active area in the northern portion of Washington and southeastern British Columbia. This steepening occurred where the shelf edge deflected cratonward to join the Kicking Horse Rim, a feature which trended sub-parallel to the dominant strike of the shelf edge, approximately perpendicular to the deflected portion. This interpretation is supported by mapping in northern Pend Oreille County which depicts abrupt facies and thickness changes across north-northeast striking syndepositional faults, and by mapping to the south that documents a thinning of Cambrian units and a decrease in slope deposits (Morton, 1989).

Central Montana Trough

The Central Montana Trough was a subsiding area that acted as a major cratonic depocenter during deposition of the late Proterozoic Belt Supergroup. It persisted with varying degrees of subsidence throughout the remainder of the Paleozoic (Peterson, 1986). Middle and Upper Cambrian sequences in the Central Montana Trough range from 300 m to over 600 m thick.

The Central Montana Trough was the location of intrashelf basins that developed between the algal-shoal complex and the

predominantly terrigenous shoreline. The Red
Lion Formation was deposited in such an
intrashelf basin (Hayden, 1990). Shales and
fine-grained sandstones of the Red Lion pass
downslope into deeper ramp ribbon carbonates
and limestone conglomerates. Ribbon limestones
consist of layers and lenses of trilobite
packstone, parallel and wave ripple-laminated
quartzose calcsiltite, and lime mudstone
arranged in fining-upward sequences (1-6 cm
thick) that are localy burrowed.

The seaward side of the basins consisted
of the peritidal algal-shoal complexes now
represented, in part, by the thick dolomitic
sequences of the Hasmark and Fishtrap
Formations. The Lemhi Arch was active on the
southwestern end of the basin where the
Cambrian units increase in siliciclastic
content and thin in places to zero. The trough
was also bordered by the Sweetgrass Arch in
north-central Montana where the Cambrian
section thins to less than 150 m (Peterson,
1986).

In summary, cratonic subsidence of the
Central Montana Trough was responsible for the
thick accumulaton of cratonic shallow-water
siliciclastics and carbonates during the
Middle and Late Cambrian.

Tectonics

Development of the cratonic margin in
northeastern Washington, northern Idaho, and
western Montana is similar to other Cambrian
margins around the edge of North America
described by Bond and others (1989), James and
others (1989), and Read (1989). Bond and
others (1989) noted the similarity in timing
of initiation of passive margins between three
areas in the Appalachian and Cordileran
margins. They suggested that it "...probably
reflects the abroad synchroneity in the
timing of continental breakup and initiation
of the passive margins around the edge of
North America, together with the extensive and
coeval submergence of the craton by the
eustatic sea-level rise beginning in Middle
Cambrian time" (Bond and others, 1989, p. 58).

In the study area, carbonate development
initially was inhibited by the high rates of
siliciclastic deposition in a rapidly
subsiding basin. These siliciclastics were
derived from Montania and the Lemhi Arch
adjacent to the margin. However, Montania was
buried by carbonate buildups during the Middle
Cambrian transgression, while the Lemhi Arch
continued to be intermittently emergent. Bond
and others (1989) attribute the burial of
marginal highs on passive margins to an
increase in rigidity and flexure of the
underlying lithosphere, a function of
abnormalities in the rate of decay during
rifting.

In northeastern Washington, the shoal
complexes backstepped from the outer edge of
the margin toward the craton due to the
effects of long-term eustatic sea-level rise
and the increased flexural bending of the
cratonic edge. According to Bond and others
(1989), sea-level rise and flexural bending
resulted in the retreat of the siliciclastic

shoreline and the cratonic migration of the
carbonates. This stage of tectonic activity
marked the development of the algal-shoal
complexes with associated intrashelf basins
over the Central Montana Trough.

Syndepositional faulting caused
oversteepening of the margin's edge which was,
in part, reponsible for the develpment of
sedimentary breccias in northeastern
Washington. Steepening of the margin occurred
during an overall slowing in the rate of
subsidence and drowning of the outer platform
in the Late Cambrian and Early Ordovician.
Rates of subsidence along the shelf edge of
Washington differed from those in the southern
Canadian Rockies where relatively stable
tectonics may have allowed the development of
the Kicking Horse Rim at the platform edge.
During this time, the shoal complex in western
Montana continued to migrate towards the
craton, over a ramp and away from the shelf
edge in northeastern Washington. Collectively
the Kicking Horse Rim, Montana, and the Lemhi
Arch were part of a positive belt located on
the cratonic margin that was buried on its
northern end in the Middle and Late Cambrian.

Comparisons of Grand Cycle thicknesses
from sections in the Central Montana Trough
indicate a post-rift exponential decrease in
the rate of subsidence. Bond and others (1989)
related similar decreases to the response of
slow cooling and thermal contraction of the
lithosphere that was heated during rifting
prior to the Early Cambrian. Post-rifting
deposition began with rapid subsidence in the
Early Cambrian, followed by a decrease in the
rate of subsidence and a decrease in siliciclastic deposition and cratonic transgression
in the Middle Cambrian. Slowing of the rate of
subsidence continued in the Late Cambrian.

SUMMARY AND CONCLUSIONS

Comparison of Cambrian paleogeographic
features of the passive margin in northeastern
Washington, northern Idaho, and western
Montana leads to several conclusions:

1) The Lemhi Arch, Montana, and the
Kicking Horse Rim were collectively part of an
intermittent outer shelf "positive area". The
Kicking Horse Rim acted as a persistent,
stationary rim through the Middle and Late
Cambrian (Aitken, 1971). Montania was buried
in the Middle Cambrian; however, the Lemhi
Arch was generally emergent during the Middle
and Late Cambrian.

2) The Lemhi Arch and large algal-shoal
complexes formed a partial barrier between
intrashelf basins in the Central Montana
Trough and outer basinal environments during
the Middle and Late Cambrian.

3) Montania was not as extensive as
described in earlier literature, but was
active in the Early Cambrian as a source of
post-rift siliciclastics. Once buried in the
Middle Cambrian, Montania may have indirectly
influenced the location of subsequent algal-
shoal development.

4) The shelf margin in northeastern Washington ranged from a ramp to an over-steepened shelf produced by faulting. Location of the oversteepened margin and associated debris flows were part of an east-west deflection in the cratonic margin toward an accretionary rim in southern Alberta.

5) Evolution of the cratonic margin in northeastern Washington followed a pattern for Cambrian passive margins documented elsewhere on the North American plate.

ACKNOWLEDGEMENTS

I would like to thank the numerous students who worked with me in the field and lab during the past decade. Special thanks are extended to Patrick Seward who critically read and edited the manuscript. John Cooper and Jack Morton reviewed the manuscript and provided many helpful suggestions.

REFERENCES CITED

Aadland, R. K., 1979, Cambrian stratigraphy of the Libby Trough, Montana [unpublished Ph.D. dissertation]: University of Idaho, Moscow, Idaho, 236 p.
_____, 1985, Lithofacies of the Middle-Upper Cambrian sequences, Libby Syncline, Montana: Montana Bureau of Mines and Geology Open-File Report 153, 52 p.

Apgar, J. L., 1986, Stratigraphy, sedimentology, and diagenesis of the Flathead Sandstone, Libby Trough, Montana [unpublished M. S. thesis]: University of Idaho, Moscow, Idaho, 95 p.

Aitken, J. D., 1966, Middle Cambrian to Middle Ordovician cyclic sedimentation, southern Rocky Mountains of Alberta: Bulletin of Canadian Petroleum Geology, v. 14, p. 405-411.
_____, 1971, Control of lower Paleozoic sedimentary facies by the Kicking Horse Rim, southern Rocky Mountains, Canada: Bulleton of Canadian Petroleum Geology, v. 19, 557-569.
_____, 1978, Revised model for depositional Grand Cycles, Cambrian of the southern Rocky Mountains, Canada: Bulletin of Canadian Petroleum Geology, v. 26, p. 515-542.
_____, 1981, Cambrian stratigraphy and depositional fabrics, southern Canadian Rocky Mountains, Alberta and British Columbia, in Taylor, M. E., ed., Guidebook for the Cambrian System in the Southern Canadian Rocky Mountains, Alberta and British Columbia: Second International Symposium on the Cambrian System: supplement to U. S. Geological Survey Open-File Report 81-743, p. 1-26.

Bond, G. C., and Kominz, M. A., 1984, Construction of tectonic subsidence curves for the early Paleozoic miogeocline, southern Canadian Rocky Mountains - implications for subsidence mechanisms, age of breakup and crustal thinning: Geological Society of America Bulletin, v. 95, p. 155-173.

Bond, G. C., Kominz, M. A., Steckler, M. S., and Grotzinger, J. P., 1989, Role of thermal subsidence flexure, and eustacy in evolution of early Paleozoic passive margin carbonate platforms, in Crevello, P. D., Wilson, J. L., Sarg, J. F., and Read, J. F., eds., Controls on Carbonate Platform and Basin Development: Society of Economic Paleontologists and Mineralogists Special Publication 44, p. 39-61.

Bush, J. H., 1989, The Cambrian System of northern Idaho and northwestern Montana, in Chamberlain, V. E., Breckenridge, R. M., and Bonnichsen, B., eds., Guidebook to the Geology of Northern and Western Idaho and Surrounding area: Idaho Geological Survey Bulletin 28, p. 103-121.

Bush, J. H., and Fischer, H. J., 1981, Stratigraphic depositional summary for Middle and Upper Cambrian strata in northwestern Montana, northeastern Washington, and northern Idaho, in Taylor, M. E., ed., Short Papers for the Second International Symposium on the Cambrian System: U. S. Geological Survey Open-File Report 81-743, p. 42-46.

Bush, J. H., and Hayden, L. L., 1987, Middle and Upper Cambrian platform evolution and paleogeography, northwestern Montana, northern Idaho, and northeastern Washington: America Association of Petroleum Geologists Bulletin [Abstract], v. 71, p. 1002.

Bush, J. H., Kacheck, D. C., and Webster, G. D., 1985, Newly discovered Ordovician strata, Libby Trough, northwestern Montana: Geological Society of America, Abstracts with Programs, v. 17, p. 211.

Carter, C., 1989a, A Middle Ordovician graptolite fauna from near the contact between the Ledbetter Slate and the Metaline Limestone in the Pend Orielle mine, northeastern Washington State: U. S. Geological Survey Bulletin 1860, Shorter Contributions to Paleontology and Stratigraphy, p. A1-A23.
_____, 1989b, Ordovician-Silurian graptolites from the Ledbetter Slate, northeastern Washington State: U. S. Geological Survey Bulletin 1860, Shorter Contributions to Paleontology and Stratigraphy, p. B1-B28.

Derstler, K., and McCandless, D. O., 1981, Cambrian trilobites and trace fossils from the southern Lemhi Range, Idaho: their stratigraphic and paleontologic significance: Geological Society of America, Abstracts with Programs, v. 13, p. 194.

Diess, C. F., 1939, Cambrian stratigraphy and trilobites of northwest Montana: Geological Society of America Special Paper 18, 135 p.
_____, 1941, Cambrian geography and sedimentation in the central Cordilleran region: Geological Society of America Bulletin, v. 52, p. 1085-1116.

Dings, M. G., and Whitebread, D. H., 1965, Geology and ore deposits of the Metaline zinc-lead district, Pend Oreille County, Washington: U. S. Geological Survey Professional Paper 489, 109 p.

472

Enterline, T. R., 1978, Depositional environment of the Pagoda, Pentagon, and Steamboat Formations (Middle Cambrian), northwest Montana [unpublished M. S. thesis]: University of Montana, Missoula, Montana, 105 p.

Evans, K. V., and Zartman, R. E., 1988, Early Paleozoic alkalic plutonism in east-central Idaho: Geological Society of America Bulletin, v. 100, p. 1981-1987.

Fischer, H. J., 1980, Lithology of a dolofloatstone facies in the Metaline Formation, Stevens County, Washington: Geological Society of America Abstracts with Programs, v. 12, p. 106-107.

_____, 1981, The lithology and diagensis of the Metaline Formation, northeastern Washington [unpublished Ph.D. dissertation]: University of Idaho, Moscow, Idaho, 175 p.

Fritz, W. H., and Norris, D. K., 1966, Lower Middle Cambrian correlations in the east-central Cordillera, in Report of Activities: Geological Survey of Canada Paper 66-1, p. 105-110.

Harrison, J. E., and Jobin, D. A., 1965, Geologic map of the Packsaddle Mountain quadrangle, Idaho: U. S. Geological Survey Quadrangle Map GQ-375, with 4 p. text.

Hayden, L. L., 1990, Stratigraphy and depositional environment of the Red Lion Formation, southwestern Montana [unpublished M. S. thesis]: University of Idaho, Moscow, Idaho, 103 p.

Hayden, L. L., and Bush, J. H., 1988, Description and significance of three isolated Cambrian outcrops in northern Idaho and northwestern Montana: Geological Society of America, Abstracts with Programs, Rocky Mountain Section, v. 20, p. 420.

Healy, M. P., 1986, Sedimentology and lithofacies of the Pilgrim Formation (Upper Cambrian), west-central Montana [Unpublished M. S.thesis]: University of Idaho, Moscow, Idaho, 146 p

Heckel, P. H., 1974, Carbonate buildups in the geologic record: a review, in Laporte, L. F., ed., Reefs in Time and Space: Society of Economic Paleontologists and Mineralogists Special Publication 18, p. 90-154.

Hilty, A. J., 1989, Lithofacies and depositional environments of Middle and Upper Cambrian rocks of the Dearborn River Canyon, west-central Montana [unpublished M. S. thesis]: University of Idaho, Moscow, Idaho, 164 p.

Hurley, A. J., 1989, The Metaline Formation-Ledbetter Slate contact in northeastern Washington [unpublished Ph. D. dissertation]: Washington State University, Pullman, Washington, 141 p.

James, N. P., Stevens, R. K., Barnes, C. R., and Knight, I., 1989, Evolution of a lower Paleozoic continental-margin carbonate platform, northern Canadian Appalachians, in Crevello, P. D., Wilson, J. L., Sarg, J. F., and Read, J. F., eds., Controls on Carbonate Platform and Basin Development: Society of Economic Paleontologists and Mineralogists Special Publication 44, p. 123-146.

Joseph, N. L., 1990, Geologic Map of the Colville 1:100,000 quadrangle, Washington and Idaho: Washington Division of Geology and Earth Resources Open-File Report 90-13, 78 p.

Lindsey, K. A., 1987, Character of origin of the Addy and Gypsy Quartzites, central Stevens and northern Pend Oreille counties, northeastern Washington [unpublished Ph.D. dissertation]: Washington State University, Pullman, Washington, 256 p.

Lochman-Balk, C., 1971, The Cambrian of the Craton of the United States, in Holland, C. H., ed., Cambrian of the New World, v. 1: New York, Wiley Interscience, p. 79-167.

_____, 1972, Cambrian System, in Mallory, R. V., ed., Geologic Atlas of the Rocky Mountain Region: Denver, Colorado, Rocky Mountain Association of Geologists, p. 60-75.

Mills, J. W., 1977, Zinc and lead ore deposits in carbonate rocks, Stevens County, Washington: Washington Division of Geology and Earth Resources Bulletin 70, 171 p.

Morton, J. A., 1989, Lower to early Middle Ordovician transition between Metaline and Ledbetter Formations in the Metaline mining district of northeastern Washington: unpublished manuscript, 76 p.

Motzer, M. E., 1980, Paleoenvironments of the lower Lakeview Limestone (Middle Cambrian), Bonner County, Idaho [unpublished M. S. thesis]: University of Idaho, Moscow, Idaho, 95 p.

Oriel, S. S., and Armstrong, F. C., 1971, Uppermost Precambrian and lowest Cambrian rocks in southeastern Idaho: U. S. Geological Survey Professional Paper 394, 52 p.

Palmer, A. R., 1960, Some aspects of the early Upper Cambrian stratigraphy of the White Pine district, White Pine County and vicinity: Intermountain Association of Petroleum Geologists, Eastern Nevada Geological Society Guidebook, Joint Field Conference, p. 43-52.

_____, 1971, The Cambrian of the Great Basin and adjoining areas, western United States, in Holland, C. H., ed., Cambrian of the New World, v. 1: New York, Wiley Interscience, p. 79-167.

Peterson, J. A., 1986, General stratigraphy and regional paleotectonics of the western Montana overthrust belt, in Peterson, J. A., ed., Paleotectonics and sedimentation in the Rocky Mountain Region, United States: American Association of Petroleum Geologists Memoir 41, p. 57-86.

Read, J. F., 1982, Geometry, facies, and development of Middle Ordovician carbonate buildups, Virginia Appalachians: American Association of Petroleum Geologists Bulletin, v. 66, p. 189-209.

_____, 1985, Carbonate platform facies models: American Association of Petroleum Geologists Bulletin, v. 69, p. 1-21.

_____, 1989, Controls on evolution of Cambrian-Ordovician platform margin, U. S. Appalachians, *in* Crevello, P. D., Wilson, J. L, Sarg, J. F., and Read, J. F., eds., Controls on Carbonate Platform and Basin Development: Society of Economic Paleontologists and Mineralogists Special Publication 44, p. 147-165.

Robison, R. A., 1960, Lower and Middle Cambrian stratigraphy of the eastern Great Basin: Intermountain Association of Petroleum Geologists - Eastern Nevada Geological Society Guidebook, Joint Field Conference, p. 43-52.

_____, 1964, Late Middle Cambrian faunas from western Utah:' Journal of Paleontology, v. 38, p. 510-566.

_____, 1976, Middle Cambrian trilobite biostratigraphy of the Great Basin: Brigham Young University Geology Studies, v. 23, part 2, p. 93-109.

Ruppel, E. T., 1986, The Lemhi Arch: A Late Proterozoic and early Paleozoic landmass in central Idaho, *in* Peterson, J. A., ed., Paleotectonics and Sedimentation in the Rocky Mountain Region, United States: American Association of Petroleum Geologists Memoir 41, p. 119-130.

Seward, W. P., 1990, Cambrian and Devonian carbonate lithologies and conodont biostratigraphy of the Whitefish-MacDonald Range, northwestern Montana and southeastern British Columbia [unpublished Ph. D.dissertation]: University of Idaho, Moscow, Idaho, 225 p.

Sloss, L. L., 1954, Lemhi Arch, a mid-Paleozoic positive element in south-central Idaho: Geological Society of America Bulletin, v. 65, p. 365-368.

_____, 1963, Sequences in the cratonic interior of North America: Geological Society of America Bulletin, v. 74, p. 93-114.

Stewart, J. H., 1972, Initial deposits of the Cordilleran geosyncline: evidence of a late Precambrian (858 m.y.) continental separation: Geological Society of America Bulletin, v. 83, p. 1345-1360.

Walcott, C. D., 1915, Cambrian problems in the Cordilleran region, *in* problems of American Geology: New Haven, Yale University Press, p. 162-233.

Wilson, J. L., 1975, Carbonate Facies in Geologic History: New York, Springer-Verlag, 471 p.

Yates, R. G., 1964, Geologic map and sections of the Deep Creek area, Stevens and Pend Oreille Counties, Washington: U. S. Geological Survey Miscellaneous Geological Investigations Map I-412.

UPPER PLATE ROCKS OF THE ROBERTS MOUNTAINS THRUST, NORTHERN INDEPENDENCE RANGE, NORTHEAST NEVADA: THE LATE CAMBRIAN(?) TO MIDDLE ORDOVICIAN SNOW CANYON FORMATION OF THE VALMY GROUP

Stephen A. Leslie
Department of Geological Sciences
The Ohio State University
Columbus, OH 43210

Peter E. Isaacson
Department of Geology
University of Idaho
Moscow, ID 83843

John E. Repetski
U.S. Geological Survey
970 National Center
Reston, VA 22092

Wayne L. Weideman
Independence Mining Co. Inc.
Elko, NV 89801

ABSTRACT

The Snow Canyon Formation of the Valmy Group in the Independence Range of northeastern Nevada consists of three informal units: a lower interbedded chert and shale unit with abundant greenstone, a middle quartzite unit with siltstone and shale interbeds, and an upper chert and shale unit. The base of the Snow Canyon Formation is the Roberts Mountains thrust and the top is conformably overlain by the McAfee Quartzite. Conodont data suggest a Late Cambrian(?) to Middle Ordovician age for the Snow Canyon Formation. Some of these conodont ages correlate well with previously determined graptolite ages for the lower unit of the Snow Canyon Formation; others, however, extend the age of its previously defined base into the earliest Early Ordovician and most likely into the Late Cambrian. Givetian to Frasnian conodonts have been recovered from rocks stratigraphically above the Roberts Mountains Formation, which were previously assigned to the Snow Canyon Formation. These Devonian rocks are suggested to be parautochthonous and may be correlative with the Popovich or Rodeo Creek limestones of local usage.

Greenstone geochemistry data, based upon plots of Ti, Zr, and Y on Zr-Zr/Y, Zr-TiO2, and Ti/100-Zr-Y*3 discrimination diagrams, support the interpretation that Snow Canyon Formation basalts were deposited in mid-ocean ridge and intraplate (ocean island or continental) environments. Ten of the 16 greenstone samples (8 from this study; 5 from Wrucke and others ,1978; 3 from Watkins and Browne, 1989) plot as ocean floor basalt and 6 plot as intraplate basalt.

Sediment gravity flows within the Snow Canyon Formation are classified as debris and grain flows. The presence of grain flows suggests a depositional setting adjacent to a relatively steep slope (at least 9 degrees). The purity of quartzite within the Snow Canyon Formation and overlying McAfee Quartzite supports the idea of a setting proximal to the shelf/slope,

and adjacent to the partially age-equivalent Eureka Quartzite as suggested by Miller and Larue (1983). These quartzites, associated with the "deep sea type" sediment and abundant greenstones within the Snow Canyon Formation, suggest a depositional setting marginal to the North American craton, in a relatively active, extensional tectonic regime.

INTRODUCTION

Detailed stratigraphic work on Ordovician western assemblage rocks in northern Nevada began with Merriam and Anderson's (1942) study of the Roberts Mountains in which they defined the Vinini Formation. Roberts (1951) defined the Valmy Formation, a correlative of the Vinini Formation, in the Battle Mountain area. Roberts and others (1958) show the Valmy Formation in the Battle Mountain area as Early to Middle Ordovician age, and as Early to Late Ordovician age in the Cortez and Shoshone Ranges. The Valmy Formation consists of typical western assemblage rocks as defined by Merriam and Anderson (1942), and has been suggested to be at least 12,000 feet (3660 m) and may be over 25,000 feet (7620 m) thick in the Shoshone Range (Gilluly and Gates, 1965). However, Ross (1976) suggested a total thickness for the Valmy and its correlatives in Nevada, Utah, and Idaho to be greater than 3,000 (915 m) but no more than 6,000 feet (1,830 m), and attributed seemingly greater thicknesses to unusual structural circumstances. Neither the top nor the base of the Valmy Formation are well defined stratigraphically, hence deposition may have begun in the Cambrian and continued into the Silurian.

The Valmy Formation is, at least in part, coeval with, and has similar lithology as the Ramshorn Slate, Kinnikinic Quartzite, Saturday Mountain Formation, and Phi Kappa Formation, and it is also time-equivalent, in part, with the Eureka Quartzite (Gilluly and Gates, 1965). The eastern assemblage Pogonip Group, consisting of the Antelope Valley Limestone, Ninemile Formation, and Goodwin Limestone, is also time correlative with much of the Valmy Formation (Nolan and

In Cooper, J.D., and Stevens, C.H., eds., 1991, **Paleozoic Paleogeography of the Western United States-II:** Pacific Section SEPM, Vol. 67, p. 475-486.

others, 1956; see also Ross and others, 1982).

In the northern Independence Range, Churkin and Kay (1967) elevated the Valmy Formation to group rank, and included in it their Snow Canyon Formation, McAfee Quartzite, and Jacks Peak Formation. Their revision of the Valmy stratigraphy was questioned by Miller and Larue (1983). Field work completed in this study, however, and by Watkins and Browne (1989), supports Churkin and Kay's (1967) stratigraphic nomenclature. Rocks of the Valmy Group are in low-angle, structural contact, on the Roberts Mountains thrust, with the underlying Roberts Mountains Formation (Silurian and Devonian).

Exposures of the Snow Canyon Formation in the Jerritt Canyon Mine area and along the Snow Canyon drainage, approximately 50 miles north-northwest of Elko, Nevada, are the focus of this study (Fig. 1). The goals of this study were to define a composite stratigraphy for the Snow Canyon Formation in the Jerritt Canyon Mine area, collect paleontologic data to better constrain its age, and examine possible depositional systems using sedimentological, paleontological, and geochemical data.

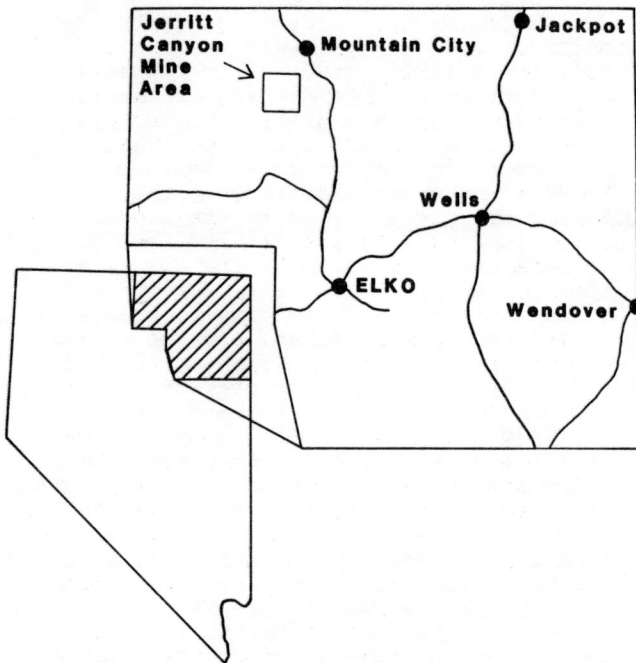

Figure 1: Index map of Nevada showing location of study area in the northern Independence Range.

STRATIGRAPHY AND PETROGRAPHY

The Snow Canyon Formation is approximately 600 meters thick and is composed of chert, clayshale, siltstone, claystone, quartzite, greenstone, and rare limestone. Structural complexity and monotonous cyclicity of lithology have inhibited systematic stratigraphic study of the Snow Canyon. Sections measured and

petrographic analyses of this study coupled with previous work, however, suggest that the Snow Canyon displays characteristic variations which can be utilized to infer relative stratigraphic position. Nine partial sections were measured from which a composite section has been assembled (Leslie, 1990). The Snow Canyon is divided into the following informal units: a lower unit, comprising interbedded chert and clayshale with abundant greenstone; a middle unit of quartzite with siltstone and clayshale interbeds; and an upper unit consisting of interbedded chert and clayshale (Fig. 2).

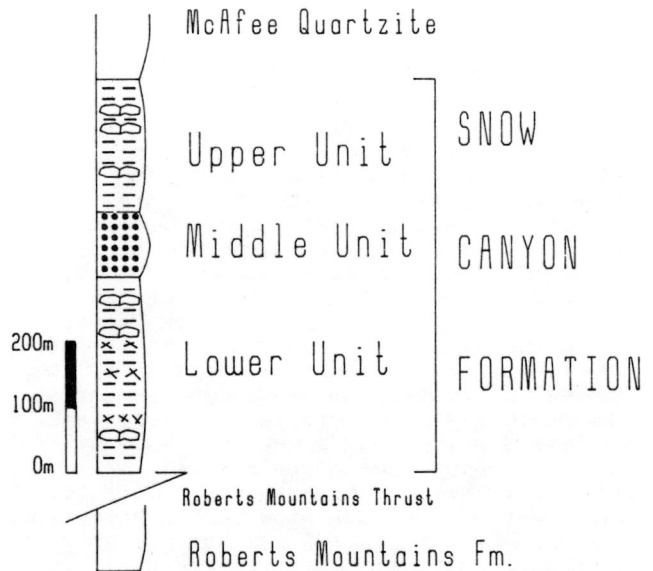

Figure 2: Composite section of the Snow Canyon Formation of the Valmy Group compiled from nine partial sections measured in the northern Independence Range.

Lower Unit

The lower unit of the Snow Canyon Formation is composed of chert, clayshale, claystone, greenstone pods, siltstone, rare limestone, and rare, discontinuous quartzite, and is approximately 300 meters thick. The base of the lower unit, and therefore the base of the Snow Canyon locally, is the Roberts Mountains thrust. Attenuated, black to tan, pyritic clayshale is the most common lithology. Clayshales contain 3-7% quartz silt and 1-3% silt size clay minerals. Chert, displaying abundant soft-sediment deformation structures, and with argillaceous partings, is the second most common lithology and it is in many places interbedded with, or present as lenses within, claystone and clayshale. Black carbonaceous staining, siliceous veins, and stylolites are typical of Snow Canyon cherts. Cherts contain 2-4% silt-size quartz grains, 1-2% silt size clay minerals, and rare dolomite rhombs. Silicispheres, which are common in some chert beds, may represent completely replaced radiolarians.

Claystone and clayshale within the lower unit are commonly silicified. In hand sample, differentiation of silicified claystone and chert is extremely difficult. However, in thin section the nearly uniform extinction of silicified claystone differentiates it from chert. This uniform extinction results from sub-parallel alignment of clay minerals.

Siltstone is interbedded with chert and clayshale. It is light greenish gray, massive-bedded to parallel and cross-laminated. Well sorted, subangular quartz grains comprise 96-97%, lath shaped clay 1-2%, and authigenic pyrite 2-3% of silt size grains. Very high-relief, high-birefringent minerals comprise less than 1% of grains and are interpreted to be either zircon or rutile. Cement makes up approximately 30% of the siltstones, of which 65% is calcite and 35% is silica. Siltstones are grain supported, with sutured and concavo-convex grain-to-grain contacts. Quartz grains are monocrystalline and commonly display undulatory extinction. Most siltstones also contain abundant anastomosing, polycrystalline, quartz veins with disseminated carbon and pyrite.

Greenstone occurs as pods within the interbedded chert and attenuated black clayshale, and is not laterally continuous over distances greater than approximately 200 meters within the study area. These pods are fault bounded and commonly display marginal brecciation and "swirling" of clayshale. Lenticular shape and attenuation of clayshale marginal to greenstone pods are the result of greater relative competency compared to the surrounding clayshale during structural deformation. Limestone is relatively rare but is commonly closely associated with greenstone throughout the Snow Canyon. Medium silt size, sub-rounded quartz grains comprise 2-4% of the limestones.

Rare, discontinuous, stylolitic quartzite of the lower unit typically displays normal grading and limonite staining after pyrite. Coarser grained material at the base of the quartzite beds is poorly sorted, parallel laminated, subrounded, and is composed of fine to coarse sand size quartz with biotite inclusions, and chert, carbonate, siltstone, and quartz aggregate grains. Finer material at the top of the quartzite beds is moderately to well sorted, cross laminated, sub-rounded and composed almost solely of medium to coarse silt size quartz grains. Quartzite beds of the lower unit are interpreted to be small-scale, submarine channel deposits.

Two conglomerate deposits in the lower unit of the Snow Canyon have a combined thickness of 14.5 meters and are 70 meters wide. The lower conglomerate overlies silty, parallel laminated limestone with rare chert interbeds. Scouring and post-depositional compaction of the underlying rocks is evidenced by angular truncation of the limestone interval and loading at the base of the conglomerate. The lower conglomerate is a 2.4 meter thick, calcite-cemented quartz arenite with rare, subrounded limestone clasts, which become more common up-section where it is a quartz-sand, matrix-supported conglomerate. The upper contact of the lower conglomerate has no apparent scoured contact.

The upper conglomerate is clast-supported, and has a coarse-grained limestone matrix containing 5-10% quartz grains. Clast composition is polymictic and includes silty, parallel laminated limestone, laminated, calcite cemented sandstone, and rare chert. The upper conglomerate is 12.1 meters thick and average clast size decreases from 30x15 centimeters in cross-section at the base to 3x1 centimeters in cross-section at the top. Clast density within the upper conglomerate is between 60% and 80%. The upper conglomerate is overlain by parallel and cross-laminated limestone interbedded with black clayshale.

Middle Unit

The middle unit of the Snow Canyon Formation is composed of quartzite, siltstone, and clayshale. This unit is approximately 100 meters thick. Quartzites are moderately sorted, subrounded to rounded, medium grained with 85-100% silica and 0-15% carbonate cement. Nearly all grains are quartz with undulatory extinction and rare inclusions of zircon or rutile. Siltstone and feldspar comprise less than 0.5% of grains. Grain contacts are sutured, concavo-convex, and long. Point contacts and floating grains comprise less than 3% of contacts. Matrix, defined as grains less than 0.065 mm, is 7-10% of the quartzites and is composed of 70-75% quartz, 25-30% pyrite, and less than 1% zircon, rutile, apatite, and clay. As apparent in plane polarized light, the cement has a distinctive orange stain, which is the result of limonite from altered pyrite. This stain imparts the distinctive orange-brown color of Snow Canyon quartzites.

Clayshale and siltstone are interbedded with the quartzites. The clayshale is orange-brown to black and composed of 90-95% clay minerals, 5-10% limonite after pyrite, and 2-5% silt size quartz grains. Varidirectional, polycrystalline quartz veins are common. Clayshale also displays variable degrees of black carbonaceous staining. Siltstone is orange-brown, massive, parallel and cross-laminated, and composed of 90-95%, moderately well sorted, subangular quartz grains, 1-2% lath shaped clay particles and 3-5% authigenic pyrite. Trace amounts of very high-relief, high-birefringent grains within the siltstone are interpreted to be either zircon or rutile. Approximately 30% of the siltstone lithology is cement, consisting of 95-100% silica and less than 5% calcite. Siltstones are grain supported, with grains in sutured and concavo-convex contact. Quartz grains are monocrystalline and commonly display undulatory extinction. Most siltstones contain abundant anastomosing, polycrystalline quartz veins.

Upper Unit

The upper unit of the Snow Canyon Formation is composed of chert, clayshale, mudstone, siltstone, and rare quartzite. This unit is approximately 200 meters thick. Chert is the most common lithology and in many places displays argillaceous partings or is interbedded with clayshale or claystone. Cherts contain 3-5% silt-size quartz grains and less than 1% silt size clay. Attenuated, black to tan, pyritic clayshale is the second most common lithology of this unit. Clayshales contain 3-7% quartz silt and 1-3% silt-size clay minerals. Claystone and clayshale are commonly silicified.

Siltstone is interbedded with chert and clayshale. It is light to dark brownish gray, massive to parallel and cross-laminated and well sorted. It consists of silt-size material, including 95-97% subangular quartz grains, 2-3% authigenic pyrite, and 1-2% lath shaped clay minerals. Cement makes up approximately 25% of the siltstones and is 80-90% silica and 10-20% carbonate. Siltstones are grain supported, with grains in sutured and concavo-convex contact. Quartz grains are monocrystalline and commonly display undulatory extinction. Most siltstones contain abundant anastomosing polycrystalline quartz veins.

Greenstone occurs as rare pods within the interbedded chert and attenuated black clayshale, and is not laterally continuous for more than 35 meters within the study area. Dark gray, parallel and trough cross-laminated, pyritic, silty limestone, with 5-10% coarse silt to fine sand size, sub-rounded quartz clasts, is rare within the upper unit. Discontinuous quartzite within the upper unit is typically poorly sorted, and has subangular to well rounded, fine to coarse sand-size grains. Nearly all grains are quartz with undulatory extinction and contain rare inclusions of zircon, rutile, and apatite. Grain contacts are sutured, concavo-convex, and long. Matrix makes up 10-15% of the quartzites and is 70-75% quartz, 25-30% pyrite, and less than 1% zircon, rutile, apatite, and clay particles. The cement is 65% silica and 35% carbonate and has a distinctive orange stain, apparent in plane polarized light, that is interpreted to be limonite after pyrite.

DEPOSITIONAL ENVIRONMENT

The lower, middle and upper units of the Snow Canyon Formation most likely had very similar depositional settings, and all units demonstrate a setting proximal to the North American craton (Fig. 3). Deposition of the lower unit is suggested to have occurred near the shelf/slope break. The abundance of black chert, claystone, and clayshale indicates a relatively deep, marine environment. Abundant greenstone suggests the area was not stable

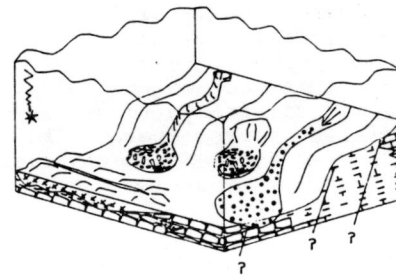

Lower unit

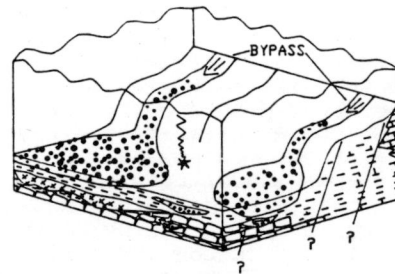

Middle unit

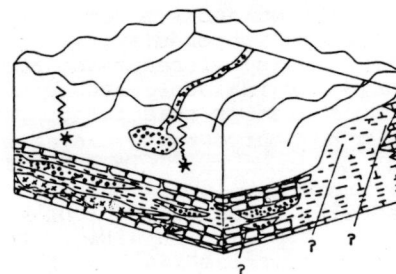

Upper unit

Figure 3: Schematic block diagrams showing hypothetical deposition of the lower, middle and upper units of the Snow Canyon Formation.

tectonically. Parallel and cross-laminated siltstone, and rare quartzites throughout the lower unit, indicate a setting close enough to the shelf, or other topographic high, for coarse clastic material to be present. Quartzites display normal grading, parallel laminations, and are capped by parallel and cross-laminated siltstone. Both the parallel and cross-laminated siltstone and quartzite deposits are interpreted to be turbidites.

The matrix-supported conglomerate in the lower unit is interpreted as a debris flow (Leslie, 1990). Debris flows, sedimentary gravity flows in which the sediment is supported above the sediment-water interface by matrix strength and buoyancy, and which "float" larger clasts in the matrix, require only slight slopes (less than 1 degree) for mobility (Cook and Mullins, 1983). However, their formation does not preclude slopes of greater steepness. The overlying clast-supported conglomerate is interpreted as a grain flow (Leslie, 1990). Grain flows are sedimentary gravity flows in which the sediment is supported above the sediment-water interface by direct grain-to-grain interactions. True grain flows require slopes of 18-30+ degrees to sustain movement; however, modified grain flows are suggested to be mobile on slopes less than 9 degrees (Cook and Mullins, 1983). Lowe (1976) discussed modified grain flows in detail.

The presence of debris flows and grain flows near the base of the lower unit of the Snow Canyon implies a slope of at least 9-18 degrees (Cook and Mullins, 1983). Although debris flows are suggested to be capable of moving great distances over gentle slopes, grain flows are "probably not mechanisms for long distance transport" (Cook and Mullins, 1983). Therefore, the presence of a grain flow within the lower unit of the Snow Canyon suggests proximity to a substantial slope.

Watkins and Browne (1989) suggested that accumulation of greenstone within the Snow Canyon represents a seamount. Although the presence of a seamount may explain the slope required for grain flow mobilization, paleontologic data suggest that the grain flow pre-dates proposed seamount development. The proposed seamount is suggested to have developed during the Llanvirnian (Watkins and Browne, 1989), while conodont ages in this study indicate sedimentary gravity flows to be of early Tremadocian age.

The absence of macrofossils within debris and grain flow clasts suggests that limestones comprising these clasts were deposited in relatively deep water, basinward of the carbonate platform, on the slope. Greenstones are indicative of an active tectonic setting which may have triggered debris and grain flows. These sediment gravity flows traveled until the slope was no longer steep enough to support mobility and were deposited within the black clayshale, chert, and greenstone facies of the Snow Canyon.

The middle unit is composed of alternating normal graded, parallel, and cross-laminated quartzite with rare dish and pillar structures, parallel and cross-laminated siltstone, and claystone. These sedimentary structures suggest that the middle unit represents cyclic deposition of turbidites. Two possible interpretations for the increase in frequency and thickness of turbidites in the middle unit are a more proximal position to the craton, or an increase in clastic material on the platform being shed into deeper water. The latter interpretation is supported by a study of Ordovician deep sea fan deposits by Miller and Larue (1983). Inundation of the carbonate platform during Middle Ordovician

time by the sand comprising the Eureka Quartzite is the suggested source for clastic material that spilled over the shelf/slope break. Miller and Larue (1983) suggested that the sand comprising the Eureka Quartzite had the same source as sand forming the McAfee Quartzite, and that these formations are age equivalents. Hence, it is not unreasonable that sand forming the cyclic turbidites within the Snow Canyon had the same source as the Eureka Quartzite, and represents bypass of the carbonate platform during the incipient stage of platform inundation.

Deposition of the upper unit represents a return to similar conditions present during deposition of the lower unit. The only marked difference between the lower and upper units is the rarity of greenstone within the upper unit. Paucity of greenstone above the lower unit may represent relaxation of tectonic extension responsible for extrusion of greenstone in the lower unit. Complete inundation of the carbonate platform by quartz sand took place after deposition of the upper unit. At this point, sand began breaching the shelf/slope break more frequently. This is manifested in the rock record in the northern Independence Range as the McAfee Quartzite (Miller and Larue, 1983), which overlies the upper unit of the Snow Canyon.

AGE OF THE SNOW CANYON FORMATION

Churkin and Kay (1967), after measuring sections in its type area, suggested that the Snow Canyon Formation ranges in age from Arenigian to Llandeilian (Early to early Middle Ordovician). These ages were based on graptolites collected from 5 sites within the Snow Canyon. Watkins and Browne (1989) suggested the Snow Canyon ranges in age from late Arenigian through Llandeilian, based upon previous work and their own graptolite data. Conodonts recovered from limestones within the Snow Canyon are used in this study to correlate with previous graptolite ages.

Limestone disaggregations have yielded poor returns for the most part. Of the 43 samples run for conodonts, fewer than 10 yielded identifiable elements (Appendix A). The diagnostic faunas found correlate well with graptolite data but they also suggest that the initial deposition of the Snow Canyon was as early as the Late Cambrian. The protoconodont _Phakelodus tenuis_ (Müller) and the paraconodont _Furnishina furnishi_ Müller, which were recovered below the lower unit debris flow, are most likely Late Cambrian in age. Both these conodonts have ranges extending from the Middle Cambrian to the lowest part of the Lower Ordovician. A Late Cambrian age for this interval is probable, based on: 1) the absence of euconodonts (true conodonts), which would be expected to be associated with primitive conodonts in the Lower Ordovician; and 2) the presence of very early Early Ordovician conodonts a short stratigraphic distance above. A sample (USGS fossil locality

10776-CO) containing Cordylodus caboti Bagnoli, Barnes and Stevens, C. intermedius Furnish, C. lindstromi Druce and Jones, Oneotodus aff. O. variabilis Lindström, and Semiacontiodus sp., recovered within the grain flow, indicates an early Early Ordovician age (early Tremadocian; early Ibexian). A paraconodont, cf. Furnishina primitiva Müller, a Late Cambrian(?) species with a poorly defined range (colln. 10780-CO), also was recovered within the grain flow. Cordylodus caboti, C. lindstromi, Oneotodus? sp., Semiacontiodus sp., and Variabiloconus sp., an assemblage of conodont species recovered above the grain flow (colln. 10778-CO), also are of early Early Ordovician age. It should be noted that the conodonts extend the known lower age limit of the lower unit of the Snow Canyon significantly. Previously, the oldest fossils reported from that unit are of late middle or early late Arenigian age (Watkins and Browne, 1989). Their stratigraphic assignment, which correlates with late Ibexian to earliest Whiterockian in North American usage, is significantly younger than the early Ibexian (early Tremadocian) and (probably) Late Cambrian ages suggested by conodonts recovered in this study low in the lower unit.

The debris and grain flows within the lower unit of the Snow Canyon are of Late Cambrian(?) through the Early Ordovician age based upon Late Cambrian(?) conodonts recovered from strata below, Early Ordovician conodonts recovered above, and conodonts of both Late Cambrian(?) and early Early Ordovician ages recovered from within the flows. Samples collected from other measured sections also yielded Early to early Middle Ordovician conodonts.

The types of conodonts present suggest deposition of the Snow Canyon in a cool, deep marine setting. For example, in the late Early to earliest Middle Ordovician faunas, Periodon? sp., Bergstroemognathus cf. B. extensus Serpagli, Juanognathus sp., J. aff. J. jaanussoni Serpagli, Tripodus laevis Bradshaw, and Protopanderodus gradatus Serpagli (collns. 10775-CO, 10779-CO, and 10781-CO) are characteristic of the Atlantic Faunal Realm, which is interpreted as a high-latitude, deep and/or cool-water region (Bergström, 1990). Likewise, in the Late Cambrian(?) and earliest Ordovician collections, all the recovered species are of cosmopolitan distribution; the absence of species characteristic of warm, shallow shelf environments is here interpreted as indicating cooler and (or) deeper depositional environments for the lower part of the lower unit of the Snow Canyon. Although northeastern Nevada was probably not located at a very high latitude during the Late Cambrian to Middle Ordovician, a cool, deep marine setting is consistent with lithologic interpretations of this study and those of previous workers (e.g. Churkin and Kay, 1967; Wrucke and others, 1978). Samples having proto- and paraconodonts to the exclusion of euconodonts suggest, but do not prove, that deposition of the Snow

Canyon, and therefore upper plate, western assemblage rocks in the Jerritt Canyon Mine area, began at least as far back as the Late Cambrian. The middle to early late Arenigian age (=late Ibexian to early Whiterockian) of the majority of conodonts collected is slightly older than the previously suggested late Arenigian to Llanvirnian-Llandeilian ages of graptolites from other Snow Canyon localities.

Devonian conodonts were recovered from a limestone approximately 60 feet (18 m) above the Roberts Mountains Formation. They include the Givetian-Frasnian age conodonts Palmatolepis sp., P. aff. P. transitans Müller, Pandorinellina sp., Polygnathus sp., P. asymmetricus ovalis Ziegler and Klapper, and P. xylus-group (colln. 12011-SD) that were identified in samples from a core drilled in the Jerritt Canyon Mine area. Age similarities between this limestone unit and the locally-named Popovich and Rodeo Creek limestones indicate that these units may be correlatives. This limestone was originally thought to be a Snow Canyon equivalent; however the latest Middle to earliest Late Devonian age of this sample indicates that it is not, and that this unit is either autochthonous or parautochthonous. Therefore, evidence is strong for the presence of a heretofore unidentified Devonian unit that overlies the Silurian and Devonian Roberts Mountains Formation in the Independence Range.

GREENSTONES OF THE SNOW CANYON FORMATION

Greenstone Petrography

Greenstones of the Snow Canyon Formation include porphyritic, microporphyritic, aphyritic, seriate, amygdoloidal, and vesicular altered basalts. Altered, phaneritic, mafic rocks also occur within the Snow Canyon, but are rare. Greenstone samples analyzed during this study are pervasively altered and contain carbonate (ankerite and calcite), chlorite, leucoxene, albite, quartz, magnetite, and clay (sericite, montmorillonite, or illite). Amygdules are composed of either calcite, chalcedonic quartz, or a combination of these minerals.

The greenstones of the Snow Canyon were affected by carbonate alteration. Carbonate alteration of mafic rocks includes the mineral assemblage quartz, albite, ankerite, dolomite, calcite, sericite, chlorite, and rutile (Siems, 1984). All these minerals (except rutile) are present in greenstones of the Snow Canyon. Silica liberated during alteration is accounted for by extensive siliceous veining and vesicle fillings. Magnetite-ilmenite (titanomagnetite), present in trace amounts and outlining relict crystal forms, is the source of titanium oxides comprising the leucoxene. Well outlined, relict crystal basal sections reflect the presence of both pyroxene and amphibole phenocrysts in the parent rock.

Plagioclase phenocrysts are typically

Major Elements (%)

Sample	SiO2	Al2O3	Fe2O3	FeO	MgO	CaO	Na2O	K2O	TiO2	P2O5	MnO	BaO	LOI	TOTAL
NGG-1	41.35	16.08	10.56	8.50	9.80	3.12	1.81	1.83	1.79	0.34	0.13	1.83	11.62	100.3
NGG-2	43.43	12.11	14.46	11.55	9.19	3.05	1.77	0.51	1.63	0.27	0.13	0.27	12.33	98.65
NGG-4	40.74	10.30	11.84	9.70	7.56	7.56	1.47	0.81	1.44	0.25	0.17	0.27	17.22	98.82
NGG-5	43.11	11.91	14.06	11.25	7.43	5.63	2.55	0.44	1.87	0.21	0.24	0.12	12.70	99.83
NGG-6	44.85	11.95	15.22	11.20	5.98	6.09	3.29	0.37	2.70	0.19	0.19	0.16	8.70	99.33
NGG-7	38.43	12.49	14.94	11.75	7.20	7.06	2.27	0.46	2.17	0.19	0.16	0.09	14.65	99.65
SCC-1	48.94	16.72	9.09	6.12	5.41	4.20	6.41	0.34	2.33	0.65	0.05	0.07	6.94	100.8
DF-4	46.24	18.53	10.63	8.39	4.06	4.26	5.73	1.00	2.95	0.33	0.13	0.12	7.24	100.2

Minor Elements (ppm)

SAMPLE	Ni	Cr	Sr	V	Nd	Sm	Y	Zr
NGG-1	74	150	58	250	17	3.9	24	105
NGG-2	49	122	38	400	11	3.0	29	73
NGG-4	40	134	64	350	5	2.4	30	71
NGG-5	41	92	54	460	5	3.1	33	78
NGG-6	24	58	114	490	11	4.9	45	140
NGG-7	46	90	102	430	6	3.6	39	100
SCC-1	11	22	44	270	37	6.1	26	180
DF-4	22	46	270	280	11	4.1	18	160

Table 1: Major oxide and trace element geochemistry of greenstones of the Snow Canyon Formation. Note: Major element analyses completed using ICP-AES; Ni, Cr, Sr, and V using AAS; Nd and Sm using NAA; and Y and Zr using XRF. All analyses Chemex Labs Inc.

altered to sericite, calcite, and albite (An 8-12). Large volumes of calcite, both within relict plagioclase phenocrysts and disseminated throughout the greenstone, suggest a calcic parent plagioclase. However, all plagioclase phenocrysts observed apparently have been albitized.

The basaltic parent rocks of the greenstones experienced at least two phases of alteration. The first was spilitic metasomatism on the sea floor. Later carbonate alteration is interpreted to have affected amphiboles and pyroxenes, commonly unaltered in spilites. The variable texture and composition of greenstones reflect different cooling rates. Some greenstones display well developed pillow texture, are aphyritic, microporphyritic or seriate, and amygdaloidal or vesicular. These greenstones represent flows on the ocean floor. Altered porphyritic and possibly seriate mafic rocks may represent hypabyssal basalts and are interpreted to be the feeder dikes to overlying flows.

Greenstone Geochemistry

Systematic trace element variations have been shown to exist between basaltic rocks deposited in different tectonic settings (Engle and others, 1965; Pearce and Cann, 1973; Floyd and Winchester, 1975; Pearce and Norry, 1979). Analyses of different magma by plots, using selected element axes or discrimination functions, may be used to characterize tectonic affinities (Pearce and Cann, 1973). Geochemical analyses for elements with high field strengths are of particular interest

as these elements are not readily mobile and most likely reflect relative abundances very close to that of original deposition, even in metamorphosed and weathered basalts (Pearce and Norry, 1979). Ti, Zr, and Y have high field strengths and have been used to suggest tectonic setting for basaltic rocks based upon correlation with basaltic rocks of known origin (Pearce and Cann, 1973; Floyd and Winchester, 1975; Pearce and Norry, 1979).

Using trace element geochemistry, Pearce and Cann (1973) demonstrated that ocean floor basalt, volcanic arc basalt, and within-plate (ocean island and continental) basalt are distinguishable. The classification suggested by Pearce and Cann (1973) has been utilized by Wrucke and others (1978), Madrid (1987), and Watkins and Browne (1989) and it is adopted in this study to suggest the tectonic setting of greenstones within the Snow Canyon Formation.

Wrucke and others (1978) conducted a study which incorporated greenstone geochemistry from the Valmy Group in the Independence Range and the Valmy Formation in the northern Shoshone Range. Ti, Zr, and Y analyses were plotted on discrimination diagrams within fields described by Pearce and Cann (1973). Based on these data, Wrucke and others (1978) suggested that Valmy Formation basalt extrusion occurred in an ocean-floor setting. Watkins and Browne (1989) analyzed greenstones from the Snow Canyon of the Valmy Group in the Independence Mountains. They suggested from plots of Ti, Zr, and Y, that greenstones of the Snow Canyon represent an Ordovician seamount formed in a within-plate setting. Madrid (1987) analyzed greenstones of the Valmy Formation in the northern Shoshone Range, north-central Nevada. He concluded from plots of Ti, Zr, and Y that these greenstones were also deposited in a within-plate setting (Madrid, 1987, fig. 54).

Greenstone samples from the northern Independence Range were analyzed in this study for major elements and selected trace elements (Table 1). Results were plotted on Zr-TiO2, Zr-Zr/Y, and Ti/100-Zr-Y*3 discrimination diagrams (Fig. 4, 5, and 6, respectively). Data plotted on Figure 4 suggest a tholeiitic composition for the majority of the basalts within the Snow Canyon. Figures 5 and 6 indicate that the majority of the basalt was deposited in an ocean-floor setting, with a minor fraction plotting in the ocean island basalt field.

Trace element plots yielding multiple tectonic settings for greenstones within western assemblage rocks are interpreted to suggest that the tectonic setting evolved from a within-plate setting to an ocean-floor setting. Further investigations documenting spatial relations of greenstone occurrences, accompanied by geochemical data, may provide further insight into the problem of greenstones which plot in both within-plate and ocean-floor fields.

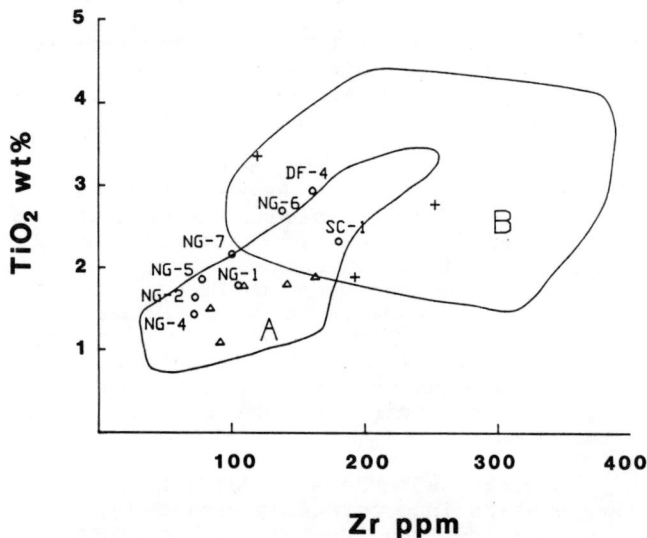

Figure 5: Discrimination plot of zirconium divided by yttrium versus zirconium of Pearce and Norry (1979). Numbered circles refer to samples in Table 1; triangles refer to samples of Wrucke and others (1978); crosses refer to samples of Watkins and Browne (1989). Field A = Island-arc basalt, Field B = Mid-ocean ridge basalt, Field C = Within-plate basalt (i.e. intraplate, ocean island, continental), Field D = Within-plate + Mid-ocean ridge basalt.

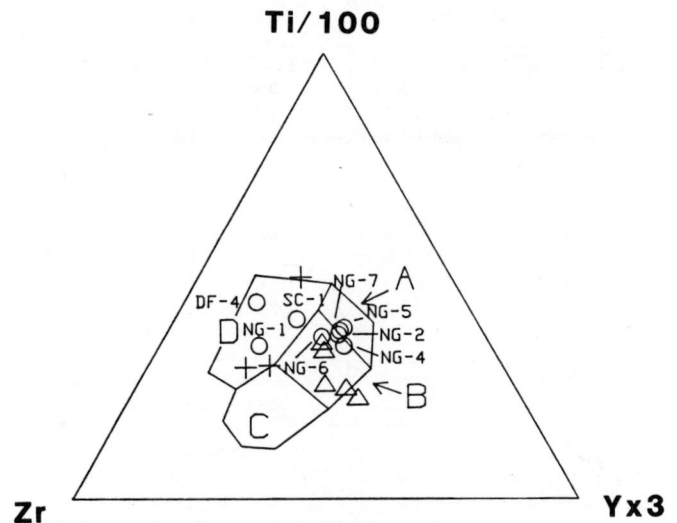

Figure 4: Titanium oxide versus zirconium diagram of Floyd and Winchester (1975). Numbered circles refer to samples in Table 1; triangles refer to samples of Wrucke and others (1978); crosses refer to samples of Watkins and Browne (1989). Field A = Oceanic tholeiite and Field B = Alkalic basalt.

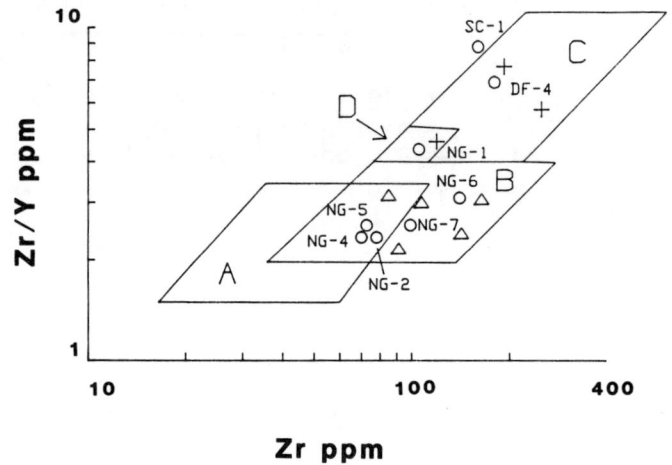

Figure 6: Titanium-zirconium-yttrium discrimination diagram of Pearce and Cann (1973). Numbered circles refer to samples in Table 1; triangles refer to samples of Wrucke and others (1978); crosses refer to samples of Watkins and Browne (1989). Field A = Low K tholeiite, Field B = Ocean-floor basalt + calcalic basalt + low K tholeiite, Field C = Calcalkaline basalt, Field D = Within-plate basalt (i.e. intraplate, ocean island, continental).

REGIONAL TECTONIC SETTING

Madrid (1987) and Turner and others (1989) have incorporated Willden's (1979) extensional theory with the back-arc thrusting theory (Burchfiel and Davis, 1975) in a detailed model describing deposition and tectonic settings through the early Paleozoic of the North American Cordillera. Western assemblage rocks of the Roberts Mountains allochthon record a depositional history of Cambrian extension, Early Ordovician subsidence, Ordovician starved-basin conditions, Middle and Late Ordovician extension, Silurian siliclastic sedimentation followed by starved-basin conditions, and Late Devonian intrabasinal extension (Turner and others, 1989). Alkalic mafic lavas and arkosic sandstone turbidites are evidence for Cambrian rifting. These deposits are overlain by deeper water pelagic rocks deposited under starved-basin conditions (Turner and others, 1989). Synvolcanic normal faults and abundant sedex (sedimentary exhalative) barite and iron-copper deposits are cited as evidence for Middle and Late Ordovician extension (Papke, 1984; Turner and others, 1989). Dolomitic siltstones and fine-grained quartzites of the Silurian Trail Creek Formation in Idaho (Dover, 1981) and the approximately coeval Elder Sandstone and Fourmile Canyon Formation in Nevada (Gilluly and Gates, 1965) record Silurian siliclastic deposition (Turner and others, 1989). Devonian extension is evidenced by alkaline basalts and sedex zinc and barite of the Slaven Chert (Papke, 1984; Madrid, 1987). Turner and others (1989) and Miller and Larue (1983) suggest the western assemblage was deposited adjacent to eastern assemblage rocks as part of the outer continental margin during the early Paleozoic and transported as a single allochthonous block eastward over the eastern and transitional assemblages. Movement of the Roberts Mountains thrust as a single allochthonous block is suggested by fold geometry studies in the Shoshone Range, Tuscarora Range, and the Battle Mountain area (Evans and Theodore, 1978; and Madrid, 1987).

The model suggested by Turner and others (1989) and Madrid (1987) appears to be the most comprehensive model that accounts for the geology discussed herein. Back-arc extension is explained by a decrease in subduction rate. As subduction rate slows, the angle of subduction increases, creating a tensional stress regime on the back-arc environment (Dube, 1988; Uyeda and Kanamori, 1979; Stewart, 1978). Although Madrid (1987) implied extension via this mechanism during the Devonian, it is plausible that it also affected Ordovician extension. The presence of alkalic (Madrid, 1987) and tholeiitic lavas (Wrucke and others, 1978; this study) may be the expression of oceanic crust developing within rifting continental crust. At the onset of extension, magma was intruded through the continental crust, which explains the alkalic signature. As extension continued, ocean crust was formed and the lavas erupted through this crust yielding a tholeiitic signature more typical of ocean-floor basalts. A schematic diagram of this tectonic evolution theory is shown in Figure 7. This model incorporates the conclusions of Madrid (1987), Turner and others (1989), and Willden (1979), and is supplemented with data from this study.

EARLY ORDOVICIAN

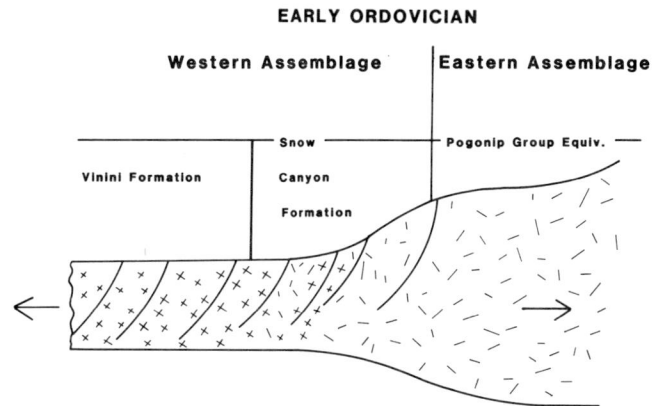

Figure 7: Reconstruction of the continental margin during Early Ordovician time in northeastern Nevada showing depositional setting of the Snow Canyon Formation (part) in a tectonically active, extensional regime, adjacent to age-equivalent eastern and western assemblage rocks.

SUMMARY AND CONCLUSIONS

The Snow Canyon Formation is the stratigraphically lowest formation of the Valmy Group, as defined by Churkin and Kay (1967), and is located in the northern Independence Range, northeast Nevada. The base of the Snow Canyon is the Roberts Mountains thrust and its top is conformably overlain by the McAfee Quartzite. The following conclusions and suggestions are the result of this study:

1) The Snow Canyon Formation is approximately 600 meters thick in the Jerritt Canyon Mine area and consists of three informal mappable units: a the lower, 300-meter-thick, unit with inter-bedded chert and clayshale and abundant greenstone; a middle, 100-meter-thick, unit of quartzite interbedded with siltstone and clayshale; and an upper, 200-meter-thick, unit of interbedded chert and clayshale.

2) Conodonts (this study) and graptolites (previous studies) support a pre-Tremadocian(?) to Arenigian-Llanvirnian age (Late Cambrian(?) to Middle Ordovician) for the Snow Canyon Formation. Conodont data also prove the presence of a late Middle to early Late Devonian unit that overlies the Roberts Mountains Formation within the Jerritt Canyon Mine area.

3) Greenstones from the Snow Canyon Formation are alkalic and tholeiitic and

484

suggest both ocean-floor and within-plate settings. Therefore, a tectonic evolution model, including an extensional regime, is suggested. Credibility of this model may be tested by mapping spatial and temporal relations between alkalic versus tholeiitic, and ocean-floor versus within-plate greenstones, within the Cordilleran margin.

4) The Snow Canyon Formation was deposited in a setting close to the shelf/slope break in an active tectonic environment. Syntectonic activity for at least the duration of deposition of the lower unit is evidenced by abundant greenstone deposits. Tectonic activity apparently persisted, although possibly not as intensively, throughout deposition of the middle and upper units, as indicated by the local presence of greenstone within these units.

In summary, deposition of the Snow Canyon Formation persisted from Late Cambrian(?) through early Middle Ordovician time. Paucity of known fossil data precludes establishment of a comprehensive conodont biostratigraphy; however, the data available support the lithostratigraphic relations asserted herein and by previous workers. Snow Canyon Formation deposition occurred in a deep, cool marine setting characterized by deposition of chert, greenstone, clayshale and submarine fan clastics. Deposition of the Snow Canyon Formation terminated, and deposition of the McAfee Quartzite began, with the inundation of the eastern assemblage carbonate platform and followed by pervasive foundering of the shelf/slope break in the late Middle Ordovician.

ACKNOWLEDGEMENTS

We wish to thank Freeport-McMoRan Gold Company (currently Independence Mining Co. Inc.) for their support of this project, in particular, Chief Mine Geologist of the Jerritt Canyon Mine, D.J. Birak. We benefitted from reviews of an earlier version of this report by S.M. Bergström, A.K. Armstrong, and J.D. Cooper. We also thank A.G. Harris, U.S. Geological Survey, for assistance in identification of Silurian through Devonian conodonts.

REFERENCES

Bergström, S.M., 1990, Relations between conodont provincialism and the changing paleogeography during the Early Paleozoic, in McKerrow, W.S., and Scotese, C.R., eds., Paleozoic Paleogeography and Biogeography: Geological Society Memoir No. 12, p. 105-121.

Burchfiel, B.C., and Davis, G.A., 1975, Nature and controls of the Cordilleran orogenesis, western United States: Extensions of an earlier synthesis: American Journal of Science, v. 275 A, p. 363-396.

Churkin, M.C., and Kay, M., 1967, Graptolite-bearing Ordovician siliceous and volcanic rocks, Northern Independence Range, Nevada: Geological Society of America Bulletin, v. 78, p. 651-668.

Cook, H.E., and Mullins, H.T., 1983, Basin margin, Scholle, P.A., Bebout, D.G., and Moore, C.H., eds., in Carbonate Depositional Environments: American Association of Petroleum Geologists Memoir 33, p. 539-618.

Dover, J.H., 1981, Geology of the Bolder-Pioneer wilderness study area, in Simons, F.S., Dover, J.H., eds., Mineral Resources of the Bolder-Pioneer Study Area, Blane and Custer Counties, Idaho: U.S. Geological Survey Bulletin 1497, p. 15-75.

Dube, T.E., 1988, Tectonic significance of Upper Devonian rocks and bedded barite, Roberts Mountains Allochthon, Nevada, U.S.A., in McMillan, N.J., Embary, A.F., Glass, D.J., eds., Devonian of the World: Proceedings of the Second International Symposium on the Devonian System, Calgary, Canada, Volume II: Sedimentation, p. 235-249.

Engle, A.E., Engle, C.G., and Havens, R.G., 1965, Chemical characteristics of ocean basalts and the upper mantle: Geologic Society of America Bulletin, v. 76, p. 719-734.

Evans, J.G., and Theodore, T.G., 1978, Deformation of the Roberts Mountains allochthon in north-central Nevada: U.S. Geological Survey Professional Paper 1060, 18 p.

Floyd, P.A. and Winchester, J.A., 1975, Magma type and tectonic setting discrimination using immobile elements: Earth and Planetary Science Letters, v. 27, p. 211-218.

Gilluly, J., and Gates, O., 1965, Tectonic and igneous geology of the northern Shoshone Range, Nevada: USGS Prof. Paper 465, 153 p.

Leslie, S.A., 1990, The Late Cambrian-Middle Ordovician Snow Canyon Formation of the Valmy Group, northeastern Nevada: [Unpublished M.S. Thesis], University of Idaho, 112 p.

Lowe, D.R., 1976, Subaqueous liquified and fluidized sediment flows and their deposits: Sedimentology, v. 23, p. 285-308.

Madrid, R.J., 1987, Stratigraphy of the Roberts Mountains allochthon in north-east Nevada: [Doctoral Dissertation], Stanford University, Stanford, California, 453 p.

Merriam, C.W., and Anderson, C.A., 1942, Reconnaissance survey of the Roberts Mountains, Nevada: Geological Society of America Bulletin, v. 5, p. 1675-1728.

Miller, E.L., and Larue, D.K., 1983, Ordovician quartzite in the Roberts Mountains Allochthon, Nevada: Deep sea fan deposits derived from cratonal North America in Stevens C.H., ed., Pre-Jurassic Rocks in western North American suspect terranes: Pacific Section Society of Economic Paleontologists and Mineralogists, p. 91-102.

Nolan, T.B., Merriam, C.W., and Williams, J.S., 1956, The stratigraphic section in the vicinity of Eureka, Nevada: U.S. Geological Survey Professional. Paper 276, 77 p.

Papke, K.G., 1984, Barite in Nevada: Nevada Bureau of Mines and Geology Bulletin, v. 98, 125 p.

Pearce, J.A., and Cann, J.R., 1973, Tectonic setting of basic volcanic rocks determined using trace element analyses: Earth and Planetary Science Letters, v. 19, p. 290-300.

Pearce, J.A. and Norry, M.J., 1979, Petrogenic implications of Ti, Zr, Y, and Nb variations in volcanic rocks: Contributions to Mineralogy and Petrology, v. 69, p. 33-47.

Roberts, R.J., 1951, Geology of the Antler Peak Quadrangle, Nevada: U.S. Geological Survey Geologic Quadrangle Map GQ-10, scale 1:62,500.

Roberts, R.J., Preston, E.H., Gilluly, J., and Ferguson, H.G., 1958, Paleozoic rocks of north-central Nevada: American Association of Petroleum Geologists Bulletin, v. 42, p. 2813-2857.

Ross, R.J., Jr., 1976, Ordovician sedimentation in the western United States, in Hill, J.G., ed., Symposium on Geology of the Cordilleran Hingeline: Rock Mountain Association of Geologists Symposium, p. 109-134.

Ross, R.J., and 27 others, 1982, The Ordovician System in the United States; Correlation chart and explanatory notes: International Union of Geological Sciences, Publication 12, 73 p., 3 charts.

Siems, P.L. 1984, Lecture Manual: Hydrothermal alteration for mineral exploration workshop: Department of Geology, College of Mines and Earth Resources, Idaho Mining and Mineralogical Research Institute and University Continuing Education, University of Idaho, 518 p.

Stewart, J.H., 1978, Basin-range structure in western North America: a review, in Smith, R.B., and Eaton, G.P., eds., Cenozoic Tectonics and Regional Geophysics of the Western Cordillera: Geological Society of America Memoir 152, p. 1-32.

Turner, R.J.W., Madrid, R.J., and Miller, E.L., 1989, Roberts Mountains allochthon: Stratigraphic comparison with lower Paleozoic outer continental margin strata of the northern Canadian Cordillera: Geology, V. 17, p. 341-344.

Uyeda, S. and Kanamori, H., 1979, Back-arc opening and the mode of subduction: Journal of Geophysical Research, v. 84, n. 3, pp. 1049-1061.

Watkins, R., and Browne, Q.J., 1989, An Ordovician continental-margin sequence of turbidite and seamount deposits in the Roberts Mountains allochthon, Independence Range, Nevada, Geological Society of America Bulletin, v. 101, p. 731-741.

Willden, R., 1979, Ruby orogeny- A major Early Paleozoic tectonic event in Newman, G.D., and Goode, H.D., eds.,

Basin and Range Symposium: Rocky Mountain Association of Geologists and Utah Geological Association, p. 55-73.

Wrucke, C.T., Churkin, M., and Heropoulos, C., 1978, Deep-sea origin of Ordovician pillow basalt and associated sedimentary rocks, northern Nevada: Geologic Society of America Bulletin, v.89, p. 1272-1280.

APPENDIX A

Identifiable conodonts recovered from 8 limestone disaggregations. Conodont specimens with locations are curated at the USGS Branch of Paleontology and Stratigraphy, Reston, VA. Conodont identification by John E. Repetski.

USGS Sample 10775-CO, CAI 4-5

Bergstroemognathus cf. B. extensus Serpagli
Juanognathus sp.
J.? sp.
Oistodus? sp.
Parapanderodus sp.
Scolopodus gracilis? Ethington and Clark
Tropodus cf. T. comptus (Branson and Mehl)
T. comptus

USGS Sample 10776-CO, CAI 4.5

Cordylodus intermedius Furnish
C. caboti Bagnoli, Barnes, & Stevens
C. lindstromi Druce & Jones
C. prion Lindström
C. proavus Müller
Cordylodus sp.
Oneotodus aff. O. variabilis Lindström
Semicontiodus sp.
Teridontus nakamurai (Nogami)

USGS Sample 12011-SD, CAI 4

Palmatolepis aff. P. transitans Müller
Palmatolepis sp.
Pandorinellina sp.
Polygnathus sp.
P. asymmetricus ovalis Ziegler & Klapper
P. xylus-group

USGS Sample 10777-CO

Furnishina furnishi Müller
Phakelodus tenuis (Müller)

USGS Sample 10778-CO, CAI 4

C. caboti Bagnoli, Barnes, & Stevens
C. lindstromi Druce & Jones
C. prion Lindström s.f.
Cordylodus sp. indet.
Oneotodus? sp.
Semicontiodus sp.
Teridontus nakamurai (Nogami)?
Variabiloconus? sp.

USGS Sample 10779-CO, CAI 4-4.5

Protopanderodus sp.
Tripodus cf. T. laevis Bradshaw

USGS Sample 10780-CO

cf. _Furnishina primitiva_ Muller

USGS Sample 10781-CO, CAI 4.5

Drepanodus sp.
Juanognathus aff. _J. jaanussoni_ Serpagli
Paracordylodus? sp.
Periodon? sp.
Protopanderodus gradatus Serpagli
Tripodus laevis Bradshaw
?_T. laevis_
Scolopodus? peselephantis Lindström

APPENDIX B

Detailed location of measured sections.

Snow Canyon-A Section

Elevation - 7240 ft., 2205 meters

U.S Public Land System location:
    The base of section SC-A is in the NW 1/4, NW 1/4, SW 1/4, Sec. 27, T. 41 N, R. 53 E of the California Mountain 7.5-minute map (1971), of the Mt. Diablo Principal Meridian System.

Spherical Coordinates:
    Longitude W 116 00'10", Latitude N 41 25'10"

Line of Section:
    Section SC-A trends N 30 W with an average dip of 40-50 NW.

Snow Canyon-B Section

Elevation - 7620 ft., 2320 meters

U.S. Public Land System location:
    The base of section SC-B is in the NW 1/4, NW 1/4, SW 1/4, Sec. 26, T. 41 N, R. 53 E of the California Mountain 7.5-minute map (1971), of the Mt. Diablo Principal Meridian System.

Spherical Coordinates:
    Longitude W 115 59'05", Latitude N 41 25'10"

Line of Section:
    Section SC-B trends N 15 W with an average dip of 25-35 NW.

Snow Canyon-C Section

Elevation - 7220 ft., 2200 meters

U.S. Public Land System location:
    The base of section SC-C is in the SW 1/4, NW 1/4, SE 1/4, Sec. 16, T. 41 N, R. 53 E of the Tuscarora 15-minute map (1956), of the Mt. Diablo Principal Meridian System.

Spherical Coordinates:
    Longitude W 116 00'45", Latitude N 41 26'51"

Line of Section:
    Section SC-C trends N 40 W with an average dip of 35-50 NW.

Snow Canyon-D Section

Elevation - 7040 ft., 2145 meters

U.S. Public Land System location:
    The base of section SC-D is in the SW 1/4, SW 1/4, SE 1/4, Sec. 16, T. 41 N, R. 53 E of the Tuscarora 15-minute map (1956), of the Mt. Diablo Principal Meridian System.

Spherical Coordinates:
    Longitude W 116 00'47", Latitude N 41 26'33"

Line of Section:
    Section SC-C trends N 50 W with an average dip of 15-25 NW.

West Generator-A Section

Elevation - 7060 ft., 2150 meters

U.S. Public Land System location:
    The base of section WG-A is in the NW 1/4, NW 1/4, SE 1/4, Sec. 33, T. 41 N, R. 53 E of the Tuscarora 15-minute map (1956), of the Mt. Diablo Principal Meridian System.

Spherical Coordinates:
    Longitude W 116 00'50", Latitude N 41 24'18"

Line of Section:
    Section WG-A trends N 50 E with an average dip of 70 NE.

West Generator-B Section

Elevation - 6680 ft., 2035 meters

U.S. Public Land System location:
    The base of section WG-B is in the SW 1/4, SE 1/4, SE 1/4, Sec. 33, T. 41 N, R. 53 E of the Tuscarora 15-minute map (1956), of the Mt. Diablo Principal Meridian System.

Spherical Coordinates:
    Longitude W 116 00'26", Latitude N 41 24'05"

Line of Section:
    Section WG-B trends N 70 W with an average dip of 20-30 NW.

North Generator Section

Elevation - 7720 ft., 2355 meters

U.S. Public Land System location:
    The base of section NG is in the NE 1/4, NE 1/4, NW 1/4, Sec. 34, T. 41 N, R. 53 E of the California Mountain 7.5-minute map (1971), of the Mt. Diablo Principal Meridian System.

Spherical Coordinates:
    Longitude W 115 59'47", Latitude N 41 24'44"

ANTLER FORELAND STRATIGRAPHY OF MONTANA AND IDAHO: THE STRATIGRAPHIC RECORD OF EUSTATIC FLUCTUATIONS AND EPISODIC TECTONIC EVENTS

S.L. Dorobek
Department of Geology
Texas A&M University
College Station, TX 77843

S.K. Reid
Department of Geology
Texas A&M University
College Station, TX 77843

M. Elrick
Department of Geology
University of New Mexico
Albuquerque, NM 87131

ABSTRACT

Devonian and Mississippian sedimentary rocks of western Montana and east-central Idaho were deposited on a cratonic platform that faced a deep basin to the west. The deep basin in Idaho probably was a northern extension of the Antler foredeep and formed as a flexural response to loading of the ancient North American continental margin by an inferred arc and thrust belt complex. Evidence for the arc and thrust belt complex is especially ambiguous in Idaho.

In order to document the timing of possible Antler convergence events, quantitative subsidence analyses were done on Devonian-Mississippian strata in Montana and Idaho. The subsidence analyses indicate that episodic subsidence events in the proximal foredeep also affected the adjacent cratonic platform, an area approximately 800 km wide (palinspastic). Isopach maps for this sequence illustrate that many depocenters and paleohighs were geographically coincident across the foreland through time. Some of these structures were tectonically inverted (i.e., paleohighs became depocenters and vice versa) during the 50-60 million years represented by this stratigraphic sequence. Many of these paleostructures were oriented at high angles to the north-south trending axis of the Antler foredeep and the inferred strike of the Antler orogenic belt. These foreland structures coincide geographically with structural trends produced during Proterozoic extension, suggesting that the Proterozoic faults were reactivated during Antler convergence.

The complex patterns of subsidence across the Montana-Idaho foreland can not fit into simple flexural models for vertical loading of unbroken elastic plates. Instead, differential subsidence of the foreland may be related to several mechanisms: 1) flexure of mechanically independent, fault-bounded segments of the foreland produced by areally limited thrust loads (subregional vertical loading); 2) transmission of compressive in-plane stresses through the foreland lithosphere (regional horizontal loading), causing reactivation of Proterozoic faults depending on the faults' orientations with respect to a varying regional stress field; and 3) waxing and waning of in-plane compressive stresses due to the episodic nature of Antler convergence.

Results from the subsidence analyses are important for understanding the development of stratigraphic sequences across the Antler foreland of Montana and Idaho. This study suggests that in settings where the foreland lithosphere is broken by ancient fault systems, the foreland may exhibit complex patterns of differential subsidence that probably reflect a composite response to both vertical and horizontal loads. Also, the simultaneous pulses of subsidence documented across large parts of the Antler foreland suggest that it may be possible to date episodes of convergence along ancient continental margins, even in cases where the ancient thrust belt complex is poorly preserved.

INTRODUCTION

The Antler orogeny and its associated foreland basin stratigraphic sequence are poorly understood outside of the "classic" Antler terranes in Nevada, despite the apparent, widespread occurrence of Late Devonian-Mississippian convergence along the western margin of North America (Roberts and Thomasson, 1964; Burchfiel and Davis, 1972; Poole, 1974; Dickinson, 1977; Speed, 1977; Dover, 1980; Schweickert and Snyder, 1981; Speed and Sleep, 1982; Miller et al., 1984; Gordey, 1988; Morrow and Geldsetzer, 1988; Oldow et al., 1989).

This paper discusses the complex patterns of differential subsidence that occurred across the Antler foreland in Montana and Idaho during Late Devonian to late Mississippian time. Quantitative subsidence analyses document the timing of episodic subsidence in the Idaho-Montana foreland. Isopach maps of various stratigraphic intervals illustrate the orientation and scale of the differential subsidence across the foreland and the tectonic inversion of paleostructural elements across the Antler foreland in Montana and Idaho. The combined results of these analyses suggest that the response of the Antler foreland was not entirely a flexural response to vertical loading, but that additional mechanism(s) were necessary to generate the observed subsidence patterns. Understanding the types of tectonic loads which produced the differential subsidence in Idaho and Montana has important implications for deciphering the development of Devonian through Mississippian stratigraphic sequences in the area.

In Cooper, J.D., and Stevens, C.H., eds., 1991, **Paleozoic Paleogeography of the Western United States-II:** Pacific Section SEPM, Vol. 67, p. 487-507.

This study illustrates how Antler tectonic events affected development of foreland stratigraphy in Idaho and Montana during the Devonian to Mississippian (also see Dorobek, this volume, and Reid and Dorobek, this volume). Differential subsidence of the distal Montana foreland was affected by poorly understood convergence events which occurred many hundreds of kilometers outboard of the craton. Finally, this study supports previous studies that suggest convergence events may be dated indirectly from subsidence histories and sedimentologic analysis of *distal* foreland stratigraphic sequences.

OVERVIEW OF FORELAND BASIN DEVELOPMENT

Foreland basins (or foredeeps) largely form as a flexural response to vertical loading of continental lithosphere by thrust sheets (Beaumont, 1981; Jordan, 1981; Turcotte and Schubert, 1982). Foreland basins typically have asymmetric cross-sectional profiles (Fig. 1); the deepest part of a foreland basin is located immediately adjacent to the thrust belt complex (or accretionary wedge) which borders one side of the basin. The basin floor progressively shallows away from the thrust load (Fig. 1). On the cratonward side of the foreland basin, a relatively low relief, uplifted area called the *forebulge* or *peripheral bulge* is formed due to isostatic compensation for the end loading of the foreland lithosphere (Fig. 1). The geometry of the foreland basin is dependent on the magnitude of applied loads, both vertical and horizontal, and on the *flexural rigidity* (or in simplistic terms, the stiffness) of the foreland lithosphere. Typically, only vertical loads (i.e., thrust loads and redistribution of thrust loads by erosion and sedimentation in the foreland basin) are considered in attempts to model foreland basin geometries. The geometry of a foreland basin (Fig. 1) can be described by: 1) the maximum *amplitude* of the deflection of the foreland lithosphere, which is dependent on the magnitude of the applied load and on the flexural rigidity of the foreland lithosphere; 2) the *wavelength* or width of the deflection, which is dependent on the flexural rigidity of the foreland lithosphere; and 3) the location, width, and height of the forebulge.

The evolution of the foreland basin is directly coupled to its adjacent fold-and-thrust belt. Lateral, cratonward migration of the fold-and-thrust belt accounts for the progressive cratonward migration and disruption observed in many foreland basins. Numerical models have been developed which relate the scale of the thrust load and rheologic properties of the loaded lithosphere to the subsidence history and development of large-scale structures in the foreland area (Quinlan and Beaumont, 1984; Schedl and Wiltschko, 1984; Stockmal et al., 1986; Beaumont et al., 1988). These theoretical models of flexural behavior generally assume that the foreland lithosphere is not broken by lithosphere-scale faults that might prevent flexure from being transmitted uniformly across the

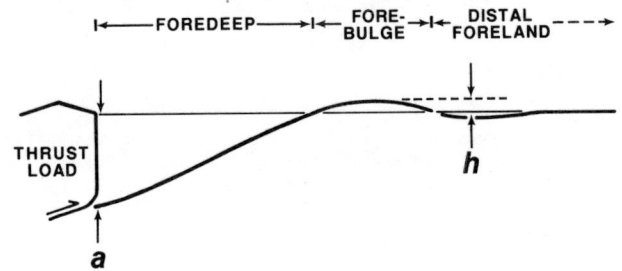

Figure 1. Schematic cross-section through hypothetical foredeep basin. *a* = amplitude or maximum deflection of foreland lithosphere produced by the adjacent thrust load; *h* = height of forebulge. Note maximum deflection occurs immediately adjacent to the thrust load and progressively shallows away from the load. Also note low amplitude secondary basin (*sensu* Flemings and Jordan, 1989) on foreland side of forebulge. No scale implied.

foreland (cf. Stockmal and Beaumont, 1987; Royden et al., 1987). In the Apennine foreland system of Italy, the foreland basin area is distinctly compartmentalized into sub-basins because the foreland lithosphere apparently is segmented by tear faults that trend at high angles to the thrust load (Royden et al., 1987). Each foreland segment is mechanically independent of adjacent segments, producing offset foreland depocenters with separate forebulge areas.

Quantitative models of foreland basin evolution are dependent on assumptions of the foreland lithosphere's rheology. Rheologies used in various approaches include uniform elastic, uniform viscoelastic (Maxwell rheology), and temperature-dependent viscosity models. In uniform elastic lithosphere models, the foreland lithosphere responds instantaneously to loading. The form and position of the deflection produced by thrust-loading does not change with time as long as the scale and position of the thrust load do not change. In uniform viscoelastic models, the lithosphere progressively softens following loading. The form of the deflection changes with time because of viscous flow in the lithosphere. Theoretically, if the foreland lithosphere behaves viscoelastically and with unchanging loads, the foredeep should deepen and the forebulge should increase in height and migrate towards the thrust load over time. Previous estimates for time constants of the lithospheric relaxation are on order of 10^6–10^7 yr (Beaumont, 1981). In temperature-dependent viscoelastic models, flow first occurs in the lower lithosphere which is hotter and less viscous than upper parts of the lithosphere. Relaxation propagates upwards into cooler, more viscous lithosphere (Quinlan and Beaumont, 1984; Beaumont et al., 1987). The time required for relaxation also increases progressively upwards because of the corresponding increase in temperature-dependent viscosity. The

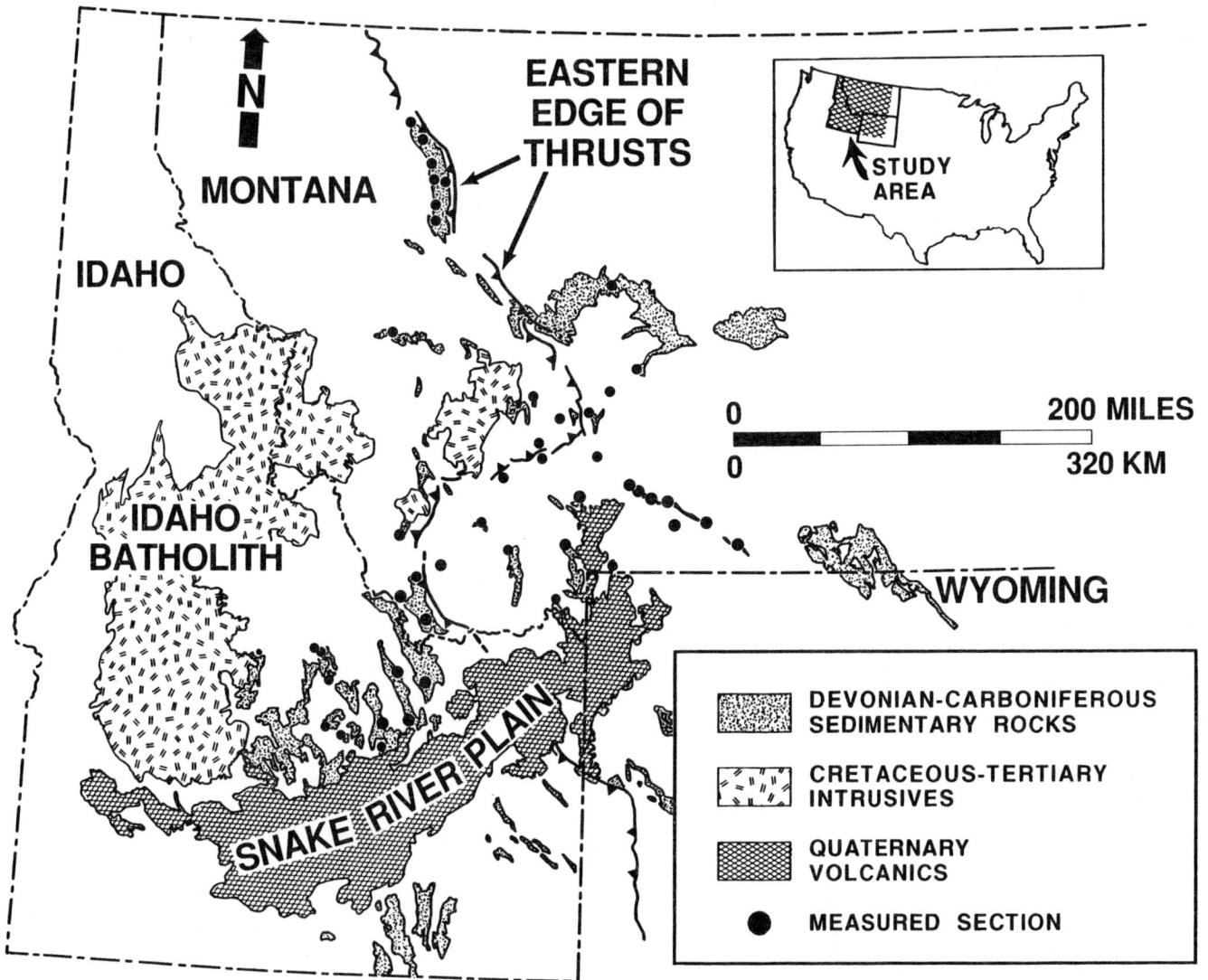

Figure 2. Map of study area. Inset map of United States in upper right corner shows location of study area.

amount of time necessary to achieve local isostatic equilibrium in the uppermost lithosphere by viscous flow is greater than the age of the Earth. Thus the temperature-dependent viscosity model can account for the true flexural response of the lithosphere as well as the component of apparent viscous relaxation after loading.

In addition to the flexural response of the foreland area, convergence along a continental margin produces horizontal, in-plane compressive stresses that can be transmitted into the interior of a foreland plate. These horizontal loads may affect foreland lithosphere many hundreds of kilometers inboard of the actual thrust load and proximal foredeep (Lambeck et al., 1984; Cloetingh, 1988). As a result, horizontal in-plane stresses might enhance deflections of the lithosphere in distal foreland or cratonic areas and therefore also affect

sedimentation far from the zone of active plate-margin convergence. However, horizontal loads generally are ignored in quantitative or conceptual models that attempt to explain the geometry of foreland basins or the development of foreland stratigraphy.

STRUCTURAL AND STRATIGRAPHIC SETTING

Regional Structural History

Devonian through Mississippian sedimentary rocks are exposed throughout central Montana and east-central Idaho (Fig. 2). In east-central Idaho and southwestern Montana, most exposures of Devonian and Mississippian rocks are allochthonous, with local windows of possible parautochthonous rocks. Eastward transport of these rocks occurred along complex thrust and tear fault systems during Pennsylvanian (?) to Early Tertiary time (Sevier and Laramide orogenies;

Skipp and Hall, 1975; Skipp et al., 1979; Ruppel and Lopez, 1984; Perry et al., 1989); structural evidence for any pre-Pennsylvanian shortening is equivocal (Dover, 1980). East of the leading edge of this fold-and-thrust belt, the amount of shortening is probably not significant enough to seriously affect palinspastic reconstructions. However, estimates for the total amount of tectonic shortening in the fold-and-thrust belt are highly variable and range from tens to several hundreds of kilometers (Skipp and Hait, 1977; Nilsen, 1977; Skipp et al., 1979; Dover, 1980; Ruppel et al., 1981; Schmidt and Hendrix, 1981; Woodward, 1981; Skipp, 1988).

Late Tertiary extension produced the present north- or northwest-trending, block-faulted mountain ranges across much of the study area (Pardee, 1950; Reynolds, 1979; DuBois, 1983). The amount of extension progressively decreases eastward to the eastern edge of the study area. Movement along some of these high-angle faults still occurs, especially in east-central Idaho.

Regional Devonian and Mississippian Stratigraphic Relationships

Details about regional stratigraphic relationships in Devonian-Mississippian strata of southwestern Montana and east-central Idaho are discussed in Dorobek and Smith (1989), Reid (1991), Dorobek (this volume), Reid and Dorobek (1989; this volume), and Dorobek et al. (in press). However, the chronostratigraphic correlation chart shown in Figure 3 illustrates the important stratigraphic correlations from the Montana platform into eastern parts of the Antler foredeep in east-central Idaho. Figure 3 also shows the variable durations and complex areal extent of regional unconformities across the study area. The significance of these complex chronostratigraphic relationships across the study area are discussed below.

DEVONIAN-MISSISSIPPIAN PALEOGEOGRAPHY AND TECTONISM

Late Proterozoic through Mississippian Tectonic History

Prior to Middle Devonian time, a passive continental margin existed across Montana and east-central Idaho. This passive margin was initiated during late Proterozoic-early Cambrian rifting (600-550 Ma; Stewart and Suczek, 1977; Armin and Mayer, 1983; Bond et al., 1983; Bond and Kominz, 1984) and persisted until Early to Middle(?) Devonian time. However, by Late Devonian-Early Mississippian time, some type of convergence occurred along the western margin of North America. This collision produced a foredeep basin that was superimposed on the underlying passive margin sequence. In general, the eastern side of this foredeep was located near the hinge zone of the antecedent lower Paleozoic passive margin; the western and deeper part of the foredeep formed above outer shelf to slope facies of the lower Paleozoic passive margin. This interval of

Late Devonian-Early Mississippian convergence is known regionally as the Antler orogeny (Roberts and Thomasson, 1964; Burchfiel and Davis, 1972; Poole, 1974; Dickinson, 1977; Speed, 1977; Dover, 1980; Speed and Sleep, 1982).

A number of tectonic models have been proposed for the Antler orogeny (Fig. 4), including arc-continent collision (Schweickert and Snyder, 1981; Speed and Sleep, 1982), back-arc thrusting adjacent to a marginal basin (Burchfiel and Davis, 1972; Miller et al., 1984) and "incipient" subduction (Johnson and Pendergast, 1981). However, most workers agree that the Antler thrust belt overrode westward-dipping continental lithosphere and that Antler foreland deposits accumulated above an earlier, mature passive margin sequence (Speed and Sleep, 1982). Similar convergence events affected much of the western margin of North America during Late Devonian-Early Mississippian time (Gordey, 1988; Morrow and Geldsetzer, 1988; Oldow et al., 1989), although unequivocal evidence for arc-continent collision is poorly preserved along much of the North American Cordillera.

Evidence for the Antler Orogeny in East-central Idaho

Geologic evidence for the Antler orogeny is especially cryptic in east-central Idaho (Nilsen, 1977; Dover, 1980), where the main evidence for an Antler orogenic highland consists of synorogenic siliciclastic sediments (Devonian Picabo and Milligen Formations and Mississippian McGowan Creek Formation, White Knob Group, and Copper Basin Formation) in the proximal foredeep (Sandberg, 1975; Sandberg et al., 1975; Skipp and Sandberg, 1975; Nilsen, 1977; Dover, 1980). Paleocurrent data, regional clast size distribution trends, and lithofacies patterns suggest a western source for most of these units (Paull et al., 1972; Nilsen, 1977; Dover, 1980; unpubl. data). In those siliciclastic units where a western source has not been documented unequivocally, a western source can be inferred based on the absence of time-equivalent siliciclastic facies in eastern parts of the Antler foredeep. Evidence for Devonian-Mississippian deformation in Idaho is equivocal (Nilsen, 1977; Dover, 1980).

Foundering of the antecedent lower Paleozoic passive margin and initiation of Antler foredeep subsidence in Montana and Idaho probably began as early as Early to Middle Devonian time, based on the presence of deep water "flysch" facies of the Lower to Upper Devonian Milligen Formation in east-central Idaho (Sandberg et al., 1975). The Milligen Formation is a 1200+ meter thick sequence of argillite and sandstone that is largely correlative with carbonate facies of the Carey Dolomite and Jefferson Formation to the east (Sandberg et al., 1975; Johnson et al., 1985). An eastern source area has been suggested for sediments in the Milligen Formation (Sandberg et al., 1975), but a western source may be more likely given that

Figure 3. A. Location map for selected
localities used to construct chronostratigraphic chart shown in Figure 3B.
M = Monarch, MT; S = Sacajawea Peak, MT;
L = Logan, MT; A = Ashbough Canyon, MT;
T = Tendoy Mountains; MC = composite
section from McGowan Creek and Grandview
Canyon, ID. These selected localities
occur along a platform-to-basin transect. Measured section locations on this
base map also were chosen for subsidence
analyses; subsidence curves for each
locality are shown in Figure 8.

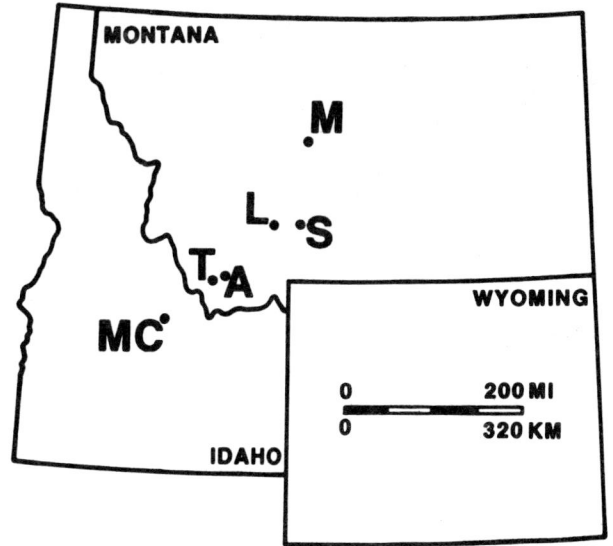

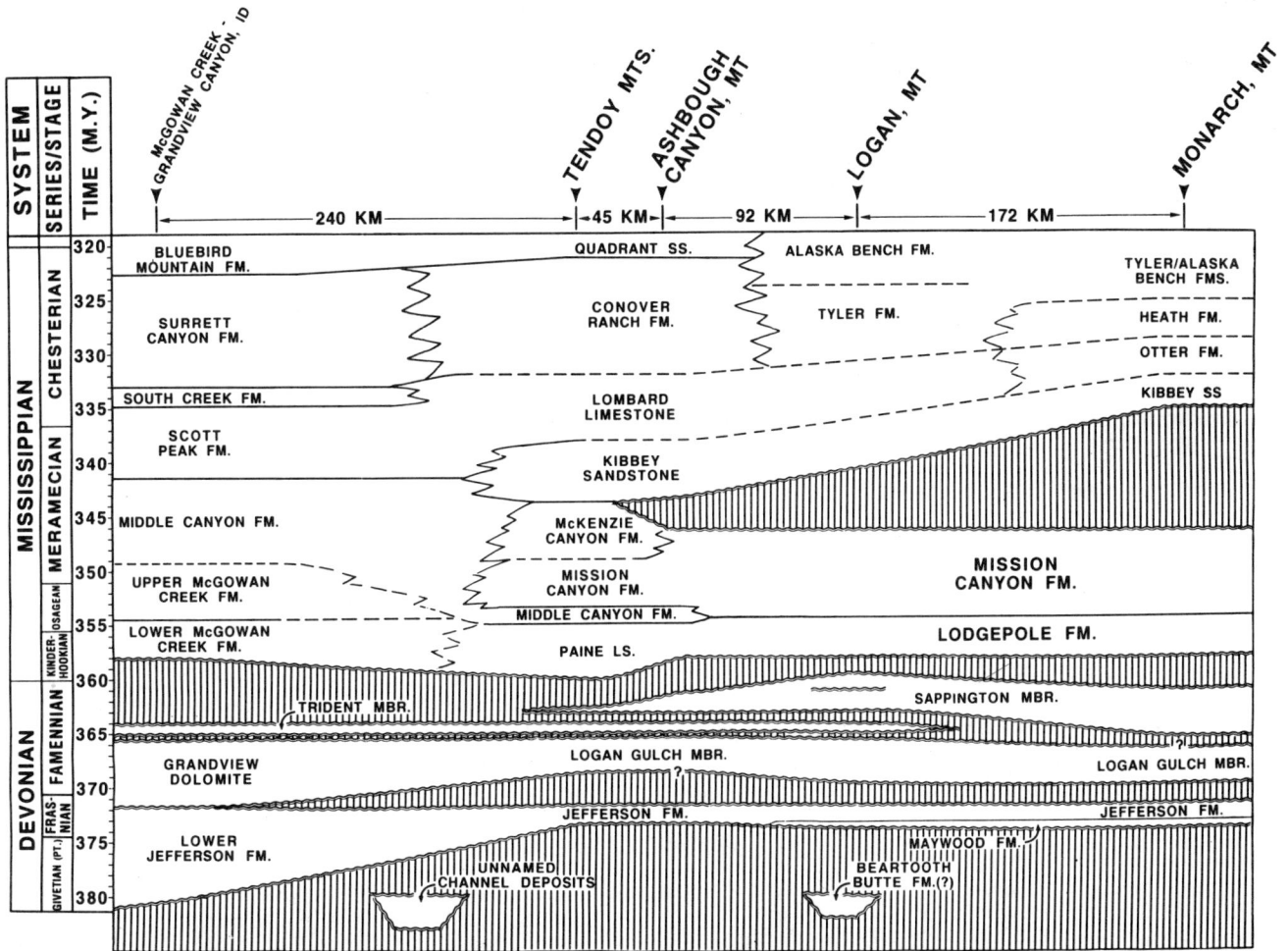

Figure 3. B. Chronostratigraphic chart for Upper Devonian to Upper Mississippian strata across
the study area. Distances between localities used to construct the chronostratigraphic
chart are palinspastic (palinspastic base from Peterson, 1986); distances on the base map
shown in Figure 3A are nonpalinspastic. Unconformities indicated by vertically ruled
areas. Note highly variable durations of hiatuses for many unconformities at a regional
scale, especially for many of the Late Devonian unconformities. Lateral contacts between
formations commonly are dashed because stratigraphic interfingering between many units has
not been observed in this structurally complex area.

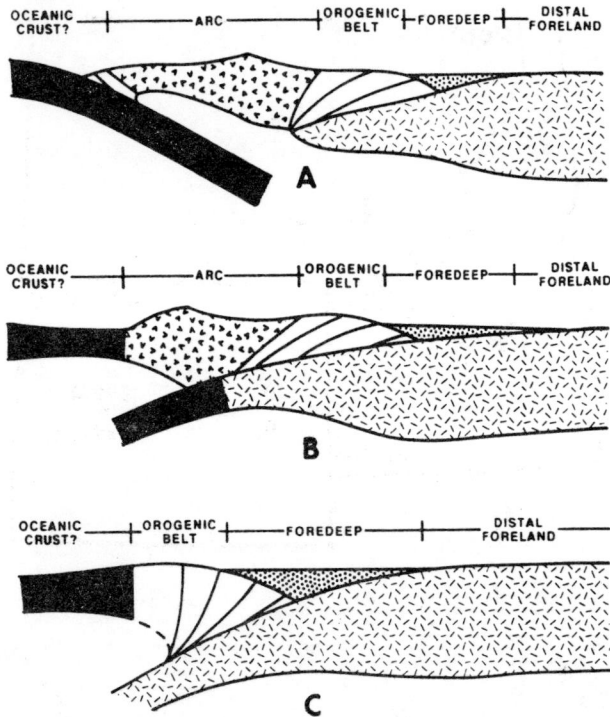

Figure 4. Some suggested tectonic models for the Antler orogeny. Diagrams not to scale. *A*. Back-arc thrusting (Burchfiel and Davis, 1972). *B*. Arc-continent collision (Speed and Sleep, 1982). *C*. "Incipient" subduction without a magmatic arc (Johnson and Pendergast, 1981).

equivalent parts of the Carey Dolomite and Jefferson Formation to the east of the Milligen Formation are nearly all dolomite and contain very little shale. These stratigraphic relationships and the results from our subsidence analyses (discussed below) suggest that subsidence in the foredeep began earlier, in the Early to Middle Devonian, than on the cratonic platform, and may reflect the initial response of the foredeep to the encroaching Antler accretionary wedge.

By late Frasnian-early Famennian time, a subregional unconformity developed between the top of the Milligen Formation (earliest Frasnian) and an overlying unnamed limestone unit of Famennian age in the Wood River area of east-central Idaho (Sandberg et al., 1975; Johnson et al., 1985). Basinal facies of the Milligen Formation must have been uplifted to produce this unconformity. In addition, sandstone and restricted peritidal facies of the "Grandview Dolomite" (uppermost part of the Jefferson Formation in the Lost River Range of east-central Idaho) were deposited to the east during the time represented by this unconformity. "Grandview Dolomite" facies apparently prograded from west to east (P.E. Isaacson, pers. comm.); sandstone lithofacies in the "Grandview Dolomite" thicken westward and sandstone paleocurrent

data indicate west-to-east transport directions (Dorobek, unpubl. data). These Late Devonian events precede deposition of conglomerate and sandstone of the upper Famennian Picabo Formation. Picabo siliciclastics also probably were derived from local uplifted areas associated with the Antler orogeny (Skipp and Sandberg, 1975; Isaacson et al., 1983).

These stratigraphic relationships suggest that an uplifted sediment source area existed to the west of the cratonic platform possibly as early as the Early Devonian, but clearly by Late Devonian time (Isaacson et al., 1983). A western source terrane also is well documented for the various Mississippian siliciclastic units (McGowan Creek Formation, White Knob Group, and Copper Basin Formation) that filled the Antler foredeep (Sandberg, 1975; Nilsen, 1977; Dover, 1980; Sandberg et al., 1983).

Preliminary provenance analyses of various Devonian through Mississippian siliciclastic units in the "proximal" Antler foredeep indicate that the sand- to pebble-size fraction in these siliciclastic units consists largely of quartz, chert, and various sedimentary rock fragments. Rare quartz grains have earlier generations of quartz overgrowths that were abraded during erosion and reworking of older quartz-cemented sandstones. Rare metamorphic(?) rock fragments and mica flakes occur in some basinal siliciclastic deposits. These petrographic data suggest that Antler deformation in east-central Idaho probably involved parts of the lower Paleozoic passive margin sequence. The deformed lower Paleozoic strata may have been the dominant source for Antler synorogenic sediments in the Idaho foredeep.

Devonian-Mississippian Differential Subsidence and Tectonic Inversion Across the Montana-Idaho Foreland

During Late Devonian through Late Mississippian time, a number of low-amplitude paleohighs and troughs extended across the Antler foreland in Montana and eastern parts of Idaho (Fig. 5). Many of these Devonian-Mississippian foreland structures had orientations that were very similar to older, Precambrian structural trends (Peterson, 1986; Tonnsen, 1986). Some of these foreland structures were tectonically active during Devonian-Mississippian time as far east as the Williston Basin (McCabe, 1954; Clement, 1986; Gerhard et al., 1987; LeFever et al., 1987). Many of the foreland structures underwent several episodes of "tectonic inversion" (*sensu* Visser, 1980; Ziegler, 1987a, 1987b) where subsidence in trough areas was interrupted by periods of uplift, or conversely, paleohighs were inverted and became depocenters.

The Precambrian structural grain which was reactivated during Antler time probably formed during a Middle Proterozoic extensional event that produced several depocenters across Montana and Idaho (Fig. 5A; Peterson, 1986; Tonnsen, 1986). These

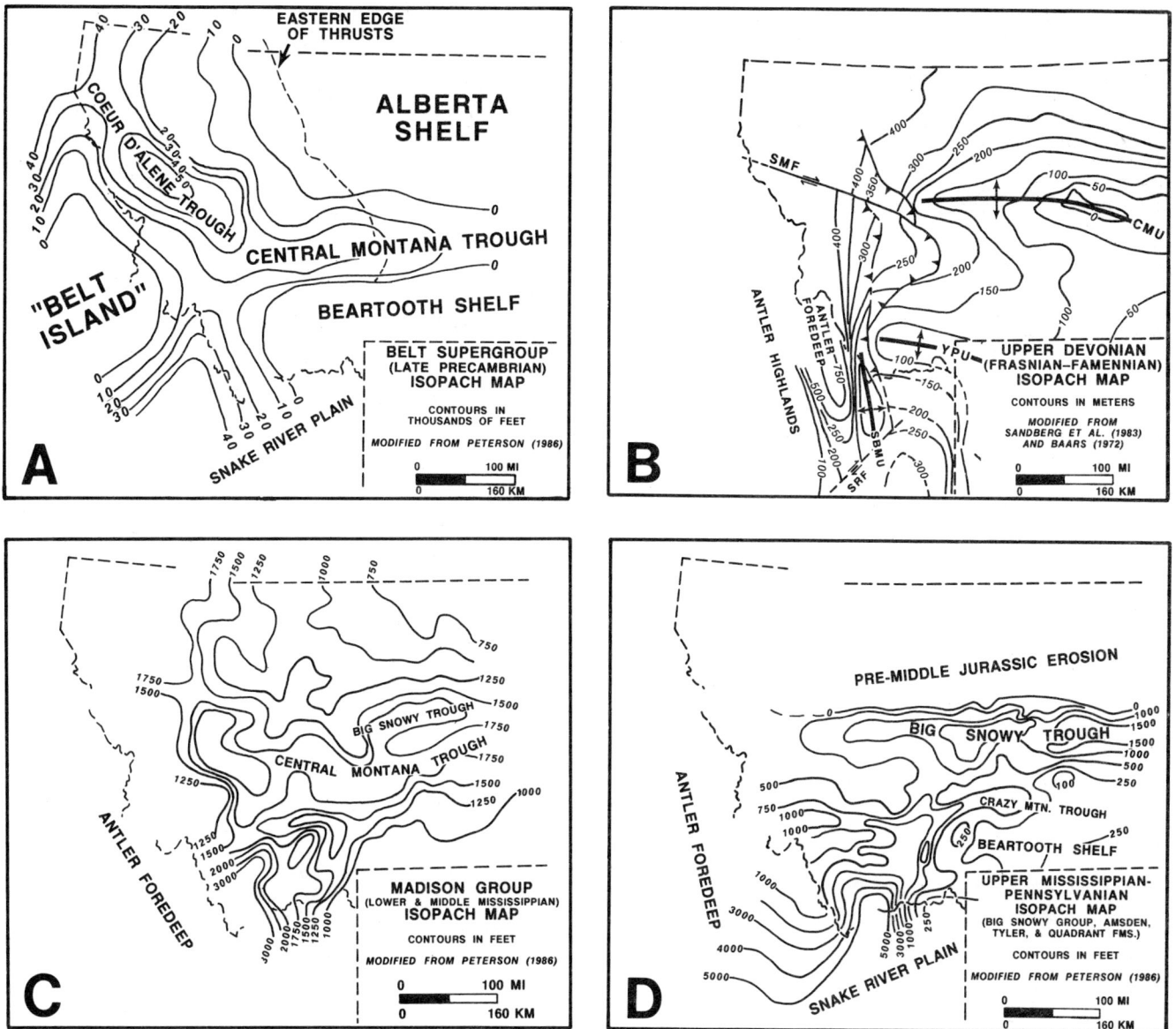

Figure 5. Isopach maps for specific time intervals; partly restored base. Figures 5A, 5C, 5D modified from Peterson (1986); Figure 5B modified from Sandberg et al. (1983). Major paleotopographic features are also indicated. Isopach thicknesses are reasonably accurate because Antler deformation never advanced far enough eastward during the Devonian-Mississippian to significantly deform and/or erode much of the sedimentary section that has been contoured in these isopach maps.

A. Isopach map for Proterozoic Belt Supergroup. Several depocenters comprise the "Belt Basin"; note especially the east-west trending Central Montana Trough.

B. Isopach map for Frasnian and lower Famennian deposits. SRF = Snake River Fault Zone; SMF = St. Marys Fault Zone; CMU = Central Montana Uplift; YPU = Yellowstone Park Uplift; SBMU = Southern Beaverhead Mountains Uplift. Note Late Devonian Central Montana Uplift geographically overlaps the Proterozoic Central Montana Trough depocenter shown in Figure 5A. Also note location of Late Devonian foredeep depocenter. Thrust traces (with sawteeth) delineate major Laramide thrusts; these thrusts are shown here in order to be consistent with the original isopach map of Sandberg et al. (1983).

C. Isopach map for Lower and Middle Mississippian Madison Group. Note Central Montana Uplift, which was present across Montana in Late Devonian time (Fig. 5B), underwent tectonic inversion and became the Central Montana Trough in Early to Middle Mississippian time.

D. Isopach map for Upper Mississippian and Pennsylvanian rocks. Note that foreland area in Montana still was dissected by numerous paleostructures that were oriented at high angles to the Antler foredeep axis. Also note that Late Mississippian-Early Pennsylvanian foredeep depocenter was located much further to the southeast than the Late Devonian foredeep depocenter.

depocenters comprise the Middle to Late Proterozoic Belt basin. As in other extensional basins, many of the depocenters in the Belt basin probably were bounded by normal faults that formed during Middle Proterozoic extension. Prior to Antler time, these inferred Precambrian faults were reactivated again during the 600-550 Ma rifting event that initiated the lower Paleozoic passive margin. Reactivation of the inferred Proterozoic faults is suggested by the geographic coincidence of Proterozoic and early Paleozoic depocenters across central Montana.

The Central Montana Trough was a prominent depocenter from the Middle Proterozoic until the beginning of the Ordovician. However, during the Middle Devonian, much of the former Central Montana Trough was inverted and became the Central Montana Uplift. By Late Devonian time, the Central Montana Uplift was onlapped by Upper Devonian units (Dorobek and Smith, 1989; Dorobek, this volume). Various parts of the Idaho-Montana foreland continued to experience differential subsidence and/or tectonic inversion during Mississippian time (Craig, 1972; Mallory, 1972; Smith, 1977; Sando et al., 1975; Peterson, 1986; Sando, 1988), producing complex linear patterns of troughs and depocenters with intervening low-relief paleohighs.

Cross-sections of stratigraphic thicknesses for several 10-20 million year intervals (Figs. 6, 7) show that the paleohighs and troughs across the Montana-Idaho foreland had (sediment-filled) cross-sectional amplitudes of about 50-350 m and wavelengths of 50-200 km. These generally east-west trending paleostructures also were oriented at high angles to the north-south trending axis of the Antler foredeep and the inferred strike of the Antler orogenic belt. Antler foredeep depocenters also appear to have migrated progressively southeastward from Devonian to Early Pennsylvanian time (Fig. 5), suggesting that the maximum thrust load also moved progressively southeastward (cf. Jordan, 1981; Quinlan and Beaumont, 1984; Stockmal et al., 1986).

SUBSIDENCE ANALYSES ACROSS THE ANTLER FORELAND

The subsidence history of the Antler foreland in Idaho and adjacent cratonic platform in Montana was highly variable, even over short distances. Quantitative subsidence analyses (or "backstripping") were conducted in order to understand the timing and relative rates of subsidence across the study area. This approach allows indirect correlation of thrust movement with coeval events in the adjacent foreland (cf. Jordan et al., 1988).

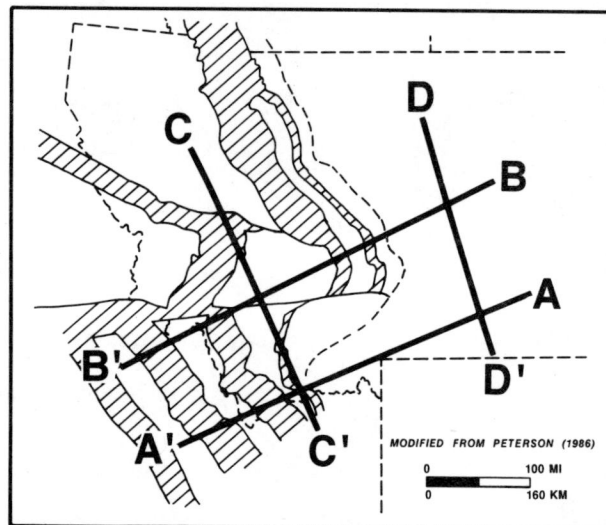

Figure 6. Palinspastic base map with locations of thickness cross-sections shown in Figure 7. Hachured regions show amount of palinspastic restoration in overthrust belt. Cross-sections A-A' and B-B' are dip-sections; cross-sections C-C' and D-D' trend approximately parallel to regional depositional strike. Palinspastic base from Peterson (1986).

Methods for subsidence analysis used in this study are modified from backstripping procedures developed by Sleep (1971) and Steckler and Watts (1978). Measured stratigraphic sections are used to construct cumulative stratigraphic thickness curves from which the effects of sediment loading and lithification (compaction and cementation) are removed. The resulting curves reflect subsidence due to tectonic forces and eustatic sea level changes. Input for the analyses includes measured stratigraphic thicknesses and percentages of lithologies, percentages of early cements (from thin section study), estimates of maximum and minimum paleobathymetries (based on facies analyses), and absolute ages of stratigraphic units within measured sections. Data used to construct subsidence curves were compiled by us and submitted to M.A. Kominz (University of Texas, Austin) for analysis. Details of subsidence analysis methods used in this study are discussed by Bond and Kominz (1984) and Bond et al. (1988, 1989).

Input Parameters

Stratigraphic Data

Subsidence curves were constructed for six stratigraphic sections which form a NE-SW

Figure 7. Cross-sections that illustrate compacted stratigraphic thicknesses (shown as shaded areas) across the study area for specific time intervals. 375-360 Ma = Late Devonian to Devonian-Mississippian boundary; 360-340 Ma = Devonian-Mississippian boundary to Middle Mississippian; 340-310 Ma = Middle Mississippian to Early Pennsylvanian. Note tectonic inversion across the foreland area, which is especially apparent on cross-sections C-C' and D-D'.

495

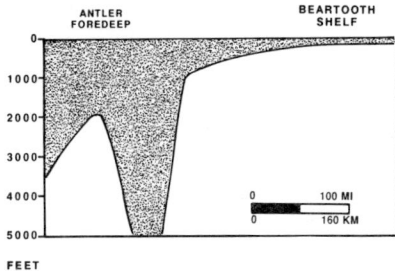 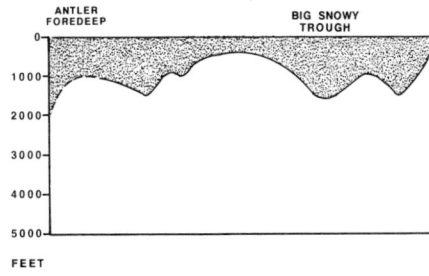

~340 - ~310 Ma

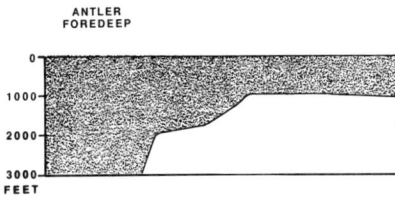 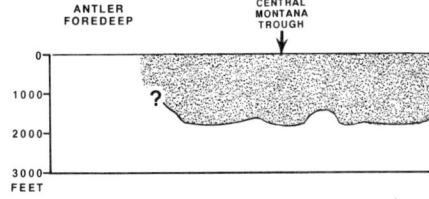

360 - ~340 Ma

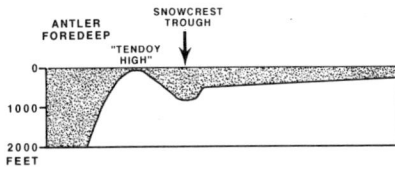 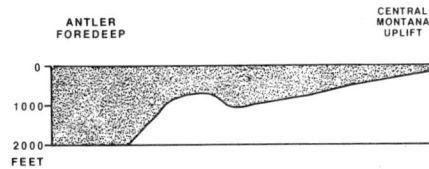

375 - 360 Ma

A'⟵⟶A B'⟵⟶B

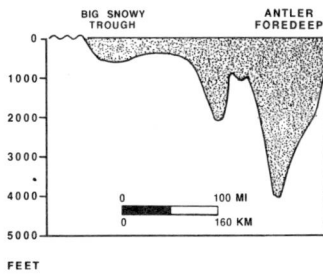 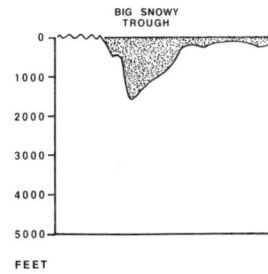

~340 - ~310 Ma

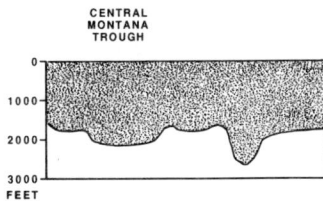 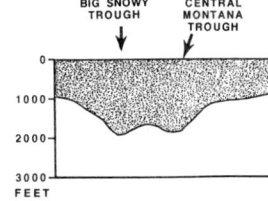

360 - ~340 Ma

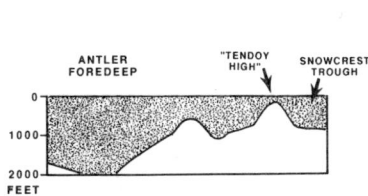 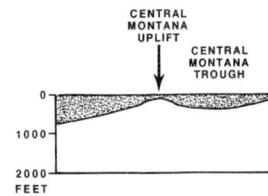

375 - 360 Ma

C⟵⟶C' D⟵⟶D'

transect across the Lower Mississippian platform-to-basin transition (Fig. 3A). This transect is approximately perpendicular to the inferred strike of the Antler foredeep and extends over a palinspastic distance of ~500 km (Peterson, 1986). Each stratigraphic section consists of Middle or Upper Devonian to uppermost Mississippian strata and records ~50 to 60 million years of Antler foredeep/distal foreland evolution (Fig. 8). Stratigraphic data used in the subsidence analyses come from detailed measured sections of Devonian through lower Mississippian strata which were completed by the authors and from published measured sections of middle to upper Mississippian strata (Walton, 1946; Mamet et al., 1971; Sando and Dutro, 1974; Sandberg, 1975; Sandberg and Poole, 1977; Sando et al., 1985; Wardlaw and Pecora, 1985). Biostratigraphic ages for these strata are from Huh (1967, 1968), Sandberg et al. (1967), Sando et al. (1969, 1976, 1985), Mamet et al. (1971), Sando and Dutro (1974), Sandberg (1975), Sandberg and Poole (1977), Gutschick et al. (1980), Hildreth (1981), Davis and Webster (1984), Johnson et al. (1985), Wardlaw and Pecora (1985), and Reid (1991).

Time Scale

Absolute ages of strata included in subsidence analyses are from dates assigned to Devonian and Mississippian biozone boundaries by Sandberg and Poole (1977), Sandberg et al. (1983), Johnson et al. (1985), Sando (1985), and Sandberg et al. (1988). Dates reported by Sando (1985) are based on the time scale of Harland et al. (1982). Sandberg and Poole (1977), Sandberg et al. (1983), and Sandberg et al. (1988) use a duration of ~13-15 million years for the Late Devonian in order to calculate conodont biozone durations. Harland et al. (1982) suggest that the Late Devonian lasted ~14 million years. Therefore, the ages of Devonian through Mississippian strata shown in Figure 8 are consistent with the same absolute time scale. Ages do not include the ranges of uncertainty given by Harland et al. (1982). Using a different time scale would slightly change the slope of segments on the subsidence curves and produce small shifts in the position of inflections; however, the *form* of the subsidence curves is not significantly affected by the choice of time scale (cf. Bond et al., 1988).

Delithification

The delithification procedure used in this study is the "maximum difference" method of Bond et al. (1989). This method generates a single subsidence curve which maximizes the differences in delithification factors for fine- and coarse-grained, calcareous and non-calcareous lithologies. Lithification of shales is assumed to be entirely due to mechanical compaction. In carbonate strata and in siliciclastic siltstones and sandstones, mechanical compaction is suppressed by an amount proportional to the percentage of early cement. Percentage of early (syndepositional to pre-

TABLE 1. ESTIMATED PALEOBATHYMETRIC RANGES FOR LITHOFACIES IN UPPER DEVONIAN THROUGH UPPERMOST MISSISSIPPIAN STRATA

Lithofacies	Water Depths (m)
Evaporitic mudstone, cryptalgalaminite, fenestral limestone	0 – 5
Skeletal-ooid grainstone, ooid grainstone *Amphipora* grainstone, massive to cross-bedded quartz sandstone	1 – 10
Bioturbated skeletal-peloid grainstone/packstone	1 – 20
Massive skeletal grainstone, stromatoporoid framestone	5 – 20
Peloid-skeletal wackestone/packstone	10 – 60
Bioturbated cherty limestone	50 – 100
Laminated cherty limestone	75 – 150
Mixed carbonate/siliciclastic rocks	100 – 250
Calcareous siltstone/sandstone	100 – 300

stylolitization) cement was estimated visually from petrographic thin sections. Total volume loss due to pressure solution and/or extensive stylolitization appears to be less than 10% in platform lithofacies, based on field and petrographic observations; pressure solution features also are more or less evenly distributed throughout all lithofacies. Therefore, major inflection points on the subsidence curves will not change with the addition of any rock volume lost to pressure solution processes. However, slopes on segments of the curves may increase slightly.

Water Depth Estimates

Subsidence curves incorporate maximum and minimum paleobathymetric estimates for each lithologic unit within the measured stratigraphic sections (cf. Bond et al., 1988; Bond et al., 1989). Ranges of water depths for each lithofacies in Devonian through Mississippian strata are listed in Table 1. Relative paleobathymetries probably are correct but overlap between these ranges reflects the uncertainty in absolute water depths for individual lithofacies. Water depth estimates are based on modern facies analogs (Ball, 1967; Purser, 1973; Hine, 1977; Halley et al., 1983), studies of similar ancient platforms and basins (Smith, 1977; Yurewicz, 1977; Wilson and Jordan, 1983), and previous paleobathymetric interpretations of Devonian through Mississippian strata in the region (Sando, 1980; Harbaugh and Dickinson, 1981; Gutschick and Sandberg, 1983).

Addition of maximum and minimum water depths to the delithified and isostatically adjusted cumulative stratigraphic thickness curve results in two subsidence curves labeled "ts+wdmax" and "ts+wdmin", respectively (Fig. 8 A-F). Short-term variation in the magnitude of subsidence ("high frequency" noise on Fig. 8 A-F) is

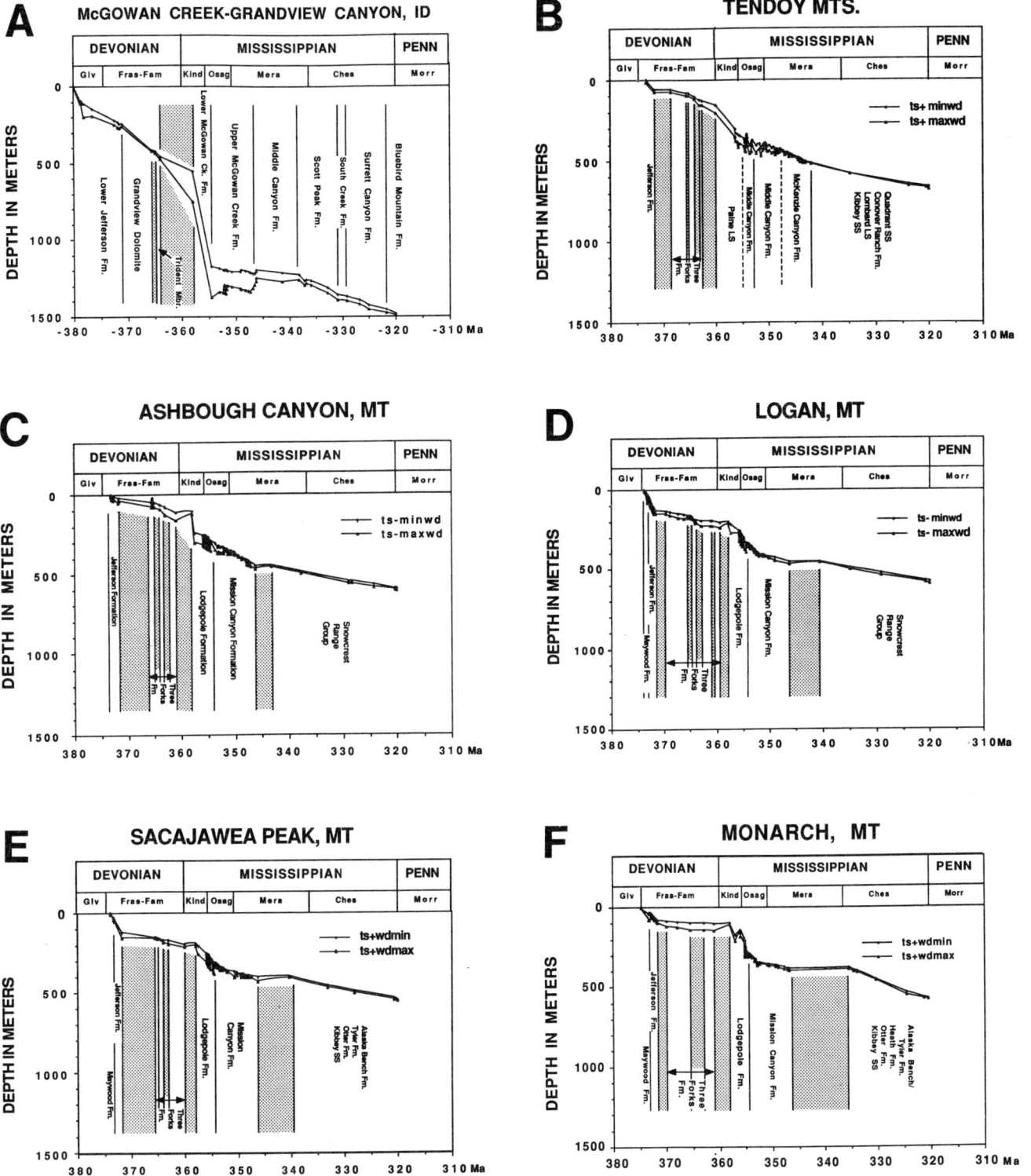

Figure 8. Subsidence curves generated for selected localities shown in Figure 3A. Unconformities indicated by stippled intervals. Two subsidence curves are shown, one generated using corrections for minimum water depth estimates for each stratigraphic unit ("ts-minwd"), and the other generated using maximum water depth estimates ("ts-maxwd"). Note that most facies were deposited in water depths less than 50 m; therefore, differences in curves due to water depth corrections are negligible. See text for discussion on significance of inflection points on subsidence curves. Monarch, Sacajawea Peak, Logan, and Ashbough Canyon are platform localities. Tendoy Mountains section is transitional area located near transition from shallow platform to deeper slope environments on eastern side of Antler foredeep. McGowan Creek-Grandview Canyon section is the most basinal locality in the Antler foredeep which has most of the Devonian to Upper Mississippian stratigraphy exposed.

largely a function of our estimated water depths. However, the uncertainties in the estimated water depths do not significantly affect the *form* of the subsidence curves (Bond et al., 1988).

Ductile Strain

The sedimentary rocks from platform locations used for the subsidence analyses do not contain extensive strain microfabrics, other than calcite twins. Mesoscale and microscale folds are not abundant at the selected platform locations. Ooids (probably the best strain gauges in these rocks) are common throughout the Devonian and Mississippian section in Montana; these grains do not exhibit any measurable ductile strain. Therefore, the measured stratigraphic thicknesses from platform locations that were used in this study are considered to be very reliable.

However, foredeep deposits in Idaho typically have been subjected to more intense deformation than any platform rocks in Montana. At the single foredeep section used for the subsidence analyses (McGowan Creek-Grandview Canyon), a large portion of the total stratigraphic thickness consists of shales and siltstones. These rocks have been subjected to several episodes of compressional deformation and have well developed cleavage. Therefore, rocks at this section have had their thicknesses changed due to ductile strain. We can not estimate the total change in stratigraphic thickness due to deformation because of poor exposure and lack of strain gauges in these very fine-grained rocks. However, the position of major inflection points should not vary on the subsidence curve shown in Figure 8A as long as the strain is distributed fairly equally through the entire section.

Removal of Loads

The subsidence analysis procedure removes sediment loads by assuming Airy compensation (Bond and Kominz, 1984; Bond et al., 1988). Methods that remove sediment loads by assuming flexural compensation require 2-dimensional (seismic profiles or balanced cross-sections) or preferably, 3-dimensional representations of regional stratigraphic thicknesses (cf. Jordan et al., 1988). Lack of balanced regional cross-sections in the Northern Rocky Mountains and complex isopach patterns for each Devonian through Mississippian stratigraphic interval prevent this approach.

Airy isostasy implies that the lithosphere possesses no strength and that all subsidence occurs directly beneath a "point load" represented by measured stratigraphic thicknesses at individual localities. By assuming local Airy compensation for measured stratigraphic thicknesses at individual localities, subsidence due to sediment loading in the foredeep is overestimated during backstripping because thrust loads (vertical loads) are actually partially supported by adjacent lithosphere. Too much subsidence due to sediment loading is removed during backstripping because this support is ignored. The amount of subsidence due to sediment loading is overestimated across the entire foredeep. Therefore, the assumption of Airy isostasy results in subsidence curves that portray slightly inaccurate magnitudes and rates of subsidence. However, the assumption of Airy compensation does not produce artificial inflections nor change the absolute ages (horizontal position) of inflection points on subsidence curves. The same inflections exist on measured and delithified cumulative stratigraphic thickness curves and on isostatically adjusted curves. Only the slopes of curve segments are affected by removal of loads (Fig. 8 A-F).

Eustatic Component in the Subsidence Curves

It is important to note that there is an eustatic component in our subsidence curves from Montana and Idaho. The simultaneous nature of inflection points on the subsidence curves may be due partially to eustatic sea level fluctuations. Previous studies have suggested time-equivalent eustatic rises that *approximately* coincide with pulses of increased subsidence on the curves from Montana and Idaho (Hallam, 1984; Johnson et al., 1985; Ross and Ross, 1987; Bond et al., 1989). However, Bond and Kominz (in press) have estimated the *total* amount of sea level rise for Late Devonian (base of Frasnian) to middle Mississippian (lower Chesterian) time was on the order of 180 m; this estimate is derived from subsidence analyses of mid-continent strata in Iowa. The Iowa section was chosen to identify the magnitude of Frasnian to Chesterian sea level change because it was deposited in an area that closely approximates a stable cratonic interior.

The 180 m eustatic rise during this time interval is distributed as follows: 1) Frasnian to top of Famennian - ~100 m; 2) base of Kinderhookian to top of Osagean - ~55 m; and 3) base of Meramecian to lower Chesterian - ~25 m. Generally, the short-term, incremental pulses of subsidence in the curves from Montana and Idaho (which include eustatic fluctuations) greatly exceed these estimates of incremental sea level rise, which suggests that the pulses of subsidence probably reflect a true tectonic signal. Exceptions to this generalization are discussed below.

In addition, the magnitude of subsidence during any interval of subsidence varies across the study area. Subsidence also is always greater in the proximal foredeep than on the adjacent platform. However, subsidence does not appear to decrease monotonically away from the foredeep as a simple flexural model would predict. Therefore, the variations in subsidence cannot entirely reflect variations in eustatic sea level but instead must be related to tectonic loading which produced differential subsidence across the foreland.

Subsidence Curves from the Antler Foreland and Their Interpretation

Six localities were selected for subsidence analyses (Fig. 3A). These localities were selected because: 1) they define a platform-to-basin transect; 2) the Devonian-Mississippian stratigraphy is well-exposed at these localities; and 3) biostratigraphic boundaries are reasonably well constrained. Subsidence analyses were not attempted for the deepest part of the Antler foredeep in east-central Idaho because age determinations and stratigraphic thicknesses are poorly constrained. The most basinward locality with enough reliable data to perform the subsidence analyses is a composite section from McGowan Creek-Grandview Canyon (Fig. 3A).

Important similarities and differences exist between the platform and foredeep basin subsidence histories. Subsidence began at McGowan Creek-Grandview Canyon at about the Middle Devonian (Fig. 8A). However, further west, subsidence began even earlier as indicated by an Early Devonian age for the base of the Milligen Formation (inferred by Sandberg et al., 1975, based on early Middle Devonian conodonts that were found 300 m from the base of the Milligen Formation). The Milligen Formation was deposited near western parts of the incipient Antler foredeep, suggesting that subsidence in the foredeep began in the Early to Middle Devonian. Unfortunately, structural complications and poor exposure make quantitative subsidence analyses impossible in this part of the Antler foredeep. In contrast to the Early to Middle Devonian onset of foredeep subsidence, subsidence of the Montana platform apparently began in the Late Devonian (Frasnian; Figs. 8 B-F).

There are several possible explanations why subsidence apparently began at different times in the foredeep and adjacent platform. First, the earlier onset of subsidence in the proximal foredeep may reflect the initial response of the foredeep to the encroaching Antler accretionary wedge. This might explain the relatively high rate and magnitude of subsidence in the incipient Antler foredeep and the apparent lack of coeval subsidence on the platform.

Alternatively, initial vertical loading by the Antler accretionary wedge may have been located above continental lithosphere that had been thinned during late Proterozoic-early Paleozoic extension. It is possible that the region of initial vertical loading had relatively low flexural rigidity, thus preventing flexure of regions far inboard from the Antler accretionary prism. However, Antler convergence occurred about 150-200 million years after the 600-550 Ma rifting event. Continental lithosphere in Idaho and Montana should have cooled and re-thickened by Antler time, making this second explanation for the earlier onset of foredeep subsidence less likely than the first.

It also is possible that the apparent pulse of Frasnian-Famennian tectonic subsidence on the Montana platform is actually due to a rise in sea level. Bond and Kominz (in press) estimated from subsidence analyses of midcontinent strata in Iowa that Frasnian-Famennian sea level rise was ~100 m. The magnitude of Frasnian-Famennian "subsidence" also is on order of 100 m or less for several platform locations (Tendoy Mountains, Ashbough Canyon, and Monarch). At the other platform locations (Logan and Sacajawea Peak), the Frasnian-Famennian "subsidence" pulse is very close to Bond and Kominz's (in press) calculated upper limit of maximum sea level rise, if all reasonable sources of error are considered in the subsidence analyses from Iowa. Therefore, the apparent increase in platform subsidence during Frasnian-Famennian time may actually record a Late Devonian sea level rise and *not* the coupled tectonic response of the Antler accretionary wedge, foredeep, and distal foreland platform. However, while a sea level rise may explain the apparent Late Devonian pulse of platform subsidence, the magnitude, rate, and timing of Devonian subsidence in the foredeep can not be explained by a sea level rise and must be attributed to tectonic processes. In addition, the remaining inflection points on the subsidence curves from all localities greatly exceed the estimated sea level rise from Early Mississippian to Late Mississippian time and must be a response to tectonic loads.

The apparent onset of platform subsidence in the Frasnian occurred during deposition of cyclic platform carbonates of the Jefferson Formation. These peritidal cyclic facies prograded away from the paleotopography produced by differential subsidence across Montana during the Frasnian. Differential subsidence across the platform also produced very complex lithostratigraphic relationships within the Jefferson Formation, which makes lithostratigraphic correlation of cycles virtually impossible, even over short distances (Dorobek and Smith, 1989; Dorobek, this volume). The differential subsidence across the distal Antler foreland in Montana and the long distance (over 400 km palinspastic) from the Antler foredeep suggest that additional mechanisms, other than flexural due to surficial end loads, are required to produce the observed subsidence patterns.

Subsidence curves for platform locations (Monarch, Sacajawea Peak, Logan, and Ashbough Canyon) are very similar in general form (Fig. 8 C-F). Except for late Frasnian to middle Kinderhookian (early Mississippian) time, the position of inflection points on the subsidence curves is remarkably consistent across the platform. Inflection points on Mississippian parts of the platform subsidence curves are well correlated with inflection points on the subsidence curve from the most basinal locality, McGowan Creek-Grandview Canyon. The relationships between platform and foredeep subsidence suggest that the proximal foredeep and distal foreland were responding to the same episodic tectonic events. Pulses of subsidence in both areas apparently reflected episodes of tectonic activity in the Antler accretionary wedge, whereas intervals of slow subsidence suggest relative tectonic quiescence.

The episodic nature of subsidence on the platform played a large role in the resultant platform stratigraphy. Platform strata that were deposited immediately following a pulse of increased subsidence typically formed onlapping or transgressive packages, at least initially (e.g., Maywood-Jefferson Formations and Lodgepole Formation; Elrick, 1990; Dorobek, this volume). Platform strata that were deposited during intervals of decreased subsidence formed offlapping or regressive packages (e.g., Mission Canyon Formation and stratigraphic equivalents; Reid and Dorobek, this volume).

CONCEPTUAL TECTONIC MODELS FOR THE ANTLER FORELAND OF MONTANA AND IDAHO

Episodic, differential subsidence of the Antler foreland occurred over a vast region (over 800 km palinspastic distance). However, this probably was not a purely flexural response to vertical loading along the western continental margin of North America. Flexural wavelengths on the order of 800-1000 km (palinspastic distance from proximal foredeep in Idaho to distal foreland platform in central Montana) are not likely without assuming unrealistically large flexural rigidities for foreland lithosphere. Many of the paleotopographic elements across the foreland also were oriented at high angles to the axis of the Antler foredeep which would not be expected if these elements had been produced solely by flexure. The axes of the foreland structures would have been subparallel to the foredeep axis if they were entirely a flexural response to the vertical load of the Antler accretionary wedge.

These observations suggest that alternative models are necessary to explain the differential subsidence across the broad area of the distal Montana foreland. The coincidence of Phanerozoic depocenters and paleohighs with older, Precambrian structures in Montana and Idaho has been recognized in previous studies (Peterson, 1986; Tonnsen, 1986; Winston, 1986; Schwartz and DeCelles, 1988). The geographic coincidence of Antler foreland structures with Precambrian structures suggests that inferred Precambrian faults or zones of weakness were somehow reactivated during Antler convergence.

Continental lithosphere beneath the Devonian-Mississippian cratonic platform across Montana most likely was thick and cold, but had been "segmented" during at least two major, pre-Antler extensional events in the Middle Proterozoic and Late Proterozoic-Early Cambrian (Hoffman, 1989; Oldlow et al., 1989). Incipient Antler convergence in the Early to Middle Devonian apparently began far outboard of the lower Paleozoic passive margin hinge zone and flexural bending may not have been transmitted very far inboard because of the distal position of the load and/or the possible low flexural rigidity of the loaded lithosphere. Frasnian subsidence in Montana

apparently began only after the Antler thrust load moved far enough eastward so that the lithosphere beneath westernmost parts of the Jefferson platform may have been within reasonable flexural wavelengths. However, distal parts of the Antler foreland require additional tectonic mechanisms to explain their subsidence histories.

Flexural bending also could explain some of the Late Devonian-Early Pennsylvanian differential subsidence across the foreland platform if the Antler accretionary wedge was areally limited throughout the course of the Antler orogeny and individual, mechanically independent segments of foreland lithosphere were loaded (Fig. 9A). Antler collision along the continental margin in Idaho probably was diachronous during Late Devonian to Early Pennsylvanian time as suggested by the southeastward migration of Antler foredeep depocenters (Fig. 5). Diachronous collision also is suggested by the apparent southeastward migration of western source areas for synorogenic siliciclastic sediment from Devonian to Pennsylvanian time (Nilsen, 1977; Skipp et al., 1979). Therefore, thrust loads probably were not distributed evenly along the ancient continental margin through time. This uneven distribution produced *subregional vertical loads* which changed position as Antler collision progressed along the margin. Individual blocks of the segmented Antler foreland, especially those bounded by east-west trending Proterozoic faults, may have responded separately and sequentially to the migrating thrust load maxima.

However, the differential subsidence of the Montana platform can not be attributed entirely to lithospheric flexure in response to vertical loads, even if the lithosphere beneath the platform was segmented by high angle faults. Much of the Montana platform was located 500-1000 km (palinspastic distance) inboard from the deepest part of the Antler foredeep. These distances are far greater than observed or calculated flexural wavelengths, assuming geologically reasonable rheologies for loaded plates. Therefore, the potential for flexural response of the segmented foreland lithosphere would have been confined to Idaho and westernmost Montana (based on maximum flexural wavelengths of 400 km and palinspastic reconstructions of Peterson, 1986 and Sandberg et al., 1983). In addition, many of the paleostructures across the cratonic platform were oriented at high angles to the axis of the Antler foredeep and the inferred strike of the Antler thrust belt; the strike of flexural features on the cratonic platform should be subparallel to the Antler foredeep/thrust belt. Finally, many of the foreland structures had relatively short wavelengths (often less than 100 km). Multiple, short wavelength foreland structures which are oriented at high angles to the axis of a foredeep are not predicted for distal foreland areas by simple flexural models.

An alternative explanation for the differential subsidence across the foreland

A.

HIGH

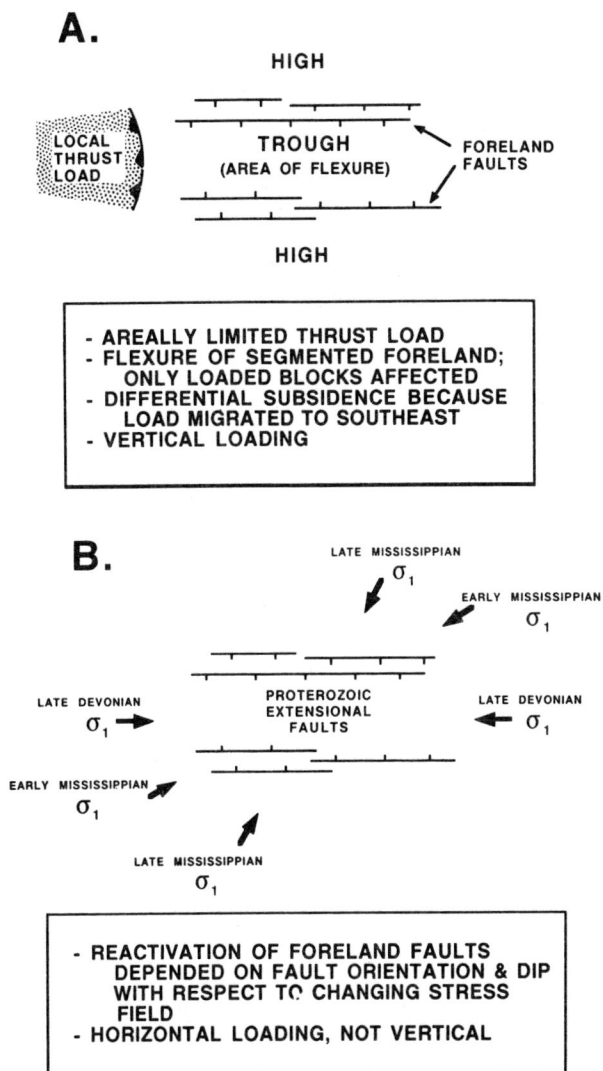

LOCAL THRUST LOAD

TROUGH
(AREA OF FLEXURE)

FORELAND FAULTS

HIGH

- AREALLY LIMITED THRUST LOAD
- FLEXURE OF SEGMENTED FORELAND; ONLY LOADED BLOCKS AFFECTED
- DIFFERENTIAL SUBSIDENCE BECAUSE LOAD MIGRATED TO SOUTHEAST
- VERTICAL LOADING

B.

LATE MISSISSIPPIAN
σ_1

EARLY MISSISSIPPIAN
σ_1

LATE DEVONIAN
σ_1

PROTEROZOIC EXTENSIONAL FAULTS

LATE DEVONIAN
σ_1

EARLY MISSISSIPPIAN
σ_1

LATE MISSISSIPPIAN
σ_1

- REACTIVATION OF FORELAND FAULTS DEPENDED ON FAULT ORIENTATION & DIP WITH RESPECT TO CHANGING STRESS FIELD
- HORIZONTAL LOADING, NOT VERTICAL

Figure 9. Generalized diagrams illustrating conceptual kinematic models which may explain the differential subsidence across the Antler foreland in Montana and Idaho.
A. Differential subsidence due to *subregional vertical loading*. Emplacement of areally limited thrust loads on broken foreland lithosphere may have caused flexural response of mechanically independent foreland blocks. See text for more complete discussion of model.

B. Differential subsidence due to *regional horizontal loading*. Transmission of in-plane compressive stress associated with Antler convergence may have caused movement on old Proterozoic faults. Amount of movement and sense of shear along ancient fault zones would have depended on orientation of faults with respect to regional stress field. Migration of Antler foredeep depocenters from Late Devonian to Late Mississippian time suggests that orientation of principal stress direction (indicated by arrows) also may have changed through time. However, actual stress field orientation at any time would have depended on the geometries along the margins and relative motions of the North American plate and colliding Antler terrane. Such data are not available at this time. Nonetheless, individual faults and fault zones apparently had different responses through time as Antler convergence progressed.

is the reactivation of Proterozoic faults by in-plane compressive stress during Antler convergence. Stress can be transmitted laterally over hundreds of kilometers inboard from modern plate margins (Gay, 1980; Zoback and Zoback, 1980; Zoback et al., 1985; Cloetingh and Wortel, 1986). Inversion tectonics, involving reversals of movement along high angle faults, are well documented in the Alpine foreland, more than 1000 km from the leading edge of Alpine thrusts (Ziegler, 1987a, 1987b). These movements have been attributed to transmission of in-plane stresses. Similar reversals in dip-slip movement on high angle faults that cut through Devonian and Mississippian strata in eastern Montana have been documented from borehole data and on seismic lines (Plawman, 1983; Clement, 1986; Nelson, pers. comm., 1990). It is difficult to explain this deformation without calling upon a regional stress field produced by tectonic loading along boundaries of the North American plate.

The actual response of Proterozoic faults across the Antler foreland would have depended on fault orientation, mechanical properties along fault surfaces, and the

magnitude and orientation of the regional stress field through time (Fig. 9B). As the Antler foredeep depocenters and inferred location of thrust load maxima migrated to the southeast from Late Devonian to Early Pennsylvanian time, the orientation of the principal stress direction also may have changed orientation with respect to the east-west trending foreland faults. As the principal stress direction changed orientation through time with respect to the pre-existing foreland faults, the amount and direction strike-slip and/or dip-slip movement also would have varied along the faults (Fig. 5B). However, the actual orientations of any possible changes in Devonian to Mississippian regional stress fields cannot be resolved more accurately without better constraints on the actual geometries of the colliding plate boundaries and possible changes in relative plate motions *during* Antler convergence.

Therefore, fault reactivation by *regional horizontal loading* (i.e., transmission of in-plane compressive stress) also might account for some of the short wavelength, variably oriented structures

502

across the Antler foreland. Relative motion along fault surfaces would have depended on fault orientation with respect to a changing regional stress field.

Finally, the subsidence curves for Montana and Idaho also provide indirect evidence for the timing and episodic nature of Late Devonian to Late Mississippian deformation in the study area. The essentially simultaneous nature of inflection points on the subsidence curves across a broad region of the Idaho-Montana foreland suggests a common tectonic mechanism for the subsidence. Given the regional tectonic setting, the subsidence events most likely reflect episodic loading events in the Antler orogenic belt. Previous studies have pointed out that subsidence analyses of foreland stratigraphy may indirectly record the history of deformation in accretionary wedges (cf. Jordan et al., 1988). This study suggests that episodic convergence events might even be identified through analyses of distal foreland strata which were deposited many hundreds of kilometers inboard from foredeep depocenters, as long as most of the accommodation space in the distal foreland was produced by tectonic subsidence and not just by eustatic sea level rise. Subtracting the estimated increments of eustatic sea level rise during the Frasnian to Meramecian (Bond and Kominz, in press) from the Idaho-Montana subsidence curves indicates that most of the pulses of apparent subsidence in the Antler foreland have a large component of remaining tectonic subsidence. This is true at least for Mississippian parts of our subsidence curves, but may not be true for Frasnian-Famennian segments of platform subsidence curves, where some of the apparent subsidence may be due to eustatic sea level rise. The episodic foreland subsidence therefore probably is a true indicator of episodic convergence and/or episodic changes in the regional stress field during much of Antler time.

SUMMARY

This study documents the response of a foredeep and adjacent cratonic platform to episodic convergence events along a continental margin that was located hundreds of kilometers outboard of the foreland area. Our approach has been to examine the response of the distal foreland area to episodic convergence through detailed study of the foreland stratigraphy. The complex chronostratigraphic and lithofacies relationships across the Antler foreland of Montana and Idaho provide a good example where models that incorporate both flexure and the effects of horizontal in-plane stress are necessary to explain the actual foreland stratigraphy.

Pre-existing basement structures apparently had a significant influence on the response of the distal Antler foreland. These pre-existing zones of weakness in the distal foreland lithosphere promoted differential subsidence which influenced foreland sedimentation. Flexural models which only incorporate vertical loading can not account for all of the differential subsidence that occurred across the Montana-Idaho foreland area during Antler convergence. The transmission of in-plane stress associated with convergence and the effect of in-plane stress on pre-existing foreland faults may explain part of the differential subsidence in the foreland.

Finally it may be possible to accurately constrain the timing of emplacement of major thrust loads by examining the subsidence history of *distal* foreland areas, if the magnitude of eustatic sea level variation can be estimated independently. This approach may be useful in structurally complex terranes where the accretionary wedge which records the deformation history of the convergence events has been destroyed by erosion or obscured by later deformation.

ACKNOWLEDGMENTS

This research was supported by U.S. Department of Energy Grant DE-FG05-87ER13767, Basic Energy Sciences to S.L. Dorobek. Such support does not constitute an endorsement by DOE of the views expressed in this paper. Acknowledgment also is made to the Donors of The Petroleum Research Fund, administered by the American Chemical Society, for partial support of this research (Grant #19519-G2 to S.L. Dorobek). Additional support was provided by the Geological Society of America Grant-in-Aid to S.K. Reid. W.J. Sando allowed us to examine unpublished data and provided comments about stratigraphy which helped to constrain our interpretations. However, views expressed herein are the responsibility of the authors. Michelle Kominz performed the subsidence analyses and, along with Gerard Bond, provided useful comments that added greatly to this study. John Cooper, Gary Girty, John Grotzinger, and Glen Stockmal also provided very useful reviews that improved the manuscript.

REFERENCES CITED

Armin, R.A., and Mayer, L., 1983, Subsidence analysis of the Cordilleran miogeocline: Implications for timing of late Proterozoic rifting and amount of extension: Geology, v. 11, p. 702-705.

Ball, M.M., 1967, Carbonate sand bodies of Florida and the Bahamas: Journal of Sedimentary Petrology, v. 37, p. 556-591.

Beaumont, C., 1981, Foreland basins: Geophysical Journal of the Royal Astronomical Society, v. 65, p. 291-329.

Beaumont, C., Quinlan, G., and Hamilton, J., 1987, The Alleghanian orogeny and its relationship to the evolution of the Eastern Interior, North America, in Beaumont, C., and Tankard, A.J., eds., Sedimentary Basins and Basin-forming Mechanisms: Canadian Society of Petroleum Geologists, Memoir 12, p. 425-445.

Beaumont, C., Quinlan, G., and Hamilton, J., 1988, Orogeny and stratigraphy: numerical models of the Paleozoic in the eastern interior of North America: Tectonics, v. 7, p. 389-416.

Bond, G.C., and Kominz, M.A., 1984, Construction of tectonic subsidence curves for the early Paleozoic miogeocline, southern Canadian Rocky Mountains: Implications for subsidence mechanisms, age of breakup, and crustal thinning: Geological Society of America Bulletin, v. 95, p. 155-173.

Bond, G.C., and Kominz, M.A., in press, Disentangling Middle Paleozoic sea level and tectonic events in cratonic margins and cratonic basins of North America: Journal of Geophysical Research.

Bond, G.C., Kominz, M.A., and Devlin, W.J., 1983, Thermal subsidence and eustasy in the lower Paleozoic miogeocline of western North America: Nature, v. 306, p. 775-779.

Bond, G.C., Kominz, M.A., and Grotzinger, J.P., 1988, Cambro-Ordovician eustasy: Evidence from geophysical modelling of subsidence in Cordilleran and Appalachian passive margins, in Kleinspehn, K.L., and Paola, C., eds., New Perspectives in Basin Analysis, Springer-Verlag, New York, p. 129-160.

Bond, G.C., Kominz, M.A., Steckler, M.S., and Grotzinger, J.P., 1989, Role of thermal subsidence, flexure, and eustasy in the evolution of early Paleozoic passive-margin carbonate platforms, in Crevello, P.D., Wilson, J.L., Sarg, J.F., and Read, J.F., eds., Controls on Carbonate Platform and Basin Development: Society of Economic Paleontologists and Mineralogists Special Publication 44, p. 39-61.

Burchfiel, B.C. and Davis, G.A., 1972, Structural framework and evolution of the southern part of the Cordilleran orogen, western United States: American Journal of Science, v. 272, p. 97-118.

Clement, J.H., 1986, Cedar Creek: A significant paleotectonic feature of the Williston Basin, in Peterson, J.A., ed., Paleotectonics and Sedimentation: American Association of Petroleum Geologists Memoir 41, p. 213-240.

Cloetingh, S., 1988, Intraplate stresses: a new element in basin analysis, in Kleinspehn, K.L., and Paola, C., eds., New Perspectives in Basin Analysis: Springer-Verlag, New York, p. 205-230.

Cloetingh, S., and Wortel, R., 1986, Stress in the Indo-Australian Plate: Tectonophysics, v. 132, p. 49-67.

Craig, L.C., 1972, Mississippian System, in Mallory, W.W., ed., Geologic Atlas of the Rocky Mountain Region: Rocky Mountain Association of Geologists, Denver, Colorado, p. 100-110.

Davis, L.E., and Webster, G.D., 1984, The age of the Alaska Bench Formation, Central Montana: Northwest Geology, v. 13, p. 1-4.

Dickinson, W.R., 1977, Paleozoic plate tectonics and the evolution of the Cordilleran continental margin, in Stewart, J.H., Stevens, C.H., and Fritsche, A.E., eds., Paleozoic Paleogeography of the Western United States: Society of Economic Paleontologists and Mineralogists, Pacific Section, Pacific Coast Paleogeography Symposium 1, p. 137-155.

Dorobek, S.L., and Smith, T.M., 1989, Cyclic sedimentation and dolomitization history of the Devonian Jefferson Formation, southwestern Montana, in French, D.E., and Grabb, R.F., eds., 1989 Field Conference Guidebook: Montana Centennial Edition, Geologic Resources of Montana, v. 1, p. 31-46.

Dorobek, S.L., Reid, S.K., Elrick, M., Bond, G.C., and Kominz, M.A., in press, Foreland response to episodic convergence: Subsidence history of the Antler foredeep and adjacent cratonic platform areas, Montana and Idaho, in Franseen, E., and Watney, L., eds., Sedimentary Modeling: Computer Simulation and Methods for Improved Parameter Definition: Kansas Geological Society Special Volume.

Dover, J.H., 1980, Status of the Antler orogeny in central Idaho - Clarification and constraints from the Pioneer Mountains, in Fouch, T.D. and Magathan, E.R., eds., Paleozoic paleogeography of the west-central United States, Rocky Mountain Paleogeography Symposium 1: Society of Economic Paleontologists and Mineralogists, Rocky Mountain Section, p. 371-386.

DuBois, D.P., 1983, Tectonic framework of basement thrust terrane, northern Tendoy Range, southwest Montana, in Powers, R.B., ed., Geologic Studies of the Cordilleran Thrust Belt, Rocky Mountain Association of Geologists, v. 1, p. 145-158.

Elrick, M., 1990, Development of cyclic ramp-to-basin carbonate deposits, Lower Mississippian, Wyoming and Montana [unpublished Ph.D. dissertation]: Virginia Polytechnic Institute and State University, Blacksburg, Virginia, 169 p.

Flemings, P.B. and Jordan, T.E., 1989, A synthetic stratigraphic model of foreland basin development: Journal of

504

Geophysical Research, v. 94, p. 3851–3866.

Gay, N.C., 1980, The state of stress in the plates, *in* Bally, A.W., Pender, P.L., McGetchin, T.R., and Walcott, R.I., eds., Dynamics of Plate Interiors: Geodynamics Series Volume 1: American Geophysical Union, Washington, D.C., p. 145–153.

Gerhard, L.C., Anderson, S.B., and Lefever, J.A., 1987, Structural history of the Nesson Anticline, North Dakota, *in* Longman, M.W., ed., Williston Basin: Anatomy of a Cratonic Oil Province: Rocky Mountain Association of Geologists, Denver, Colorado, p. 337–353.

Gordey, S.P., 1988, Devono-Mississippian clastic sedimentation and tectonism in the Canadian Cordilleran miogeocline, *in* McMillan, N.J., Embry, A.F., and Glass, D.J., eds., Devonian of the World: Volume 2: Sedimentation: Canadian Society of Petroleum Geologists, Memoir 14, p. 1–14.

Gutschick, R.C., and Sandberg, C.A., 1983, Mississippian continental margins of the conterminous United States, *in* Stanley, D.J., and Moore, G.T., eds., The Shelfbreak: Critical Interface on Continental Margins: Society of Economic Paleontologists and Mineralogists Special Publication No. 33, p. 79–96.

Gutschick, R.C., Sandberg, C.A., and Sando, W.J., 1980, Mississippian shelf margin and carbonate platform from Montana to Nevada, *in* Fouch, T.D., and Magathan, E.R., eds., Paleozoic Paleogeography of the West-central United States, Rocky Mountain Paleogeography Symposium 1: Society of Economic Paleontologists and Mineralogists, Rocky Mountain Section, p. 111–128.

Hallam, A., 1984, Pre-Quaternary sea-level changes: Annual Reviews Earth and Planetary Sciences, v. 12, p. 205–243.

Halley, R.B., Harris, P.M., and Hine, A.C., 1983, Bank margin environment, *in* Scholle, P.A., Bebout, D.G., and Moore, C.H., eds., Carbonate Depositional Environments: American Association of Petroleum Geologists Memoir 33, p. 463–506.

Harbaugh, D.W. and Dickinson, W.R., 1981, Depositional facies of Mississippian clastics, Antler foreland basin, central Diamond Mountains, Nevada: Journal of Sedimentary Petrology, v. 51, p. 1223–1234.

Harland, W.B., Cox, A.V., Llewellyn, P.G., Pichon, C.A.G., Smith, A.G., and Walters, R., 1982, A Geologic Time Scale: Cambridge, Cambridge University Press, 131 p.

Harrison, J.E., Griggs, A.B., and Wells, J.D., 1974, Tectonic features of the Precambrian Belt basin and their influence on post-Belt structures: U.S. Geological Survey Professional Paper 866, 15 p.

Hildreth, G.D., 1981, Stratigraphy of the Mississippian Big Snowy Formation of the Armstead Anticline, Beaverhead County, Montana: Montana Geological Society 1981 Field Conference, Southwest **Montana**, p. 49–57.

Hine, A.C., 1977, Lily Bank, Bahamas: History of an active oolite sand shoal: Journal of Sedimentary Petrology, v. 47, p. 1554–1581.

Hoffman, P.F., 1989, Precambrian geology and tectonic history of North America, *in* Bally, A.W., and Palmer, A.R., eds., The Geology of North America – An Overview: The Geology of North America, v. A, Geological Society of America, p. 447–512.

Huh, O.K., 1967, The Mississippian System across the Wasatch Line east central Idaho, extreme southwestern Montana, *in* Centennial Basin of Southwest Montana: Montana Geological Society, 18th Annual Field Conference Guidebook, p. 31–62.

Huh, O.K., 1968, Mississippian stratigraphy and sedimentology across the Wasatch Line, east-central Idaho and extreme southwestern Montana [unpublished Ph.D. dissertation]: University Park, Pennsylvania State University, 176 p.

Isaacson, P.E., Simpson, K.R., and McFaddan, M.D., 1983, Depositional setting and carbonate buildup succession in Jefferson Formation (Devonian), central Idaho – a harbinger of Antler uplift?: Geological Society of America, Cordilleran-Rocky Mountain Sections Meeting, Abstracts with Programs, v. 15, no. 5, p. 306.

Johnson, J.G., Klapper, G., and Sandberg, C.A., 1985, Devonian eustatic fluctuations in Euramerica: Geological Society of America Bulletin, v. 96, p. 567–587.

Johnson, J.G. and Pendergast, A., 1981, Timing and mode of emplacement of the Roberts Mountain allochthon, Antler orogeny: Geological Society of America Bulletin, v. 92, p. 648–658.

Jordan, T.E., 1981, Thrust loads and foreland basin development, Cretaceous, western United States: American Association of Petroleum Geologists Bulletin, v. 65, p. 2506–2520.

Jordan, T.E., Flemings, P.B., and Beer, J.A., 1988, Dating thrust-fault activity by use of foreland-basin strata, *in* Kleinspehn, K.L., and Paola, C., eds., New Perspectives in Basin Analysis: Springer-Verlag, New York, p. 307–330.

Lambeck, K., McQueen, H.W.S., Stephenson, R.A., and Denham, D., 1984, The state of stress within the Australian continent: Annales Geophysicae, v. 2, p. 723–742.

LeFever, J.A., LeFever, R.D., and Anderson, S.B., 1987, Structural evolution of the central and southern portions of the Nesson Anticline, North Dakota, *in* Carlson, C.G., and Christopher, J.E., eds., Fifth International Williston Basin Symposium: Saskatchewan Geological Society Special Publication Number 9, p. 147–156.

Mallory, W.W., 1972, Regional synthesis of the Pennsylvanian System, *in* Mallory, W.W., ed., Geologic Atlas of the Rocky Mountain Region: Rocky Mountain Association of Geologists, Denver, Colorado, p. 111–127.

Mamet, B.L., Skipp, B., Sando, W.J., and Mapel, W.J., 1971, Biostratigraphy of Upper Mississippian and associated Carboniferous rocks in south-central Idaho: American Association of Petroleum Geologists Bulletin, v. 55, p. 20-33.

McCabe, W.S., 1954, Williston basin Paleozoic unconformities: American Association of Petroleum Geologists Bulletin, v. 51, p. 883-917.

Miller, E.L., Holdsworth, B.K., Whiteford, W.B., and Rodgers, D., 1984, Stratigraphy and structure of the Schoonover sequence, northeastern Nevada: Implications for Paleozoic plate-margin tectonics: Geological Society of America Bulletin, v. 95, p. 1063-1076.

Morrow, D.W., and Geldsetzer, H.H.J., 1988, Devonian of the eastern Canadian Cordillera, in McMillan, N.J., Embry, A.F., and Glass, D.J., eds., Devonian of the World: Volume 1: Regional Syntheses: Canadian Society of Petroleum Geologists, Memoir 14, p. 85-121.

Nilsen, T.H., 1977, Paleogeography of Mississippian turbidites in south-central Idaho, in Stewart, J.H., Stevens, C.H., and Fritsche, A.E., eds., Paleozoic Paleogeography of the West-central United States, Pacific Coast Paleogeography Symposium 1: Society of Economic Paleontologists and Mineralogists Pacific Section, p. 275-299.

Oldow, J.S., Bally, A.W., Avé Lallemant, H.G., and Leeman, W.P., 1989, Phanerozoic evolution of the North American Cordillera; United States and Canada, in Bally, A.W., and Palmer, A.R., eds., The Geology of North America; An Overview: Boulder, Colorado, Geological Society of America, v. A, p. 139-232.

Pardee, J.T., 1950, Late Cenozoic block faulting in western Montana: Geological Society of America Bulletin, v. 61, p. 359-406.

Paull, R.A., Wolbrink, M.A., Volkmann, R.G., and Grover, R.L., 1972, Stratigraphy of the Copper Basin Group, Pioneer Mountains, south-central Idaho: American Association of Petroleum Geologists Bulletin, v. 56, p. 1370-1401.

Perry, W.J., Jr., Dyman, T.S., and Sando, W.J., 1989, Southwestern Montana recess of Cordilleran thrust belt, in French, D.E., and Grabb, R.F., eds., Geological Resources of Montana, Vol. 1: Montana Geological Society, 1989 Field Conference Guidebook: Montana Centennial Edition, p. 261-270.

Peterson, J.A., 1986, General stratigraphy and regional paleotectonics of the western Montana overthrust belt, in Peterson, J.A., ed., Paleotectonics and Sedimentation in the Rocky Mountain Region, United States: American Association of Petroleum Geologists, Memoir 41, p. 57-86.

Plawman, T.L., 1983, Fault with reversal of displacement, central Montana, in Bally, A.W., ed., Seismic Expression of Structural Styles, v. 3: American Association of Petroleum Geologists Studies in Geology Series No. 15, p. 3.3-1 – 3.3-2.

Poole, F.G., 1974, Flysch deposits of the Antler foreland basin, western United States, in Dickinson, W.R., ed., Tectonics and Sedimentation: Society of Economic Paleontologists and Mineralogists Special Publication 22, p. 58-82.

Purser, B.H., 1973, The Persian Gulf - Holocene carbonate sedimentation and diagenesis in a shallow water epicontinental sea: New York, Springer-Verlag, 471 p.

Quinlan, G.M., and Beaumont, C., 1984, Appalachian thrusting, lithospheric flexure and the Paleozoic stratigraphy of the eastern interior of North America: Canadian Journal of Earth Sciences, v. 21, p. 973-996.

Reid, S.K., 1991, Evolution of the Lower Mississippian Mission Canyon platform and distal Antler foredeep, Montana and Idaho [unpublished Ph.D. dissertation]: Texas A&M University, College Station, Texas, 115 p.

Reid, S.K., and Dorobek, S.L., 1989, Application of sequence stratigraphy in the Mississippian Mission Canyon Formation and stratigraphic equivalents, southwestern Montana and east-central Idaho, in French, D.E., and Grabb, R.F., eds., 1989 Field Conference Guidebook: Montana Centennial Edition, Geologic Resources of Montana, v. 1, p. 47-53.

Reynolds, M.W., 1979, Character and extent of basin-range faulting, western Montana and east-central Idaho, in Newman, G.W., and Goode, H.D., eds., 1979 Basin and Range Symposium: Rocky Mountain Association of Geologists, p. 185-193.

Roberts, R.J., and Thomasson, M.R., 1964, Comparison of Late Paleozoic depositional history of northern Nevada and central Idaho: U.S. Geological Survey Professional Paper 475-D, Article 122, p. D1-D6.

Ross, C.A., and Ross, J.P., 1987, Late Paleozoic sea levels and depositional sequences, in Ross, C.A., and Haman, D., eds., Timing and Depositional History of Eustatic Sequences: Constraints on Seismic Stratigraphy: Cushman Foundation for Foraminiferal Research, Special Publication No. 24, p. 137-149.

Royden, L., Patacca, E., and Scandone, P., 1987, Segmentation and configuration of subducted lithosphere in Italy: An important control on thrust-belt and foredeep-basin evolution: Geology, v. 15, p. 714-717.

Ruppel, E.T., and Lopez, D.A., 1984, The thrust belt in southwest Montana and east-central Idaho: U.S. Geological Survey Professional Paper 1278, 41 p.

Ruppel, E.T., Wallace, C.A., Schmidt, R.G., and Lopez, D.A., 1981, Preliminary interpretation of the thrust belt in southwest and west-central Montana and east central Idaho, in Southwest Montana: Montana Geological Society, 1981 Field Conference and Symposium Guidebook, p. 139-159.

506

Sandberg, C.A., 1975, McGowan Creek Formation, new name for Lower Mississippian flysch sequence in east-central Idaho: U.S. Geological Survey Bulletin 1405-E, 11 p.

Sandberg, C.A., and Poole, F.G., 1977, Conodont biostratigraphy and depositional complexes of Upper Devonian cratonic-platform and continental shelf rocks in the Western United States, in Murphy, M.A., Berry, W.B.N., and Sandberg, C.A., eds., Western North America: Devonian: California University, Riverside, Campus Museum Contributions 4, p. 144-182.

Sandberg, C.A., Mapel, W.J., and Huddle, J.W., 1967, Age and regional significance of basal part of Milligen Formation, Lost River Range, Idaho: United States Geological Survey Professional Paper 575-C, p. C127-C131.

Sandberg, C.A., Poole, F.G., and Johnson, J.G., 1988, Upper Devonian of western United States, in McMillan, N.J., Embry, A.F., and Glass, D.J., eds., Devonian of the World: Canadian Society of Petroleum Geologists, Memoir 14, p. 183-220.

Sandberg, C.A., Hall, W.E., Batchelder, J.N., and Axelsen, C., 1975, Stratigraphy, conodont dating, and paleotectonic interpretation of the type Milligen Formation (Devonian), Wood River area, Idaho: U.S. Geological Survey Journal of Research, v. 3, no. 6, p. 707-720.

Sandberg, C.A., Gutschick, R.C., Johnson, J.G., Poole, F.G., and Sando, W.J., 1983, Middle Devonian to Late Mississippian geologic history of the Overthrust Belt region, western United States, in Powers, R.B., ed., Geologic Studies of the Cordilleran Thrust Belt: Rocky Mountain Association of Geologists, v. 2, p. 691-719.

Sando, W.J., 1980, The paleoecology of Mississippian corals in the western conterminous United States: Acta Palaeontologica Polonica, v. 25, p. 619-631.

Sando, W.J., 1985, Revised Mississippian time scale, western interior region, conterminous United States: United States Geological Survey Bulletin, v. 1605-A, p. A15-A26.

Sando, W.J., 1988, Madison Limestone (Mississippian) paleokarst: A geologic synthesis, in James, N.P. and Choquette, P.W., eds., Paleokarst: New York, Springer-Verlag, p. 256-277.

Sando, W.J., and Dutro, J.T., Jr., 1974, Type sections of the Madison Group (Mississippian) and its subdivisions in Montana: United States Geological Survey Professional Paper 842, 22 p.

Sando, W.J., Gordon, M., Jr., and Dutro, J.T., Jr., 1975, Stratigraphy and geologic history of the Amsden Formation (Mississippian and Pennsylvanian) of Wyoming: U.S. Geological Survey Professional Paper 848-A, p. A1-A83.

Sando, W.J., Sandberg, C.A., and Perry, W.J., Jr., 1985, Revision of Mississippian Stratigraphy, Northern Tendoy Mountains, southwest Montana, in Sando, W.J., ed., Mississippian and Pennsylvanian Stratigraphy in Southwest Montana and Adjacent Idaho: U.S. Geological Survey Bulletin, v. 1657, p. A1-A10.

Sando, W.J., Dutro, J.T., Jr., Sandberg, C.A., and Mamet, B.L., 1976, Revision of Mississippian stratigraphy, eastern Idaho and northeastern Utah: United States Geological Survey Journal of Research, v. 4, p. 467-479.

Schedl, A., and Wiltschko, D.V., 1984, Sedimentological effects of a moving terrain: Journal of Geology, v. 92, p. 273-287.

Schmidt, C.J., and Hendrix, T.E., 1981, Tectonic controls for thrust belt and Rocky Mountain foreland structures in the northern Tobacco Root Mountains - Jefferson Canyon area, southwestern Montana: Montana Geological Society, 1981 Field Conference and Symposium Guidebook, p. 167-180.

Schwartz, R.K., and DeCelles, P.G., 1988, Cordilleran foreland basin evolution in response to interactive Cretaceous thrusting and foreland partitioning, southwestern Montana, in Schmidt, C.J., and Perry, W.J., Jr., eds., Interaction of the Rocky Mountain Foreland and the Cordilleran Thrust Belt: Geological Society of America Memoir 171, p. 489-513.

Schweickert, R.A., and Snyder, W.S., 1981, Paleozoic plate tectonics of the Sierra Nevada and adjacent areas, in Ernst, W.G., ed., The Geotectonic Evolution of California (Rubey Volume 1): Englewood Cliffs, New Jersey, Prentice-Hall, p. 182-201.

Skipp, B., 1988, Cordilleran thrust belt and faulted foreland in the Beaverhead Mountains, Idaho and Montana, in Schmidt, C.J., and Perry, W.J., Jr., eds., Interaction of the Rocky Mountain Foreland and the Cordilleran Thrust Belt: Geological Society of America Memoir 171, p. 237-266.

Skipp, B., and Hait, M.H., Jr., 1977, Allochthons along the northeast margin of the Snake River Plain, Idaho: 29th Annual Field Conference, 1977, Wyoming Geological Association Guidebook, p. 499-515.

Skipp, B., and Hall, W.E., 1975, Structure and Paleozoic stratigraphy of a complex of thrust plates in the Fish Creek Reservoir area, south-central Idaho: U.S. Geological Survey Journal of Research, v. 3, no. 6, p. 671-689.

Skipp, B., and Sandberg, C.A., 1975, Silurian and Devonian miogeosynclinal and transitional rocks of the Fish Creek Reservoir units, central Idaho: U.S. Geological Survey Journal of Research, v. 3, no. 6, p. 691-706.

Skipp, B., Sando, W.J., and Hall, W.E., 1979, The Mississippian and Pennsylvanian (Carboniferous) Systems in the United States - Idaho: United States Geological Survey Professional Paper 1110-AA, 42 p.

Sleep, N.H., 1971, Thermal effects of the formation of Atlantic continental margins by continental breakup: Geophysical Journal of the Royal Astronomical Society, v. 24, p. 325-350.

Smith, D.L., 1977, Transition from deep- to shallow-water carbonates, Paine Member, Lodgepole Formation, central Montana, *in* Cook, H.E., and Enos, P., eds., Deep-water Carbonate Environments: Society of Economic Paleontologists and Mineralogists Special Publication 25, p. 187-201.

Speed, R.C., 1977, Island-arc and other paleogeographic terranes of the late Paleozoic age of the western Great Basin, *in* Stewart, J.H., Stevens, C.H., and Fritsch, A.E., eds., Paleozoic Paleogeography of the Western United States, Pacific Coast Paleogeography Symposium 1: Society of Economic Paleontologists and Mineralogists, Pacific Section, p. 349-362.

Speed R.C. and Sleep, N.H., 1982, Antler orogeny and foreland basin: A model: Geological Society of America Bulletin, v. 93, p. 815-828.

Steckler, M.S., and Watts, A.B., 1978, Subsidence of the Atlantic-type continental margin off New York: Earth and Planetary Science Letters, v. 41, p. 1-13.

Stewart, J.H., and Suczek, C.A., 1977, Cambrian and latest Precambrian paleogeography and tectonics in the western United States, *in* Stewart, J.H., Stevens, C.H., and Fritsche, A.G., eds., Paleozoic Paleogeography of the Western United States, Pacific Coast Paleogeography Symposium 1: Society of Economic Paleontologists and Mineralogists, Pacific Section, p. 1-17.

Stockmal, G.S., and Beaumont, C., 1987, Geodynamic models of convergent margin tectonics: the southern Canadian Cordillera and the Swiss Alps, *in* Beaumont, C., and Tankard, A.J., eds., Sedimentary Basins and Basin-Forming Mechanisms: Canadian Society of Petroleum Geologists Memoir 12, p. 393-411.

Stockmal, G.S., Beaumont, C., and Boutilier, R., 1986, Geodynamic models of convergent tectonics: the transition from rifted margin to overthrust belt and consequences for foreland-basin development: American Association of Petroleum Geologists Bulletin, v. 70, p. 181-190.

Tonnsen, J.J., 1986, Influence of tectonic terranes adjacent to the Precambrian Wyoming Province on Phanerozoic stratigraphy in the Rocky Mountain region, *in* Peterson, J.A., ed., Paleotectonics and Sedimentation: American Association of Petroleum Geologists Memoir 41, p. 21-39.

Turcotte, D.L., and Schubert, G., 1982, Geodynamics - Applications of Continuum Physics to Geological Problems: John Wiley and Sons, New York, 450 p.

Visser, W.A., ed., 1980, Geological Nomenclature: Royal Geological and Mining Society of the Netherlands-Bohn, Scheltema, and Holkema, Utrecht; Martinus Nijhoff, The Hague, 540 p.

Walton, P.T., 1946, Ellis, Amsden, and Big Snowy Group, Judith Basin, Montana: American Association of Petroleum Geologists Bulletin, v. 30, p. 1294-1305.

Wardlaw, B.R. and Pecora, W.C., 1985, New Mississippian-Pennsylvanian stratigraphic units in southwest Montana and adjacent Idaho, *in* Sando, W.J., ed., Mississippian and Pennsylvanian Stratigraphy in Southwest Montana and Adjacent Idaho: United States Geological Survey Bulletin, v. 1656, p. B1-B9.

Wilson, J.L. and Jordan, C., 1983, Middle shelf environment, *in* Scholle, P.A., Bebout, D.G., and Moore, C.H., eds., Carbonate Depositional Environments: American Association of Petroleum Geologists Memoir 33, p. 297-343.

Winston, D., 1986, Sedimentation and tectonics of the Middle Proterozoic Belt Basin and their influence on Phanerozoic compression and extension in western Montana and northern Idaho, *in* Peterson, J.A., ed., Paleotectonics and Sedimentation: American Association of Petroleum Geologists Memoir 41, p. 87-118.

Woodward, L.A., 1981, Tectonic framework of disturbed belt of west-central Montana: American Association of Petroleum Geologists Bulletin, v. 65, p. 291-302.

Yurewicz, D.A., 1977, Sedimentology of Mississippian basin carbonates, New Mexico and west Texas - the Rancheria Formation, *in* Cook, H.E., and Enos, P., 1977, eds., Deep-water Carbonate Environments: Society of Economic Paleontologists and Mineralogists Special Publication 25, p. 203-219.

Ziegler, P.A., 1987a, Compressional intraplate deformations in the Alpine foreland - an introduction: Tectonophysics, v. 137, p. 1-5.

Ziegler, P.A., 1987b, Late Cretaceous and Cenozoic intra-plate compressional deformations in the Alpine foreland - a geodynamic model: Tectonophysics, v. 137, p. 389-420.

Zoback, M.D., Moos, D., and Mastin, L., 1985, Well bore breakouts and in situ stress: Journal of Geophysical Research, v. 90, p. 5523-5530.

Zoback, M.L., and Zoback, M.D., 1980, State of stress in the conterminous United States: Journal of Geophysical Research, v. 85, P. 6113-6156.

CYCLIC PLATFORM CARBONATES OF THE DEVONIAN JEFFERSON FORMATION, SOUTHWESTERN MONTANA

S.L. Dorobek
Department of Geology
Texas A&M University
College Station, TX 77843

ABSTRACT

The Jefferson Formation (Upper Devonian) in southwestern Montana consists of cyclic shallow marine carbonate deposits that grade westward into a thick sequence of shallow water to deep ramp, slope, and basinal facies in central Idaho. Jefferson lithofacies were deposited on a 700-800 km (nonpalinspastic) wide ramp that graded eastward into the Williston Basin, westward into the Antler foredeep, northward into Upper Devonian carbonate platform complexes and shale basins of western Canada, and southward onto the craton and extensions of the ramp. Several paleostructural features extended across Montana and affected depositional patterns on the Jefferson ramp in Late Devonian time. Differential subsidence across the Jefferson ramp affected depositional patterns and most likely was a response to Antler convergence many 100's of kilometers (palinspastic) outboard of the ramp.

Individual shallowing-upward Jefferson cycles in Montana are tens of meters to less than 1 m thick. Cycle lithofacies include, in ascending stratigraphic order: local, thin, basal siliciclastic facies; basal, thin bedded, fine grained, shaly dolostone/limestone with closely spaced hardgrounds that grade upward into burrow-homogenized, irregularly bedded dolostone/limestone; coarsely crystalline dolostone with abundant laminar to domal stromatoporoids; thin bedded *Amphipora* dolowackestone; rare ooid dolograinstone/packstone; cryptalgalaminated/stromatolitic/thrombolitic dolomudstone; and local solution-collapse breccia caps. All lithofacies rarely occur together within a single shallowing-upward cycle. Overall, subtidal facies are dominated by poorly sorted and/or muddy lithologies with low diversity faunas, attesting to the generally restricted, low energy conditions across this broad region. Quartz sand beds only occur adjacent to a paleohigh in southwestern Montana (Southern Beaverhead Mountains Uplift). Cycles either are capped by subaerial erosional disconformities with variable relief or grade upward over a few centimeters into basal units of the overlying cycle.

Inner ramp localities contain thinner cycles (<5 m thick) with fewer basal, subtidal facies. Outer ramp localities contain thicker individual cycles; outer ramp cycles also are dominated by thicker subtidal facies. Middle parts of the ramp generally have greater numbers of cycles than on either the inner or outer ramp. This probably results because: 1) thin cycles on the middle ramp are recorded as very thin sequences or nondepositional surfaces on the inner ramp; 2) lowermost Jefferson deposits on the inner ramp may be younger than deposits on downdip parts of the ramp; and 3) the outer ramp may not have become shallow enough or was not subaerially exposed following aggradation during many cycle intervals. Individual outer ramp cycles may be amalgamated and therefore are not easily recognized.

The Jefferson Formation may be an important analogue for other cyclic carbonate sequences which form on restricted, low energy ramps. The Jefferson Formation also illustrates cyclic depositional processes on a carbonate ramp that developed during transition from a passive margin to a convergent margin. Even though the Jefferson ramp was far from the active zone of convergence (>500 km palinspastic distance), tectonic loads associated with the convergence events caused differential subsidence across the ramp, as suggested by differences in cycle thickness, facies composition, and subsidence histories. Therefore, the Jefferson Formation is an example of cyclic platform sedimentation in a tectonically active setting, unlike many other cyclic platform sequences which presumably formed on relatively stable, linearly subsiding platforms.

INTRODUCTION

Shallowing-upward carbonate sequences have been documented on many continental platforms throughout the geologic record. Cyclic platform carbonate sequences of Devonian age are also very common worldwide (Wilson, 1967; Read, 1973; Wong and Oldershaw, 1980; Cutler, 1983; Viau, 1983; Goodwin and Anderson, 1985; Dorobek and Read, 1986; Wilson and Pilatzke, 1987). Well-documented examples of fourth- and fifth-order (i.e., 105 to 104 yr duration; *sensu* Vail et al., 1977) shallowing-upward cycles of Late Devonian age are remarkably similar

In Cooper, J.D., and Stevens, C.H., eds., 1991, **Paleozoic Paleogeography of the Western United States-II:** Pacific Section SEPM, Vol. 67, p. 509-526.

509

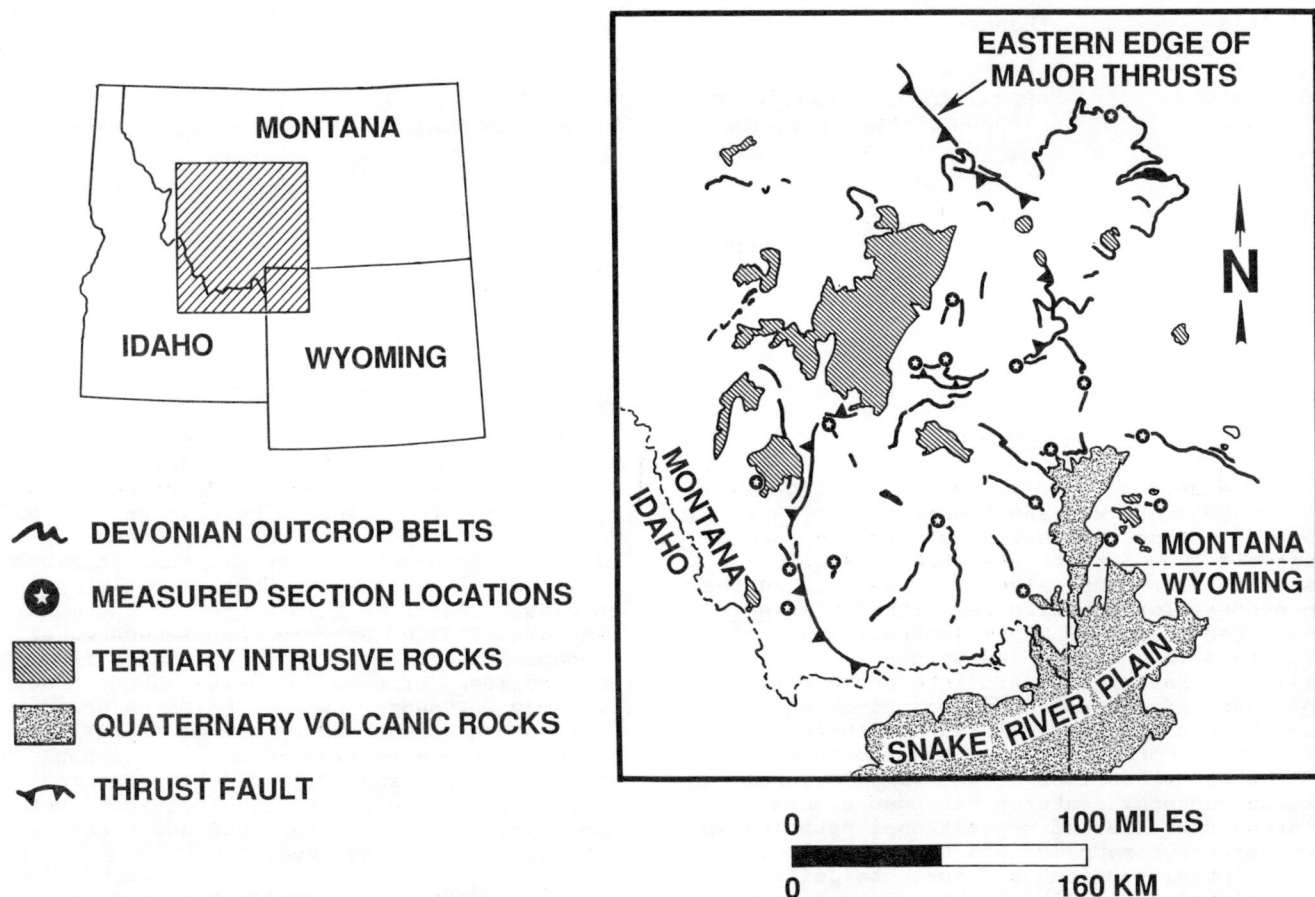

Figure 1. Map of study area showing locations of measured sections and Devonian outcrop belts.

worldwide, yet factors controlling their development are not fully understood.

The Upper Devonian (mostly Frasnian) Jefferson Formation of southwestern Montana consists of numerous thin to thick (<1 to 40 m) shallowing-upward cycles. Lithofacies composition and thicknesses of individual cycles vary vertically within a measured section and laterally between sections. While cycles in equivalent rocks from the Williston Basin and eastern Montana (Duperow Formation) have been examined previously (Wilson, 1967; Wilson and Pilatzke, 1987), no detailed studies have been done on cyclic sequences farther to the west in southwestern Montana.

This paper describes cycles in the Jefferson Formation of southwestern Montana and discusses the role of differential subsidence on cycle generation. The Jefferson Formation may be an important analogue for other cyclic carbonate sequences which form during early stages of plate convergence on the outer (marginal) parts of formerly stable platforms. Because of the tectonic setting, Jefferson cycles differ from other cyclic carbonate sequences that formed on linearly subsiding passive margins or isolated platforms. Analysis of the Jefferson cycles also may help to determine the causes of Upper Devonian cyclic sequences elsewhere.

STRUCTURAL AND STRATIGRAPHIC SETTING

Regional Structural History

The Upper Devonian Jefferson Formation is exposed throughout central and southwestern Montana (Fig. 1). In western Montana, most exposures of Devonian rocks are allochthonous and occur within the north-south trending Montana Disturbed Belt. Eastward transport of these rocks occurred along complex thrust and tear fault systems during Pennsylvanian (?) to Early Tertiary time (Sevier and Laramide orogenies; Skipp and Hall, 1975; Skipp et al., 1979; Ruppel and Lopez, 1984; Perry et al., 1989). East of the leading edge of this fold-and-thrust belt, the amount of shortening is minimal and does not affect palinspastic reconstructions. However, estimates for the total amount of tectonic shortening in the fold-and-thrust belt are highly variable and range from tens to several hundreds of kilometers (Skipp and Hait, 1977; Nilsen, 1977; Skipp et al., 1979; Dover, 1980; Ruppel et al., 1981; Schmidt and Hendrix, 1981; Woodward, 1981; Skipp, 1988).

Regional Devonian Stratigraphic Relationships

Devonian deposits in the study area unconformably overlie much older rocks, from Precambrian siliciclastics to lower Paleozoic siliciclastics and carbonates (Sloss and Moritz, 1951; Scholten, 1957, 1960; Scholten and Hait, 1962; Sandberg, 1961; Churkin, 1962; Loucks, 1977; Ruppel, 1986). Pre-Devonian subcrop maps (Sandberg and Mapel, 1967; Baars, 1972; Peterson, 1986) and field observations (Scholten, 1957, 1960; Churkin, 1962; Ruppel, 1986) show that on a regional scale, the oldest subcrops of Precambrian and Cambrian sedimentary rocks occur near the present Montana-Idaho border, just north of the Snake River Plain (Fig. 2). Progressively younger Paleozoic sedimentary rocks underlie Devonian deposits to the east and west of this subcrop trend.

The Jefferson Formation conformably overlies the Devonian Maywood Formation (Middle Frasnian; Sandberg and Poole, 1977; Johnson et al., 1985) across most of the study area (Fig. 3). The Maywood Formation (10 to 90 m thick) consists of dolomitic siltstone and argillaceous dolostone with less common mudstone, limestone, and dolostone layers (Benson, 1966). Large-scale incised valleys also occur locally beneath the Jefferson Formation in western Montana and east-central Idaho. The strata that fill these features are Early to Middle Devonian in age and are either: 1) included with the Jefferson Formation; 2) named separately (Beartooth Butte Formation); 3) or are unnamed (Churkin, 1962; Scholten and Hait, 1962; Sandberg and Mapel, 1967; Mapel and Sandberg, 1968; Hoggan, 1981).

Stratigraphic and sedimentologic relationships for the Jefferson Formation of southwestern Montana (generally 90 to 170 m thick) have been described previously (Sloss and Laird, 1947; Sloss, 1950; Sloss and Moritz, 1951; Andrichuck, 1962; Sandberg, 1962, 1965; McMannis, 1962, 1965; Benson, 1966; Mapel and Sandberg, 1968; Loucks, 1977; Sandberg and Poole, 1977; Sandberg et al., 1988). Two separate units are recognized within the Jefferson Formation: a thick,

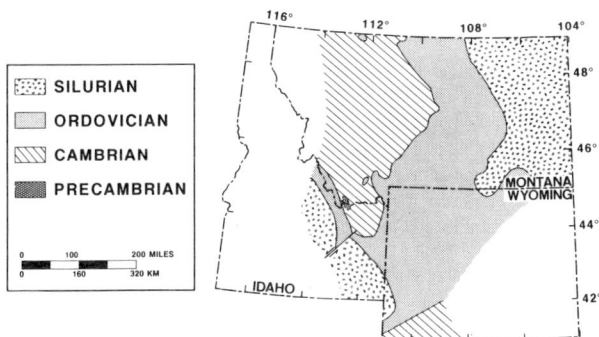

Figure 2. Pre-Devonian paleogeologic map. Modified from Sandberg and Mapel (1967) and Baars (1972) using data from Churkin (1962), Scholten (1957; 1960), and unpublished field data. Note oldest rocks beneath Devonian deposits are located in southwestern Montana near the present Montana-Idaho border. Lower Paleozoic subcrop patterns are highly generalized across the entire area.

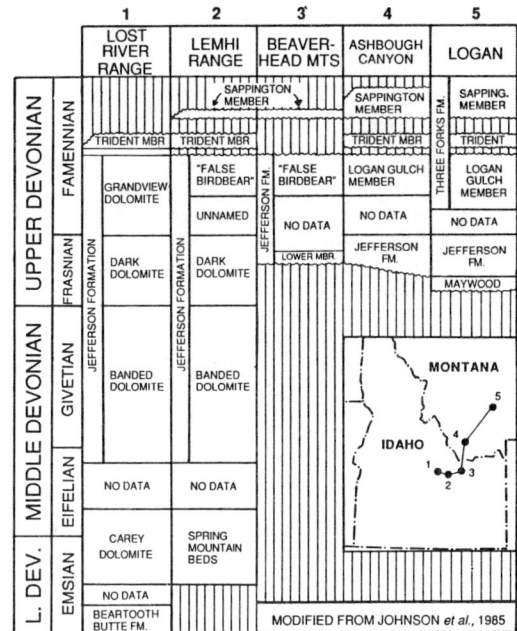

Figure 3. Generalized stratigraphic correlation chart for Upper Devonian strata in southwestern Montana and east-central Idaho. Modified from Johnson et al. (1985).

unnamed lower unit (80-150 m thick) and a thinner, upper unit called the Birdbear Member (15-25 m thick), which locally rests disconformably on the lower member (Sandberg and Poole, 1977). The lower unit consists mostly of interbedded dolostone and limestone. The Birdbear Member consists mostly of massively bedded dolostone (Sandberg, 1965). In the study area, the amount of limestone in the entire Jefferson Formation generally increases to the east (Smith and Dorobek, 1989).

The Jefferson Formation of central and southwestern Montana is mostly Frasnian in age. However, the upper part of the Birdbear Member locally may be Famennian (Sandberg and Poole, 1977). In contrast, the Jefferson Formation in east-central Idaho was deposited over a much longer time span, from at least Givetian through Famennian time (Fig. 3; Johnson et al., 1985).

JEFFERSON PALEOGEOGRAPHY AND PALEOTECTONICS

The Jefferson Formation was deposited on a 700-800 km wide platform that graded eastward into the Williston Basin, westward into the Antler foredeep, northward into Upper Devonian isolated platform complexes and shale basins of western Canada, and southward onto the craton and southern extensions of the platform. The Antler orogeny was a regional orogenic episode that occurred during deposition of the Jefferson Formation and probably was related to arc-continent collision along the western margin of North America (Burchfiel and Davis, 1972; Poole, 1974; Dickinson, 1977; Speed, 1977; Speed and Sleep, 1982). Therefore, Upper

Devonian strata in western North America record the transition from an earlier passive margin to a convergent margin. The Jefferson Formation in Montana was deposited on the broad cratonic platform located east of the incipient Antler foredeep in Idaho.

Published isopach maps and regional lithofacies patterns indicate that Late Devonian sedimentation in the study area was affected by a number of paleotopographic highs. These include the east-west trending Central Montana Uplift to the north and the Yellowstone Park and Southern Beaverhead Mountains Uplifts to the south (Fig. 4). These paleostructures formed during early stages of the Antler orogeny (Sandberg and Poole, 1977) and they coincide with Precambrian structural trends (cf. Peterson, 1986; Tonnsen, 1986; Winston, 1986). Dorobek et al. (in press and this volume) suggest that these Late Devonian structures across Montana reflect reactivation of Precambrian zones of weakness. Reactivation and tectonic inversion across Montana may have been a response to both vertical and horizontal loads associated with Antler convergence events.

These Late Devonian structures across Montana affected deposition of the Jefferson Formation. The Jefferson Formation is thinner and contains fewer total cycles around paleohighs than on other parts of the ramp. Shallow subtidal to peritidal facies also dominate the facies within individual cycles around paleohighs, while deeper subtidal facies dominate cycles on the distal ramp in west-central Montana. The higher bathymetric gradients around paleohighs resulted in abrupt facies changes across the Jefferson ramp, which may partially explain the lateral lithofacies and thickness variation within cycles.

The Jefferson ramp, to the north and east of the Southern Beaverhead Mountains and Yellowstone Park Uplifts, was a broad, relatively shallow, depositional surface with a regional westward slope (Sandberg et al., 1983; Dorobek, 1987; Dorobek and Smith, 1989). Paleogeographic reconstructions, lithofacies distribution, and published isopach maps indicate that a tectonic hinge line or abrupt line of flexure existed near the present Montana-Idaho border (Fig. 4). West of this hinge line, the Jefferson Formation thickens considerably and consists of a 500-800 m thick sequence of shallow peritidal to deep water deposits (Dorobek, 1987; Isaacson and Dorobek, 1988). This abrupt increase in thickness may reflect the easternmost effects of lithospheric flexure by the encroaching Antler accretionary wedge in Late Devonian time (Dorobek et al., in press and this volume).

Throughout the rest of this paper, "inner ramp" refers to locations which are from eastern parts of the study area (i.e., closer to regional paleostrandline positions to the east) or sections which are proximal to paleotopographic highs on the platform. "Outer ramp" refers to locations in western Montana which are distal from any

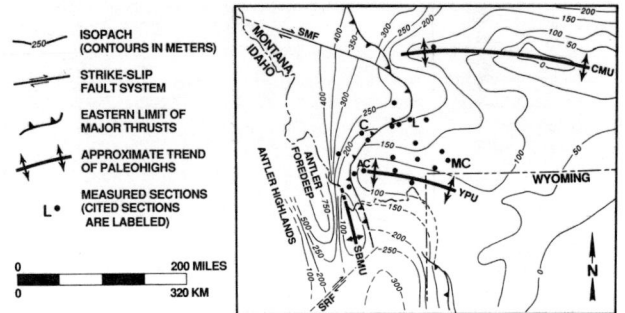

Figure 4. Isopach map for Frasnian and lower Famennian rocks of Montana, northern Wyoming, and Idaho. Major paleostructures also labeled. SRF = Snake River Fault Zone; SMF = St. Marys Fault Zone; CMU = Central Montana Uplift; YPU = Yellowstone Park Uplift; SBMU = Southern Beaverhead Mountains Uplift. Eastern limit of thrust sheets indicated by sawteeth on thrust traces. Labeled measured sections: C = Camp Creek; AC = Ashbough Canyon; L = Logan; MC = Mill Creek. Modified slightly from Sandberg et al. (1983).

paleotopographic high and which are located near the line of flexure into the Antler foredeep.

LITHOFACIES DESCRIPTION AND INTERPRETATION

Shallowing-upward cycles in the Jefferson Formation consist of basal subtidal facies which grade upward into progressively shallower marine facies. Individual cycles either are capped by subaerial erosional disconformities with variable relief or grade upward over a few centimeters into basal units of the overlying cycle. Lithofacies are described in order of ascending stratigraphic succession.

Basal Siliciclastic Facies

Description

Thin beds (2-30 cm thickness) of mudstone, siltstone, or quartz arenite occur at the bases of some cyclic sequences that formed near the Southern Beaverhead Mountains Uplift in southwestern Montana. Basal facies in most other cycles in the study area are bioturbated, muddy carbonates (see description below). Thin basal siliciclastic beds typically overlie sharp erosional surfaces and locally contain intraclastic horizons derived from the top of the underlying cycle. Mudstone and siltstone beds are dark red-brown to dark gray and slightly nodular to thin, irregularly bedded. Rare brachiopods occur locally. Quartz arenite beds are white to yellowish gray and structureless to vaguely laminated. Quartz sand is well rounded, well sorted, and fine to coarse grained; rare quartz grains have abraded quartz overgrowths. Scattered brachiopod, crinoid, and bryozoan fragments

Figure 5. Jefferson cycle lithofacies.
  a) Nodular to irregularly thin bedded, bioturbated dolowackestone/mudstone facies. Bioturbation is generally intense. Hammer for scale.
  b) Laminar stromatoporoid subfacies. Polished slab photograph. Note sheetlike growth morphology of stromatoporoids.
  c) *Amphipora* dolowackestone subfacies. Polished slab photograph. Mostly oblique and cross-sectional views of *Amphipora*.
  d) Photomicrograph of ooid dolograinstone/packstone facies. Ooids/coated grains replaced by coarse crystalline dolomite; lime mud or other fine-grained carbonate matrix between ooids is replaced by fine crystalline dolomite.

occur locally.

Interpretation

Basal siliciclastic beds represent transgressive facies which were deposited and/or reworked during initial submergence of the underlying cycle. Quartz sand and mud may have been transported out onto parts of the ramp which were adjacent to paleohighs. Paleoclimate during Jefferson deposition was arid to semi-arid, as indicated by the presence of evaporites (sulfates, halite) in the upper parts of cycles (see description below). Fine grained sand may have been transported by eolian and/or fluvial processes across subaerially exposed parts of the Jefferson ramp and subsequently reworked by marine processes (cf. Driese and Dott, 1984; Chan and Kocurek, 1988). The dark red color of some basal mudstones also may indicate initial deposition in subaerial environments. Conversely, the structureless nature, thin irregular bedding, presence of marine faunas, and coarse grained nature of some sands suggest that the basal siliciclastic beds were transported solely by marine currents during early stages of marine transgression. The abraded quartz overgrowths on rare quartz sand grains and regional subcrop patterns beneath the Jefferson Formation (Fig. 2) suggest that the sand probably was eroded and recycled from Lower Paleozoic sedimentary units that were uplifted and exposed in southwesternmost Montana prior to Jefferson deposition.

Bioturbated Mudstone/wackestone/dolostone

Description

Fine crystalline dolomudstone/wackestone, lime mudstone, and skeletal lime wackestone comprise the basal facies of most Jefferson cycles or overlie thin basal siliciclastic deposits in some cycles. These dark-colored basal carbonates (1-15 m thick) are massive to irregularly medium bedded and are thicker and more common on outer parts of the Jefferson ramp. In general, faunal abundance and diversity are low. Scattered skeletal grains include disarticulated brachiopods, crinoid and stromatoporoid fragments, calcispheres, rare favositid corals, calcareous algae, and bryozoans. Closely spaced hardgrounds (2-5 cm spacing) are common in the basal few meters of the lowermost cycles from the outer ramp (Fig. 5a). Hardgrounds commonly are bored (<1 cm diameter borings) and locally are stained by finely disseminated, microcrystalline, iron-bearing authigenic minerals; hardgrounds commonly are reddish orange on weathered surfaces. Most depositional layering is disrupted by anastomosing burrow networks, but gently undulatory to horizontal planar laminations are preserved locally near the transition into overlying shallower facies. Primary depositional textures in dolomud-stone/wackestone are identical to those observed in lime mudstone and skeletal wackestone, but they are slightly obscured by dolomitization. Bioturbated limestones form the thickest accumulations of undolomitized sediment in the entire Jefferson Formation; overlying shallower water facies generally are massively dolomitized.

Interpretation

Bioturbated mudstone/wackestone/ dolostone is the deepest water carbonate facies in the Jefferson cycles. Evidence for deeper water deposition includes the lack of abundant current-generated sedimentary structures, dark color, poor sorting, intense bioturbation, and sparse open marine faunas. Closely spaced hardgrounds at the bases of some outer ramp cycles probably record periods of very slow deposition during initial transgression of the outer ramp. Sediment accumulation rates probably were outpaced briefly during relative sea level rise, leading to formation of hardground surfaces. Rare planar to gently undulatory laminations are interpreted as storm layers that were not disrupted by burrowing infauna. These layers may reflect relative shallowing into storm wave base where the effects of storm activity on the sea floor were more frequent and burrowing organisms were less likely to destroy storm-generated layering.

Stromatoporoid Facies

Description

Massive to medium, irregularly bedded stromatoporoid layers conformably overlie basal bioturbated facies on the outer ramp or form the basal deposits of some inner ramp cycles. Three subfacies are recognizable based on stromatoporoid growth morphology. These subfacies are described below, in order of common stratigraphic succession.

Laminar Stromatoporoid Dolostone Subfacies - Fine to medium crystalline, dark-colored, medium to thick irregularly bedded dolomudstone/wackestone with laminar to sheet-like stromatoporoids occur locally above basal bioturbated beds in some outer ramp cycles (Fig. 5b). Individual stromatoporoids are <1-20 cm thick and 5 cm-2 m wide and locally contain borings. Some stromatoporoids are partially leached. Dolomudstone/wackestone matrix between stromatoporoids is bioturbated to locally laminated and contains scattered brachiopods, crinoids, corals, and fragments of laminar stromatoporoids.

Nodular-Hemispherical Stromatoporoid Dolostone Subfacies - Lenticular stromatoporoids often grade upward into beds containing stromatoporoids with nodular to hemispherical growth morphologies. In some beds, nodular stromatoporoids encrust laminar stromatoporoids. Laterally discontinuous hardgrounds/firmgrounds also form substrates for some nodular to hemispherical stromatoporoids. Individual stromatoporoids usually are no closer than 15 cm from one another; they often are overturned and bored. Matrix between stromatoporoids consists of bioturbated to locally laminated skeletal dolowackestone with poorly sorted and abraded brachiopod, coral, crinoid, and fragments of nodular-hemispherical stromatoporoids.

Amphipora Dolowackestone - Fine to coarsely crystalline *Amphipora* (stick-like stromatoporoids) dolowackestone layers occur as the basal deposits of inner ramp cycles or overlie beds with larger stromatoporoids in outer ramp cycles (Fig. 5c). *Amphipora* dolowackestone layers are dark gray to brown, 2-40 cm thick, and even to irregularly bedded. They commonly are interbedded with medium to coarsely crystalline skeletal dolostones. *Amphipora* typically are flat-lying, but are not highly abraded. Scattered bryozoans, brachiopods, and corals are intermixed with the *Amphipora*.

Interpretation

Stromatoporoid beds probably were deposited near fairweather wave base as indicated by the presence of current-generated horizontal laminations. Minor abrasion of stromatoporoid fragments and reorientation from growth position may be due to bioerosion or to reworking by fairweather and storm waves. Stromatoporoid growth morphologies also may be indicative of depositional conditions. Sheetlike, laminar growth morphologies may have been: 1) a strategy to evenly distribute the weight of the organism and prevent sinking into bottom muds; 2) a strategy to maximize the amount of available light across the surface of the entire stromatoporoid; or 3) a growth response to low sedimentation rates (Stearn, 1982; Hallock and Schlager, 1986). Nodular to hemispherical growth forms typically overlie and locally encrust sheetlike stromatoporoids and may reflect continued shallowing into more current-agitated environments. Firm substrates (i.e., laminar stromatoporoids, hardgrounds/firmgrounds, or less muddy substrates) may have been important for nodular to hemispherical stromatoporoids. Nodular/hemispherical stromatoporoids, with their high mass-to-basal surface area ratio, probably would have sunk into the bottom muds colonized by sheetlike stromatoporoids. *Amphipora* dolowackestone beds are the shallowest stromatoporoid facies in the Jefferson Formation and indicate relatively quiet water, "lagoonal" environments similar to interpretations for *Amphipora* facies from other Upper Devonian carbonate sequences (Wilson, 1967; Read, 1973; Wong and Oldershaw, 1980; Burchette, 1981). The flat-lying, unabraded nature of the *Amphipora* and lenticular, pod-like bedding characteristics of some *Amphipora* dolowackestone beds suggest that the *Amphipora* were not transported far prior to deposition and that some beds may be death accumulations of local, monospecific *Amphipora* communities.

Ooid Dolograinstone/packstone

Description

Medium to coarsely crystalline ooid dolograinstone/packstone (Fig. 5d) occurs as rare 10-50 cm thick layers in inner ramp cycles. Where present, this lithofacies forms the basal deposits of individual cycles that are less than 2 m thick. Most oolitic beds are structureless, but bidirectional tabular cross-bed sets occur locally. Ooid dolograinstone/packstone (with scattered peloids and intraclasts) also locally occurs as matrix between stromatolitic/thrombolitic heads in upper parts of cycles.

Interpretation

Ooid sands from modern tropical to sub-tropical marine environments reflect shallow, high energy settings. In many modern settings, ooid sands form on top of or between preexisting submarine topographic highs, leeward of islands, or where tidal currents are amplified by basin shape or sea floor topography (Ball, 1967; Loreau and Purser, 1973; Hine, 1977; Hine et al., 1981; Halley et al., 1983).

Ooid dolograinstone/packstone facies from Jefferson cycles probably reflect similar current-agitated, shallow marine environments. Restriction of oolitic sediments to basal facies in inner ramp cycles may be related to maximum water depth attained at the beginning of a cycle when the ramp initially was flooded by a relative rise in sea level. Oolitic deposits formed only on the inner ramp where maximum water depth would have been only on the order of a few meters.

Laminated Dolostone

Description

Fine to medium crystalline, planar to ripple laminated dolostones occur as thin layers (<30 cm thick) in the upper parts of some cycles. Planar laminations typically are about 1 cm thick and grade laterally and vertically into 1-2 cm thick current ripple cross-laminae. Rare skeletal grains include articulated ostracodes, calcispheres, and fragments of calcareous algae, brachiopods, and crinoids.

Some laminated dolostones grade laterally into stromatolitic heads. Stromatolites and thrombolites are relatively rare in Jefferson cycles. They are most common near the tops of outer ramp cycles. Individual heads are 5-40 cm wide and up to 50 cm high. Stromatolites have smooth, well laminated, domal heads ("SH" heads) to knobby, slightly digitate heads ("LLH" heads; Logan et al., 1964). Thrombolites (*sensu* Aitken, 1967) have similar dimensions as well laminated stromatolites, but layering is highly irregular and sediment microfabrics are clotted. Thrombolite heads are irregularly digitate to domal (Fig. 6a) and are highly bored locally. Some thrombolites contain abundant skeletal fragments trapped within their clotted internal fabric. Desiccation features are not present in laminated dolostones.

Interpretation

Laminated, fine crystalline dolostone and stromatolitic/thrombolitic dolostone were deposited in shallow subtidal to possibly lower intertidal environments. Current-ripple and planar laminations indicate deposition by directional currents. Low diversity faunas in some beds suggest somewhat restricted environmental conditions or high energy mobile sand substrates; abraded skeletal grains may have been transported from less restricted environments by storm or fairweather currents. Stromatolite/ thrombolite morphologies also suggest formation in current-agitated subtidal to lower intertidal environments (cf. Logan et al., 1964; Logan et al., 1974). Absence of desiccation features suggests that exposure was, at most, intermittent and/or areally limited.

Cryptalgalaminite-Evaporite Facies

Description

Thin cryptalgalaminated dolomudstone (<10 cm to 1.5 m thick) forms the capping facies in many Jefferson cycles (Fig. 6b). Individual planar laminations are smooth to crinkly and submillimeter to a few millimeters thick. Shallow mudcracks, prism cracks, and tepee structures occur locally. Dolomite and calcite cements occur locally as pseudomorphs after evaporite minerals (halite, gypsum, anhydrite) in surface exposures of the cryptalgalaminite facies (Fig. 6c). In subsurface cores from boreholes to the north of the study area (Cascade and Teton Counties, Montana), bedded to nodular evaporites are interbedded with cryptalgalaminite facies in the upper parts of cycles.

Interpretation

Cryptalgalaminite-evaporite lithofacies formed after shallowing of depositional surfaces to intertidal and supratidal levels. Abundant desiccation features indicate extended periods of subaerial exposure. Preserved evaporite deposits in subsurface cores and evaporite pseudomorphs in surface

516

Figure 6. Jefferson cycle lithofacies.
 a) Thrombolitic heads. Note irregular digitate thrombolite structure in upper left
    (arrow) and crudely laminated domal head, outlined by dashed line, near 15 cm long
    scale.
 b) Sharp erosional cycle contact. Light gray cryptalgalaminated dolomudstone
    (intertidal to supratidal) is cap facies of underlying cycle. Basal facies of
    overlying cycle is dark gray, highly bioturbated dolowackestone (deep subtidal).
    Note laminations which drape over local microtopography on underlying cycle cap
    (arrow). Black part of staff on left side of photograph is 30 cm long.
 c) Halite pseudomorphs in surface exposure of cryptalgalaminite facies. Individual
    halite pseudomorphs 2-5 mm across.
 d) Solution collapse breccia. Note angular breccia clasts of mostly cryptalgalaminated
    dolomudstone (randomly oriented arrows). Scale divisions in centimeters.
    Stratigraphic up, as indicated by arrow labelled "UP", is toward right side of
    photograph.

exposures indicate arid climatic conditions. Evaporite pseudomorphs formed either during early, near-surface, meteoric diagenesis when tidal flat facies shallowed to sea level and subsequently were subaerially exposed, or during very late meteoric diagenesis after Late Cretaceous-Tertiary uplift of the Jefferson Formation (Smith and Dorobek, 1989).

Solution Collapse Breccia

Description

Solution collapse breccia frequently occurs in the upper parts of cycles from surface exposures of the Jefferson Formation. These breccias are not common in cores from the subsurface. A major solution breccia horizon also occurs in some areas at the very top of the Jefferson Formation, beneath the overlying Three Forks Formation. Breccia clasts consist of very angular sand- to boulder-size host dolostone with fine-grained, dark to yellow-gray dolomitic mud matrix (Fig. 6d). Breccia clasts include all cycle lithologies, but cryptalgalaminite and laminated dolostone are the most abundant clast types. Some breccia clasts have truncated stylolites and/or tectonic fractures at their margins and some breccia horizons are crosscut by later tectonic fractures. Rarely, sediments from overlying beds locally drape over the irregular upper surface of some brecciated or eroded cryptalgalaminite layers (Fig. 6b).

Interpretation

Most solution collapse breccias appear

to have formed after uplift of Jefferson rocks in Late Cretaceous-Tertiary time. Truncated stylolites and truncated tectonic fractures at clast margins suggest many solution breccias formed after deep burial pressure solution of host dolostone and after some phases of tectonic fracturing. Dissolution of evaporites by near-surface meteoric waters may explain many of the breccias in the Jefferson Formation. Meteoric dissolution probably occurred when evaporite horizons were uplifted in Late Cretaceous-Tertiary time and entered near surface meteoric environments. This also would explain the less frequent occurrence of solution breccias in the Jefferson Formation from the subsurface.

However, some thin solution breccia horizons formed soon after deposition of intertidal/supratidal facies which cap individual Jefferson cycles. These breccia horizons occur entirely within intertidal/supratidal facies. Subtidal sediments that drape over the irregular topography of underlying brecciated cycle caps (Fig. 6b) indicate that these thin breccias formed soon after deposition of underlying cycle caps. Intertidal/supratidal caps on some cycles apparently were subaerially exposed during relative sea level falls. The cycle tops became recharge surfaces for meteoric waters during subaerial exposure. Meteoric waters dissolved evaporite nodules and layers, producing local solution collapse breccias.

GENERAL LITHOFACIES TRENDS AND LATERAL VARIATION IN CYCLE BOUNDARIES

Lithofacies composition of individual Jefferson cycles varies regionally along transects normal to strandline positions on the Jefferson ramp. Most outer ramp cycles are thicker than inner ramp cycles (Fig. 7). Deep subtidal facies (with hardground horizons) also are thicker and more common as the basal facies of outer ramp cycles. In contrast, basal facies of inner ramp cycles are dominated by stromatoporoid facies, oolitic beds, and various laminated dolostones.

The greater thickness of individual cycles and the dominance of basal, deep subtidal facies on the outer ramp is the result of two factors: 1) regional paleoslope and 2) incomplete aggradation of the outer ramp during some cycle intervals before an ensuing sea level rise. Because of the regional paleoslope, the outer ramp would have been deeper than the inner ramp during any relative sea level rise, regardless of the mechanism for rise. Therefore, subtidal sedimentation would have been greater on the outer ramp during most cycle intervals (cf. Hardie, 1986). Hardgrounds in basal facies of outer ramp cycles indicate slow sedimentation rates and suggest rapid, high amplitude relative sea level rises caused incipient drowning (Kendall and Schlager, 1981) of the outer ramp. Basal lithofacies of inner ramp cycles (stromatoporoid facies, ooid dolograinstone/packstone, laminated

dolostones) also reflect maximum depth of submergence, which generally would have been much shallower than on the outer ramp.

Boundaries between outer ramp cycles often are gradational, while most inner ramp cycles are bounded by subaerial erosion surfaces. For some outer ramp locations, thin cryptalgalaminite caps (which lack subaerial exposure features) or shallow subtidal facies grade upward over a few centimeters into deep subtidal facies of the overlying cycle (Fig. 7). This indicates that prograding tidal flats barely reached the outer ramp before being submerged by the next relative sea level rise. In fact, cycle boundaries on the outer ramp are often difficult to determine as they may be cryptic surfaces *within* subtidal facies (e.g., Camp Creek section, Fig. 7). In contrast, disconformities frequently cap inner ramp cycles and indicate periods of subaerial exposure prior to deposition of the overlying cycle.

CYCLE ANALYSIS

Quantitative analysis of Jefferson cycles is difficult because of: 1) the highly variable thicknesses and lithofacies composition of individual Jefferson cycles; 2) the inability to correlate individual cycles across the entire ramp; 3) the lack of very high resolution time lines (<105 yr spacing) across the Jefferson ramp; and 4) the effects of differential subsidence across the Jefferson ramp. However, understanding these limitations provides some insight into the mechanisms which produced Jefferson cycles.

Missing Barrier Facies and Tidal Flat Progradation

Well developed barrier facies (e.g., reefs or carbonate sand belts) are absent from the Jefferson ramp of southwestern Montana and adjacent parts of east-central Idaho. Low energy conditions across the Jefferson ramp may have prevented accumulation of reefs or shoals. The dominance of fine-grained subtidal facies which lack abundant current-generated sedimentary structures also indicates ramp-wide, generally low energy conditions. Dampening of waves and deflection of oceanic circulation systems may have been accomplished by the areally extensive, low relief uplifts which extended across the Jefferson platform (Fig. 4). Other workers have suggested that dominant wind patterns and oceanic circulation systems during the Late Devonian moved from roughly northeast to southwest across western Canada and western Montana (Klovan, 1974; Heckel and Witzke, 1979; Witzke and Heckel, 1988). Therefore, the east-west trending Central Montana Uplift (Fig. 4) would have been an effective obstruction which generated a broad low energy shadow on its southern, leeward side. All of southwestern Montana would have been included in this low energy shadow.

Tidal flat location and progradation directions also were affected by

518

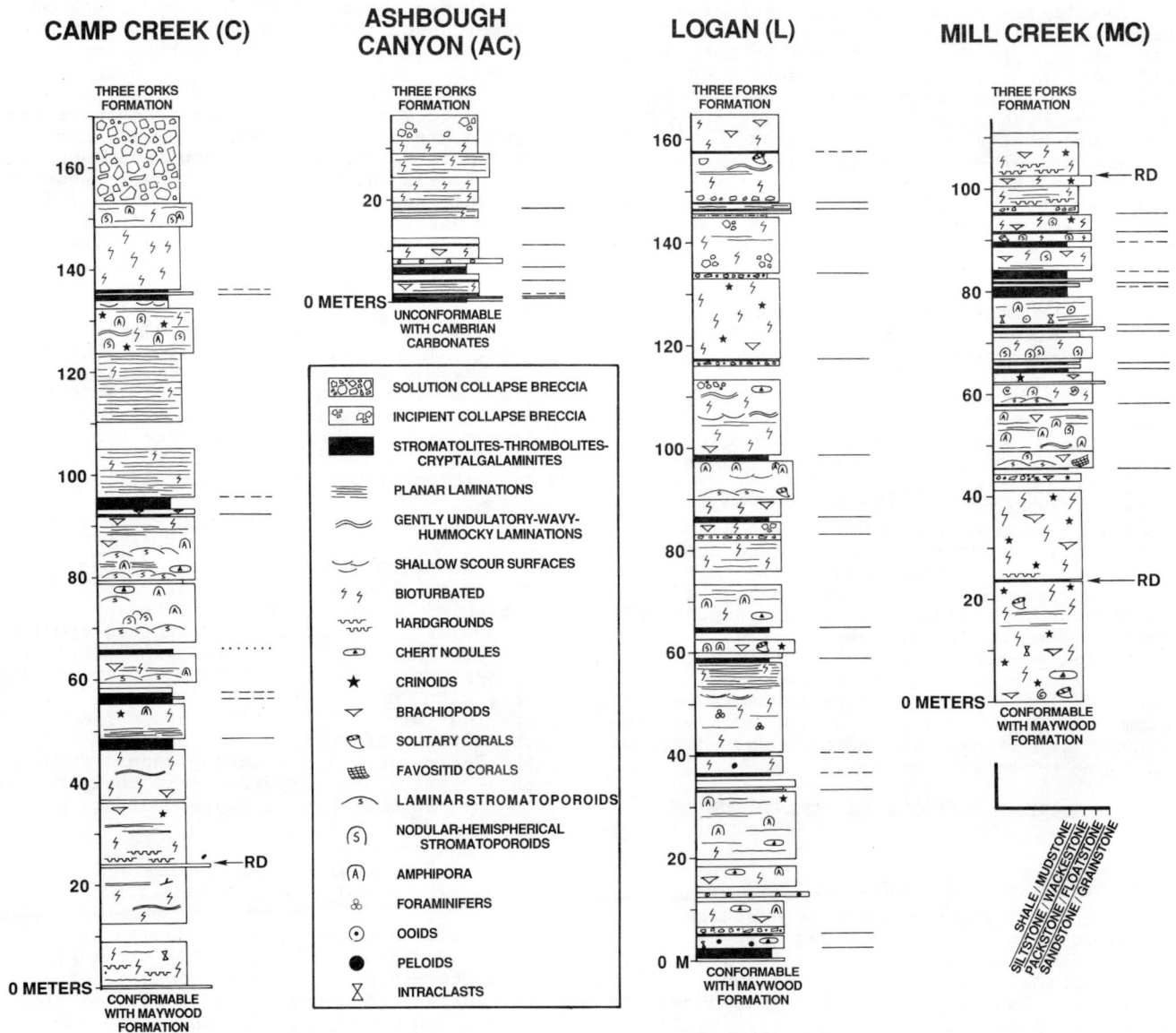

Figure 7. Selected measured sections which illustrate variation in lithofacies composition and thickness of cycles. Section locations shown on Figure 4. Cycle boundaries indicated to right of stratigraphic columns: solid line = sharp, typically erosional contact; dashed line = gradational contact; dotted line = approximate contact (actual contact may be covered); "RD" = rapid deepening surface. Rapid deepening surfaces interpreted as incipient drowning events and probably represent cycle boundaries. Lower "RD" at Mill Creek manifested as 5 cm thick black shale, overlain by deep subtidal deposits with multiple hardgrounds.

paleotopography across the Jefferson platform. Tidal flats probably prograded away from regional paleohighs (like the Yellowstone Park, Southern Beaverhead Mountains, and Central Montana Uplifts) or from the regional strandline to the southeast (Fig. 4). Cycles from the middle ramp are halfway between paleotopographic highs and the eastern paleostrandline. Cryptalgalaminite caps on some of these cycles require that tidal flats prograded 200 km from the nearest paleostrandline position on the platform during development of a single cycle. Tidal flats would have prograded at

average rates of 0.8 to 20 km/1000 yr, assuming 200 km progradation distance, 20,000 to 500,000 yr cycle periods, and tidal flat progradation occurred during the regressive phases of cycles, arbitrarily chosen to be one-half of the total cycle period. Similar progradation rates have been demonstrated for Holocene tidal flats from the Persian Gulf and the Bahamas (Kinsman, 1969; Patterson and Kinsman, 1977; Hardie, 1986). However, these estimates may represent maximum progradation rates for Jefferson tidal flats. Lower progradation rates are necessary if tidal flats built away from local islands or shoals

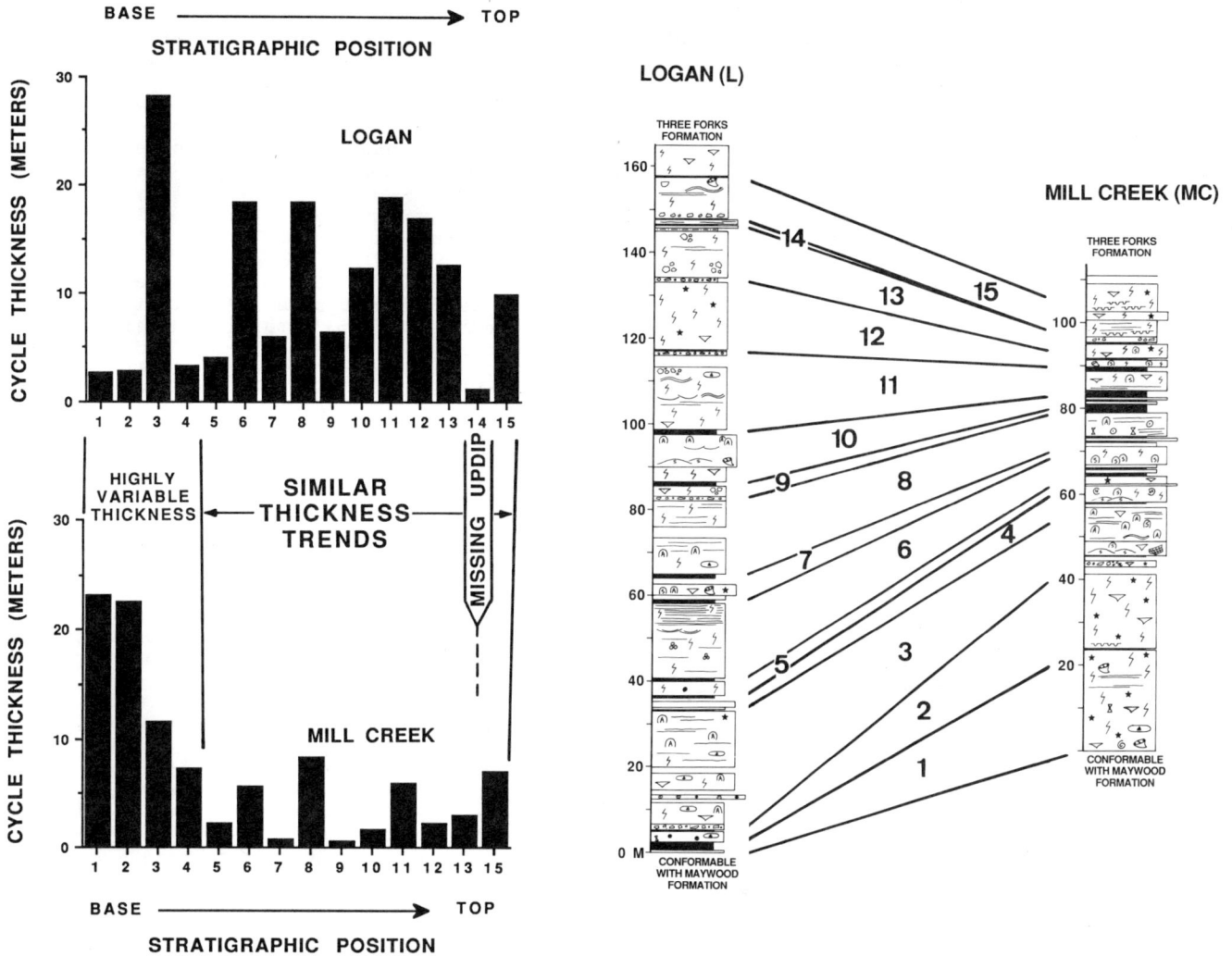

Figure 8. Method used for correlating middle ramp cycles at Logan and Mill Creek. Cycle thicknesses plotted according to stratigraphic occurrence on left side of figure. Comparison of thickness trends between sections shows that the first four cycles at both Logan and Mill Creek have highly variable thicknesses that are not comparable between these sections. This variablity is attributed to differential subsidence between Logan and Mill Creek and/or to antecedent topography prior to Jefferson deposition. However, cycles 5 through 15 show very similar changes in cycle thickness, that is, a thin (or thick) cycle at Logan is recorded as a thin (or thick) cycle at Mill Creek; cycle correlations (shown on right side of figure) are based on recognition of similar thickness trends between sections. Cycle 14 at Logan is very thin (~1.5 m) and was not recorded on updip parts of the middle ramp (e.g., at Mill Creek). This is the only technique which provides some measure of certainty to cycle correlations and may be useful on other platforms where cycle bounding surfaces cannot be traced laterally.

across middle parts of the Jefferson ramp. These shoal facies may have built up to shallow subtidal or intertidal depths and then acted as local paleohighs from which tidal flats prograded during relative stillstands or lowstands of sea level.

Cycle Correlations

It is difficult to lithostratigraphically correlate individual cycles across the Jefferson ramp. Cycle boundaries cannot be traced with certainty between measured sections spaced less than 30 km apart. Attempts to simply connect cycle

boundaries results in uncertain correlations because the same number of cycles do not occur even in adjacent measured sections. Cycle thicknesses between adjacent or widely spaced sections also are extremely variable. Therefore, larger scale correlation of stratal packages is not obvious as it is not clear which groups of cycles should be bundled together to form larger scale depositional sequences (cf., Goldhammer et al., 1987; Sarg, 1988).

An alternative method for correlation is shown in Figure 8, where cycle thicknesses for two middle ramp sections are plotted relative to stratigraphic occurrence of each cycle. These "thickness distribution" plots were used to lithostratigraphically correlate individual cycles (Fig. 8). The two localities used are Logan and Mill Creek; Mill Creek is closer to the eastern paleostrandline (Fig. 4).

The thicknesses of the first four cycles at both locations are extremely variable (Fig. 8). The Mill Creek cycles show a decreasing trend in thickness from the basal cycle (23.1 m) to cycle 4 (7.3 m); these first four cycles comprise 64.5 m (63%) of the total section. In contrast, the first four cycles at Logan show no apparent trends in thickness and comprise 36.9 m (23%) of the total section. This extreme variation in cycle thickness between sections which are 90 km apart (palinspastic distance) is attributed possibly to the inherited topography on the platform prior to the start of Jefferson deposition or to local effects of differential subsidence (discussed below).

More importantly, the thicknesses of cycles 5 through 15 indicate the same exact thickness trends at both sections. Only cycle 14 at Logan does not fit this trend. Cycle 14 at Logan is only 1.2 m thick and apparently is not recorded at the updip Mill Creek section, hence Mill Creek has one less cycle than Logan (Fig. 8).

Clearly this correlation technique is tenuous. Confidence in cycle correlations hinges on the confidence level placed on the stratigraphic distribution of cycle thickness. In addition, this technique only has been successful for correlating Jefferson cycles from middle ramp locations where cycle boundaries are easily defined and where subsidence was great enough so that most cycles were recorded. The technique is not successful at outer ramp locations where cycle boundaries are often difficult to identify. Lack of apparent stratigraphic trends in cycle thicknesses at many sections, regardless of paleogeographic location, and numerous missing cycles at inner ramp locations also limits the usefulness of this technique. This technique also is useful for illustrating the highly variable thicknesses of individual cycles across the Jefferson ramp and it might be useful for cycle correlation on other platforms where cycle bounding surfaces can not be traced laterally because of inadequate exposure.

Effects of Late Devonian Eustasy on Cycle Development

Jefferson cyclic sedimentation most likely was affected by eustatic sea level fluctuations (Dorobek and Smith, 1989). Compelling evidence which suggests that eustatic sea level fluctuations may have been the dominant driving force for cyclicity in the Jefferson Formation is the similar number of cycles found in time equivalent strata of the Williston Basin. Sixteen Frasnian cycles were deposited in the Williston Basin (Wilson, 1967, 1985; Wilson and Pilatzke, 1987), while fifteen to seventeen cycles are recorded on middle parts of the Jefferson

ramp in southwestern Montana.

Eustatic sea level fluctuations caused by tectonic processes probably do not occur at rates fast enough nor with the short term periodicities calculated for individual Jefferson cycles (20 to 400 kyr, estimated by dividing the decompacted thicknesses of individual Jefferson cycles by average rate of subsidence at the same locality). Other mechanisms which may account for Late Devonian eustasy include:

1) Glacio-eustasy: Calculated durations for most Jefferson cycles fall within the Milankovitch band of astronomical cycles (Berger et al., 1984; Arthur and Garrison, 1986). Therefore, Jefferson cycles might be attributable to eustatic sea level fluctuations related to growth and melting of continental glaciers. Sedimentologic evidence for major continental glaciation during the Frasnian is poorly constrained (Hambrey and Kluyver, 1981; Rocha-Campos, 1981a, 1981b; Caputo and Crowell, 1985), but major continental glaciation seems well-established by mid-Famennian time (Caputo and Crowell, 1985). Frasnian glacio-eustatic fluctuations might have occurred during incipient stages of major continental glaciation which culminated in extensive Famennian glaciation.

2) Desiccation of isolated ocean basins: Eustatic sea level rise caused by desiccation of isolated ocean basins may outpace platform carbonate sedimentation (Berger and Winterer, 1974; Hsü and Winterer, 1980; Schlager, 1981). This seems unlikely to have caused Jefferson cyclicity because global sea level probably was at a highstand during the Frasnian (House, 1975; Vail et al., 1977; Hallam, 1984) and multiple basin flooding/desiccation cycles would have had to occur with frequencies similar to those calculated for Jefferson platform cycles.

3) Sea level fluctuations caused by geoidal variations: Mörner (1976, 1981) has argued that global sea level is affected regionally and locally by sinusoidal differences in elevation of geodetic sea level. The magnitude of this effect on intrabasinal sea level may be as much as 10 to 100 m. This difference in elevation also may migrate across the earth's surface as "wavelike oscillations" and the concomitant sea level fluctuations can occur over similar time scales as those estimated for Jefferson cycle durations (Mörner, 1976). It is impossible to evaluate geoidal variations as a cause for "regional eustasy" across the Jefferson platform, but it can not be ruled out as another possible mechanism for generating Jefferson cycles. If geoidal variations did produce Jefferson cycles, then they had to occur at frequencies similar to those calculated for Jefferson cycles.

Effects of Differential Subsidence and Topography Across the Jefferson Ramp

Differential subsidence and topography across the Jefferson ramp influenced sedimentation patterns and partially account for the highly variable thicknesses of Jefferson cycles. Several paleohighs and troughs extended across Montana and adjacent parts of Idaho prior to and during Jefferson deposition (Sandberg and Mapel, 1967; Sandberg et al., 1983; Peterson, 1986). These paleostructures coincide with Precambrian structural trends (Peterson, 1986; Tonnsen, 1986; Winston, 1986) and underwent tectonic inversion (i.e., paleohighs became troughs and vice versa) from Late Devonian to Early Mississippian time (Dorobek et al., in press and this volume). These paleostructures across the Jefferson ramp had wavelengths of 50-200 km and amplitudes of about 50-350 m (Dorobek et al., in press and this volume). The coincidence of these paleostructures with Precambrian structural trends suggests that ancient zones of weakness in the lithosphere were reactivated during the Antler orogeny (Dorobek et al., in press and this volume). The topography produced by these paleostructures influenced Jefferson cycle development by producing local and regional depositional gradients across the Jefferson ramp.

Topography across the platform also may have developed following deposition of individual cycles. Shallow water, subtidal areas in the interiors of modern carbonate platforms may have several meters (generally less than 10 m) of local relief. This relief generally is produced by variable sediment accumulation rates across the inner platform. Higher accumulation rates might occur where local carbonate sand shoals or bioherms/biostromes are located; intervening subtidal areas have lower accumulation rates and form low areas. Similar variations in sediment accumulation rates might have occurred across the Jefferson platform and produced irregular, local topography. The antecedent irregular topography produced during one cycle might have been filled in during deposition of the next successive cycle, which in turn, produced its own irregular topography along its upper bounding surface. These patterns of variable sediment accumulation across the Jefferson platform might explain some of the variation in cycle thicknesses. However, highly prograded Holocene tidal flats produce relatively flat, smooth depositional surfaces. By analogy, prograded tidal flats on the Jefferson ramp probably also produced a relatively flat surface across the middle ramp. If it is accepted that tidal flat caps on middle ramp cycles reflect regional tidal flat progradation, then relative sea level rises at the beginning of each cycle would have flooded flat, smooth surfaces. Changes in cycle thickness across the middle ramp can then only be attributed to compaction or to variable subsidence across the platform. Thickness differences which might be due to compaction probably are negligible in that the same lithofacies occur at both sections, and burial histories are nearly identical.

Table 1. Changes in slope/cycle between Logan and Mill Creek, MT (90 km apart)

CYCLE NO.	THICKNESS AT LOGAN	THICKNESS AT MILL CREEK	DIFFERENCE IN THICKNESS	SLOPE/ CYCLE
1	2.7 m	23.1 m	-20.4 m	-22.7 cm/km
2	2.9	22.5	-19.6	-21.8
3	28.2	11.6	16.6	18.4
4	3.3	7.3	-4.0	-4.4
5	4.0	2.3	1.7	1.9
6	18.4	5.7	12.7	14.1
7	6.0	0.8	5.2	5.8
8	18.3	8.3	10.0	11.1
9	6.4	0.6	5.8	6.4
10	12.2	1.6	10.6	11.8
11	18.8	6.0	12.8	14.2
12	16.8	2.2	14.6	16.2
13	12.6	3.0	9.6	10.7
14	1.2	0	1.2	1.3
15	9.9	7.0	2.9	3.2

However, differential subsidence may have had a more dominant effect on Jefferson deposition. Although time-slice isopach maps were not generated for the Jefferson Formation in the study area, Hurley (1962) demonstrated that significant differential subsidence and tectonic inversion occurred over very short distances in the Sweetgrass Arch area of northwestern Montana during Jefferson time. In addition, isopach maps and stratigraphic cross-sections of Upper Devonian, Lower-Middle Mississippian, and Upper Mississippian-Lower Pennsylvanian strata also show significant differential subsidence and tectonic inversion across Montana (Dorobek et al., in press and this volume). Reactivation of the Precambrian structures and the tectonic inversion probably was a complex response to both vertical and horizontal tectonic loads related to the Antler orogeny (Dorobek et al., in press and this volume). Quantitative subsidence curves also show that Frasnian subsidence rates and magnitudes varied significantly at several locations on the Jefferson platform (Dorobek et al., in press and this volume). It should be noted that some of this variation in apparent subsidence rates and magnitudes may be a record of eustatic sea level rise; Jefferson facies onlapped the irregular antecedent topography that existed across Montana at the start of the Frasnian and what appears to be variable subsidence on subsidence curves, may actually be variation in the amount of transgression and/or onlap (Dorobek et al., in press and this volume).

If we accept that the cycle correlations shown in Figure 8, it is possible to calculate changes in slope across the Jefferson platform through time. Only sections from the middle ramp provide an estimate of the amount of differential subsidence. Local slopes on the middle ramp were calculated for individual cycles by taking the difference in thickness for each correlated cycle at Logan and Mill Creek (Fig. 8) and dividing this by the palinspastic distance between the locations (Table 1). Slopes on the ramp vary between <2 to 16 cm/km (Table 1). While these still are low slopes typical of carbonate ramps, the order of magnitude difference in the range of slopes and the rapid changes in slope between some cycles suggests significant changes in topography.

These data suggest that differential subsidence and/or irregular topography produced by variable sedimentation across the Jefferson platform affected cycle development. The extreme variation in cycle thickness and apparently random bundling of stratal sequences across the ramp might be explained by these mechanisms. Neither of these mechanisms can be unequivocally demonstrated at the short time scales represented by individual Jefferson cycles. However, isopach maps and subsidence curves for Devonian through Mississippian strata (Dorobek et al., this volume and in press) demonstrate that differential subsidence and tectonic inversion of paleostructures clearly occurred across Montana over longer time scales. Thus, differential subsidence is inferred to have occurred during the Frasnian across Montana and probably influenced Jefferson cycle development.

SUMMARY

Identifying the exact mechanism(s) that caused cyclical sedimentation on the Jefferson ramp is difficult. Eustatic sea level fluctuations may be the dominant cause of cyclic sedimentation on linearly subsiding platforms (mature passive margins, some isolated platforms), but in convergent settings, both differential subsidence and eustatic sea level fluctuations may play key roles in cycle development and facies architecture. The Jefferson Formation may record both eustatic and differential subsidence "signals" because it was deposited on a platform that was adjacent to an incipient foredeep.

In summary, regional study of the Frasnian Jefferson Formation of southwestern Montana suggests that:

1) The Jefferson Formation was deposited on a broad, shallow marine ramp that graded basinward into an incipient(?) foreland basin. Therefore, the Jefferson Formation in Montana contains the critical stratigraphic record of a carbonate ramp's response during foundering of an earlier passive margin and its transition to a convergent margin. A number of paleotopographic highs greatly influenced deposition on the ramp.

2) Jefferson ramp facies consist of shallowing-upward carbonate cycles. Cycles may have been caused by eustatic sea level fluctuations, but other factors influenced cycle thickness and lithofacies composition. The cycles can not be correlated confidently across the ramp using standard lithostratigraphic techniques. Variability in cycle thickness and lithofacies composition possibly reflects: a) large-scale topography produced by regional structures across Montana; b) small-scale topography produced by variable sediment accumulation rates; and c) differential subsidence across the ramp. The patterns of differential subsidence appear to be related to reactivation of Precambrian structural

trends during the Antler orogeny.

3) The cycles developed on a ramp which lacked laterally extensive barrier facies (at least within the study area). The absence of extensive barrier facies in the study area suggest relatively low energy conditions across the ramp.

4) Jefferson tidal flats may have prograded from paleohighs and the regional strandline to the southeast. Tidal flats prograded at average rates of 0.8 to 20 km/1000 yr, which fall within the range of progradation rates for some Holocene tidal flats. Lower progradation rates are necessary if progradation was initiated from local shoals across the Jefferson ramp.

ACKNOWLEDGMENTS

Supported by a grant from Texaco U.S.A. and partially by a faculty summer stipend during tenure at Washington State University. Acknowledgment also is made to J. Fred Read for numerous discussions concerning cyclicity in the stratigraphic record and for comments on an early version of this manuscript. J. Dravis, N. Hurley, S. Reid, T. Smith, and P. Wong provided constructive criticisms of an early version of the manuscript. D. Zenger and J. Cooper provided useful reviews of the final manuscript. J. Harris, D. Quigley, S. Reid, T. Smith, and P. Whitsitt served as field assistants at various stages of the research.

BIBLIOGRAPHY

Aitken, J.D., 1967, Classification and environmental significance of cryptalgal limestones and dolomites, with illustrations from the Cambrian and Ordovician of southwestern Alberta: Journal of Sedimentary Petrology, v. 37, p. 1163-1178.

Andrichuck, J.M., 1962, Regional stratigraphic analysis of Devonian System in Wyoming, Montana, southern Saskatchewan, and Alberta: American Association of Petroleum Geologists Bulletin, v. 35, p. 2368-2408.

Arthur, M.A., and Garrison, R.E., 1986, Cyclicity in the Milankovitch band through geologic time: an introduction: Paleoceanography, v. 1, p. 369-372.

Baars, D.L., 1972, Devonian System, in Mallory, W.W., ed., Geologic Atlas of the Rocky Mountain Region: Rocky Mountain Association of Geologists, Denver, Colorado, p. 90-99.

Ball, M.M., 1967, Carbonate sand bodies of Florida and the Bahamas: Journal of Sedimentary Petrology, v. 37, p. 556-591.

Benson, A.L., 1966, Devonian stratigraphy of western Wyoming and adjacent areas: American Association of Petroleum Geologists Bulletin, v. 50, p. 2566-2603.

Berger, A., Imbrie, J., Hays, J., Kukla, G., and Saltzman, B., 1984, Milankovitch and Climate: D. Reidel Publishing, Boston, 895 p.

Berger, W.H., and Winterer, E.L., 1974, Plate stratigraphy and the fluctuating carbonate line, in Hsü, K.J., and Jenkyns, H.C., eds., Pelagic Sediments on Land and Under the Sea: International Association of Sedimentologists Special Publication 1, p. 11-48.

Burchette, T.P., 1981, European Devonian reefs: a review of current concepts and models, in Toomey, D.F., ed., European Fossil Reef Models: Society of Economic Paleontologists and Mineralogists Special Publication No. 30, p. 85-142.

Burchfiel, B.C., and Davis, G.A., 1972, Structural framework and evolution of the southern part of the Cordilleran orogen, western United States: American Journal of Science, v. 272, p. 97-118.

Caputo, M.V., and Crowell, J.C., 1985, Migration of glacial centers across Gondwana during Paleozoic Era: Geological Society of America Bulletin, v. 96, p. 1020-1036.

Chan, M.A., and Kocurek, G., 1988, Complexities in eolian and marine interactions: Processes and eustatic controls on erg development: Sedimentary Geology, v. 56, p. 283-300.

Churkin, M., Jr., 1962, Facies across Paleozoic miogeosynclinal margin of central Idaho: American Association of Petroleum Geologists Bulletin, v. 46, p. 569-591.

Cutler, W.G., 1983, Stratigraphy and sedimentology of the Upper Devonian Grosmont Formation, northern Alberta: Canadian Petroleum Geology, v. 31, p. 282-325.

Dickinson, W.R., 1977, Paleozoic plate tectonics and the evolution of the Cordilleran continental margin, in Stewart, J.H., Stevens, C.H., and Fritsche, A.E., eds., Paleozoic Paleogeography of the Western United States, Pacific Coast Paleogeography Symposium 1: Society of Economic Paleontologists and Mineralogists, Pacific Section, p. 137-155.

Dorobek, S.L., 1987, Cyclic platform dolomites and platform-to-basin transition of the Devonian Jefferson Formation, Montana and Idaho [abs.]: Fourth Annual Midyear Meeting, Society of Economic Paleontologists and Mineralogists, Abstracts with Programs, v. IV, p. 22.

Dorobek, S.L., and Read, J.F., 1986, Sedimentology and basin evolution of the Siluro-Devonian Helderberg Group, central Appalachians: Journal of Sedimentary Petrology, v. 56, p. 601-613.

Dorobek, S.L., and Smith, T.M., 1989, Cyclic sedimentation and dolomitization history of the Devonian Jefferson Formation, southwestern Montana, in French, D.E., and Grabb, R.F., eds., Geological Resources of Montana, 1989 Field Conference Guidebook, Montana Centennial Edition: Montana Geological Society, v. 1, p. 31-46.

Dorobek, S.L., Reid, S.K., Elrick, M., Bond, G.C., and Kominz, M.A., in press, Foreland response to episodic convergence: Subsidence history of the Antler foredeep and adjacent cratonic platform areas, Montana and Idaho, in Franseen, E., and Watney, L., eds., Sedimentary Modeling: Computer Simulation and Methods for Improved Parameter Definition: Kansas Geological Society Special Volume.

Driese, S.G., and Dott, R.H., Jr., 1984, Model for sandstone-carbonate "cyclothems" based on upper member of Morgan Formation (Middle Pennsylvanian) of northern Utah and Colorado: American Association of Petroleum Geologists Bulletin, v. 68, p. 574-597.

Dover, J.H., 1980, Status of the Antler orogeny in central Idaho - Clarification and constraints from the Pioneer Mountains, in Fouch, T.D. and Magathan, E.R., eds., Paleozoic Paleogeography of the West-Central United States, Rocky Mountains Paleogeography Symposium 1: Society of Economic Paleontologists and Mineralogists, Rocky Mountain Section, p. 371-386.

Goldhammer, R.K., Dunn, P.A., and Hardie, L.A., 1987, High frequency glacio-eustatic sea level oscillations with Milankovitch characteristics recorded in the Middle Triassic platform carbonates in northern Italy: American Journal of Science, v. 287, p. 853-892.

Goodwin, P.W., and Anderson, E.J., 1985, Punctuated aggradational cycles: A general hypothesis of episodic stratigraphic accumulation: Journal of Geology, v. 93, p. 515-533.

Hallam, A., 1984, Pre-Quaternary sea-level changes: Annual Reviews Earth and Planetary Sciences, v. 12, p. 205-243.

Halley, R.B., Harris, P.M., and Hine, A.C., 1983, Bank margin environment, in Scholle, P.A., Bebout, D.G., and Moore, C.H., eds., Carbonate Depositional Environments: American Association of Petroleum Geologists Memoir 33, p. 463-506.

Hallock, P., and Schlager, W., 1986, Nutrient excess and the demise of coral reefs and carbonate platforms: Palaios, v. 1, p. 389-398.

Hambrey, M.J., and Kluyver, H.M., 1981, Evidence of Devonian or Early Carboniferous glaciation in the Agades region of Niger, in Hambrey, M.J., and Harland, W.B., eds., Earth's Pre-Pleistocene Glacial Record, International Geological Correlation Programme, Project 38: Pre-Pleistocene Tillites: Cambridge University Press, Cambridge, p. 188-190.

Hardie, L.A., 1986, Stratigraphic models for carbonate tidal-flat deposition, in Hardie, L.A., and Shinn, E.A., Carbonate Depositional Environments, Modern and Ancient. Part 3: Tidal Flats: Colorado School of Mines Quarterly, v. 81, no. 1, p. 59-74.

Heckel, P.H., and Witzke, B.J., 1979, Devonian world paleogeography determined from distribution of carbonates and related lithic paleoclimatic indicators, in House, M.R., Scrutton, C.T., and Bassett, M.G., eds., The Devonian System: Palaeontological Association, London, Special Papers in Palaeontology,

524

no. 23, p. 99-123.

Hine, A.C., 1977, Lily Bank, Bahamas: history of an active oolite sand shoal: Journal of Sedimentary Petrology, v. 47, p. 1554-1581.

Hine, A.C., Wilber, R.J., and Neumann, A.C., 1981, Carbonate sand bodies along contrasting shallow bank margins facing open seaways in northern Bahamas: American Association of Petroleum Geologists Bulletin, v. 65, p. 261-290.

Hoggan, R., 1981, Devonian channels in the southern Lemhi Range, Idaho: Montana Geological Society, 1981 Field Conference, p. 45-47.

House, M.R., 1975, Facies and time in Devonian tropical areas: Proceedings Yorkshire Geological Society, v. 40, p. 233-287.

Hsü, K.J., and Winterer, E.L., 1980, Discussion on causes of world-wide changes in sea level: Journal of the Geological Society of London, v. 137, p. 509-510.

Hurley, G.W., 1962, Distribution and correlation of Upper Devonian formations, Sweetgrass Arch area, northwestern Montana, in Billings Geological Society Guidebook, 13th Annual Field Conference, p. 23-32.

Isaacson, P.E., and Dorobek, S.L., 1988, Regional significance and interpretation of a coral-stromatoporoid carbonate buildup succession, Jefferson Formation (Upper Devonian), east-central Idaho, in Devonian of the World: Canadian Society of Petroleum Geologists Memoir 14, p. 581-589.

James, N.P., 1984, Shallowing-upward sequences in carbonates, in R.G. Walker, ed., Facies Models, 2nd Edition: Geoscience Canada Reprint Series 1, p. 213-228.

Johnson, J.G., Klapper, G., and Sandberg, C.A., 1985, Devonian eustatic fluctuations in Euramerica: Geological Society of America Bulletin: v. 96, p. 567-587.

Kendall, C.G.St.C., and Schlager, W., 1981, Carbonates and relative changes in sea level: Marine Geology, v. 44, p. 181-212.

Kinsman, D.J.J., 1969, Modes of formation, sedimentary associations and diagnostic features of shallow-water and supratidal evaporites: American Association of Petroleum Geologists Bulletin, v. 53, p. 830-840.

Klovan, J.E., 1974, Development of western Canadian Devonian reefs and comparison with Holocene analogues: American Association of Petroleum Geologists Bulletin, v. 58, p. 787-799.

Logan, B.W., Hoffman, P., and Gebelein, C.D., 1974, Algal mats, cryptalgal fabrics, and structures, Hamelin Pool, Western Australia, in Logan, B.W., Read, J.F., Hagan, G.M., Hoffman, P., Brown, R.G., Woods, P.J., and Gebelein, C.D., Evolution and Diagenesis of Quaternary Carbonate Sequences, Shark Bay, Western Australia: American Association of Petroleum Geologists Memoir 22, p. 140-194.

Logan, B.W., Rezak, R., and Ginsburg, R.N., 1964, Classification and environmental significance of algal stromatolites: Journal of Geology, v. 72, p. 68-83.

Loreau, J.P., and Purser, B.H., 1973, Distribution and ultrastructure of Holocene ooids in the Persian Gulf, in Purser, B.H., ed., The Persian Gulf, Holocene Carbonate Sedimentation and Diagenesis in a Shallow Epicontinental Sea: Springer-Verlag, New York, p. 279-328.

Loucks, G.G., 1977, Geologic history of the Devonian northern Alberta to southwest Arizona, in Twenty-ninth Annual Field Conference, Wyoming Geological Association Guidebook, p. 119-134.

Mapel W.J., and Sandberg, C.A., 1968, Devonian paleotectonics in east-central Idaho and southwestern Montana: U.S. Geological Survey Professional Paper 600D, p. D115-D125.

McMannis, W.J., 1962, Devonian stratigraphy between Three Forks, Montana, and Yellowstone Park, in The Devonian System of Montana and Adjacent Areas: Billings Geological Society 13th Annual Field Conference, p. 4-12.

McMannis, W.J., 1965, Resume of depositional and structural history of western Montana: American Association of Petroleum Geologists Bulletin, v. 49, p. 1801-1823.

Mörner, N-A, 1976, Eustasy and geoid changes: Journal of Geology, v. 84, p. 123-151.

Mörner, N-A, 1981, Revolution in Cretaceous sea-level analysis: Geology, v. 9, p. 344-346.

Nilsen, T.H., 1977, Paleogeography of Mississippian turbidites in south-central Idaho, in Stewart, J.H., Stevens, C.H., and Fritsche, A.E., eds., Paleozoic Paleogeography of the West-central United States, Pacific Coast Paleogeography Symposium 1: Society of Economic Paleontologists and Mineralogists Pacific Section, p. 275-299.

Patterson, R.J., and Kinsman, D.J.J., 1977, Marine and continental groundwater sources in a Persian Gulf coastal sabkha: American Association of Petroleum Geologists, Studies in Geology, v. 4, p. 381-397.

Perry, W.J., Jr., Dyman, T.S., and Sando, W.J., 1989, Southwestern Montana recess of Cordilleran thrust belt, in French, D.E., and Grabb, R.F., eds., Geological Resources of Montana, 1989 Field Conference Guidebook, Montana Centennial Edition: Montana Geological Society, v. 1, p. 261-270.

Peterson, J.A., 1986, General stratigraphy and regional paleotectonics of the western Montana overthrust belt, in Peterson, J.A., ed., Paleotectonics and Sedimentation in the Rocky Mountain Region, United States: American Association of Petroleum Geologists, Memoir 41, p. 57-86.

Poole, F.G., 1974, Flysch deposits of the Antler foreland basin, western United States, in, Dickinson, W.R., ed., Tectonics and Sedimentation: Society of Economic Paleontologists and Mineralogists, Special Publication 22,

p. 58-82.

Read, J.F., 1973, Carbonate cycles, Pillara Formation (Devonian), Canning Basin, Western Australia: Bulletin of Canadian Petroleum Geology, v. 21, p. 38-51.

Rocha-Campos, A.C., 1981a, Late Devonian Curuá Formation, Amazon Basin, Brazil, *in* Hambrey, M.J., and Harland, W.B., eds., Earth's Pre-Pleistocene Glacial Record, International Geological Correlation Programme, Project 38: Pre-Pleistocene Tillites: Cambridge University Press, Cambridge, p. 888-891.

Rocha-Campos, A.C., 1981b, Middle-Late Devonian Cabeças Formation, Parnaíba Basin, Brazil, *in* Hambrey, M.J., and Harland, W.B., eds., Earth's Pre-Pleistocene Glacial Record, International Geological Correlation Programme, Project 38: Pre-Pleistocene Tillites: Cambridge University Press, Cambridge, p. 892-895.

Ruppel, E.T., 1986, The Lemhi Arch: a Late Proterozoic and early Paleozoic landmass in central Idaho, *in* Peterson, J.A., ed., Paleotectonics and Sedimentation: American Association of Petroleum Geologists Memoir 41, p. 119-130.

Ruppel, E.T., and Lopez, D.A., 1984, The thrust belt in southwest Montana and east-central Idaho: U.S. Geological Survey Professional Paper 1278, 41 p.

Ruppel, E.T., Wallace, C.A., Schmidt, R.G., and Lopez, D.A., 1981, Preliminary interpretation of the thrust belt in southwest and west-central Montana and east central Idaho, *in* Southwest Montana: Montana Geological Society, 1981 Field Conference and Symposium Guidebook, p. 139-159.

Sandberg C.A., 1961, Widespread Beartooth Butte Formation of Early Devonian age in Montana and Wyoming and its Paleogeographic significance: American Association of Petroleum Geologists Bulletin, v. 45, p. 1301-1309.

Sandberg, C.A., 1962, Stratigraphic section of type Three Forks and Jefferson Formations at Logan Montana, *in* Billings Geological Society Guidebook, 13th Annual Field Conference, p. 47-50.

Sandberg, C.A., 1965, Nomenclature and lithologic subdivisions of the Jefferson and Three Forks Formations of southern Montana and northern Wyoming: U.S. Geological Survey Bulletin, 1194-N, 18 p.

Sandberg, C.A., Gutschick, R.C., Johnson, J.G., Poole, F.G., and Sando, W.J., 1983, Middle Devonian to Late Mississippian geologic history of the Overthrust Belt region, western United States: Rocky Mountain Association of Geologists, Geologic Studies of the Cordilleran Thrust Belt, v. 2, p. 691-719.

Sandberg, C.A., and Mapel, W.J., 1967, Devonian of the northern Rocky Mountains and Plains, *in* Oswald, D.H., ed., International Symposium on the Devonian System, Alberta Society of Petroleum Geologists, v. 1, p. 843-877.

Sandberg, C.A., and Poole, F.G., 1977, Conodont biostratigraphy and depositional complexes of Upper-Devonian cratonic-platform and continental shelf rocks in the Western United States, *in* Murphy, M.A., Berry, W.B.N., and Sandberg, C.A., eds., Western North America; Devonian: California University, Riverside, Campus Museum Contributions 4, p. 144-182.

Sandberg, C.A., Poole, F.G., and Johnson, J.G., 1988, Upper Devonian of western United States, *in* McMillan, N.J., Embry, A.F., and Glass, D.J., eds., Devonian of the World: Canadian Society of Petroleum Geologists Memoir 14, p. 183-220.

Sarg, J.F., 1988, Carbonate sequence stratigraphy, *in* Wilgus, C.K., Hastings, B.S., Kendall, C.G.St.C., Posamentier, H.W., Ross, C.A., and Van Wagoner, J.C., eds., Sea-level Changes: An Integrated Approach: Society of Economic Paleontologists and Mineralogists Special Publication 42, p. 155-181.

Schlager, W., 1981, The paradox of drowned reefs and carbonate platforms: Geological Society of America Bulletin, v. 92, p. 197-211.

Schmidt, C.J., and Hendrix, T.E., 1981, Tectonic controls for thrust belt and Rocky Mountain foreland structures in the northern Tobacco Root Mountains - Jefferson Canyon area, southwestern Montana, *in* Southwest Montana: Montana Geological Society, 1981 Field Conference and Symposium Guidebook, p. 167-180.

Scholten, R., 1957, Paleozoic evolution of the geosynclinal margin north of the Snake River Plain, Idaho-Montana: Geological Society of America Bulletin, v. 68, p. 151-170.

Scholten, R., 1960, Sedimentation and tectonism in the thrust belt of southwestern Montana and east-central Idaho: Wyoming Geological Association 15th Annual Field Conference Guidebook, p. 77-84.

Scholten, R., and Hait, M.H., 1962, Devonian system from shelf edge to geosyncline, southwestern Montana-central Idaho: Billings Geological Society, 13th Annual Field Conference Guidebook, p. 13-22.

Skipp, B., 1988, Cordilleran thrust belt and faulted foreland in the Beaverhead Mountains, Idaho and Montana, *in* Schmidt, C.J., and Perry, W.J., Jr., eds., Interaction of the Rocky Mountain Foreland and the Cordilleran Thrust Belt: Geological Society of America Memoir 171, p. 237-266.

Skipp, B., and Hait, M.H., Jr., 1977, Allochthons along the northeast margin of the Snake River Plain, Idaho: 29th Annual Field Conference, 1977, Wyoming Geological Association Guidebook, p. 499-515.

Skipp, B., and Hall, W.E., 1975, Structure and Paleozoic stratigraphy of a complex of thrust plates in the Fish Creek Reservoir area, south-central Idaho: U.S. Geological Survey Journal of Research, v. 3, no. 6, p. 671-689.

Skipp, B., Sando, W.J., and Hall, W.E., 1979, The Mississippian and Pennsylvanian (Carboniferous) Systems in the United States - Idaho: United States Geological Survey Professional Paper 1110-AA, 42 p.

526

Sloss, L.L., 1950, Paleozoic sedimentation in Montana area: American Association of Petroleum Geologists Bulletin, v. 34, p. 423-451.

Sloss, L.L., and Laird, W.M., 1947, Devonian system in central and northwestern Montana: American Association of Petroleum Geologists Bulletin, v. 31, p. 1404-1430.

Sloss, L.L., and Moritz, C.A., 1951, Paleozoic stratigraphy of southwestern Montana: American Association of Petroleum Geologists Bulletin, v. 35, p. 2135-2169.

Smith, T.M., and Dorobek, S.L., 1989, Dolomitization of the Devonian Jefferson Formation in south-central Montana: Mountain Geologist, v. 26, p. 81-96.

Speed, R.C., 1977, Island-arc and other paleogeographic terranes of late Paleozoic age in the western Great Basin, in Stewart, J.H., Stevens, C.H., and Fritsche, A.E., eds., Paleozoic paleogeography of the western United States, Pacific Coast Paleogeography Symposium 1: Society of Economic Paleontologists and Mineralogists, Pacific Section, p. 349-362.

Speed, R.C., and Sleep, N.H., 1982, Antler orogeny and foreland basin: A model: Geological Society of America Bulletin, v. 93, p. 815-828.

Stearn, C.W., 1982, The shapes of Paleozoic and modern reef-builders: a critical review: Paleobiology, v. 8, p. 228-241.

Tonnsen, J.J., 1986, Influence of tectonic terranes adjacent to the Precambrian Wyoming Province on Phanerozoic stratigraphy in the Rocky Mountain region, in Peterson, J.A., ed., Paleotectonics and Sedimentation: American Association of Petroleum Geologists Memoir 41, p. 21-39.

Vail, P.R., Mitchum, R.M., Jr., and Thompson, S., III, 1977, Seismic stratigraphy and global changes of sea level, Part 4: Global cycles of relative changes of sea level, in Payton, C.E., ed., Seismic Stratigraphy -- Applications to Hydrocarbon Exploration: American Association of Petroleum Geologists, Memoir 26, p. 83-97.

Viau, C., 1983, Depositional sequences, facies and evolution of the Upper Devonian Swan Hills reef buildup, central Alberta, Canada, in Harris, P.M., ed., Carbonate Buildups - A Core Workshop: Society of Economic Paleontologists and Mineralogists, Core Workshop No. 4, p. 112-143.

Wilson, J.L., 1967, Carbonate-evaporite cycles in lower Duperow Formation of Williston Basin: Bulletin of Canadian Petroleum Geology, v. 15, p. 230-312.

Wilson, J.L., 1985, Tectonic controls of carbonate platforms, in Harris, P.M., Moore, C.H., and Wilson, J.L., Carbonate Depositional Environments, Modern and Ancient, Part 2: Carbonate Platforms: Colorado School of Mines Quarterly, v. 80, no. 4, p. 9-29.

Wilson, J.L., and Pilatzke, R.H., 1987, Carbonate-evaporite cycles in lower Duperow Formation of the Williston Basin, in Longman, M.W., ed., Williston Basin: Anatomy of a Cratonic Oil Province: Rocky Mountain Association of Geologists, Denver, Colorado, p. 119-146.

Winston, D., 1986, Sedimentation and tectonics of the Middle Proterozoic Belt Basin and their influence on Phanerozoic compression and extension in western Montana and northern Idaho, in Peterson, J.A., ed., Paleotectonics and Sedimentation: American Association of Petroleum Geologists Memoir 41, p. 87-118.

Witzke, B.J., and Heckel, P.H., 1988, Paleoclimatic indicators and inferred Devonian paleolatitudes of Euramerica, in McMillan, N.J., Embry, A.F., and Glass, D.J., eds., Devonian of the World: Canadian Society of Petroleum Geologists Memoir 14, p. 49-63.

Wong, P.K., and Oldershaw, A.E., 1980, Causes of cyclicity in reef interior sediments, Kaybob Reef, Alberta: Bulletin of Canadian Petroleum Geology, v. 28, p. 411-424.

Woodward, L.A., 1981, Tectonic framework of Disturbed Belt of west-central Montana: American Association of Petroleum Geologists Bulletin, v. 65, p. 291-302.

DEVELOPMENT OF THIRD- AND FOURTH-ORDER DEPOSITIONAL SEQUENCES IN THE LOWER MISSISSIPPIAN MISSION CANYON FORMATION AND STRATIGRAPHIC EQUIVALENTS, IDAHO AND MONTANA

S.K. Reid and S.L. Dorobek
Department of Geology
Texas A&M University
College Station, Texas 77843

ABSTRACT

The Mission Canyon Formation and stratigraphic equivalents were deposited on a ramp-like carbonate platform which developed on the foreland side of the Mississippian Antler foredeep in southwestern Montana and east-central Idaho. In individual outcrops, these strata can be subdivided into 20 to 60 m thick shallowing-upward units which are the local expressions of depositional sequences. Regional biostratigraphic and lithostratigraphic correlations of these shallowing-upward units allows reconstruction of the former laterally continuous and genetically related set of facies that characterized each sequence prior to deformation.

Twelve, perhaps up to fourteen, third- to fourth-order depositional sequences are recognized in this study area. These sequences can be correlated with subintervals of the Mission Canyon and Charles Formations in the central Williston Basin. Sequences in southwestern Montana and subintervals in the Williston Basin have similar stacking patterns. Sequence sets are bounded by correlative major marine flooding surfaces. These similarities suggest that the same mechanism(s) influenced evolution of the Mission Canyon Formation and stratigraphic equivalents in both areas.

Correlation of sequences with subintervals in the Williston Basin, deposition of sequences during inferred relative quiescence in the Antler thrust belt, and the presence of an Early Carboniferous ice cap suggest that third- and fourth-order glacioeustatic sea level fluctuations probably controlled sequence development. A lack of evidence for drowning of the Mission Canyon platform suggests that the rates and magnitudes of sea level oscillations were on the order of ten to, at most, a few tens of meters. Results of this study should be compared to other detailed studies of Lower Mississippian strata around the world in order to construct a more accurate sea level curve for the Early Mississippian.

INTRODUCTION

Sequence stratigraphy is a technique which allows assessment of the influence of relative sea level fluctuations on facies geometries in sedimentary basins (*sensu* Van Wagoner et al., 1988). However, sequence stratigraphy alone cannot differentiate between variations in sediment accumulation, eustasy and tectonic subsidence which interact simultaneously to produce changes in relative sea level. In fact, no single approach (e.g., hypsometry, backstripping, forward modeling or sequence stratigraphic analysis) allows unequivocal isolation of the individual components of relative sea level change (Kendall and Lerche, 1988).

Numerous workers have applied sequence stratigraphy to seismic profiles across carbonate shelf-to-basin transitions from tectonically stable, passive margin settings (e.g., Austin et al., 1986; Mullins et al., 1988; Wilgus et al., 1988; Rudolph and Lehmann, 1989). A few outcrop-oriented studies have successfully utilized sequence stratigraphy to characterize carbonate platform-to-basin transitions (e.g., Sarg, 1988; Franseen et al., 1989). However, these outcrop studies have been limited to well exposed, undeformed sequences where stratal boundaries can be traced physically through mapping or on photomosaics. In structurally complex regions, such as the North American Cordillera, undisturbed, laterally continuous outcrops are rare and stratal boundaries cannot be traced physically over great distances.

This study documents the influence of relative sea level fluctuations on the evolution of a carbonate ramp which developed on the foreland side of the Mississippian Antler foredeep in southwestern Montana and east-central Idaho. Existing regional stratigraphic relationships are significantly improved by the sequence stratigraphic analysis developed in this study. Sequence correlations also allow reconstruction of facies geometries across the tectonically shortened ramp-to-basin transition. Comparison of depositional sequences identified in this study with other detailed studies of Lower Mississippian strata may lead to refinement of existing Early Mississippian sea level curves.

Study Area and Approach

Stratigraphic sections of the Lower

In Cooper, J.D., and Stevens, C.H., eds., 1991, **Paleozoic Paleogeography of the Western United States-II:** Pacific Section SEPM, Vol. 67, p. 527-541.

527

528

Mississippian Mission Canyon Formation and its stratigraphic equivalents were measured and described in detail along a NE-SW transect across southwestern Montana and east-central Idaho (Fig. 1). Depositional sequences were identified at widely separated outcrops and correlated using published and new biostratigraphic data and lithostratigraphic correlation techniques. Reconstructed facies geometries were compared to coeval strata in the Williston Basin of northwestern North Dakota. In addition, quantitative subsidence analyses of measured sections within the study area were used to assess the relative influence of subsidence on sequence development (Dorobek et al., in press; Reid, 1991). Sequence stratigraphic relationships and subsidence curves then were compared with published relative sea level curves for the Carboniferous.

REGIONAL GEOLOGY

Paleogeography and Stratigraphy

During Early Mississippian time, a broad, shallow marine carbonate platform extended across Montana. The platform was located approximately 5°–10° north of the paleoequator (Sando, 1976; Gutschick et al., 1980; Gutschick and Sandberg, 1983). This platform developed on the foreland side of an extension of the Antler foreland basin located in east-central Idaho (Sando, 1976; Dover, 1980; Gutschick et al., 1980; Sandberg et al., 1983). Both the axis of the foredeep and the platform-to-basin transition trended approximately north-south relative to Mississippian plate reconstructions.

The Antler foredeep formed in response to Late Devonian-Early Mississippian convergence along the western margin of North America (Roberts and Thomasson, 1964; Burchfiel and Davis, 1972; Poole, 1974; Dickinson, 1977; Speed, 1977; Dover, 1980; Speed and Sleep, 1982). A number of tectonic models have been proposed for the Antler orogeny, including arc-continent collision (Speed and Sleep, 1982), back-arc thrusting adjacent to a marginal basin (Burchfiel and Davis, 1972) and "incipient" subduction (Johnson and Pendergast, 1981). However, most workers agree that the Antler thrust belt (or accretionary prism; Speed and Sleep, 1982) overrode westward-dipping continental lithosphere and that the foredeep formed above an earlier, mature passive margin. Foundering of this antecedent passive margin and initiation of Antler foredeep subsidence in Montana and Idaho probably began as early as the Middle Devonian (Dorobek et al., in press).

Synorogenic siliciclastic conglomerates and turbidites of the lower Copper Basin Formation (>2000 m thick) and the McGowan Creek Formation (100–1100 m thick) were deposited in the proximal Antler foredeep during the Early Mississippian (Sandberg, 1975; Nilsen, 1977). Correlative subtidal ramp facies of the Lodgepole Formation (150–300 m thick) were deposited across the Antler foreland (Sandberg, 1975; Nilsen, 1977; Skipp et al., 1979; Gutschick and Sandberg, 1983).

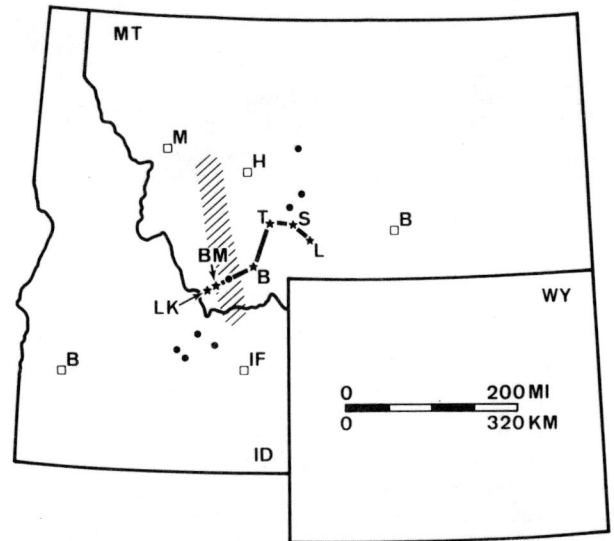

Figure 1. Study area. Filled circles are measured section locations. Labelled stars are locations discussed in the text. Open boxes are major cities: B = Billings, H = Helena, M = Missoula, IF = Idaho Falls, B = Boise. Cross-section from L to LK is shown in Figure 6. Diagonally ruled area is the approximate location and trend of the Lower Mississippian platform-to-basin transition.

The Mission Canyon Formation (200–400 m thick) and stratigraphic equivalents conformably overlie the McGowan Creek and Lodgepole Formations and predominantly consist of shallow subtidal to peritidal platform facies throughout central and southwestern Montana (Fig. 2, Reid and Dorobek, 1989). These facies grade westward into a relatively narrow belt of thick skeletal grainstone/packstone in southwestern Montana which also is included in the Mission Canyon Formation (Huh, 1967, 1968; Rose, 1976; Nichols, 1980; Peterson, 1986; Reid and Dorobek, 1989). Near the Idaho-Montana border, grainstones of the Mission Canyon Formation interfinger with cherty limestones of the Middle Canyon Formation (Sando et al., 1985; this study). Farther west, the upper member of the McGowan Creek Formation and the Middle Canyon Formation are correlative with the Mission Canyon Formation and consist of silty, spicular limestones, spiculites and calcareous siltstones (Fig. 2; Huh, 1967, 1968; Sandberg, 1975; Nilsen, 1977; Skipp et al., 1979; Gutschick and Sandberg, 1983).

The top of the Mission Canyon Formation is a regional unconformity which represents from 9 to 14 m.y. of subaerial exposure in central Montana (Fig. 2; Sando et al., 1976; Skipp et al., 1979; Gutschick et al., 1980; Sandberg et al., 1983; Sando, 1988). The Mission Canyon Formation was extensively karstified across most of Montana during this time (Middleton, 1961; Roberts, 1966; Sando, 1974, 1988). Farther west however, deposition was continuous and peritidal facies of the

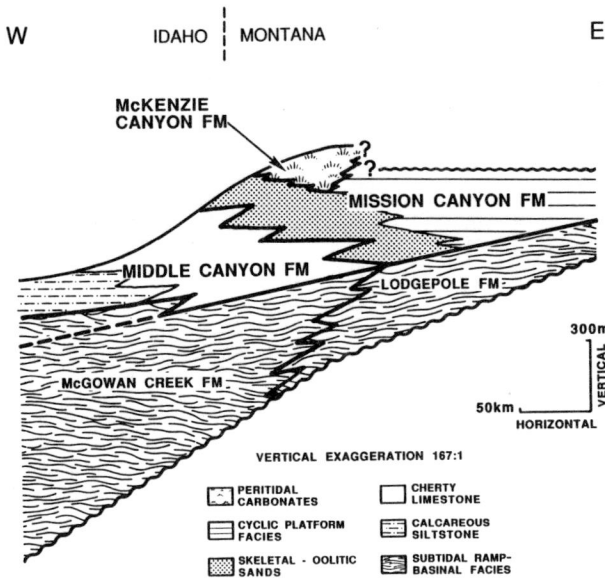

W   IDAHO | MONTANA   E

McKENZIE CANYON FM

MISSION CANYON FM

MIDDLE CANYON FM   LODGEPOLE FM

McGOWAN CREEK FM

300m VERTICAL

50km HORIZONTAL

VERTICAL EXAGGERATION 167:1

- PERITIDAL CARBONATES
- CHERTY LIMESTONE
- CYCLIC PLATFORM FACIES
- CALCAREOUS SILTSTONE
- SKELETAL - OOLITIC SANDS
- SUBTIDAL RAMP-BASINAL FACIES

Figure 2. Highly schematic NE-SW stratigraphic cross-section and platform model for the Mission Canyon Formation and stratigraphic equivalents. Cross-section extends from central Montana to east-central Idaho. Strata formed a ramp profile (note extreme vertical exaggeration). Lodgepole Formation and stratigraphic equivalents onlapped the Mississippian/Devonian unconformity. Mission Canyon Formation and stratigraphic equivalents prograded into distal Antler foredeep. Unconformity on top of the platform does not extend beyond the platform-to-basin transition.

McKenzie Cànyon Formation (Fig. 2) and equivalent deep-water facies of the upper Middle Canyon Formation were deposited in middle to late Meramecian time during development of the unconformity surface (Sando et al., 1985).

Platform Model

The Mission Canyon platform has been characterized as a ramp (*sensu* Ahr, 1973) with slopes of less than a few degrees (Gutschick et al., 1980) and, conversely, as a platform with 200-400 m relief at its margin (Rose, 1976). Thin carbonate gravity flow deposits and soft sediment deformation (folding and sedimentary boudinage structures; Reid, 1991; W. Perry, pers. comm.) in slope facies indicate that somewhat greater slopes existed west of the platform-to-basin transition. More rapid sedimentation on the Mission Canyon platform than in the distal Antler foredeep also must have produced some relief across the platform-to-basin transition (Sandberg, 1975; Sando, 1976; Skipp et al., 1979; Gutschick et al., 1980; Sandberg et al., 1983).

Diverse assemblages of reef-building organisms apparently did not exist during the Early Mississippian (Heckel, 1974; James, 1984) and a reefal rim did not develop along the Mission Canyon platform margin. Instead, wave-agitated skeletal banks formed a relatively narrow facies belt along the platform margin (Huh, 1968; Rose, 1976; Nichols, 1980; Peterson, 1986; Reid, 1991). Therefore, the Lower Mississippian carbonate platform that extended across Montana and Idaho might be best characterized as a progradational carbonate ramp with a narrow belt of skeletal banks along the transition from shallow platform settings to slope environments (Fig. 2).

SEQUENCE STRATIGRAPHY OF THE MISSION CANYON FORMATION AND STRATIGRAPHIC EQUIVALENTS

Correlations of Lower Mississippian strata are reasonably well-constrained in Montana and east-central Idaho at the formation and member scale. However, smaller scale lithofacies relationships are poorly documented because the time-stratigraphic resolution of biozones used for regional correlations is relatively low. Lack of detailed regional correlations has hampered previous efforts to reconstruct the tectonically shortened Mission Canyon platform-to-foredeep transition. A full assessment of mechanisms which affected deposition of the Mission Canyon Formation and stratigraphic equivalents depends on reconstruction of this critical transition. This study uses sequence stratigraphic concepts to improve the existing regional stratigraphic framework and to provide the detail necessary to assess the influence of relative sea level fluctuations on sedimentation patterns across the Antler foreland.

Regional Biostratigraphic Zonation

Foraminifera, conodonts, and corals have been used to define biozones within Mississippian strata of the Northern Rocky Mountains (Fig. 3; Sando et al., 1969; Sandberg et al., 1983; Sando, 1985; Sando and Bamber, 1985). Because all three zonations have been used to correlate these strata, this study follows the composite biozonation of Sando (1985) as a matter of convenience. Composite biochronozones range in duration from 0.75 to 3.4 m.y., depending on the absolute time scale used to calculate biozone durations (Sandberg and Poole, 1977; Sandberg et al., 1983; Sando, 1985). Published and new biostratigraphic data and formation-scale correlations between the stratigraphic sections examined in this study are summarized in Reid (1991, see Plates 1, 2, Appendix B).

Basic Sequence Stratigraphic Concepts

Sequence stratigraphy, originally introduced as seismic stratigraphy (Payton, 1977), "is the study of rock relationships within a chronostratigraphic framework of repetitive, genetically related strata bounded by surfaces of erosion or nondeposition, or their correlative conformities" (Van Wagoner et al., 1988, p. 39). Sequence stratigraphic units largely are defined and identified by the nature of the bounding surfaces between stratigraphic units and the lateral geometry of the strata within these units (Van Wagoner et al., 1988). Definition of these units does not depend on absolute thickness, the amount of time during which they form, or interpretation of regional or global origin (Van Wagoner et al., 1988).

Basic sequence stratigraphic concepts can be used to identify sequences and sequence boundaries of different spatial and temporal scales. Recognition of scale variations is essential for better interpretation of mechanisms which control sequence development. Unfortunately, no nomenclatural hierarchy clearly accommodates relative magnitudes of sequences and sequence boundaries. Vail et al. (1977) attempted to rectify this problem by proposing a hierarchical ordering of *sechrons* (or cycles; Haq et al., 1988), which are the geochronologic equivalents of stratigraphic sequences. Magnitudes of unconformities (sequence boundaries) are ranked according to hiatus durations but still are described by ambiguous terms such as "minor, medium and major" (e.g., Haq et al., 1988).

The smallest sequence stratigraphic unit is the *parasequence* which essentially is equivalent to fourth and fifth order (10^5 to 10^4 year duration), shallowing-upward, depositional cycles (Sarg, 1988; Haq et al., 1988; Van Wagoner et al., 1988). The fundamental sequence stratigraphic unit is the *sequence*, which is of third to fourth order scale (10^6 to 10^5 year duration). Sequences are relatively conformable successions of genetically related strata bounded by unconformities and their correlative conformities (Mitchum, 1977; Vail et al., 1977; Haq et al., 1988). The definition of unconformity is restricted to a surface which separates stratigraphic units and which represents a significant hiatus (Van Wagoner, et al., 1988). Evidence of subaerial erosional truncation (and, in some areas, correlative submarine erosion) or subaerial exposure is typical (Van Wagoner, et al., 1988). A *supersequence* (second order, 10^7 year duration) is a stacked set of sequences which is bounded by major unconformities (Vail et al., 1977; Haq et al., 1988). Stacked sets of supersequences form *megasequences* (first order, 10^8 year duration; Vail et al., 1977; Haq et al., 1988). Sequences identified by Sloss (1963) are examples of supersequences and megasequences (Mitchum et al., 1977; Haq et al., 1988).

Approximate Durations of Sequences within the Mission Canyon Formation and Stratigraphic Equivalents

Approximate durations of sequences recognized in this study were calculated for several stratigraphic sections using measured thicknesses and published biostratigraphic

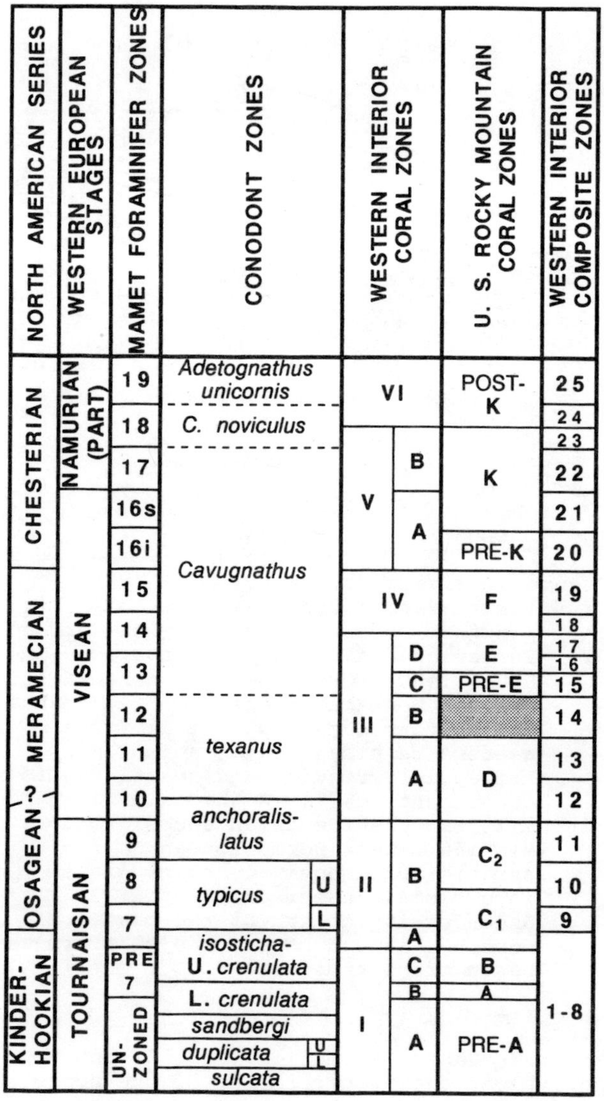

Figure 3. Biostratigraphic zonation used in the region. Compiled from Sando et al. (1969), Sando (1985), and Sando and Bamber (1985).

data (Gutschick et al., 1980; Sandberg et al., 1983; Sando et al., 1985). Absolute ages of biozone boundaries were determined using the Mississippian time scales of Sando (1985), Ross and Ross (1987), and Sandberg et al. (1983). Average accumulation rates for the Mission Canyon Formation and stratigraphic equivalents were calculated in order to estimate sequence durations. For example, deposition of the Mission Canyon Formation and stratigraphic equivalents (525 m thick) at location BM (Fig. 1) took 10.9 to 12.4 m.y. (depending on the time scale used). This is the amount of time between the composite zone 9/10 and composite zone 14/15 boundaries which occur at the base and the top of the section respectively. A minimum average accumulation rate of 4.2 cm/1000 yr was calculated by dividing measured thickness by 12.4 m.y. A maximum average accumulation rate of 6.7 cm/1000 yr was calculated using a decompacted thickness of 730 m, assuming 40% compaction throughout the section, divided by 10.9 m.y. Sequence durations were calculated by dividing measured sequence thicknesses by minimum average accumulation rate (minimum durations) or by dividing decompacted sequence thicknesses by maximum average accumulation rate (maximum durations). Excluding errors associated with the absolute time scales used for these calculations, sequences range from 0.4 to 1.3 m.y. in duration (third to fourth order, *sensu* Vail et al., 1977) and appear to provide a degree of chronostratigraphic resolution that is significantly higher than most Osagean to Meramecian biochronozones.

Recognition of Sequences in Outcrops

Recognition of sequences and sequence boundaries in widely separated outcrops of Lower Mississippian rocks from different thrust sheets depends on lithofacies interpretations, especially interpretations of relative paleobathymetry for each lithofacies. In many cases, particularly in deeper water settings, sequences are defined by the stacking order of individual facies within the sequences. Sequence boundaries in deeper water sediments often are not unconformities, but surfaces of abrupt deepening.

Relative paleobathymetries of lithofacies recognized in the Mission Canyon Formation and stratigraphic equivalents are, from deepest to shallowest (see Reid, 1991 for a complete discussion of relative paleobathymetric estimates): calcareous siltstone/sandstone; mixed carbonate/siliciclastic facies; laminated cherty limestone; bioturbated cherty limestone; peloid-skeletal wackestone/packstone; massive skeletal grainstone; skeletal-ooid grainstone; skeletal-peloid grainstone/packstone; fenestral limestone; restricted peritidal facies. In individual outcrops, sequences in the Mission Canyon Formation and its stratigraphic equivalents occur as 18 to 62 m thick, shallowing-upward units composed of these lithofacies. The set of lithofacies which locally defines each sequence varies depending on stratigraphic position of the sequence and on paleogeographic location of the measured section.

Platform Settings

In shallow-water platform sections, sequences consist of, in ascending order: 1) peloid-skeletal wackestone/packstone; 2) skeletal-ooid grainstone; 3) skeletal-peloid grainstone/packstone; and 4) thin, stacked parasequences of peritidal facies (Figs. 4, 6). Not all facies within this generalized sequence are present within each individual sequence. Facies which cap parasequences that occur at the tops of platform sequences often show evidence of subaerial exposure such as desiccation features, evaporite pseudomorphs and vadose diagenetic fabrics. Sequence

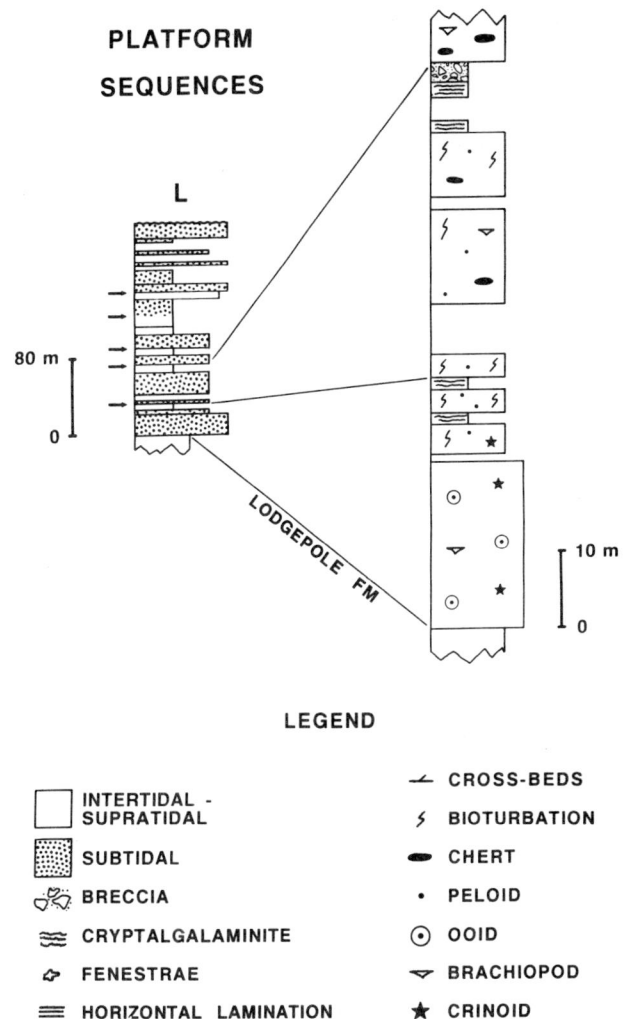

Figure 4. Typical platform sequences from location L (see Figure 1). Left column is the entire thickness of the Mission Canyon Formation (224 m). Arrows alongside left column indicate sequence boundaries. Enlargement of basal two sequences shown in right column.

boundaries in shallow-water platform settings are placed along the highest parasequence cap (1-10 m thick interval of peritidal facies) which is overlain by relatively thick (~3 to 10 m) subtidal facies (Figs. 4, 6). Careful observation of clast lithologies within solution collapse breccias allows some sequence boundaries to be placed within brecciated horizons because overall shallowing-upward characteristics frequently are preserved within the breccias. Lithofacies patterns within individual sequences are discussed in more detail in Reid (1991).

Outer Ramp and Slope Settings

Sequences can be identified in outer ramp and slope settings but meter-scale parasequences are difficult to recognize. Outer ramp and upper slope sequences are

DEEPER WATER SEQUENCES

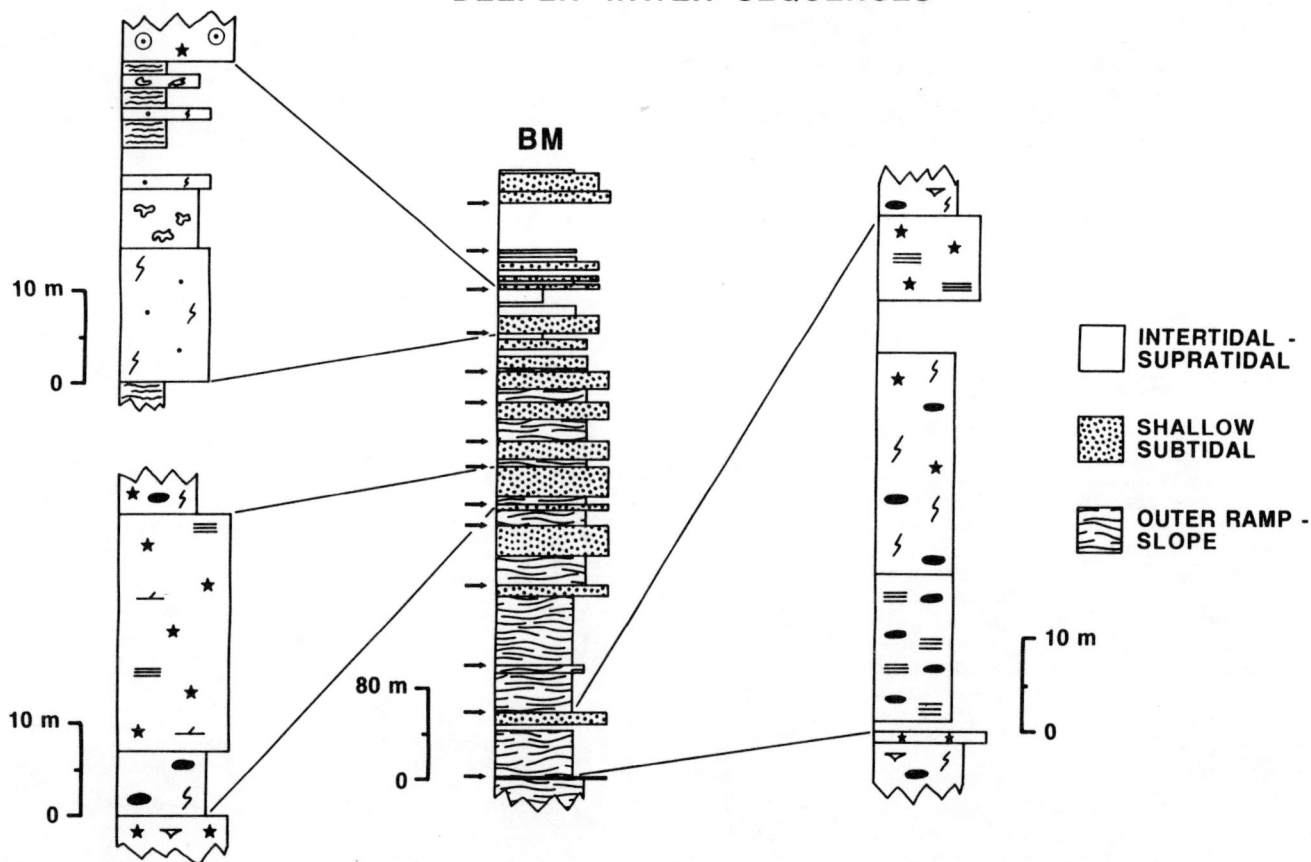

Figure 5. Sequences from deeper water settings at location BM (see Figure 1). Center column is entire thickness of Middle Canyon, Mission Canyon and McKenzie Canyon Formations (522 m). Arrows indicate sequence boundaries. Enlargements of specific sequences shown on left and right. Lithologic symbols same as in Figure 4.

characterized by repetitive intervals of bioturbated cherty limestone overlain by massive skeletal grainstone (Figs. 5, 6). Sequences in deeper slope settings consist of thick intervals of laminated cherty limestone overlain by bioturbated cherty limestone (Fig. 6, locations BM, LK). These relationships suggest that sequences shallow upward (see Reid, 1991 for detailed discussion of lithofacies and interpretations). In both settings, deeper-water slope facies (bioturbated cherty limestone or laminated cherty limestone) abruptly overlie shallower-water deposits (grainstone or bioturbated cherty limestone). The contact of deeper-water facies with underlying shallower-water facies is a marine-flooding surface (*sensu* Van Wagoner et al., 1988).

Reconstruction of the Platform-to-Basin Transition

Local sequence expressions must be correlated in order to reconstruct the former laterally continuous and genetically related set of facies that characterized each sequence prior to deformation (Fig. 6; detailed measured sections used to construct

this diagram are available from the authors upon request). Some datum must be established in order to correlate individual stratigraphic sequences because, unlike biozones, these lithostratigraphic units are repetitive. Therefore, at least one sequence must be biostratigraphically correlated between outcrops in order to lithostratigraphically correlate overlying and underlying sequences.

The datum used in this study was the basal sequence of the Mission Canyon Formation and stratigraphic equivalents (Fig. 6) because isolated exposures of this sequence can be regionally correlated along the C_1/C_2 megafaunal zone boundary and the composite zone 9/10 boundary (Figs. 3, 6; Reid, 1991). Sequence boundaries, as previously defined, were correlated lithostratigraphically.

Figure 6 shows reconstructions of sequences along a transect across the Mission Canyon platform to Antler foredeep transition (see Fig. 1 for section locations). This figure illustrates regional lithofacies and

SEQUENCE RECONSTRUCTION

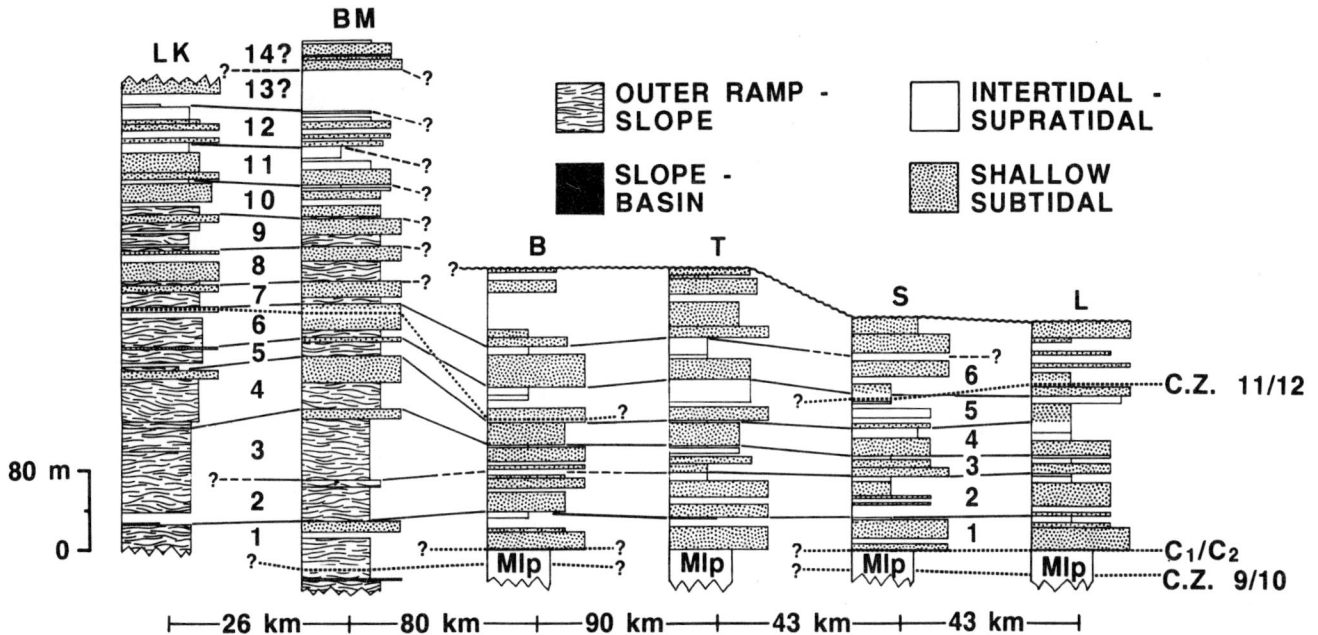

Figure 6. Reconstructed sequences across the platform-to-basin transition. Line of section shown on Figure 1. Solid lines between sections are sequence boundaries (dashed where inferred). Dotted lines labelled C.Z. 9/10, C_1/C_2, and C.Z. 11/12 are composite zone and western interior megafaunal biozone boundaries (Figure 3). Large numbers refer to numbered sequences discussed in text. Distances between sections are palinspastic.

measured thickness variations within individual sequences. Sequences 7 through 14 from near the platform-to-basin transition cannot be identified in shallow water platform areas because of poor exposure and because of removal or nondeposition of these sequences during middle Meramecian subaerial exposure and karstification.

Relationship of Sequences to Informal Subintervals of the Mission Canyon and Charles Formations in the Williston Basin

Larger-scale, shallowing-upward units composed of carbonates and evaporites (~10 to 40 m thick) are used to subdivide the Mission Canyon and Charles Formations in the Williston Basin into informal subintervals (Harrison and Flood, 1956; Ballard, 1963; Harris et al., 1966; Carlson and Lefever, 1987; Hendricks, 1988). Boundaries of subintervals are placed at gamma-ray marker horizons which are regionally extensive but which cannot be traced to the center of the Williston Basin (Harris et al., 1966). Biostratigraphic data indicate that these markers are time lines (Waters and Sando, 1987a).

Disagreement still exists concerning the position and origin of siliciclastic-rich marker horizons within subintervals. Originally, markers were interpreted as argillaceous, transgressive deposits at the bases of informal subintervals (Harris et al., 1966). Other workers have suggested that marker beds occur at the tops of subintervals (Malek-Aslani, 1971; Shanley, 1983; Elliott, 1987). However, the original interpretation of Harris et al. (1966) probably applies to markers which separate the Rival and Midale subintervals and the Tilston and Frobisher-Alida intervals. Both markers, referred to as "transgressive discontinuities" by Petty (1988), consist of argillaceous and/or cherty limestone with abundant crinoids, solitary rugose corals, and brachiopods (Fuller, 1956; Waters and Sando, 1987a, 1987b; Lindsay, 1988).

Informal subintervals of the Mission Canyon and Charles Formations in the central Williston Basin can be correlated with the depositional sequences recognized in this study (Figs. 6, 7). The correlation datum is the composite zone 11/12 boundary, which coincides with the top of the *Stelechophyllum banffense* coral zonule in the Williston Basin (Waters and Sando, 1987a, 1987b). The 4.5 to 6 m thick *Stelechophyllum banffense* coral zonule is parallel to and occurs within 6 m of the K-1 (Fryburg) gamma ray marker which separates the Sherwood and Mohall subintervals (Harris et al., 1966; Waters and Sando, 1987a; Petty, 1988). Biostratigraphic correlations of subsurface rocks in the Williston Basin with the type section of the

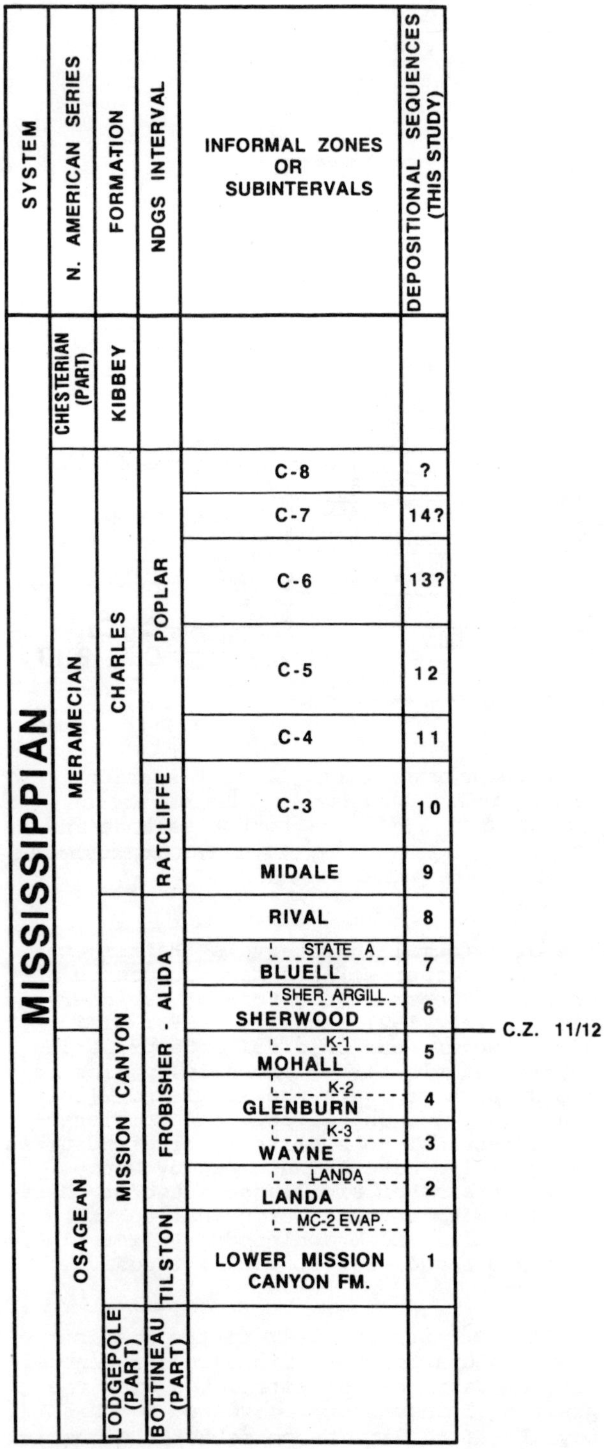

SYSTEM	N. AMERICAN SERIES	FORMATION	NDGS INTERVAL	INFORMAL ZONES OR SUBINTERVALS	DEPOSITIONAL SEQUENCES (THIS STUDY)
MISSISSIPPIAN	CHESTERIAN (PART)	KIBBEY			
	MERAMECIAN	CHARLES	POPLAR	C-8	?
				C-7	14?
				C-6	13?
				C-5	12
				C-4	11
			RATCLIFFE	C-3	10
				MIDALE	9
	OSAGEAN	MISSION CANYON	FROBISHER - ALIDA	RIVAL	8
				STATE A / BLUELL	7
				SHER. ARGILL / SHERWOOD	6
				K-1 / MOHALL	5
				K-2 / GLENBURN	4
				K-3 / WAYNE	3
				LANDA / LANDA	2
			TILSTON	MC-2 EVAP / LOWER MISSION CANYON FM.	1
		LODGEPOLE (PART)			
		BOTTINEAU (PART)			

(C.Z. 11/12 — labelled at right of chart at the Sherwood level)

Figure 7. Correlation of sequences recognized in this study with informal subintervals of Harrison and Flood (1956) and Harris et al. (1966) from the Williston Basin. Position of composite zone 11/12 boundary (C.Z. 11/12) labelled on right of chart. The composite zone 11/12 boundary is the primary datum for correlation between the two regions.

Mission Canyon Formation (Sando, 1960; Sando, 1978; Sando and Mamet, 1981) further support the correlations shown in Figure 7.

Comparison of Stacking Patterns in Montana and the Williston Basin

In Lower Mississippian strata of southwestern Montana, progradational sequence stacking is indicated by a progressive upward increase in the thickness of shallow-water facies (Fig. 6, locations LK, BM). In contrast, retrogradational sequence stacking is expressed in some parts of the Lower Mississippian stratigraphy as an upward increase in the thickness of deeper-water facies (Fig. 6, locations LK, BM). Harrison and Flood (1956) and Harris et al. (1966) have documented stacking patterns of correlative subintervals in the Williston Basin (Fig. 8). Sequence 1 and the unnamed sequence below it (equivalent to the uppermost Lodgepole Formation) show an upward increase in the thickness of skeletal grainstone (Fig. 6, location BM). Skeletal grainstone horizons indicate shallower water depths and define sequence tops. However, the top of Sequence 2 lacks grainstone and instead consists of bioturbated cherty limestone. Farther west (location LK), the Sequence 2/3 boundary cannot be identified but appears to occur within a thick interval of laminated cherty limestone (Fig. 6). Therefore, the Sequence 1/2 boundary is interpreted as a major marine flooding surface (*sensu* Van Wagoner et al., 1988). This boundary correlates with the Tilston/ Frobisher-Alida boundary in the Williston Basin, which Petty (1988) refers to as a "transgressive discontinuity" (Figs. 7, 8). Interpretation of this surface as a major marine flooding surface would explain the lack of skeletal grainstone at the top of Sequence 2 in slope environments and coeval retrogradation of the Landa subinterval in the Williston Basin (Figs. 7, 8).

Between Sequences 2 and 8 in southwestern Montana (Fig. 6, locations LK and BM) and the Landa and Rival subintervals in the Williston Basin (Fig. 8) sequences/subintervals are dominated by progressively thicker intervals of shallow water lithofacies up-section. Sequence 9 and the Midale subinterval consist of thick intervals of deeper-water lithofacies which sharply overlie thick grainstones (Fig. 6, Sequence 8, location LK) and evaporites (Fig. 8, Rival subinterval) at the top of the underlying progradational sequence sets. The correlative bases of Sequence 9 and the Midale subinterval are interpreted as a second major marine flooding surface. Above this surface sequences again become progradational in both areas. Therefore, sequences in southwestern Montana and subintervals in the Williston Basin comprise two progradational sequence sets which are bounded by correlative major marine flooding surfaces. These relationships suggest that the same mechanism(s) influenced evolution of the Mission Canyon Formation and stratigraphic equivalents in both areas.

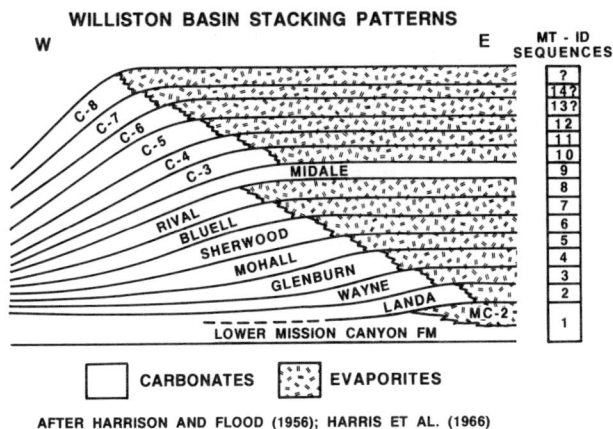

WILLISTON BASIN STACKING PATTERNS

CARBONATES   EVAPORITES

AFTER HARRISON AND FLOOD (1956); HARRIS ET AL. (1966)

Figure 8. Schematic stratigraphic cross-section illustrating stacking patterns of informal subintervals in the Williston Basin (based on well log and core analyses of Harrison and Flood, 1956; Harris et al., 1966). Correlative sequences from this study shown to the right of the cross-section. Not to scale.

EXTERNAL CONTROLS ON SEQUENCE DEVELOPMENT

The Mission Canyon Formation and stratigraphic equivalents were deposited during relatively slow Osagean to Meramecian subsidence across the entire Antler foreland of Idaho and Montana (Dorobek et al., this volume; Reid, 1991). Slow subsidence appears to represent a time of relative quiescence in the Antler accretionary wedge. A gradual second-order eustatic sea level fall, inferred from the onlap-offlap curve of Ross and Ross (1987), also may have occurred during Osagean to Meramecian subsidence (Fig. 9). Relatively slow subsidence during this time (Dorobek et al., this volume) suggests that this gradual eustatic sea level fall may have counteracted tectonic subsidence to produce a slow relative sea level rise.

Episodic thrusting events with periods of the same duration as the sequences discussed above (0.4 to 1.3 m.y.) cannot be unequivocally identified within the Osagean to Meramecian segments of subsidence curves from Dorobek et al. (this volume and in press) and Reid (1991) because such short-term events are well below the resolution of Mississippian biochronozones. Therefore, short-term episodic thrusting events, with concomitant pulses of subsidence, cannot be eliminated as a possible cause for the sequences documented in this study. However, simultaneous flexure across a palinspastic distance of >1000 km would be required in order to produce coeval relative sea level fluctuations in southwesternmost Montana and the Williston Basin. Widths of modern and ancient foredeeps generally are less than 250 km (cf. Jordan, 1981; Karner and Watts, 1983; Stockmal and Beaumont, 1987). Therefore, it is unlikely that sequences in the Mission

Canyon Formation and stratigraphic equivalents formed as a response to flexure alone.

Alternatively, the depositional sequences recognized in this study could have formed in response to relative sea level oscillations produced by in-plane compressive stresses (cf. Cloetingh, 1988). This mechanism could affect the entire Antler foreland and the Williston Basin simultaneously but cannot explain the periodicity suggested by the sequences. However, waxing and waning of in-plane compressive stresses and concomitant relative sea level fluctuations could be related to episodic convergence events in the Antler accretionary wedge (Dorobek et al., this volume).

Although a tectonic origin for the sequences in the Mission Canyon Formation and stratigraphic equivalents cannot be eliminated completely, they more likely formed in response to third- and fourth-order eustatic sea level fluctuations. Eustatic sea level fluctuations are preferred over "yo-yo" tectonics as the dominant mechanism for forming the Lower Mississippian sequences because: 1) the area over which these sequences can be correlated is immense (over 1000 km) and it is difficult to imagine

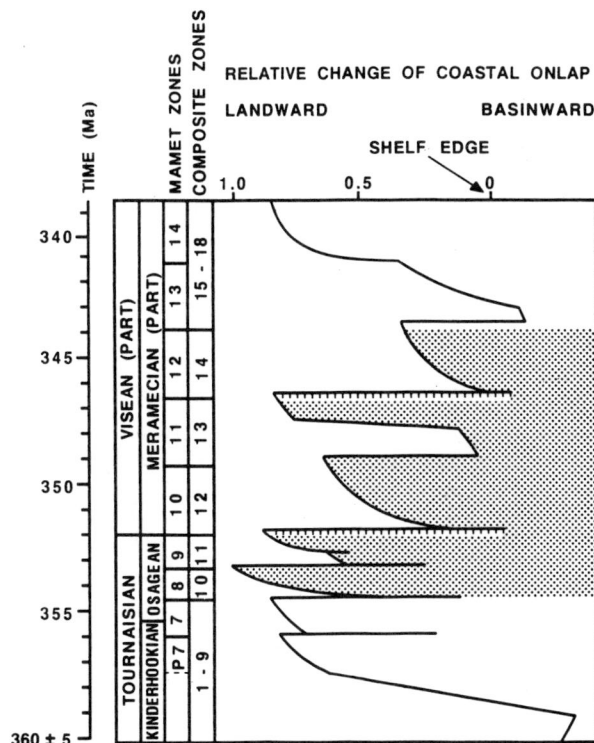

Figure 9. Coastal onlap-offlap curve of Ross and Ross (1987), modified to fit time scale of Sando (1985). Shaded portion corresponds with time of deposition of the Mission Canyon Formation and stratigraphic equivalents. Hachured lines are "exposed lowstand surfaces" of Ross and Ross (1987).

tectonic processes which could affect two different tectonic provinces at exactly the same time; 2) the sequences have somewhat regular periodicities which are difficult to explain by some type of regular, periodic tectonic mechanism; and 3) sequences from southwestern Montana and the central Williston Basin have very similar stacking patterns and have coeval major marine flooding surfaces which are more easily explained by eustatic sea level change than by periodic pulses of subsidence.

Sea level fluctuations during the Early Mississippian may have been glacioeustatic (cf. Vail et al., 1977; Ross and Ross, 1987). Late Paleozoic (Gondwanan) glaciation apparently began in the Early Carboniferous (Kinderhookian to Osagean) and culminated in the Permo-Carboniferous (Frakes and Crowell, 1969; Crowell, 1978; Caputo and Crowell, 1985). Caputo and Crowell (1985) suggested that Gondwanaland was never covered by a single large ice cap. Instead, smaller ice centers migrated across Gondwanaland from South America (Early Carboniferous) to eastern Australia and Antarctica (early Late Permian). Volume changes in the relatively small Early Carboniferous ice cap would have produced relatively low amplitude sea level fluctuations, perhaps on the order of ten to, at most, a few tens of meters.

Low amplitude sea level oscillations are consistent with a lack of evidence for drowning (*sensu* Schlager, 1981) of the Mission Canyon platform (Reid, 1991). Rapid sea level rise at the beginning of each third- to fourth-order sea level oscillation probably flooded regionally extensive tidal flat and supratidal areas (i.e., sequence tops). However, the predominance of cross-bedded skeletal-ooid and peloidal grainstones at the bases of platform sequences suggests that the rates and magnitudes of sea level rises were neither great enough to cause drowning of the Mission Canyon platform nor to submerge the platform surface below fairweather wave base.

Correlation of platform sequences with coeval strata deposited in tens of meters of water suggests that a purely autocyclic origin for sequences in the Mission Canyon Formation and stratigraphic equivalents is unlikely. Autocyclic mechanisms (Ginsburg, 1971) require water depths well within the zone of optimum carbonate sediment production (<10 m) in order to promote tidal flat progradation. Autocyclicity also requires prograding tidal flats to advance to the platform margin, which decreases sediment supply to the tidal flats by decreasing the area of the carbonate mud "factory". However, tidal flat progradation would cease well before reaching the platform margin because *larger* amounts of sediment are required to aggrade deeper water platform areas to sea level (Hardie, 1986). Therefore, only meter-scale tidal flat parasequences near the tops of sequences in the Mission Canyon Formation (Figs. 4, 5) might have been affected by autocyclic processes.

The coastal onlap-offlap curve of Ross and Ross (1987) suggests six third-order eustatic cycles occurred during deposition of the Mission Canyon Formation and stratigraphic equivalents (Fig. 9). At least twelve third- to fourth-order cycles are suggested by the sequence stratigraphic relationships established in this study (Figs. 6, 7, 8). Results from this study and other detailed studies of Lower Mississippian strata around the world may allow construction of a more refined sea level curve.

Subaerial Exposure of the Mission Canyon Platform

Subaerial exposure of the Mission Canyon platform probably began during a significant fall in eustatic sea level at the Sequence 7/8 boundary. Relatively "instantaneous" subaerial exposure of the flat-topped Mission Canyon platform at this time would explain both the near coincidence of the Sequence 7/8 boundary with the tops of platform sections and the relatively small variation in thicknesses (224 to 286 m) of complete measured sections of platform strata at many localities (Fig. 6). Thickness variations probably reflect the karst topography (with up to ~60 m of local relief) which developed on the exposed platform (cf. Sando, 1988). Exposure of the Mission Canyon platform at the Sequence 7/8 boundary also is suggested by the coeval onset of skeletal bank deposition in former slope settings (e.g., at location LK) and widespread deposition of the Frobisher evaporite (Rival subinterval) across the Williston Basin (Figs. 6, 7, 8). In addition, platform sediment supply to lower slope and basinal environments appears to have decreased at this time (Reid, 1991).

Lack of biostratigraphic control prevents accurate dating of the correlative conformity of the Sequence 7/8 exposure surface (Fig. 6, locations BM, LK). The youngest biostratigraphic date at the top of the Mission Canyon Formation in shallow platform areas suggests that this surface is of late early Meramecian age (composite zone 13/14 boundary; Sando, 1976, 1985; Gutschick et al., 1980; Sandberg et al., 1983). If this age is correct, the Sequence 7/8 boundary coincides with an "exposed lowstand surface" on the Ross and Ross (1987) onlap-offlap curve (Fig. 9).

Sea level must have fallen to or just below the platform-to-basin transition during subaerial exposure of the platform. Third- to fourth-order sea level fluctuations which controlled deposition of Sequences 8 through 14 primarily affected former upper slope and platform margin areas. The major flooding event which occurred at the beginning of Sequence 9 apparently only affected areas near the platform-to-basin transition (and near the center of the Williston Basin). The Mission Canyon platform remained emergent until the middle to late Meramecian because it did not subside rapidly enough to allow flooding by relatively low amplitude sea level fluctuations. Therefore, the unconformity between locations B and BM (where strata become conformable) may have been the site of local onlap of the platform by the McKenzie Canyon Formation over former upper slope and platform margin areas (Figs. 6). The McKenzie Canyon Formation probably formed as a platform margin carbonate wedge (*sensu* Sarg, 1988; Reid, 1991). Biostratigraphy of all strata overlying the unconformity must be better constrained before subaerial exposure of the Mission Canyon platform and onlap of the McKenzie Canyon platform margin carbonate wedge can be fully understood.

CONCLUSIONS

The Mission Canyon Formation and stratigraphic equivalents contain at least twelve, perhaps up to fourteen, third- to fourth-order depositional sequences. These sequences are expressed as 20 to 60 m thick, shallowing upward units in isolated outcrops. Biostratigraphic and lithostratigraphic correlation of these units has allowed reconstruction of facies geometries across the tectonically shortened platform-to-basin transition. Depositional sequences in southwestern Montana and east-central Idaho correlate with informal subintervals of the Mission Canyon and Charles Formations in northwestern North Dakota. Sequences and subintervals in both areas have nearly identical stacking patterns. Sequence/subinterval sets are bounded by correlative major marine flooding surfaces.

The Mission Canyon Formation and stratigraphic equivalents were deposited during relatively slow Osagean to Meramecian subsidence of the Antler foreland. Published coastal onlap-offlap curves suggest that a gradual second-order eustatic sea level fall also occurred at this time. Gradual eustatic sea level fall may have counteracted tectonic subsidence to produce a slow relative sea level rise but tectonic subsidence still must have outpaced long-term eustatic sea level fall. Sequences within the Mission Canyon Formation and stratigraphic equivalents may have formed in response to third- and fourth-order eustatic sea level fluctuations superimposed on the second-order eustatic sea level fall. However, the onlap-offlap curve of Ross and Ross (1987) indicates that six third-order sea level fluctuations occurred during the Osagean to middle Meramecian while this study suggests the occurrence of at least twelve oscillations. Results of this study should be compared to other detailed studies of Lower Mississippian strata around the world in order to construct a more accurate sea level curve.

The Mission Canyon platform was subaerially exposed in the late early Meramecian. A platform margin carbonate wedge (the McKenzie Canyon Formation) developed over former upper slope and platform margin areas in southwesternmost Montana where subsidence of the Antler foreland outpaced falling sea level. Onlap of these strata over the Mission Canyon karst plain cannot be

fully assessed until the biostratigraphic time lines in strata overlying the Mission Canyon Formation are improved.

ACKNOWLEDGMENTS

This research was supported by U.S. Department of Energy Grant DE-FG05-87ER13767, Basic Energy Sciences to S.L. Dorobek. Such support does not constitute an endorsement by DOE of the views expressed in this paper. Acknowledgment also is made to the Donors of The Petroleum Research Fund, administered by the American Chemical Society, for partial support of this research (Grant #19519-G2 to S.L. Dorobek). Additional support was provided by the Geological Society of America Grants-in-Aid to S.K. Reid. T. Smith and J. Harris provided assistance in the field. W.J. Sando identified corals collected by Reid, allowed us to examine unpublished data, and provided comments about stratigraphy which helped to constrain our interpretations. A.G. Harris kindly identified conodonts in samples collected at location LK by Reid. However, views expressed herein are the responsibility of the authors. This paper was improved by the insightful reviews of John Cooper and Mitch Harris.

REFERENCES CITED

Ahr, W.M., 1973, The carbonate ramp: An alternative to the shelf model: Transactions, Gulf Coast Association of Geological Societies, v. 23, p. 221-225.

Austin, J.A., Schlager, W., Palmer, A.A., et al., 1986, Proceedings of the Ocean Drilling Program: Part A, Initial Reports, v. 101: Ocean Drilling Program, College Station, TX, Texas A&M University, p. 569.

Ballard, F.V., 1963, Structural and stratigraphic relationships in the Paleozoic rocks of eastern North Dakota: North Dakota Geological Survey Bulletin 40, 42 p.

Burchfiel, B.C. and Davis, G.A., 1972, Structural framework and evolution of the southern part of the Cordilleran orogen, western United States: American Journal of Science, v. 272, p. 97-118.

Caputo, M.V. and Crowell, J.C., 1985, Migration of glacial centers across Gondwana during Paleozoic Era: Geological Society of America Bulletin, v. 96, p. 1020-1036.

Carlson, C.G. and Lefever, J.A., 1987, The Madison Group, nomenclatural review with a look to the future, *in* Carlson, C.G. and Christopher, J.E., eds., Fifth International Williston Basin Symposium: Saskatchewan Geological Society Special Publication No. 9, p. 77-82.

Cloetingh, S., 1988, Intraplate stresses: a new element in basin analysis, *in* Kleinspehn, K.L., and Paola, C., eds., New Perspectives in Basin Analysis: Springer-Verlag, New York, p. 205-230.

Crowell, J.C., 1978, Gondwanan glaciation, cyclothems, continental positioning, and climate change: American Journal of Science, v. 278, p. 1345-1372.

Dickinson, W.R., 1977, Paleozoic plate tectonics and the evolution of the

538

Cordilleran continental margin, *in* Stewart, J.H., Stevens, C.H., and Fritsche, A.E., eds., Paleozoic Paleogeography of the Western United States, Pacific Coast Paleogeography Symposium 1: Society of Economic Paleontologists and Mineralogists, Pacific Section, p. 137-155.

Dorobek, S.L., Reid, S.K., Elrick, M., Bond, G.C., and Kominz, M.A., *in press*, Foreland response to episodic convergence: Subsidence history of the Antler foredeep and adjacent cratonic platform areas, Montana and Idaho, *in* Sedimentary Modeling: Computer Simulation and Methods for Improved Parameter Definition: Kansas Geological Society Bulletin.

Dover, J.H., 1980, Status of the Antler orogeny in central Idaho - Clarification and constraints from the Pioneer Mountains, *in* Fouch, T.D. and Magathan, E.R., eds., Paleozoic Paleogeography of the West-Central United States, Rocky Mountains Paleogeography Symposium 1: Society of Economic Paleontologists and Mineralogists, Rocky Mountain Section, p. 371-386.

Elliott, T.L., 1987, Carbonate facies, depositional cycles, and the development of secondary porosity during burial diagenesis: Mission Canyon Formation, Haas Field, North Dakota, *in* Peterson, J.A., Kent, D.M., Anderson, S.B., Pilatzke, R.H., and Longman, M.W., eds., Williston Basin: Anatomy of a Cratonic Oil Province: Rocky Mountain Association of Geologists, Denver, Colorado, p. 385-405.

Frakes, L.A. and Crowell, J.C., 1969, Late Paleozoic glaciation - Part I, South America: Geological Society of America Bulletin, v. 80, p. 1007-1042.

Franseen, E.K., Fekete, T.E., and Pray, L.C., 1989, Evolution and destruction of a carbonate bank at the shelf margin: Grayburg Formation (Permian), Western Escarpment, Guadalupe Mountains, Texas, *in* Crevello, P.D., Wilson, J.L., Sarg, J.F., and Read, J.F., eds., Controls on Carbonate Platform and Basin Development: Society of Economic Paleontologists and Mineralogists Special Publication 44, p. 289-304.

Fuller, J.G.C.M., 1956, Mississippian in the Saskatchewan portion of the Williston Basin: A review, *in* First International Williston Basin Symposium: North Dakota and Saskatchewan Geological Societies, Conrad Publishing, Bismark, North Dakota, p. 29-35.

Ginsburg, R.N., 1971, Landward movement of carbonate mud: New model for regressive cycles in carbonates [Abs.]: American Association of Petroleum Geologists Bulletin, v. 55, p. 340.

Gutschick, R.C., and Sandberg, C.A., 1983, Mississippian continental margins of the conterminous United States, *in* Stanley, D.J., and Moore, G.T., eds., The Shelfbreak: Critical Interface on Continental Margins: Society of Economic Paleontologists and Mineralogists Special Publication 33, p. 79-96.

Gutschick, R.C., Sandberg, C.A., and Sando, W.J., 1980, Mississippian shelf margin and carbonate platform from Montana to Nevada, *in* Fouch, T.D., and Magathan, E.R., eds., Paleozoic Paleogeography of the West-Central United States, Rocky Mountain Paleogeography Symposium 1: Society of Economic Paleontologists and Mineralogists, Rocky Mountain Section, p. 111-128.

Hardie, L.A., 1986, Stratigraphic models for carbonate tidal-flat deposition, *in* Hardie, L.A. and Shinn, E.A., eds., Carbonate Depositional Environments Modern and Ancient, Part 3: Tidal flats: Colorado School of Mines Quarterly, v. 81, p. 59-74.

Harris, S.H., Land, C.B., and McKeever, J.H., 1966, Relation of Mission Canyon stratigraphy to oil production in north-central North Dakota: American Association of Petroleum Geologists Bulletin, v. 50, p. 2269-2276.

Harrison, R.L. and Flood, A.L., 1956, Mississippian correlation in the international boundary areas, *in* First International Williston Basin Symposium: North Dakota and Saskatchewan Geological Societies, Conrad Publishing, Bismark, North Dakota, p. 36-51.

Heckel, P.H., 1974, Carbonate buildups in the geologic record: A review, *in* Laporte, L.F., ed., Reefs in Time and Space: Society of Economic Paleontologists and Mineralogists Special Publication 18, p. 90-155.

Hendricks, M.L., 1988, Shallowing-upward cyclic carbonate reservoirs in the lower Ratcliffe interval (Mississippian), Williams and McKenzie Counties, North Dakota, *in* Goolsby, S.M. and Longman, M.W., eds., Occurrence and Petrophysical Properties of Carbonate Reservoirs in the Rocky Mountain Region: Rocky Mountain Association of Geologists, p. 371-380.

Huh, O.K., 1967, The Mississippian System across the Wasatch Line east central Idaho, extreme southwestern Montana, *in* Centennial Basin of Southwest Montana: Montana Geological Society Guidebook, 18th Annual Field Conference, Billings, Montana, p. 31-62.

Huh, O.K., 1968, Mississippian stratigraphy and sedimentology, across the Wasatch Line, east-central Idaho and extreme southwestern Montana [unpublished Ph.D. dissertation]: University Park, The Pennsylvania State University, 176 p.

James, N.P., 1984, Reefs, *in* Walker, R.G., ed., Facies Models: Geoscience Canada, Reprint Series 1, p. 229-244.

Johnson, J.G. and Pendergast, A., 1981, Timing and mode of emplacement of the Roberts Mountain allochthon, Antler orogeny: Geological Society of America Bulletin, v. 92, p. 648-658.

Jordan, T.E., 1981, Thrust loads and foreland basin development, Cretaceous, western United States: American Association of Petroleum Geologists Bulletin, v. 65, p. 2506-2520.

Karner, G.D. and Watts, A.B., 1983, Gravity anomalies and flexure of the lithosphere at mountain ranges: Journal of

Geophysical Research, v. 88, p. 10449-10477.

Kendall, C.G.St.C. and Lerche, I., 1988, The rise and fall of eustasy, in Wilgus, C.K., Hastings, B.S., Kendall, C.G.St.C., Posamentier, H.W., Ross, C.A., and Van Wagoner, J.C., eds., Sea-level Changes: An Integrated Approach: Society of Economic Paleontologists and Mineralogists Special Publication 42, p. 1-17.

Lindsay, R.F., 1988, Mission Canyon Formation reservoir characteristics in North Dakota, in Goolsby, S.M. and Longman, M.W., eds., Occurrence and Petrophysical Properties of Carbonate Reservoirs in the Rocky Mountain Region: Rocky Mountain Association of Geologists, p. 317-346.

Malek-Aslani, M., 1971, Depositional environments of Mission Canyon (Mississippian) oil fields in north-central North Dakota [Abs.]: American Association of Petroleum Geologists Bulletin, v. 55, p. 351.

Mamet, B.L., Skipp, B., Sando, W.J., and Mapel, W.J., 1971, Biostratigraphy of Upper Mississippian and associated Carboniferous rocks in south-central Idaho: American Association of Petroleum Geologists Bulletin, v. 55, p. 20-33.

Middleton, G.V., 1961, Evaporite solution breccias from the Mississippian of southwest Montana: Journal of Sedimentary Petrology, v. 31, p. 189-195.

Mullins, H.T., Gardulski, A.F., Hine, A.C., Melillo, A.J., Wise, S.W., Jr., and Applegate, J., 1988, Three-dimensional sedimentary framework of the carbonate ramp slope of central west Florida: A sequential seismic stratigraphic perspective: Geological Society of America Bulletin, v. 100, p. 514-533.

Nichols, K.M., 1980, Depositional and diagenetic history of porous dolomitized grainstones at the top of the Madison Group, Disturbed Belt, Montana, in Fouch, T.D., and Magathan, E.R., eds., Paleozoic Paleogeography of the West-Central United States, Rocky Mountain Paleogeography Symposium 1: Society of Economic Paleontologists and Mineralogists, Rocky Mountain Section, p. 163-173.

Nilsen, T.H., 1977, Paleogeography of Mississippian turbidites in south-central Idaho, in Stewart, J.H., Stevens, C.H., and Fritsche, A.E., eds., Paleozoic Paleogeography of the Western United States, Pacific Coast Paleo-geography Symposium 1: Society of Economic Paleontologists and Mineralogists Pacific Section, p. 275-299.

Peterson, J.A., 1986, General stratigraphy and regional paleotectonics of the western Montana overthrust belt, in Peterson, J.A., ed., Paleotectonics and Sedimentation in the Rocky Mountain Region, Unites States: American Association of Petroleum Geologists Memoir 41, p. 57-86.

Petty, D.M., 1988, Depositional facies, textural characteristics, and reservoir

properties of dolomites in Frobisher-Alida interval in southwest North Dakota: American Association of Petroleum Geologists Bulletin, v. 72, p. 1229-1253.

Poole, F.G., 1974, Flysch deposits of the Antler foreland basin, western United States, in Dickinson, W.R., ed., Tectonics and Sedimentation: Society of Economic Paleontologists and Mineralogists Special Publication 22, p. 58-82.

Read, J.F., 1985, Carbonate platform facies models: American Association of Petroleum Geologists Bulletin, v. 69, p. 1-21.

Reid, S.K., 1991, Evolution of the Lower Mississippian Mission Canyon platform and distal Antler foredeep, Montana and Idaho [unpublished Ph.D. dissertation]: College Station, Texas A&M University, 115 p.

Reid, S.K. and Dorobek, S.L., 1989, Application of sequence stratigraphy in the Mississippian Mission Canyon Formation and stratigraphic equivalents, southwestern Montana and east-central Idaho, in French, D.E, and Grabb, R.F., eds., 1989 Field Conference Guidebook: Montana Centennial Edition, Geologic Resources of Montana, v. 1, p. 47-53.

Roberts, A.E., 1966, Stratigraphy of Madison Group near Livingston, Montana, and discussion of karst and solution-breccia features: United States Geological Survey Professional Paper 526-B, p. B1-B23.

Roberts, R.J., and Thomasson, M.R., 1964, Comparison of Late Paleozoic depositional history of northern Nevada and central Idaho: United States Geological Survey Professional Paper 475-D, p. D1-D6.

Rose, P.R., 1976, Mississippian carbonate shelf margins, western United States: United States Geological Survey Journal of Research, v. 4, p. 449-466.

Ross, C.A., and Ross, J.P., 1987, Late Paleozoic sea levels and depositional sequences, in Ross, C.A., and Haman, D., eds., Timing and Depositional History of Eustatic Sequences: Constraints on Seismic Stratigraphy: Cushman Foundation for Foraminiferal Research Special Publication No. 24, p. 137-149.

Rudolph, K.W. and Lehmann, P.J., 1989, Platform evolution and sequence stratigraphy of the Natuna platform, South China Sea, in Crevello, P.D., Wilson, J.L., Sarg, J.F., and Read, J.F., eds., Controls on Carbonate Platform and Basin Development: Society of Economic Paleontologists and Mineralogists Special Publication 44, p. 353-361.

Sandberg, C.A., 1975, McGowan Creek Formation, new name for Lower Mississippian flysch sequence in east-central Idaho: United States Geological Survey Bulletin, v. 1405-E, 11 p.

Sandberg, C.A. and Poole, F.G., 1977, Conodont biostratigraphy and depositional complexes of Upper-Devonian cratonic-platform and continental shelf rocks in the western United States, *in* Murphy, M.A., Berry, W.B.N., and Sandberg, C.A., eds., Western North America, Devonian: California University, Riverside, Campus Museum Contributions 4, p. 144-182.

Sandberg, C.A., Gutschick, R.C., Johnson, J.G., Poole, F.G., and Sando, W.J., 1983, Middle Devonian to Late Mississippian geologic history of the Overthrust Belt region, western United States: Rocky Mountain Association of Geologists, Geologic Studies of the Cordilleran Thrust Belt, v. 2, p. 691-719.

Sando, W.J., 1960, Corals from well cores of Madison Group, Williston Basin: United States Geological Survey Bulletin 1071-F, 190 p.

Sando, W.J., 1974, Ancient solution phenomena in the Madison Limestone (Mississippian) in north-central Wyoming: United States Geological Survey Journal of Research, v. 2, p. 133-141.

Sando, W.J., 1976, Mississippian history of the northern Rocky Mountains Region: United States Geological Survey Journal of Research, v. 4, p. 317-338.

Sando, W.J., 1978, Coral zones and problems of Mississippian stratigraphy in the Williston Basin: Montana Geological Society Williston Basin Symposium, Billings, Montana, p. 231-237.

Sando, W.J., 1985, Revised Mississippian time scale, western interior region, conterminous United States: United States Geological Survey Bulletin, v. 1605-A, p. A15-A26.

Sando, W.J., 1988, Madison Limestone (Mississippian) paleokarst: A geologic synthesis, *in* James, N.P., and Choquette, P.W., eds., Paleokarst: New York, Springer-Verlag, p. 256-277.

Sando, W.J., and Bamber, E.W., 1985, Coral zonation of the Mississippian System in the western interior province of North America: United States Geological Survey Professional Paper 1334, 61 p.

Sando, W.J. and Mamet, B.L., 1981, Distribution and stratigraphic significance of foraminifera and algae in well cores from Madison Group (Mississippian), Williston Basin, Montana: United States Geological Survey Bulletin 1529-F, 12 p.

Sando, W.J., Mamet, B.L., and Dutro, J.T., Jr., 1969, Carboniferous megafaunal and microfaunal zonation in the Northern Cordillera of the United States: United States Geological Survey Professional Paper 613-E, 29 p.

Sando, W.J., Sandberg, C.A., and Perry, W.J., Jr., 1985, Revision of Mississippian Stratigraphy, Northern Tendoy Mountains, southwest Montana, *in* Sando, W.J., ed., Mississippian and Pennsylvanian Stratigraphy in Southwest Montana and Adjacent Idaho: United States Geological Survey Bulletin, v. 1656, p. A1-A10.

Sando, W.J., Dutro, J.T., Jr., Sandberg, C.A., and Mamet, B.L., 1976, Revision of Mississippian stratigraphy, eastern Idaho and northeastern Utah: United States Geological Survey Journal of Research, v. 4, p. 467-479.

Sarg, J.F., 1988, Carbonate sequence stratigraphy, *in* Wilgus, C.K., Hastings, B.S., Kendall, C.G.St.C., Posamentier, H.W., Ross, C.A., and Van Wagoner, J.C., eds., Sea-Level Changes: An Integrated Approach: Society of Economic Paleontologists and Mineralogists Special Publication 42, p. 155-181.

Shanley, K.W., 1983, Stratigraphy and depositional model, upper Mission Canyon formation (Mississippian) northeast Williston Basin, North Dakota [unpublished M.S. thesis]: Golden, Colorado, Colorado School of Mines, 172 p.

Skipp, B., Sando, W.J., and Hall, W.E., 1979, The Mississippian and Pennsylvanian (Carboniferous) Systems in the United States - Idaho: United States Geological Survey Professional Paper 1110-AA, 42 p.

Speed, R.C., 1977, Island-arc and other paleogeographic terranes of the late Paleozoic age of the western Great Basin, *in* Stewart, J.H., Stevens, C.H., and Fritsch, A.E., eds., Paleozoic Paleogeography of the Western United States, Pacific Coast Paleogeography Symposium 1: Society of Economic Paleontologists and Mineralogists, Pacific Section, p. 349-362.

Speed R.C. and Sleep, N.H., 1982, Antler orogeny and foreland basin: A model: Geological Society of America Bulletin, v. 93, p. 815-828.

Stockmal, G.S., and Beaumont, C., 1987, Geodynamic models of convergent margin tectonics: the southern Canadian Cordillera and the Swiss Alps, *in* Beaumont, C., and Tankard, A.J., eds., Sedimentary Basins and Basin-Forming Mechanisms: Canadian Society of Petroleum Geologists Memoir 12, p. 393-411.

Vail, P.R., Mitchum, R.M., Jr., Thompson, S., III, 1977, Seismic stratigraphy and global changes of sea level, Part 4: Global cycles and relative changes of sea level, *in* Payton, C.E., ed., Seismic Stratigraphy - Application to Hydrocarbon Exploration: American Association of Petroleum Geologists Memoir 26, p. 49-212.

Van Wagoner, J.C., Posamentier, H.W., Mitchum, R.M., Vail, P.R., Sarg, J.F., Loutit, T.S., and Hardenbol, J., 1988, An overview of the fundamentals of sequence stratigraphy and key definitions, *in* Wilgus, C.K., Hastings, B.S., Kendall, C.G.St.C., Posamentier, H.W., Ross, C.A., and Van Wagoner, J.C., eds., Sea-Level Changes: An Integrated Approach: Society of Economic Paleontologists and Mineralogists Special Publication 42, p. 39-45.

Waters, D.L. and Sando, W.J., 1987a, Coral zonules: New tools for petroleum exploration in the Mission Canyon Limestone and Charles Formation, Williston Basin, North Dakota: Rocky Mountain Association of Geologists 1987 Symposium, p. 193-208.

Waters, D.L. and Sando, W.J., 1987b, Corals from the Madison Group, Williston Basin, North Dakota, *in* Carlson, C.G. and Christopher, J.E., eds., Fifth

International Williston Basin Symposium: Saskatchewan Geological Society Special Publication 9, p. 83-97.

Waters, D.L. and Sando, W.J., 1987c, Depositional cycles in the Mississippian Mission Canyon Limestone and Charles Formation, Williston Basin, North Dakota, *in* Carlson, C.G. and Christopher, J.E., eds., Fifth International Williston Basin Symposium: Saskatchewan Geological Society Special Publication 9, p. 123-133.

Wilgus, C.K., Hastings, B.S., Kendall, C.G.St.C., Posamentier, H.W., Ross, C.A., and Van Wagoner, J.C., 1988, Sea-Level Changes: An Integrated Approach: Society of Economic Paleontologists and Mineralogists Special Publication 42, 407 p.

DEPOSITIONAL HISTORY AND PALEOGEOGRAPHIC SETTING OF THE JUNIPER GULCH MEMBER OF THE MIDDLE PENNSYLVANIAN-LOWER PERMIAN SNAKY CANYON FORMATION OF CENTRAL IDAHO

K.L. CANTER
Coyote Geologic Services
2595 Spruce St., Ste A.
Boulder, Colorado 80302

and

P.E. ISAACSON
Department of Geology
University of Idaho
Moscow, Idaho 83843

ABSTRACT

The Juniper Gulch Member of the Pennsylvanian-Permian Snaky Canyon Formation is characterized by interbedded carbonate and sandstone cycles deposited as progradational shallowing-upward carbonate platform to eolian dune sequences. Carbonate facies include skeletal and oolitic shoals, phylloid algal and Palaeoaplysina mound-shaped carbonate buildups, open platform/ lagoon, restricted platform/peritidal, and supratidal sequences. Eolian dune deposits are characterized by planar-tabular to wedge-planar crossbedded quartz sandstones. These interbedded carbonate and sandstone cycles record the spreading of the Upper Pennsylvanian-Lower Permian erg into south-central Idaho from the northeast. Shallow-water carbonate platform and tidal flat sedimentation was terminated by prograding eolian dunes derived from the coeval Quadrant-Tensleep sand sea to the north and east.

INTRODUCTION

The Snaky Canyon Formation (Skipp and others, 1979) is a thick sequence of platform-dominated late Paleozoic interbedded sandstone and carbonate rocks composed of three members; the Bloom, Gallagher Peak Sandstone, and Juniper Gulch Members. The focus of this paper is the sedimentology and stratigraphy of the Upper Pennsylvanian to Lower Permian Juniper Gulch Member.

Within the Juniper Gulch Member of the Snaky Canyon Formation are phylloid algae and Palaeoaplysina-dominated buildups that are interbedded with quartz sandstones. The paleontology of these buildups in the area north of Howe, Idaho was first noted by Breuninger (1971, 1976), who described the upper portions of these mounds. This paper focuses on the sedimentology and depositional facies of three sections of the member emphasizing the lateral changes.

The principal section for the study, the Arco Hills section, is located at the southern end of the Lost River Range (Fig. 1). Other locations considered here are the Cedar Canyon section in the southern Lemhi Range and the Snaky Canyon type locality in the southern Beaverhead Range (Fig. 1). The original distance between localities of the Juniper Gulch member in these three ranges could have been quite different than that at present, owing to extensive post-depositional thrusting in the region (Skipp and Hait, 1977). Assuming that much of this deformation was eastward-directed, actual depositional sites for each section would have been west of the present localities.

STRATIGRAPHIC SETTING

Pennsylvanian and Permian strata presently located east of the White Knob Mountains in south- to east-central Idaho are composed of a thick sequence of dominantly shallow-water, siliciclastic-rich platform to platform margin carbonate rocks. Investigations by Skipp and others (1979 and 1980) document the deposition of over 1,400m of carbonate and sandstone in this area. This carbonate- and siliciclastic-rich interval is divided into three formations, from oldest to youngest: Bluebird Mountain, Snaky Canyon, and Phosphoria formations.

To the east of the study area, on the Wyoming platform, equivalent carbonate and siliciclastic rocks interfinger with quartzose sandstone of the Pennsylvanian Quadrant Formation and dolomite of the Permian Phosphoria Formation (Fig. 2). The interbedded sandstones and dolomites in the Quadrant Formation are interpreted as representing progradational cycles of peri-tidal to supratidal carbonates capped by eolian dunes (Saperstone and Ethridge, 1984). The carbonates of the Phosphoria Formation generally are considered to represent shallow, extremely restricted lagoonal deposition.

Approximately 25km west of the Lost River Range in the White Knob Mountains, some of these units interfinger with coarse, orogenic highland-derived clastic material where medium- to coarse-grained, cherty, quartz sandstones and conglomerates occur within the lower member (Bloom) of the Snaky Canyon Formation. The relationship of the younger carbonates of the Snaky Canyon Formation (Gallagher Peak Sandstone and Juniper Gulch members) and the Phosphoria Formation to these orogenic deposits has not been established.

The thick, alternating sandstones and carbonates of the Pennsylvanian-Permian

In Cooper, J.D., and Stevens, C.H., eds., 1991, **Paleozoic Paleogeography of the Western United States-II:** Pacific Section SEPM, Vol. 67, p. 543-550.

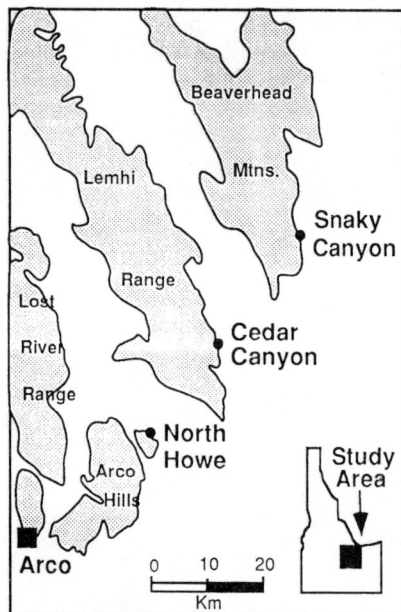

Fig. 1. Locations of measured sections of the Juniper Gulch Member of the Snaky Canyon Formation used in this study.

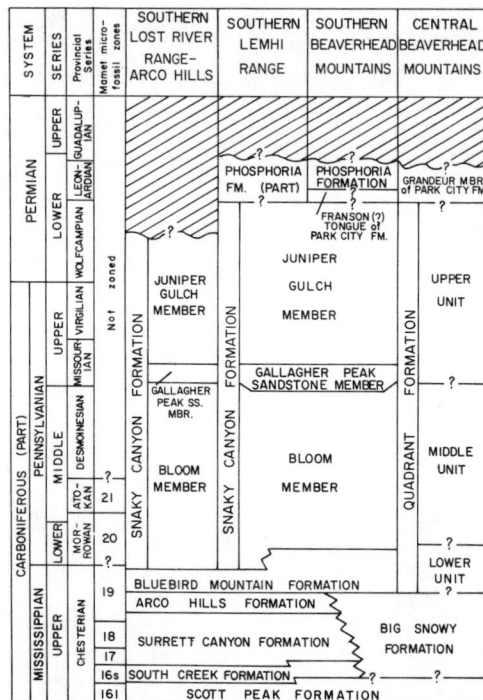

Fig. 2. Generalized correlation chart illustrating the relationship of the Juniper Gulch Member of the Snaky Canyon Formation to coeval units in adjacent regions.

Oquirrh Formation and the calcareous siltstones of the Cassia Basin (Morgan, 1980) lie south of the study area and south of the basalts of the Snake River Plain. Although these units are lithologically similar to the rocks of the Snaky Canyon Formation, their relationship to similar rocks north of the Snake River Plain is poorly understood and is in need of additional study.

STRATIGRAPHY OF THE JUNIPER GULCH MEMBER

Introduction

Data and interpretations from three measured sections taken at Snaky Canyon, Cedar Canyon, and north Howe were used in this study. Locations of these sections are listed in the Appendix and a detailed description of each section is presented below. Phylloid algae and hydrozoan mounds are common in both the North Howe and Cedar Canyon sections, but were rarely observed in the type section at Snaky Canyon. Silici-clastic rocks are common in all of these sections but are particularly abundant in the type section.

Snaky Canyon Type Section

At the Snaky Canyon locality in the southern Beaverhead Range, the Juniper Gulch Member is approximately 600m thick. Here, the member can be divided into four units based on facies associations. The lowermost unit, consisting of dolomitic, sandy, fossiliferous wackestone and packstone, cherty dolomite wackestone and packstone, and non-fossiliferous, sandy and cherty dolomite mudstone is 125m thick. Sandy limestone float is common in the covered interval at the top of the unit.

The overlying unit, approximately 200m

thick, is rich in pelmatozoan debris. Chert nodules and lenses are common at the base and top of this dolomite-rich wackestone and packstone interval. This fossiliferous interval is overlain by 160m of cherty, dolomitic mudstone; sandy, oolitic grainstone; and dolomitic to calcareous sandstone. Fossils are limited to sparse, abraded bioclastic debris.

The uppermost 155m is characterized by abundant phylloid algae (Archaeolithophyllum) and minor Palaeoaplysina debris. Quartz sand also is common and other rock types present include calcareous sandstone; sandy dolomite breccia; sandy wackestone and packstone; sandy dolomitic mudstone; and bedded chert. The contact with the overlying Phosphoria Formation is covered and abrupt.

The Snaky Canyon section represents peritidal to supratidal marine environments that were periodically inundated by floods of siliciclastic material. This restricted, inner-platform to coastal setting is indicated by an allochem assemblage limited to stromatolitic algae, encrusting forms of phylloid algae, such as Archaeolithophyllum, ooids, and fine-grained bioclastic debris. Most carbonate rocks are quartz sand-rich. Sedimentary features indicative of peritidal to supratidal environments include desiccation breccia; flat pebble conglomerate; and white, brecciated chert beds that are interpreted to represent silcretes. Segregated sandstone units contain high-angle wedge to planar-tabular cross beds and contain few marine fossils, and are bimodally sorted. These features suggest that many of these sands were deposited as eolian dunes.

Cedar Canyon Section

At Cedar Canyon in the southern Lemhi Range (Fig. 1), a partial section of Juniper Gulch Member is at least 335m thick. The lowermost unit, located above a thrust fault, is 110m thick. This unit is characterized by sandstone, sandy mudstone breccia, sandy packstone rich in grapestone aggregates and arenaceous foraminifera, and peloidal wackestone/ packstone. The overlying 110m consists of coral and hydrozoan bafflestone, bioclastic wackestone/packstone, _Eugono-phyllum_ and _Palaeoaplysina_ bindstone, and recrystallized limestone. The carbonates become increasingly more quartz sand-rich up section.

Sandstone, sandy calcareous or dolomitic mudstone, and algal-mat bindstone compose the next 57m of section. Less common rocks include fossiliferous packstone and grain-stone. This unit is overlain by approximately 68m of sand-rich, vuggy dolomite mudstone; sandy mudstone; and dolomitic algal-mat bindstone. Allochems are dominated by encrusting phylloid algae and stromato-litic algae.

The depositional sequence at Cedar Canyon appears similar to that observed at Snaky Canyon, and also represents restricted peritidal deposition. An increase in phylloid algae and other reef-dwelling organisms, and composite grains in the middle portion of this section, such as grapestones, indicate a marine-influenced lagoonal paleo-environment. Overlying lithologies suggest a return to restricted, eolian-influenced peritidal conditions.

Arco Hills Section

North of the town of Howe (Fig. 1), a section approximately 700m thick is characterized by five depositional cycles, each containing at least one carbonate buildup (Canter and Isaacson, 1983). Cycle 1 (265m thick) contains alternating beds of well-sorted, crossbedded, medium-grained sandstone and bioclastic packstone/ grainstone. Coral and/or bryozoan bafflestone, skeletal wackestone/packstone and sandy, dolomitic mudstone are less common (Fig. 3). The buildup interval in this cycle was terminated by a flood of siliciclastic sediment. Cycle 2 (165m thick) is litho-logically similar to the first cycle, but it also includes fenestrate bryozoan lime floatstone. Its contact with the overlying cycle is abrupt and may be erosional.

Cycle 3 (70m thick) contains slightly sandy, skeletal wackestones and packstones rich in phylloid algae and _Palaeoaplysina_ debris, and algal and encrusting bryozoan bindstone. Its contact with the overlying cycle is erosional, and this erosional surface can be recognized throughout the Arco Hills area (Breuninger, 1971).

Cycle 4 (71m thick) is lithologically

similar to Cycles 1 and 2. The lowermost strata consist of sandstone-rich cycles that contain thick interbeds of skeletal wacke-stones and packstone overlain by coralline bafflestone. Diversely fossiliferous wacke-stone and packstone, _Archaeolithophyllum_(?) bindstone, and sandy and/or dolomitic mudstone are present at the top of the cycle.

Cycle 5 (140m thick) is similar to Cycles 1, 2, and 4, with the exception of the presence of numerous beds of stromatolitic and _Archaeolithophyllum_(?), sandy bindstone at the top of the unit. This section is terminated by a fault.

The lower two cycles at the north Howe stratigraphic section are characterized by many smaller scale cycles of interbedded carbonates and sandstones. A general small-scale cycle begins with a basal unit of skeletal packstone or grainstone deposited above a sharp, planar contact with the underlying unit, typically a sandstone. Bedding within these 1 to 3 meters-thick grain-supported beds varies from crossbedded to massive. These packstones/grainstones represent transgressive shoal deposits that grade upward into lagoonal deposits that contain small patch reefs comprised of ramose and fenestrate bryozoans and/or syringoporoid or colonial rugose corals. Sandy, dolomitic mudstones and/or planar-tabular to trough cross-stratified sandstones cap these small-scale progradational cycles. These sand-rich units record the progradation of eolian dunes over shallow-water carbonate sediments.

A deepening event is indicated by the rocks of Cycles 3 and 4. Siliciclastic interbeds are less common in these cycles that are instead characterized by diversely fossiliferous packstones and phylloid algal and _Palaeoaplysina_ boundstones. These lithologies indicate well-oxygenated, shallow marine conditions. The deposition of these shallow marine carbonate cycles over the interbedded siliciclastics and carbonates of Cycles 1 and 2 results in a backstepping facies pattern, occurring in response to a slight rise in sea level.

A return to restricted peritidal marine conditions occurred with the deposition of Cycle 5. This cycle is composed of sandy limestone and dolomite and interbedded quartz sandstone. Carbonate buildups in this cycle are dominated by encrusting growth morphol-ogies capable of withstanding more intense wave energy than those in the earlier cycles.

DEPOSITIONAL HISTORY AND PALEOGEOGRAPHY

Introduction

Correlation of the cycles within the Juniper Gulch Member to Upper Pennsylvanian and Lower Permian rocks in the adjacent areas of southwestern Montana and western Wyoming is difficult because of unclear age relation-ships and lack of detailed stratigraphic data (Saperstone and Ethridge, 1984; Fig. 4). Additionally, because of tectonic complexi-ties, correlations between the sections of the member are tentative and based on marker

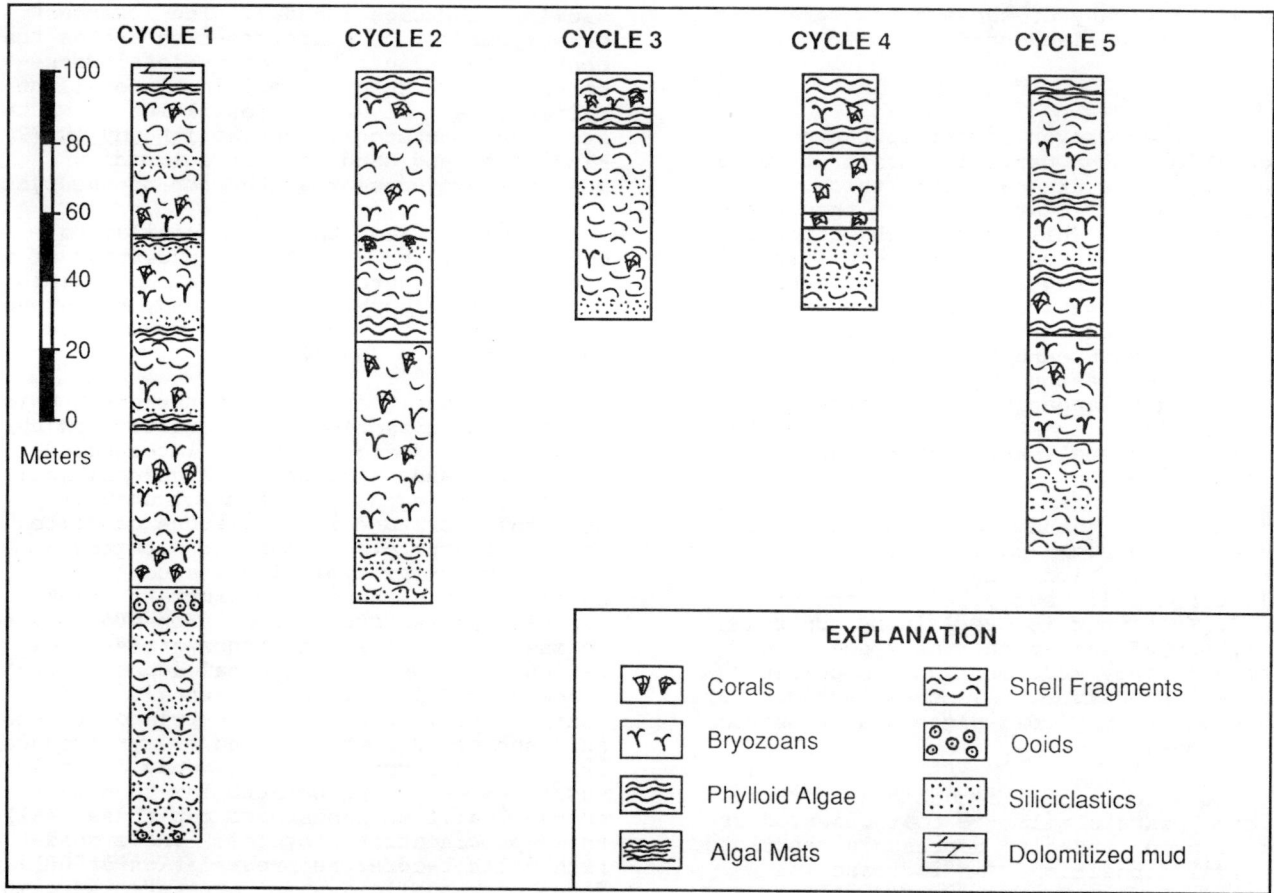

Fig. 3. Idealized vertical sequence of the five depositional cycles described from the north Howe section (Fig. 1). Cycle 1 is oldest; Cycle 5 is youngest. The lower part of Cycle 1 is characterized by interbedded sandstone and packstone/grainstone. Dark-gray, mud-rich bryozoan and coral bafflestone and skeletal wackestone and packstone are abundant in the upper part of Cycle 1 and throughout most of Cycle 2. Cycles 3 an 4 contain well-established hydrozoan and phylloid algal buildups. Interbedded sandstone and packstone/grainstone, and thin phylloid algal and algal mat boundstones are common in Cycle 5.

beds, such as oolitic grainstones and key sandstones because detailed biostratigraphic data are lacking. In order to better understand the relationship of the member to the Pennsylvanian sandstones to the north and east, and to the Permian Phosphoria Formation, a more regionally comprehensive detailed stratigraphic study is needed. Within the study area, however, a depositional history for these rocks can be constructed. A preliminary paleogeographic model that incorporates the importance of the interbedded quartz sandstones in the Juniper Gulch Member to the interpretation of the depositional history of the member as well as the regional relationship of the Juniper Gulch Member to the eolian sandstones to the north and east is presented below.

Facies Descriptions

Five carbonate facies as defined by Wilson (1975) occur repeatedly in these sections (Canter, 1984). An idealized carbonate-dominated depositional cycle consists of (from base to top): carbonate-shoal (Facies Belt 6), carbonate-buildup (Facies Belt 5), lagoon (Facies Belt 7), restricted platform/peritidal (Facies Belt 8), and supratidal/marginal marine (Facies Belt 9). Representative photographs of significant lithologies are presented in Figure 5.

Carbonate-shoal facies (Facies Belt 5) are typically composed of skeletal and peloidal, often echinoderm-rich packstones and grainstones that are generally located near the base of a cycle. Oolitic grainstones are present in some cycles. The grain-supported lithologies were deposited as sand waves in an agitated shoal or tidal-bar environment.

Two facies belts represent the normal-marine open platform environment, open lagoon (Facies Belt 7) and carbonate buildup (Facies Belt 5). Whole-fossil brachiopod-rich wackestones and extensively bioturbated packstones were deposited in low energy open lagoons. Associated with these mud-rich lithologies are phylloid algae and Palaeoaplysina bafflestones and bindstones (Facies Belt 5). These buildups are initially colonized by ramose

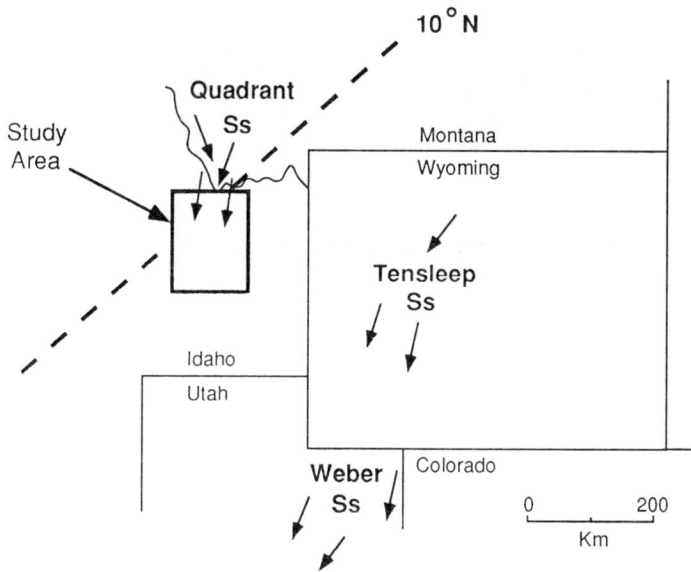

Fig. 4. Paleowind directions from Pennsylvanian and Permian sandstone deposits of the Wyoming shelf (from Saperstone and Ethridge, 1984).

bryozoans or colonial corals (Canter, 1984). Phylloid algae, such as Eugonophyllum and Archaeolithophyllum and the hydrozoan (?) Palaeoaplysina then constructed mound-shaped buildups on the colonized substrate (Fig. 6). Some of these buildups grew upward into increasing more shallow and agitated waters, resulting in the deposition of platy algae- or Palaeoaplysina-rich grainstones as capping and flanking beds adjacent to these mounds.

Restricted inner platform environments (Facies Belt 8) are represented by the typically sandy and dolomitic lithologies of the upper part of a cycle. Common allochems include peloids, intraclasts, coated grains, and abraded skeletal material. Supratidal conditions (Facies Belt 9) are indicated by the occurrence of stromatolitic algal bound-stones and vuggy, sandy dolomite mudstones. These rocks are typically overlain by quartz sandstones.

The quartz sandstones are fine- to medium-grained and typically trough to planar-tabular cross-stratified. The grains are well-sorted and well-rounded. Thickness of the sand units averages 3 meters. These texturally mature sands are virtually free of any skeletal material, and marine-derived allochems are limited to minor (<10%) amounts of ooids admixed with the detrital quartz in the trough cross-stratified sandstones. Upper bed boundaries between these sandstones and overlying carbonates are sharp; no gradational contacts were observed in the Arco Hills and Cedar Canyon sections. Gradational contacts between these two lithologies were observed only at the type section. The sedimentary features and textures of these sandstones indicate that they were deposited by eolian dunes that prograded across a carbonate tidal flat. Some dunes may have been re-worked into trough crossbedded sand units in tidal

channels or in adjacent subtidal marine environments.

Paleogeography

Interbedded sandstones and carbonates indicative of shallow-water deposition characterize the lower part of the member in the study area. At the Arco Hills section, these carbonate rocks include skeletal lime packstones and grainstone, sparsely fossil-iferous, brachiopod-rich lime wackestone, and coral and/or bryozoan lime bafflestone. Most are mottled from extensive bioturbation. Grapestones and depositional breccias (including flat-pebble conglomerates) are abundant in the Cedar Canyon section, whereas sandy lime wackestone and packstone and cherty dolomite characterize the type section. Collectively, this suite of carbonate rocks represents restricted, shallow-water lagoon (Arco Hills and Cedar Canyon localities) to peritidal deposition (Snaky Canyon section; Fig. 6).

At about the time that Quadrant Sandstone deposition apparently ceased in southwestern Montana (Blakey and others, 1988, Fig. 2), the Gallagher Peak Sandstone, which underlies the Juniper Gulch Member, was deposited in south-central Idaho. One possible explanation for the cessation in Quadrant deposition is that the Upper Pennsylvanian-Lower Permian erg spread to the south due to a widespread regressive event and is recorded by the deposition of the middle member of the Snaky Canyon Formation. Similarly, the source area that provided sand to the Middle to Upper Pennsylvanian Tensleep Formation in western Wyoming may have also contributed quartz sand to this area.

The continued influence of this erg in Lower Permian strata is manifested in the numerous clean, well-sorted sandstone interbeds in the Juniper Gulch Member. In the lower units of this member, the inter-bedded high-angle trough cross-stratified sands are interpreted to represent subtidal sand wave deposition. Planar-tabular and wedge-planar crossbedded sandstones may have been deposited as prograding eolian dunes.

Slightly deeper water, more marine-dominated facies characterize the rocks that make up the middle part of the Juniper Gulch Member. Sandstone interbeds are less common and carbonate lithologies contain a diverse fossil assemblage. Carbonate buildups constructed by phylloid algae and/or Palaeoaplysina are common in the Arco Hills and Cedar Canyon sections, while less extensively developed phylloid algal buildups occur in coeval strata in the type section at Snaky Canyon.

Cyclic, restricted peritidal carbonate and siliciclastic eolian coastal dune deposition was re-established during the accumulation of the upper part of the Juniper Gulch Member. This is manifested by an increase in the number of interbedded sand-stones and in the amount of sandy dolomite. Carbonate buildups are sandy and dominated by encrusting growth forms that adapted to

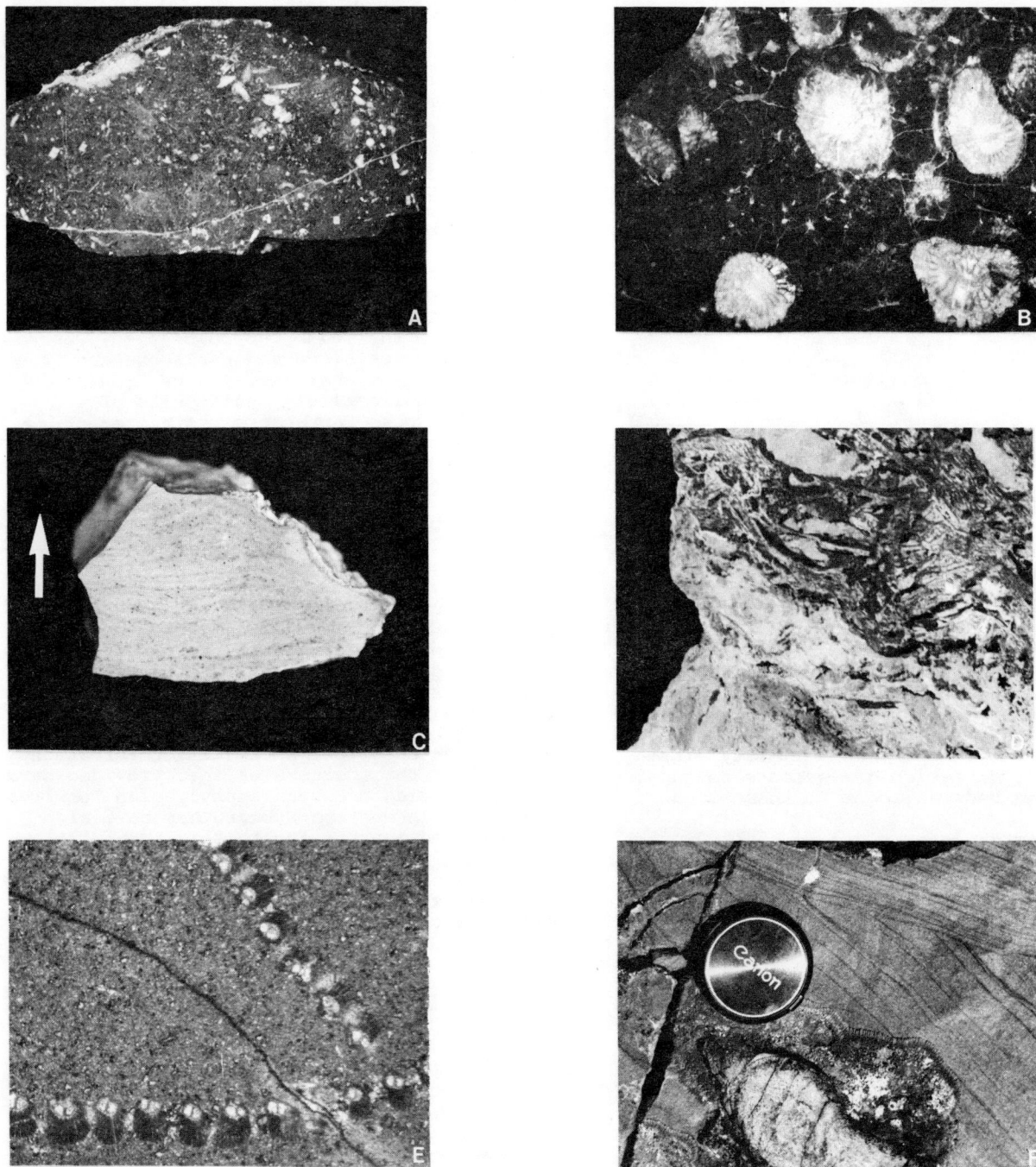

Fig. 5. Representative lithologies from the Juniper Gulch Member with interpreted facies belt (after Wilson, 1975). 5a - bioturbated packstone (Facies Belt 6/7), 5b - rugose coral bafflestone (Facies Belt 7), 5c - phylloid algal and foraminifera bindstone (Facies Belt 5 - high energy), 5d - phylloid algal bafflestone with mud-filled umbrella structures (Facies Belt 5 -low energy), 5e -fenestrate bryozoan floatstone (Facies Belt 5 - buildup core), 5f - crossbedded sandstone (Eolian Facies). All photographs X 1 except as indicated.

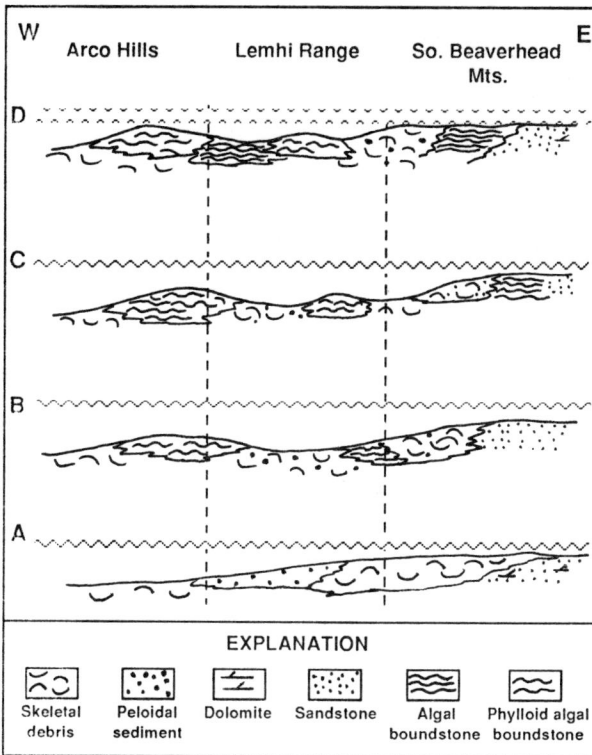

Fig. 6. Cross sections illustrating facies development of the Juniper Gulch Member through time, with A being oldest. A - character of initial inner platform; B - buildup growth results in the establishment of a barrier, causing infilling shoreward of buildup and lagoonal sediments; C -with continued buildup growth and infilling of lagoon, intertidal conditions develop behind buildup; and D - lagoon is filled with carbonate and siliciclastic material derived from the east.

relatively high energy environments. Although Skipp and others (1979) state that the contact between the Juniper Gulch Member of the Snaky Canyon and the lower member of the Phosphoria Formation is abrupt but conformable, it is likely that subaerial exposure may have affected the upper portion of the member in the study area, and that the contact between the formations is marked by a sequence boundary that is overlain by the transgressive carbonates of the Phosphoria Formation.

CONCLUSIONS

The Juniper Gulch Member of the Snaky Canyon Formation is characterized by repetitious interbeds of marine carbonate and quartz sandstone that were deposited as shallowing-upward cycles in an environment where siliciclastic eolian dunes prograded over shallow-water, dominantly peritidal carbonates. A typical carbonate to siliciclastic cycle is represented by the following (from bottom to top): (a) sharp, low-angle contact between the underlying sandstone and the overlying carbonate; (b) skeletal packstone and grainstone or, more rarely, cross-bedded oolitic grainstone; (c)

skeletal wackestone and coral and/or bryozoan boundstone; (d) sandy and dolomitic wackestone and packstone; and (e) cross-stratified sandstone.

During marine transgressive events, sandstone beds were truncated by erosion associated with the high-energy transgressive sea. Carbonate shoals were deposited after the initial transgression. A progradational carbonate-dominated depositional sequence overlies the basal shoal deposits. A variety of shallow-water carbonate facies occur within this progradational sequence, including carbonate buildup (Facies Belt 5), open lagoon (Facies Belt 7), restricted platform and peritidal (Facies Belt 8), and supratidal (Facies Belt 9). Eolian dunes prograded across the carbonate tidal flat, terminating carbonate sedimentation in most cycles.

ACKNOWLEDGEMENTS

This paper is based on an M.S. thesis supervised by Dr. P.E. Isaacson at the University of Idaho. Funds for the study were provided by a Featherstone Grant, Idaho Mining and Minerals Research Grant, and a grant from the Idaho DOE for continued research. Coyote Geologic Services defrayed the cost of manuscript preparation.

This study was improved by the helpful comments and criticisms of many people, including R.C. Geesaman, V.J. Coringrato, and Jane Estes. Excellent suggestions for improving the manuscript were provided by Dr. Calvin Stevens.

REFERENCES CITED

Blakey, R.C., Peterson, F., and Kocurek, G., 1988, Synthesis of late Paleozoic and Mesozoic eolian deposits of the Western Interior of the United States. Sedimentary Geology, v. 56, p. 3-125.

Breuninger, R.H., 1971, Late Pennsylvanian and Early Permian carbonate mounds in the Arco Hills and southern Lemhi Range, Idaho [PhD thesis]: Missoula, University of Montana, 216 p.

Breuninger, R.H., 1976, Palaeoaplysina (Hydrozoan?) carbonate buildups from upper Paleozoic of Idaho. American Association of Petroleum Geology Bulletin, v. 60, p. 584-607.

Canter, K.L., 1984, Geology, stratigraphy, lithologic descriptions, depositional environment, and successional stages of the Juniper Gulch Member of the Snaky Canyon Formation, east-central Idaho [M.S. thesis]: Moscow, University of Idaho, 108 p.

Canter, K.L., and Isaacson, P.E., 1983, Successional stages in recurrent (cyclic?) carbonate buildups, Snaky Canyon Formation (Pennsylvanian-Permian), south-central Idaho. Geological Society of America Abstracts with Programs, v. 15, p. 306.

550

Morgan, W.A., 1980, Euxinic Early Permian
    sedimentation in the Cassia Basin of
    southern Idaho, in Fouch, T.D., and
    Magathan, E.R., eds., Paleozoic
    Paleogeography of the West-Central
    United States: Rocky Mountain Section,
    Society Economic Paleontologists and
    Mineralogists, Rocky Mountain
    Paleography Symposium 1, p. 305-326.

Saperstone, H.I., and Ethridge, F.G., 1984,
    Origin and paleotectonic setting of the
    Pennsylvanian Quadrant Sandstone,
    Southwestern Montana: Thirty-fifth
    Annual Field Conference, Wyoming
    Geological Association Guidebook, p.
    309-331.

Skipp, B.A., and Hait, M.H., Jr., 1977,
    Allochthons along the northeast margin
    of the Snake River Plain, Idaho, in
    Heisey, E.L., and others, eds., Rocky
    Mountain Thrust Belt, Geology and
    Resources: 29th Annual Field Conference,
    Wyoming-Montana-Utah Geological
    Association Guidebook, p. 499-515.

Skipp, B.A., Hoggan, R.D., Schleicher, D.L.,
    and Douglas, R.C., 1979, Upper Paleozoic
    carbonate bank in east-central Idaho -
    Snaky Canyon, Bluebird Mountain, and
    Arco Hills formation, and their
    paleotectonic significance: U.S.
    Geological Survey Bulletin 1486, 78 p.

Wilson, J.L., 1975, Carbonate facies in
    geologic history: Berlin, Springer-
    Verlag Publications, 471 p.

APPENDIX

Detailed locations of stratigraphic
sections, Juniper Gulch Member of the Snaky
Canyon Formation, south-central and east-
central Idaho:

Snaky Canyon: S 1/2, Sec. 21, T 9 N, R
32 E, Snaky Canyon quadrangle, Clark
County;

Cedar Canyon: NE 1/4, SE 1/4, Sec. 17,
T 7 N, R 30 E, Deer Canyon quadrangle,
Butte County; and

North Howe: SE 1/4, SE 1/4, Sec. 31,
T 6 N, R 29 E, Howe quadrangle, Butte
County.

PENNSYLVANIAN AND PERMIAN SUN VALLEY GROUP, WOOD RIVER BASIN, SOUTH-CENTRAL IDAHO

J. Brian Mahoney
Department of Geological Sciences
University of British Columbia
Vancouver, B.C., Canada V6T 2B4
U.S. Geological Survey
Branch of Western Mineral Resources
Spokane, Washington 99204

Paul Karl Link
Department of Geology
Idaho State University
Pocatello, Idaho 83209
U.S. Geological Survey
Branch of Western Mineral Resources
Spokane, Washington 99204

Bradford R. Burton
Norcen Energy Resources
715 Fifth Avenue S.W.
Calgary, Alberta, Canada T2P 2X7

Jeffrey K. Geslin
Department of Earth and Space Sciences
University of California, Los Angeles
Los Angeles, California, 90024

J.P. O'Brien
Department of Geology
Idaho State University
Pocatello, Idaho 83209

ABSTRACT

We define the Sun Valley Group of south-central Idaho to include the Desmoinesian to Leonardian Wood River, Dollarhide, and Grand Prize Formations, which consist of 2000 to 3400 m of mixed carbonate-siliciclastic strata deposited in the epicratonic Wood River basin. In the Wood River Formation we revise the basal Hailey Member to make it more easily mappable, and define overlying Eagle Creek and Wilson Creek Members.

The Middle Pennsylvanian part of the Sun Valley Group contains proximally-derived conglomerate and interbedded biostromal limestone deposited on a braid-delta and adjacent slope. The Late Pennsylvanian to Early Permian part of the group comprises a south-sloping fine-grained mixed carbonate-siliciclastic ramp-apron system. This system received cratonal quartz sand from a northern source and comminuted bioclastic debris and micrite from the coeval Snaky Canyon Formation to the east.

The Sun Valley Group contains parts of three cycles of sedimentation which become coarser grained and then finer grained upward. Though biostratigraphic control is limited, it appears that siliciclastic grain size of the Sun Valley Group reflects second order eustatic changes.

Publication approved by Director, U. S. Geological Survey, January 23, 1991

The Wood River basin may have been a northern continuation of the Oquirrh basin to the south; the basins share similar styles of sedimentation and general times of subsidence. Subsidence of Wood River basin was influenced by Middle Pennsylvanian transpression related to the Ancestral Rockies orogeny and in the Late Pennsylvanian and Early Permian by crustal loading related to tectonism along the Late Paleozoic convergent western margin of North America.

INTRODUCTION

Upper Pennsylvanian and Lower Permian sedimentary rocks in the Boulder, White Cloud, Smoky, and Pioneer Mountains of south central Idaho (Fig. 1, inset) were deposited in the Wood River basin, one of several depocenters that developed in a narrow sinuous tract along the western margin of North America during the late Paleozoic (Skipp and Hall, 1980; Kluth, 1986; Mahoney and others, 1990; Snyder and others, 1991). In this paper we formally define the stratigraphy and describe the sedimentology of the Sun Valley Group. The Sun Valley Group includes all rocks recognized as deposited in the Wood River basin and contained within the previously defined Wood River, Grand Prize, and Dollarhide Formations, and the informally named Carrietown sequence (Lindgren, 1900; Umpleby and others, 1930; Hall and others, 1974, 1978a; Skipp and Hall, 1980; Hall, 1985).

In Cooper, J.D., and Stevens, C.H., eds., 1991, **Paleozoic Paleogeography of the Western United States-II:** Pacific Section SEPM, Vol. 67, p. 551-579.

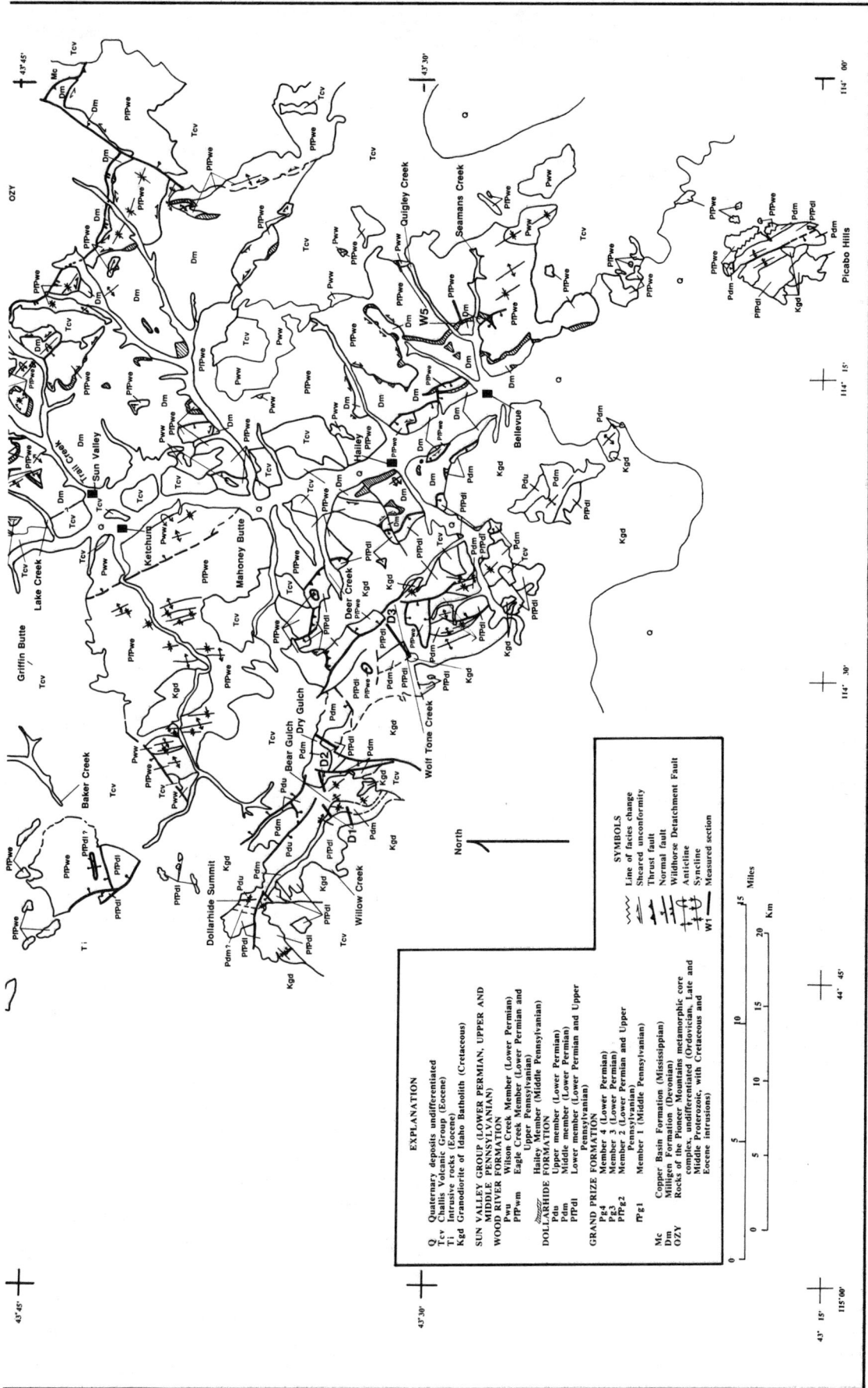

Figure 1. Location and geologic map of the Sun Valley Group on the Hailey and Challis 2-degree quadrangles, south-central Idaho. Inset map shown regional relations. Locations of stratigraphic reference sections are shown. Geologic mapping was completed as part of the Hailey CUSMAP project of the U.S. Geological Survey.

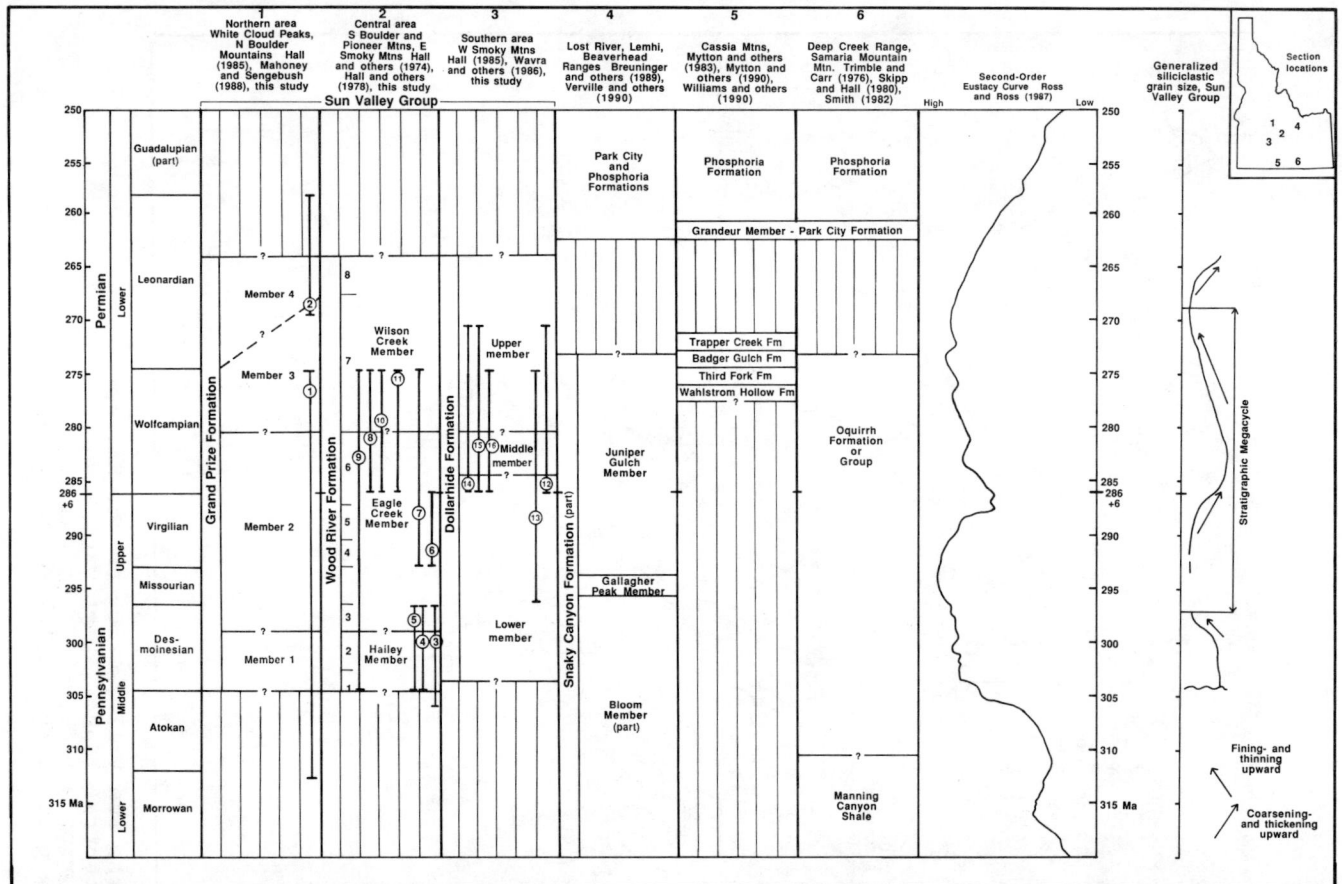

Figure 2. Correlation chart for Middle Pennsylvanian to Lower Permian rocks of southern Idaho. Locations of stratigraphic columns are shown in index map at upper right. Sea-level chart from Ross and Ross (1987). Parts of three generalized siliciclastic grain-size cycles of Sun Valley Group strata are shown at right. Within Sun Valley Group bars with numbers indicate ranges of previously published and new biostratigraphic collections. Ranges are based on Loeblich and Tappan (1988). Circled numbers indicate stratigraphic position of the collection. Identified taxa and name of biostratigrapher are as follows.

GRAND PRIZE FORMATION:
1. conodont elements of _Adetognathus_ sp., elements of _Hindeodus_ cf. _H. Ninutus_ (Ellison), elements of _Idiognathosus_ sp., (late Morrowan to Wolfcampian), Member 2? or 3?, Peach Creek, identified by A.G. Harris (Hall, 1985, p. 125);
2. conodont _Neogondelella idahoensis_ (middle-late Leonardian), Member 3 or 4, Pole Creek, identified by B.R. Wardlaw (Hall, 1985, p. 125).

WOOD RIVER FORMATION:
3. _Wedekindellina_ (late Atokan (?) to Desmoinesian), Hailey Member limestone, north of Seamans Creek in Principal Reference Section, (Bostwick, 1955) and _Pseudozaphrentoides_ (Middle Pennsylvanian brachiopod), Eagle Creek Member (unit 3), Wilson Creek section (Thomasson, 1959);

4. _Beedeina_ and numerous brachiopods (Desmoinesian), Hailey Member limestone, north of Seamans Creek in principal reference section, (Bostwick, 1955; Hall and others, 1974, p. 91);
5. _Beedeina_ (Desmoinesian), Eagle Creek Member (unit 3), north of Seamans Creek in principal reference section, Hall and others (1974, p. 91);
6. _Triticites_, _Pseudofusulinella_ sp.and _Triticites_ sp. aff _cullomensis_ Dunbar and Condra (Virgilian), Eagle Creek Member (unit 4), Seamans Creek in principal reference section, Hall and others (1974, p. 92);
7. _Triticites_ sp. (Missourian to Wolfcampian), Eagle Creek Member (unit 5), north of Seamans Creek in principal reference section, Hall and others (1974, p. 93.);
8. _Schubertella_ and _Staffella_ (unit 6 north of Seamans Creek in principal reference section, Hall and others (1974, p. 94)) plus _Triticites cullomensis_ (Wolfcampian), Eagle Creek Member, mid-upper unit 6, Wilson Creek Ridge (type section) Burton (1988), identified by C.A. Ross;
9. _Pseudofusulinella_ (Desmoinesian to Wolfcampian), Eagle Creek Member, limestone at top of unit 6, Wilson Creek Ridge type section (Burton, 1988), identified by C.A. Ross;
10. _Triticites confertus_, _T. pinguis_, _T. meeki_, _T. cellamagnus_, _Pseudofusulinella utahensis_, _P. grandensis_, _P. elkoensis_, (Wolfcampian), Wilson Creek Member, lower unit 7, Basin Gulch type section (Burton, 1988), identified by C.A. Ross;

11. _Pseudofusulina grandensis_, _P._
elkoensis, _P. wellsensis_, _Schwagerina_,
Paraschwagerina (Wolfcampian), Wilson Creek
Member, upper unit 7 or 8, Basin Gulch type
section (Burton, 1988), identified by C.A.
Ross; this collection is thought to be from
upper limestone bed tentatively assigned to
Leonardian-Guadalupian(?) by Hall and
others (1978, p. 581.);

DOLLARHIDE FORMATION (12 through 16 are
reported for the first time here, and in
O'Brien (1991):
12. _Schwagerina sp._ (Wolfcampian to mid-
Leonardian), lower member, Sky Ranch flat,
west of Bellevue, two collections, by P.K.
Link and R.S. Lewis, identified by, C.A.
Ross and D.A. Myers;
13. possibly _Triticites_ (Missourian to
Wolfcampian), lower member?, Sky Ranch
flat, west of Bellevue, collected by P.K.
Link, identified by R.C. Douglass;
14. possibly _Pseudofusulina_ (Wolfcampian
to Leonardian), lower member, Wolf Tone
Creek, R.C. Douglass written communication
to W. Hall, 1978 (Hall, 1985, p. 124);
15. Suggestive of _Schwagerina_ (Virgilian-
Wolfcampian), elevation 8640' on ridge at
northern headwaters of Deer Creek (SE1/4
Sec. 29, T. 3 N., R. 16 E., Buttercup
Mountain quadrangle), collected by J.P.
O'Brien, identified by D.A. Myers;
16. _Bartramella_ sp. (elevation 8640' on
ridge at northern headwaters of Deer Creek,
SE1/4 Sec. 29, T. 3 N., R. 16 E., Buttercup
Mountain quadrangle), collected by R.S.
Lewis, identified by D.A. Myers. Since this
locality yielded both _Bartramella_ and
Schwagerina (#15 above), the collection may
be _B. heglarensis_, which is associated with
Schwagerina sublettensis (Wolfcampian) in
the Sublett Range of southern Idaho
(Thompson and others, 1958).

Sedimentary rocks of the Wood River,
Grand Prize, and Dollarhide Formations were
named severally because they were thought
to be allochthonous tectonostratigraphic
units bounded by regional thrust faults
(Hall, 1985). Our work demonstrates that
the contacts between these formations are
in most cases changes in sedimentary or
metamorphic facies (Fig. 1). The
previously defined "structural" distinction
between these sedimentary units is not
valid. The three formations are mappable
lateral variations of originally
contiguous strata deposited in one
sedimentary basin.

We propose the name Sun Valley Group
(derived from the prosperous resort town
just east of Ketchum, Idaho) to include the
Wood River, Grand Prize, and Dollarhide
Formations (Fig. 2). The type area of the
Sun Valley Group is the area occupied by
these formations on Figure 1. We retain
the names Wood River, Grand Prize, and
Dollarhide Formations, but formally modify
previously published definitions of these
formations. Within the Wood River
Formation we redefine the basal Hailey
Member (revised Hailey Conglomerate Member

(Hall and others, 1974)) and propose formal
Eagle Creek (middle) and Wilson Creek
(upper) Members. The Dollarhide Formation
is expanded to include the "Carrietown
sequence" (Skipp and Hall, 1980; Link and
others, 1988), which is a metamorphic
equivalent of the lower member of the
Dollarhide Formation adjacent to the Idaho
batholith (Whitman, 1990). The term Sun
Valley Group will supplant previously used
informal terms such as "Wood River basin
strata", and will facilitate collective
discussion of three genetically related
stratigraphic units.

REGIONAL GEOLOGIC SETTING

In south-central Idaho, rocks of the
Sun Valley Group are exposed in a belt
bounded on the west and north by the
Cretaceous Idaho batholith, on the east by
the Pioneer Mountains metamorphic core
Complex and Eocene Challis Volcanic Group
and associated intrusive rocks, and on the
south by the Neogene Snake River Plain
magmatic downwarp (Fig. 1, inset). Age
equivalent rocks are assigned to the
Oquirrh Group on the south side of the
Snake River Plain and to the Snaky Canyon
Formation east of the Pioneer thrust (Fig.
2).

The Sun Valley Group unconformably
overlies the Devonian Milligen Formation
(Sandberg and others, 1975; Otto and
Turner, 1987; Turner and Otto, 1988;
Burton, 1988) and the Paleozoic Salmon
River assemblage (Sengebush, 1984; Hall,
1985; Hall and Hobbs, 1987; Mahoney and
Sengebush, 1988), which strongly resembles
the Milligen (Fig. 1). The Sun Valley
Group and these underlying lower Paleozoic
strata comprise the "Ketchum sequence" of
Roberts and Thomasson (1964), which
approximates the "central Idaho black-shale
mineral belt" of Hall (1985). The Milligen
Formation and part of the Salmon River
assemblage were deformed by the Late
Devonian to Mississippian Antler orogeny
and display tight recumbent folds with an
axial-planar cleavage that is not present
in the Sun Valley Group (Roberts and
Thomasson, 1964; Dover, 1980; Turner and
Otto, 1988).

The Late Devonian to Mississippian
Antler orogeny produced an elongate, north-
trending orogenic belt along the western
margin of North America. In south-central
Idaho, this tectonically positive area
formed the western margin of a flysch
trough in which accumulated the
Mississippian Copper Basin Formation (Ross,
1960; Nilsen, 1977; Skipp and others,
1979b). Deposition associated with the
Antler orogenic belt ceased in the Late
Mississippian, and a hiatus exists between
the Mississippian Copper Basin Formation
and the Sun Valley Group (Skipp and Hall,
1980). Uplift of the former flysch trough
to form the Middle Pennsylvanian Copper
Basin highland was coeval with the initial
subsidence of the Wood River basin to the
west (Skipp and Hall, 1980). The existence
of an Antler-age fabric in subjacent

formations, the unconformity between these lower Paleozoic strata and the Sun Valley Group, and the absence of Mississippian strata beneath the Sun Valley Group suggest that the Wood River basin developed on the site of the Late Devonian to Mississippian positive area (Skipp and Hall, 1980).

In the Boulder and White Cloud Mountains the contact between the Sun Valley Group and subjacent Paleozoic strata is a locally sheared unconformity (Sengebush, 1984; Burton, 1988; Mahoney and Sengebush, 1988; Burton and others, 1989) and not a regional thrust fault (Wood River thrust of Dover (1983) and Hall (1985)). The youngest age and maximum thickness of the Sun Valley Group is not known since the top of the Sun Valley Group is the present day erosional surface, or an unconformity below Eocene Challis Volcanic Group.

The original character of much of the Sun Valley Group has been altered by structural deformation, metamorphism, and intrusion. The Sun Valley Group was deformed by compression associated with the Late Cretaceous Sevier orogeny which produced generally east-vergent, locally overturned folds with wavelengths of several kilometres (Fig. 1). These folds may be cored by blind (not surfacing) thrust faults of small displacement or may be a decollement fold train above the underlying Pioneer thrust (Dover, 1981; 1983; Fig. 1). Only one thrust fault, west of Hailey, has been mapped within Sun Valley Group strata (Skipp, written communication, 1990; Fig. 1). At least two thrust plates (the Pioneer and Copper Basin plates) separate the Sun Valley Group from the coeval Snaky Canyon Formation to the east (Skipp and Hait, 1977; Link and others, 1988).

Although much of the Sun Valley Group is unmetamorphosed, near the Idaho batholith the Grand Prize Formation is silicified and bleached white, and the Dollarhide Formation is locally metamorphosed to sillimanite-grade (Mahoney and Sengebush, 1988; Whitman, 1990). The regional but largely static metamorphism of the Dollarhide Formation has been dated at 83.9 +/- 3.4 Ma (whole rock K-Ar date from upper member Dollarhide Formation at the head of Deer Creek, R.L. Armstrong, written communication, 1990).

Paleogene extension which resulted in uplift of the Pioneer Mountains metamorphic core complex produced low-angle normal faults with top-to-the-northwest translation that cut the Sun Valley Group (Wust, 1986, Burton and others, 1989; Burton and Link, 1989). These faults were formerly mapped as thrust faults (Umpleby and others, 1930; Dover, 1981; 1983; Hall, 1985). Neogene "Basin and Range" steeply dipping normal faults cut the Sun Valley Group and younger rocks.

SUN VALLEY GROUP STRATIGRAPHY

Introduction

Much of the Sun Valley Group consists of mixed carbonate-siliciclastic strata, a sediment-type which is less well studied than end-member carbonate or siliciclastic sediments. In this paper, lithology is described using the scheme of Mount (1985), which is a first-order textural and compositional classification based on the relative abundance of four end-member components: sand or silt, terrigenous mud, micrite (lime mud), and allochems. End-members of this scheme are further classified by the schemes of Dunham (1962) for carbonate rocks and Folk (1960) for siliciclastic rocks.

The Sun Valley Group can be discussed in terms of eight lithofacies which are described in Table 1. These include conglomerate, bioclastic limestone, micritic sandstone, banded micritic sandstone/siltstone, graded silty micrite, silty micrite, sandy micrite, and carbonaceous siltstone. These lithofacies are laterally and vertically gradational, and the vertical arrangement of lithofacies is similar in each formation, allowing the group to be discussed as a whole.

Silicification and carbon content overprint these lithofacies, defining the distinction of one formation from another. The Grand Prize Formation, in addition to being sandier and coarser-grained than the other formations, has undergone metamorphic silicification. The Dollarhide Formation is characteristically dark-colored and finer-grained than the Wood River or Grand Prize Formations.

Wood River Formation

Review of Nomenclature

The Wood River Formation, as named by Lindgren (1900), included all Paleozoic rocks exposed near the Wood River Valley. Umpleby and others (1930) restricted the formation to light-colored rocks of Pennsylvanian and Permian age. The formation is now known to comprise rocks of Desmoinesian to Leonardian age (Bostwick, 1955; Ross, 1960; Hall and others, 1974) but may include rocks as young as Guadalupian (Hall and others, 1978a; Skipp and Hall, 1980). Thomasson (1959) completed a regional stratigraphic study and subdivided the Wood River Formation into four members, of which only the Hailey Conglomerate Member was subsequently formally defined (Hall and others, 1974).

Hall and others (1974, 1978a) divided the Wood River Formation east of Bellevue into eight informal units totalling approximately 3000 m. The measured sections of Hall and others (1974) are here designated the composite principal reference section of the Wood River Formation. Goodman (1983) determined that

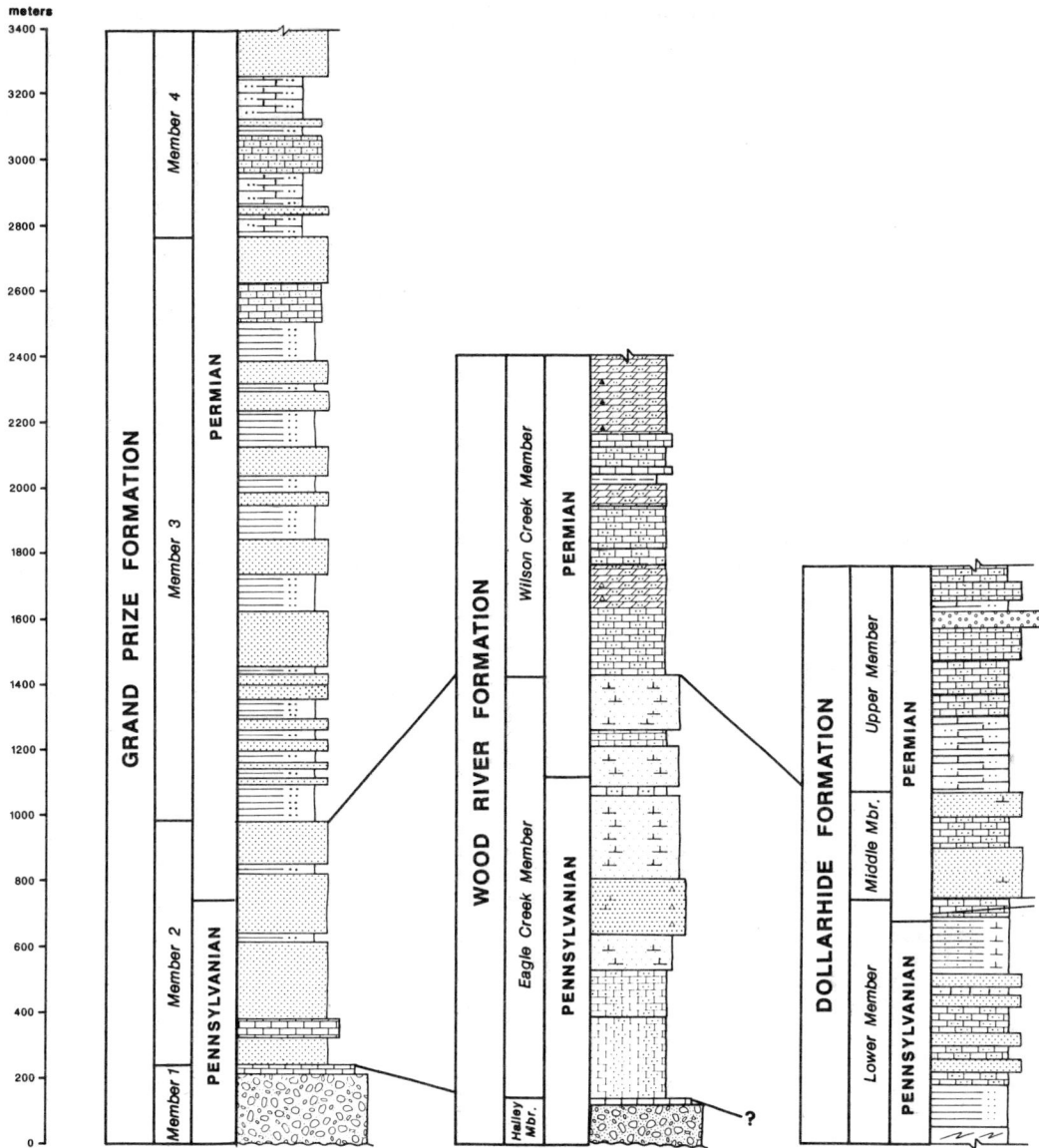

Figure 3. Generalized stratigraphic columns of the Sun Valley Group, south-central Idaho.

Table 1: LITHOFACIES PRESENT IN THE SUN VALLEY GROUP

LITHOFACIES	SEDIMENTOLOGIC DESCRIPTION	INTERPRETATION	OCCURRENCE
Facies most abundant in lower part of Sun Valley Group.			
Conglomerate	Medium-bedded to massive matrix- and clast-supported pebble to boulder conglomerate containing subangular to rounded clasts of quartzite, argillite, sandstone, chert, and rare limestone; beds are tabular and laterally continuous, intercalated with subordinate thin- to medium-bedded, parallel laminated, fine- to coarse-grained sandstone.	Unchannelized sediment gravity flow deposits	Hailey Member, Wood River Fm.; member 1, Grand Prize Fm., thin beds in lower and upper members, Dollarhide Fm.
Bioclastic limestone	Thin-bedded to massive blue-grey bioclastic grainstone and packstone, locally conglomeratic or sandy; intercalated with *in situ* biostromes, micritic sandstone, siltstone, and thin-bedded pebble and granule conglomerate.	Bioclastic debris flows and biostromes	Hailey Member, Wood River Fm. member 1, Grand Prize Fm; lower and upper members, Dollarhide Fm.
Facies most abundant in lower and upper fine-grained part of megacycle in middle and upper Sun Valley Group.			
Banded micritic sandstone and siltstone	Thin-bedded couplets of very fine- to fine-grained light grey to brown micritic sandstone containing convolute- and cross-laminations gradationally overlain by parallel laminated carbonaceous micritic siltstone; couplets are tabular, laterally continuous, and display sharp upper and lower contacts; rhythmically interbedded couplets give lithofacies a distinct banded appearence; contains base-cut-out Bouma sequences (T(bcde), T(cde), T(de)) intercalated with amalgamated top-cut-out T(aaa) sequences; and minor thin-bedded blue-grey silty to sandy micrite containing convolute laminae and synsedimentary folds.	Silt to sand turbidites and calc-turbidites	Member 3, Grand Prize Fm; upper and lower members, Dollarhide Fm.
Carbonaceous siltstone	Thin-bedded dark-grey to black carbonaceous, locally calcareous, siltstone with parallel laminations; locally aggregates >70m; contains base-cut-out (T(cde), T(de), T(eee)) Bouma sequences; intercalated with thin- to medium-bedded bluish-grey silty to sandy micrite and micritic siltstone containing parallel and convolute laminae, graded bedding, synsedimentary folds, and bioturbation; beds are tabular and laterally continuous.	Silt turbidites	Lower and upper members, Dollarhide Fm.; member 4, Grand Prize Fm.; Wilson Creek Member, Wood River Fm.
Graded silty micrite	Carbonaceous and calcareous banded siltstone arranged in fining and thinning upward cyclic packages (10-30 cm) of micritic sandstone and sandy to silty micrite; complete and base-cut-out Bouma sequences (T(cdef), T(def), T(ef); sedimentary structures include parallel, cross, and convolute laminae, graded beds, soft-sediment deformation; beds are tabular and laterally continuous with sharp upper and lower contacts between packages, diagenetic chert nodules and beds occur locally, especially in Wilson Creek Member.	Mixed carbonate and siliciclastic silty turbidites	Lower Eagle Creek Member and Wilson Creek Member, Wood River Fm., lower and upper members, Dollarhide Fm.
Silty micrite	Thin to medium-bedded blue-grey silty micrite (locally argillaceous or dolomitic) with thin mudstone laminae; intense bioturbation locally destroys bedding character; contains base-cut-out Bouma sequences ((T(def), T(ef)); intercalated with mudstone and micritic siltstone.	Mixed carbonate and siliciclastic distal turbidites and hemi-pelagic deposits	Wilson Creek Member, Wood River Formation; upper member, Dollarhide Formation.
Sandy micrite	Thin- to medium-bedded blue-grey sandy micrite; bioclastic in part, with parallel and convolute laminae; contains partial base- and top-cut-out turbidites (T(abc); T(bcde)); interbedded with black carbonaceous siltstone and silty micrite.	Mixed carbonate-siliciclastic silt to sand turbidites, bioclastic debris flows	Members 3 and 4, Grand Prize Fm., Eagle and Wilson Creek Members, Wood River Fm., lower and upper members, Dollarhide Fm.
Facies most abundant in coarse-grained middle part of megacycle			
Micritic sandstone	Medium-bedded to massive, medium to fine-grained, grey, brown-weathering micritic sandstone, locally siliceous, with interbeds of medium-bedded sandy micrite; local conglomerate; medium-bedded intervals contain parallel laminae and trough cross-laminae, graded bedding, and flute casts; contains T(aaa), T(ab), T(abc) and T(abcde) Bouma sequences; sedimentary structures are rare in massive intervals.	Sediment gravity flow deposits, amalgamated top-cut-out turbidites and liquified sediment gravity flow deposits.	Member 2 Grand Prize Fm., Eagle Creek Member Wood River Fm., middle member, Dollarhide Dollarhide Fm.

the lower portion of the principal reference section is repeated by faulting and that Hall's informal units are laterally discontinuous. Skipp and Hall (1975) measured a partial section of the Wood River Formation at Fish Creek Reservoir, 35 km east of Bellevue (Section W6, Figs. 1, 4). Our work (Link and others, 1988; Burton, 1988; Link and others, 1988; Burton and Link, 1989; Link and Mahoney, 1989) subdivides the formation into three regionally mappable members, which are formalized in this report.

Stratotype

In the principal reference section, Hall and others (1974) measured units 1-6 of the Wood River Formation 4.3 km east of Bellevue, on the ridge north of Seamans Creek, and measured units 7-8 in the Quigley Creek drainage (Section W5; Figs. 1, 4). We designate two additional supplemental reference sections which are more completely exposed, especially in the upper part of the formation: 1) Basin Gulch section (Section W3), exposed on the ridgecrest northeast of the head of Lake Creek; 2) Wilson Creek section (Section W4), measured on the ridgecrest north of Wilson Creek (Figs. 1, 4). Additional measured sections of the Wood River Formation include the Murdock Creek (Section W1) and Lake Creek (Section W2) sections (Figs. 1, 4).

Member Designation

The Wood River Formation is herein divided into three formal members: the Hailey Member (basal 200 m), the Eagle Creek Member (middle 880+ m), and the Wilson Creek Member (upper 800+ m). In our proposed terminology, the Hailey Member contains unit 1 (Hailey Conglomerate Member) and bioclastic limestone of unit 2 of Hall and others (1974), the Eagle Creek Member contains their units 3-6, and the Wilson Creek Member contains units 7-8 of Hall and others (1974, 1978a)(Figs. 2, 3).

Hailey Member:

We redefine the Hailey Member of the Wood River Formation to include the Hailey Conglomerate Member and interfingering thin bioclastic limestone (unit 2 of Hall and others (1974)). The type locality for the Hailey Conglomerate Member is west of Hailey (Thomasson, 1959; Hall and others, 1974). We designate the type section of the herein defined Hailey Member as the lower 180 m of the principal reference section east of Bellevue (Hall and others, 1974, p. 90; Section W5 of Fig. 4). The type section for the Hailey Member is described in Table 2. The "Hailey Member" is simplified from "Hailey Conglomerate Member" because of the inclusion of the bioclastic limestone which is vertically and laterally gradational with the conglomerate. The original Hailey Conglomerate Member was mapped at 1:24,000 by Hall and others (1978b) and Batchelder and Hall (1978), at 1:48:000 by Dover

(1983), and at 1:62,500 by Dover (1981). The Hailey Member has been mapped as "lower member, Wood River Formation" by Rember and Bennett (1979) and Link and others (1988).

The Hailey Member is 0-200 m thick, and consists of 0-180 m of light-brown to light-grey conglomerate gradationally overlain by 15-30 m of bluish-grey bioclastic limestone (Figs. 3, 4). The conglomerates of the Hailey Member form resistant ledges exposed in bold relief above the subjacent Milligen Formation. The member is primarily exposed east of the Wood River in the Pioneer and Boulder Mountains, although it is exposed in a few areas west of the Wood River near Hailey (Fig. 1).

The basal contact of the Hailey Member is a locally sheared unconformity with the underlying Devonian Milligen Formation (Burton, 1988; Burton and others, 1989; Ratchford, 1989). The Hailey Member was deposited on an irregular surface, and is locally depositionally absent (Winsor, 1981). In many areas interstratal slip along this contact has produced a shear zone in which the Hailey Member occurs as boudins or is attenuated, particularly in fold hinges. The upper contact of the Hailey Member is placed at the first appearance of distinctive light-purple silty micrite of the Eagle Creek Member (Figs. 3, 4). The age of the Hailey Member is Desmoinesian, based on coral, fusulinid, and phylloid green algae biostratigraphy (Bostwick, 1955; Hall and others, 1974; Fig. 2).

Eagle Creek Member:

The middle portion of the Wood River Formation contains 880-1300 m of strata consisting of light-purple silty micrite overlain by light-brown micritic sandstone, light-grey sandy micrite, and subordinate quartz arenite, herein designated the Eagle Creek Member (Fig. 4). The name is chosen from Eagle Creek, a tributary on the east side of the Wood River north of Ketchum (Fig. 1). The type section of the Eagle Creek Member is located approximately 7 kilometres southeast of Eagle Creek, on the ridgecrest between Trail and Wilson Creeks (section W4, Figs. 1, 4). The type section of the Eagle Creek Member is described in Table 2. Additional reference sections exist at Basin Gulch and Lake Creek (W3 and W2, Fig. 4). The Eagle Creek Member includes units 3 through 6 of Hall and others (1974), and corresponds to the rocks described at Lake Creek by Thomasson (1959).

The Eagle Creek Member is exposed in discontinuous cliffs and ledges from Bellevue north to Pole Creek, and creates conspicous talus cones. The micritic sandstone of the member has a well-developed reddish-brown weathering rind. The Eagle Creek Member forms the bulk of the outcrop area of the Wood River Formation (Fig. 1). Purple silty micrite

560

of the basal Eagle Creek Member abruptly overlies bioclastic limestone of the Hailey Member. The Eagle Creek Member is gradationally overlain by graded silty micrite of the Wilson Creek Member (Figs. 3, 4; Table 1). Fusulinid assemblages constrain the age of the Eagle Creek Member to late Desmoinesian to Wolfcampian (Hall and others, 1974; Fig. 2, this study).

Wilson Creek Member:

The upper Wood River Formation consists of over 800 m of thin-bedded light-brown graded silty micrite, light-brown silty micrite, light-grey sandy micrite, and dark-grey carbonaceous siltstone, with subordinate medium-bedded micritic sandstone, assigned to the Wilson Creek Member (Figs. 3, 4; Table 1). In the upper part of the member, the micrite fraction is locally diagenetically replaced by chert (Figs. 3, 4).

The name Wilson Creek is derived from a tributary to Trail Creek northeast of Ketchum (Fig. 1) and was first proposed in an unpublished thesis by Thomasson (1959). The type section of the Wilson Creek Member is located in the Basin Gulch section (section W2, figure 4), on the ridgecrest between Lake Creek and Basin Gulch, approximately 3 kilometres northwest of Wilson Creek. The type section of the Wilson Creek Member is described in Table 2. The Wilson Creek Member is generally light-brown to reddish-brown, fine-grained, thin-bedded, and weathers easily to reddish-brown regolith. The upper portion of the member is dolomitic in part, and contains diagenetic chert. The member is a slope former, and is sparsely exposed east of Bellevue and Hailey, although it is well-exposed in alpine ridges of the high Boulder Mountains northeast of Ketchum (Fig. 1).

Figure 4. Measured stratigraphic reference sections of the Pennsylvanian and Permian Wood River Formation. Symbols are shown on legend for Figure 3. Locations are as follows:
Section W1--Murdock Creek--Section measured in northeast-dipping beds on the ridge between Murdock Creek and the East Fork of the North Fork of the Wood River, Amber Lakes quadrangle, starting just southwest of peak 8635 and proceeding northeastward along long ridgeline to peak 9783. NE1/4 Sec. 27, through Sec. 23, to SW1/4 Sec. 13, T.6N. R.17E., Blaine County, Idaho. Base of section is 43°40'23" N., 114°24'22" W.
Section W2--Lake Creek section - measured southwest to northeast along ridge between Lake and Eagle Creeks in northeast-dipping beds, Rock Roll Canyon 7.5 minute quadrangle.; 9040' to 9000' elevation, across peak 9675'; NW 1/4 of Sec. 8 to NE 1/4 of Sec. 5, T.5.N., R.18E., Blaine County, Idaho. Base of section is 43°46'56"N., 114°20'34"W;
Section W3--Basin Gulch Section; includes type section of Wilson Creek Member - measured southwest to northeast along ridge northeast of the head of Lake Creek in northeast-dipping beds; Rock Roll Canyon 7.5 minute quadrangle; measured from 7960' elevation to peak 9677', across 10,200' ridge crest; SW 1/4 of Sec. 3, T.6N., R.18E. and NE 1/4 of Sec. 34, T.6.N., R.18E., Blaine County and Custer Counties, Idaho. Base of section is 43°47'58"N., 114°18'45"W.
Section W4--Wilson Creek Ridge; includes type section of Eagle Creek Member. Section measured from southwest to northeast on the ridge between Wilson Creek and Trail Creek, in northeast-dipping beds, Phi Kappa Mountain and Rock Roll Canyon quadrangles. Hailey Member measured about 500 m east of the base of the ridge on the west bank of Wilson Creek, NW1/4 Sec. 13, T.5 N., R.19 E.. Eagle Creek and Wilson Creek Members measured along the ridge from elevation 6600 feet to swale at elevation 9000 ft. in the NE1/4 Sec. 14 and SW1/4, Sec. 12, T.5N., R.18E. Blaine County Idaho. Base of section is 43°46'13" N., 114°16'18"W.
Section W5--Bellevue (Principal Reference Section and type section of Hailey Member, described by Hall and others (1974); composite section; units 1-6 measured southwest to northeast along ridge north of Seamans Creek, Seamans Creek quadrangle, from 6390' elevation to top of ridge; base of section is 43°29'30", 114°14'00"; unit 7 measured on Quigley Creek in Sec. 21, T.3 N., R. 19E, starting 2300 ft north of the SW corner of section 21 and measured toward the east up the ridge from an altitude of 6,640 to 7,300 ft elevation, Blaine County, Idaho. Base of section is 43°34'30"N.,114°12'30"W.
Section W6- Fish Creek Reservoir (Hailey Member and Eagle Creek Member), from Skipp and Hall (1975, Fig. 8); Section measured along ridge in southwest-dipping beds from 5400' to about 6000', Fish Creek Reservoir quadrangle, SE1/4 Sec. 15, NE1/4 Sec. 22, NW1/4 sec. 23, T. 1 N., R. 22 E., Blaine County Idaho. Base of section is 43°24'45"N., 113°49'00"W.

WOOD RIVER FORMATION

Table 2: Descriptions of type sections for three members of the Wood River Formation

Type Section of Hailey Member Wood River Formation, measured by Hall and others, 1974) east of Bellevue on the ridge north of Seamans Creek in northeast-dipping beds from 6390' to 6600' elevation, SW1/4 Sec. 28, T.2N., R.19E., Seamans Creek quadrangle (Section W5, Figure 4)

(meters)

Conformable Contact to Eagle Creek
    Member (unit 3)
Top of Hailey Member
Unit 2 of Hall and others (1974)
    Limestone, bluish-grey, medium- to thick-   15.0
    bedded, fine-grained with locally abun-
    dant crinoidal debris, bryozoa, and
    brachiopod fragments. Contains 5-10
    percent detrital quartz grains.
Conformable contact to unit 2
Unit 1
    Chert pebble conglomerate, light grey,   1.5
    siliceous cement
    Limestone, brown, fine-grained, algal;   12.5
    and chert pebble conglomerate, beds
    0.3-0.6 m thick.
    Chert pebble conglomerate, as above   0.3
    Limestone, brown, fine-grained, algal,   16.2
    with 10 percent fine-grained quartzite
    beds 1.3 cm to 2.5 cm thick.
    Chert pebble conglomerate, light-grey and   64.0
    brownish-grey, as above
    Limestone, micritic, brown, silty, in part   0.6
    silicified.
    Chert pebble conglomerate, light-gray,   3.7
    chalcedonic matrix.
    Limestone, brown, fine-grained, sandy.   1.5
    Chert-pebble conglomerate, siliceous   16.8
    matrix, chert and quartzite clasts; and
    quartzite, light-green, fine-grained,
    thick-bedded.
    Limestone, sandy, brown.   0.3
    Chert pebble conglomerate, light-grey,   31.0
    thick-bedded, siliceous cement, a few
    well-rounded light-brown limestone and
    white quartzite clasts. Subrounded chert
    pebbles 1 cm to 2.5 cm long.
    Total thickness Hailey Member   **135.5 m**

Type Section of Eagle Creek Member, Wood River Formation, Measured on ridge north of Wilson Creek, in northeast-dipping beds from 6600' to 8800' elevation, NE1/4 Sec. 14 and SW1/4 Sec. 12, T.5N. R.18E., Rock Roll Canyon Quadrangle (measured section W4 of Figure 4, Burton, 1988)

Wilson Creek Member (gradational contact)
Top of Eagle Creek Member
Unit 6 of Hall and others, 1974)
    Micritic sandstone medium grey,   20
    weathering brown, fine-grained trough
    cross laminated, syndepositional
    convolute laminae, thin bedded inter-
    bedded with silty micrite.
    Micritic sandstone (quartz arenite); grey,   483
    weathering brown, thick-bedded to mas-
    sive, fine- to coarse grained, fractured;
    sparse trough cross lamination, load
    casts and flute casts; *Arenicolites*.
    Section is cut by fault at 470 m.
Unit 5 of Hall and others (1974)
    Fine sandstone (quartz arenite); siliceous   160
    or partly calcareous, light brown, well
    indurated, thickly bedded to massive;
    crude parallel laminae, dish structures,
    intensely fractured.
Unit 4 of Hall and others (1974)
    Silty micrite to fine micritic sandstone;   160
    brown, thin- to medium bedded, trough
    cross- and convolute laminated, weakly
    graded beds.
    Silty micrite; dark grey, weathering grey   40
    brown, wavy parallel laminae; interbeds
    of dark grey micritic mudshale.
    Very fine micritic sandstone; dark brown,   20
    thin-bedded, convolute laminae, load
    casts, micritic mudstone partings;
    Scalarituba

Unit 3 of Hall and others (1974)
    Silty micrite and mudshale; dark brown to   175
    purple-grey, weathers pink-grey, moderately
    bioturbated, crudely fissle; *Neonereites,*
    Scalarituba, Phycosiphon, Zoophycos.
    Silty micrite; grey brown and dark brown,   85
    thinbedded, wavy parallel laminae, burrowed
    to intensely bioturbated; contains interbeds
    up to 1.5 m thick of silty allochem lime-
    stone containing crinoid, brachiopod, and
    rugose coral bioclasts, *Neonereites,*
    Scalarituba, Muensteria?, Spirophycos?
Top of Hailey Member (bioclastic packstone of
    Unit 2 of Hall and others, 1974).
Total thickness Eagle Creek Member   **1143m**

Type Section of the Wilson Creek Member, Wood River Formation, measured southwest to northeast across ridge northeast of the head of Lake Creek; from 9200' elevation across 10,200' ridgecrest to 9200' knob on ridge southwest of 9677', north of Basin Gulch; starts in NW1/4 Sec. 3, T.5N., R.18E. and continues into NE 1/4 of Sec. 34, T.6.N., R.18E.; Rock Roll Canyon quadrangle (Section W3 of Figure 4, Burton, 1988)

Top of measured section: hinge of tight
    syncline
Unit 7 of Hall and others (1974).
    Fine sandy micrite, medium brown,   175
    weathering dark brown, thin-bedded,
    partly silicified and dolomitic; arranged
    in fining-upward cyclic packets 8 to 17 m
    thick with massive siliceous micrite in
    lower 2-3 m, overlain by thin-bedded
    dolomicrite with trough cross-and convo-
    lute laminae, overlain by intensely
    bioturbated carbonaceous silty
    micrite.
    Sandy and coarse silty micrite, thin-,   180
    bedded, siliceous, trough cross- and
    convolute laminated; mottled grey/orange-
    brown/black; *Neonereites*
    Allochemic sandy micrite: medium to dark   30
    grey, bioclasts include crinoid columnals,
    scaphopods?, cephalopods, bryozoa,
    fusulinids
    Micritic mudshale; very dark brown, silty,   20
    bioturbated; *Neonereites, Phycosiphon,*
    Paleophycus?
    Fine micritic sandstone (quartz arenite);   20
    medium grey, weathering brown, calcareous;
    thick bedded to massive.
    Interbedded coarse silty micrite and fine   67
    micritic sandstone; medium grey to brown,
    silty micrite portions bioturbated.
    Micritic mudshale; very dark brown, silty,   8
    bioturbated; *Neonereites, Phycosiphon,*
    Paleophycus?
    Silty micrite and very fine sandy micrite   252
    medium grey, weathering light grey to
    yellow-brown; arranged in thinning and
    fining upward sequences (15 to 25 m thick)
    containing complete turbidites at the base
    passing upward to partial (base-cut-out)
    T(ace) and T(cde) turbidites and to thin
    very dark brown intensely silicified T(def)
    turbidites at the top; basal sequences are
    intensely bioturbated.
    Silty micrite and micritic sandstone;   35
    medium grey to light brown; medium beds
    contain T(acde) silt turbidite sequences;
    Neonereites, Scalarituba, Phycosiphon,
    Zoophycos, Planolites.
    Fine sandy micrite; medium grey, weather-   8
    ing medium brown, thin to medium bedded,
    brown micritic mudstone partings.
    Fine micritic sandstone; medium grey, thin   5
    bedded with micritic mudstone partings,
    load casts, *Scalarituba*
Conformable contact, top of Eagle Creek member
    (micritic sandstone, Unit 6 of Hall and others,
    1974)
Total thickness Wilson Creek Member   **800+ m**

Graded silty micrite at the base of the Wilson Creek Member gradationally overlies micritic sandstone of the Eagle Creek Member. The upper contact of the Wilson Creek Member is the modern erosion surface or an unconformity below Eocene Challis Volcanic Group. Fusulinid biostratigraphy from the Boulder Mountains constrains the age of the Wilson Creek Member at Wolfcampian to Leonardian (Hall and others, 1974, 1978; this study; Fig. 2). The member may be as young as Guadalupian, based on schaphopod and gastropod biostratigraphy reported by J. Dover (Hall and others, 1978a). However, a collection of scaphopods, phylloid green algae, and fusulinids from the same bed described by Dover (1983) (the "upper cherty limestone") yielded fusulinids ascribed to the Wolfcampian (C.A. Ross, written communication, 1988), so the Guadalupian assignment remains tentative (Fig. 2). Younger rocks may occur in the structurally complex area to the north of this collection, but cannot accurately be assigned a stratigraphic position in the Wood River Formation.

Grand Prize Formation

Review of Nomenclature

The Grand Prize Formation comprises the northern part of the Sun Valley Group and is exposed primarily north of Pole Creek in the White Cloud Peaks (Fig. 1). The formation was named by Hall (1985) for a sequence of fine-grained quartzite, limy siltstone, banded siltite, and dark carbonaceous silty limestone exposed north of Grand Prize Gulch, a tributary to Pole Creek. It has previously been included in the Pole Creek sequence (Skipp and Hall, 1980; Sengebush, 1984) or the Wood River Formation (Tschanz and Kiilsgaard, 1986). Mahoney and Sengebush (1988) expanded and revised Hall's definition of the Grand Prize Formation, and subdivided the formation into four informal members.

Stratotype

Hall (1985) established a 1450 m type section of the Grand Prize Formation on the north side of Pole Creek, and subdivided it into four informal units (Sections G5a,b, Figs. 1, 5). Mahoney and Sengebush (1988) recognized that the type section of the Grand Prize Formation was both incomplete and structurally repeated (Fig. 5). They established a composite supplemental reference section for the Grand Prize Formation, with member 1 measured in Strawberry Basin, and members 2-4 measured on the ridge north of Fourth of July Creek (Section G2, Figs. 1, 5). We designate an additional reference section for members 2, 3 and 4 on the ridge south of Champion Lakes (Section G3, Figs. 1, 5). Partial sections of the Grand Prize Formation were also measured at the HooDoo Mine (members 2, 3; section G1, Fig. 5), in Washington Basin (members 1, 2; section G3, Fig. 5), and in the Salmon River Headwaters (members 2, 3; section G6, Fig. 5).

Member Designation

Member 1:

Member 1 of the Grand Prize Formation is 0-400 m thick, and consists of 0-350 m of light-brown to light-grey polymict conglomerate and sandstone overlain by 30-50 m of dark grey carbonaceous bioclastic limestone (Figs. 3, 5). This designation is restricted from that of Mahoney and Sengebush (1988), who included approximately 400 m of graded micritic sandstone in the upper portion of member 1. We recognize this graded micritic sandstone to be petrologically and stratigraphically similar to member 2 and include it there. Member 1 is resistant, and forms conspicuous ledges above the Paleozoic Salmon River assemblage in the central and northern White Cloud Peaks.

The lower contact of member 1 is an erosional unconformity with the underlying Paleozoic Salmon River assemblage. This contact has been strongly sheared during Mesozoic compression, and original stratigraphic thicknesses of the basal conglomerate are generally not preserved. The conglomerate is locally depositionally absent or structurally removed. Sheared conglomerate along this contact has been previously interpreted as a thick tectonic breccia (Winsor, 1981; Hall, 1985; Fisher and Johnson, 1987), but is now recognized as a brecciated conglomerate (Sengebush, 1984; Mahoney and Sengebush, 1988). The upper contact of member 1 is gradational with overlying member 2, and is placed above the carbonaceous bioclastic limestone of member 1. Member 1 is undated, but is the lithostratigraphic correlative of the Hailey Member of the Wood River Formation, which is of Desmoinesian age (Figs. 2, 3).

Member 2:

Member 2 of the Grand Prize Formation consists of 500-1100 m of thick-bedded to massive light-brown micritic sandstone, with subordinate light-grey sandy micrite, and dark-grey carbonaceous siltstone. Member 2 is medium-bedded to massive, and forms discontinuous cliffs and ledges in the northern Smoky Mountains and in the White Cloud Peaks. The lower contact of member 2 is placed at the base of graded micritic sandstone that overlies carbonaceous bioclastic limestone of member 1. The upper contact of member 2 is gradational with the banded micritic sandstone and siltstone of member 3 (Figs. 3, 5, Table 1). The age of member 2 is thought to be Desmoinesian to Wolfcampian, based on Wolfcampian conodonts recovered from overlying members 3 and 4 in the type section (Hall, 1985), and on lithostratigraphic correlations with the Eagle Creek Member of the Wood River Formation (Figs. 2, 3).

564

Member 3:

Member 3 of the Grand Prize Formation consists of 650-1700 m of rhythmically interbedded couplets of light-grey, fine-grained banded micritic sandstone and overlying carbonaceous siltstone interbedded with thick-bedded light-brown micritic sandstone and light-grey sandy micrite (Figs. 3, 5; Table 1). Member 3 includes the banded lithologies described by Hall (1985) from the type section on Pole Creek. Banded sandstone amd siltstone of member 3 interfingers with micritic sandstone of member 2, such that the thickness and relative amount of each member varies between sections, although the aggregate thickness of the two members remains constant.

Member 3 is distinctly banded, and has an irregular weathering pattern, with the sandier intervals exposed in bold relief against the more easily weathered finer grained intervals. The member forms light-grey cliffs throughout the White Cloud Peaks (Fig. 1). The lower contact of member 3 is gradational with underlying member 2, and the upper contact is gradational into the carbonaceous siltstone of member 4. The age of member 3 is thought to be Wolfcampian to Leonardian, based on conodonts collected from members 3 and 4 in the type section, and stretched and corroded conodonts collected from limestone interbeds in member 3(?) on Peach Creek (Hall, 1985). Member 3 is lithologically correlative with the Wilson Creek Member of the Wood River Formation which contains Wolfcampian and Leonardian strata.

Member 4:

Member 4 of the Grand Prize Formation consists of over 450 m of thin- to medium-bedded carbonaceous siltstone, sandy to silty micrite, and minor micritic sandstone (Figs. 3, 5; Table 1). Member 4 contains the upper carbonaceous siltstone and limestone described by Hall (1985) from the type section on Pole Creek. The member weathers to a dark-grey regolith, and is a slope former in the southern White Cloud Peaks (Fig. 1). Member 4 is not recognized north of Strawberry Basin (Fig. 1).

Figure 5. Measured stratigraphic reference sections of the Pennsylvanian and Permian Grand Prize Formation. Locations are as follows:
Section G1 - Hoodoo Mine section; measured on east flank of peak 10,050', west of Hoodoo Lake in west-dipping beds, Robinson Bar quadrangle; Section measured from 8820'-10,050', Custer County, Idaho. Base of section is 44°10'35"N., 114°38'15"W.
Section G2 - Fourth of July Creek section; composite section; 0-500 m interval measured in east-dipping beds, Strawberry Basin, north of Blackman Peak, 9800'-10,111' elevation, Washington Peak quadrangle; base of section is 44°04'28"N, 114°39'32". Interval from 550-2500 m measured from west to east in steeply west-dipping overturned beds on ridge north of Fourth of July Creek, 7800'-9200' elevation, Obsidian and Washington Peak quadrangles, Custer County, Idaho. Base of section is 44°03'18"N, 114°43'48"W..
Section G3 - Champion Lakes section; measured on ridge south of Champion Lakes Basin, Horton Peak and Washington Peak quadrangles.; meausured from east to west om steeply west-dipping beds; section starts at 9920', southeast of peak 10,167', and continues along ridge to west side of Champion Lakes Basin, to north of peak 10,081', Custer County Idaho. Base of section is 43°58'53"N, 114°40'00"W.
Section G4 - Washington Basin section; measured from east to west in steeply west-dipping beds on east flank of peak 10,519', Washington Peak quadrangle; section starts at 10,200' and goes along sawtooth ridge to top of peak 10,519, Custer County, Idaho. Base of section is 44°15'00", 114°39'38".
Section G5a - Pole Creek section 1 (type section of Hall, 1985); measured in northwest-dipping beds on north side of Pole Creek, near its confluence with Grand Prize Gulch, on south flank of peak 10,166', Horton Peak quadrangle, 7970' to 9560' elevation, Custer County, Idaho. Base of section is 43°56'27"N., longitude 114°41'00"W.
Section G5b - Pole Creek section 2, separated from Pole Creek section 1 by low-angle fault. The type section of Hall (1985) crosses this fault. Section continues from 9560' to 10,166' elevation above section G5a, Horton Peak 7.5 minute quadrangle, Custer County, Idaho. Base of section is 43°56'28"N., 114°41'00"W.
Section G6 - Salmon River Headwaters section; measured on west side of Salmon river, on east flank of peak 9423' in northwest-dipping beds, Frenchman Creek quadrangle, 7800' to 9423' feet elevation, Camas County, Idaho. Base of section is 43°48'45"N., 114°46'33"W.

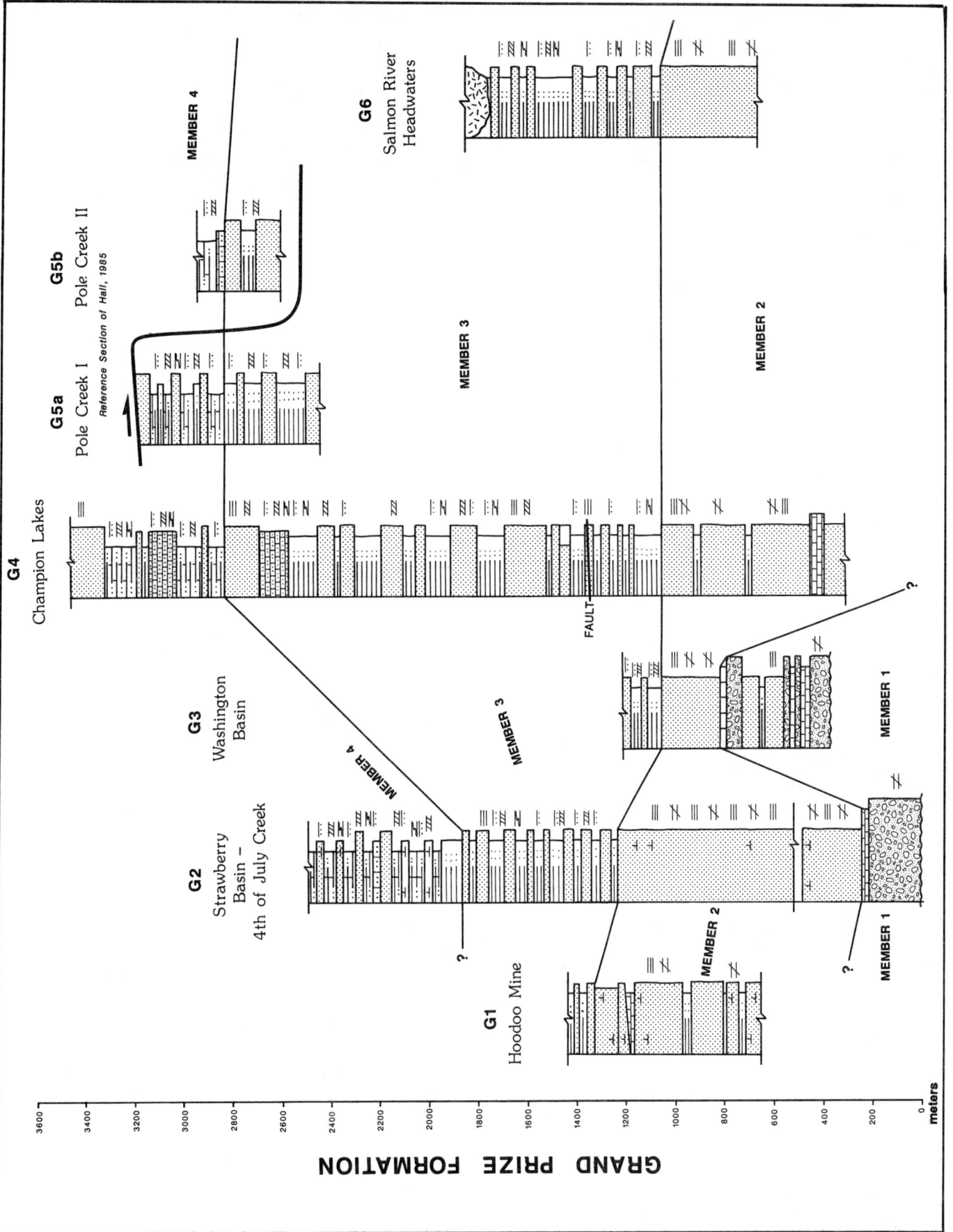

GRAND PRIZE FORMATION

566

The lower contact of member 4 is
gradational with underlying member 3, and
the upper contact is not recognized due to
erosion. Member 4 is believed to be
Wolfcampian to Leonardian age, based on
conodonts recovered from the type section,
and lithostratigraphic correlations with
the Wilson Creek Member of the Wood River
Formation and the upper member of the
Dollarhide Formation (Hall, 1985; Figs. 2,
3).

Dollarhide Formation

Review of Nomenclature

The Dollarhide Formation comprises the
southern and southwestern part of the Sun
Valley Group, and is exposed in a complex
series of folds intruded by the Idaho
batholith and associated stocks in the
Smoky Mountains, west of the Wood River
(Fig. 1). It was named by Hall (1985) for
a sequence of dark-colored ("sooty")
carbonaceous limestone, siltite, fine-
grained quartzite and granule conglomerate
exposed on Dollarhide Summit, west of
Ketchum (Fig. 1). Wavra (1985), Geslin
(1986), and Wavra and others (1986)
subdivided the Dollarhide into three
informal members which we adopt here with
minor modifications (O'Brien, 1991; Figs.
3, 6).

Skipp and Hall (1980), Geslin (1986),
Darling (1988) and Link and others, (1988)
described the metasedimentary "Carrietown
sequence" as being in thrust contact below
the lower member of the Dollarhide
Formation near the edge of the Idaho
batholith (Fig. 1). Whitman (1990)
recognized that this contact is not
structural, but is a lithologic change
accentuated by metamorphism. These
metasedimentary rocks are now interpreted
as a marginal tectonite to the Idaho
batholith, and are stratigraphically
contained within the lower member of the
Dollarhide Formation. The name "Carrietown
sequence" is abandoned.

Stratotype

Hall (1985) established the type
locality for the Dollarhide Formation on
Dollarhide Summit, but did not
differentiate units. Hall (1985)
designated a partial reference section on
the east side of Wolf Tone Creek, because
it yielded the only fossil data for the
formation (Early Permian fusulinids)(Fig.
2). Wavra and others (1986) describe a
2300 m partial reference section west of
Wolf Tone Creek (Section D3, Figs. 1, 6).
We herein designate the principal reference
section for the Dollarhide Formation east
of Willow Creek (Section D1, Figs. 1, 6)
which is the most complete section through
all three members. Section D1 is modified
from that of Geslin (1986) to eliminate
structural repetition in the middle and
upper members (O'Brien, 1991). Additional
partial reference sections are located in
Dry Gulch (lower and middle members) and
north of Bear Gulch (middle and upper
members)(composite Section D2; Figs. 1, 6).

Member Designation

Lower member:

The lower member of the Dollarhide
Formation is at least 800 m thick, and
consists of rhythmically interbedded, very
fine-grained, light-grey to light-brown
micritic sandstone and dark-grey
carbonaceous silty micrite, with
subordinate medium- to thick-bedded light
brown micritic sandstone and light-grey
lenticular conglomerate containing both
extra- and intrabasinal clasts. The lower
member contains a minimum of 175 m of
quartzite, phyllite and calc-silicate
hornfels formerly assigned to the
"Carrietown sequence". The lower member
designated here corresponds with the lower
member of Wavra (1985), Geslin (1986), and
Wavra and others (1986) with the addition
of the Carrietown rocks at the base and
carbonaceous limestone at the top. The
lower member is exposed in the southern
Smoky Mountains, where it forms slopes
punctuated by thin ridges of resistant
micritic sandstone.

The stratigraphic base of the
Dollarhide Formation is not recognized due
to intrusion of the Idaho batholith, and
the upper contact is placed above a
mappable carbonaceous limestone and below
thick micritic sandstone of the middle
member (O'Brien, 1991). Hall (1985)
assigned the entire formation to the Early
Permian, based on fusulinid collections
from the lower member along Wolf Tone
Creek, but a new fusulinid collection that
includes Triticites sp., and a conodont
collection from the lower member (B.R.
Wardlaw, oral communication, 1990) suggest
the member includes Desmoinesian to
Virgilian rocks as well.

Middle member:

The middle member of the Dollarhide
contains approximately 300 m of fine-
grained light-brown micritic sandstone and
light-grey sandy micrite, with subordinate
dark-grey to black carbonaceous siltstone
and lenticular conglomerate. The lower
contact of the middle member is placed at
the lowest thick, micritic sandstone. This
placement of the base of the middle member
is higher than that of Geslin (1986) and
Wavra and others (1986), who included about
150 m of dark limestone in the base of the
middle member. The reassignment is more
practicable for purposes of geologic
mapping because it restricts the middle
member to only light-colored and sandy
strata (O'Brien, 1991). The middle member
becomes finer grained and bedding becomes
thinner upward into carbonaceous siltstone
of the upper member.

The middle member forms prominent
light-colored cliffs throughout the central
and southern Smoky Mountains. Fusulinids
collected near the head of Deer Creek,
including Schwagerina sp. and Bartramella
sp., suggest the middle member is early
Wolfcampian in age.

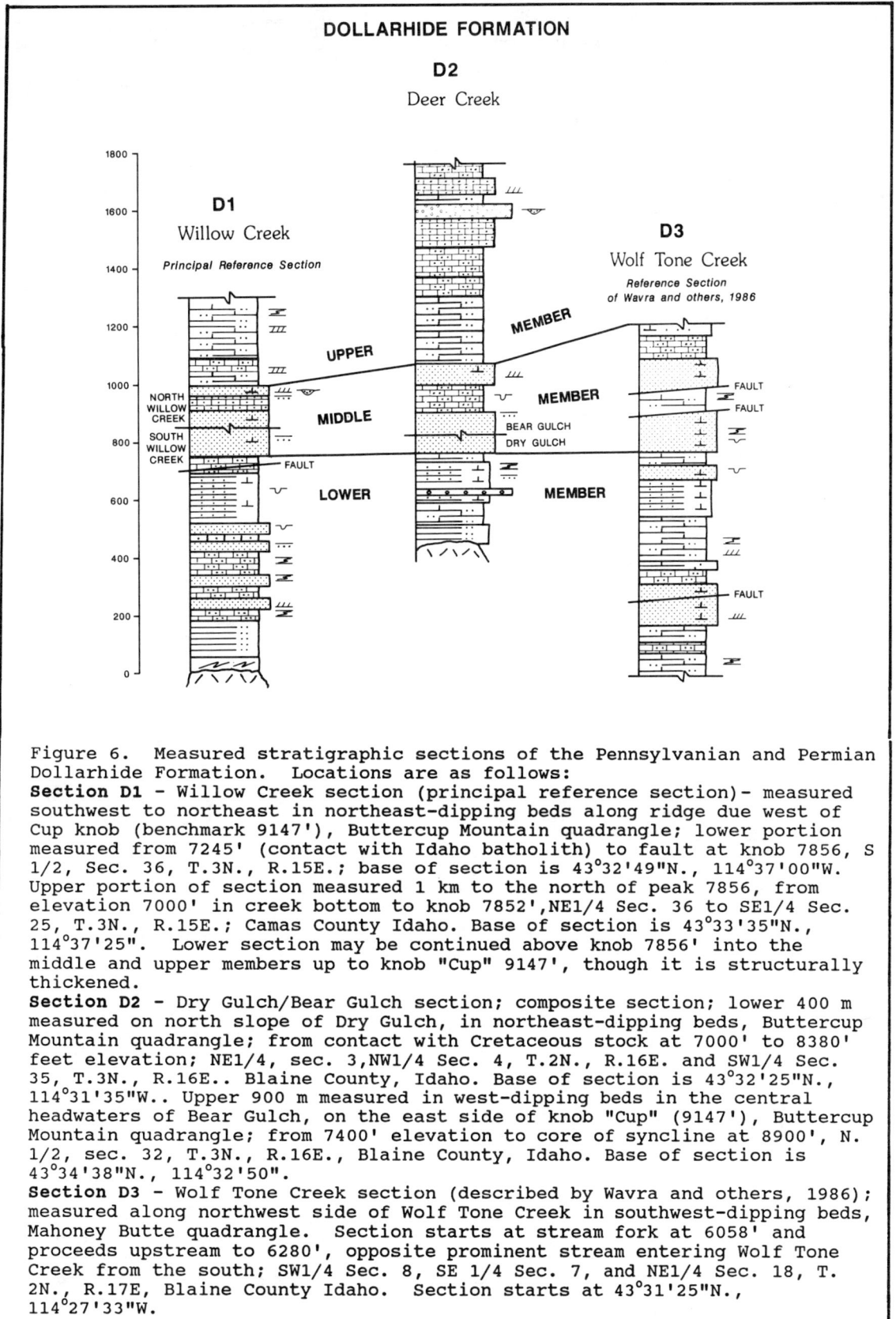

Figure 6. Measured stratigraphic sections of the Pennsylvanian and Permian Dollarhide Formation. Locations are as follows:

Section D1 - Willow Creek section (principal reference section)- measured southwest to northeast in northeast-dipping beds along ridge due west of Cup knob (benchmark 9147'), Buttercup Mountain quadrangle; lower portion measured from 7245' (contact with Idaho batholith) to fault at knob 7856, S 1/2, Sec. 36, T.3N., R.15E.; base of section is 43°32'49"N., 114°37'00"W. Upper portion of section measured 1 km to the north of peak 7856, from elevation 7000' in creek bottom to knob 7852',NE1/4 Sec. 36 to SE1/4 Sec. 25, T.3N., R.15E.; Camas County Idaho. Base of section is 43°33'35"N., 114°37'25". Lower section may be continued above knob 7856' into the middle and upper members up to knob "Cup" 9147', though it is structurally thickened.

Section D2 - Dry Gulch/Bear Gulch section; composite section; lower 400 m measured on north slope of Dry Gulch, in northeast-dipping beds, Buttercup Mountain quadrangle; from contact with Cretaceous stock at 7000' to 8380' feet elevation; NE1/4, sec. 3,NW1/4 Sec. 4, T.2N., R.16E. and SW1/4 Sec. 35, T.3N., R.16E.. Blaine County, Idaho. Base of section is 43°32'25"N., 114°31'35"W.. Upper 900 m measured in west-dipping beds in the central headwaters of Bear Gulch, on the east side of knob "Cup" (9147'), Buttercup Mountain quadrangle; from 7400' elevation to core of syncline at 8900', N. 1/2, sec. 32, T.3N., R.16E., Blaine County, Idaho. Base of section is 43°34'38"N., 114°32'50".

Section D3 - Wolf Tone Creek section (described by Wavra and others, 1986); measured along northwest side of Wolf Tone Creek in southwest-dipping beds, Mahoney Butte quadrangle. Section starts at stream fork at 6058' and proceeds upstream to 6280', opposite prominent stream entering Wolf Tone Creek from the south; SW1/4 Sec. 8, SE 1/4 Sec. 7, and NE1/4 Sec. 18, T. 2N., R.17E, Blaine County Idaho. Section starts at 43°31'25"N., 114°27'33"W.

Upper Member:

The upper member of the Dollarhide Formation is approximately 900 m thick, and is composed of thin-bedded to laminated dark-grey carbonaceous siltstone, light-grey silty micrite, minor light- brown micritic sandstone and conglomerate (Table 1). The upper member gradationally overlies the middle member, and the upper boundary is not recognized due to erosion. The member weathers to a dark-grey regolith, and is a poorly exposed slope former throughout the Smoky Mountains (Fig. 1). The age of the upper member is poorly constrained, but must be at least Wolfcampian to Leonardian, based on fossil control lower in the section.

RELATIONS BETWEEN FORMATIONS OF THE SUN VALLEY GROUP

Our geologic mapping indicates that the contacts between formations of the Sun Valley Group include the following relations, from south to north:

A) The Dollarhide-Wood River contact occurs in four localities:
  1) In the Picabo Hills the lower Dollarhide Formation passes eastward into the Eagle Creek Member of the Wood River Formation across a set of southwest-vergent folds. Facies change is indicated on Figure 1.
  2) Along Deer Creek (B. Skipp, written communication, 1990), the lower member of the Dollarhide Formation is thrust over the Eagle Creek Member of the Wood River Formation, and silicified quartz arenite of the Eagle Creek Member lies above middle and lower members of the Dollarhide Formation on a low-angle normal fault.
  3) North of Deer Creek, the middle member of the Dollarhide Formation is mapped south of the Mahoney Butte volcanic field, but Eagle Creek Member of the Wood River Formation is mapped in structurally continuous strata to the north. In this area, the middle member of the Dollarhide Formation is a tongue of partially silicified micritic sandstone, petrographically identical to the Eagle Creek Member (Fig. 7).
  4) At the western headwaters of Deer Creek what was mapped as a flat fault contact of the middle member (Eagle Creek Member) of the Wood River Formation above upper member of the Dollarhide Formation (Geslin, 1986; Link and others, 1988) is now known to be a continuous sequence across a syncline into the upper part of the upper member of the Dollarhide Formation. The uppermost Dollarhide Formation at this locale is light-colored and sandy, resembling the Eagle Creek Member of the Wood River Formation.
  5) North of Baker Creek, the Eagle Creek Member overlies the lower member of the Dollarhide along a low-angle normal fault (Stewart, 1987).

B) The contact between lithologically identical carbonaceous siltstones of the upper member of the Dollarhide Formation and member 4 of the Grand Prize Formation is intruded by Eocene dacite porphyry west of the headwaters of the Salmon River. The upper member of the Dollarhide and member 4 of the Grand Prize are interpreted as lateral equivalents (Figs. 1, 2, 3, 7).

C) The Wood River Formation passes northward into the Grand Prize Formation in two places. These contacts are shown as lines of facies change on Figure 1.
  1) North of Pole Creek massive micritic sandstone of the Eagle Creek Member intertongues with thick-bedded member 2 of the Grand Prize Formation. This relation was shown as a thrust fault by Hall (1985).

  2) In Galena Gulch medium-bedded micritic sandstone and thin-bedded sandy micrite and silty micrite of the Wilson Creek Member, Wood River Formation are gradationally overlain by black carbonaceous siltstone and sandy to silty micrite of member 4 of the Grand Prize Formation (Figs. 1, 7).

STRATIGRAPHIC RELATIONS OF LITHOFACIES

The sequential distribution of lithofacies in the Sun Valley Group (listed in Table 1) defines parts of three cycles of sedimentation (Fig. 2). Each of these cycles may be related to a change in tectonic regime or source area in the Wood River basin.

Lower part of Sun Valley Group

The basal contact of the Wood River and Grand Prize Formations is an unconformity with underlying lower Paleozoic rocks. This unconformity is overlain by conglomerate lithofacies of the Hailey Member of the Wood River Formation, and member 1 of the Grand Prize Formation (Figs. 3, 4, 5, 7). The stratigraphic base of the Dollarhide Formation is not recognized. The Hailey Member contains graded beds of pebble to boulder conglomerate with generally tabular bed geometry. The conglomerate lithofacies of the Grand Prize Formation is thicker than that of the Hailey Member, and contains a higher proportion of matrix-supported conglomerate and sandstone interbeds with complete and partial turbidite sequences. Within the Wood River Formation, the conglomerate lithofacies thickens to the southwest, but clast size coarsens to the northeast, with largest clasts 1 m in diameter (Winsor, 1981; Burton, 1988). Thomasson (1959) reports southeast-directed paleocurrents from the Hailey conglomerate. The polymict assemblage of chert, quartzite, siliceous argillite, and minor limestone reflects a provenance of older Paleozoic rocks that occur in the region. With the exception of angular siliceous argillite rip-up clasts derived from the subjacent Milligen Formation at the base of the Hailey member, conglomerate clasts are well-rounded to sub-rounded, implying moderate transport distance.

The conglomerate lithofacies grades laterally and vertically into bioclastic limestone, and becomes finer grained and thinner bedded upward. The bioclastic limestone includes both biostromal limestone containing in situ rugose corals, crinoids, gastropods, and spinose brachiopods (Pseudozaphrentoides) (top of the Hailey Member) and bioclastic limestone debris flows and limestone turbidites. The conglomerate lithofacies of the Grand Prize Formation has a smaller proportion of limestone interbeds than the Hailey Member, and contains thin-bedded, non-fossiliferous limestone intercalated with medium- to thick-bedded sandstone. The top of member 1 of the Grand Prize Formation contains 10-50 m of thick-bedded to massive carbonaceous bioclastic limestone containing abundant fossil debris.

The Hailey Member records the only period of photic zone deposition in the Wood River Formation. Deposition of the overlying deep-water fine-grained mixed carbonate-siliciclastic strata of the middle and upper Sun Valley Group marks a major increase in subsidence rate and a change in provience for the Wood River basin.

Stratigraphic Megacycle
Middle and Upper Parts of the
Sun Valley Group

The conglomerate and bioclastic limestone lithofacies of the lower part of the Sun Valley Group are overlain by fine-grained silty micrite and micritic sandstone lithofacies of the middle part (Figs. 3, 5, 7; Table 1). This contact is placed at the Hailey-Eagle Creek boundary in the Wood River Formation and the member 1-2 contact in the Grand Prize Formation. In the Dollarhide Formation, the placement of this contact is uncertain and it may be stratigraphically below the exposed portion of the lower member. This transition marks the start of a 1300-2000 m stratigraphic megacycle, initiated in the late Desmoinesian, which first becomes coarser-grained and then finer-grained upward (Figs. 2, 7). Within the megacycle, strata are laterally continuous, thin-bedded to massive, and arranged in 50-120 m thick cyclic packages that become finer grained and thinner bedded upward.

In the Wood River Formation, the coarsening-upward part of the megacycle begins in the Eagle Creek Member, which consists of approximately 250 m of silty micrite lithofacies gradationally overlain by more than 800 m of micritic sandstone lithofacies (Figs. 3, 4). In the Grand Prize Formation, the lower half of the megacycle contains about 400 m of graded micritic sandstone lithofacies overlain by more than 800 m of thick-bedded to massive micritic sandstone (Figs. 3, 5). In the lower member of the Dollarhide Formation this portion of the megacycle contains thick carbonaceous sandy micrite and banded micritic sandstone and siltstone, with sparse beds of matrix-supported

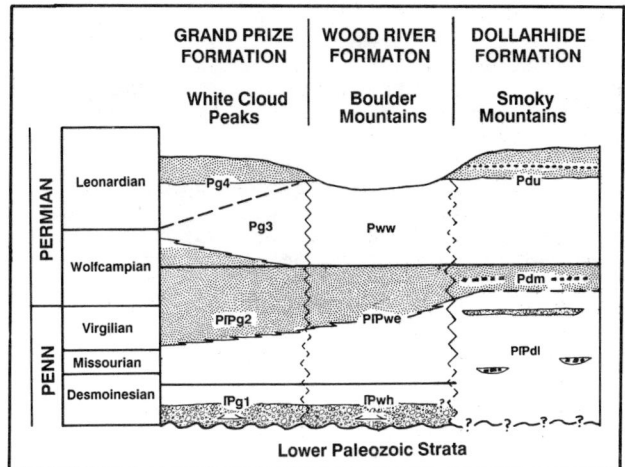

Figure 7. Schematic diagram showing facies change within the Sun Valley Group across the Wood River basin. Conglomerate symbol = conglomerate, dots = quartz sand wedge, solid rectangles = intrabasinal clasts. Solid sawtooth line indicates observed transition between formations, dashed line indicates inferred transition. Double-sided arrow indicates sheared unconformity at base of Sun Valley Group. Letter symbols: Grand Prize Formation, IPg1-member 1, PIPg2-member 2, Pg3-member 3, Pg4-member 4; Wood River Formation: IPwh-Hailey Member, PIPwe-Eagle Creek Member, Pww-Wilson Creek Member; Dollarhide Formation: PIPdl-lower member, Pdm-middle member, Pdu-upper member.

conglomerate containing extra- and intrabasinal clasts and limestone turbidites similar to those in the Hailey Member to the east. The megacycle continues in graded silty micrite of the upper part of the lower member and becomes coarsest in the 300 m thick micritic sandstone of the middle member (Figs. 3, 6).

The coarsest part of the megacycle is a northward-thickening wedge of micritic sandstone which comprises the middle part of the Eagle Creek Member of the Wood River Formation, member 2 of the Grand Prize Formation, and the middle member of the Dollarhide Formation (Figs. 3, 7). The age of this coarsest part of the megacycle is constrained to the late Virgilian and early Wolfcampian by biostratigraphy in the Wood River and Dollarhide Formations. The base of the sandstone tongue appears to be younger in the basinal Dollarhide Formation (early Wolfcampian) than in the Wood River Formation (late Virgilian), and the top is stratigraphically highest in member 3 of the Grand Prize Formation, suggesting the sandstone tongue prograded from the north (Figs. 2, 7).

The fining-upward portion of the megacycle begins in the micritic sandstone lithofacies of the middle part of the Sun Valley Group and passes upward in all three formations to thin-bedded silty micrite of the upper part of the group (Figs. 3, 7).

The start of this fining-upward trend is correlative between formations, and is constrained as early Wolfcampian by fusulinid biostratigraphy in the Wood River and Dollarhide Formations. In the Wood River Formation, this contact occurs where micritic sandstone of the Eagle Creek Member becomes finer-grained upward into 200 m of graded silty micrite at the base of the Wilson Creek Member, which is overlain by more than 650 m of intensely bioturbated silty micrite (Figs. 3, 4). In the Grand Prize Formation the contact is present where micritic sandstone of member 2 grades upward into banded micritic sandstone and siltstone and micritic sandstone of member 3 (Figs. 3, 5). These strata generally become finer-grained and thinner-bedded upward into carbonaceous siltstone and silty micrite in member 4. At the top of the middle member of the Dollarhide Formation, micritic sandstone becomers finer-grained and thinner-bedded upward into graded silty micrite, silty micrite, and carbonaceous siltstone of the upper member of the Dollarhide (Figs. 3, 6; Table 1).

The upper parts of the megacycle are generally characterized by a decrease in grain size, increase in organic content, and fining and thinning upward cyclicity. The silty micrite and carbonaceous siltstone lithofacies represent the top of the fining-upward part of the megacycle.

Uppermost Sun Valley Group

Carbonaceous siltstones of the lower part of member 4 of the Grand Prize and the lower part of the upper member of the Dollarhide are gradationally overlain by medium- to thick-bedded graded micritic sandstone in the uppermost portion of both formations. In the Dollarhide Formation, the upper 200 m of the upper member are coarser grained and more feldspathic than the rest of the member and contain extrabasinal pebble conglomerate and intrabasinal boulder conglomerate (Figs. 3, 6). This increase in grain size signifies the end of the stratigraphic megacycle of the middle and upper Sun Valley Group, and the beginning of another coarsening-upward cycle. The age of this transition is poorly constrained at late Wolfcampian to Leonardian (Figs. 3, 7).

PALEOSLOPES

Synsedimentary folds and convolute laminae are abundant in the middle and upper portion of the Sun Valley Group. In such folds, the axes provide the strike of the paleoslope, with slope direction determined by sense of vergence. Figure 8 shows paleoslope direction rose diagrams from the Virgilian to Leonardian stratigraphic megacycle in the middle and upper part of the Sun Valley Group.

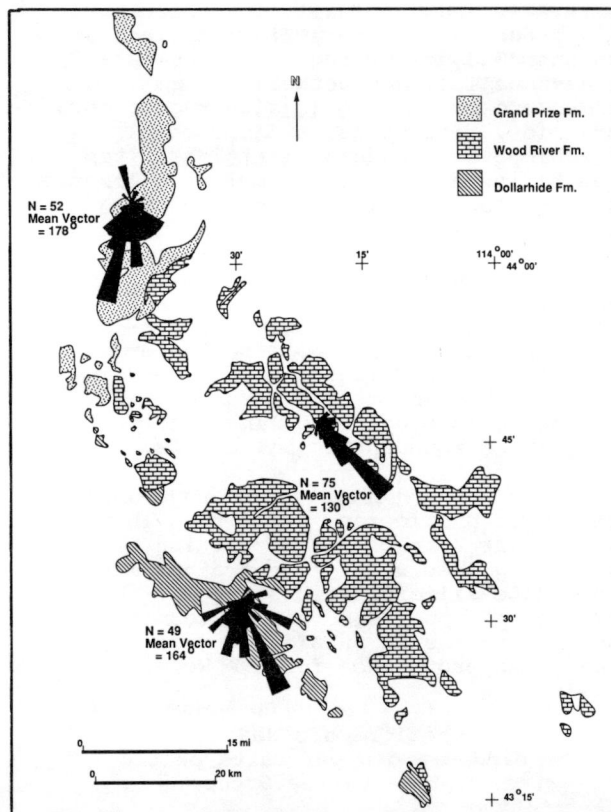

Figure 8. Paleoslope rose diagram for the Virgilian to Leonardian middle and upper parts of the three formations of the Sun Valley Group (compiled by J.K. Geslin). Roses represent paleoslope directions as indicated by the orientation of synsedimentary slump fold axes and convolute laminae and corrected for tectonic tilt. Data is derived from members 2, 3, and 4 of the Grand Prize Formation, the Eagle Creek and Wilson Creek Members of the Wood River Formation, and the lower and middle members of the Dollarhide Formation.

Paleocurrent indicators are not plotted on Figure 8, as they show significant dispersion at any one locality, reflecting the radial distribution characteristic of low-gradient turbidites (Burton, 1988). The uniformity of paleoslope indicators throughout the basin indicates that they are more reliable than paleocurrent indicators. The consistency of 176 data points measured in thirteen stratigraphic sections through a vertical range of 1500 m is striking, and suggests that the Wood River basin was characterized by a south-southeast dipping paleoslope from at least Virgilian to Leonardian time.

DEPOSITIONAL ENVIRONMENT

The Sun Valley Group was deposited in two distinct depositional environments: 1) subaqueous photic zone braid-delta and adjacent slope (lower portion of Sun Valley Group, Figs. 3, 7, 9); 2) sub-wavebase distal submarine ramp to mixed carbonate-siliciclastic apron and basin (medial and upper parts of the Group, Figs. 3, 7, 10). We use the term "ramp" in the sense of Heller and Dickinson (1985) for a sub-wavebase unchannelized siliciclastic depositional construct lacking an identifiable point source. We use the term "carbonate apron" in the sense of Mullins and Cook (1986) for an unchannelized base-of-slope construct fed by a line-source carbonate platform or shelf.

Two distinct mixing processes of carbonate-siliciclastic particles are demonstrated in the different depositional environments. In the Hailey Member, mixing occurred by growth and degradation of biostromes produced by robust carbonate-secreting organisms in an otherwise siliciclastic environment. This is a variant of the facies mixing model of Mount (1984) who describes a similar process on rimmed platforms. In the middle and upper parts of the group, mixing of carbonate and siliciclastic sediments may have occurred either during sediment gravity flow transport within Wood River basin or during transport through a postulated east or northeast shelf area. There is a general decrease in siliciclastic grain size and increase in carbonate content from the north into the central and southern parts of the group, suggesting mixing of sediment from two source areas, with the carbonate source south of the siliciclastic source.

The Desmoinesian conglomerate lithofacies of the lower Sun Valley Group contains tabular, laterally continuous beds, both clast- and matrix-supported conglomerate, graded bedding, weak pebble imbrication, and both partial and complete Bouma turbidite sequences. The conglomerate lithofacies are interpreted as the product of sediment gravity flow, including high concentration turbulent flow, debris flow, and minor turbidity currents, with sediment derived from a proximal metasedimentary source (Fig. 9, Table 1). Local biostromes in the Hailey Member developed on a conglomeratic substrate between active depositional lobes. These carbonate buildups supplied debris to bioclastic limestone debris flows and limestone turbidites in deeper portions of the basin.

A major change in provenance and depositional environment is marked by the contact between the Hailey and Eagle Creek Members of the Wood River Formation and members 1 and 2 of the Grand Prize Formation. Above this contact the siliciclastic fraction is craton-derived compositionally supermature quartz sand. The carbonate fraction is predominantly micrite and comminuted bioclastic debris.

The terrigenous clay fraction is minimal, suggesting it was separated from the sand fraction during transport [aeolian?] and did not settle in the Wood River basin.

The micritic sandstone, banded micritic sandstone and siltstone, graded silty micrite, and silty micrite lithofacies of the middle and upper portions of the Sun Valley Group lack shallow water bedforms or fauna, contain complete and partial Bouma sequences and hemipelagic sediment, and host bathyal trace fossil assemblages. Strata of the middle and upper portions of the Sun Valley Group are interpreted as the product of sub-wavebase sediment gravity flows, primarily turbidity currents, with minor liquified sediment gravity flows and debris flows. The majority of upper Sun Valley Group strata is fine-grained sand or silt, and the strata bear the characteristics of silt turbidites which display a characteristic set of sedimentary structures, but do not have a wide range of grain size (Stow and Piper, 1984). The carbonaceous siltstone and silty micrite lithofacies are the product of base-cut-out turbidites and hemipelagic sedimentation in a quiet water setting removed from significant sediment influx.

The middle and upper parts of the Sun Valley Group contain tabular, laterally continuous, thin- to thick-bedded strata, abundant sand and silt turbidite sequences, bioclastic debris flows, and hemipelagic sediments. The strata are arranged in 50 to 120 m cyclic packages that become finer-grained and thinner bedded upward, and the sandstone/siltstone ratio varies laterally along strike, reflecting turbidite deposition by migrating distributary systems. The lateral continuity of Sun Valley Group strata and the lack of distinct lobes or channel features suggest the cyclic packages were fed by a line, rather than a point, source. However, the gradual transition from siliciclastic-dominated sediments in the north (Grand Prize Formation) to carbonate-dominated sediments in the south (Wood River Formation) precludes sediment derivation from a homogeneous source. We suggest that the middle and upper parts of the Sun Valley Group are the product of sediment gravity flows in both a siliciclastic-dominated submarine ramp and a mixed carbonate-siliciclastic apron (Fig. 10).

Spatial variation between formations of the Sun Valley Group is evident within the combined ramp-apron system. Members 2, 3 and 4 of the Grand Prize Formation contain a higher proportion of sandstone and a smaller proportion of carbonate than the Wood River or Dollarhide Formations, and were deposited on the medial to distal portions of a northerly-derived siliciclastic-dominated ramp. The Eagle Creek and Wilson Creek Members of the Wood River Formation received a higher influx of micrite and bioclastic debris, and a smaller proportion of siliciclastic sediment than the Grand Prize Formation, and were deposited on the distal portion of

572

a mixed carbonate-siliciclastic apron fed from the east and northeast. The entire Dollarhide Formation was deposited in a slope to basinal setting, in more distal parts of the ramp-apron system than the Wood River and Grand Prize Formations.

Bottom-water conditions were dysaerobic in the deeper parts of the Wood River basin. Silt turbidites of the silty micrite lithofacies, Wilson Creek Member, contain a diverse bathyal ichnofossil assemblage, in which the distribution and abundance of ichnofossils were controlled by the delivery of oxygenated bottom water via turbidity currents (Burton and Link, 1990). Abundant organic material in the lower and upper members of the Dollarhide Formation, Wilson Creek Member and member 4 of the Grand Prize Formation indicates that more dysaerobic conditions prevailed in the deeper (Dollarhide) portion of the basin and in the upper part of the megacycle.

BASIN ANALYSIS

Sun Valley Group strata represent coeval, laterally contiguous facies of the late Paleozoic Wood River basin. Northeast-directed Mesozoic compression resulted in less than 50% shortening across map-scale fold trains so that present-day lateral facies relationships represent a foreshortened view of the original basin configuration.

Initial deposition in Wood River basin began in the Desmoinesian and produced the Hailey Member of the Wood River Formation, member 1 of the Grand Prize Formation, and possibly a portion of the lower member of the Dollarhide Formation. These rocks were deposited in a subaqueous braid-delta unconformably above deformed subjacent lower Paleozoic rocks of the Antler orogenic belt. The conglomerate lithofacies is coeval with coarse clastic debris in the Bloom Member of the Snaky Canyon Formation to the east, and in the Oquirrh Formation to the south (Skipp and others, 1989a). The Middle Pennsylvanian Copper Basin highland east of the Wood River basin has been proposed to be the source of these conglomerate clasts (Skipp and Hall, 1980). Provenance, thickness, and grain-size trends suggests the proximal source of metasedimentary clasts was northeast of the Wood River basin. Our work suggests that the Copper Basin highland was an active sediment source during the Desmoinesian, after which the highland became subdued, and was no longer a major sediment source. In the Fish Creek Reservoir area (Section W6, Figs. 1, 4) pebble conglomerate clasts persist into probable early Virgilian age strata, suggesting that there the Copper Basin highland persisted as a source area until early Virgilian (Skipp and Hall, 1975).

In late Desmoinesian time, the character of the Wood River basin changed markedly. Locally derived conglomerate and intrabasinal carbonate deposition was superceded by fine-grained mixed carbonate-siliciclastic sedimentation. Texturally and compositionally mature craton-derived fine-grained sand and silt, mixed with micrite and reworked bioclastic material, was deposited in a sub-wavebase ramp-apron system. By Virgilian to Wolfcampian time, the Wood River basin had developed a northern, siliciclastic submarine ramp facies (Grand Prize Formation), a central mixed carbonate-siliciclastic apron facies (Wood River Formation), and a southwestern slope to basinal facies (Dollarhide Formation) (Figs. 1, 10).

The increase in micrite and bioclastic debris in the central and southwestern facies requires a carbonate source to the south of the northern facies. The coeval Snaky Canyon Formation to the east records carbonate platform sedimentation from Early Pennsylvanian to Early Permian time, and may have been the source of the carbonate fraction of Sun Valley Group strata (Skipp and others, 1979a; Breuninger and others, 1988; Fig. 2).

The middle and upper parts of the Sun Valley Group contain a stratigraphic megacycle which first becomes coarser-grained and then finer-grained upward and which may reflect 2nd order eustatic changes (Fig. 2). The coarse portion of the megacycle in each formation is broadly associated with a second-order eustatic lowstand near the Virgilian-Wolfcampian boundary, as reported by Ross and Ross (1987). This sand pulse, though it appears diachronous (younger to the south and west), resembles the facies of a lowstand systems tract (Vail, 1987) with sand delivered to the basinal depocenter (middle member of the Dollarhide Formation). The fining-upward part of the megacycle may represent a eustatic rise and a highstand systems tract with coarser sediment retained in upslope sites (Wilson Creek Member and member 3 of the Grand Prize Formation). The increase in sand content in the uppermost portion of the Grand Prize and Dollarhide Formations may represent the beginning of a second-order lowstand in the late Wolfcampian (Figs. 2, 7). Correlation of sedimentation rates and eustatic changes is tentative due to insufficient biostratigraphic control in the Sun Valley Group.

In summary, the Wood River basin was an elongate northwest-trending basin flanked on the east/northeast by the emergent Copper Basin highland through Desmoinesian time, and by the Snaky Canyon carbonate platform from late Desmoinesian to Leonardian time. The Wood River basin evolved from a Desmoinesian subaqueous braid-delta in which accumulated proximally-derived coarse grained clastic debris and intrabasinal carbonate to a late Desmoinesian to Leonardian depocenter accumulating fine-grained cratonal sand derived from the north and carbonate debris derived from the east and northeast in a mixed siliciclastic-carbonate sub-wavebase ramp-apron system.

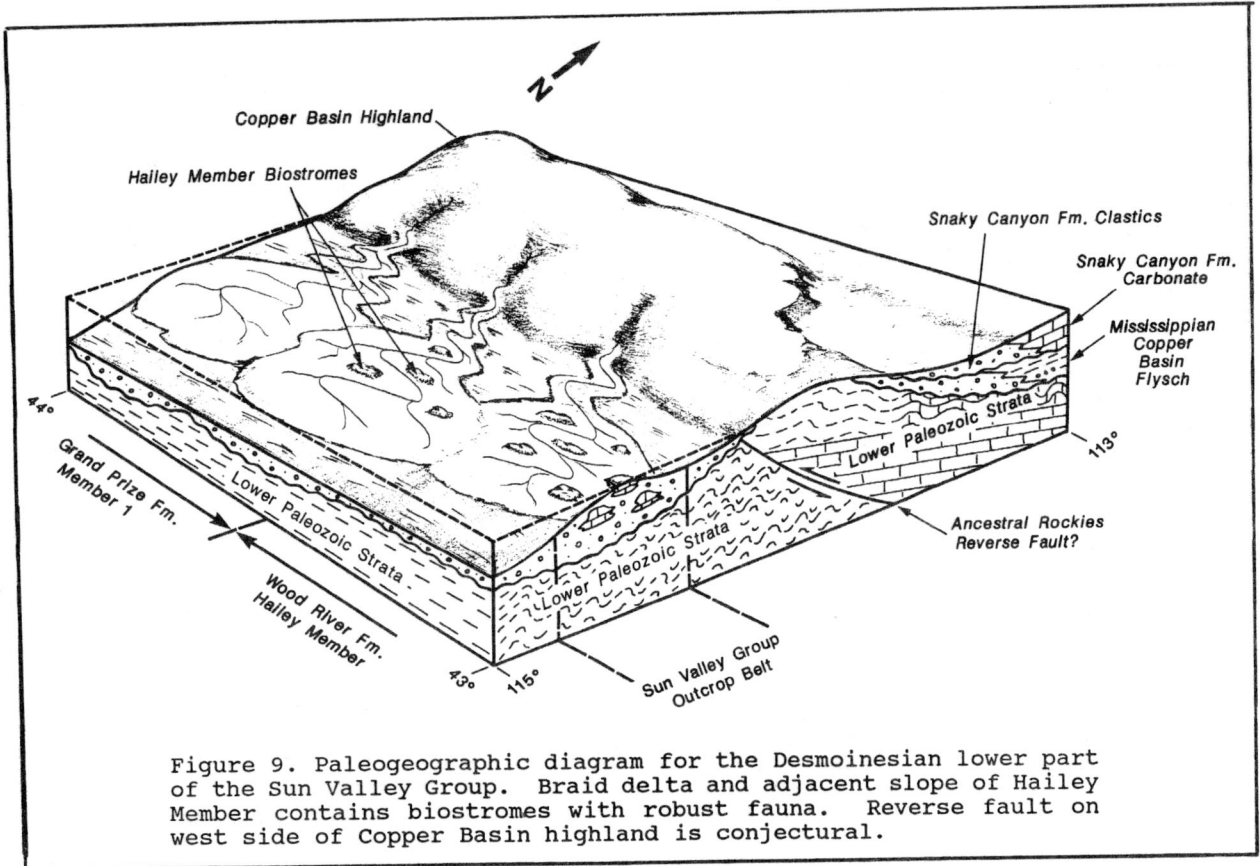

Figure 9. Paleogeographic diagram for the Desmoinesian lower part of the Sun Valley Group. Braid delta and adjacent slope of Hailey Member contains biostromes with robust fauna. Reverse fault on west side of Copper Basin highland is conjectural.

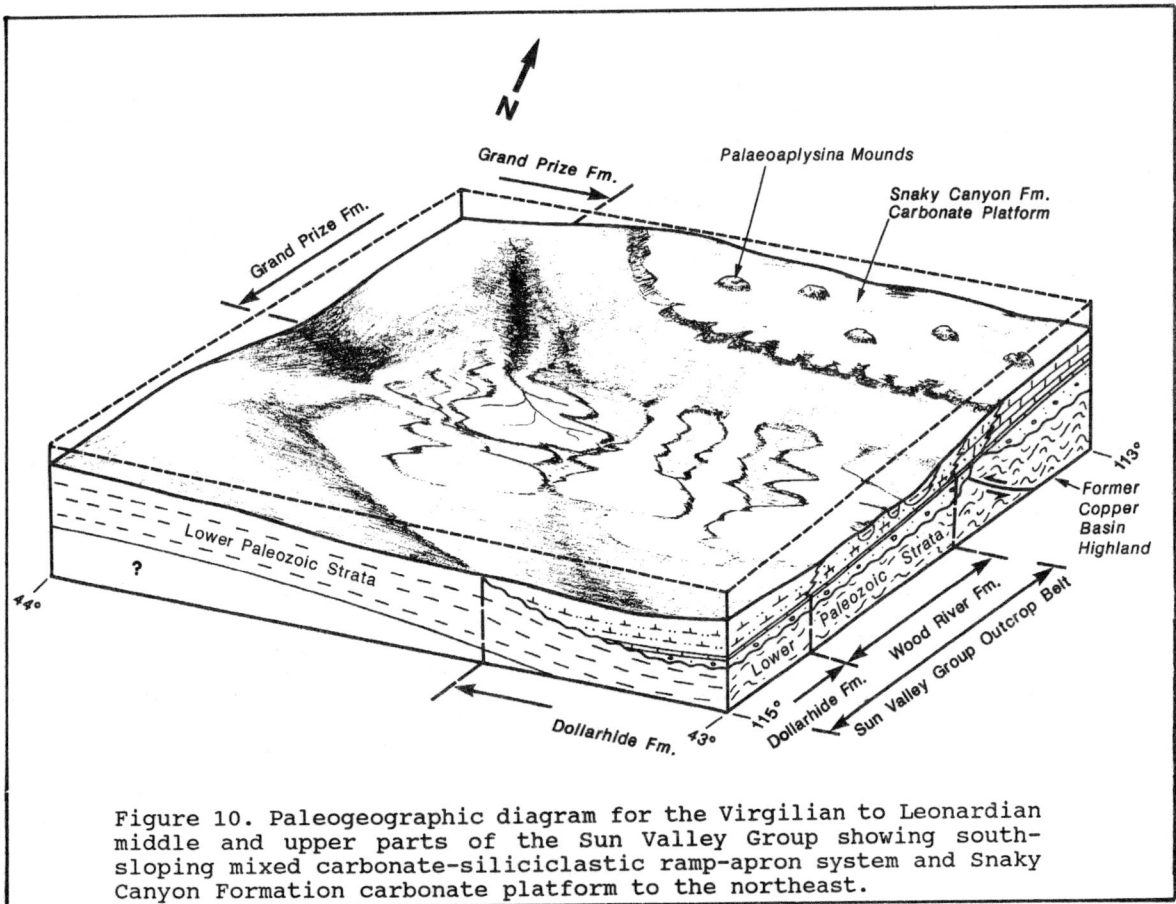

Figure 10. Paleogeographic diagram for the Virgilian to Leonardian middle and upper parts of the Sun Valley Group showing south-sloping mixed carbonate-siliciclastic ramp-apron system and Snaky Canyon Formation carbonate platform to the northeast.

REGIONAL RELATIONS

The Wood River basin, the Oquirrh basin (south), the Dry Mountain trough, and the Prophet/Ishbel trough (north) formed a narrow, sinuous depositional tract inboard of the tectonically active Late Pennsylvanian-Early Permian western margin of North America (Jordan and Douglass, 1980; Richards, 1989; Henderson, 1989; Mahoney and others, 1990; Snyder and others, 1991; Fig. 11).

Relations Between
Wood River and Oquirrh Basins

The evolution of the Wood River and Oquirrh basins are strongly parallel. The Oquirrh basin was an elongate northwest-trending basin that deepened to the west, and was flanked on the east and northeast by a shallow water shelf and carbonate platform (Jordan and Douglass, 1980; Fig. 11). Initial subsidence of the Oquirrh basin predated that of the Wood River basin, and may have been related to Ancestral Rockies tectonism in Middle Pennsylvanian time (Yancey and others, 1980; Jordan and Douglass, 1980; Kluth, 1986). The Oquirrh basin evolved from a mixed carbonate-siliciclastic shallow marine environment in the Desmoinesian to a Virgilian to Leonardian deep water trough in which deposits of mixed carbonate-siliciclastic sediment gravity flows accumulated (Jordan and Douglass, 1980).

The similarities between the Wood River and Oquirrh basins indicate the basins may have formed a continuous northwest-trending depocenter from Desmoinesian to Leonardian time. Our paleoslope data indicate that the Wood River basin may have been the northern arm of the much larger Oquirrh basin. Both basins display contemporaneous stratigraphic megacycles which first become coarser and then become finer upward (Jordan and Douglass, 1980). Both basins were filled with large volumes of craton-derived fine-grained sand and easterly derived carbonate debris.

Differences between the Oquirrh and Wood River basins include details of timing and subsidence history. The majority of the Oquirrh basin shallowed to a restricted shelf during the late Wolfcampian to early Leonardian, with continued subsidence recorded only in the northwestern part (Cassia basin of Mytton and others, 1983; Fig. 2, column 5). The Wood River basin remained a sub-wavebase depocenter from late Desmoinesian to Leonardian time. Wolfcampian subsidence along basin-bounding faults is postulated for the Oquirrh basin (Jordan and Douglass, 1980). Basin-bounding faults have not been proven in the Wood River basin.

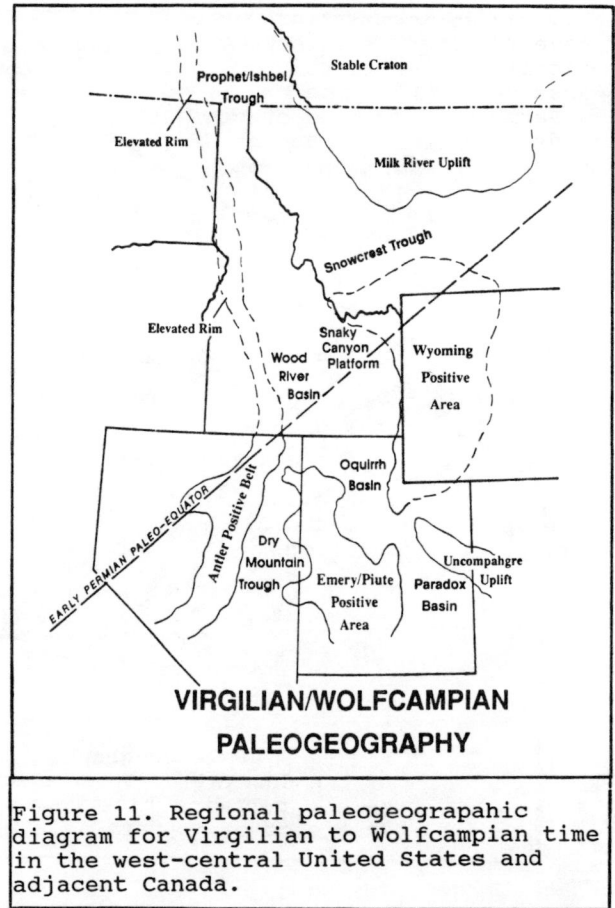

Figure 11. Regional paleogeograpahic diagram for Virgilian to Wolfcampian time in the west-central United States and adjacent Canada.

Coeval Basins

The Wood River basin is located south of the Prophet/Ishbel trough and northeast of the Dry Mountain trough (Fig. 11). The Prophet/Ishbel trough and Dry Mountain trough were asymmetric, two-sided, north-trending elongate basins flanked on the east by a carbonate ramp/platform, and flanked on the west by a neutral to low relief sediment barrier or "elevated rim" (terminology of Richards, 1989). This western barrier was an intermittently reactivated remnant of the Antler orogenic belt (Schwarz, 1987; Richards, 1989; Snyder and others, 1991; Fig. 11). This sediment barrier served to separate continentally derived sediment from volcanogenic sediment with western provenance (Stewart and others, 1977; Richards, 1989). The absence of volcanogenic sediment in the Wood River basin suggests the basin was removed from the volcanogenic depocenter to the west, perhaps by a subdued remnant of the Antler orogenic belt (Fig. 11).

Snyder and others (1991) propose that localized reactivation of the Antler orogenic belt was responsible for episodic subsidence in the Dry Mountain Trough during early Wolfcampian time. Richards (1989) documents differential uplift and episodic subsidence in the Prophet/Ishbel trough in Desmoinesian to Leonardian time.

Reactivation of the Antler orogenic belt may have resulted from interaction with tecnically active western volcanic terranes during the Late Pennsylvanian and Early Permian (Oldow and others, 1989; Richards, 1989; Snyder and others, 1991).

Strata of the Wood River and Oquirrh basins record deposition of fine-grained, cratonally derived, mixed carbonate-siliciclastic sediment from late Desmoinesian to Leonardian time. In contrast, late Paleozoic Ancestral Rockies basins to the east, such as the Paradox basin, were filled with detritus derived from proximal block uplifts by Wolfcampian time (Kluth, 1986). We propose the late Desmoinesian to Leonardian evolution of the Wood River basin differed from that of Ancestral Rockies basins to the east, and that subsidence was tectonically controlled in Late Pennsylvanian and Early Permian time by plate interactions to the west.

CONCLUSIONS

1. The Wood River, Grand Prize, and Dollarhide Formations of south-central Idaho comprise the herein designated Sun Valley Group. Sun Valley Group strata are coeval, laterally contiguous, genetically related sedimentary rocks deposited in the Desmoinesian to Leonardian Wood River basin.

2. Sun Valley Group strata record two distinct stages of deposition: 1) Desmoinesian mixed coarse-grained carbonate-siliciclastic sedimentation in a subaqueous braid-delta and deep-water slope. Conglomerate clasts may have been derived from the Copper Basin highland to the northeast. 2) Late Desmoinesian to Leonardian mixed fine-grained carbonate-siliciclastic sedimentation in a sub-wavebase submarine ramp-apron system. This system received siliciclastic sediment from a northern cratonal source and carbonate sediment from an eastern source (possibly the Snaky Canyon carbonate platform).

3. The Wood River basin was epicratonic, and not a west-facing continental margin, as suggested by Wavra and others (1986), for the following reasons: 1) facies distribution and paleoslope data indicate that the basin had a south to southeast-dipping paleoslope, not a west-dipping slope as would be expected on the Cordilleran continental margin; 2) biostratigraphic control suggests sedimentation was irregular, indicative of tectonically-controlled local subsidence and not exponentially-decaying thermally-controlled subsidence; 3) there is no evidence of connection between the Wood River basin and late Paleozoic volcanogenic basins on the western margin of North America. The absence of west-derived volcanogenic sediment in the Sun Valley Group suggests the Wood River basin was separated from this subduction-influenced western margin by a barrier or "elevated rim"; 4) Wood River basin contains fine-grained mixed carbonate-siliciclastic

strata comparable to that deposited in other epicratonic basins inboard of the western edge of North America in late Paleozoic time.

5. The late Desmoinesian to Leonardian part of the Sun Valley Group contain a 1300-2000 m megacycle which first becomes coarser-grained then finer-grained upward. Timing of this large-scale cyclicity is coincident with second-order eustatic changes, although biostratigraphic control is tentative. Sedimentary facies suggest the Wood River basin may have contained highstand systems tracts in the early Virgilian and middle to late Wolfcampian and a lowstand systems tract near the Virgilian-Wolfcampian boundary.

6. The two-stage evolution of the Wood River basin is the result of two distinct tectonic influences on basin subsidence and sediment provenance. Initial subsidence of the Wood River basin, and associated uplift of the Copper Basin highland, began in the Desmoinesian in response to transpression associated with Ancestral Rockies tectonism (Kluth, 1986). The change in provenance, depositional environment, water depth, paleotransport direction and biologic component in late Desmoinesian-early Virgilian time indicates basin readjustment to a different tectonic regime. Late Pennsylvanian-Early Permian tectonic loading of the western continental margin influenced basin subsidence in the Dry Mountain trough to the south, in the Prophet trough to the north, and, we propose, in the Wood River basin.

ACKNOWLEDGEMENTS

Our study of the Sun Valley Group from 1985-1990 comprised 1:24,000 scale mapping, plus stratigraphic, petrologic, geochemical, mineral resource, and paleontologic study sponsored by the U.S. Geological Survey Hailey CUSMAP project (R.G. Worl, director). The majority of the research was conducted by M.S. students at Idaho State University: Geslin (1986), Stewart (1987), Darling (1987), Mahoney (1987), Burton (1989), Whitman (1990), O'Brien (1991). This work was supervised by P.K. Link, D.W. Rodgers, W.R. Hackett, H.T. Ore, C.W. Blount and B. Dog. Mahoney and Link were supported from 1986-1990 by the U.S. Geological Survey, and Burton was supported by Norcen Energy Resources, Ltd. Additional financial support was obtained from the Idaho State Board of Education, Idaho State University Faculty Research Council, Idaho Geological Survey, Everett Community College, and Sigma Xi.

We owe our background in the geology of south-central Idaho to Dr. Betty Skipp and the late W.E. Hall of the U.S. Geological Survey, to whom we are deeply indebted. Wayne Hall studied Upper Paleozoic strata in south-central Idaho for 20 years (1965-1985) and established the foundation upon which our work is based. In the course of our work we were forced to reject Hall's concepts of thrust-bounded tectonostratigraphic units in the Wood River area, but we understand how the concepts arose inasmuch as we followed them ourselves for several years.

Fusulinid biostratigraphy was provided by C.A. Ross and C.F. Kluth of Chevron Inc., and Raymond C. Douglass and Donald A Myers of the U.S. Geological Survey. We would like to thank Anna Goldberg of Norcen Energy Resources, Ltd. and Ryan McDermott of Idaho State University for help with figure preparation. The assistance of the Idaho State University Geology Alumni Mapping Project and the B.U.L.L.S.H.I.T. team is gratefully acknowledged. The manuscript was improved by reviews from G.R. Winkler, B. Skipp, W.S. Snyder, M. Maclachlan, H. Lyatsky, and C. Stevens.

REFERENCES CITED

Batchelder, J.N., and Hall, W.E., 1978, Preliminary geologic map of the Hailey 7.5-minute quadrangle, Idaho: U.S. Geological Survey Open-file report 78-545, scale 1:24,000.

Bostwick, D.A., 1955, Stratigraphy of the Wood River Formation, south-central Idaho: Journal of Paleontology, v. 29, p. 941-951.

Breuninger, R.H., Canter, K.L., and Isaacson, P.E., 1988, Pennsylvanian-Permian Palaeoaplysina and algal buildups, Snaky Canyon Formation, south-central Idaho, U.S.A., in Geldsetzer, H.H.J., James, N.P., and Tebbut, G.E., eds., Reefs - Canada and Adjacent areas: Canadian Society of Petroleum Geologists Memoir 13, p. 631-637.

Burton, B.R., 1988, Stratigraphy of the Wood River Formation in the eastern Boulder Mountains, Blaine and Custer Counties, south-central Idaho [M.S. Thesis]: Pocatello, Idaho State University, 165 p.

Burton, B.R., and Link, P.K., 1989, Lake Creek mineralized area, Blaine County, Idaho: in Winkler, G.R., Soulliere, S.J., Worl, R.G., and Johnson, K.M., eds., Geology and mineral deposits of the Hailey and western Idaho Falls 1°x2° quadrangles, Idaho: U.S. Geological Survey Open-File Report 89-639, p. 74-85.

_____, and Link, P.K., 1990, Ichnology of fine-grained mixed carbonate-siliciclastic turbidite facies of the upper member, Wood River Formation (Lower Permian), south-central Idaho: Abstracts volume, 13th International Sedimentological Congress, Nottingham England, p. 313

_____, Link, P.K., and Rodgers, D.W., 1989, Death of the Wood River thrust: Structural relations in the Pioneer and Boulder Mountains, south-central Idaho: Geological Society of America Abstracts with Programs, v. 21, no. 5, p. 62.

Darling, R.S., 1987, The geology and ore deposits of the Carrietown silver-lead-zinc district, Blaine and Camas Counties, Idaho [M.S. thesis]: Pocatello, Idaho State University, 168 p.

_____, 1988, Ore deposits of the Carrietown silver-lead-zinc District, Blaine and Camas Counties, Idaho: in Link, P.K., and Hackett, W. R., editors, Guidebook to the geology of central and southern Idaho: Idaho Geological Survey Bulletin 27, p. 193-198.

Dover, J.H., 1980, Status of Antler orogeny in central Idaho, clarification and constraints from the Pioneer Mountains, in Fouch, T.D. and Magathan, E.R., eds., Paleozoic paleogeography of the west-central United States: Denver, Rocky Mountain Section, Society of Economic Paleontologists and Mineralogists, West-central United States Paleogeography Symposium I, p. 371-381.

_____, 1981, Geology of the Boulder-Pioneer wilderness study area, Blaine and Custer Counties, Idaho, in Mineral resources of the Boulder-Pioneer wilderness study area, Blaine and Custer Counties, Idaho: U.S. Geological Survey Bulletin 1497, p. 1-75.

_____, 1983, Geologic map and sections of the central Pioneer Mountains, Blaine and Custer Counties, central Idaho: U.S. Geological Survey Miscellaneous Investigations Series, Map I-1319, scale 1:48,000.

Dunham, R.J., 1962, Classification of carbonate rocks according to depositional texture, in Ham, W.E., ed., Classification of carbonate rocks: American Association of Petroleum Geologists Memoir 1, p. 108-121.

Fisher, F. S., and Johnson, K.M., eds., 1987, Preliminary manuscript for Mineral-resource assessment and geology of the Challis 1° x 2° quadrangle, Idaho: U.S. Geological Survey Open-File Report 87-840, 251 p., 23 plates, scale 1:250,000.

Folk, R.L., 1980, Petrology of Sedimentary Rocks: Austin, Tx., Hemphill Pub. Co., 184 p.

Geslin, J.K., 1986, The Permian Dollarhide Formation and Paleozoic Carrietown sequence in the SW 1/4 of the Buttercup Mountain quadrangle, Blaine and Camas Counties, Idaho [M.S. Thesis]: Pocatello, Idaho State University, 116 pp.

Goodman, H.M., 1983, Comparison of three stratigraphic sections of the Wood River Formation in the vicinity of Bellevue, Idaho [M.S. thesis]: Pullman, Washington State University, 92 p.

Hall, W.E., 1985, Stratigraphy and mineral deposits in middle and upper Paleozoic rocks of the black-shale mineral belt, central Idaho, in McIntyre, D.H., ed., Symposium on the geology and mineral resources of the Challis 1°x2° quadrangle, Idaho: U.S. Geological Survey Bulletin 1658-J, p. 117-132.

_____, and Hobbs, S.W., 1987, Black shale terrane, in Fisher, F. S., and Johnson, K.M., eds., Preliminary manuscript for Mineral-resource assessment and geology of the Challis 1°x2° degree quadrangle, Idaho: U.S. Geological Survey Open-File Report 87-840, p. 29-42.

_____, Batchelder, J., and Douglass, R.C., 1974, Stratigraphic section of the Wood River Formation, Blaine County, Idaho: U.S. Geological Survey Journal of Research, v. 2, p. 89-95.

_____, Rye, R.O., and Doe, B.R., 1978a, Wood River mining district, Idaho - intrusion-related lead-silver deposits derived from country rock source: U.S. Geological Survey Journal of Research, v. 6,, p. 579-592.

_____, Batchelder, J.H., and Tschanz, C.M., 1978b, Geologic map of the Sun Valley 7.5-minute quadrangle, Blaine County Idaho: U.S. Geological Survey Open-file Report 78-1058, scale 1:24,000.

Heller, P.L., and Dickinson, W.R., 1985, Submarine ramp facies model for delta-fed, sand-rich turbidite systems: American Association of Petroleum Geologists Bulletin, vol. 69,, p. 960-976.

Henderson, C.M., 1989, The Lower Absaroka Sequence: Upper Carboniferous and Permian, in Ricketts, B.D., ed., Western Canada Sedimentary Basin - A Case History; p. 203-214.

Jordan, T.E., and Douglass, R.C., 1980, Paleogeography and structural development of the Late Pennsylvanian to Early Permian Oquirrh Basin, northwestern Utah, in Fouch, T.D. and Magathan, E.R., eds., Paleozoic paleogeography of the west-central United States: Denver, Rocky Mountain Section, Society of Economic Paleontologists and Mineralogists, West-central United States Paleogeography Symposium I, p. 217-238.

Kluth, C.F., 1986, Plate tectonics of the Ancestral Rocky Mountains, in Peterson, J.A., ed., Paleotectonics and sedimentation in the Rocky Mountain region, United States: American Association of Petroleum Geologists Memoir 41, p. 353-369.

Lindgren, W., 1900, The gold and silver veins of Silver City, DeLamar, and other mining districts in Idaho: U.S. Geological Survey Twentieth Annual Report, part 3, p. 65-256.

Link, P.K., and Mahoney, J.B., 1989, Stratigraphic setting of sediment-hosted mineralization in the eastern Hailey 1°x2° quadrangle, Blaine, Custer, and Camas Counties, south-central Idaho, in Winkler, G.R., Soulliere, S.J., Worl, R.G., and Johnson, K.M., eds., Geology and mineral deposits of the Hailey and western Idaho Falls 1°x2° Quadrangles, Idaho: U.S. Geological Survey Open-File Report 89-639, p. 53-69.

_____, Mahoney, J.B., Burton, B.R., Ratchford, M.E., Turner, R.J.W., and Otto, B.R., 1987, Introduction to the geology of the central Idaho black shale mineral belt: Northwest Geology, v. 16, p. 61-84.

_____, Skipp, B., Hait, M.H., Jr., Janecke, S., and Burton, B.R., 1988, Structural and stratigraphic transect of south-central Idaho: A field guide to the Lost River, White Knob, Pioneer, and Smoky Mountains, in Link, P.K., and Hackett, W.R., eds., Guidebook to the geology of central and southern Idaho: Idaho Geological Survey Bulletin 27, p. 5-42.

Loeblich, A.R., Jr,. and Tappan, H., 1988, Foraminiferal genera and their classification: New York, Van Nostrand Reinhold Co., 970 p.

Mahoney, J.B., 1987, Geology of the northern Smoky Mountains and stratigraphy of a portion of the Permian Grand Prize Formation, Blaine and Camas Counties, Idaho [M.S. thesis]: Pocatello, Idaho State University, 98 p.

_____, and Sengebush, R.M., 1988, Stratigraphy of the Lower Permian Grand Prize Formation, south-central Idaho, in Link, P.K., and Hackett, W.R., eds., Guidebook to the geology of central and southern Idaho: Idaho Geological Survey Bulletin 27, p. 169-179.

_____, Link, P.K., Burton, B.R., 1990, Tectonic setting and sedimentologic history of an epicratonic basin in south-central Idaho: tentative correlations of late Paleozoic basins on the western edge of North America: Geological Association of Canada-Mineralogical Association of Canada Abstracts with Programs, v. 15, p. A82.

Mount, J.F., 1984, Mixing of siliciclastic and carbonate sediments in shallow shelf environments: Geology, v. 12, p. 432-435.

_____, 1985, Mixed carbonate and siliciclastic sediments: a proposed first order textural and compositional classification: Sedimentology, v. 32, p. 435-442.

Mullins, H.T., and Cook, H.E., 1986, Carbonate apron models: Alternatives to the submarine fan model for paleoenvironmental analysis and hydrocarbon exploration: Sedimentary Geology, v. 48, p. 37-79.

Mytton, J.W., Morgan, W.A., and Wardlaw, B.R., 1983, Stratigraphic relations of Permian units, Cassia Mountains, Idaho, in Miller, D.M, Todd, V.A, and Howard, K.A., eds., Tectonic and stratigraphic studies in the eastern Great Basin: Geological Society of America Memoir 157, p. 281-303.

_____, Williams, P.L., and Morgan, W.A., 1990, Geologic map of the Stricker 4 Quadrangle, Cassia County, Idaho: U.S. Geological Survey Miscellaneous Investigations Series, Map I-2052, scale 1:48,000.

Nilsen, T.H., 1977, Paleogeography of Mississippian turbidites in south-central Idaho, in Stewart, J.H., Stevens, C.H., and Fritsche, A.E., eds., Paleozoic paleogeography of the western United States: Los Angeles, Pacific Section, Society of Economic Paleontologists and Mineralogists, Pacific Coast Paleogeography Symposium I, p. 275-299.

O'Brien, J.P., 1991, Stratigraphy of the Pennsylvanian-Permian Dollarhide Formation in the Buttercup Mountain quadrangle, Blaine County, Idaho [M.S. thesis]: Pocatello, Idaho State University.

Oldow, J.S., Balley, A.W., Ave' Lallemant, H.G., and Leeman, W.P., 1989, Phanerozoic evolution of the North American Cordillera; United States and Canada, Chapter 8, in Balley, A.W., and Palmer, A.R., eds., The Geology of North America; An Overview: Geological Society of America, The Geology of North America, Volume A, p. 139-232.

Otto, B.R., and Turner, R.J.W., 1987, Stratigraphy and structure of the Milligen Formation, Sun Valley area, Idaho: Northwest Geology, v. 16, p. 95-103.

Ratchford, M.E., 1989, Geology fo the Boulder Basin, Blaine and Custer Counties, Idaho [M.S. thesis]: Moscow, University of Idaho, 154p.

Rember, W.C. and Bennett, E.H., 1979, Geologic map of the Hailey quadrangle, Idaho: Idaho Bureau of Mines and Geology Geologic Map Series, scale 1:250,000.

Richards, B.C., 1989, Upper Kaskaskia Sequence - Uppermost Devonian and Lower Carboniferous, in Ricketts, B.D., ed., Western Canada Sedimentary Basin - A Case History; p. 165-196.

Roberts, R.J., and Thomasson, M.R., 1964, Comparison of late Paleozoic depositional history of northern Nevada and central Idaho: U.S. Geological Survey Professional Paper 475-D, p. D1-D6.

Ross, C.P., 1960, Diverse interfingering Carboniferous strata in the Mackay quadrangle, Idaho: U.S. Geological Survey Professional Paper 400-B, p. B232-B233.

Ross, C.A., and Ross, J.P., 1987, Late Paleozoic sea levels and depositional sequences, in Ross, C.A., and Haman, D., eds., Timing and depositional history of eustatic sequences: Constraints on seismic stratigraphy: Cushman Foundation for Foraminiferal Research Special Publication 24, p. 137-150.

Sandberg, C.A., Hall, W.E., Batchelder, J.N., and Axelson, C., 1975, Stratigraphy, conodont dating, and paleotectonic interpretation of the type Milligen Formation (Devonian), Wood River area, Idaho: U.S. Geological Survey Journal of Research, v. 3, p. 707-720.

Schwarz, D.L., 1987, Geology of the lower Permian Dry Mountain Trough, Buck Mountain, Limestone Peak, and Secret Canyon areas, east-central Nevada [M.S. thesis]: Pocatello, Idaho State University and Boise State University, 149p.

Sengebush, R.M., 1984, The geology and tectonic history of the Fourth of July Creek area, White Cloud Peaks, Custer County, Idaho [M.S. thesis]: Missoula, University of Montana, 79 p.

Skipp, Betty, and Hall, W.E., 1975, Structure and stratigraphy of a complex of thrust plates in the Fish Creek Reservoir area, south-central Idaho: U.S. Geological Survey Journal of Research, v. 3, p. 671-689.

_____, and Hait, M.H., Jr., 1977, Allochthons along the northeast margin of the Snake River Plain, Idaho: Casper, Wyoming Geological Association Twenty-Ninth Annual Field Conference Guidebook, p. 499-515.

_____, and Hall, W.E., 1980, Upper Paleozoic paleotectonics and paleogeography of Idaho, in Fouch, T.D. and Magathan, E.R., eds., Paleozoic paleogeography of the west-central United States: Denver, Rocky Mountain Section, Society of Economic Paleontologists and Mineralogists, West-central United States Paleogeography Symposium I, p. 387-422.

_____, Hoggan, R.D., Schleicher, D.L., and Douglass, R.C., 1979a, Upper Paleozoic carbonate bank in east-central Idaho--Snaky Canyon, Bluebird Mountain, and Arco Hills Formations, and their paleotectonic significance: U.S. Geological Survey Bulletin 1496, 78 p.

_____, Sando, W.J., and Hall, W.E., 1979b, The Mississippian and Pennsylvanian Systems in the United States, Idaho: U.S. Geological Survey Professional Paper 1110, p. AA1-AA42.

Smith, J.F., Jr., 1982, Geologic Map of the Strevell 15-minute Quadrangle, Cassia County, Idaho: U.S. Geological Survey Miscellaneous investiagtions Series Map I-1403, scale 1:62,500.

Snyder, W.S., Spinosa, C., and Gallegos, D.M., 1991, Pennsylvanian-Permian tectonism on the western U.S. continental margin, in Rains, G.L., Leslie, R.E., Shaffer, R.W., and Wilkinson, W.H., eds., Geology and Ore Deposits of the Great Basin, Reno, Geological Society of Nevada Symposium.

Stewart, D.E., 1987, Geology of the northern half of the Baker Peak quadrangle, Blaine and Camas Counties, Idaho [M.S. thesis]: Pocatello, Idaho State University, 113 p.

Stewart, J.H., MacMillan, J.R., Nichols, K.M., Stevens, C.H., 1980, Deep-water Upper Paleozoic rocks in north-central Nevada - a study of the type area of the Havallah Formation, in Stewart, J.H., Stevens, C.H., and Fritsche, A.E., eds., Paleozoic paleogeography of the western United States: Los Angeles, Pacific Section, Society of Economic Paleontologists and Mineralogists, Pacific Coast Paleogeography Symposium I, p. 337-347.

Stow, D.A.V., and Piper, D.J.W., 1984, Deep-water fine-grained sediments: facies models, in Stow, D.A.V., and Piper, D.J.W., eds., Fine-grained sediments: Deep-water processes and facies: Boston, Blackwell Scientific Publications, p. 611-646.

Thomasson, M.R., 1959, Late Paleozoic stratigraphy and paleotectonics of central and eastern Idaho [Ph.D. Thesis]: Madison, University of Wisconsin, 224 p.

Thompson, M.L., Dodge, H.W., and Youngquist, W., 1958, Fusulinids from the Sublett Range, Idaho: Journal of Paleontology, v. 32, p. 113-125.

Trimble, D.E., and Carr, W.J., 1976, Geology of the Rockland and Arbaon quadrangles, Power Countie, Idaho: U.S. Geological Survey Bulletin 1399, 115p.

Tschanz, C.M., and Kiilsgaard, T.H., 1986, Geologic appraisal of mineral resources of the eastern part of the Sawtooth National Recreation Area, Idaho: U.S. Geological Survey Bulletin 1545, p. 53-208.

Turner, R.J.W., and Otto, B.R., 1988, Stratigraphy and structure of the Milligen Formation, Sun Valley area, Idaho: in Link, P.K., and Hackett, W.R., eds., Guidebook to the geology of central and southern Idaho: Idaho Geological Survey Bulletin 27, p. 153-167.

Umpleby, J.B., Westgate, L.G., and Ross, C.P., 1930, Geology and ore deposits of the Wood River region, Idaho: U.S. Geological Survey Bulletin 27, p. 153-167.

Verville, G.J., Sanderson, G.A., Baesemann, J.F., and Hampton, G.L., III, 1990, Pennsylvanian fusulinids from the Beaverhead Mountains, Morrison Lake area, Beaverhead County, Montana: The Mountain Geologist, v. 27, p. 47-55.

Wavra, C.S., 1985, Structural geology and petrology of the SW1/4 of the Mahoney Butte quadrangle, Blaine County, Idaho [M.S. thesis]: Moscow, University of Idaho, 103 p.

_____, Isaacson, P.E., and Hall, W.E., 1986, Studies of the Idaho black shale belt: Stratigraphy, depositional environment, and economic geology of the Permian Dollarhide Formation: Geological Society of America Bulletin, v. 97, p. 1504-1511.

Whitman, S.K., 1990, Metamorphic petrology and structural geology of the Pennsylvanian-Permian Dollarhide Formation, Blaine and Camas Counties, south-central Idaho [M.S. thesis]: Pocatello, Idaho State University, 108 p.

Williams, P.L., Mytton, J.W., and Covington, H.R., 1990, Geologic map of the Stricker 1 Quadrangle, Cassia, Twin Falls, and Jerome Counties, Idaho: U.S. Geological Survey Miscellaneous Investigations Series, Map I-2078, scale 1:48,000.

Winsor, H.C., 1981, The paleogeography and paleoenvironments of the Middle Pennsylvanian part of the Wood River Formation in south-central Idaho [M.S. Thesis]: San Jose, San Jose State University, 81 p.

Wust, S.L., 1986, Extensional deformation with northwest vergence, Pioneer core complex, central Idaho: Geology, v. 14, p. 712-714.

Yancey, T.E., Ishibashi, G.D., and Bingham, P.T., 1980, Carboniferous and Permian stratigraphy of the Sublett Range, south-central Idaho, in Fouch, T.D. and Magathan, E.R., eds., Paleozoic paleogeography of the west-central United States: Denver, Rocky Mountain Section, Society of Economic Paleontologists and Mineralogists, West-central United States Paleogeography Symposium I, p. 259-269.

OVERVIEW OF EARLY PALEOZOIC MAGMATISM IN THE EASTERN KLAMATH MOUNTAINS, CALIFORNIA: AN ISOTOPIC PERSPECTIVE

E.T. Wallin
Department of Geoscience
University of Nevada
Las Vegas, NV 89154

N. Lindsley-Griffin
Department of Geology
University of Nebraska
Lincoln, NE 68588

J.R. Griffin
Department of Geology
University of Nebraska
Lincoln, NE 68588

ABSTRACT

Igneous rocks of the Yreka and Trinity terranes record a complex history of magmatic events that occurred discontinuously over a 200 m.y. period between the Early Cambrian and the Middle Devonian. Lower Cambrian rocks occur as fault-bounded units within the northwestern part of the Trinity terrane, and as tectonic blocks in melange of the Yreka terrane. The textures and structures of the strongly foliated, ductilely deformed Cambrian rocks indicate a complicated tectonic history prior to their juxtaposition with younger rocks of the Trinity and Yreka terranes. Samples dated as Early Cambrian (565 Ma) consist of plagiogranites and metagabbro; one plagiogranite contains inherited Precambrian zircon.

Ordovician ultramafic and mafic rocks of the Trinity terrane range in age from 472 to 435 Ma. Presumably, two plagiogranites of the Trinity complex dated at 475 and 469 Ma are also related genetically to the oceanic lithosphere represented by the ultramafic rocks. A fault that places Ordovician mantle tectonite of the Trinity terrane against Cambrian metagabbro is intruded by Late Silurian pegmatitic gabbro of the China Mountain pluton, which indicates that the Cambrian and Ordovician rocks were juxtaposed before or during the Late Silurian. Isotopic dating and our recent mapping indicate Late Silurian magmatism was voluminous and widespread throughout the Trinity terrane.

Well-studied Devonian rocks of the Redding terrane appear to represent an immature, submerged to island arc setting. Basaltic to andesitic pillow lavas, flows and dikes occur locally within the Yreka and Trinity terranes. Within melange, these poorly dated volcanic rocks may be Cambrian, Ordovician, or Silurian in age. However, volcanic rocks that overlie the melange are constrained to be Early to Middle Devonian in age and may be a minor northern manifestation of the well developed Devonian magmatic arc in the Redding terrane to the south. These minor Lower to Middle Devonian volcanics and the dikes that presumably fed them are undeformed, which indicates they were erupted after the Yreka and Trinity terranes were amalgamated. Small swarms of sheeted dikes related to this volcanism suggest eruption during post-amalgamation extension.

INTRODUCTION

Until recently, conventional wisdom held that early Paleozoic magmatism in the Klamath Mountains consisted of two principal events, namely the generation of Ordovician oceanic lithosphere of the Trinity complex, and development of the Devonian volcanic arc of the Redding terrane. Irwin (1985) compiled existing data concerning the age and origin of plutonic belts throughout the Klamath Mountains. Since then, new isotopic studies have revealed areally extensive Cambrian and Silurian igneous rocks within the Trinity complex as well as Cambrian rocks in the Yreka terrane. Thus, it seems appropriate to review this recent progress from the perspective of isotope geology and to highlight uncertainties that remain regarding the timing and tectonic significance of early Paleozoic magmatic events in the Klamath Mountains.

CAMBRIAN

Lower Cambrian rocks occur within melange of the Yreka terrane and within the northwestern part of the Trinity complex (Fig. 1). Most of the Cambrian rocks that have been dated are plagiogranites, although U-Pb zircon ages for one large body of strongly foliated amphibolitic metagabbro of the Trinity complex reveal it is coeval with the plagiogranites. The plagiogranites range in composition from tonalite to quartz diorite to trondhjemite, and have been subjected to varying degrees of very low- to low-grade metamorphism.

The strain exhibited by Cambrian plagiogranitic rocks of the region is highly variable. The Skookum Butte block in melange of the Yreka terrane (Wallin et al., 1988) is relatively undeformed and exhibits only millimeter-scale zones of cataclasis and bent plagioclase twin lamellae. In contrast, several plagiogranites within the Trinity terrane are strongly foliated and display well-developed features indicative of both ductile and brittle deformation, including dynamically recrystallized quartz and feldspar porphyroblasts and zones of cataclasis (Lindsley-Griffin, this volume). The plagiogranites were mapped provisionally as Ordovician by previous workers (e.g., Hotz, 1977, 1978; Wagner and Saucedo, 1987). Wallin et al. (1988) demonstrated that many plagiogranitic

In Cooper, J.D., and Stevens, C.H., eds., 1991, **Paleozoic Paleogeography of the Western United States-II:** Pacific Section SEPM, Vol. 67, p. 581-588.

582

Cretaceous and younger deposits

Mesozoic plutonic rocks (Irwin, 1985).

Undivided western Paleozoic/Triassic belt and central metamorphic belt
(Devonian to Triassic)

Metamorphic rocks with minor carbonate blocks.
(Wright, 1982; Irwin, 1989; Goodge, 1990).

Redding terrane (Devonian to Jurassic)
Primarily volcanic and sedimentary rocks with minor plutonic rocks.
Includes Copley Greenstone, Balaklala Rhyolite, Kennett Fm., Baird
Fm., Bragdon Fm., McCloud Limestone, Nosoni Fm., Dekkas Andesite,
Pit Fm., Hosselkus Limestone, and numerous Jurassic units (Burchfiel
and Davis, 1981; Miller, 1989).

Yreka terrane (Ordovician to Devonian)
Terrigenous clastic metasedimentary rocks with minor carbonate rocks.
Includes Sissel Gulch Graywacke, Duzel Phyllite, Moffett Creek Fm.,
Antelope Mountain Quartzite, Gazelle Fm., and various disrupted
melange units, including the schist of Skookum Gulch which contains
blocks of Lower Cambrian tonalite (Hotz, 1977; Potter et al., 1977;
Lindsley-Griffin and Griffin, 1983; Wallin, 1989).

Trinity complex (lower Paleozoic)
Ultramafic rocks and gabbros with subordinate plagiogranite and
basalt. Lower Cambrian rocks represented by lower pattern;
Ordovician rocks represented by upper pattern; Silurian rocks shown in
black (Lindsley-Griffin, 1982; Irwin, 1985; Jacobsen et al., 1984; Wallin
et al., 1988; Wallin, 1990).

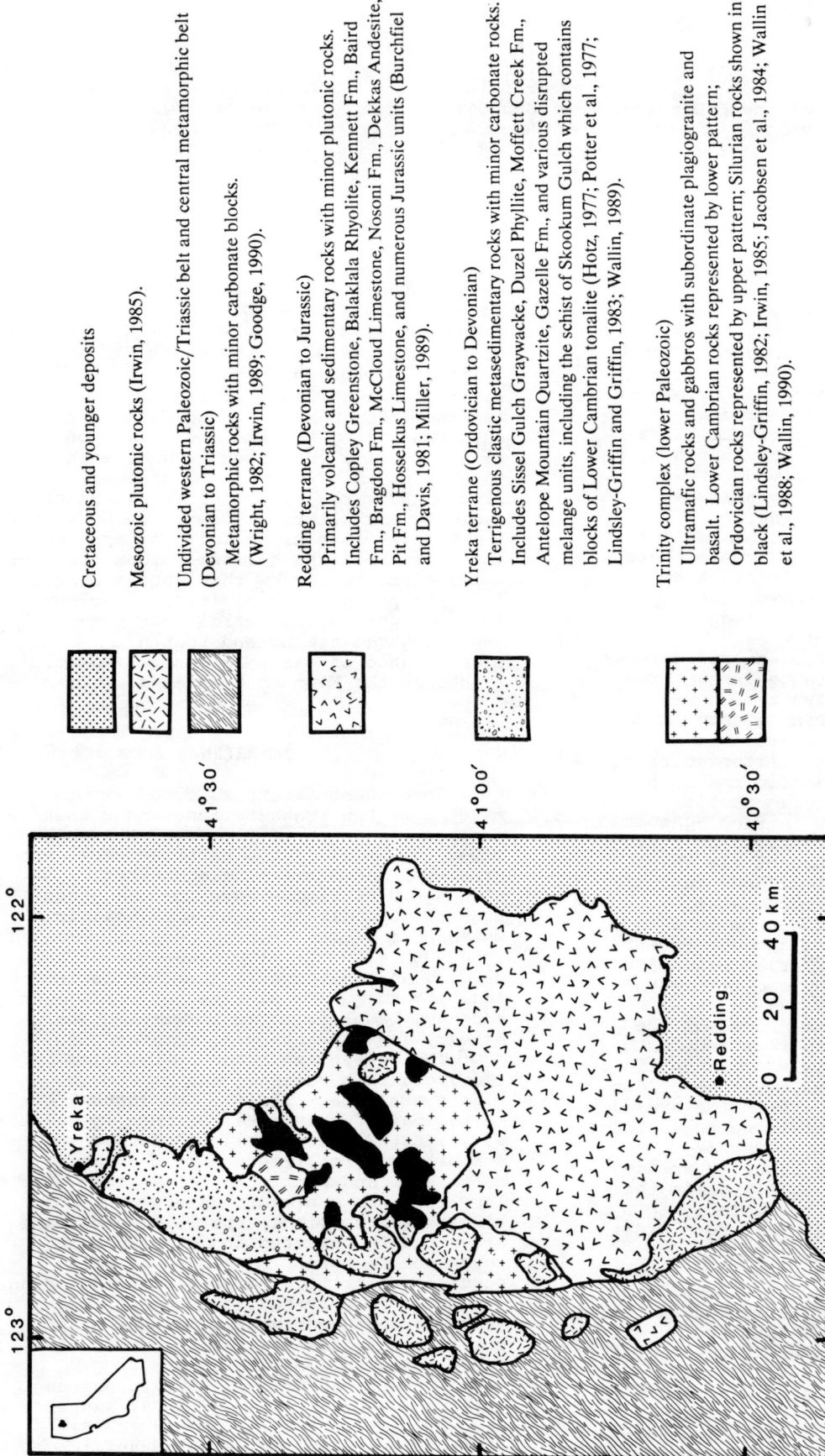

Figure 1: Generalized geologic map of the eastern Klamath Mountains, California (modified after Irwin, 1977).

bodies in the area are Early Cambrian (565 Ma) in age, although lithologically similar rocks of Middle Ordovician age occur as well. A detailed map of the Cambrian portion of the Trinity complex is presented by Lindsley-Griffin (this volume).

The structural context of plagiogranites within the Yreka terrane is fairly well known. The Trail Gulch and Skookum Butte bodies (Wallin, 1990) occur as tectonic blocks within metamorphosed tectonic melange of the Skookum Gulch schist, which has yielded Ordovician-Silurian isotopic ages of metamorphism (Hotz, 1977; Potter et al., 1981; Cotkin and Armstrong, 1987). Interpretation of these bodies as tectonic blocks is generally accepted because no intrusive contacts have been observed between the plagiogranites and other components of the Skookum Gulch schist.

Cambrian rocks of the Trinity complex include plagiogranite, metagabbro, and possibly spilitic basalt and harzburgite (Lindsley-Griffin, this volume). The geometry and structural position of these Cambrian rocks remain enigmatic. They may be large blocks in a matrix-poor, block-on-block tectonic melange or a fragment of a crustal-scale volcanoplutonic edifice of Early Cambrian age (Wallin, 1990). The partial preservation of what is possibly ophiolitic stratigraphy within Cambrian rocks of the Trinity complex is consistent with the latter hypothesis (Lindsley-Griffin, this volume).

The juxtaposition of Cambrian and Ordovician rocks appears to be tectonic rather than intrusive because only faulted contacts have been observed among them. The uncertainty as to whether emplacement of the Cambrian assemblage involved a thin- or thick-skinned deformation is discussed by Wallin et al. (in press). Although plagiogranitic rocks in both the Yreka and Trinity terranes may have formed in different settings, they do exhibit broad lithologic similarity and are intimately associated with coeval mafic volcanic rocks.

Several lines of evidence indicate the petrotectonic setting of the Cambrian rocks was quite different than that of younger rocks in the Trinity complex. Ordovician and Silurian rocks of the Trinity complex appear to be more primitive than the Cambrian rocks. Based on sparse data, Ordovician and Silurian rocks appear to exhibit higher E_{Nd} and lower initial $^{87}Sr/^{86}Sr$ values than those of the Cambrian rocks (Table 1). In addition, the Trail Gulch block in melange of the Yreka terrane has been shown to contain inherited Precambrian zircon, which indicates some involvement of cratonic material in its petrogenesis (Wallin, 1990). Scatter in the $^{207}Pb/^{206}Pb$ ages of zircon (556 - 579 Ma) from the metagabbro body within the Trinity complex may also reflect a minor component of xenocrystic Precambrian zircon (Wallin et al., in press). Despite these new constraints, the structure and petrogenesis of the Lower Cambrian rocks remain highly uncertain and are important topics for future research.

ORDOVICIAN

Ordovician(?) rocks of the Trinity complex consist mainly of lherzolite, plagioclase lherzolite, harzburgite, dunite, and pyroxenite; with minor gabbro and plagiogranite in fault slices (Fig. 1). Prior to widespread application of the concepts of plate tectonics, these rocks were interpreted as an intrusion that had been emplaced along a thrust fault (Lipman, 1964). K-Ar whole rock ages (439-333 Ma) reported for several gabbros within the Trinity complex provided the first isotopic evidence that the complex was early Paleozoic in age (Lanphere et al., 1968). Minor Middle Ordovician plagiogranites (475 and 469 Ma) are tentatively included within the Trinity complex although they occur in a poorly exposed area adjacent to melange of the newly defined Gregg Ranch Complex (Wallin et al., 1988; Lindsley-Griffin et al., this volume).

Lindsley-Griffin (1977, 1982) interpreted part of the northwesternmost Trinity complex as a remnant of an ophiolite. Many of the rocks related to the sequence studied by Lindsley-Griffin have been shown to be Early Cambrian in age and thus are unrelated genetically to Ordovician ultramafic rocks of the Trinity complex (Wallin et al., 1988). The tectonic significance of the apparent 100 m.y. magmatic gap between Lower Cambrian and Middle Ordovician rocks of the Trinity complex remains enigmatic (Wallin et al., in press). Lindsley-Griffin (this volume) reexamines the confusing evolution of the nomenclature that has been applied to the Trinity complex and has incorporated recent advances in field geology and geochronology into a model for its evolution.

Quick (1981a, 1981b) emphasized that many features of the Trinity complex are atypical of ophiolites that were generated at spreading-ridges and argued that Ordovician ultramafic rocks of the Trinity complex formed by the rise of an upper mantle diapir in the vicinity of an island-arc or back-arc setting. Jacobsen et al. (1984) reported a Sm-Nd isochron age of 472 ±32 Ma for plagioclase lherzolite that remains the best estimate of the age of crystallization for ultramafic rocks of the Trinity complex (Table 1). This plagioclase lherzolite yielded a highly positive E_{Nd} signature (+10.4) that is identical within uncertainty to that of the Ordovician Kings River ophiolite (+10.7) in the northern Sierra Nevada terrane (Jacobsen et al., 1984; Shaw et al., 1987). This similarity strengthens tectonostratigraphic correlations that have been proposed between the eastern Klamaths and the northern Sierra Nevada (Davis, 1969; Schweickert and Snyder, 1981, Wallin et al., 1988).

Table 1. Isotopic Ages and Petrogenetic Indicators for Lower Paleozoic Rocks

Rock Unit	Isotopic Age	Initial $^{87}Sr/^{86}Sr$	Epsilon Nd(T)
Cambrian plagiogranite of Skookum Butte melange block	565 ±5 (7)	0.70636 (9)	+4.4 (9)
Cambrian plagiogranite of Trail Gulch melange block with inherited zircon	567 ±78 (6)	0.70612 (9)	--
Cambrian metagabbro of Trinity complex	556-579 (8)	0.70527 (9)	--
Ordovician plagiogranite of Trinity complex	475 ±10 (7)	--	--
Ordovician plagioclase lherzolite of Trinity complex	472 ±32 (2)	<0.70246 (2)	+10.4 (2)
Ordovician gabbro of Trinity complex	470 (4,7)	--	--
Ordovician plagiogranite of Trinity complex	469 ±21 (7)	--	--
Ordovician or Silurian microgabbro of Trinity complex	435 ±21 (2)	0.70337-0.70376 (2)	+6.7 (2)
Ordovician or Silurian clinopyroxenite dike of Trinity complex	--	0.7029 (2)	+7.3 (2)
Silurian pegmatitic gabbro of Trinity complex (China Mtn.)	415 ±3 (8)	0.70471 (9)	--
Silurian pegmatitic trondhjemite dikes of Trinity complex	412 ±10 (7)	--	--
Devonian Copley Greenstone	--	0.7049-0.7050 (3)	--
Devonian Copley Greenstone	--	0.70345-0.71221 (1)	-3.8 - +8.0 (1)
Devonian Mule Mountain Stock	400 ±3 (3)	0.7041-0.7055 (3)	+7.0 - +7.5 (3)
Devonian Balaklala Rhyolite	--	0.7029-0.7064 (1)	--
Devonian Balaklala Rhyolite	--	0.70453-0.70586 (5)	+7.5 - +7.9 (5)
Devonian Balaklala Rhyolite	--	0.70750-0.71412 (1)	+4.7 - +7.9 (1)

Numbers in parentheses indicate reference cited. Isotopic ages are U-Pb zircon ages except for Sm-Nd isochron ages published by Jacobsen et al., 1984. References cited: 1) Brouxel et al., 1987; 2) Jacobsen et al., 1984; 3) Kistler et al., 1985; 4) Mattinson and Hopson, 1972; 5) Rouer et al., 1989; 6) Wallin, 1990; 7) Wallin et al., 1988; 8) Wallin et al., in press; 9) Wallin, Martin, and Lindsley-Griffin, unpub. data.

Ordovician mantle tectonites of the Trinity complex are cut by a plethora of younger, undeformed mafic dikes. Undeformed microgabbro and pyroxenite dikes yielded distinctly lower E_{Nd} values (+6.6 to +7.3) than those for the plagioclase lherzolite (Jacobsen et al., 1984). This difference indicates the younger dikes were not derived from melting of the plagioclase lherzolite and must have been derived from a distinctly different mantle source (Jacobsen et al., 1984). We believe these younger dikes probably formed during the Silurian magmatism discussed in the next section.

Boudier et al. (1989) interpreted ultramafic and mafic rocks of the Trinity complex as a single fragment of Ordovician oceanic lithosphere albeit with the caveat that it is an atypical example. The isotopic data reported by Jacobsen et al. (1984) and Wallin et al. (1990, in press) preclude such an interpretation. We concur with Jacobsen et al. (1984) who stated " ... we infer that some ophiolitic bodies represent many distinct stages of evolution of the depleted mantle covering substantial segments of geologic time. In general, it appears that we may not assume both a contemporaneous and cogenetic origin for the depleted and enriched parts of oceanic crust that are juxtaposed in a single geologic section". Available isotopic data indicate unequivocally that the Trinity complex is such a composite terrane.

SILURIAN

The isotopic age of Silurian igneous rocks in the Trinity complex was first determined by Mattinson and Hopson (1972) for dikes of pegmatitic gabbro and pegmatitic trondhjemite. These ages were recalculated using newly accepted decay constants to 412 ±10 Ma (Wallin et al., 1988). The magnitude of this Late Silurian magmatism has been recognized only recently.

The sizable China Mountain pluton consists of pegmatitic gabbro that has yielded a concordant U-Pb zircon age of 415 ±3 Ma (Wallin et al., in press). The China Mountain pluton is undeformed and stitches together deformed Cambrian and Ordovician rocks of the Trinity complex (see Fig. 2 of Lindsley-Griffin, this volume). Undeformed plutonic rocks similar in texture, composition, and structural context to those of the China Mountain pluton occur throughout the Trinity complex (Schwindinger and Anderson, 1987; Lindsley-Griffin, this volume). We interpret these undeformed rocks to be Silurian in age and postulate that the Late Silurian magmatism represents a widespread and significant tectonic event (Fig. 1). The nature of the tectonic event that produced these magmas is unknown and remains an important topic for future research.

The precise isotopic ages that are available suggest the existence of a magmatic gap of approximately 50 m.y. in the Trinity terrane between the Middle Ordovician and Late Silurian. An important question that remains is whether the Late Silurian magmatism represents a discrete tectonic event or whether it is simply a late stage of continuous magmatism that began in the Middle Ordovician and extended into the Early Devonian.

DEVONIAN

The immature Devonian volcanic arc represented by the West Shasta massive sulfide district has been studied extensively as a result of its economic importance (Albers and Bain, 1985). These Devonian volcanics have been interpreted variously as a submerged volcanic arc (Doe et al., 1985) or an immature island arc (Lapierre et al., 1985a). Watkins and Flory (1986) discussed evidence for the existence of islands during deposition of the Middle Devonian Kennett Formation.

Magmatism in the Redding terrane commenced with the eruption of basalt and basaltic andesite of the Copley Greenstone. The Copley has not been dated directly, therefore its age has been determined from cross-cutting relationships. In its uppermost portion, the Copley is interbedded with the Balaklala Rhyolite which in turn is intruded by trondhjemite of the Mule Mountain stock that has been dated at 400 Ma (Albers et al., 1981; Kistler et al., 1985). Furthermore, the Copley is overlain locally by the Middle Devonian Kennett Formation. Thus, the uppermost Copley must be Early Devonian and the Copley may be as old as Silurian near its base.

The interfingering of the Copley Greenstone and the Balaklala Rhyolite indicates that they are at least in part coeval. Both units exhibit remarkably similar REE patterns which suggests that they were derived from the same depleted source in the lower crust or mantle (Bence and Taylor, 1985). Barker et al. (1979) noted the similarity between the REE patterns of the Balaklala and those of the Mule Mountain stock and concluded that they were cogenetic. Lapierre et al. (1985b) argued that the development of bimodal volcanism in the Devonian volcanic arc sequence was the result of a phase of intra-arc extensional tectonics.

Isotopic studies of the petrogenesis of these Devonian volcanic rocks generally support the interpretation that they represent a submerged or immature island-arc system (Table 1). Sm-Nd data indicate that the Copley, the Balaklala, and the Mule Mountain stock were derived from a normal MORB-type depleted mantle source (Kistler et al., 1985; Brouxel et al., 1987; Rouer et al., 1989). Initial $^{87}Sr/^{86}Sr$ values calculated for these rocks exhibit a considerable range because of extensive hydrothermal alteration and thus are less reliable petrogenetic indicators than the Sm-Nd data (Table 1). Despite the elevated initial $^{87}Sr/^{86}Sr$ values of some of the altered rocks, the lowest values are consistent with such a tectonic setting (Kistler et al., 1985; Brouxel et al., 1987; Rouer et al., 1989).

Although this proposed tectonic setting is widely accepted, there are generally unappreciated bits of isotopic evidence that suggest minor contamination of the Devonian igneous rocks by pre-existing crustal material. Study of the Pb isotopic systematics of whole rocks and ores from the Devonian of the West Shasta district revealed some $^{207}Pb/^{204}Pb$ and $^{208}Pb/^{204}Pb$ values that were sufficiently radiogenic as to require contamination from either a considerable amount of subducted sediment or pre-existing crust (Doe et al., 1985). In addition, Brouxel et al. (1987) reported negative E_{Nd} values for some samples of the Copley Greenstone that may similarly reflect minor contamination. Kistler et al. (1985) reported a 575 ±75 Ma Sm-Nd isochron age for rocks of the Copley, Balaklala, and Mule Mountain stock which they interpreted as too old and thus geologically meaningless. In light of the new data from the Trinity-Yreka composite terrane discussed above, this Sm-Nd age raises the intriguing possibility that the Devonian magmas were derived from Cambrian igneous rocks in the basement of the island arc. If true, it implies the Cambrian terrane may have been of regional extent and that the basement of the island arc was complex. Taken a step further, it implies coevolution of the Redding terrane and the Trinity-Yreka composite terrane since the Devonian.

The Devonian record in the Trinity-Yreka composite terrane consists solely of dikes, pillow basalts, and flows. Although these basalts have not been dated isotopically, their age may be inferred from stratigraphic and intrusive contacts. Pillow basalts are interstratified with the Early Devonian Gazelle Formation (as redefined by Lindsley-Griffin et al., this volume) and also overlie melange that contains an Early Devonian fauna (Wallin, unpub. data; Lindsley-Griffin and Griffin, 1983). Some exposures of these basalts exhibit small swarms of sheeted dikes which suggest at least localized extension during their emplacement.

Available geochemical data permit the interpretation that these Devonian basalts are cogenetic with the Copley Greenstone to the south, but they are not compelling (Lindsley-Griffin, 1982; Brouxel et al., 1989). The volume of basalt in the Trinity-Yreka composite terrane is minor relative to that of the Copley, and the basalts do not represent a well-developed magmatic arc edifice. Given our model of the Gazelle Formation as a trench-related basin (Lindsley-Griffin et al., this volume) associated with eastward-dipping subduction represented by the central metamorphic belt (Peacock, 1987; Peacock and Norris, 1989), the basalts appear to represent near-trench volcanism. A carefully conducted comparison of Devonian volcanic rocks in the Trinity-Yreka composite terrane to those of the Redding terrane would be an important contribution to our understanding of Devonian paleogeography and tectonics.

ACKNOWLEDGEMENTS

We thank G.H. Girty for his helpful comments on the manuscript. Wallin has been supported by the University Research Council of the University of Nevada, Las Vegas.

REFERENCES CITED

Albers, J.P. and Bain, J.H.C., 1985, Regional setting and new information on some critical geologic features of the West Shasta District, California: Economic Geology, v. 80, p. 2072-2091.

Albers, J.P., Kistler, R.W. and Kwak, L., 1981, The Mule Mountain stock, an early Middle Devonian pluton in northern California: Isochron West, v. 31, p. 17.

Barker, F., Millard, H.T., Jr., and Knight, R.J., 1979, Reconnaissance geochemistry of Devonian island-arc volcanic and intrusive rocks, West Shasta district, California; in F. Barker, ed., Trondhjemites, Dacites, and Related Rocks: Amsterdam, Elsevier, p. 531-545.

Bence, A.E. and Taylor, B.E., 1985, Rare earth element systematics of West Shasta metavolcanic rocks: petrogenesis and hydrothermal alteration: Economic Geology, v. 80, p. 2164-2176.

Boudier, F., Le Sueur, E. and Nicolas, A., 1989, Structure of an atypical ophiolite: The Trinity complex, eastern Klamath Mountains, California: Geological Society of America Bulletin, v. 101, p. 820-833.

Brouxel, M., LaPierre, H., Michard, A., and Albarede, F., 1987, The deep layers of a Paleozoic arc: geochemistry of the Copley-Balaklala series, northern California: Earth and Planetary Science Letters, v. 85, p. 386-400.

Brouxel, M., Lecuyer, C. and LaPierre, H., 1989, Diversity of magma types in a lower Paleozoic island arc-marginal basin system (eastern Klamath Mountains, California, U.S.A.: Chemical Geology, v. 77, p. 251-274.

Burchfiel, B.C. and Davis, G.A., 1981, Triassic and Jurassic tectonic evolution of the Klamath Mountains-Sierra Nevada geologic terrane; in W.G. Ernst, ed., The Geotectonic Development of California, Rubey Volume 1: Prentice-Hall, Englewood Cliffs, New Jersey, p. 50-70.

Cotkin, S.J. and Armstrong, R.L., 1987, Rb/Sr age, geochemistry, and tectonic significance of blueschist from the schist of Skookum Gulch, eastern Klamath Mountains, California - introducing the Callahan event: Geological Society of America Abstracts with Programs v. 19, p. 367-368.

Davis, G.A., 1969, Tectonic correlations, Klamath Mountains and western Sierra Nevada, California: Geological Society of America Bulletin, v. 94, p. 3709-3716.

Doe, B.R., Delevaux, M.H. and Albers, 1985, The plumbotectonics of the West Shasta mining district, eastern Klamath Mountains, California: Economic Geology, v. 80, p. 2136-2148.

Goodge, J.W., 1990, Tectonic evolution of a coherent Late Triassic subduction complex, Stuart Fork terrane, Klamath Mountains, northern California: Geological Society of America Bulletin, v. 102, p. 86-101.

Hotz, P.E., 1977, Geology of the Yreka quadrangle, Siskiyou County, California: U.S. Geological Survey Bulletin 1436, 72 p.

----------, 1978, Geologic map of the Yreka quadrangle and parts of the Fort Jones, Etna, and China Mountain quadrangles, California: U.S. Geological Survey Open-File Report 78-12.

Irwin, W.P., 1985, Age and tectonics of plutonic belts in accreted terranes of the Klamath Mountains, California and Oregon; in D.G. Howell, ed., Tectonostratigraphic terranes of the Circum-Pacific Region, American Association of Petroleum Geologists, Earth Sciences Series Volume 1, Tulsa, Oklahoma, p. 187-199.

----------, 1989, Terranes of the Klamath Mountains, California and Oregon; in M.C. Blake and D.S. Harwood, eds., Tectonic evolution of northern California: International Geological Congress, 28th, Field Trip Guidebook T108, p. 19-32.

Jacobsen, S.B., Quick, J.E., and Wasserburg, G.E., 1984, A Nd and Sr isotopic study of the Trinity peridotite - Implications for mantle evolution: Earth and Planetary Science Letters, v. 68, p. 361-378.

Kistler, R.W., McKee, E.H., Futa, K., Peterman, Z.E. and Zartman, R.E., 1985, A reconnaissance Rb-Sr, Sm-Nd, U-Pb, and K-Ar study of some host rocks and ore minerals in the West Shasta Cu-Zn district, California: Economic Geology, v. 80, p. 2128-2135.

Lanphere, M.A., Irwin, W.P., and Hotz, P.E., 1968, Isotopic age of the Nevadan orogeny and older plutonic and metamorphic events in the Klamath Mountains, California: Geological Society of America Bulletin, v. 79, p. 1027-1052.

LaPierre, H., Albarede, F., Albers, J., Cabanis, B. and Coulon, C., 1985a, Early Devonian volcanism in the eastern Klamath Mountains, California: Evidence for an immature island arc: Canadian Journal of Earth Sciences, v. 22, p. 214-227.

LaPierre, H., Cabanis, B., Coulon, C., Brouxel, M. and Albarede, F., 1985b, Geodynamic setting of Early Devonian Kuroko-type sulfide deposits in the eastern Klamath Mountains (northern California) inferred by the petrological and geochemical characteristics of the associated island-arc volcanic rocks: Economic Geology, v. 80, p. 2100-2113.

Lindsley-Griffin, N., 1977, Paleogeographic implications of ophiolites: The Ordovician Trinity complex, Klamath Mountains, California: in J.H. Stewart, C.H. Stevens, and A.E. Fritsche, eds., Paleozoic Paleogeography of the Western United States: Society of Economic Paleontologists and Mineralogists, Pacific Section, Pacific Coast Paleozoic Paleogeography Symposium, v. 1, p. 409-420.

--------------------, 1982, Structure, stratigraphy, petrology, and regional relationships of the Trinity ophiolite, eastern Klamath Mountains, California [unpublished Ph.D. dissertation]: University of California, Davis, 445 p.

Lindsley-Griffin, N. and Griffin, J.R., 1983, The Trinity terrane - an early Paleozoic microplate assemblage; in C.H. Stevens, ed., Pre-Jurassic rocks in western North American suspect terranes: Pacific Section, Society of Economic Paleontologists and Mineralogists, Los Angeles, California, p. 63-76.

Lipman, P. W., 1964, Structure and origin of an ultramafic pluton in the Klamath Mountains, California: American Journal of Science, v. 262, p. 199-222.

Mattinson, J.M. and Hopson, C.A., 1972, Paleozoic ophiolitic complexes in Washington and northern California: Carnegie Institute of Washington Yearbook, v. 71, p. 578-583.

588

Miller, M.M., 1989, Intra-arc sedimentation and tectonism: Late Paleozoic evolution of the eastern Klamath terrane, California: Geological Society of America Bulletin, v. 101, p. 170-187.

Peacock, S.M., 1987, Serpentinization and infiltration metasomatism in the Trinity peridotite, Klamath province, northern California: implications for subduction zones: Contributions to Mineralogy and Petrology, v. 95, p. 55-70.

Peacock, S.M. and Norris, P.J., 1989, Metamorphic evolution of the Central Metamorphic Belt, Klamath Province, California: an inverted metamorphic gradient beneath the Trinity peridotite: Journal of Metamorphic Geology, v. 7, p. 191-209.

Potter, A.W., Hotz, P.E., and Rohr, D.M., 1977, Stratigraphy and inferred tectonic framework of lower Paleozoic rocks in the eastern Klamath mountains, northern California: in J.H. Stewart, C.H. Stevens, and A.E. Fritsche, eds., Paleozoic paleogeography of the western United States: Pacific Section, Society of Economic Paleontologists and Mineralogists, Paleozoic Paleogeography Symposium Volume 1, 421-440.

Potter, A.W., Hotz, P.E., and Lanphere, M., 1981, Evidence of Ordovician-Silurian subduction and Silurian or older igneous units of possible magmatic arc origin, eastern Klamath Mountains, northern California: Geological Society of America Abstracts with Programs, v. 13, p. 101.

Quick, J.E., 1981a, The origin and significance of large, tabular dunite bodies in the Trinity Peridotite, northern California: Contributions to Mineralogy and Petrology, v. 78, p. 413-422.

Quick, J.E., 1981b, Petrology and petrogenesis of the Trinity peridotite, an upper mantle diapir in the eastern Klamath Mountains, California: Journal of Geophysical Research, v. 86, B12, p. 11837-11863.

Rouer, O., LaPierre, H., Mascle, G., Coulon, C., and Albers, J., 1989, Geodynamic implications of Devonian silicic arc magmatism in the Sierra Nevada and Klamath Mountains, California: Geology, v. 17, p. 177-180.

Schweickert, R.A. and Snyder, W.S., 1981, Paleozoic plate tectonics of the Sierra Nevada and adjacent regions; in W.G. Ernst, ed., The Geotectonic Development of California: Prentice Hall, Englewood Cliffs, New Jersey, p. 182-201.

Schwindinger, K.R. and Anderson, A.T., Jr., 1987, Probable low-pressure intrusion of gabbro into serpentinized peridotite, northern California: Geological Society of America Bulletin, v. 98, p. 364-372.

Shaw, H.F., Chen, J.H., Saleeby, J.B., and Wasserburg, G.J., 1987, Nd-Sr-Pb systematics and age of the Kings River ophiolite, California: Implications for depleted mantle evolution: Contributions to Mineralogy and Petrology, v. 96, p. 281-290.

Wagner, D.L. and Saucedo, G.J., 1987, Geologic map of the Weed quadrangle: California Division of Mines and Geology Regional Geologic Map Series N. 4A, 1:250,000.

Wallin, E.T., 1989, Provenance of lower Paleozoic sandstones in the eastern Klamath Mountains and the Roberts Mountains allochthon, California and Nevada [unpublished Ph.D. thesis]: University of Kansas, Lawrence, 152 p.

Wallin, E.T., 1990, Petrogenetic and tectonic significance of xenocrystic Precambrian zircon in Lower Cambrian tonalite, eastern Klamath Mountains, California: Geology, v. 18, p. 1057-1060.

Wallin, E.T., Lindsley-Griffin, N., Griffin, J.R., and Potter, A.W., in press, Pre-Late Silurian amalgamation of Cambrian and Ordovician terranes in the Trinity complex, eastern Klamath Mountains, California: Geology.

Wallin, E.T., Mattinson, J.M. and Potter, A.W., 1988, Early Paleozoic magmatic events in the eastern Klamath Mountains, northern California: Geology, v. 16, p. 144-148.

Watkins, R. and Flory, R.A., 1986, Island arc sedimentation in the Middle Devonian Kennett Formation, eastern Klamath Mountains, California: Journal of Geology, v. 94, p. 753-761.

Wright, J.E., 1982, Permo-Triassic accretionary subduction complex, southwestern Klamath Mountains, northern California: Journal of Geophysical Research, v. 87, p. 3805-3818.

THE TRINITY COMPLEX: A POLYGENETIC OPHIOLITIC ASSEMBLAGE

Nancy Lindsley-Griffin
Department of Geology, 214 Bessey Hall
University of Nebraska, Lincoln, NE 68588

ABSTRACT

Rocks within the Trinity terrane formed over a time span of nearly 200 million years, from earliest Cambrian (565 Ma) to Middle Devonian (about 380 Ma). The Trinity complex is composed of three main components: 1) the Lower Cambrian Trinity ophiolite, which consists of a partially dismembered sequence of spilitic basalt, plagiogranite, metagabbro, and harzburgite; 2) an ophiolitic assemblage of Ordovician age that comprise the main mass of Trinity peridotite plus minor fragments of plagiogranite and gabbro; and 3) pegmatitic gabbros, pyroxenites, and related rocks that intruded the Cambrian and Ordovician rocks in the Late Silurian. Middle Devonian? basaltic dike swarms that penetrate the Trinity terrane are not part of the Trinity complex because they postdate its amalgamation with the Yreka terrane. During formation of this early Paleozoic oceanic terrane, repeated magmatism was interspersed with deformational events that juxtaposed blocks of different ages along wide, vertical fault zones, some of which were later intruded by Silurian plutons. From Late Silurian to Early Devonian, the Trinity complex formed the leading edge of a tectonic plate that was overriding the subducting slab represented by the central metamorphic terrane. By Middle Devonian, the Trinity complex had amalgamated with the Yreka and central metamorphic terranes. This magmatic and structural history demonstrates the Trinity complex is not a single slice of normal oceanic lithosphere formed at a mid-ocean ridge, but instead consists of separate and distinct components that were tectonically amalgamated before the Silurian magmatism. Thus, the Trinity composite terrane represents atypical oceanic lithosphere such as might be found in an oceanic plateau, a transform fault zone, or a long-lived back-arc basin.

INTRODUCTION

The recent recognition of Cambrian and Silurian rocks within the ophiolitic Trinity complex (Wallin *et al.*, 1988; 1990), in addition to the previously known Ordovician rocks, indicates that it is not a single slice of lithosphere formed at a typical oceanic ridge. This paper summarizes data on the Trinity complex in light of the new isotopic ages and recent field work, and presents a new model for its origin and history.

Regional Geologic Setting

The lower Paleozoic Trinity terrane is a fault-bounded, subhorizontal sheet of mafic and ultramafic rocks, over 1900 km^2 in extent, within the eastern Klamath terrane or eastern Klamath plate of Irwin (1981, 1985). As shown in Figure 1, it lies between the Yreka and central metamorphic terranes to the northwest and the Redding terrane to the southeast.

The Yreka terrane consists of fault-bounded sheets of lower Paleozoic marine sedimentary and metasedimentary rocks and melange that have been thrust over the Trinity complex (Hotz, 1977, 1978; Lindsley-Griffin and Griffin, 1983). Geophysical data suggest the Yreka terrane overlies a basement of dense rocks of probable ultra-mafic composition, generally interpreted to be a subsurface continuation of the Trinity complex (Irwin and Bath, 1962; LaFehr, 1966; Griscom, 1977; Fuis and Zucca, 1984).

West of the Yreka terrane, the central metamorphic terrane (Fig. 1) consists of an elongate, steeply dipping ultramafic slab, with minor amphibolite and metasedimentary melange, of approximately 375 km^2 extent. The north end of the central metamorphic terrane bounds the Yreka terrane and its south end abuts the Trinity terrane. To the south, the central metamorphic terrane widens and consists of Devonian schists (Irwin, 1981).

The Redding terrane (Fig. 1) comprises a sequence of marine metavolcanics and metasediments that ranges in age from Early Devonian to Middle Jurassic; it is roughly homoclinal with a regional dip to the east (Irwin, 1981). Although rocks of the Redding terrane are generally regarded as having been deposited on a basement consisting of Trinity complex (Irwin, 1981), the contact between Redding terrane and Trinity terrane is now a fault.

Ophiolites

Formerly regarded as igneous intrusions, most mafic-ultramafic complexes like the Trinity complex now are considered to represent fragments of oceanic lithosphere that have been tectonically accreted to the continental margin. Where a complete succession of peridotite, gabbro, diabase, and basalt is preserved, the assemblage may be called an ophiolite (Anonymous, 1972;

In Cooper, J.D., and Stevens, C.H., eds., 1991, **Paleozoic Paleogeography of the Western United States-II:** Pacific Section SEPM, Vol. 67, p. 589-607.

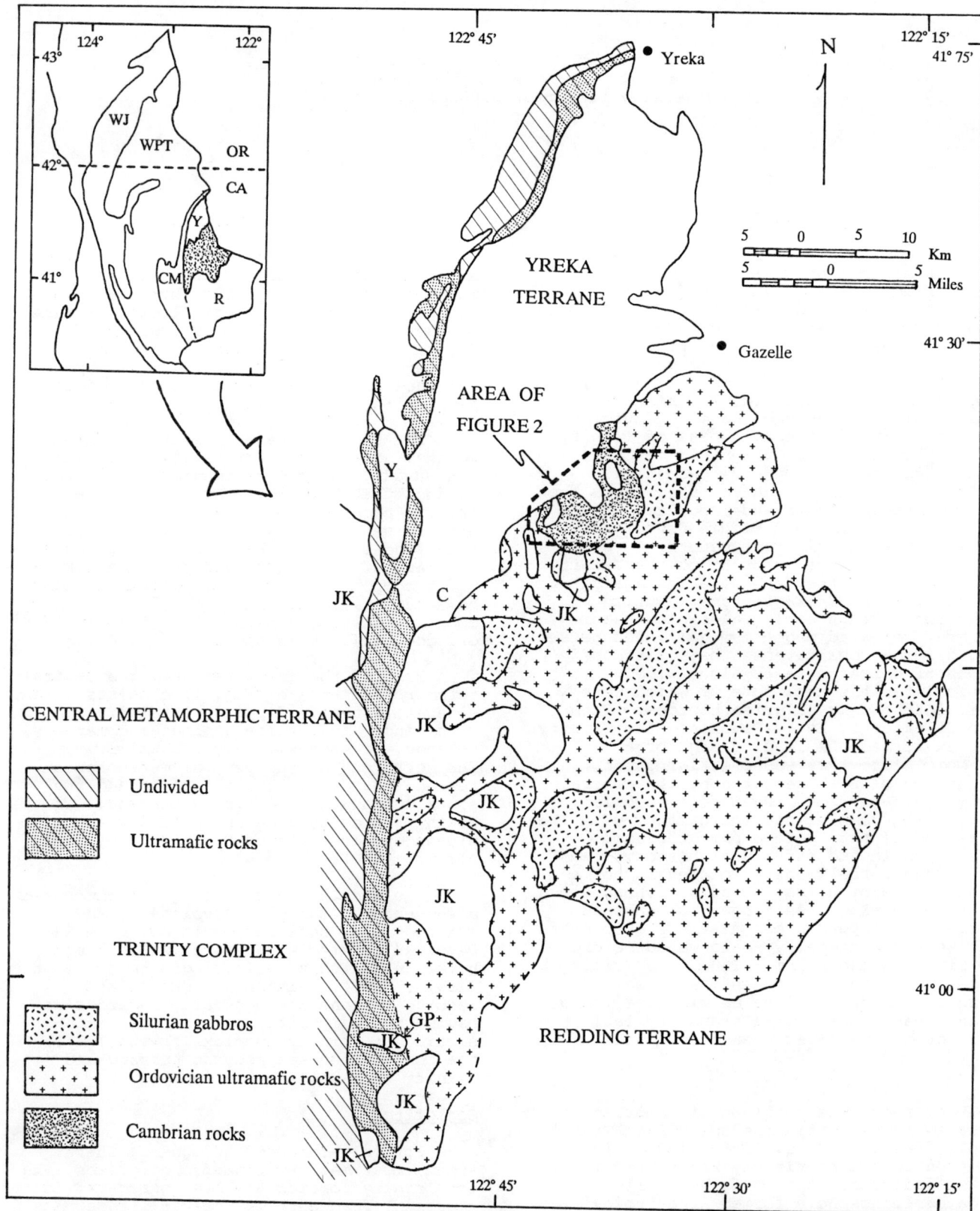

Figure 1. Geologic map of Trinity terrane. Inset shows relationship of Trinity terrane (stippled) to other terranes of the Klamath Mountains. Symbols include: R, Redding terrane; Y, Yreka terrane; CM, central metamorphic terrane; WPT, western Paleozoic and Triassic terrane; WJ, western Jurassic terrane; JK, Jurassic and Cretaceous plutons; GP, Gibson Peak. Geology modified after Lindsley-Griffin and Griffin (1983), Strand (1962), Wagner and Saucedo (1987).

Coleman, 1977). Use of the term ophiolite does not imply a specific genesis, because ophiolitic assemblages may form in a wide variety of tectonic settings, ranging from mid-ocean ridge to volcanic arc to the edge of a continental margin (Coleman, 1977; Moores, 1982). Thus, it is incorrect to use the term ophiolite as a synonym for an oceanic ridge assemblage; the origin of each ophiolite must be interpreted individually.

The Trinity complex is ophiolitic in nature because it consists of the appropriate mafic and ultramafic rocks, but the ophiolite stratigraphy is preserved only locally in fault blocks along the northwestern margin of the terrane. The Trinity complex differs from the ideal ophiolite of Coleman (1977) in that it formed over a time range of nearly 200 Ma; its older components are faulted together and its younger components intrude the older ones. The precise origin of the various components within the Trinity complex is at present poorly understood. The atypical character of this ophiolitic complex was recognized by Lindsley-Griffin (1975, 1977a) and by Quick (1981b).

Terrane or Complex?

A complex is a body of rocks "so intricately involved...that the rocks cannot be readily differentiated in mapping" (Bates and Jackson, 1987, p. 135). The Trinity complex consists of three major components: 1) the Trinity ophiolite, a sequence of Cambrian metagabbro and plagiogranite, and possibly Cambrian spilitic basalt, faulted against Cambrian? harzburgite; 2) ophiolitic Ordovician peridotites and minor Ordovician plagiogranites and gabbros; and 3) Silurian pegmatitic gabbros and related rocks. The numerous small mafic dikes of unknown age that cut the mafic-ultramafic rocks are part of the Trinity complex but are not part of the Trinity ophiolite.

A terrane is "a fault-bounded body of rock of regional extent, characterized by a geologic history different from that of contiguous terranes" (Bates and Jackson, 1987, p. 679). In this paper, the term is used for allochthonous bodies accreted to the continent at an active margin (Irwin, 1972; Jones et al., 1983). The Trinity terrane consists of igneous and metaigneous rocks of Cambrian, Ordovician, and Silurian age that appear to have a unique history not shared by rocks in adjoining terranes until after the end of the Silurian. Thus, as used in this paper, Trinity terrane is virtually synonymous with Trinity complex. The Yreka terrane includes igneous, metamorphic, and sedimentary rocks that range in age from Early Cambrian to Middle Devonian, and whose geologic history is different from that of the Trinity terrane. The Trinity-Yreka composite terrane includes both the Trinity and the Yreka terranes, along with rocks of Middle Devonian and later age that intruded or were deposited on them after their juxtaposition in Early (to Middle?) Devonian.

Previous Work

The name "Trinity" was first used by Hershey (1901) when he defined seven formations of pre-Cretaceous age in the southeastern Klamath Mountains; he named the "serpentine" of this region the Trinity Formation. Hinds (1932) applied the name "Trinity Alps batholith" to serpentinites and gabbroic rocks of the eastern Klamaths, and tentatively assigned a Jurassic age to the body based on the ages of similar rocks elsewhere in California. Since then, it has been variously referred to as "Trinity ultramafic pluton" (Lipman, 1964; Davis, 1966, 1968, 1969; Hotz, 1971), "Trinity ultramafic intrusion" (Davis et al., 1965), "Trinity ultramafic sheet" (Lanphere et al., 1968; Irwin, 1977, 1981), "Trinity complex" (Mattinson and Hopson, 1972a; Goullaud, 1977), and "Trinity ophiolite" (Hopson and Mattinson, 1973; Lindsley-Griffin, 1973). The term "Trinity complex" is most appropriate, based on our current understanding of the body.

Brewer (1955) first recognized the close relationship between serpentinized peridotite and hornblende diorite, hornblende gabbro, and dikes in the China Mountain area (Fig. 2). Recognition of the Trinity complex as an ophiolitic assemblage came in the 1970's (Mattinson and Hopson, 1972a, 1972b; Irwin, 1973; Lindsley-Griffin, 1973) and resulted in a number of detailed field studies of the mafic and ultramafic rocks (Goullaud, 1977; Lindsley-Griffin, 1977a, 1977b, 1982; Throckmorton, 1978; Quick, 1981a, 1981b, 1981c; Schwindinger and Anderson, 1987).

Regional relationships between the Trinity complex and surrounding areas have been extensively discussed in the literature. The contributions of W.P. Irwin (1960, 1964, 1966, 1973, 1977, 1981, 1985) are of special note because he recognized that the Trinity terrane is the oldest part of the Klamath Mountains, and that the various subprovinces or terranes become progressively younger to the west and represent sequential accretion of terranes against the Trinity terrane.

Isotopic Dates

Isotopic dates have been reported for the Trinity complex by a number of authors, but the significance of many is uncertain because the samples were collected without the benefit of detailed structural maps. Newly published isotopic data (Wallin et al., 1988, and this volume), combined with my detailed geologic and structural surveys of the sample sites during the summers of 1989 and 1990, permit new insights into the origin and history of the Trinity complex. Sample sites of previously published isotopic dates (Mattinson and Hopson, 1972b; Jacobsen et al., 1984; Wallin et al., 1988) were investigated in order to evaluate their relationship to the Trinity complex, but some could not be located. Wallin et al. (this volume) summarize isotopic data for

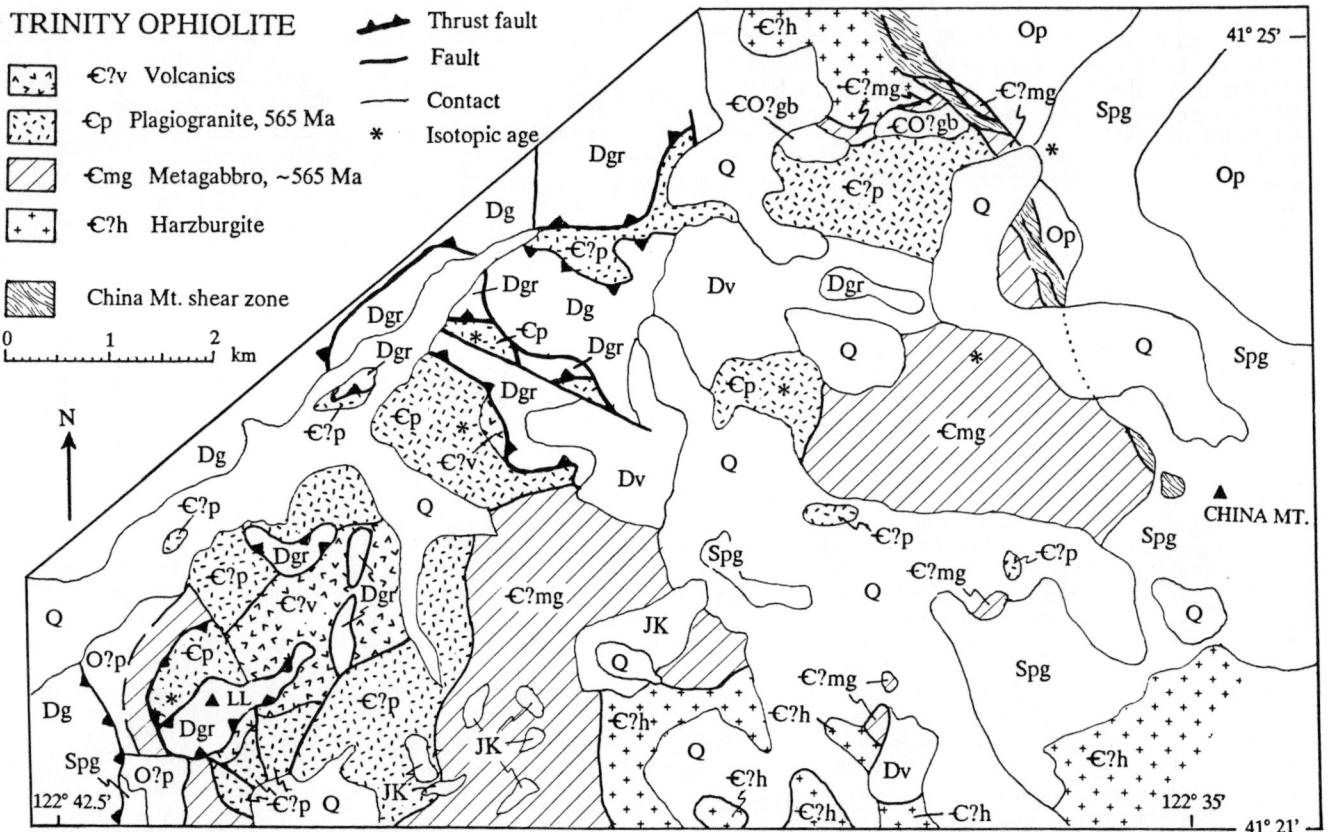

Figure 2. Geologic map of northwestern edge of Trinity complex, showing Cambrian rocks of the Trinity ophiolite. Other units of the Trinity complex: €O?gb, gabbros of Cambrian or Ordovician age; Op, Ordovician peridotite; Spg, Silurian pegmatitic gabbros. Yreka terrane units: Dg, Gazelle Formation; Dgr, Gregg Ranch Complex. Dv, Devonian volcanic rocks; JK, Jurassic-Cretaceous intrusions; Q, Quaternary sediments. LL, Lovers Leap.

the Trinity complex and other igneous rocks in the eastern Klamath region.

The difficulty of interpreting isotopic dates from the eastern Klamaths is illustrated by geological relationships at Lovers Leap (Fig. 2). Here, plagiogranites have been found to be both Ordovician (Mattinson and Hopson, 1972b; Wallin et al., 1988) and Cambrian (Wallin et al., 1988). The Cambrian sample site lies structurally above at least one of the Ordovician sites, which suggests a fault, but poor exposure prevents a clear understanding of the relationship. Mattinson and Hopson (1972b) describe one of their Ordovician sites as within a fault slice. It is probable that Lovers Leap consists of a tectonic melange of the Cambrian and Ordovician basement rocks overlain by polygenetic melange of the Lower Devonian Gregg Ranch Complex (Fig. 2; see Lindsley-Griffin et al., this volume). Both melanges are penetrated by several different ages of dikes and dike swarms, which yield at least one Late Jurassic date (Brouxel, Lapierre, Zimmerman, 1989). Other dikes at Lovers Leap are lithologically similar to rocks known to be Silurian, Devonian, and Cretaceous.

The Trinity complex is of early Paleozoic age, but is intruded or overlain by igneous rocks of middle Paleozoic to late Mesozoic age. Thus, within the areal extent of the Trinity terrane (Figs. 1, 2), there may be found gabbroic rocks of Cambrian, Ordovician, Silurian, Devonian, and probable Jura-Cretaceous age. Plagiogranites, granitoids and other intermediate rock types have been dated as Cambrian, Ordovician, and Cretaceous; geologic relationships suggest that some intermediate composition rocks in the terrane are also Silurian, Jurassic, and possibly Devonian.

The only isotopic date available for ultramafic rocks of the complex is Ordovician (Jacobsen et al., 1984). However, at least five distinct map units of ultramafic rocks can be recognized within the Trinity complex; geologic relationships with gabbros whose ages are known suggest that both Cambrian and Silurian ultramafic rocks may be present in addition to the Ordovician unit. Basaltic rocks of two ages are present within the boundaries of the Trinity terrane. Although neither volcanic unit has been isotopically dated, geological restrictions on their age are discussed below.

Because each major rock type within the Trinity terrane (ultramafics, gabbros, plagiogranites, basalts) includes several different ages of similar rock compositions, age and geochemical data not linked to a detailed geologic map are of limited utility. Interpretations based on geochemical data from melange zones, such as the melanges and dike complexes at Lovers Leap and Gregg Ranch (Brouxel and Lapierre, 1988; Brouxel, Lecuyer, Lapierre, 1989), are misleading because they do not correctly distinguish between suites of different ages and origins.

CAMBRIAN AND ORDOVICIAN OPHIOLITIC ROCKS

Cambrian Trinity Ophiolite

The Trinity ophiolite consists of a partially dismembered sequence of Cambrian metagabbros and plagiogranites and associated Cambrian? volcanic and ultramafic rocks that occur along the northwestern edge of the Trinity complex (Figs. 1, 2). This sequence, previously considered to be Ordovician (Lindsley-Griffin, 1977a, 1977b, 1982), has yielded isotopic ages of 565 Ma for plagiogranites of the dike complex unit and for the "amphibolitic gabbro" unit, a deformed metagabbro (Wallin et al., 1988, 1990, and this volume). Isotopic dates from these two units confirm that much of the ophiolite sequence is actually Cambrian in age. Geologic relationships suggest the Cambrian ophiolite includes harzburgite, metagabbro, plagiogranite, and spilitic pillow basalts. These highly deformed rocks are faulted against less deformed gabbro and peridotite fragments of Ordovician age, and are intruded by dikes and small stocks of Cambro-Ordovician (?), Silurian, Devonian, Jurassic, and Cretaceous age.

Wallin (1990, p. 1057) concluded that "Lower Cambrian plutonic rocks in the eastern Klamath Mountains are probably not ophiolitic in origin." However, he was using the term "ophiolitic" as a synonym for typical oceanic crust, rather than as generally defined and as used in this paper (Anonymous, 1972; Coleman, 1977; Bates and Jackson, 1987). Although Wallin (1990) found xenocrystic Precambrian zircon in tonalite blocks within melange of the Yreka terrane, no evidence has yet been found of Precambrian zircon in rocks of the Trinity complex.

Cambrian Ophiolite Stratigraphy

Stratigraphic relationships within the Cambrian ophiolite sequence are poorly preserved, and most contacts within this assemblage appear to be faults. East of Lovers Leap, a stratigraphic sequence from metagabbro to plagiogranite to basaltic lavas appears to be undisrupted by faulting, but no isotopic ages have been obtained for the metagabbro and exposures are poor. This metagabbro (Fig. 3) is correlated with the Cambrian ophiolite because its structural style is similar to the structural style of the isotopically dated Cambrian metagabbro.

The stratigraphic relationship between Cambrian plagiogranites and the overlying Cambrian? basaltic dikes and lavas is well established by field relationships and map pattern (Fig. 2). The lavas overlie the plagiogranites; basaltic feeder dikes intrude the plagiogranites and locally can be traced upward into the lavas (Lindsley-Griffin, 1977a, 1977b, 1982). It is significant that no dikes of the Cambrian? basalts cut harzburgite or Cambrian metagabbro. Thus, the Cambrian? basalts must have been erupted through the Cambrian plagiogranites before the ophiolite sequence was dismembered. The assignment of these basalts to the Cambrian ophiolite also is supported by their extensive deformation. The similarity of their structural style to that of the Cambrian metagabbro and plagiogranite units suggests they have shared the same structural history.

Sheeted dike complexes related to eruption of the Devonian basalts are not part of the Cambrian Trinity ophiolite, although such dikes intrude the Cambrian rocks at many localities. Although these sheeted dike complexes and volcanics have been included in the "Trinity ophiolite" by some authors (Brouxel and Lapierre, 1988; Brouxel et al., 1988; Brouxel, Lecuyer, Lapierre, 1989), they post-date the juxtaposition of both the Trinity and the Yreka terranes and therefore cannot be part of either the Trinity ophiolite or the Trinity complex. Ultramafic and mafic dikes related to the Silurian pegmatitic gabbros are not part of the Cambrian Trinity ophiolite, but they are part of the Trinity complex.

Harzburgite

No isotopic dates have been obtained for the fault-bounded block of harzburgite that I correlate with the Trinity ophiolite. It is tentatively assigned to the Cambrian ophiolite because: 1) it occurs along the northwestern edge of the Trinity terrane near the Cambrian blocks (Fig. 2); 2) its textures and structures suggest that it is more intensely deformed than most of the Ordovician peridotites and thus is likely to be pre-Ordovician; 3) it exhibits a structural style similar to that of rocks known to be Cambrian; and 4) unlike the Ordovician peridotite, it lacks feldspar clots formed by partial fusion, suggesting a different history.

Composition

Harzburgite is peridotite composed of large orthopyroxene grains in a matrix of olivine, with accessory amounts of chrome spinel and magnetite (ol + opx + sp + mag) and trace amounts of clinopyroxene (cpx). The Trinity harzburgite unit contains an average of 30%-40% orthopyroxene (enstatite), but the orthopyroxene content ranges from 10% to 75%. Layers and lenses of dunite are present where orthopyroxene content is low, and local concentrations of spinels form podiform chromite bodies and thin layers. The harzburgite is extensively

Figure 3. Cambrian? metagabbro unit of the Trinity ophiolite. Undated schistose hornblende gabbro and amphibolite schlieren resemble the 570 m.y. Cambrian metagabbro.

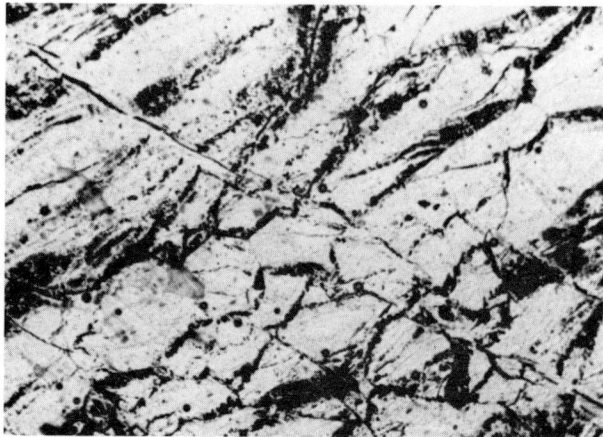

Figure 6. Schlieren of amphibolite in Cambrian metagabbro of the Trinity ophiolite. Pencil is 10.5 cm long.

Figure 4. Cambrian? harzburgite of the Trinity ophiolite. Photomicrograph of planar fabric defined by olivine "ghosts" outlined by magnetite, and trains of dismembered orthopyroxenes. Width of view is 3mm.

Figure 7. Mylonitic Cambrian metagabbro of the Trinity ophiolite; individual layers have been ductilely deformed into boudins.

Figure 5. Hornblende gneiss and amphibolite of the Cambrian metagabbro, Trinity ophiolite. Scale is 15 cm.

Figure 8. Cambrian plagiogranite of the Trinity ophiolite. Photomicrograph shows quartz and plagioclase boudins in fragmental matrix of quartz, plagioclase, and chlorite after hornblende. View is 10mm across.

serpentinized (50%-95%); the olivines are completely destroyed and the orthopyroxenes extensively altered with serpentine cores.

Textures and Structures

The original texture of the harzburgite is recognizable because secondary magnetite grains that result from the serpentinization process outline the olivine grains. The olivine "ghosts", magnetite, and chrome spinel grains are elongated into trains that produce a planar fabric faintly visible in outcrop and readily visible in thin section (Fig. 4). The large flattened orthopyroxene oikocrysts are aligned in trains parallel to this fabric; typically grains within the same train have nearly the same optical orientation, suggesting they once may have been part of a single large grain which has been dismembered with little or no rotation. Strain lamellae within orthopyroxene grains also indicate the extensive deformation that characterizes this unit.

In addition to the mineral foliation, compositional banding or layering is locally present, although much of the harzburgite unit is massive. The layers, typically 20-50 cm thick, are caused by changes in the proportions of pyroxene to olivine. Compositional layering is subparallel to the mineral foliation where both can be observed in the same outcrop.

Metagabbro

The Cambrian age of 556-579 Ma (Wallin *et al.*, 1988; 1990; and this volume) was determined for the large fault-bounded block of metagabbro that comprises the western slope of China Mountain (Fig. 2). This fault block is bounded on the east by a broad, steep fault zone; both the fault zone and the metagabbro are intruded by undeformed Silurian pegmatitic gabbro of the China Mountain pluton.

Composition

The Cambrian metagabbro consists of amphibolite and amphibole gneiss, locally interlayered with leucocratic gneiss and clinopyroxenite. Less deformed parts of the unit are uralitized layered gabbro. Amphibolite layers consist of up to 95% hornblende, whereas leucocratic layers may contain as little as 20% hornblende. Leucocratic layers contain albite and locally common quartz. The clinopyroxenite consists of about 97% augitic diopside and a few percent magnetite. The metagabbro unit is hydrothermally altered; epidote coats joints and fracture surfaces and occurs as veins up to 5 cm thick.

Textures and Structures

The metagabbro is strongly deformed. Although it contains some zones of nearly undeformed layered gabbro with graded bedding and relict cumulate textures, much of the metagabbro is characterized by ductilely necked and thinned bands and schlieren derived presumably from an original cumulate layering (Figs. 5, 6). Near the boundaries of the Cambrian metagabbro block, the rock is mylonitized, with more leucocratic layers pulled apart into boudin (Fig. 7).

In clinopyroxenite, textural evidence supports the interpretation of a history of intense deformation. The recrystallized matrix consists of small, clean clinopyroxene grains with a uniformly even grain size and equiangular grain boundaries, that surround sparse relict clinopyroxene grains. The large relict grains exhibit internal strain lamellae and fractures, and are aligned in trains with nearly the same optical orientation. This suggests they are dismembered parts of the same original grain.

Along fault contacts between metagabbro and harzburgite (Fig. 2), blocks of ductilely deformed clinopyroxenite interlayered with amphibolite occur in serpentinite and serpentine schist matrix. The larger blocks of clinopyroxenite are locally intercalated with boudins and stringers of amphibolite and amphibolitic metagabbro, which suggests the clinopyroxenite may have originated as the basal ultramafic part of a cumulate gabbro.

Plagiogranite

Cambrian isotopic ages of 565 Ma (Wallin *et al.*, 1988; 1990; and this volume) have been obtained for most of the fault-bounded blocks of plagiogranite that form a substrate for basaltic dikes and lavas of the Trinity ophiolite sequence (Fig. 2).

Composition

The Cambrian plagiogranites consist of plagioclase feldspar (40%-60%), quartz (15%-35%), and altered mafic minerals (trace to 25%). Proportions of these minerals are highly variable, hence the rocks range from quartz-rich to quartz-poor, and from leucocratic to locally mafic. The feldspar is extensively saussuritized and typically consists of a felted mass of zoisite, actinolite, and chlorite. Unaltered plagioclase ranges in composition from An_{25} to An_{40}, well within the reported range for oceanic plagiogranites (Coleman, 1977). No alkali feldspar has been observed in these rocks. The mafic minerals were probably hornblende originally; some of the plagiogranites still contain up to 20% recognizable hornblende. However, in most samples the mafic minerals have been altered to chlorite and actinolite.

Textures and Structures

Like the metagabbro unit, the plagiogranites are intensely deformed. In these rocks, evidence of extensive ductile to brittle deformation is seen in stretched and dismembered clots of feldspar and quartz grains, and the streaked foliated appearance of the rock in outcrop. In thin section

(Fig. 8), clots of quartz and plagioclase grains exhibit ductile stretching and boudinage, overprinted by brittle tension fractures oriented perpendicular to the stretching direction. Large grains are broken into subgrains that have been rotated slightly relative to each other, with small recrystallized grains along subgrain boundaries. The deformed quartzofeldspathic clots are surrounded by a fragmental matrix composed of quartz, plagioclase, and altered hornblende. Some of the larger matrix fragments would fit back together if the intervening matrix were removed (Fig. 8). The finest grain sizes within the matrix exhibit the clean, fresh appearance and equiangular grain junctions typical of recrystallization textures. This evidence suggests the intense ductile deformation was followed by brittle deformation and partial recrystallization. If so, the plagiogranites experienced long-continued deformation under changing P-T conditions.

Spilitic Dikes and Pillow Lavas

In several localities, altered basaltic dikes penetrate the Cambrian plagiogranites and can be traced stratigraphically upward into Cambrian? altered basaltic lavas. Some dikes are sheared into phacoids with plagiogranite wrapping around them, and the volcanic rocks are also fractured and sheared in appearance, although spilitic pillow lavas are locally preserved at Lovers Leap. Their structural style and their intimate spatial relationship to the Cambrian plagiogranites support the assignment of these basaltic rocks to the Cambrian ophiolite sequence. Dikes of microgabbro and plagiogranite appear to be associated with the spilitic dikes where they intrude the plagiogranite. The lack of well developed sheeted structure in the dikes is evidence that the complex did not form at a typical spreading ridge.

As noted by Lindsley-Griffin (1977a, 1982), plagiogranites are spatially associated with these Cambrian? basaltic volcanics at Lovers Leap, Gregg Ranch, and Crater Creek (Figs. 1, 2). At all three localities the ophiolitic rocks are overlain by a thrust sheet of melange containing Middle Ordovician through Middle Devonian blocks (Gregg Ranch Complex of Lindsley-Griffin et al., this volume). At Gregg Ranch and Crater Creek the plagiogranite-basalt-melange sequence also is overlain by a second, post-Lower Devonian basalt sequence, a circumstance which has caused much confusion in the literature. The Devonian basalt (discussed below) is not part of the Trinity ophiolite, nor is it part of the Trinity complex; instead, it overlaps the Trinity-Yreka composite terrane.

Composition

The Cambrian? volcanic unit consists of basaltic spilites and keratophyres. The most typical rock type contains 1-5 mm euhedral to subhedral phenocrysts of feld-spar (partially saussuritized albite) and sparse hornblende in an aphanitic matrix. Phenocrysts comprise 5%-50% of the rock. Rocks of this unit are extensively altered to an assemblage of albite, saussurite, red and green chert, calcite, chlorite, and epidote. The rocks also contain ubiquitous pyrite, chalcopyrite, bornite, and a variety of iron and copper oxides. Calcite veins are common. Locally, so much calcite has replaced the original matrix that the rock fizzes on contact with HCl: a useful technique for distinguishing it from the younger pillow lavas of Devonian age.

The REE and trace element data that are unquestionably from the Cambrian? basalt unit (Lindsley-Griffin, 1982) exhibit nearly flat REE patterns that are slightly depleted in LREE, similar to the patterns exhibited by oceanic tholeiites. The Ti-Zr values fall within the overlap field of Pearce and Cann (1971, 1973), near the average value for ophiolitic extrusives (Coleman, 1977).

Brouxel (Brouxel et al., 1988; Brouxel, Lecuyer, Lapierre, 1989) analyzed samples from dikes and volcanics at Lovers Leap and found two geochemical suites, one LREE-depleted and the other LREE-enriched. The LREE-depleted suite undoubtedly represents the Cambrian? volcanic unit, as the geochemistry is very similar to that reported by Lindsley-Griffin (1982) for this unit. Brouxel's LREE-enriched suite from Lovers Leap probably represents dikes of Devonian or younger age. Brouxel, Lecuyer, and Lapierre (1989, p. 260) interpreted the LREE-depleted suite as low-K tholeiites typical of "immature island-arc volcanics". The data for the Cambrian? volcanic unit also could be interpreted as a backarc marginal basin origin. Because of extensive alteration in this unit, geochemical data should be interpreted with caution, but the evidence does support an oceanic origin for these rocks.

Textures and Structures

The Cambrian? basalts range from a feldspar-rich porphyritic pillow lava to massive aphanitic lava flows to breccia. The pillows are difficult to recognize because of extensive fracturing and shearing that is characteristic of the unit, but are best preserved at Lovers Leap (Fig. 2). The pillow structures exhibit both concentric and radial fractures, aphanitic chilled margins, and a coarser grained porphyritic interior. Locally, clots of chert and siliceous mudstone occur in the interstices between pillows. Basaltic dikes crosscut the pillows. The pillow structures are clear evidence of a submarine origin for the basaltic lavas of the Trinity ophiolite.

Ordovician Ophiolitic Rocks

A complete ophiolite sequence of Ordovician age does not appear to be present within the Trinity complex. In the area shown on Figure 2, where the best ophiolite sequence is displayed (Lindsley-Griffin,

1977a, b; 1982), most of the plagiogranites and all of the volcanic rocks are now believed to be Cambrian. Deformed gabbros of unknown age are probably Cambrian as well, although further isotopic dating is necessary to confirm this hypothesis. Thus, ophiolitic rocks of Ordovician age consist only of fault slivers and melange blocks along the northwestern edge of the Trinity complex and the broad expanse of ultramafic rocks that make up the bulk of the Trinity terrane (Fig. 1).

Ordovician Ultramafic Rocks

East of the China Mountain fault zone (Figs. 1, 2) lies a unit of relatively unserpentinized peridotites and dunites that were mapped by Quick (1981) and by Lindsley-Griffin (1982); most of the northeastern and north-central part of the Trinity complex consists of this unit, but its extent to the south is not known. Jacobsen *et al.* (1984) obtained a Sm-Nd mineral isochron age of 472 ± 32 Ma for a feldspathic pocket within this peridotite. They interpreted this age as the time when small pockets of basaltic magma formed by adiabatic partial melting. Thus, this is a minimum age for the perido-tite, rather than the time of crystalli-zation. The description of composition, textures, and structures given below is drawn from Quick (1981a, 1981b, 1981c) and Lindsley-Griffin (1977a, 1977b, 1982) and is supplemented by Lindsley-Griffin's field and lab work in 1988-1990.

Composition

The Ordovician ultramafic rocks of the Trinity complex are unusual because of their high plagioclase content. Peridotite com-prises 60-70% of the unit (Quick, 1981b) with dunite and pyroxenite comprising the remainder. Lithologic types include lherzo-lite (ol + opx + cpx) and plagioclase lherzolite, harzburgite (ol + opx) and plagioclase harzburgite, dunite, and spinel pyroxenite (cpx + opx + sp). Chrome spinel is a ubiquitous accessory, which locally forms small layers or pods. The ortho-pyroxene is enstatite; the clinopyroxene is diopside or chrome diopside (Lindsley-Griffin, 1982). Serpentinization is extensive near the Silurian gabbroic plutons and other intrusions and along fault zones, but elsewhere serpentine represents only about 20-50% of the rock (Quick, 1981b). Contacts between different rock types are commonly gradational, caused by changes in proportions of pyroxene to olivine, or abundance of feldspar (Fig. 9).

Textures and Structures

Pyroxene grains within the Ordovician lherzolite, harzburgite, and pyroxenite have been stretched and partially dismembered into a planar fabric by ductile deformation. Mineral foliation and compositional layering are characteristic of most of the Ordovician ultramafic rocks; such structures are typi-cal of mantle tectonites and may have formed by cumulate processes or by pressure

solution during mantle creep (Coleman, 1977; Dick and Sinton, 1979). Some dunite bodies appear to crosscut peridotite structures; these bodies appear less deformed than the surrounding peridotite (Fig. 10). Quick (1981a, 1981b) interpreted these as having formed after most of the ductile deformation by a complex process related to the genera-tion and expulsion of basaltic magma from suboceanic mantle.

Crosscutting the tectonic fabric are stringers and pockets of feldspar, now altered to albite or saussurite, about which are clustered tiny recrystallized grains of clinopyroxene and orthopyroxene (Fig. 11). Several authors have concluded that these distinctive textures represent a partial melting event in which small amounts of basaltic melt formed during pressure-release melting during or after ductile deformation (Menzies and Allen, 1974; Lindsley-Griffin, 1976, 1977a, 1982; Quick, 1981b, 1981c).

Figure 9. Ordovician feldspathic lherzolite of the Trinity complex. Clots of plagio-clase that crosscut mineral foliation and compositional layering may have formed by partial fusion. Scale is 15 cm.

Figure 10. Ordovician peridotite of the Trinity complex. Dunite "dikes" crosscut strongly foliated, hackly appearing, peridotites. Scale is 15 cm.

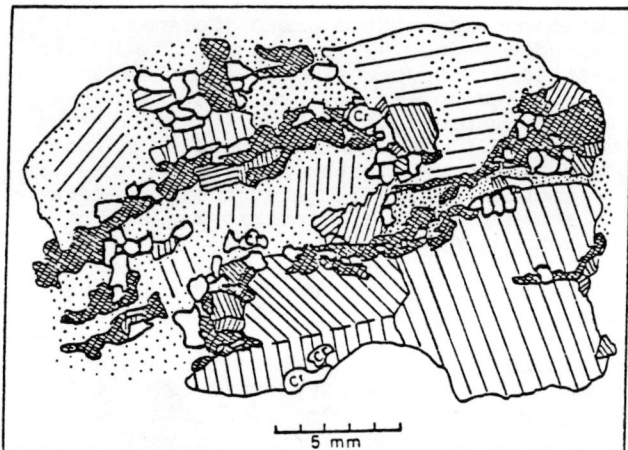

Figure 11. Sketch of partial fusion texture in Ordovician feldspathic lherzolite of the Trinity complex. Relict melt pockets consist of irregular plagioclase grains (cross-hatched) surrounded by unstrained fresh clinopyroxenes (unpatterned) and unstrained fresh orthopyroxenes (closely spaced lines); these are elongated parallel to layering and mineral foliation and crosscut the large kinked orthopyroxenes (widely spaced lines) that are largely altered to serpentine (stippled). Chromite (cr) is primary.

Ordovician Gabbro and Plagiogranite

Mattinson and Hopson (1972b, p. 582) reported two Ordovician isotopic ages from the Trinity-Yreka terranes: 1) 455 Ma for a "trondhjemite" boulder in graywacke conglomerate at Lovers Leap, and 2) 480 Ma for "xenomorphic hornblende gabbro...in a fault slice at the northern margin of the Trinity sheet", also at Lovers Leap. These ages were recalculated using new constants and yielded 440 Ma for the boulder and 470 Ma for the hornblende gabbro (Wallin *et al.*, 1988). The 440 Ma boulder lies within a conglomeratic melange block; the 470 Ma fault slice may also represent melange, as it is in fault contact with Cambrian plagiogranites. Wallin *et al.* (1988) reported two new isotopic ages of 469 and 475 Ma for Ordovician "hornblende tonalite of the Trinity ophiolite". Both these samples are also from Lovers Leap, but their structural setting is uncertain.

Mattinson and Hopson (1972b) reported initial Sr^{87}/Sr^{86} ratios for the two samples from Lovers Leap. Although such sparse data should be interpreted with caution, it is significant that the data fall within ranges for Mesozoic ophiolites in Greece (Pindos) and Cyprus (Troodos), both of which were interpreted to be ocean ridge assemblages. Thus, the ophiolitic Ordovician rocks are of probable ocean-ridge origin, even though the ophiolite is dismembered and incomplete.

Composition

The Ordovician plagiogranites are highly variable in composition, accounting

for the variety of rock names applied to them in the literature: "tonalite", "hornblende gabbro", "trondhjemite". Because of this range of compositional types, the best term for the rocks is "plagiogranite" (Coleman and Peterman, 1975; Coleman, 1977). These quartz-bearing rocks consist of plagioclase (albite), with hornblende ranging from less than 10% (trondhjemitic) to as much as 30% (tonalitic if quartz-rich; gabbroic if quartz-poor). Quartz abundances vary from 15% to 35%, and alkali feldspar is absent or present only in trace amounts.

Textures and Structures

Ordovician plagiogranites are characterized in part by the absence of textures and fabric caused by deformation. In contrast to the highly deformed Cambrian plagiogranites and metagabbros, the Ordovician rocks exhibit a nonfoliated appearance, with a well preserved hypidiomorphic-granular texture (Fig. 12).

Ordovician Stratigraphic Relationships

The Ordovician ophiolitic rocks at Lovers Leap consist of hornblende gabbro and plagiogranite faulted against serpentinized lherzolite and plagioclase lherzolite. Although the lherzolite at Lovers Leap has not been isotopically dated, its composition is similar to that of the Ordovician lherzolite and dunite east of the China Mountain fault zone (Fig. 2), and it includes partial melt pockets like those of the Ordovician unit. Thus, even though the Lovers Leap lherzolite is more serpentinized and much more intensely deformed than the Ordovician lherzolite, I believe it to be Ordovician.

Numerous small intrusions of undeformed, nearly massive, fresh-looking gabbro and plagiogranite (Fig. 13) occur within the Cambrian? metagabbro unit that has not been isotopically dated. These small dikes resemble Ordovician gabbro and plagiogranite in composition and structural style, and may also be Ordovician. In addition, two small stocks of hornblende gabbro intrude Cambrian plagiogranite west of the China Mountain shear zone (Fig. 2). Although no isotopic data are available for these two bodies, they are probably Ordovician or Late Cambrian, because they post-date deformation within the Cambrian rocks and are cut off by the pre-Late Silurian China Mountain shear zone.

SILURIAN PEGMATITIC GABBRO SUITE

New isotopic data establish the age of the China Mountain pluton (Fig. 2) as 415 Ma, Late Silurian (Wallin *et al.*, 1988, 1990). This pluton, a large stock that intrudes ophiolitic Cambrian and Ordovician rocks on the northwestern edge of the Trinity complex, is part of a distinctive suite of mafic and ultramafic dikes and stocks that is widespread throughout the Trinity complex (Wallin *et al.*, 1990). The intrusive nature of these rocks has long

Figure 12. Massive Ordovician hornblende gabbro boulder in conglomerate block at Lovers Leap. Correlated with Trinity complex because of 440 m.y. age (Mattinson and Hopson, 1972b). View is 25 cm across.

Figure 15. Silurian pegmatitic gabbro suite of the Trinity complex: medium to coarse grained and pegmatitic hornblende-pyroxene gabbro.

Figure 13. Nonfoliated leucogabbro dike cutting foliated hornblende metagabbro of the Cambrian? metagabbro unit, Trinity ophiolite.

Figure 16. Silurian pegmatitic gabbro suite of the Trinity complex: pegmatitic clinopyroxenite dike intruding foliated Ordovician lherzolite. Scale is 15 cm.

Figure 14. Silurian pegmatitic gabbro suite of the Trinity complex, showing typical appearance of the pyroxene gabbro.

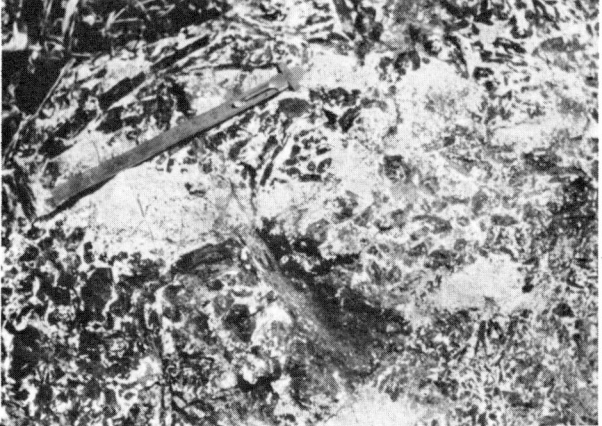

Figure 17. Silurian pegmatitic gabbro suite of the Trinity complex: hornblende diorite consisting of skeletal, euhedral hornblende crystals up to 10 cm long in a matrix of altered plagioclase. Scale is 15 cm.

been recognized (Lipman, 1964; Lanphere *et al.*, 1968; Lindsley-Griffin, 1975, 1977, 1982; Goullaud, 1977; Throckmorton, 1978; Irwin, 1985; Schwindinger and Anderson, 1987).

Although included in the Trinity ophiolite by some authors (Brouxel and Lapierre, 1988; Brouxel *et al.*, 1988; Brouxel, Lecuyer, Lapierre, 1989), the Silurian gabbroic suite is not part of an ophiolite sequence because it does not comprise a continuous sheet above the peridotite or below the dike complex-basalt. Instead, it consists of individual stocks and dikes, as well as dike swarms, that crosscut the older deformed ophiolitic assemblages (Lindsley-Griffin, 1977a, 1977b, 1982). The Silurian plutons are, however, included in the Trinity complex and the Trinity terrane because they intrude only rocks of the Trinity terrane; they do not intrude rocks of the Yreka terrane or the central metamorphic terrane.

Composition

The dominant rock types in the Silurian suite are pyroxene gabbro (Fig. 14) and hornblende pyroxene gabbro (Fig. 15). The pyroxene is typically a green clinopyroxene (diopside) that weathers bronze or reddish brown and comprises 30-40% of the rock. Plagioclase comprises 40-60% of the rock and is typically saussuritized; fresh plagioclase is calcic (An_{60} to An_{96}, according to Schwindinger and Anderson, 1987). Black hornblende is present as interstitial material or as rims on pyroxene grains (Fig. 15); it comprises 5-15% of the hornblende pyroxene gabbros. Quartz is locally present as an accessory in all rock types of the Silurian suite.

Subordinate rock types in the Silurian suite include pyroxenites, hornblende diorites, quartz diorites, and aplite dikes. The pyroxenites include clinopyroxenite (green diopside), websterite (cpx + opx + ol), and wehrlite (cpx + ol). Cumulate pyroxenites occur as individual stocks and as parts of gabbroic stocks; pyroxenite dikes intrude rocks of the Cambro-Ordovician basement complex (Fig. 16). Diorites range from hornblende diorite (Fig. 17) to quartz diorite and quartz-feldspar aplite, and typically occur as dikes and dike swarms that intrude the basement complex, the pyroxenites, and the gabbros.

Textures and Structures

The Silurian gabbroic suite is termed "pegmatitic gabbro" because the abundance of pegmatitic grain sizes is the distinguishing feature of the unit (Fig. 14); however, regions of fine to medium grained rocks are common (Fig. 15). Grain size in the gabbros varies abruptly; these variations commonly define a crude layering that Schwindinger and Anderson (1987) attributed to episodic water saturation of the magma. The larger gabbro and clinopyroxenite plutons exhibit cumulate textures. Most of the pyroxenite,

gabbro, and hornblende diorite dikes are pegmatitic, with an average grain size of 2-5 cm; locally, individual crystals attain a size of >5 cm by 20 cm. (Figs. 16, 17).

The lack of a penetrative deformational fabric in rocks of the Silurian suite confirms that it intruded after deformation ceased in the Cambro-Ordovician rocks. Mutually intrusive relationships allow the following intrusive sequence to be established: pyroxenite dikes and stocks, gabbro dikes and stocks, hornblende diorite dikes, diorite and quartz diorite dikes, aplite dikes.

POST-EARLY DEVONIAN ROCKS
AND THE TRINITY COMPLEX

Devonian Lavas and Dike Swarms

A suite of altered basaltic pillow lavas (Fig. 18), pillow breccias, hyaloclastites, and massive lava-flow rocks overlies both the Trinity and Yreka terranes (Fig. 2). The extrusive rocks are associated with basaltic dikes and dike swarms that contain xenoliths of the Late Silurian pegmatitic gabbro suite (Fig. 19), hence, they must be post-Late Silurian in age. These dikes intrude the Cambrian Trinity ophiolite, the Ordovician ophiolitic rocks of the Trinity complex, and the stocks and dikes of the Silurian pegmatitic gabbro suite. At a number of localities along the northwestern edge of the Trinity complex, the dikes can be traced upward into the extrusive suite, and the volcanic rocks unconformably overlie Lower Devonian Gazelle Formation or melange containing fossiliferous rocks of Early Devonian (Emsian) age (Gregg Ranch Complex of Lindsley-Griffin *et al.*, this volume). Thus, the dike swarms and lavas are probably Middle (to Late?) Devonian in age.

Figure 18. Devonian pillow lavas that overlie the Trinity-Yreka composite terrane. The hammer handle at lower edge of view is 4 cm wide and 30 cm long.

Figure 19. Devonian pyroxene basalt dike containing xenoliths of Silurian pegmatitic gabbro. Scale is 15 cm.

Devonian dikes and lavas range in composition from green pyroxene (diopside) basalt, to basaltic andesite with hornblende phenocrysts, to green aphanitic spilite. The green pyroxene basalt dikes represent the earliest intrusion in the dike assemblage (Peter H. Masson, pers. comm., 1990). At many localities along the northwestern edge of the Trinity terrane, the Devonian dikes form small sheeted dike complexes, as well as large individual dikes. In the central and eastern part of the Trinity complex, Devonian sheeted-dike suites and small sheeted packets, consisting of 3-10 dikes, intrude the Ordovician peridotites (Peter H. Masson, pers. comm., 1990). These sheeted complexes are evidence of a Middle Devonian rifting event, but their limited extent suggests the rifting was minor and total extension was not great.

Although included in the Trinity ophiolite by some authors (Brouxel and Lapierre, 1988; Brouxel et al., 1988; Brouxel, Lecuyer, Lapierre, 1989; Lapierre et al., 1987), the Devonian dike swarms and lavas are not part of the Cambrian Trinity ophiolite, the lower Paleozoic Trinity complex, nor the Trinity terrane. However, because they postdate the juxtaposition of the Trinity and Yreka terranes, the Devonian basalts may be considered part of the Trinity-Yreka composite terrane.

Jurassic-Cretaceous Intrusions

Plutonic rocks of Jurassic and Cretaceous age (Lanphere et al., 1968; Wagner and Saucedo, 1987) intrude the Trinity complex (Figs. 1, 2) and stitch together the Trinity-Yreka composite terrane, the central metamorphic terrane, and other terranes of the Klamath Mountains (Irwin, 1985). These plutons are intermediate in composition, ranging from granodiorite to quartz diorite and diorite to trondhjemite.

ULTRAMAFIC ROCKS OF THE CENTRAL METAMORPHIC TERRANE

The Yreka terrane is bounded on the west by ultramafic and minor mafic rocks that extend from Yreka south to Callahan where they appear to merge with the Trinity complex (Figs. 1, 2). These ultramafic rocks are part of the central metamorphic terrane because they differ texturally and structurally from the Trinity complex, and because they have experienced a different geologic history than the Trinity complex (Lindsley-Griffin, 1977a, 1982; Lindsley-Griffin and Griffin, 1983). However, because the contact between ultramafic rocks of two adjacent terranes is difficult to recognize, they have been included in the Trinity terrane by many workers (Irwin and Lipman, 1962; Lipman, 1964; Davis et al., 1965; Goullaud, 1977). The contact between the two terranes lies south of Callahan (Fig. 1), but is well camouflaged by Mesozoic plutons and associated contact metamorphism.

Unlike the subhorizontal sheet of the Trinity complex, this ophiolitic fragment appears to dip steeply eastward beneath the Trinity-Yreka composite terrane (Lindsley-Griffin and Griffin, 1983), which suggests it may be a structural entity separate from the subhorizontal sheet of the Trinity terrane. The amphibolites reported by Hotz (1977, 1978) that overlie this ultramafic slab, and separate it from the Trinity-Yreka terrane, have K-Ar dates of 398 and 407 Ma (Kelley et al., 1987), suggesting Devonian metamorphism.

South of Callahan (Fig. 1), where the ultramafic slab is continuous with the Trinity complex, the boundary between the central metamorphic terrane and the Trinity terrane is more difficult to recognize. The most likely location for this suture is east of Gibson Peak (Fig. 1) in the Trinity Alps. The area is occupied by several Mesozoic plutons that obscure relationships in the basement rocks. The dashed trace of this suture shown on Figure 1 is based on the detailed study of the Trinity Alps area by Lipman (1964). Lipman described ultramafic rocks along the western edge of this belt that are very unlike those of the Trinity complex and which I consider to be part of the central metamorphic terrane. Although peridotites of the central metamorphic terrane exhibit mineral foliation, compositional layering, and deformational textures like those of the Trinity complex peridotites, their primary composition and metamorphic mineralogy are different from those of the Trinity complex, especially the Ordovician feldspathic lherzolite and dunite.

According to Lipman (1964), peridotite in the central metamorphic terrane is an olivine-rich harzburgite with 10-15% enstatite associated with minor dunite and enstatite pyroxenite. No feldspar has been

602

observed in the ultramafic rocks, unlike lherzolite and dunite of the Trinity terrane in which feldspar is a ubiquitous, though minor component. Thus, peridotites of the central metamorphic terrane apparently did not experience adiabatic partial melting. In harzburgite of the central metamorphic terrane, pyroxene is preferentially altered, but most of the olivine is fresh (Lipman, 1964). In contrast, harzburgite of the Trinity complex contains totally altered olivine, and partially altered pyroxene. These observations suggest different origins and histories for peridotites in the two terranes.

Lipman (1964) also described pegmatites of pyroxene gabbro, hornblende diorite, and hornblende tonalite that are correlative with the distinctive suite of Silurian pegmatitic gabbros and related rocks of the Trinity terrane. These undeformed gabbroic dikes and veins occur only east of Gibson Peak pluton (Fig. 1), and are separated from the nonfeldspathic harzburgites by a zone of tremolite schist. I interpret the zone of schist as a fault between two different bodies of peridotite. Dikes of Silurian rocks are abundant and ubiquitous throughout the Trinity terrane, but are absent in the central metamorphic terrane. Thus, I interpret the western boundary of Silurian pegmatites as the western boundary of the Trinity terrane, where it is faulted against the central metamorphic terrane (Fig. 1). The two terranes had different geologic histories until they were juxtaposed in post-Late Silurian time.

ORIGIN OF THE TRINITY COMPLEX

Cambrian Ophiolite

Most components of the Cambrian ophiolite assemblage are probably of oceanic origin. The harzburgite is typical of suboceanic mantle tectonites, metagabbro is typical of deformed oceanic gabbros, and plagiogranites meet the petrographic criteria for oceanic plagiogranites (Coleman and Peterman, 1975; Coleman, 1977). However, the lack of geochemical data for these rocks makes interpretation of their origin difficult. The exact type of oceanic setting cannot be determined from currently available data, but it was probably not a typical oceanic spreading center. Some rocks now correlated with the ophiolite may be blocks within melange of the overlying Yreka terrane; some may be continental in origin as suggested by Wallin (1990).

Previously published hypotheses (Brouxel et al., 1988; Brouxel, Lecuyer, Lapierre, 1989; Lapierre et al., 1985, 1987) that advocated an island-arc origin for the Trinity ophiolite apparently were based on samples from the Devonian basalts and related dikes rather than samples from the Cambrian? basalts of the ophiolite sequence. However, the low-K tholeiites reported from Lovers Leap (Brouxel, Lecuyer, Lapierre, 1989) are probably Cambrian? basalts of the Trinity ophiolite.

For samples that I collected within the Cambrian? basaltic rocks, the Zr-Ti-Y values (Lindsley-Griffin, 1982) plot within the overlap field of Pearce and Cann (1971, 1973) and thus are permissive of either island-arc or ocean-ridge origin. However, the REE patterns exhibit the relatively flat profile typical of oceanic basalts, although with more variation than normal MORB.

A reasonable model based on available data is that the Cambrian ophiolite assemblage formed as a "normal" ophiolite sequence of harzburgite, layered gabbro, plagiogranite, and basalt that was subsequently dismembered, deformed, and juxtaposed against the fragments of Ordovician ophiolitic rocks. The data could be interpreted as permissive of a number of different tectonic settings, such as: a marginal basin undergoing backarc rifting, a rifting continent undergoing transition into an oceanic spreading center, or localized rifting along a curvilinear transform fault.

Ordovician Ophiolitic Rocks

The Ordovician rocks interpreted as ophiolitic are clearly of oceanic origin. Quick (1981b) interpreted the Ordovician ultramafic rocks as a mantle diapir; such an origin is consistent with formation in a marginal basin beneath a backarc spreading center as suggested by Lindsley-Griffin (1977a, b). Quick (1981b) hypothesized that the mantle diapir, because of its feldspar-rich composition, may have been located near a continental margin. He also postulated a tectonic setting within either a volcanic arc or back-arc basin.

The initial Sr^{87}/Sr^{86} ratios that Mattinson and Hopson (1972b) report for Ordovician gabbroic rocks are consistent with an origin at an oceanic ridge. No other Ordovician rocks have been identified, and a complete ophiolite succession of Ordovician age does not appear to be present in the Trinity complex.

The Ordovician peridotites, dunites, and minor gabbros of the Trinity complex formed in an oceanic setting which included mantle diapirism, partial fusion and generation of gabbroic magmas. Ordovician gabbros in fault slices at Lovers Leap may have formed at an oceanic ridge. Undeformed gabbros of Ordovician? (or Cambrian?) age may represent intrusion of gabbro into older deformed ocean crust during an abortive rifting episode. Alternatively, they could have intruded along a "leaky" transform fault, as older Cambrian oceanic crust passed an Ordovician ridge segment.

Pre-Silurian Amalgamation

Map relationships demonstrate that the Cambrian plagiogranites and the Cambrian metagabbro are faulted against each other and against adjoining ultramafic rock units of both Ordovician and unknown ages (Fig. 2). These faults are intruded by undeformed Silurian gabbro of the China Mountain pluton. Thus, the Cambrian and Ordovician assemblages were juxtaposed before intrusion in the Late Silurian. The present attitude of this fault is nearly vertical, but its original orientation is uncertain.

The China Mountain fault zone (Fig. 2) is a broad, sinuous fault zone characterized by phacoids and slivers of harzburgite, Cambrian metagabbro, and clinopyroxenite in serpentine schist. These fault blocks exhibit textural and structural evidence of a repeated ductile to brittle deformation, recrystallization, intrusion of mafic dikelets, and continued deformation (Lindsley-Griffin, 1982). These features suggest a major fault in oceanic crust along which deformation was associated with minor magmatism. A likely tectonic setting for such a sequence of events would be along an oceanic fracture zone or transform fault. Alternatively, the China Mountain fault zone may represent a suture between two oceanic terranes, one Cambrian and the other Ordovician, that amalgamated in Late Ordovician to Silurian time.

Intrusion of Silurian Gabbros

Map relationships (Fig. 2) demonstrate that the Silurian pegmatitic gabbro suite intruded after the juxtaposition of the Cambrian and the Ordovician oceanic terranes. The association of pyroxene gabbro, hornblende pyroxene gabbro, and pyroxenite with only minor components of intermediate composition suggests these Silurian magmas were produced in an oceanic setting.

Jacobsen *et al.* (1984) studied a post-deformation intrusive pyroxenite dike that I consider to be part of the Silurian suite. They concluded that the pyroxenite was not derived from partial melting of the Ordovician peridotites, but from a different source area in the mantle. Schwindinger and Anderson (1987) concluded that the pegmatitic gabbros at Castle Lake crystallized at a relatively low pressure. They hypothesized that the gabbros may have formed "in a shallow, oceanic body of serpentinizing peridotite and were metamorphosed by a sodium-poor hydrothermal fluid...derived either from deserpentinization of peridotite or by natural boiling (distillation) of sea water at a pressure <800 bars" (p. 372).

Thus, the Silurian gabbroic suite formed in an oceanic setting at relatively shallow depths, and was derived from a source other than the Ordovician peridotites of the Trinity complex. Without additional geochemical data, little can be deduced about the tectonic setting except that it was not a site of active crustal deformation, and it was probably not a typical oceanic spreading center.

Post-Silurian Events

The Trinity complex was juxtaposed with melange of the Yreka terrane in late Early to early Middle Devonian, possibly as the culmination of the subduction that emplaced

rocks of the central metamorphic terrane along the western edge of the Yreka and Trinity terranes. Middle Devonian basaltic lavas were erupted through the Trinity-Yreka composite terrane during a minor rifting episode that may have been related to development of the volcanic arc represented by the Copley Greenstone of the Redding terrane. Some fault zones that cut the Trinity complex may post-date the Devonian lavas (Peter H. Masson, pers. comm., 1990). In Middle to Late Jurassic, the Trinity-Yreka-central metamorphic composite terrane was amalgamated with successively younger ocean crust and island arc terranes to the west, and the entire mass was accreted to the North American continental margin in latest Jurassic to earliest Cretaceous time (Irwin, 1981, 1985).

EARLY PALEOZOIC PALEOGEOGRAPHY OF THE TRINITY TERRANE

The Trinity terrane formed in an oceanic setting over a time span of nearly 200 Ma, from earliest Cambrian (565 Ma) to Middle Devonian time (post-Emsian, or Eifelian, about 380 Ma). Such a long-lived oceanic terrane could not represent a single slice of oceanic crust and upper mantle. A likely modern analog for its early Paleozoic paleogeographic setting would be a long-lived backarc marginal basin like the Japan Sea. In the middle Paleozoic, the Trinity complex may have comprised a thick crustal block analogous to an oceanic plateau.

The Trinity ophiolite formed in the Early Cambrian by slow rifting in an oceanic setting. Ordovician components of the Trinity complex formed at an ocean ridge of unknown character. In the Silurian, both the Ordovician and the Cambrian assemblages were faulted together, possibly within an oceanic transform fault complex, but the fault had become inactive by Late Silurian when voluminous gabbroic magmatism occurred.

From Late Silurian to early Middle Devonian, the Trinity complex formed the leading edge of a plate that was overriding the east-dipping subducting slab of the central metamorphic terrane. This produced the accretionary complex known as the Yreka terrane, which consists of the Ordovician assemblage of Duzel Phyllite, Sissel Gulch Graywacke, and Antelope Mountain Quartzite; the Silurian Moffett Creek Formation; the Devonian Gazelle Formation; the Silurian tectonic melange of Skookum Gulch (Hotz, 1977, 1978); and the Early to Middle Devonian polygenetic melange of the Gregg Ranch Complex (Lindsley-Griffin *et al.*, this volume). During the amalgamation of these terranes, isolated near-trench volcanic centers developed that erupted minor pillow lavas over the composite Trinity-Yreka terrane. The association of small sheeted-dike swarms with the Devonian volcanic rocks suggests minor extension.

It is important to remember that this paleogeographic setting was not two-dimensional, and that it may have included a

604

number of microplates and triple junctions. For example, both the Devonian volcanism and the cessation of Devonian subduction could have been caused by migration of a triple junction along the active subduction zone. A triple junction containing at least one divergent or convergent plate margin component would have provided a heat source for the near-trench extension and volcanism, and passage of the triple junction would have changed plate motions so that the subduction zone became inactive.

By Middle Devonian, the Trinity terrane had amalgamated with the Yreka and central metamorphic terranes to form an oceanic plateau. This oceanic plateau was only marginally affected by development of island arc volcanism and volcaniclastic sedimentation in the Redding terrane to the east, but it served as a high-standing nucleus against which younger terranes to the west were accreted in the Jurassic.

Acknowledgements

I thank G.F. Brem, J.D. Cooper, J.R. Griffin, and E.T. Wallin for their helpful reviews of the manuscript. Thanks also go to the many field geologists who have shared constructive criticism and their best outcrops with me, especially P.H. Masson, Joanne Danielson, A.W. Potter, J.R. Griffin, and E.T. Wallin. Field and laboratory work during 1987-1990 were supported by the University of Nebraska-Lincoln Research Council and the UNL Department of Geology.

REFERENCES CITED

Anonymous, 1972, Ophiolites--Penrose Conference report: Geotimes, v. 17, n. 12, p. 24-25.

Bates, R.L., and Jackson, J.A., eds., 1987, Glossary of geology, third edition: Alexandria, Virginia, American Geological Institute, 788 p.

Brewer, W.A., III, 1955, The geology of a portion of the China Mountain quadrangle, California [unpublished M.A. thesis]: Berkeley, California, University of California, 47 p.

Brouxel, M., and Lapierre, H., 1988, Geochemical study of an early Paleozoic island-arc--back-arc basin system. Part 1: The Trinity ophiolite (northern California): Geological Society of America Bulletin, v. 100, p. 1111-1119.

Brouxel, M., Lapierre, H., Michard, A., and Albarede, F., 1988, Geochemical study of an early Paleozoic island-arc--back-arc basin system. Part 2: Eastern Klamath, early to middle Paleozoic island-arc volcanic rocks (northern California): Geological Society of America Bulletin, v. 100, p. 1120-1130.

Brouxel, M., Lapierre, H., and Zimmerman, J., 1989, Upper Jurassic mafic magmatic rocks of the eastern Klamath Mountains, northern California: Remnant of a volcanic arc built on young continental crust: Geology, v. 17, p. 273-276.

Brouxel, M., Lecuyer, C., and Lapierre, H., 1989, Diversity of magma types in a lower Paleozoic island arc--marginal basin system (eastern Klamath Mountains, California, U.S.A.): Chemical Geology, v. 77, p. 251-264.

Coleman, R.G., 1977, Ophiolites: Ancient oceanic lithosphere?: New York, NY, Springer-Verlag, 229 p.

Coleman, R.G., and Peterman, Z.E., 1975, Oceanic plagiogranites: Journal of Geophysical Research, v. 80, p. 1099-1108.

Davis, G.A., 1966, Metamorphic and granitic history of the Klamath Mountains, California: California Division of Mines and Geology Bulletin 190, p. 39-50.

_____1968, Westward thrust faulting in the south-central Klamath Mountains, California: Geological Society of America Bulletin, v. 79, p. 911-934.

_____1969, Tectonic correlations, Klamath Mountains and western Sierra Nevada, California: Geological Society of America Bulletin, v. 80, p. 1095-1108.

Davis, G.A., Holdaway, M.J., Lipman, P.W., and Romey, W.D., 1965, Structure, metamorphism, and plutonism in the south-central Klamath Mountains, California: Geological Society of America Bulletin, v. 76, p. 933-966.

Dick, H.J.B., and Sinton, J.M., 1979, Compositional layering in alpine peridotites: Evidence for pressure solution creep in the mantle: Journal of Geology, v. 87, p. 403-416.

Fuis, G.S., and Zucca, J.J., 1984, A geologic cross section of northeastern California from seismic refraction results, in Nilsen, Tor H., ed., Geology of the Upper Cretaceous Hornbrook Formation, Oregon and California: Society of Economic Paleontologists and Mineralogists, Pacific Section Publication No. 42, p. 203-209.

Goullaud, L., 1977, Structure and petrology in the Trinity mafic-ultramafic complex, Klamath Mountains, northern California, in Lindsley-Griffin, N., and Kramer, J.C., eds., Geology of the Klamath Mountains, northern California: Geological Society of America Cordilleran Section Guidebook, p. 112-133.

Griscom, A., 1977, Aeromagnetic and gravity interpretation of the Trinity ophiolite complex, northern California (abs.): Geological Society of America Abstracts with Programs, v. 9, p. 426-427.

Hershey, O.H., 1901, Metamorphic formations of northwestern California: The American Geologist, v. 27, p. 225-245.

Hinds, N.E.A., 1932, Paleozoic eruptive rocks of southern Klamath Mountains: University of California Publications in Geological Science, v. 23, p. 375-410.

Hopson, C.A., and Mattinson, J.M., 1973, Ordovician and Late Jurassic ophiolitic assemblages in the Pacific northwest (abs.): Geological Society of America Abstracts with Programs, v. 5, p. 57.

Hotz, P.E., 1971, Geology of lode gold districts in the Klamath Mountains, California and Oregon: U.S. Geological Survey Bulletin 1290, 91 p.

_____ 1977, Geology of the Yreka Quadrangle, Siskiyou County, California: U.S. Geological Survey Bulletin 1436, 72 p.

_____ 1978, Geologic map of the Yreka Quadrangle and parts of the Fort Jones, Etna, and China Mountain Quadrangles, California: U.S. Geological Survey Open File Report 78-12, 1:62,500.

Irwin, W. P., 1960, Geologic reconnaissance of the northern Coast Ranges and Klamath Mountains, California, with a summary of the mineral resources: California Division of Mines Bulletin 179, 80 p.

_____ 1964, Late Mesozoic orogenies in the ultramafic belts of northwestern California and southwestern Oregon: U.S. Geological Survey Professional Paper 501-C, p. 1-9.

_____ 1966, Geology of the Klamath Mountains province: California Division of Mines and Geology Bulletin 190, p. 19-38.

_____ 1972, Terranes of the western Paleozoic and Triassic belt in the southern Klamath Mountains, California, in Geological Survey Research, 1972: U.S. Geological Survey Professional Paper 800-C, p. C103-C111.

_____ 1973, Sequential minimum ages of oceanic crust in accreted tectonic plates of northern California and southern Oregon (abs.): Geological Society of America Abstracts with Programs, v. 5, p. 62-63.

_____ 1977, Review of Paleozoic rocks of the Klamath Mountains, in Stewart, J.H., Stevens, C.H., and Fritsche, A.E., eds., Paleozoic paleogeography of the western United States: Pacific Coast Paleogeography Symposium I: Society of Economic Paleontologists and Mineralogists, Pacific Section Publication No. 7, p. 441-454.

_____ 1981, Tectonic accretion of the Klamath Mountains, in Ernst, W.G., ed., The geotectonic development of California: Rubey Volume 1: Englewood Cliffs, NJ, Prentice-Hall, p. 29-49.

_____ 1985, Age and tectonics of plutonic belts in accreted terranes of the Klamath Mountains, California and Oregon, in Howell, D.G., ed., Tectonostratigraphic terranes of the circum-Pacific region: Circum-Pacific Council for Energy and Mineral Resources, Earth Science Series Number 1, p. 187-199.

Irwin, W.P., and Bath, G.D., 1962, Magnetic anomalies and ultramafic rock in northern California: U.S. Geological Survey Professional Paper 450-B, p. 65-67.

Irwin, W.P., and Lipman, P.W., 1962, A regional ultramafic sheet in the eastern Klamath Mountains, California: U.S. Geological Survey Professional Paper 450-C, p. 18-21.

Jacobsen, S.B., Quick, J.E., and Wasserburg, G.J., 1984, A Nd and Sr isotopic study of the Trinity peridotite; implications for mantle evolution: Earth and Planetary Science Letters, v. 68, p. 361-378.

Jones, D.L., Howell, D.G., Coney, P.J., and Monger, J.W.H., 1983, Recognition, character, and analysis of tectono-stratigraphic terranes in western North America, in Hashimoto, M., and Uyeda, S., eds., Accretion tectonics in the Circum-Pacific regions: Tokyo, Terra Scientific Publishing Co., and Dordrecht, D. Reidel, p. 21-35.

Kelley, F.R., Wagner, D.L., and Saucedo, G.J., 1987, Radiometric ages of rocks in the Weed Quadrangle, California, in Wagner, D.L., and Saucedo, G.J., compilers: California Division of Mines and Geology, Regional Geologic Map Series, Weed Quadrangle, Map No. 4A (Geology), 15 p.

LaFehr, T.R., 1966, Gravity in the eastern Klamath Mountains, California: Geological Society of America Bulletin, v. 77, p. 1177-1190.

Lanphere, M.A., Irwin, W.P., and Hotz, P.E., 1968, Isotopic age of the Nevadan orogeny and older plutonic and metamorphic events in the Klamath Mountains, California: Geological Society of America Bulletin, v. 79, p. 1027-1052.

Lapierre, H., Albarede, F., Albers, J., Cabanis, B., and Coulon, C., 1985, Early Devonian volcanism in the eastern Klamath Mountains, California: evidence for an immature island arc: Canadian Journal of Earth Science, v. 22, p. 214-226.

606

Lapierre, H., Brouxel, M., Albarede, F., Coulon, C., Lecuyer, C., Martin, P., Mascle, G., and Rouer, O., 1987, Paleozoic and lower Mesozoic magmas from the eastern Klamath Mountains (North California) and the geodynamic evolution of northwestern America: Tectonophysics, v. 140, 155-177.

Lindsley-Griffin, N., 1973, Lower Paleozoic ophiolite of the Scott Mountains, eastern Klamath Mountains, California (abs.): Geological Society of America Abstracts with Programs, v. 5, p. 71-72.

_____1975, Geology of the northwestern edge of the Trinity ophiolite, eastern Klamath Mountains, California (abs.): EOS, Transactions of the American Geophysical Union, v. 56, p. 1079.

_____1976, Feldspathic lherzolites of the Trinity ophiolite complex, eastern Klamath Mountains, California (abs.): EOS, Transactions of the American Geophysical Union, v. 57, p. 1025.

_____1977a, Paleogeographic implications of ophiolites: the Ordovician Trinity complex of the Klamath Mountains, northern California, in Stewart, J.H., Stevens, C.H., and Fritsche, A.E., eds., Paleozoic paleogeography of the western United States: Pacific Coast Paleogeography Symposium I: Society of Economic Paleontologists and Mineralogists, Pacific Section Publication No. 7, p. 409-420.

_____1977b, The Trinity ophiolite, Klamath Mountains, California, in Coleman, R.G., and Irwin, W.P., eds., North American ophiolites: Oregon Department of Geology and Mineral Industries Bulletin 95, p. 107-120.

_____1982, Structure, stratigraphy, petrology and regional relationships of the Trinity ophiolite, eastern Klamath Mountains, California [unpublished Ph.D. dissertation]: Davis, California, University of California, 453 p.

Lindsley-Griffin, N., and Griffin, J.R., 1983, The Trinity terrane: An early Paleozoic microplate assemblage, in Stevens, C.H., ed., Pre-Jurassic rocks in western North American suspect terranes: Society of Economic Paleontologists and Mineralogists, Pacific Section Publication No. 32, p. 63-75.

Lipman, P.W., 1964, Structure and origin of an ultramafic pluton in the Klamath Mountains, California: American Journal of Science, v. 262, p. 199-222.

Mattinson, J.M., and Hopson, C.A., 1972a, Paleozoic ages of rocks from ophiolitic complexes in Washington and northern California (abs.): EOS, Transactions of the American Geophysical Union, v. 53,

p. 543.

_____1972b, Paleozoic ophiolitic complexes in Washington and northern California: Carnegie Institution Yearbook, v. 71, p. 578-583.

Menzies, M., and Allen, C., 1974, Plagioclase lherzolite--residual mantle relationships within two eastern Mediterranean ophiolites: Contributions to Mineralogy and Petrology, v. 45, p. 197-213.

Moores, E.M., 1982, Origin and emplacement of ophiolites: Reviews of Geophysics and Space Physic, v. 20, p. 735-760.

Pearce, J.A., and Cann, J.R., 1971, Ophiolite origin investigated by discriminant analysis using Ti, Zr, and Y: Earth and Planetary Science Letters, v. 19, p. 290-300.

_____1973, Tectonic setting of basic volcanic rocks determined using trace element analyses: Earth and Planetary Science Letters, v. 24, p. 419-426.

Quick, J.E., 1981a, The origin and significance of large, tabular dunite bodies in the Trinity peridotite, northern California: Contributions to Mineralogy and Petrology, v. 78, p. 413-422.

_____1981b, Petrology and petrogenesis of the Trinity peridotite, an upper mantle diapir in the eastern Klamath Mountains, northern California: Journal of Geophysical Research, v. 86, p. 11837-11863.

_____1981c, Petrology and petrogenesis of the Trinity peridotite, northern California, part I [unpublished Ph.D. dissertation]: Pasadena, California, California Institute of Technology, 288 p.

Schwindinger, K.R., and Anderson, A.T., Jr., 1987, Probable low-pressure intrusion of gabbro into serpentinized peridotite, northern California: Geological Society of America Bulletin, v. 98, p. 364-372.

Strand, R.G., 1962, Geologic atlas of California, Redding sheet: California Division of Mines and Geology, 1:250,000.

Throckmorton, M.L., 1978, Petrology of the Castle Lake peridotite-gabbro mass, eastern Klamath Mountains, California [unpublished M.S. thesis]: Santa Barbara, California, University of California, 109 p.

Wagner, D.L., and Saucedo, G.J., compilers, 1987, Geologic map of the Weed quadrangle, 1:250,000, California: California Division of Mines and Geology Regional Geologic Map Series,

607

Map 4A.

Wallin, E.T., 1990, Petrogenetic and tectonic significance of xenocrystic Precambrian zircon in Lower Cambrian tonalite, eastern Klamath Mountains, California: Geology, v. 18, p. 1057.

Wallin, E.T., Mattinson, J.M., and Potter, A.W., 1988, Early Paleozoic magmatic events in the eastern Klamath Mountains, northern California: Geology, v. 16, p. 144-148.

Wallin, E.T., Lindsley-Griffin, N., and Potter, A.W., 1990, Trinity ultramafic complex, Klamath Mountains, California: Cambro-Ordovician ophiolite or composite terrane? (abs.): Geological Society of America Abstracts with Programs, v. 22, p. 91-92.

REDEFINITION OF THE GAZELLE FORMATION OF THE YREKA TERRANE, KLAMATH MOUNTAINS, CALIFORNIA: PALEOGEOGRAPHIC IMPLICATIONS

Nancy Lindsley-Griffin
Department of Geology
University of Nebraska
Lincoln NE 68588-0340

John R. Griffin
Department of Geology
University of Nebraska
Lincoln NE 68588-0340

E. Timothy Wallin
Department of Geoscience
University of Nevada
Las Vegas NV 89154-4010

ABSTRACT

The structural and lithologic complexity of the lower Paleozoic rocks of the Yreka terrane requires a redefinition of the stratigraphic nomenclature in order to clarify interpretation of the paleogeographic setting. We restrict the name Gazelle Formation to the Lower Devonian siliciclastic sediments that exhibit continuous bedding and an internal stratigraphy that is consistent from one locality to the next. We propose the name Gregg Ranch Complex for the subjacent, structurally disrupted rocks that consist of a mixture of Middle Ordovician through Lower Devonian sedimentary rocks and exotic blocks of igneous and metamorphic rocks in a sheared, scaly mudstone matrix. The redefined Gazelle Formation overlies the Gregg Ranch Complex along a sheared contact that is typically marked by tectonic interleaving of the two units. The Gregg Ranch Complex is interpreted as melange formed in an accretionary complex at an active convergent plate margin; the Gazelle Formation is interpreted as trench or trench-slope basin fill that was deposited during the Early Devonian convergence.

INTRODUCTION

The lower Paleozoic sedimentary rocks of the Yreka terrane in the eastern Klamath Mountains (Figs. 1, 2) are lithologically diverse and structurally complex. Numerous previous attempts to define a usable stratigraphy have resulted in a perplexing variety of usages of the name Gazelle Formation (e.g.: Wells et al., 1959; Griffin, 1971; Merriam, 1972; Rohr, 1972; Lindsley-Griffin, 1977b; Potter et al., 1977; Olson, 1978; Lindsley-Griffin and Griffin, 1983; Potter, 1990a, 1990b; Cram-Barry and Wallin, 1990). This confusion has hindered understanding of the early Paleozoic stratigraphy, structure, and paleogeography of the Yreka terrane.

In order to provide a more rational stratigraphic framework to facilitate interpretation of the paleogeography, we propose the following two formation names for lower Paleozoic rocks of the Yreka terrane.

1. The name Gazelle Formation is restricted to the siliceous shales, mudstones, and sandstones of the Gazelle as defined by Wells et al. (1959). The redefined Gazelle Formation includes all the continuous beds that form a stratigraphic succession and that lack extensive structural disruption and dismemberment. The type locality is Gazelle Mountain (Fig. 2), for which the unit was originally named (Wells et al., 1959).

2. The name Gregg Ranch Complex is proposed herein for the laterally discontinuous mixed lithologies that were included in the Gazelle Formation as it was previously defined (Wells et al., 1959; Merriam, 1961, 1972). This newly named formation includes the structurally disrupted blocks of limestone, graywacke, chert, conglomerate, shale, sandstone, serpentinite, meta-peridotite, amphibolite, actinolite schist, pillow basalt, marble, and quartzite that are surrounded by a sheared, scaly, muddy to silty matrix. The Gregg Ranch Complex is a melange composed of olistoliths and fault slices that have been modified and overprinted by a succession of tectonic events. The type locality is the historic Gregg Ranch (Fig. 2).

The Gregg Ranch Complex now includes the Payton Ranch Limestone Member of Churkin and Langenheim (1960), as well as the following informally named units: Grouse Creek formation (Merriam, 1961; Olson, 1978), Grouse Creek unit (Boucot et al., 1974), Lovers Leap formation (Rohr, 1972), Kangaroo Creek formation (Rohr, in Lindsley-Griffin and Rohr, 1977), Potter's "Member 1" of the Gazelle Formation (Potter, 1977; Potter et al., 1977), portions of Potter's "Member 3" of the Gazelle Formation (Potter, 1977; Potter et al., 1977), Horseshoe Gulch unit (Potter et al., 1977; Potter, 1990a, 1990b), Gregg Ranch unit (Boucot et al., 1974; Potter, 1990a, 1990b). Historical development of the nomenclature is discussed in detail in Appendix I.

Melange

According to Bates and Jackson (1987, p. 410), melange is defined as:

A body of rock mappable at a scale of 1:24000 or smaller, characterized by a lack of internal continuity of contacts or strata and by the inclusion of fragments and blocks of all sizes, both exotic and native, embedded in a fragmental matrix of finer grained material.

In Cooper, J.D., and Stevens, C.H., eds., 1991, **Paleozoic Paleogeography of the Western United States-II**: Pacific Section SEPM, Vol. 67, p. 609-624.

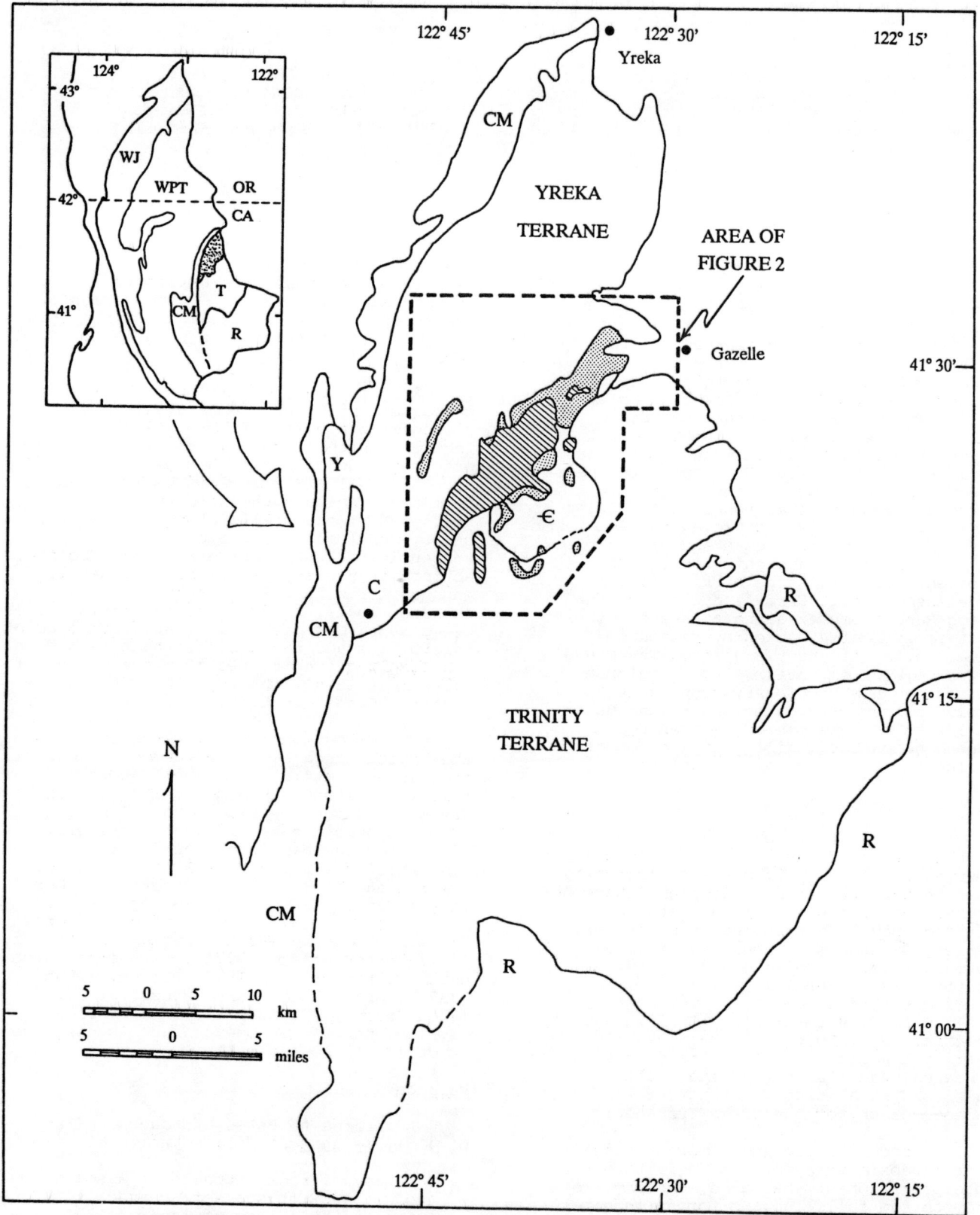

Figure 1. Location of the Yreka-Trinity composite terrane. Stipples: Gregg Ranch Complex, diagonal lines: Gazelle Formation. €, Cambrian rocks of the Trinity complex; C, Callahan. Inset shows Klamath Mountain subprovinces. Pattern: Yreka terrane; R, Redding terrane; T, Trinity terrane; CM, central metamorphic terrane; WPT, western Paleozoic-Triassic terrane; WJ, western Jurassic terrane.

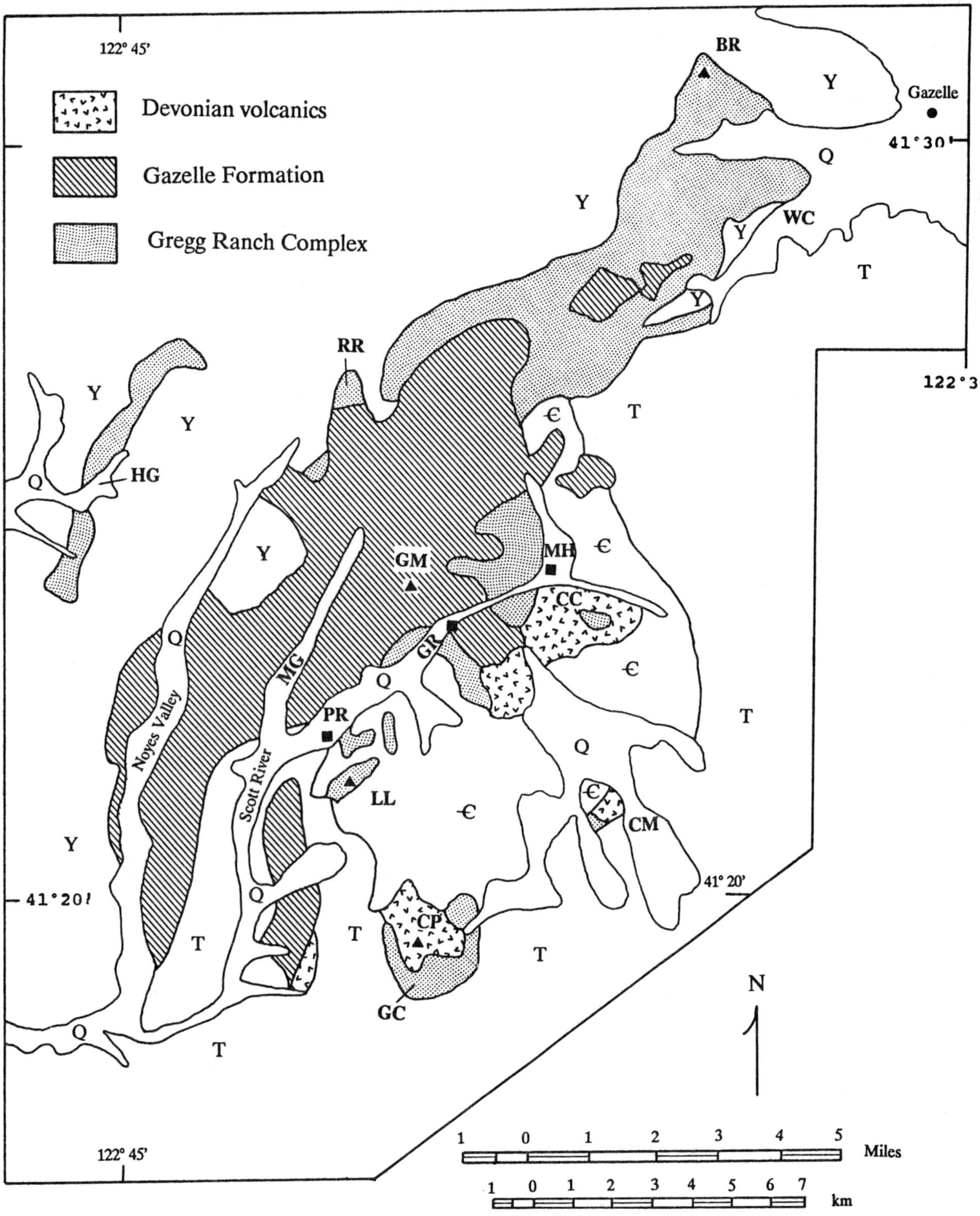

Figure 2. Geologic map of the Gazelle Formation and the Gregg Ranch Complex. Unpatterned map units: €, Cambrian rocks of the Trinity complex; T, Trinity complex undifferentiated; Y, Yreka terrane undifferentiated; Q, Quaternary sediments. Localities named in text include, from north to south: BR, Bonnet Rock; WC, Willow Creek; RR, Robbers Rock; HG, Horseshoe Gulch; GM, Gazelle Mountain; MH, Mountain House; CC, Crater Creek; GR, Gregg Ranch; PR, Parker Ranch; LL, Lovers Leap; CM, Cabin Meadow; CP, Cooper Peak; GC, Grouse Creek.

The melange concept was brought to the attention of American geologists by K.J. Hsu (1968, 1974), who applied the idea to interpretation of the Franciscan Complex. Since then, a number of different definitions have been proposed, but the usage above is the most widely employed. The three major processes that contribute to melange formation are: downslope sliding and slumping of unconsolidated material or fragments, diapirism of overpressured water-saturated sediment, and tectonic slicing and dismemberment (Maxwell, 1974; Cowan, 1985). Because melanges may form in a variety of tectonic settings, the term should be used in a descriptive sense that is not intended to connote a particular genesis (Silver and Beutner, 1980; Raymond, 1984).

Most melanges are characterized by blocks of native and exotic lithologies in a muddy matrix. Native blocks form by disruption of brittle layers that were originally interbedded with the matrix; exotic blocks are derived from rock-stratigraphic units foreign to the main body of the melange. Blocks of all sizes may be present, and the larger blocks commonly preserve a primary internal stratigraphy. The melange matrix may be structureless, but commonly is characterized by scaly foliation, produced by shearing along anasto-mosing foliation surfaces defined by flattened, stretched, or partially recrystallized clay or mica minerals; the matrix defined by these folia typically weathers out as tiny chips or scales.

Comparison of Gazelle Formation and Gregg Ranch Complex

The need for redefinition of these units becomes evident when the criteria used previously to define the Gazelle Formation are re-examined in the light of recent detailed structural and stratigraphic studies.

Previous Criteria for Defining Units

Previous authors have used different criteria for defining stratigraphic units, resulting in confusion in correlating units and interpreting their history. Wells *et al.* (1959) and Merriam (1961, 1972) used topographic and areal location combined with lithologic type. These criteria did not take block faulting into account and resulted in some structurally uplifted units being placed stratigraphically high in the section. Thus, the upper informal units shown on the Composite Columnar Section of Wells *et al.* (1959, p.647) are now part of our proposed Gregg Ranch Complex, as is the redefined type section of the Gazelle Formation of Merriam (1972).

Potter *et al.* (1977) and Potter (1987) used the presence of feldspar to distinguish between members of the Gazelle Formation. Based on this criterion, the boundary between Potter's "member 1" and "member 2" occurs within the lower portion of our redefined Gazelle Formation, so that the

formation boundaries proposed by the several workers do not agree. Potter's "member 3" (Potter *et al.*, 1977) does not appear to be useful in mapping because it contains characteristics of both his other units.

A number of workers have used the criteria of mixed lithologies and disrupted strata to separate map units from the Gazelle Formation or to define units within the Gregg Ranch Complex (Lindsley-Griffin, 1977a, 1977b, 1982; Lindsley-Griffin and Griffin, 1983; Cram-Barry and Wallin, 1990). This lithostratigraphic approach is most consistent with that recommended by the North American Commission on Stratigraphic Nomenclature (1983), and results in mappable units that are easily recognized in the field. We have used these lithostrati-graphic criteria in redefining the two units.

Lithologic character

Our redefined Gazelle Formation is a sequence consisting of interbedded terrigenous and hemipelagic eugeoclinal sedimentary rocks that were deposited together in a single deep marine setting; as restricted in this paper, the Gazelle contains no exotic lithologies. The Gregg Ranch Complex is an assemblage of native and exotic blocks that formed in a wide variety of environments, ranging from carbonate reefs to shallow and deep marine. The Gregg Ranch Complex also includes exotic blocks of igneous and metamorphic rocks; all these blocks are contained within a muddy matrix that exhibits scaly foliation.

Stratigraphy and Structural Style

The redefined Gazelle Formation consists of continuous beds which can be traced for hundreds of meters; it exhibits a stratigraphic sequence at least 125 m thick that is consistent from one part of the area to another. Structural disruption in the Gazelle Formation is restricted to tectonic slicing along the basal contact and to displacement along high angle block faults.

The Gregg Ranch Complex consists of laterally discontinuous blocks in a fine-grained scaly foliated matrix. The blocks range from cobble-sized to a maximum long dimension of nearly 2 km. Although a coherent internal stratigraphy may be preserved in larger blocks, none of these local stratigraphies can be traced outside its home block nor correlated with other local stratigraphies in the complex. The smaller blocks are flattened and elongated parallel to the semipenetrative scaly foliation of the matrix, forming wedge-shaped phacoids.

Age and Contact Relationships

The Gazelle Formation structurally overlies the Gregg Ranch Complex both at Gazelle Mountain and at Gregg Ranch. However, the Gregg Ranch Complex is approximately the same age as the Gazelle

Formation, or may be younger, even though many of its constituent blocks formed before the strata of the Gazelle Formation were deposited. Although the original contact between the two units may have been depositional, the present contact is sheared, and we consider the relationship between the two units to be tectonic rather than stratigraphic in nature.

DESCRIPTION OF THE NEWLY DEFINED FORMATIONS

Gazelle Formation

The location of the Gazelle Formation within the Trinity-Yreka composite terrane is shown on Figure 1; the areal extent of the Gazelle Formation and its relationship to the Gregg Ranch Complex are shown on Figure 2. Most of the Gazelle Formation exposures lie within the uppermost drainage of the Scott River in the vicinity of Meadow Gulch, but small outliers of the formation occur east of Gazelle Mountain and southwest of Lovers Leap (Fig. 2).

Type Locality

The type locality for the redefined Gazelle Formation is established as the slopes of Gazelle Mountain (Fig. 2). Although no type locality was specified by Wells *et al.* (1959), the formation name was originally chosen because: "Some of the most extensive exposures of the formation are found on the flanks of Gazelle Mountain from which this sedimentary sequence takes its name" (Wells *et al.*, 1959, p. 646).

Lithology

The dominant lithologies in the Gazelle Formation are siliceous shale and siliceous mudstone, with interbeds of siltstone and sandstone, and rare interbeds of black chert and black rusty weathering limestone. The shale and mudstone are characterized by intervals of medium- to thick-bedded, structureless black mudstone and siltstone (Fig. 3), alternating with intervals of thin- to medium-bedded, thinly laminated, blue-black to gray shale and siltstone (Fig. 4). Siliceous shale grades into siliceous mudstone by an increase in grain size and a loss of fissility; both grade into chert by an increase in silica until the rock is too hard to be scratched by a geologic pick. The siliceous shale and mudstone are composed of abundant radiolarians and sparse sponge spicules in a dark organic-rich matrix (Fig. 5). Cherts in the Gazelle Formation are typically black or dark gray, with yellowish brown to gray laminae, and poorly preserved, recrystallized radiolarians.

Thick beds of sandstone and fine conglomerate are intercalated with the pelites (Fig. 6). In the lower part of the Gazelle Formation, sandstone and granule to pebble conglomerate layers up to 0.5 m thick are composed of angular mudstone, chert, and volcanic fragments in a medium- to coarse-grained, quartzofeldspathic sandstone matrix (Lindsley-Griffin, 1982). The upper part of the Gazelle Formation includes four major sandstone units in thick (1 to 3 m), amalgamated, mostly structureless beds that form prominent outcrops, some of which can be traced continuously for several kilometers.

Sedimentary Structures

The siltstones, sandstones, and granule-pebble conglomerates of the Gazelle Formation exhibit sedimentary structures typical of turbidites, including partial Bouma sequences, graded bedding, sharp bases with thin intraformational breccias that scour downward into underlying beds, and laminated tops (Fig. 4). Soft-sediment deformation in the form of slump folds and brittle-ductile microfaults is common near the tops of thinly laminated siltstones and mudstones, indicating early downslope sliding and slumping.

In the laminated Gazelle mudstones, the small normal microfaults that die out in both directions record extensional strain. These probably formed during layer-parallel extension during downslope sliding such as characterizes most modern marine sediments in trench-slope environments (Kemp *et al.*, 1990). Mud-filled veins that occupy some of these microfaults may represent dewatering veins that are characteristic of modern diatomaceous oozes and muds (Lindsley-Griffin *et al.*, 1990); if so, they are evidence of episodically rapid deposition of the Gazelle sediments.

Stratigraphy and Contact Relationships

The Gazelle Formation rests depositionally on melange of the Gregg Ranch Complex, although its lower contact with the melange is locally sheared. At Parker Ranch (AP Ranch) (Fig. 2), the Gazelle Formation rests on, and is partially interbedded with, altered pillow lavas of uncertain age. The upper contact of the Gazelle Formation is eroded off in most localities, but at Cooper Peak (Fig. 2), the Gazelle Formation is overlain by post-Lower Devonian basalt and keratophyre. The Gazelle Formation is at least 200 meters thick.

Age

The age of the Gazelle Formation is poorly constrained as late Siegenian to middle Emsian or younger, that is, Early (to Middle?) Devonian.

At present, only three conodont localities have been described from limy bodies that were interbedded with undisrupted strata. These localities include: 1) a calcareous nodule in shale, identified by N. M. Savage as late Siegenian to middle Emsian; 2) calcareous sandstone or sandy limestone, identified by N.M. Savage as at least as young as late Emsian; 3) a limestone float block, identified by N. M. Savage as Emsian to Middle Devonian (Boucot *et al.*, 1974, p. 694-695; Savage, 1977, p. 57; Lindsley-

614

Figure 3. Massive siliceous mudstone and siltstone beds of the Gazelle Formation (hammer for scale).

Figure 6. Sandstone and granule conglomerate beds intercalated with siliceous pelites of the Gazelle Formation (hammer for scale).

Figure 4. Thinly laminated blue-gray to gray siliceous mudstone of the Gazelle Formation.

Figure 7. Phacoidal blocks of graywacke in sheared scaly mudstone matrix of the Gregg Ranch Complex (scale is in cm).

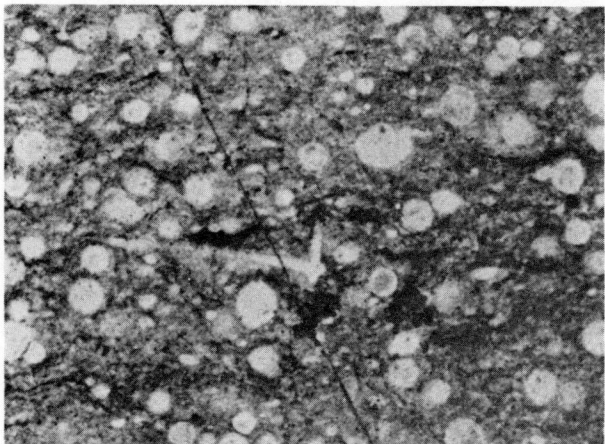

Figure 5. Photomicrograph of the siliceous mudstone of the Gazelle Formation, containing abundant radiolarians and a sponge spicule (width of view is 3 mm).

Figure 8. Prominent knockers and block trains of cherts and reef carbonates of the Gregg Ranch Complex in the Willow Creek area.

Griffin, 1982, p. 402).

Two localities of probable plant material have been observed within the Gazelle Formation, one south of Lovers Leap that is probably Devonian (A.J. Boucot, oral comm., 1974, in Lindsley-Griffin, 1982, p. 396.), and in Meadow Gulch (collected by Wallin during the 1990 field season).

Thinly laminated siliceous shales of Gazelle Formation have abundant radiolarians and other microfossils (Fig. 5; Lindsley-Griffin and Fisher, 1989). H. Y. Ling (written comm., 1990) reports after "...cursory observation under a stereo-microscope..." that the radiolarians have an age of Devonian or older.

Sandstone Composition and Provenance

Petrographic analyses of volcaniclastic sandstones from the Gazelle Formation reveal that they are texturally and compositionally immature, very fine- to coarse-grained feldspatholithic wackes (Cram-Barry and Wallin, 1990). These wackes contain a variety of mafic and intermediate(?) volcanic rock fragments, angular plagioclase, and subordinate angular to subangular monocrystalline quartz. The textural and compositional immaturity of these rocks indicates they are monocyclic in origin.

Wallin (1989a, b) used geochronology of detrital zircon to demonstrate that sandstones of the Gazelle Formation were derived, at least in part, from a pre-existing Ordovician-Silurian volcanoplutonic terrane rather than from coeval Early Devonian volcanism. An important aspect of current research is whether this source terrane was a cryptic volcanic arc complex that has now been displaced tectonically or whether it was the hanging wall of the Devonian arc-trench system (cf. Peacock, 1987; Peacock and Norris, 1989).

Environment of deposition

The abundance of siliceous microfossils in the finely laminated shales of the Gazelle Formation, as well as the organic-rich matrix revealed by concentrations of dead oil (Fig. 5) are evidence of high organic content. These shales are analogous to the organic concentrations found in modern diatomaceous muds and oozes that have been recovered by ocean drilling from areas of marine upwelling and high organic productivity (Lindsley-Griffin and Fisher, 1989). A suitable modern analog for the Gazelle Formation is found on the Peru margin, where diatomaceous sediments exhibit the same alternation between massive and laminated muds, and many of the same sedimentary structures (Lindsley-Griffin *et al.*, 1990; Kemp *et al.*, 1990).

Gregg Ranch Complex

The areal extent of the Gregg Ranch Complex is shown on Figure 2. All of its upper and lower contacts are sheared, except where it is unconformably overlain by Devonian volcanics at Crater Creek, Gregg Ranch, Parker Ranch, and Cooper Peak (Fig. 2).

Previous use of the name Gregg Ranch

The informal name Gregg Ranch unit was used by Boucot *et al.* (1974, p. 695) for rocks of the Gregg Ranch area (Fig. 2) that consist of mixed clastic rocks, Silurian and Devonian limestones, and volcanic lithologies. Potter (1990a, 1990b) used the informal name Gregg Ranch unit for limestone and sedimentary rocks exposed in the Gregg Ranch area that he considered to be Ordovician.

Based on detailed mapping by Lindsley-Griffin, summer 1973, rocks of the Gregg Ranch area were determined to be highly sheared and structurally disrupted. She suggested that the "...lower Paleozoic rocks at Gregg Ranch may comprise a submarine slide breccia or melange" (Lindsley-Griffin *et al.*, 1974). Lindsley-Griffin later used the informal name Gregg Ranch melange unit for these rocks (1977b, 1982).

Type locality

The type locality of the Gregg Ranch Complex is the region from the historic Gregg Ranch to the hill west of Mountain House (Fig. 2). It was at this ranch, one of the two earliest homesteads in the valley, that the melange character of this unit was first recognized and mapped. The hill west of Mountain House exhibits many exotic blocks in scaly matrix, as do roadcuts along the county road from Gregg Ranch to the pass north of Mountain House. The sheared contact between the Gregg Ranch Complex and the Gazelle Formation is well displayed in the valley between Mountain House and Gazelle Mountain.

Both the lithologic assemblage and the structural style of the Gregg Ranch Complex vary from place to place; thus the type locality is intended to typify only the general characteristics of the complex. In the future, as our understanding of the detailed nature of the Gregg Ranch Complex increases, it may be helpful to define formal melange units: tectono-stratigraphic units within the larger body of melange based on the different compositions and structural style of blocks and matrix. However, the melange units used below are informal groupings only, to facilitate discussion. More detailed descriptions of the informal melange units and their characteristics may be found in Lindsley-Griffin (1977b, 1982).

Lithology

The most distinctive characteristic of the Gregg Ranch Complex is the ubiquitous presence of a variety of exotic blocks that consist of sedimentary, igneous, and metamorphic rocks, mixed with the sedimentary native blocks and matrix. The native lithologies consist of feldspar-rich, volcaniclastic siltstone and sandstone blocks suspended in a matrix of dark greenish-black foliated mudstone (Fig. 7). The greenish color of muddy matrix in the Gregg Ranch Complex is distinctly different from the bluish color of mudstones in the Gazelle Formation. Locally preserved partial Bouma sequences, graded bedding, laminated tops, and slump folds suggest these native lithologies originated as turbidites interbedded with hemipelagic muds.

Exotic lithologies consist of shallow and deep marine, igneous, and metamorphic rocks. The most prominent of these exotic rocks are the light gray carbonates and varicolored cherts that are strung into block trains which form prominent knockers projecting above the poorly exposed matrix (Fig. 8). Many of the carbonates formed in reef environments, as evidenced by their fossilized framework organisms. The cherts contain recrystallized radiolarians and other siliceous microfossils; they are typically thin bedded to finely laminated and occur in a range of colors from white and yellow to dark blue and red. The varicolored layered cherts of the Gregg Ranch Complex are distinctly different from the structureless black cherts of the Gazelle Formation.

Exotic lithologies of igneous and metamorphic origin include diorite and plagiogranite, spilitic pillow basalt, greenstones, serpentinite and serpentinite breccia, harzburgite, talcose metaperidotite, amphibolite and gneissic metagabbro, actinolite schist, pyroxenite, gabbro, phyllite, marble, and quartzite. The quartzite blocks are petrographically indistinguishable from calcareous metaquartzite of the Moffett Creek Formation of Hotz (1977, 1978), and may have been derived from it.

Numerous conglomeratic blocks with sheared margins are characterized by clasts of both native lithologies (shales, mudstones, graywackes) and exotic lithologies (reef carbonates, radiolarian cherts, phyllites, quartzites, and diorites). Many of these conglomeratic blocks contain clasts of native lithologies, and locally appear to be channeled into the surrounding melange matrix. Thus, the conglomerates probably formed as channel deposits and debris flows.

Stratigraphy and Structure

Evidence of deformation within rocks of the Gregg Ranch Complex is abundant. Boundaries of blocks may be recognized by an increase in shearing and the development of foliation parallel to the edge of the block; internal bedding, if present, is cut off abruptly at block boundaries. Blocks of more competent lithologies, like the cherts, are undeformed away from their edges, or else have responded to deformation by breaking along numerous healed microfaults. The carbonates commonly have responded to deformation by ductilely necking and

breaking into phacoidal blocks (Fig. 8); many shelly fossils contained within these carbonate blocks have been stretched and dismembered parallel to the long dimension of the block. Most conglomeratic blocks have been extensively modified by tectonic flattening of clasts and the development of a semipenetrative scaly foliation in the conglomerate matrix.

Large olistoliths and fault-bounded blocks in which original stratigraphies are preserved have been recognized at Bonnet Rock (Fig. 9), Robbers Rock, Horseshoe Gulch, Gregg Ranch, and Lovers Leap (Fig. 2). Several large olistoliths at Lovers Leap are 0.5 to 1.0 km long (Lindsley-Griffin, 1977b, 1982). Potter (1977, 1990a, 1990b; Potter et al., 1977) used the sequences preserved within these olistoliths and tectonic blocks to develop his excellent detailed biostratigraphy.

Tectonic breccia zones are readily recognized where they consist of exotic lithologies. Typically, a large exotic block of reef carbonate, radiolarian chert, or greenstone is surrounded by swarms of progressively smaller fragments that have been tectonically sliced from the original block and transported along shear surfaces within the scaly matrix. With continued transport and slicing, the fragments were progressively reduced in size until they merged with the muddy matrix.

Melange units

Informal melange units for rocks now included in the Gregg Ranch Complex were tabulated by Lindsley-Griffin (1977b, 1982), based on block assemblages and distinct faunas. Such units are useful because they permit the recognition of different components in a polygenetic melange, and may help to interpret its origin. For more detailed descriptions, see Lindsley-Griffin (1982).

The Bonnet Rock melange unit is characterized by numerous exotic blocks of chert and Silurian limestone, with sparse serpentinite (Churkin, 1958), red and green mudstone, and pillow lava; the poorly exposed matrix is mudstone. The Silurian fauna includes brachiopods, conodonts, and graptolites. The brachiopods are similar to those of Alaska and the Urals (Merriam, 1961; Boucot et al., 1973a).

Along Willow Creek and westward to Robbers Rock is a melange unit that contains limestone and chert blocks in matrix of mudstone and graywacke siltstone; sparse serpentinite blocks are also present (Churkin, 1958). Conodonts from limestones are Silurian to Early Devonian; the brachiopods, corals, and trilobites are Silurian (Wenlockian-Ludlovian). The trilobites are similar to specimens from the Silurian of Bohemia, and brachiopods are similar to those in Silurian rocks in Alaska (Churkin, 1961).

The Mountain House melange unit

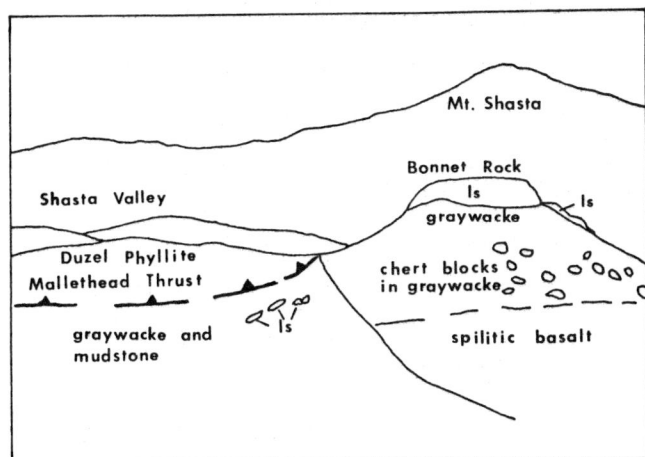

Figure 9. Bonnet Rock, a large olistolith within melange of the Gregg Ranch Complex. Sketch shows geologic relationships.

contains sparse coral-bearing limestone blocks of probable Early Devonian age, along with numerous blocks of radiolarian chert and chert breccia, and igneous and metamorphic rocks, in a strongly foliated mudstone, sandstone, and conglomerate matrix. Igneous and metamorphic rocks include actinolite schist, serpentinized harzburgite, talcose meta-peridotite, pyroxenite, amphibolite, gneissic metagabbro, marble, greenstone, and sparse phyllite and metaquartzite. Radiolarian cherts are of Devonian, and Ordovician to Silurian age (Irwin et al., 1978). Silurian brachiopods have been found within a block of metaquartzite (Lindsley-Griffin, 1982) that is lithologically similar to the Moffett Creek Formation.

Small slivers of Gregg Ranch Complex are exposed along Crater Creek and in Cabin Meadow (Fig. 2); these contain Lower Devonian coral-bearing limestone blocks, as well as volcanogenic graywackes and

conglomerates in matrix of mudstone and siltstone.

The Gregg Ranch melange unit contains numerous blocks of limestone, limestone conglomerate, recrystallized limestone, and graywacke conglomerate in a mudstone matrix. These limestones have yielded a suite of corals different from those at Lovers Leap (Merriam, 1972), Ordovician gastropods, Early Devonian conodonts, and brachiopods of Late Ordovician, Silurian(?), and Early Devonian age. A single block of thin-bedded carbonaceous shale contains Late Ordovician brachiopods and trilobites; the trilobites are of the Siberian-Western Europe Province (Boucot et al., 1973a).

The Lovers Leap melange unit consists of several large olistoliths of limestone, graywacke, and conglomerate in mudstone and graywacke matrix; limestones are of Middle Ordovician and Late Silurian age (Rohr, 1972), and one "trondhjemite" boulder in a conglomerate has an isotopic age of 440 m.y. (Mattinson and Hopson, 1972; Wallin et al., 1988). A single Silurian gastropod was found in sheared graywacke sandstone matrix (Lindsley-Griffin, 1982). The fauna is of Old World type (Potter and Boucot, 1971; Boucot et al., 1973a, 1973b).

In the Grouse Creek melange unit (Fig. 2), the sparse coral-bearing limestones are of Early Devonian age, based on a conodont fauna that is unique in the eastern Klamath Mountains (Savage, 1976, 1977). Shelly faunas are similar to the Tasman Subprovince (Boucot et al., 1973a). Other exotic lithologies include mafic and ultramafic rocks, greenstones, and a conglomerate composed of phyllite clasts that are very different in composition and structural style from other phyllites of the Yreka terrane.

The Horseshoe Gulch melange unit includes blocks of Ordovician, Silurian, and Siluro-Devonian limestone (Zdanowicz, 1971; Potter, 1987, 1990a, 1990b) and Middle Ordovician shale (Berry et al., 1973) in a scaly mudstone matrix. Other blocks include spilitic basalt, volcanolithic graywacke, and calcareous quartzite and phyllitic quartzite (Moffett Creek Formation?). The brachiopods and trilobites are typical of the Pacific Province (Potter and Boucot, 1971; Boucot et al., 1973a); the graptolites are similar to both Yukon and Nevada faunas (Berry et al., 1973).

Age and Fauna

Most of the lower Paleozoic fossils that have been found within the Yreka terrane are from melange blocks of the Gregg Ranch Complex. These blocks range in age from Middle Ordovician to Early Devonian, based on a rich fauna that includes brachiopods, gastropods, trilobites, graptolites, and conodonts. The age of a melange is not given by the age of the blocks contained within it, because the blocks must have formed before being

618

incorporated into the melange. The age of the youngest blocks within the melange provide only a maximum age of formation of the melange. Thus, the age of the Gregg Ranch Complex is Early to Middle Devonian.

The rich fauna contained within the blocks is characterized by many similarities to Pacific Rim provinces, especially among the brachiopods and trilobites; thus, it falls within the peripheral biofacies of Johnson (1983). This suggests a Pacific source area for most of the blocks in the Gregg Ranch Complex. However, the tenuous faunal links to North American (Nevada) may suggest that the basin in which these rocks were deposited was marginal to North America as well as to the Pacific.

Origin

The Gregg Ranch Complex is a classic example of an accretionary complex formed at an active convergent margin, as evidenced by its mixture of exotic and native litho-logies, structural style, and peripheral biofacies fauna. The greatly varied lithologic and paleontologic character of the melange units described above highlights the polygenetic nature of the Gregg Ranch Complex, with its fragments of many ages and origins. Olistoliths of shallow marine sediments probably were incorporated into the complex by submarine sliding; converse-ly, some may represent fragments of allochthonous terranes that were incor-porated into the complex by offscraping during subduction. The igneous and metamorphic lithologies are dominated by rock types characteristic of oceanic lithosphere (Lindsley-Griffin, 1977a, b); these may have been incorporated into the complex by structural slicing and infolding or by diapiric rise of fluid-saturated muds.

Contact Relationships

Where the Gazelle Formation rests upon melange of the Gregg Ranch Formation, local shearing and tectonic slicing disrupts bedding and produces a zone of imbricate fault-bounded slices of both units. Detailed mapping (Lindsley-Griffin and Fisher, 1989; Cram-Barry and Wallin, 1990) indicates the scaly foliation, which is characteristic of the structurally lower Gregg Ranch Complex, disappears rapidly above the contact by passing upward into an incipient fissility that rapidly changes into relatively undeformed strata. From the contact zone downward into the structurally lower Gregg Ranch Complex, the degree of shearing increases over a short distance (< 10 m) as fissility grades rapidly into scaly foliation, and the tectonic slices of Gazelle Formation become progressively smaller and fewer until they are completely replaced by the phacoidal blocks of the Gregg Ranch Complex. These relationships are similar to those observed in modern accretionary complexes such as the Peru Margin (Lindsley-Griffin and Fisher, 1989; Kemp *et al.*, 1990).

Some blocks within melange of the Gregg Ranch Complex are lithologically indistin-guishable from the thinly laminated, radiolarian-rich, siliceous shales of the Gazelle Formation. We interpret this as evidence that part of the Gazelle Formation may have formed prior to the Gregg Ranch Complex, and was incorporated into the melange either as olistoliths or as tectonic slices. However, structural evidence supports the interpretation that portions of the two units may be coeval as suggested by Lindsley-Griffin and Fisher (1989). The contact between the two units is a fault characterized by both syndepositional and postdepositional shearing and tectonic slicing. Thus, part of the Gazelle Formation appears to have been deposited over the Gregg Ranch Complex while the accretionary complex was actively forming.

EARLY DEVONIAN PALEOGEOGRAPHIC SETTING

The Gazelle Formation most likely was deposited in an Early Devonian trench or trench-slope basin, and was derived at least in part from an older Ordovician-Silurian volcanic terrane (Wallin, 1989a, b). The abundance of siliceous microfossils suggests the basin may have been located on a west-facing active margin in a zone of marine upwelling (Lindsley-Griffin and Fisher, 1989; Cram-Barry and Wallin, 1990). The Gregg Ranch Complex consists of trench sediments and exotic blocks of olistoliths and tectonic slices that comprised the accretionary complex at this same Early Devonian convergent margin.

The Gazelle-Gregg Ranch assemblage comprises one component of the Yreka terrane, elements of which began forming in the Middle Ordovician. From Late Silurian to early Middle Devonian, the east-dipping central metamorphic terrane was subducted beneath a plate of which the Trinity complex formed the leading edge (Lindsley-Griffin, this volume). Culmination of this tectonic interaction occurred in the Early (to Middle?) Devonian with the formation of the Gazelle-Gregg Ranch assemblage, thrusting of the Yreka terrane over the Trinity complex, and amalgamation of the central metamorphic terrane to the Trinity-Yreka terrane (Lindsley-Griffin, this volume).

Acknowledgements

This paper is dedicated to Rodney Gregg, in acknowledgement of the many decades of assistance and encouragement he has given to geologists and students working in the eastern Klamath Mountains. We thank our students Marc Fisher (University of Nebraska-Lincoln) and Christi Cram-Barry (University of Nevada-Las Vegas) for their assistance with field work. We thank A.W. Potter for lively discussion about field relationships in the region. Field and lab work by Lindsley-Griffin and Griffin was supported by the UNL Research Council and Department of Geology; field and lab work by Wallin was supported by NSF Grant #EAR89-16161 and the UNLV University Research

Council. We thank K.R. Aalto and J.D. Cooper for reviewing the manuscript.

REFERENCES CITED

Averill, C.V., 1931, Preliminary report on economic geology of the Shasta quadrangle: California Division of Mines Report XXVII of the State Mineralogist, v. 27, p. 2-65.

Bates, R.L., and Jackson, J.A., eds., 1987, Glossary of geology, third edition: Alexandria, VA, American Geological Institute, 788 p.

Bergstrom, S.M., Potter, A.W., Porter, R.W., Boucot, A.J., and Rohr, D.W., 1980, Biostratigraphic and biogeographic significance of Ordovician conodonts in the eastern Klamath Mountains, northern California (abs.): Geological Society of America Abstracts with Programs, v. 12, n. 5, p. 219.

Berry, W.B.N., and Boucot, A.J., 1970, Correlation of North American Silurian rocks: Geological Society of America Special Paper 102, 153 p.

Berry, W.B.N., Lindsley-Griffin, N., Potter, A.W., and Rohr, D.M., 1973, Early Middle Ordovician graptolites from the eastern Klamath Mountains, Siskiyou County, California (abs.): Geological Society of America Abstracts with Programs, v. 5, p. 11.

Boucot, A.J., 1971, Aenigmastrophia, new genus, a difficult Silurian brachiopod: Smithsonian Contributions to Paleobiology Number 3, p. 155-158.

Boucot, A.J., Dean, W.T., Martinsson, A., Potter, A.W., Rexroad, C.B., Rohr, D.M., Savage, N.M., and Wright, A.J., 1973a, Biogeographic relations of the pre-late Middle Devonian of the eastern Klamath belt, northern California (abs.): Geological Society of America Abstracts with Programs, v. 5, p. 14.

Boucot, A.J., Dean, W.T., Martinsson, A., Potter, A.W., Rexroad, C.B., Rohr, D.M., Savage, N.M., and Wright, A.J., 1973b, Pre-late Middle Devonian biostratigraphy of the Eastern Klamath belt, northern California (abs.): Geological Society of America Abstracts with Programs, v. 5, p. 15.

Boucot, A.J., Dunkle, D.H., Potter, A., Savage, N.M., and Rohr, D., 1974, Middle Devonian orogeny in western North America?: A fish and other fossils: Journal of Geology, v. 82, p. 691-708.

Brewer, W.A., III, 1955, The geology of a portion of the China Mountain quadrangle, California [unpublished M.A. thesis]: Berkeley, California, University of California, 47 p.

Churkin, M., Jr., 1958, Silurian stratigraphy of part of the Yreka and China Mountain quadrangles, Siskiyou County, California [unpublished M.A. thesis]: Berkeley, California, University of California, 85 p.

_____1960, Silurian strata of the Klamath Mountains, California: American Journal of Science, v. 258, p. 258-273.

_____1961, Silurian trilobites from the Klamath Mountains, California: Journal of Paleontology, v. 35, p. 168-175.

_____1965, First occurrence of graptolites in the Klamath Mountains, California: United States Geological Survey Professional Paper 525-C, p. C72-C73.

Churkin, M., Jr., and Langenheim, R.L., Jr., 1960, Silurian strata of the Klamath Mountains, California: American Journal of Science, v. 258, p. 258-273.

Cowan, D.S., 1985, Structural styles in Mesozoic and Cenozoic melanges in the western Cordillera of North America: Geological Society of America Bulletin, v. 96, p. 451-462.

Cram-Barry, C.A., and Wallin, E.T., 1990, Stratigraphy and sedimentology of the Devonian Gazelle Formation, eastern Klamath Mountains, N. California (abs.): Geological Society of America Abstracts with Programs, v. 22, p. 15-16.

Diller, J.S., 1886, Notes on the geology of northern California: United States Geological Survey Bulletin 33, 32 p.

_____1902, Topographic development of the Klamath Mountains: United States Geological Survey Bulletin 196, 69 p.

_____1903, Klamath Mountains section: American Journal of Science, Fourth Series, v. 15, p. 342-362.

_____1906, Description of the Redding quadrangle, California: United States Geological Survey Geologic Atlas, Folio 138, 14 p.

Diller, J.S., and Schuchert, C., 1894, Discovery of Devonian rocks in California: American Journal of Science, Third Series, v. 37, p. 416-422.

Elias, R.J., and Potter, A.W., 1984, Late Ordovician solitary rugose corals of the eastern Klamath Mountains, northern California: Journal of Paleontology, v. 58, 5, p. 1203-1214.

Griffin, J.R., 1971, Lower Paleozoic geosyncline of the Klamath Mountains: possible relationship to continental margin: Report to Geological Society of America Research Committee on Penrose Grant Number 1392-70: Boulder,

Colorado, Geological Society of America, 17 p.

Hershey, O.H., 1901, Metamorphic formations of northwestern California: The American Geologist, v. 27, p. 225-245.

Heyl, G.R., and Walker, G.W., 1949, Geology of limestone near Gazelle, Siskiyou County, California: California Journal of Mines and Geology, v. 45, p. 514-520.

Hopson, C.A., and Mattinson, J.M., 1973, Ordovician and Late Jurassic ophiolitic assemblages in the Pacific northwest (abs.): Geological Society of America Abstracts with Programs, v. 5, p. 57.

Hotz, P.E., 1977, Geology of the Yreka Quadrangle, Siskiyou County, California: U.S. Geological Survey Bulletin 1436, 72 p.

_____1978, Geologic map of the Yreka Quadrangle and parts of the Fort Jones, Etna, and China Mountain Quadrangles, California: U.S. Geological Survey Open File Report 78-12, 1:62,500.

Hsu, K.J., 1968, Principles of melanges and their bearing on the Franciscan-Knoxville paradox: Geological Society of America Bulletin, v. 79, p. 1063-1074.

_____1974, Melanges and their distinction from olistostromes, in Dott, R.H., Jr., and Shaver, R.H., eds., Modern and ancient geosynclinal sedimentation: Society of Economic Paleontologists and Mineralogists Special Publication No. 19, p. 321-333.

Irwin, W. P., 1960, Geologic reconnaissance of the northern Coast Ranges and Klamath Mountains, California, with a summary of the mineral resources: California Division of Mines Bulletin 179, 80 p.

_____1981, Tectonic accretion of the Klamath Mountains, in Ernst, W.G., ed., The geotectonic development of California: Rubey Volume 1: Englewood Cliffs, NJ, Prentice-Hall, p. 29-49.

Irwin, W.P., Jones, D.L., and Kaplan, T.A., 1978, Radiolarians from pre-Nevadan rocks of the Klamath Mountains, California and Oregon, in Howell, D.G., and McDougall, K.A., eds., Mesozoic paleogeography of the western United States: Pacific Coast Paleogeography Symposium 2: Society of Economic Paleontologists and Mineralogists, Pacific Section Publication No. 8, p. 303-310.

Johnson, J.H., and Konishi, K., 1959, Some Silurian calcareous algae from northern California and Japan, in Johnson, J.H., Konishi, K., and Rezak, R., eds., Studies of Silurian (Gotlandian) algae:

Colorado School of Mines Quarterly, v. 54, p. 131-158.

Johnson, J.G., 1983, Early Paleozoic Cordilleran biofacies, provinces, and realms, in Stevens, C.H. ed., Pre-Jurassic rocks in western North America suspect terranes: Society of Economic Paleontologists and Mineralogists, Pacific Section Publication No. 32, p. 1-5.

Kemp, A., and Lindsley-Griffin, N., 1990, Variations in structural style within Peruvian forearc sediments, in Suess, E., von Huene, R., et al., Proceedings of the Ocean Drilling Program, Scientific Results, 112: College Station, TX, (Ocean Drilling Program), p. 17-31.

Lindsley-Griffin, N., 1973, Lower Paleozoic ophiolite of the Scott Mountains, eastern Klamath Mountains, California (abs.): Geological Society of America Abstracts with Programs, v. 5, p. 71-72.

_____1975, Geology of the Northwestern edge of the Trinity ophiolite, eastern Klamath Mountains, California (abs.): EOS, Transactions of the American Geophysical Union, v. 56, n. 12, p. 1079.

_____1977a, Early Paleozoic subduction complex in the eastern Klamath Mountains: Implications for Cordilleran tectonic models (abs.): Geological Society of America Abstracts with Programs, v. 9, n. 4, p. 454.

_____1977b, Paleogeographic implications of ophiolites: the Ordovician Trinity complex of the Klamath Mountains, northern California, in Stewart, J.H., Stevens, C.H., and Fritsche, A.E., eds., Paleozoic paleogeography of the western United States: Pacific Coast Paleogeography Symposium 1: Society of Economic Paleontologists and Mineralogists, Pacific Section Publication No. 7, p. 409-420.

_____1982, Structure, stratigraphy, petrology and regional relationships of the Trinity ophiolite, eastern Klamath Mountains, California [unpublished Ph.D. dissertation]: Davis, California, University of California, 453 p.

_____1983, Lower Paleozoic rocks of the eastern Klamath Mountains, Northern California; constraints on early Paleozoic paleogeography (abs.): Geological Society of America Abstracts with Programs, v. 15, p. 426-427.

Lindsley-Griffin, N., and Rohr, D.M., 1977, Lovers Leap: A geologic puzzle, in Lindsley-Griffin, N., and Kramer, J.C. eds., Geology of the Klamath Mountains, northern California: Geological Society of America, Cordilleran Section

Guidebook, p. 47-69.

Lindsley-Griffin, N., and Griffin, J.R., 1983, The Trinity terrane: An early Paleozoic microplate assemblage, in Stevens, C.H., ed., Pre-Jurassic rocks in western North American suspect terranes: Society of Economic Paleontologists and Mineralogists, Pacific Section Publication No. 32, p. 63-75.

Lindsley-Griffin, N., and Fisher, M., 1989, Significance of the tectonic contact between lower Paleozoic melange and shales of the Gazelle Formation, Yreka terrane, Klamath Mountains, California (abs.): Geological Society of America Abstracts with Programs, v. 21, p. 107.

Lindsley-Griffin, N., Rohr, D.M., Griffin, J.R., and Moores, E.M., 1974, Geology of the Lovers Leap area, eastern Klamath Mountains, California, in McGeary, D.F.R., ed., Geological Guide to the Klamath Mountains, Annual Field Trip Guidebook, Sacramento, California, Geological Society of Sacramento, p. 82- 100.

Lindsley-Griffin, N., Kemp, A., and Swartz, J., 1990, Vein structures of the Peru margin, Leg 112, in Suess, E., von Huene, R., et al., Proceedings of the Ocean Drilling Program, Scientific Results, 112: College Station, TX, (Ocean Drilling Program), p. 3-16.

Mattinson, J.M., and Hopson, C.A., 1972, Paleozoic ophiolitic complexes in Washington and northern California: Carnegie Institution Yearbook, v. 71, p. 578-583.

Maxwell, J.C., 1974, Anatomy of an orogen: Geological Society of America Bulletin, v. 85, p. 1195-1204.

Merriam, C.W., 1961, Silurian and Devonian rocks of the Klamath Mountains, California: United States Geological Survey Professional Paper 424-C, p. 188-190.

_____1972, Silurian rugose corals of the Klamath Mountains region, California: United States Geological Survey Professional Paper 738, 50 p.

North American Commission on Stratigraphic Nomenclature, 1983, North American stratigraphic code: American Association of Petroleum Geologists Bulletin, v. 67, p. 841-875.

Oliver, W.A., Jr., 1975, Age of corals from northern California: Journal of Paleontology, v. 49, p. 424.

Olson, G.A., 1978, Geology of the Grouse Creek area, China Mountain Quadrangle, California [unpublished M.S. thesis]: Corvallis, Oregon, Oregon State University, 129 p.

Peacock, S.M., 1987, Serpentinization and infiltration metasomatism in the Trinity peridotite, Klamath province, northern California: Implications for subduction zones: Contributions to Mineralogy and Petrology, v. 95, p. 55-70.

Peacock, S.M., and Norris, P.J., 1989, Metamorphic evolution of the central metamorphic belt, Klamath province, California: an inverted metamorphic gradient beneath the Trinity peridotite: Journal of Metamorphic Geology, v. 7, p. 191-209.

Potter, A.W., 1977, Stratigraphy of Moffett Creek and Gazelle Formation northeast of Gregg Ranch, eastern Klamath Mountains, northern California, in Lindsley-Griffin, N., and Kramer, J.C., eds., Geology of the Klamath Mountains, northern California: Geological Society of America Cordilleran Section, Field Trip Guidebook, p. 80-103.

Potter, A.W., 1987, Stratigraphy and selected Ordovician brachiopods from the Horseshoe Gulch and Gregg Ranch areas, eastern Klamath Mountains, northern California [unpublished Ph.D. dissertation]: Corvallis, Oregon, Oregon State University, 367 p.

Potter, A.W., 1990a, Middle and Late Ordovician brachiopods from the eastern Klamath Mountains, northern California, Part I: Palaeontographica abt A, v. 212, p. 31-158.

Potter, A.W., 1990b, Middle and Late Ordovician brachiopods from the eastern Klamath Mountains, northern California, Part II: Palaeontographica abt A, v. 213, p. 1-114.

Potter, A.W., and Boucot, A.J., 1971, Ashgillian, Late Ordovician brachiopods from the eastern Klamath Mountains of northern California (abs.): Geological Society of America Abstracts with Programs, v. 3, p. 180-181.

Potter, A.W., Hotz, P.E., and Rohr, D.M., 1977, Stratigraphy and inferred tectonic framework of lower Paleozoic rocks in the eastern Klamath Mountains, northern California, in Stewart, C.H., and Fritsche, A.E., eds., Paleozoic paleogeography of the western United States: Pacific Coast Paleogeography Symposium 1: Society of Economic Paleontologists and Mineralogists, Pacific Section Publication No. 7, p. 421-440.

Raymond, L.A., ed., 1984, Melanges: Their nature, origin, and significance: Geological Society of America Special Paper 198, 170 p.

Rigby, J. Keith, and Potter, Alfred W., 1980, Ordovician sphinctozoan sponges from the Klamath Mountains, California,

The first known early Paleozoic occurrence (abs.): Geological Society of America Abstracts with Programs, v. 12, p. 509-510.

Rohr, D.M., 1972, Geology of the Lovers Leap area, China Mountain Quadrangle, California [unpublished M.S. thesis]: Corvallis, Oregon, Oregon State University, 95 p.

_____1978, Stratigraphy, structure, and early Paleozoic gastropoda of the Callahan area, Klamath Mountains, California [unpublished Ph.D. dissertation]: Corvallis, Oregon, Oregon State University, 182 p.

_____1980, Ordovician-Devonian gastropoda from the Klamath Mountains, California: Paleontographica abt. A., v. 171, p. 141-199.

Rohr, D.M., and Boucot, A.J., 1971, Northern California (Klamath Mountains) pre-Late Silurian igneous complex (abs.): Geological Society of America Abstracts with Programs, v. 3, p. 186.

Rohr, D.M., and Potter, A.W., 1973, Paleozoic rocks of the Callahan-Gazelle area, Klamath Mountains, northern California (abs.): Geological Society of America Abstracts with Programs, v. 5, p. 97.

Rohr, D.M., Boucot, A.J., and Potter, A.W., 1971, Northern California (Klamath Mountains) pre-Late Silurian igneous complex (abs.): Geological Society of America Abstracts with Programs, v. 3, p. 186.

Savage, N.M., 1976, Lower Devonian (Gedinnian) conodonts from the Grouse Creek area, Klamath Mountains, northern California: Journal of Paleontology, v. 50, p. 1180-1190.

Savage, N.M., 1977, Lower Devonian conodonts from the Gazelle Formation, Klamath Mountains, northern California: Journal of Paleontology, v. 51, p. 57-62.

Silver, E.A., and Beutner, E.C., 1980, Melanges--Penrose conference report: Geology, v. 8, p. 32-34.

Stauffer, C.R., 1930, The Devonian of California: California University Department of Geological Sciences Bulletin, v. 19, p. 81-118.

Strand, R.G., 1962, Geologic atlas of California, Redding sheet: California Division of Mines and Geology, 1:250,000.

Wagner, D.L., and Saucedo, G.J., compilers, 1987, Geologic map of the Weed quadrangle, 1:250,000, California: California Division of Mines and Geology Regional Geologic Map Series, Map 4A.

Wallin, E.T., 1988, Evolution of early Paleozoic terrigenous clastic sedimentation in the Yreka-Callahan terrane, eastern Klamath Mountains, California (abs.): Geological Society of America Abstracts with Programs, v. 20, p. 231.

_____1989a, Provenance of lower Paleozoic sandstones in the eastern Klamath Mountains and the Roberts Mountains allochthon, California and Nevada [unpublished Ph.D. dissertation]: Lawrence, Kansas, University of Kansas, 152 p.

_____1989b, Reconstruction of the timing and sequence of early Paleozoic sedimentation in the Yreka-Callahan terrane, eastern Klamath Mountains, California (abs.): Geological Society of America Abstracts with Programs, v. 21, p. 155.

Wallin, E.T., Mattinson, J.M., and Potter, A.W., 1988, Early Paleozoic magmatic events in the eastern Klamath Mountains, northern California: Geology, v. 16, p. 144-148.

Wells, F.G., Walker, G.W., and Merriam, C.W., 1959, Upper Ordovician (?) and Upper Silurian formations of the northern Klamath Mountains, California: Geological Society of America Bulletin, v. 70, p. 645-649.

Westman, B.J., 1947, Silurian of the Klamath Mountains province, California (abs.): Geological Society of America Bulletin, v. 58, p. 1263.

Zdanowicz, T.A., 1971, Stratigraphy and structure of the Horseshoe Gulch area, Etna and China Mountain quadrangles, California [unpublished M.S. thesis]: Corvallis, Oregon, Oregon State University, 88 p.

APPENDIX I. PREVIOUS WORK

Fossils of the Gregg Ranch Complex

The majority of paleontological localities within the Yreka terrane (Yreka-Callahan terrane of Irwin, 1981) occur within the Gregg Ranch Complex. The first geologist to study these fossils was Diller (1886, 1902, 1903; Diller and Schuchert, 1894), who wrote about the fossiliferous limestones that lie within the drainages of Willow Creek and the Lovers Leap--Gregg Ranch area (Fig. 2). Other early workers who described fossils from the Gregg Ranch Complex are Hershey (1901), Stauffer (1930), Averill (1931), Merriam (1961, 1972), Westman (1947), Churkin (1958, 1960, 1961, 1965; Churkin and Langenheim, 1960), Johnson and Konishi (1959), and Wells *et al.* (1959).

Beginning in the early 1970s, A.J. Boucot, his students, and their colleagues began a major project to understand the paleontology of the Callahan-Gazelle area. As a result of this work, many new paleontological localities have been discovered and old localities have been recollected from rocks within our newly defined Gregg Ranch Complex. Papers describing these fossils, their ages, and the rocks that contain them include Boucot (1971), Zdanowicz (1971), Rohr (1972, 1978, 1980), Berry *et al.* (1973), Boucot *et al.* (1973a, 1973b), Oliver (1975), Savage (1976), Olson (1978), Potter (1987, 1990a, 1990b), Bergstrom *et al.* (1980), Rigby and Potter (1980), Elias and Potter (1984).

Definitions of the Gazelle Formation

Wells, Walker, and Merriam (Wells *et al.*, 1959) divided the rocks between Shasta and Scott Valleys (Yreka terrane) into two units and named them the Duzel Formation and the Gazelle Formation. They (p. 646) described the Gazelle Formation as consisting of "hard, fine-grained, siliceous graywacke, dark-gray to black siltstone and mudstone, and siliceous and feldspathic grit. It also contains appreciable quantities of chert conglomerate, limestone, and limestone conglomerate, and is overlain by volcanic flow, agglomerates and pyroclastic rocks."

Merriam (1961, p. 189) described the Gazelle Formation as consisting of "gray greenish and reddish shale, gritty sandstone and graywacke, siliceous conglomerate, bedded chert and massive gray limestone.... tuffaceous interbeds are present." This description includes melange units (Gregg Ranch Complex) of the Bonnet Rock and Willow Creek area as well as the shales of Gazelle Mountain (restricted Gazelle Formation of this paper). In 1972, Merriam redefined the Gazelle Formation and established a type locality at Bonnet Rock, within melange of our Gregg Ranch Complex.

Potter (1977; also Potter *et al.*, 1977) developed a detailed chronostratigraphy of the Gregg Ranch area, based partly on the intact stratigraphic relationships preserved within some melange blocks. These rocks all belong within the newly defined Gregg Ranch Complex.

Lindsley-Griffin and Griffin (1983) proposed that part of the Gazelle Formation consists of melange and suggested separation of Gazelle into two distinct units.

Stratigraphy

As early as 1949, Heyl and Walker recognized the discontinuous nature of the limestones of the Gregg Ranch Complex. They pointed out that the stratigraphic section below the limestone at Bonnet Rock (Fig. 2) varies from place to place, and concluded that "the chert, the conglomerate, and in places the limestone occur as lenses or discontinuous beds in the shale and sandstone" (p. 515, 517). They described limestone bedding as being discordant with underlying rocks, and suggested that "the limestone apparently was thrust eastward relative to the underlying crumpled shale and chert" (p. 517).

In the Bonnet Rock-Willow Creek area, Churkin (1958) described isolated occurrences of serpentinite within sedimentary rock by saying "These rocks are composed of green flaky antigorite serpentine and are strongly sheared, having developed under stress along faults and fractures in the sedimentary rocks" (p. 27). He also noted the discordant nature of bedding in the main limestone block relative to bedding in the shale and sandstone immediately below. Although his work predates the introduction of the melange concept to North America by Hsu (1968, 1974), Churkin recognized that the disruption in these rocks was tectonic in origin.

Wells *et al.* (1959, p. 648) pointed out the discontinuous nature of some of their informal members, particularly the limestone, in their original definition of the Gazelle Formation.

Zdanowicz (1971, p. 62), in his M.S thesis on Horseshoe Gulch, observed the shattered characteristic and lack of stratal continuity of the rocks of Horseshoe Gulch: "Nowhere in the area has the mapping shown any sedimentary contacts."

The first person to propose that rocks of the Gazelle Ranch Complex represent tectonic melange was J.R. Griffin (1971); structural studies confirming the disrupted nature of the assemblage were completed by Lindsley-Griffin (1977, 1982). Papers supporting the polygenetic melange interpretation include Lindsley-Griffin and Griffin (1983), Lindsley-Griffin and Fisher (1989), Wallin (1989a, 1989b), and Cram-Barry and Wallin (1990).

Numerous other geologists have mapped within the Gregg Ranch Complex, many of whom also recognized its disrupted character and struggled with the problem of its stratigraphic relationships. They include Brewer (1955), Churkin (1958, 1961; Churkin and Langenheim, 1960), California Institute of Technology students under the direction of A. J. Boucot (unpublished maps, 1964), Zdanowicz (1971), Rohr (1972, 1978; Rohr and Boucot, 1971; Rohr and Potter 1973; Rohr *et al.*, 1971), Lindsley-Griffin (1973, 1975, 1977a, 1977b, 1982, 1983; Lindsley-Griffin *et al.*, 1974; Lindsley-Griffin and Rohr 1977; Lindsley-Griffin and Griffin, 1983), Hotz (1977, 1978), Olson (1978), Potter (1977, 1990; Potter *et al.*, 1977), Fisher (Lindsley-Griffin and Fisher, 1989), Wallin (1988b), and Cram-Barry (Cram-Barry and Wallin, 1990).

Correlations

The fossiliferous limestones of the Willow Creek area (Fig. 2) were initially considered to be Carboniferous by J.S.

Diller (1886). However, Diller later recognized the age of the fossils to be Devonian (Diller, 1902, 1903, 1906; Diller and Schuchert, 1894), and correlated the rocks with the Kennett Formation. Averill (1931) correlated rocks of the Willow Creek area with the Kennett and Bragdon Formations of Diller (1906). Brewer (1955) also considered the Gazelle Formation to be correlative with the Kennett Formation. In 1960, Irwin pointed out that the correlation of Gazelle area rocks with the Kennett Formation was inaccurate. Merriam (1961) correlated Silurian rocks of his Gazelle Formation with rocks of the Taylorsville area in the Sierra Nevada. He also stated that the Gazelle no longer should be correlated with the Kennett, but suggested instead: "it is not, however, unlikely that the Gazelle formation with its tuffaceous interbeds is correlative with the Copley greenstone and Balaklala rhyolite which underlie the Kennett" (p. 189).

As paleontological data accumulated and a better understanding of the age of the Gazelle Formation and Gregg Ranch Complex was achieved, early attempts to correlate these rocks with other units were discarded, and these lower Paleozoic rocks were recognized as a distinct assemblage. However, most of the existing correlations of Gazelle Formation are based on fossils from the Gregg Ranch Complex. For example, Silurian localities which supposedly establish the age of the Gazelle Formation of Berry and Boucot (1970) are actually from melange blocks rather than shales and siltstones of the Gazelle Formation.

Merriam (1961) considered the rocks of the Willow Creek-Bonnet rock area to be of Silurian and Early Devonian age. He also placed rocks of the Lovers Leap area (Parker Rock), Grouse Creek area, and McConnaughy Creek (Horseshoe Gulch) in the Gazelle Formation. We consider these rocks to be part of the Gregg Ranch Complex.

THE BILLY'S PEAK MAFIC COMPLEX OF THE TRINITY SHEET, CALIFORNIA:
ROOTS OF A PALEOZOIC ISLAND ARC

Scott W. Petersen
Westinghouse Hanford
P.O. Box 1970
Richland, WA 99352

Calvin G. Barnes
Department of Geosciences
Texas Tech University
Lubbock, TX 79409

James D. Hoover
Westinghouse Hanford
P.O. Box 1970
Richland, WA 99352

ABSTRACT

The Billy's Peak mafic complex is located in the Trinity sheet of the eastern Klamath belt. It consists of a complex assemblage of gabbro, diabase, and trondhjemite, which intruded ultramafic rocks of the Trinity sheet. At least five episodes of intrusion occurred, beginning with gabbro and ending with the last phase of trondhjemite injection. The western half of this complex contains diabase dikes with east-west strike and steep southward dip.

Many workers have assigned the Trinity sheet an ophiolitic origin, although it is anomalous in many respects. Our mineralogical, structural, and geochemical work suggest that the Billy's Peak mafic complex was formed in an island-arc setting, and is not a classic ophiolite. Similarities with other mafic intrusions in the Trinity sheet indicate that these intrusions may have formed in an oceanic island-arc setting.

INTRODUCTION

The Billy's Peak mafic complex is located in the west-central portion of the Trinity mafic-ultramafic sheet (Fig. 1). The Trinity sheet forms the westernmost, basal portion of the eastern Klamath belt, which is composed of Lower Paleozoic metavolcanic and metasedimentary rocks in the north, and Silurian to Jurassic arc-related rocks to the southeast. This entire sequence has also been called the eastern Klamath plate (Irwin, 1981) and the eastern Klamath terrane (Irwin, 1985).

The Trinity sheet is a structurally complex unit which consists predominantly of ultramafic and gabbroic rocks. Many of these rocks have been metamorphosed to at least lower greenschist facies, but in some areas they retain their original igneous textures and, locally, their primary mineralogy. Those ultramafic rocks unaffected by structural or metamorphic disruption indicate an upper mantle/lower crust origin in an oceanic tectonic regime. In addition to ultramafite and gabbro,

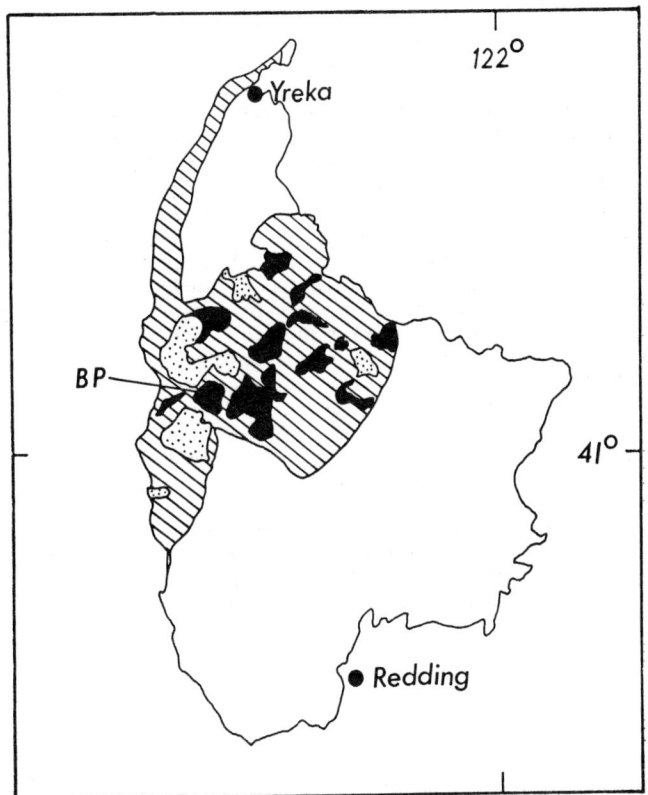

Figure 1. General geology of the Trinity sheet within the eastern Klamath belt (outline). Ultramafic rocks of the Trinity sheet are shown in a diagonal pattern. Black bodies are mafic plutons and associated rocks; stippled pattern represents Mesozoic plutons. BP=Billy's Peak mafic complex.

outcrops of basaltic rocks and subordinate amounts of intermediate and felsic igneous rocks are present throughout the Trinity sheet.

The Trinity sheet contains some of the oldest rocks in the Klamath Mountains, and possibly the entire western Cordillera. Plagioclase peridotite has yielded a Nd/Sm

In Cooper, J.D., and Stevens, C.H., eds., 1991, **Paleozoic Paleogeography of the Western United States-II**: Pacific Section SEPM, Vol. 67, p. 625-633.

date of 472 ± 32 Ma (Jacobsen and others, 1984). Leucocratic rocks associated with some of these gabbroic complexes have uranium-lead zircon ages of 440-475 Ma (Mattinson and Hopson, 1972; Wallin and others, 1988). Gabbros near the Billy's Peak area have hornblende K/Ar dates of 418 ± 17 and 439 ± 18 Ma (Lanphere and others, 1968), but these are now generally considered as minimum ages. Too few samples have been dated to determine the temporal relationship between ultramafite and gabbro, or the age distribution of the gabbroic intrusions of the Trinity sheet. It seems reasonable to assume that most or all of the gabbro was intruded into a relatively cool host sometime before the eastern Klamath belt was accreted to the continental margin, in late Devonian time.

The structural style of the Trinity sheet varies greatly. In the Yreka-Callahan area (northern portion of the Trinity sheet), Lindsley-Griffin (1977) described the Trinity sheet as an ophiolitic melange, containing all of the lithologies present in ophiolites. Most of the rest of the Trinity sheet appears to be more coherent, consisting of peridotite (typically serpentinized) and gabbro. It is difficult to determine the degree of structural disruption in the peridotite, owing to poor exposures; however, the relative structural integrity seen throughout much of the area (see, for example, Goulloud, 1977; Quick, 1981), and clear evidence that most of the gabbros intrude serpentinite (Goulloud, 1977; Petersen, 1987; Boudier and others, 1989), indicate that the bulk of the Trinity sheet is not melanged.

The Billy's Peak mafic complex is a small body of igneous rocks that intrude the western part of the Trinity sheet (Fig. 2). Most maps show this area as a southern lobe

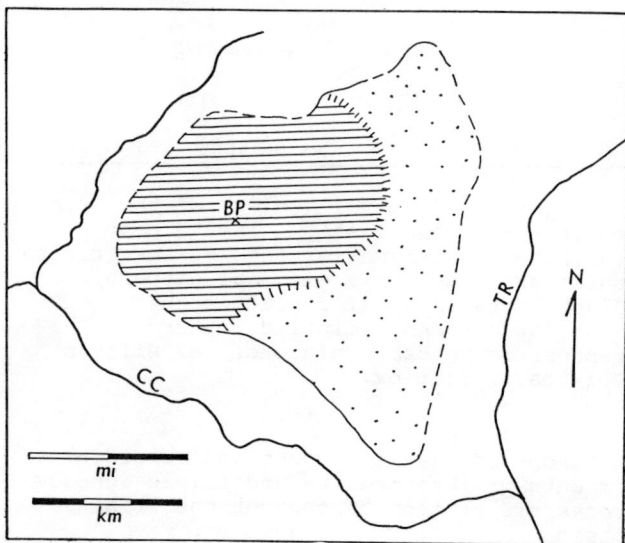

Figure 2. Schematic of geology of the Billy's Peak mafic complex. Horizontal pattern is the dike complex, stippled area represents gabbroic rocks. CC=Coffee Creek; TR=Trinity river; BP=Billy's Peak.

of the Craggy Peak pluton. The Billy's Peak complex covers about 50 km^2, and consists predominantly of gabbro and basaltic dikes, with minor amounts of siliceous dikes and sills, whereas the Craggy Peak pluton is a late-Jurassic dioritic to trondhjemitic pluton. The mineralogy and intrusive nature of the Billy's Peak mafic complex is very similar to several neighboring gabbroic plutons (Fig. 1), which is suggestive of a shared geologic history.

LITHOLOGY AND MINERALOGY

Over 90% of the rocks in the Billy's Peak mafic complex are basaltic in composition. Textures of these basic rocks range widely, from very coarse-grained pegmatitic gabbro to aphanitic basalt and andesite. The remaining <10% of the complex is composed of leucocratic dikes and sills, most of which are tonalitic and trondhjemitic in composition. Most of these igneous rocks have been metamorphosed to lower greenschist facies. Xenoliths of the host serpentinite are present throughout the complex, but are most abundant toward the margins.

The eastern portion of the Billy's Peak mafic complex consists almost exclusively of hornblende gabbro, whereas the western half consists of basaltic dikes with subordinate dikes and pods of tonalite-trondhjemite and screens of gabbro. It is this western part that makes the complex unique in the Trinity sheet: very few dike complexes have been found in the Trinity sheet, and the Billy's Peak area contains by far the largest and best exposed area of these structures yet mapped.

Gabbro

Gabbro intrudes the ultramafic rock surrounding the Billy's Peak area, and occurs as screens between basaltic dikes. The gabbro consists mainly of Ca-amphibole, plagioclase, and opaque minerals. Nearly all of the rocks have been metamorphosed to lower greenschist facies, but patches have escaped textural and chemical alteration. The primary assemblage is in many cases altered to actinolitic amphibole and Na-plagioclase + epidote. Grain size varies from medium-grained (2-5 mm) to very coarse. Locally, pegmatitic gabbro contains individual hornblende crystals up to 20 cm in length. Most of the gabbroic rocks are melagabbro with color index as high as 75, but gabbroic lithologies range from rare hornblendeite to felsic pegmatites with a color index of 35. Primary textures are predominantly cumulate; primary plagioclase, olivine, and pyroxene are present. One sample of medium-grained gabbro (sample 434) contained unzoned plagioclase crystals which ranged from An_{73} to An_{94} in composition. Primary amphibole in this same sample contained high Al_2O_3 and TiO_2 contents (9.81 and 1.41 wt. % respectively).

Figure 3. Size layering in pegmatitic hornblende gabbro. Hammer handle is 40 cm long.

Most of the gabbro is isotropic, but layering is locally present (Fig. 3). The rare layering by phase or grain size is discontinuous over even a few meters, and is typically truncated by other intrusions (see discussion in *Structure*, below).

Basalt

Basaltic dikes are the predominate lithology in the western half of the Billy's Peak area (Fig. 2). The two main types are aphanitic and porphyritic; both cross-cut the gabbro. As will be discussed later, compositions range from basaltic to andesitic, but will be termed basaltic for simplicity. These dikes are composed predominantly of Ca-amphibole and plagioclase, with varying amounts of quartz and opaque minerals. Relict igneous textures such as plagioclase phenocrysts and flow alignment of primary amphibole are preserved.

As with the gabbros, many of the basaltic dikes are metamorphosed; they exhibit sausseritized plagioclase and recrystallized, fibrous amphibole. However, many samples were apparently not affected by metamorphism, and single samples display a large amount of textural heterogeneity. The anorthite content in plagioclase of an aphanitic basaltic dike (sample 425B) ranges from An_{38} to An_{84}. No systematic zoning is evident in the plagioclase. Rare pyroxene phenocrysts in an otherwise aphanitic dike (404C) are magnesian augite; they are rimmed by fine-grained Ca-amphibole.

Trondhjemite

Felsic rocks occur as discordant pods, dikes, and sills in the gabbro and basaltic dikes. All but a very few are true trondhjemites, with color index <10% and extremely low K_2O contents. Most of the felsic rocks contain <5% mafic minerals. Rocks of this composition and association have elsewhere been termed plagiogranite (e.g., Coleman and Donato, 1979). Less than 10% of the Billy's Peak mafic complex is composed of this rock type.

Plagioclase (~An_{39}) and quartz are the main mineral phases, with small amounts of Ca-amphibole, Fe-Ti oxides, epidote, and apatite. Polygonal boundaries on quartz grains and fibrous amphibole were observed in all of the samples, consistent with low-grade metamorphism of the trondhjemite.

STRUCTURE

The Billy's Peak mafic complex affords the rare opportunity to study structural relationships among intrusive rocks in a large, recently glaciated area. The structural relationship between the various rock types records a complex history of igneous activity, involving at least five episodes of intrusion. These are illustrated in Figure 4.

The gabbroic unit was the first to intrude the ultramafic rocks of the Trinity sheet. It is commonly in direct contact with the host ultramafic rock along the margin of the Billy's Peak mafic complex. Gabbro also occurs as screens between the basaltic dikes. The screens vary from <1 meter to >30 meters in width. Chilled

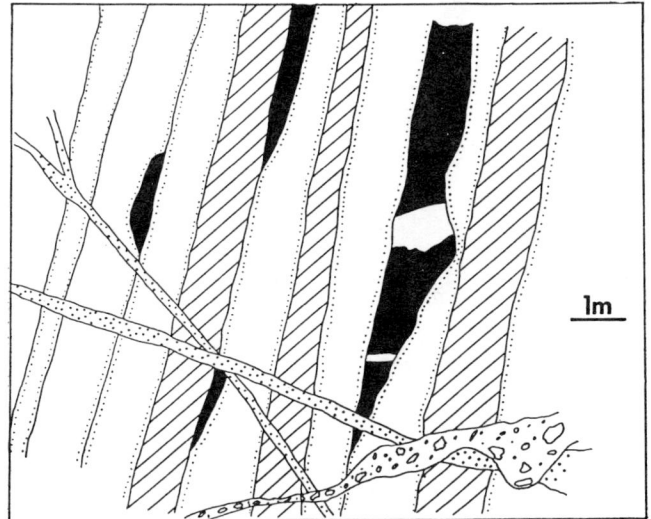

Figure 4. Schematic of intrusive relations in the Billy's Peak area. Solid black is gabbro screens, unpatterned is Trondhjemite I intruding gabbro. Diagonally ruled unit is porphyritic basalt; light stipple shows chilled margins in aphanitic basalt. Late-stage Trondhjemite II and igneous breccia are shown in heavy stipple.

margins of diabasic dikes against peridotite, the absence of shearing at the ultramafite/gabbro contact, and xenoliths of peridotite all indicate that the gabbro was intrusive, not tectonically juxtaposed against the ultramafite.

The structural relationships between various gabbroic lithologies is ambiguous in many cases, but layered gabbro appears to be the oldest unit. The other gabbroic rocks were apparently intruded in random order. For example, hornblendeite both intrudes and is intruded by isotropic gabbro. The absence of chilled margins and presence of plastic deformation between these types of gabbro indicate that all were intruded in a hot environment, and intrusion was penecontemporaneous.

Small pods and dikes of trondhjemite cut all of the gabbroic rock types. These felsic rocks are volumetrically minor, accounting for a few percent of the gabbro outcrops. Because of the existence of a later episode of felsic intrusion, these rocks are referred to as Trondhjemite I.

Intrusion of the basaltic dikes was volumetrically the most important event in the Billy's Peak mafic complex. These are of two types: porphyritic and diabasic. They are very similar in mineralogy and composition, but texturally and temporally distinct. The porphyritic dikes were intruded first, and exhibit chilled margins against gabbroic rocks and Trondhjemite I. The diabasic dikes have chilled margins against all the earlier rock types. Both the diabasic and porphyritic dike types display sheeting (Fig. 5), but this is very rare.

Figure 6. Intrusive relationships in late-stage Trondhjemite II and igneous breccia. View is about 60 cm in width.

The last intrusive episodes were volumetrically minor, and consisted of emplacement of narrow (usually <20 cm) dikes of a second pulse of trondhjemite (Trondhjemite II), porphyritic basalt, and veins of brecciated mafic rock in a felsic matrix (Fig. 6).

Orientations of the basaltic dikes, presented in the stereographic projection of Figure 7, reveal a strong maxima corresponding to an east-west strike and a dip of 70° to the south. Both the porphyritic and diabasic dikes have east-west strike, but the diabasic dikes dip more steeply (maxima=80°).

Figure 5. Sheeted dikes in the Billy's Peak mafic complex. View is toward the west. Hammer handle is 40 cm long.

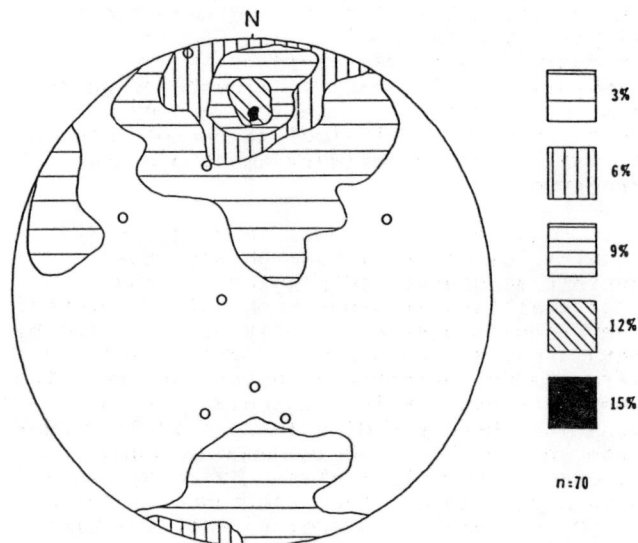

Figure 7. Lower hemisphere stereographic projection of poles to mafic dikes. Open circles are poles to layering in gabbro.

Structural data on the other intrusive phases in the area were difficult to obtain because of their scarcity and generally nonplanar nature. On the basis of a limited number of measurements of variable quality, Trondhjemite II dikes are apparently subparallel to the basaltic dikes. The dip of layering in gabbro varies widely, but strikes roughly east-west (Fig. 7).

GEOCHEMISTRY

Major and trace element analyses are listed in Table 1, along with the rock type. Major element abundances were determined by flame atomic absorption spectrophotometry at Texas Tech University. Trace elements were determined by INAA at University of Texas at El Paso (rare earth elements) and inductively coupled plasma at Texas Tech University.

Most of the rocks in the Billy's Peak mafic complex have been affected by lower greenschist facies metamorphism. This presents the possibility that some samples may have been selectively enriched or depleted in the more mobile elements (e.g., Na_2O, K_2O, Ba). All of the samples discussed here were examined petrographically before analysis, and only those showing the least alteration were selected.

Major Elements

Major element compositions (Table 1) reveal that mafic rocks of the Billy's Peak area are mostly low-K tholeiitic basalts. They have relatively high Mg-numbers (molar $Mg/(Mg+Fe^{2+})$), as shown in Figure 8. The lack of iron enrichment indicates a calc-alkaline affinity (Fig. 9). Trondhjemites are SiO_2-rich and extremely low in K_2O.

Trace Elements

Chondrite-normalized analyses of samples from the Billy's Peak area are shown on Figure 10, along with representative average patterns for island arc and mid-ocean ridge basalts. Mid-ocean ridge basalt can be distinguished from island-arc

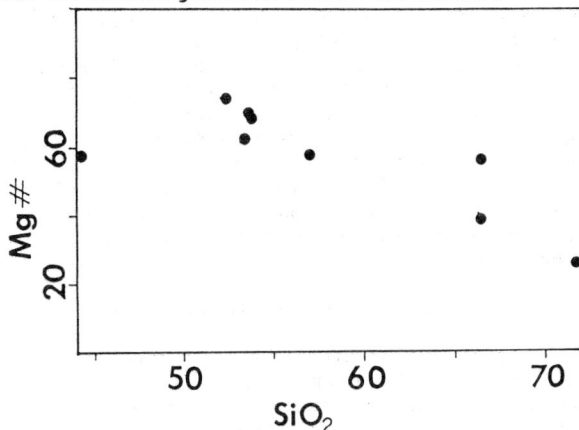

Figure 8. Variation diagram of SiO_2 vs Mg# (mol. $Mg/(Mg+Fe^{2+})$).

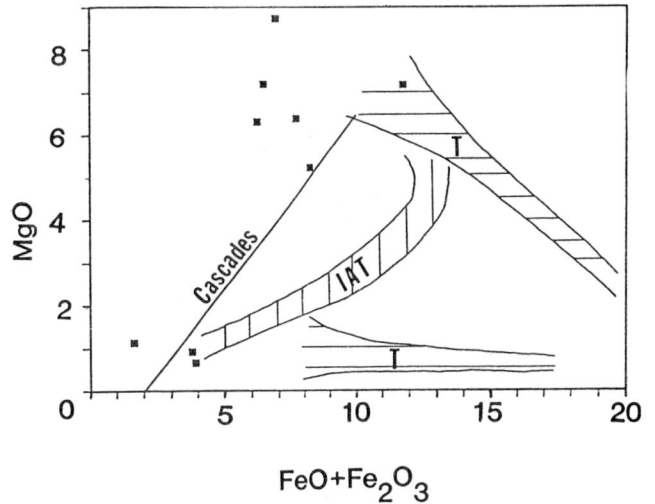

Figure 9. Plot of MgO vs $FeO+Fe_2O_3$, showing calc-alkaline affinity of the Billy's Peak mafic complex. T=oceanic tholeiite, IAT=island-arc tholeiite. Cascades=typical trend of Cascades calc-alkaline series rocks. After Jakes and Gill (1970).

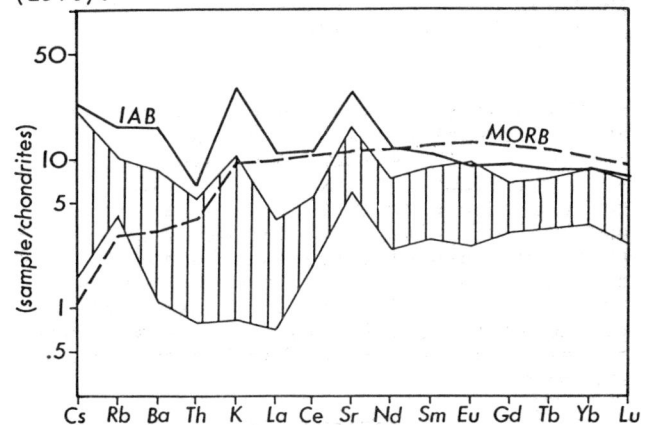

Figure 10. Chondrite-normalized abundances of trace elements. Billy's Peak samples are represented by vertical shading. Heavy dashed and solid lines are average values for mid-ocean ridge basalts and island arc basalts, respectively.

tholeiite because of the relative depletion of large ion lithophile elements in the mid-ocean ridge basalt. Figure 10 indicates that the Billy's Peak rocks do not show this depletion, and that their patterns are similar to those of island-arc tholeiite. Enrichment of K and Sr relative to mid-ocean ridge basalt is also typical of island-arc basalts and is present in the Billy's Peak samples.

The Th-Hf-Ta diagram developed by Wood (1980; Fig. 11) can also be used to discriminate tectonic environment. Data from Billy's Peak plot in both the mid-ocean ridge basalt field (A) and primitive arc tholeiite field (D), with most of the samples in the latter category. Samples of Paleozoic basalts from the southern portion

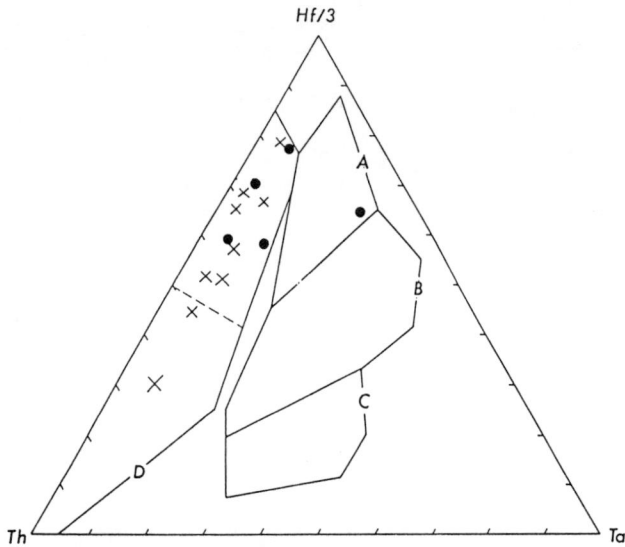

Figure 11. Trace element discrimination diagram of Wood (1980). Dots are Billy's Peak samples, X's are samples of Copley greenstone (Lapierre et al., 1985a). A=NMORB, B=EMORB, C=within-plate basalts, D=destructive plate-margin basalts.

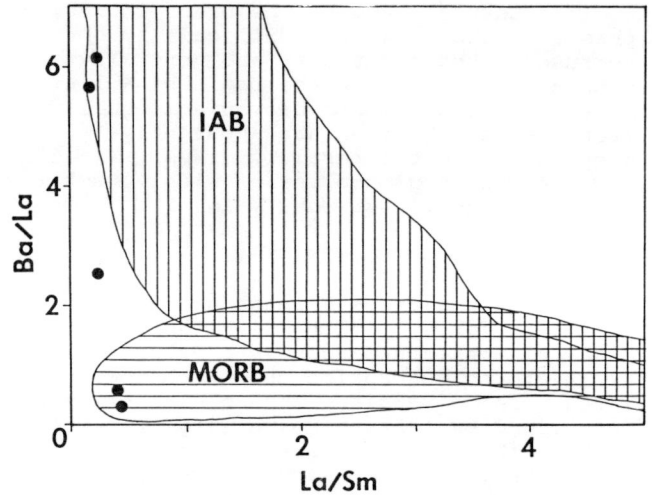

Figure 12. Variation diagram of chondrite-normalized La/Sm vs Ba/La, after Perfit and others (1980). IAB=island-arc basalts, MORB=mid-ocean ridge basalts.

of the eastern Klamath belt (Lapierre and others, 1985a) are also represented on Figure 11. These rocks may have erupted in a similar tectonic environment. The relationship between Ba/La and La/Sm is another sensitive indicator of environment (Fig. 12), assuming that Ba was not mobilized during metamorphism. If this is true, then the Billy's Peak rocks have a geochemical character that is transitional between mid-ocean ridge basalts and island arc basalts.

DISCUSSION

Trinity Sheet

The tectonic setting of the Trinity sheet has been problematic for many years. Recent researchers have recognized the oceanic affinity of all rocks in the Trinity sheet, but interpretations of the specific tectonic environment vary. Lindsley-Griffin (1977) postulated that it is a true ophiolite, formed at an oceanic ridge spreading center, and was dismembered and melanged during accretion to North America. The work of Boudier and others (1989) also assigned the Trinity sheet to an origin as oceanic lithosphere, possibly a back-arc basin.

The difficulty in calling the Trinity sheet a true ophiolite involves several factors. First, the structural disruption found in some areas, especially in the north, has effectively destroyed original stratigraphy. Although most of the lithologies present in true ophiolites are exposed, their original structural-stratigraphic relationships can only be surmised. Second, the volume of gabbro and

especially hypabyssal basaltic rocks is much less than is typical of oceanic crust or most ophiolites. The reconstructed pseudo-stratigraphic column of Lindsley-Griffin (1977, figure 2) shows the mafic portion to be about 2 km thick, compared to 6 km in typical ophiolites (Coleman, 1977). It is possible that these stratigraphically higher sections have been removed, either during emplacement or by erosion, as suggested by Boudier and others (1989). Third, the occurrence of discrete mafic plutons throughout the ultramafic section is atypical of ophiolites, in which gabbro is present as sheet-like masses and basaltic rocks are present as diabasic dikes and pillow lavas.

It is possible that the Trinity sheet was formed in an oceanic island-arc environment. In this case, the gabbroic plutons would represent the roots of arc volcanoes and the diabasic dikes would be feeders to these volcanoes. Evidence that the Trinity peridotite was emplaced as a diapir (Quick, 1981) is consistent with this interpretation, as are tectonic inferences made on the basis of trace element discrimination diagrams. One problem with this scenario is the paucity of volcanic rocks associated with the gabbro and diabase. Some lavas with geochemical and structural features suggestive of an island-arc origin (Lapierre and others, 1985b) are present in the area, but most were probably removed by erosional and/or tectonic processes.

Billy's Peak Mafic Complex

Although the Billy's Peak mafic complex constitutes only a small part of the Trinity sheet, its good exposure and variety of lithologies provide a means for testing several theories concerning this area. With the exception of the well-developed dike complex, structures and lithologies are

SAMPLE	404C	427A	427B	428	433B	434	438	441B	467
SiO_2	56.97	52.31	53.32	53.58	66.41	44.23	71.60	53.70	66.40
TiO_2	0.71	0.37	0.66	0.44	0.43	1.97	0.36	0.43	0.41
Al_2O_3	16.01	14.46	17.68	16.63	18.02	17.44	14.09	18.82	14.61
FeO	6.82	5.52	6.74	5.53	1.57	9.40	3.35	5.08	2.56
Fe_2O_3	1.39	1.35	0.93	0.90	0.07	2.38	0.57	1.12	1.21
MnO	0.16	0.14	0.11	0.12	0.03	0.15	0.03	0.11	0.07
MgO	5.24	8.73	6.39	7.20	1.14	7.18	0.66	6.32	0.92
CaO	9.53	10.45	5.84	9.80	5.84	12.44	3.33	11.38	4.91
Na_2O	1.67	2.26	5.09	2.92	5.74	1.61	4.80	2.07	3.69
K_2O	0.09	0.01	0.11	0.09	0.16	0.13	0.13	0.15	0.12
H_2O+	0.50	1.61	1.17	0.94	0.52	0.94	0.35	0.61	0.27
TOTAL	99.20	97.31	98.11	98.23	100.06	97.92	99.36	99.90	95.34
Rock Type	cpx+ hbld diab	hbld diab	hbld bas	hbld gabb	Tron I	hbld gabb	Tron II	hbld gabb	Tron I
Mg#	58	74	63	70	56	58	26	69	39
La	1.270	0.227	0.305		1.113	0.820			
Ce	4.60	1.57	1.72		3.90	2.06			
Nd	4.5	1.5	2.3		3.6	3.1			
Sm	1.724	0.557	1.015		1.347	1.213			
Eu	0.568	0.229	0.320		0.856	0.486			
Gd	0.00	0.00	0.00		0.00	0.00			
Tb	0.340	0.178	0.260		0.330	0.390			
Tm	0.400	0.103	0.130		0.168	0.120			
Yb	1.920	0.800	1.510		1.100	1.180			
Lu	0.278	0.104	0.211		0.170	0.156			
Sc	38.90	45.00	46.90		9.26	67.50			
Cr	80.30	738.00	43.00		1.03	5.81			
Co	24.48	26.88	20.93		4.46	37.60			
Ni	31.50	203.00	27.20		4.74	14.80			
Rb	3.8	2.6	1.5		3.5	3.0			
Sr	72	67	97		183	109			
Zr	33.40	8.94	17.90		47.10	6.55			
Nb	22.8	12.9	23.9		0.0	50.3			
Ba	4.18	16.80	20.60		33.30	5.09			
Cs	0.02	0.17	0.13		0.21	0.26			
Hf	1.36	0.32	0.69		1.11	0.53			
U						0.52			
Th	0.28	0.05	0.01		0.08	0.07			
Ta	0.04	0.02	0.09		0.03	n.d.			
Y	22.5	13.4	24.0		14.2	13.9			

Table 1. Whole-rock major and trace element analyses.

similar to gabbroic bodies in the west (Goullaud, 1973), north (Bailey, 1980), and east (Matzner, 1986).

Lithology

The predominance of igneous hornblende as the primary mafic phase in the Billy's Peak intrusions is atypical of most ophiolitic and island-arc related rocks, and indicates that the basaltic magma had high P_{H2O}. This water could have been incorporated into the melt at the source of melting, or after emplacement into the peridotite. Work on the nearby Castle Lake gabbro (Schwindinger and Anderson, 1987), which is very similar to gabbro at Billy's Peak but contains no basaltic dikes, suggests that water was inherited from the host serpentinite. The presence of hornblende-bearing dikes within the gabbro at Billy's Peak implies that most of the water was inherited from the source. This also argues for a small degree of partial melting in the Billy's Peak source region,

consistent with the observed enrichment of alkalis and alkaline earths.

Structure

The consistent orientation of dikes in the Billy's Peak mafic complex suggests an extensional environment, and is atypical of classic ophiolites owing to the lack of regular sheeting. The field relations suggest slow spreading that would be typical of a back-arc basin or oceanic island-arc. The dominant east-west strike and steep dips of the dikes would seem to indicate a north-south spreading direction (modern coordinates). Paleomagnetic evidence for rotation in the eastern Klamath belt is ambiguous. Renne and others (1988) concluded that little evidence exists to support large-scale rotation in this area, although it may have been active on a local scale. A much younger dike complex in the mid-Jurassic Josephine ophiolite (to the northwest) also has a dominant east-west strike. Harper and others (1985) suggested

632

that these dikes are the result of a transpressional tectonic regime that resulted in transform faulting parallel to the coast and connecting spreading ridges perpendicular to the coast.

Geochemistry

Major and trace element compositions strongly argue for an immature oceanic island-arc origin for the Billy's Peak mafic complex. The high Mg-numbers and transitional tholeiitic/calc-alkaline character of the suite (Fig. 12) indicate a mantle source with minor fractionation of clinopyroxene and plagioclase. Trace element abundance patterns, especially those of the large ion lithophile elements, closely parallel those in rocks from active island-arcs (Fig. 10), and are markedly different from mid-ocean ridge basalts.

ACNOWLEDGMENTS

We thank Melanie Barnes for her assistance in the lab, and Regan Weeks for her help in the field. This work was supported by NSF grant EAR-8720141.

REFERENCES CITED

Bailey, L. E., 1980, Geology of Scott Summit, Klamath Mountains, California (unpublished M.S. thesis): Eugene, Oregon, University of Oregon, 111 p.

Boudier, F., LeSueur, E., and Nicolas A., 1989, Structure of an atypical ophiolite: The Trinity complex, eastern Klamath Mountains, California: Geological Society of America Bulletin, v. 101, p. 820-833.

Coleman, R. G., 1977, Ophiolites, Wyllie, P. J., ed.: Springer-Verlag, New York, 229 p.

_____, and Donato, M. M., 1979, Oceanic plagiogranite revisited, in Barker, F., ed., Trondhjemites, Dacites, and Related Rocks: New York, Elsevier, p. 150-168.

Goullaud, L., 1977, Structure and petrology in the Trinity mafic-ultramafic complex, Klamath Mountains, California, in Lindsley-Griffin, N., and Kramer, J. C., eds., Geology of the Klamath Mountains: Geological Society of America Cordilleran Section Field Trip Guidebook, p. 112-133.

_____, 1973, Petrology and structure of a gabbroic body in the Trinity ultramafic pluton, Klamath Mountains, California (unpublished M.S. thesis): Seattle, Washington, University of Washington, 54 p.

Harper, G. D., Saleeby, J. B., and Norman, E. A. S., 1985, Geometry and tectonic setting of seafloor spreading for the Josephine ophiolite, and implications for Jurassic accretionary events along the California margin, in Howell, D. G., ed., Tectonostatigraphic Terranes of the Circum-Pacific Region: Houston, Texas, Circum-Pacific Council of Energy and Mineral Resources, p. 239-257.

Irwin, W. P., 1981, Tectonic accretion of the Klamath Mountains, in Ernst, W. G., ed., The Geotectonic Development of California: Englewood Cliffs, New Jersey, Prentice-Hall, p. 9-49.

_____, 1985, Age and tectonics of plutonic belts in accreted terranes of the Klamath Mountains, California and Oregon, in Howell, D. G., ed., Tectonostratigraphic Terranes of the Circum-Pacific Region: Houston, Texas, Circum-Pacific Council of Energy and Minineral Resources, p. 187-199.

Jacobsen, S. B., Quick, J. E., and Wasserburg, G. J., 1984, A Nd and Sr isotopic study of the Trinity peridotite: Implications for mantle evolution: Earth and Planetary Science Letters, v. 68, p. 361-378.

Jakes, P., and Gill, J., 1970, Rare earth elements and the island arc tholeiitic series: Earth and Planetary Science Letters, v. 9, p. 17-28.

Lanphere, M. A., Irwin, W. P., and Hotz, P. E., 1968, Isotopic age of the Nevadan orogeny and older plutonic and metamorphic events in the Klamath Mountains, California: Geological Society of America Bulletin, v. 79, p. 1027-1052.

Lapierre, J., Cabanis, B., Coulon, C., Brouxel, M., and Albarede, F., 1985a, Geodynamic setting of Early Devonian Kuroko-type sulfide deposits in the Eastern Klamath Mountains (northern California) inferred by the petrological and geochemical characteristics of the associated island-arc volcanic rocks: Economic Geology, v. 80, p. 2100-2113.

_____, Albarede, F., Albers, J., Cabanis, B., and Coulon, C., 1985b, Early Devonian volcanism in the eastern Klamath Mountains, California: Evidence for an immature island arc: Canadian Journal of Earth Science, v. 22, p. 214-227.

Lindsley-Griffin, N., 1977, Paleogeographic implications of ophiolites: The Ordovician Trinity complex, Klamath Mountains, California, in Stewart, J. H., Stevens, C. H., and Fritsche, A.E., eds., Paleozoic Paleogeography of the Western United States: Pacific Coast Paleogeography Symposium 1, Pacific Section, Society of Economic Paleontologists and Mineralogists, p. 409-420.

Mattinson, J. M., and Hopson, C. A., 1972, Paleozoic ages of rocks from ophiolitic complexes in Washington and Northern California: American Geophysical Union Transactions, v. 53, p. 543.

Matzner, D. M., 1986, Metamorphism of early Paleozoic island arc and Mesozoic plutonic rocks intruding the Trinity peridotite, eastern Klamath Mountains, Northern California (unpublished M.S. thesis): Los Angeles, California, University of California Los Angeles, 172 p.

Perfit, M. R., Gust, D. A., Bence, A. E., Arculus, R. J., and Taylor, S. R., 1980, Chemical characteristics of island-arc basalts: Implications for mantle sources: Chemical Geology, v. 30, p. 227-256.

Petersen, S. W., 1987, The Billy's Peak mafic dike complex in the Trinity sheet, Klamath Mountains, California: Geological Society of America Abstracts with Programs, v. 19, p. 439.

Quick, J. E., 1981, Petrology and petrogenesis of the Trinity peridotite, an upper mantle diapir in the eastern Klamath Mountains, northern California: Journal of Geophysical Research, v. 86, p. 11837-11863.

Renne, P. R., Scott, G. R., and Bazard, D. R., 1988, Multicomponent paleomagnetic data from the Nosoni Formation, eastern Klamath Mountains, California: Cratonic Permian primary directions with Jurassic overprints: Journal of Geophysical Research, v. 93, p. 3387-3400.

Schwindinger, K. R., and Anderson, A. T., 1987, Probable low-pressure intrusion of gabbro into serpentinized peridotite, northern California: Geological Society of America Bulletin, v. 98, p. 364-372.

Wallin, E. T., Mattinson, J. M., and Potter, A. W., 1988, Early Paleozoic magmatic events in the eastern Klamath Mountains, northern California: Geology, v. 16, p. 144-148.

Wood, D. A., 1980, The application of a Th-Hf-Ta diagram to problems of tectonomagmatic classification and to establishing the nature of crustal contamination of basaltic lavas of the British Tertiary Volcanic Province: Earth and Planetary Science Letters, v. 50, p. 11-30.

THE UPPER PERMIAN FUSULINIDS REICHELINA AND PARAREICHELINA IN NORTHERN CALIFORNIA: EVIDENCE FOR LONG-DISTANCE TECTONIC TRANSPORT

Calvin H. Stevens
San Jose State University
San Jose, California 95192

Michael D. Luken
Chevron Overseas Petroleum, Inc.
San Ramon, California 94583

Merlynd K. Nestell
University of Texas
Arlington, Texas 76019

ABSTRACT

The Permian Tethyan fusulinacean genera Reichelina and Parareichelina have been recovered from limestone blocks in the eastern Hayfork terrane of the western Klamath Mountains in northern California. These fusulinaceans and other associated foraminifers constitute a Late Permian foraminiferal assemblage unlike any other known in either autochthonous North America or in accreted allochthonous blocks or terranes along the western margin of North America. The common species of the genus Reichelina in this fauna is similar to Reichelina media Miklukho-Maclay, first described from Late Permian rocks in the northern Caucasus Mountains, and subsequently reported from other Late Permian rocks of so-called Tethyan affinity at a number of localities from the Mediterranean to Japan. The very rare genus Parareichelina has been described previously only from Late Permian rocks in the northern Caucasus Mountains and the southern Sikhote Alin Mountains region of southeastern Siberia.

The presence of these and other typical Tethyan fossils in Late Permian limestone blocks in the Hayfork terrane suggest these rocks originally were deposited in the paleo-Pacific Ocean, thousands of kilometers west of cratonal North America, as parts of tropical atolls.

INTRODUCTION

Several genera of very small fusulinaceans developed quite unusual forms through partial or complete uncoiling of their later volutions in Late Permian time. Most of these genera have been considered as restricted to the Tethyan faunal province and to a narrow stratigraphic interval in the Permian Basin in West Texas. Recently, however, two of these fusulinacean genera, Reichelina and Parareichelina, have been recovered from limestone blocks in the western Klamath Mountains in northwestern California (Fig. 1). In addition to the unusual morphologies of these fusulinaceans, these and the associated fossils possess considerable significance for the interpretation of the origin of the tectonostratigraphic terrane in which they occur.

The genus Reichelina is known to occur in Permian rocks at several localities in western North America, but the species most common in the rocks of the Hayfork terrane is the most advanced and the youngest (Nestell

and others, 1981). The genus Parareichelina, recently discovered by Luken (1985), never before had been reported from the western hemisphere, and at present is documented from only two regions in the Soviet Union: the northern Caucasus Mountains east of Sochi, and the southern Sikhote Alin Mountains of southeasternmost Siberia east of Vladivostok (Fig. 2). Miklukho-Maclay (1963) also noted the presence of Parareichelina in the Crimea, but we have not been able to find any description or illustration of specimens from this area.

The objectives of this paper are to call attention to the uniqueness of these unusual fusulinaceans, to illustrate their variation, to document their presence in North America, and to note their widely separated geographic occurrences.

GEOLOGIC SETTING

The Klamath Mountains consist of four lithotectonic terranes referred to by Irwin (1960, 1966) as the eastern Klamath subprovince, the central metamorphic subprovince, the western Paleozoic and Triassic subprovince (the region of interest in this study), and the western Jurassic subprovince (Fig. 1). The western Paleozoic and Triassic belt itself was divided into three terranes by Irwin (1972): the Northfork, Hayfork, and Rattlesnake Creek terranes. The eastern part of the Hayfork terrane (as revised by Wright, 1982) has yielded Permian foraminifers typical of the Tethyan province including Reichelina and Parareichelina, the genera shown in Figures 3-5. Although probable Late Permian algae and small foraminifers have been identified from at least seven localities in the Hayfork terrane (Luken, 1985), Reichelina has been recovered from only two of the localities and Parareichelina from only one.

PERMIAN FAUNAL PROVINCES

Study of foraminifers and corals in Permian limestones in far western United States has shown that three distinct paleobiogeographic provinces are represented: North American, eastern Klamath, and Tethyan. In California, all three of these faunal provinces are represented (Luken and others, 1985; Stevens and others, 1990). The North American faunal province is present in eastern California, and fossils of the eastern Klamath province are abundant in the coherent sedimentary rocks in the eastern Klamath Mountains in northern California.

In Cooper, J.D., and Stevens, C.H., eds., 1991, **Paleozoic Paleogeography of the Western United States-II:** Pacific Section SEPM, Vol. 67, p. 635-642.

The Tethyan faunal province, to which the fossils of concern here belong, is best represented in limestone blocks in the western Klamath Mountains in northwestern California.

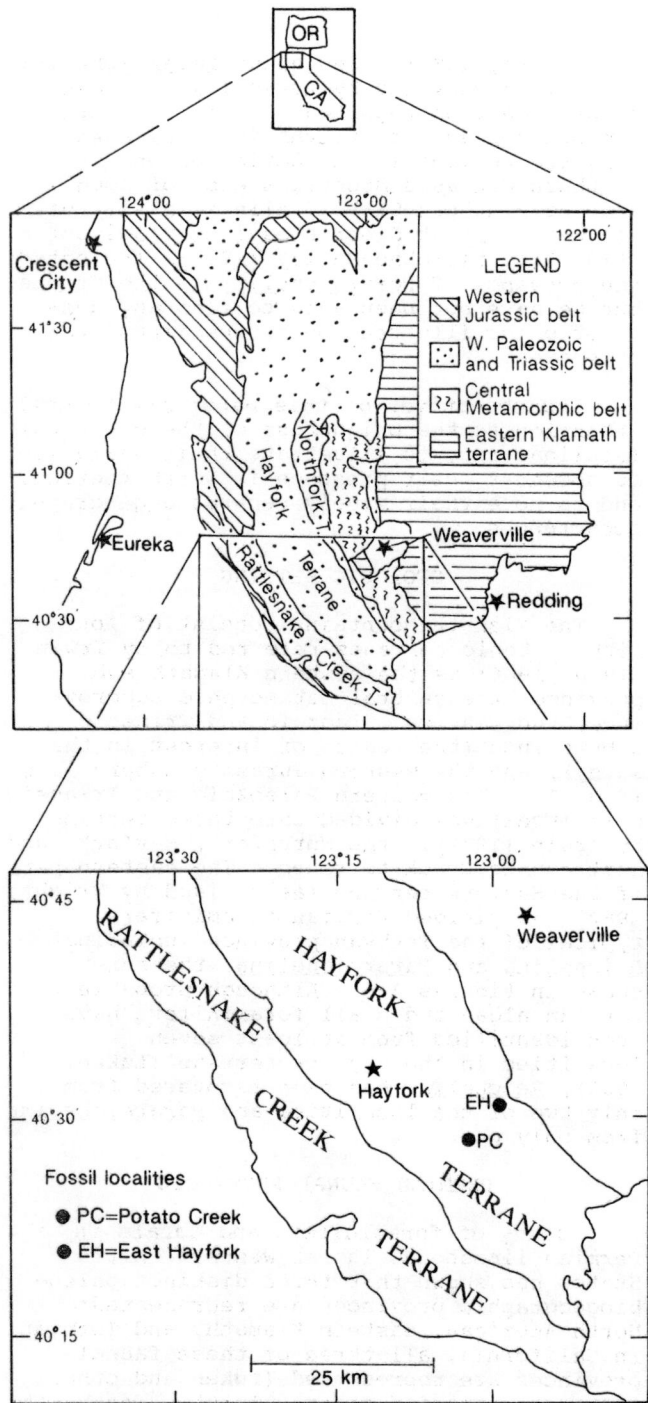

Figure 1. Tectonostratigraphic belts and terranes in the Klamath Mountains (after Irwin, 1972, and Blome and Irwin, 1983).

LATE PERMIAN LOCALITIES YIELDING FUSULINACEANS

The east Hayfork Creek locality (locality 12 of Luken, 1985; locality 18 of Irwin, 1972) consists of a large limestone mass that caps a ridge between the north and east forks of Hayfork Creek in the SE1/4, NW1/4 sec. 27, T.31N., R.10W., Trinity County, California. Samples were obtained from the extreme southeastern part of the outcrop that faces an unnamed drainage, which feeds into the east fork of Hayfork Creek. Rare specimens of Reichelina and rather uncommon specimens of Parareichelina occur in a fine- to coarse-grained, foraminiferal packstone and wackestone from the lower part of the outcrop facing the creek. Large foraminifers include Agathammina, Climacammina, Cribrogenerina, Palaeotextularia, and the fusulinacean Staffella; small foraminifers include Ammodiscus, Frondina, Geinitzina, Globivalvulina, Langella, Neoendothyra, Pachyphloia, Protonodosaria, Robuloides, and Tuberitina (Luken, 1985).

The Potato Creek locality (locality 4K1 of Nestell and others, 1981; locality 15 of Luken, 1985) consists of scattered limestone blocks exposed on a ridge west of Potato Creek in the NW1/4, NE1/4 sec. 7, T.30N, R.10W, Trinity County, California. Samples were taken from locally fossiliferous rocks on the north side of the outcrop. Reichelina

Figure 2. Worldwide distribution of Parareichelina plotted on a north-polar projection. A) Hayfork terrane, Klamath Mountains, northern California; B) southern Sikhote Alin Mountains, Siberia; C) northern Caucasus Mountains, USSR.

occurs commonly in a fine- to moderately coarse-grained, partially dolomitized limestone. Other foraminifers include Agathammina, Climacammina, Colaniella, Cornuspira, Frondina, Geinitzina, Globivalvulina, Langella, Neoendothyra, Pachyphloia, Protonodosaria, Tetrataxis, Robuloides, Tuberitina, and the fusulinaceans Nankinella, Sphaerulina and Staffella (Nestell and others, 1981; Luken, 1985).

ORIGIN OF FUSULINID-BEARING LIMESTONES

Many geologists (e.g., Irwin, 1981; Mortimer, 1985; Luken, 1985) have interpreted the terranes of the western Klamath Mountains to be allochthonous to North America, based in part on the exotic nature of the fossils in the limestone blocks. Luken (1985), in a study of many limestone blocks in the western Paleozoic and Triassic belt, recorded a wide variety of limestone textures and structures. He considered them to represent fragments of atolls formed on seamounts far from any continent because: (1) the blocks are intimately associated with other types of oceanic rocks including mafic volcanic rocks, chert, argillite, and tuff; (2) limestone has been observed in depositional contact above volcanic rock in the Rattlesnake Creek terrane and below chert in the Hayfork terrane; and (3) the limestones lack terrigeneous clastic debris other than that of volcanic origin. Davis (1968) has observed a similar relationship between a large limestone mass and volcanic rocks in the northern extension of the Northfork terrane. The fusulinacean-bearing limestones considered here consist of fine- to coarse-grained foraminiferal packstone and grainstone, which also contain dasycladacean and codiacean algae and colonial corals (Stevens and others, 1987). These limestone blocks are interpreted to have been derived from the floor of an atoll lagoon.

RELATIONSHIPS AND OCCURRENCE OF REICHELINA AND PARAREICHELINA

Reichelina and Parareichelina are morphologically somewhat similar fusulinid genera that typify Upper Permian rocks of the Tethyan realm. Both are tightly coiled for several volutions, but they then become uncoiled or partially uncoiled, giving the adult shell the shape of a cornucopia or a disk.

The genus Reichelina, first described by Erk (1941), is a relatively common genus, which occurs throughout the Middle and Upper Permian from the zone of Cancellina (Erk, 1941) into the zone of Palaeofusulina (Sheng and Chang, 1958). Geographically it occurs throughout the Tethyan region from the Mediterranean to Japan.

In North America, Reichelina was first described from the lower and middle part of the Lamar Member of the Bell Canyon Formation in West Texas (Skinner and Wilde, 1955). Species of this genus also are known from the Upper Permian La Difunta beds of Las Delicias, Coahuila, Mexico (Wardlaw and others, 1979), from allochthonous pods of limestone in the Virtue Hills of the Baker terrane of eastern Oregon (Nestell and Blome, 1988), and from rocks of the Cache Creek Group in British Columbia (Monger and Ross, 1971; Nestell, 1981).

Reichelina sp. A, from the Potato Creek locality, exhibits a pronounced "flare" or uncoiled portion and is very similar to Reichelina media described by Miklukho-Maclay (1954) from Upper Permian rocks exposed east of Sochi, USSR in the northern Caucasus Mountains. This advanced morphology is interpreted to mean that the limestone in the Hayfork terrane is Late Permian (Djulfian) in age. Several other species of Reichelina that exhibit a pronounced uncoiled portion of the test similar to R. media have been reported from Greece (Nestell and Grant, 1988), Yugoslavia (Pantic, 1963), China (Sheng, 1963), Cambodia (Tien, 1979), Japan, (Kobayashi, 1975; Igo and Igo, 1977), and eastern Siberia (Sosnina, 1981).

Reichelina sp. B, from the east Hayfork Creek locality, possibly is a new species. It appears to be at a developmental stage more or less equivalent to that of R. changhsingensis Sheng and Chang, first reported from the Palaeofusulina zone in eastern China.

Miklukho-Maclay (1954, 1960), in studies on the Late Permian foraminifers from the northern Caucasus Mountains, named the genus Parareichelina for a form that has more complex septa than Reichelina and is extensively uncoiled. Parareichelina probably evolved from a species of Reichelina, and Miklukho-Maclay (1968) stated that, "The structure of the spiral parts of Parareichelina and Reichelina are very similar, but the last volution of the test of

Figure 3. 1-12, Parareichelina sp. from east Hayfork Creek locality; 1-4, inclined equatorial sections; 5-7, 9-12, longitudinal sections; 8, equatorial section of outer volution. Number 10 from Stevens' sample; all others from Luken's samples. All figures X85.

Figure 4. 1, 2, Reichelina sp.B from east Hayfork Creek locality, equatorial sections. 3-14, Reichelina sp. A from Potato Creek locality; 3-10, 13, 14, longitudinal sections; 11, 12, equatorial sections. Numbers 3, 4, 11-14 from Nestell's samples; all others from Luken's samples. All figures X85.

Figure 5. 1-12, Reichelina sp. A from Potato Creek locality. 1-7, longitudinal sections; 8-12, equatorial sections. Numbers 2-4, 6, 7, 11 from Nestell's samples; all others from Luken's samples. All figures X85.

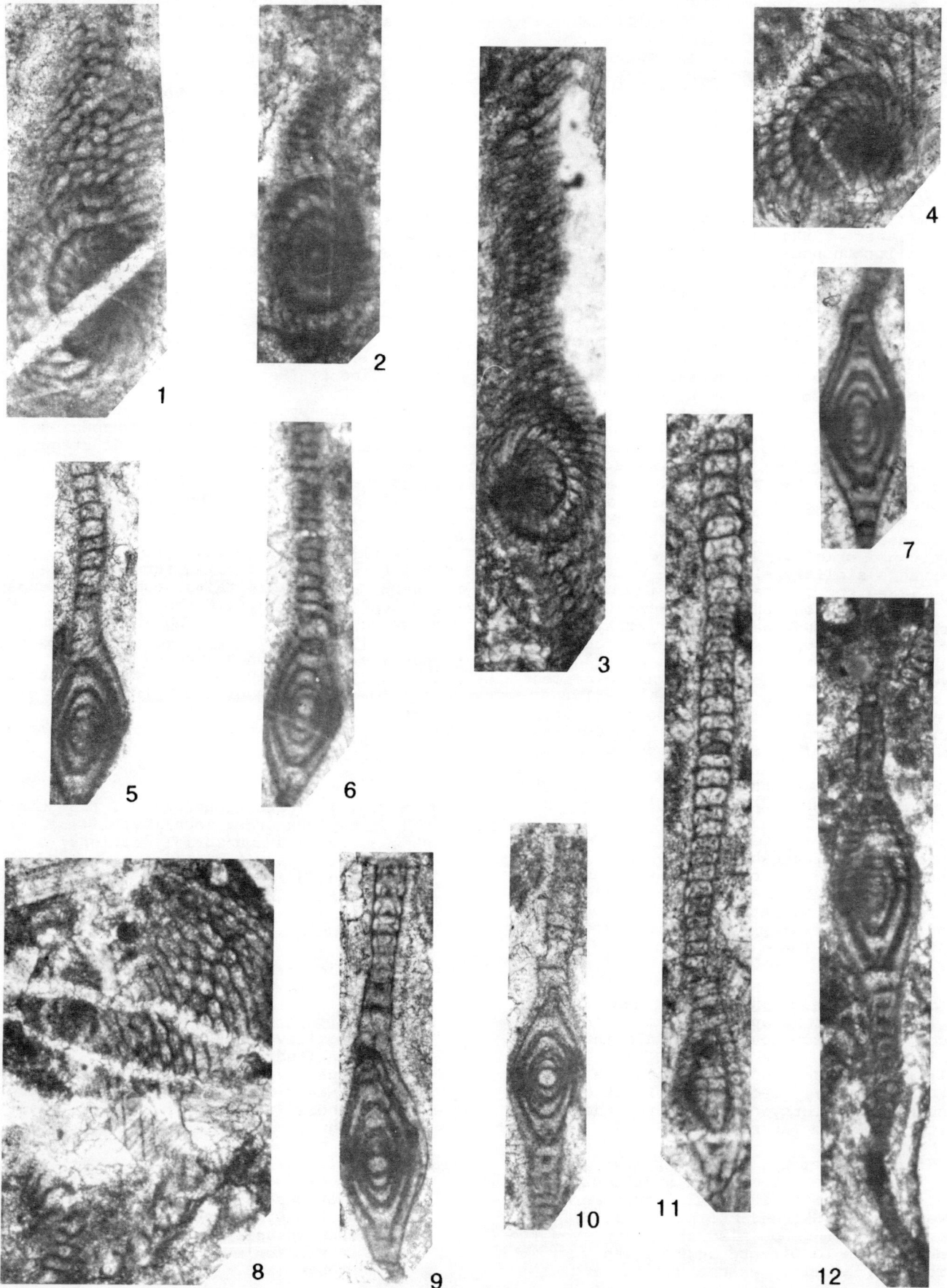

Figure 3

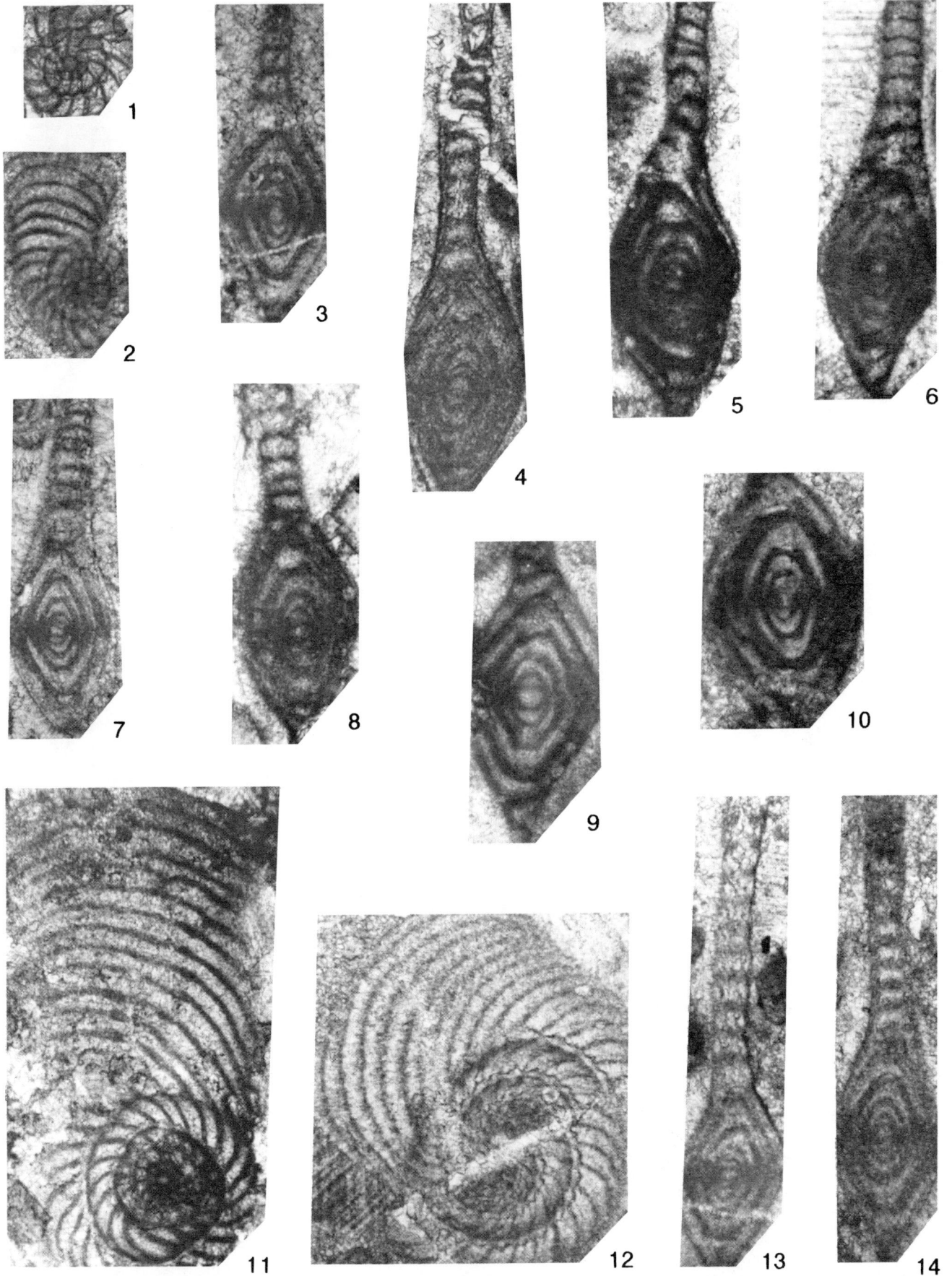

Figure 4

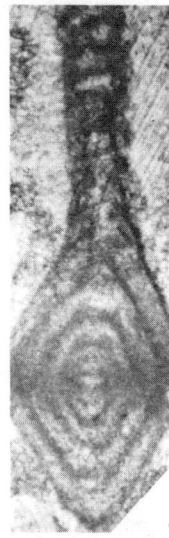

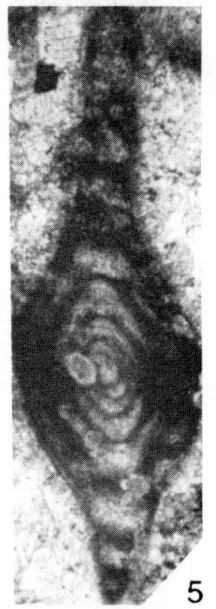

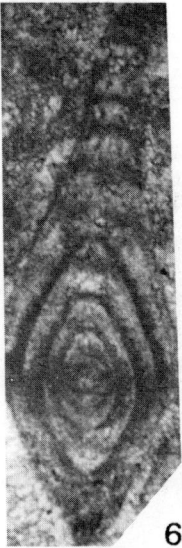

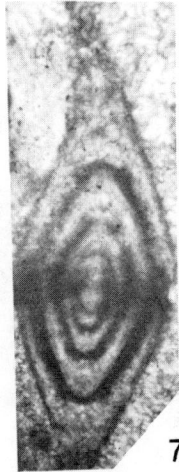

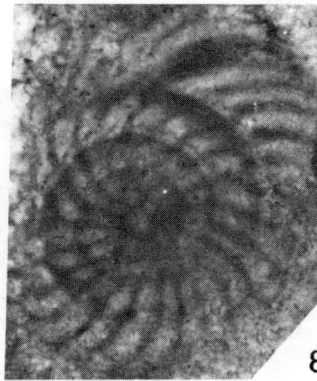

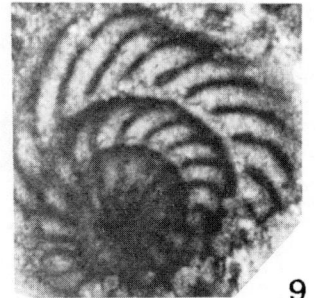

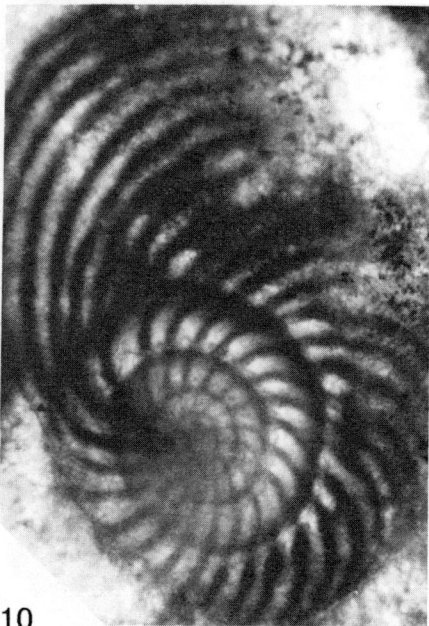

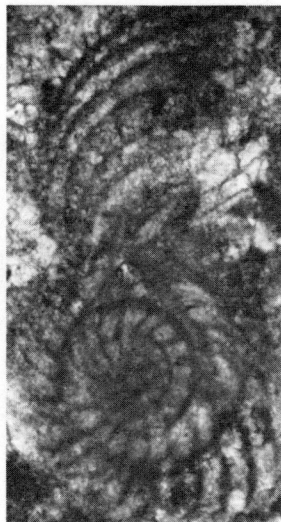

Figure 5

Parareichelina is more uncoiled and the septa in the uncoiled volution are more plicate". The early development of the test of the two genera is very similar, but the last, highly expanded volution of *Parareichelina* "wraps around" and gives the adult the form of a disk.

Miklukho-Maclay (1954, 1960) described two species of *Parareichelina*, *P. tenuissima* and *P. reticulata*, both from the northern Caucasus Mountains. Later, Sosnina (1968) described three species of *Parareichelina*, *P. mira*, *P. subangusta*, and *P. rhomboidea*, all from Upper Permian rocks exposed in the southern Sikhote Alin Mountains region east of Vladivostok, USSR. The precise geological settings of the rocks bearing these two genera are not stated with the original descriptions; Sosnina (1968) noted only that *P. mira* occurs in "organogenic fragmented and foraminiferal limestone".

The species of *Parareichelina* discovered in the rocks of the Hayfork terrane has an uncoiled portion (Fig. 3) similar to *P. reticulata* from the northern Caucasus (which has a very large uncoiled portion exceeding twice the length of the central tightly coiled part of the test). The Hayfork species also is similar to the other species from eastern Siberia. The uncoiled part of the test is similar to that of *P. mira*, whereas the inner volutions are similar to those of *P. rhomboidea*. The precise interrelationships among the five described species of *Parareichelina* have not been resolved because of their very small size and the difficulty in obtaining well oriented specimens.

PALEOGEOGRAPHIC SIGNIFICANCE

The fusulinaceans *Reichelina* and *Parareichelina* from limestone blocks in the eastern Hayfork terrane of the western Klamath Mountains, as well as the associated array of other foraminifers, form an assemblage of fossils that differs from all others in North America. Assemblages of Tethyan fossils are known from limestone blocks in tectonic belts stretching from the southern Sierra Nevada to Kodiak Island, Alaska, and in a thin stratigraphic interval in the Permian Basin of West Texas, but even these occurrences are distinctly different from and older than those described here from the western Klamath Mountains. Faunal assemblages similar to those recovered from the Klamath Mountains, however, compose the *Palaeofusulina* association of Ross (1967), which occurs in a belt of terranes that stretches from the Middle East to eastern Asia.

Reichelina is widespread in the Tethyan province and thus is not as convincingly exotic as *Parareichelina*, which is known only from the Caucasus Mountains and northeasternmost Siberia. All limestones bearing *Parareichelina* undoubtedly originated in the same paleobiogeographic province, perhaps in the same general geographic area, only to have been dispersed widely later.

Other paleobiogeographic work (Stevens and others, 1990) has suggested that the eastern Klamath terrane has been displaced thousands of kilometers from its position in Permian time. Therefore, it is not unreasonable to postulate that limestone blocks in the eastern Hayfork terrane, containing faunas that are even more exotic to North America, may have travelled considerably farther.

REFERENCES CITED

Blome, C.D., and Irwin, W.P., 1983, Tectonic significance of late Paleozoic to Jurassic radiolarians from the Northfork terrane, Klamath Mountains, California, in Stevens, C.H., ed., Pre-Jurassic rocks in western North American suspect terranes: Los Angeles, Pacific Section, Society of Economic Paleontologists and Mineralogists, p. 77-89.

Davis, G.A., 1968, Westward thrust faulting in the south-central Klamath Mountains, California: Geological Society of America Bulletin, v. 79, p. 911-934.

Erk, A.S., 1941, Sur la presence du genre *Codonofusiella* Dunbar and Skinner dans le Permien de Bursa (Turquie): Eclogae Geologique Helvetiae, v. 34, p. 234-253.

Igo, H., and Igo, H., 1977, Upper Permian fusulinaceans contained in the pebbles of the basal conglomerate of the Adoyama Formation, Kuzu, Tochigi Prefecture, Japan: Transactions of the Palaeontological Society of Japan, n.s., no. 106, p. 89-99.

Irwin, W.P., 1960, Geologic reconnaissance of the northern Coast Ranges and Klamath Mountains, California: California Division of Mines and Geology Bulletin 179, 80 p.

_____, 1966, Geology of the Klamath Mountains province: California Division of Mines and Geology Bulletin 190, p. 19-38.

_____, 1972, Terranes of the western Paleozoic and Triassic belt in the southern Klamath Mountains, California: U.S. Geological Survey Professional Paper 800-C, p. C103-C111.

_____, 1981, Tectonic accretion of the Klamath Mountains, in Ernst, W.G., ed., The geotectonic development of California (Rubey volume 1): Englewood Cliffs, New Jersey, Prentice-Hall, p. 29-49.

Kobayashi, F., 1975, *Palaeofusulina-Reichelina* fauna contained in the pebbles of intraformational conglomerate distributed in the Okutama district, west Tokyo: Transactions of the Palaeontological Society of Japan, n.s., no. 100, p. 220-229.

Luken, M.D., 1985, Petrography and origin of limestones in the western Paleozoic and Triassic belt, Klamath Mountains, California [M.S. thesis]: San Jose, California, San Jose State University, 131 p.

Luken, M.D., Stevens, C.H., and Magginetti, R., 1985, Permian faunal relationships across California: Geological Society of America Abstracts with Programs, v.

17, p. 367.

Miklukho-Maclay, A.D., 1963, Verkhnii Paleozoi Srednei Azii (Upper Paleozoic of Central Asia): Izdatel'stvo Leningradskogo Universiteta, 329 p.

Miklukho-Maclay, K.V., 1954, Foraminifera of the Upper Permian strata of the northern Caucasus: Trudy Vsesoiuznogo Nauchno-issledovatel'skogo Geologicheskogo Instituta (VSEGEI) Ministerstva Geologii i Okhrany Nedr (Moscow), 163 p.

_____, 1960, New Late Permian fusulinids of the northern Caucasus, in Markovsky, B.P., ed., Novye vidy drevnikh rastenii i bespozvonochnykh SSSR: Moscow, Gosgeodtekhizdat, pt. 1, p. 144-145.

_____, 1968, On the structure of some unique fusulinids: Ezhegodnik Vsesoyuznogo Paleontologicheskogo Obshchestva, v. 18, p. 15-18.

Monger, J.W.H., and Ross, C.A., 1971, Distribution of fusulinaceans in the western Canadian Cordillera: Canadian Journal of Earth Sciences, v. 8, p. 259-278.

Mortimer, N., 1985, Constraints on the Permian to Jurassic evolution of the Klamath Mountains: Geological Society of America Abstracts with Programs, v. 17, p. 371-372.

Nestell, M.K., 1981, Fusulinid affinities: central California to southern British Columbia: Geological Society of America Abstracts with Programs, v. 13, p. 519.

Nestell, M.K., and Blome, C.B., 1988, Paleontological constraints on the Baker terrane, eastern Oregon: Geological Society of America Abstracts with Programs, v. 20, p. A309.

Nestell, M.K., and Grant, R.E., 1988, Permian fusuline succession on Hydra, Greece: Beijing, 11th International Congress on Carboniferous Stratigraphy and Geology, Abstracts with Papers, Symposium and Miscellaneous v. 2, p. 471-472.

Nestell, M.K., Irwin, W.P., and Albers, J.P., 1981, Late Permian (early Djulfian) Tethyan foraminifera from the southern Klamath Mountains, California: Geological Society of America Abstracts with Programs, v. 13, p. 519.

Pantic, S., 1963, Upper Permian microfossils from the Anisian conglomerates of Haj-Nehaj, Montenegro: Vestnik Zavoda za Geoloska i Geofizicka Istrazivanja, ser. A, no. 21, p. 145-168.

Ross, C.A., 1967, Development of fusulinid (Foraminiferida) faunal realms: Journal of Paleontology, v. 41, p. 1341-1354.

Sheng, J.C., 1963, Permian fusulinids of Kwangsi, Kueichow, and Szechuan: Palaeontologia Sinica, no. 149, n.s. B, no. 10, 247 p.

Sheng, J.C., and Chang, L.H., 1958, Fusulinids from the type locality of the Changhsing Limestone: Acta Palaeontologica Sinica, v. 6, no.2, p. 211.

Skinner, J.W., and Wilde, G.L., 1955, New fusulinids from the Permian of west Texas: Journal of Paleontology, v. 29, p. 927-940.

Sosnina, M.I., 1968, New Late Permian fusulinids of Sikhote-Alin, in

Markovsky, B.P., ed., Novye vidy drevnikh rastenii i bespozvonochnykh SSSR: Moscow, Gosgeodtekhizdat, p. 99-128.

_____, 1981, Several Permian fusulinids of the Far East: Ezhegodnik Vsesoyuznogo Paleontologicheskogo Obshchestva, v. 24, p. 13-34.

Stevens, C.H., Miller, M.M., and Nestell, M.K., 1987, A new Permian waagenophyllid coral from the Klamath Mountains, California: Journal of Paleontology, v. 61, p. 690-699.

Stevens, C.H., Yancey, T.E., and Hanger, R.A., 1990, Significance of the provincial signature of Early Permian faunas of the eastern Klamath terrane: in Harwood, D.S., and Miller, M.M., eds., Paleozoic and Early Mesozoic paleogeographic relations: Sierra Nevada, Klamath Mountains, and related terranes: Geological Society of America Special Paper 255, p. 201-218.

Tien, N.D., 1979, Etude micropaleontologique (Foraminiferes) de materiaux du Permien du Cambodge [Ph.D thesis]; Paris, Universite de Paris, sud centre D'Orsay, 167 p.

Wardlaw, B., Furnish, W.M., and Nestell, M.K., 1979, An interpretation of the geology and paleontology of the Permian beds near Las Delicias, Coahuila, Mexico: Geological Society of America Bulletin, v. 90, p. 111-116.

Wright, J.E., 1982, Permo-Triassic accretionary subduction complex, southwestern Klamath Mountains, northern California: Journal of Geophysical Research, v. 87, p. 3805-3818.

PERMIAN AND TRIASSIC PALEOGEOGRAPHY OF THE EASTERN KLAMATH ARC AND EASTERN HAYFORK SUBDUCTION COMPLEX, KLAMATH MOUNTAINS, CALIFORNIA

M. Meghan Miller[*]
Division of Geological and
Planetary Sciences,
California Institute
of Technology,
Pasadena, California   91125

Jason B. Saleeby
Division of Geological and
Planetary Sciences,
California Institute
of Technology,
Pasadena, California   91125

ABSTRACT

Middle Permian and Middle Triassic volcanic-hypabyssal intrusive complexes form ensimatic arc deposits in the eastern Klamath terrane, northern California. Sedimentary matrix melange with blocks of sandstone, chert, and Tethyan fauna-bearing limestone compose the westward-lying eastern Hayfork terrane. Limestone olistoliths were derived from seamounts and incorporated into a subduction complex that was active during the Late Triassic and probably as early as the Permian. Geologic and biogeographic relations imply a genetic relationship between the ensimatic arc and subduction complex, constraining Permo(?)-Triassic subduction as eastward-dipping.

Disrupted quartzose sandstone from the melange yielded detrital zircon for isotopic provenance study. Nine multi-grain fractions yield $^{207}Pb/^{206}Pb$ ages that cluster between 2.046 and 2.139 Ga. Such ages suggest, but do not require, a North American source. Based on quartz and zircon provenance, the sandstone was not derived from older units in the Klamath Mountains. These data imply that the trench along which the accretionary wedge was formed intersected a continental source along strike; quartz sands were transported parallel to the length of the trench, and were then accreted during convergence between two oceanic plates. Subsequent, trench-parallel tectonic transport also may have affected these rocks. These geologic relations provide a first order constraint on Permo-Triassic paleogeography of related island arc and melange terranes in the southwestern Cordillera.

INTRODUCTION

The Klamath Mountains, northern California and southern Oregon, comprise a Paleozoic and early Mesozoic accretionary province, where ensimatic arc deposits, subduction complex, intra-arc ophiolite and transform assemblages have been structurally imbricated. A belt of similar rocks extends throughout the western Cordillera (Figure 1). Whereas most faults that currently juxtapose differing tectonostratigraphic assemblages within the Klamath Mountains (Fig. 2) were active as recently as the Late Jurassic (Harper and Wright, 1984; Wright and Fahan, 1988), many workers have argued for older paleogeographic and possible genetic relationships between some terranes

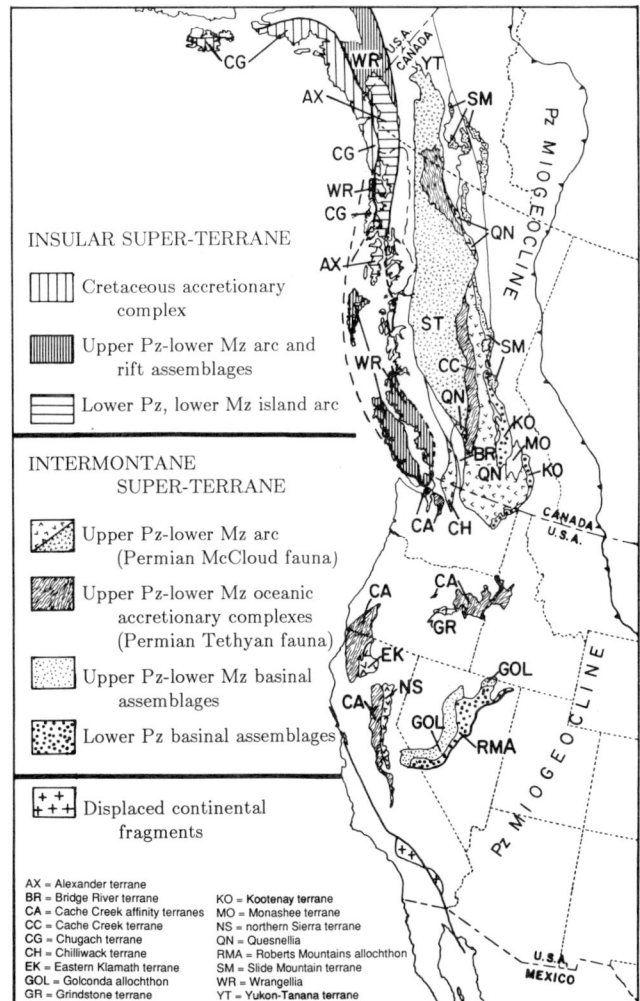

Figure 1. Paleozoic and early Mesozoic tectonic elements of the western North American Cordillera, from Miller and Saleeby (1989).

---

[*] Present address: Jet Propulsion Laboratory 183-501, California Institute of Technology, 4800 Oak Grove Drive, Pasadena, California   91109

In Cooper, J.D., and Stevens, C.H., eds., 1991, **Paleozoic Paleogeography of the Western United States-II:** Pacific Section SEPM, Vol. 67, p. 643-652.

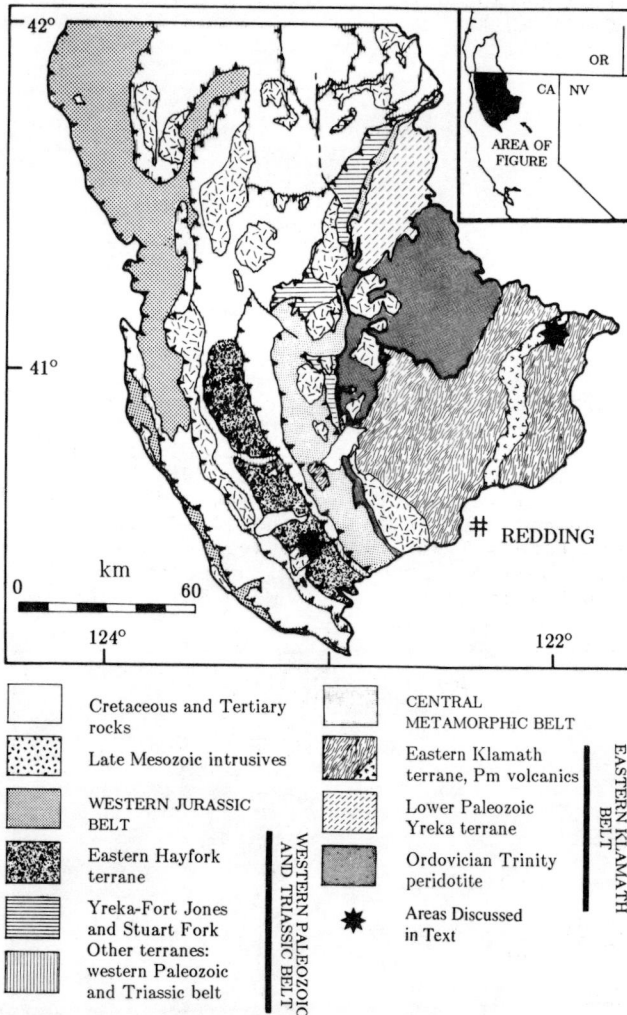

Figure 2. Lithotectonic belts of the Klamath Mountains. Simplified from Wright and Fahan (1988). Stars indicate areas discussed in the text. Area of Figure 3 corresponds to westerly star.

(e.g., Wright, 1982; Wright and Fahan, 1988). In this context, sedimentary matrix melange of the eastern Hayfork terrane has been interpreted as the subduction complex related to island arc volcanism to the east, in the eastern Klamath terrane (Fig. 2; Wright, 1982; Miller and Wright, 1987).

In order to better understand the paleogeographic setting of these related convergent margin lithotectonic assemblages, we conducted isotopic provenance studies of detrital zircon from quartzite in the eastern Hayfork melange. An early Proterozoic minimum age for detrital zircon suggests an ultimate source of older, sialic crust; the immediate source was mature, quartz-rich sediment. This paper summarizes these results and discusses the paleogeographic and tectonic setting of Permo-Triassic plate convergence.

GEOLOGIC SETTING

The Klamath Mountains consist of four regional lithotectonic belts that are bounded by east-dipping thrust faults (Fig. 2; Irwin, 1966). Collectively, these belts record a long-lived history of ensimatic convergent margin magmatism, deformation, and accretion. They record the protracted development of predominantly eastward-dipping subduction from the early or middle Paleozoic until the Middle Jurassic (Miller and Saleeby, 1989).

Island arc deposits of the easternmost and structurally highest eastern Klamath terrane record persistent Middle Devonian to Middle Jurassic volcanism, punctuated by sporadic carbonate platform development and epiclastic sedimentation (Miller, 1989). Together with coeval, biogeographically related island arc terranes elsewhere in western North America (Fig. 1), these rocks are thought to have formed a northeast Pacific fringing arc (Burchfiel and Davis, 1972, 1975; Miller 1987). Geologic relations have been used to suggest ties to marginal basin rocks of the Golconda allochthon to the east, and indirectly to western North America (Miller, 1989; Tomlinson and Wright, 1986). Biogeographic relations contradict such inferences, and have been used to suggest large geographic separations from the North American province to the east and from the Tethyan province that lay to the west (Stevens and others, 1990). Conflicting interpretations of eastern Klamath island arc paleogeography remain unresolved. Our approach stresses consideration of the stratigraphic and structural settings of epiclastic rocks and distinctive faunal associations in evaluating paleogeography.

The Permian and Triassic section of the eastern Klamath island arc terrane, of interest here, consists of Lower Permian carbonate platform deposits containing distinctive fusulinids and corals of "McCloud" or Klamath type, succeeded by a thick sequence of mid-Permian (early Guadalupian) submarine pyroclastic rocks and lava flows of the Bollibokka Group. The Permian sequence is overlain by Lower(?) and Middle Triassic tuffaceous and crystal-rich turbidites of the Pit Formation (see summary in Miller, 1990). This section attests to Permo-Triassic subduction-related magmatism.

Permian faunas from the eastern Klamath terrane differ from coeval faunas of the North American and Tethyan provinces (Stevens, 1985). Biogeographic relations suggest isolation from both of the more widespread provinces (Stevens and others, 1990). Nevertheless, regional geologic relations indicate ties to marginal basin rocks to the east, and indirectly, to western North America.

To the west of the island arc terrane lies the eastern Hayfork melange and broken formation (Fig. 2), composed of disrupted chert, argillite, and sandstone (Fig. 3;

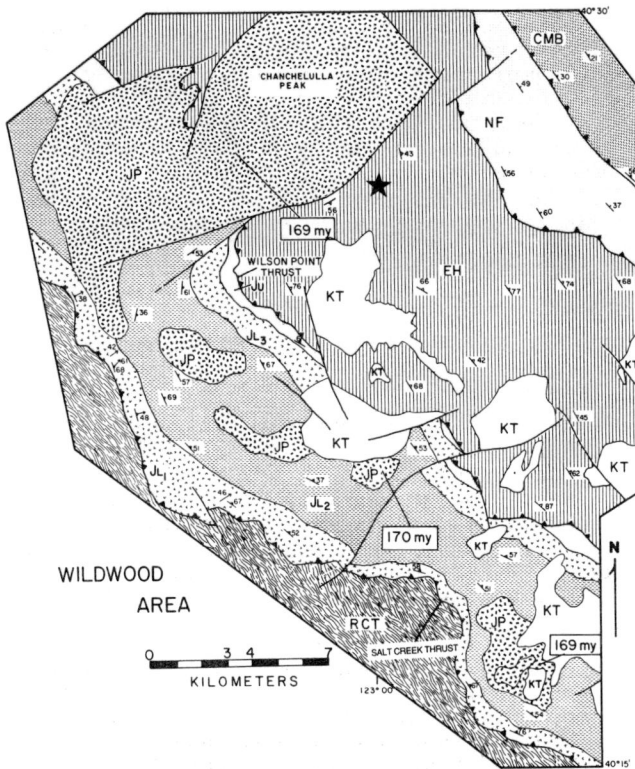

Figure 3. Geologic map of the southern part of the eastern Hayfork terrane, from Wright and Fahan (1988). Star indicates zircon sample locality, eastern Hayfork terrane.

Wright, 1982). The eastern Hayfork terrane also belongs to a regionally extensive belt of dismembered Permo-Triassic and locally older oceanic rocks associated with Late Triassic blueschist and exotic limestone blocks containing Tethyan fauna. These terranes are lithologically and biogeographically similar to the extensive Cache Creek terrane in British Columbia, and are referred to as terranes of "Cache Creek affinity" (Fig. 1; Miller, 1987).

Radiolarians and conodonts from melange-matrix chert in the eastern Hayfork terrane yield both Permian and Triassic ages (Irwin and others, 1978, 1983), implying the age-span of accretionary wedge development. Quartzose sandstone occurs both as blocks and interbedded with disrupted chert and argillite that form the melange matrix (Wright, 1982). Limestone olistoliths are mostly Permo-Carboniferous although some Silurian, Devonian and early Paleozoic ages were reported in the early literature (Irwin and Galanis, 1976). Limestone slide blocks came from seamounts or plateaus on older oceanic crust; thus, the oldest blocks give a minimum age for the oldest subducted oceanic crust.

Jurassic rocks lie between the arc and accretionary wedge deposits (Fig. 3), implying to some that the eastern Hayfork terrane may be younger (e.g. Irwin, in press). Jurassic tectonic reworking is well documented (Wright and Fahan, 1988); nevertheless, no post-Triassic rocks are known from the accretionary wedge. We follow the interpretation of Wright (1982), that existing age data strongly support a Late Triassic and older (Permo-Triassic) age of subduction-related accretion. In this context, Lower Jurassic chert that lies between the arc and accretionary complex (North Fork terrane) either formed in a younger (Early Jurassic) fore-arc setting (Saleeby and others, in press) or was trapped in an out-of-sequence paleogeographic position by middle Mesozoic strike-slip faulting.

Exotic blocks of limestone and blueschist also form important components of the eastern Hayfork terrane, and attest to a convergent margin origin. Distinctive Tethyan faunal assemblages characterize some of the limestone blocks (Irwin, 1972; Irwin and Galanis, 1976; Nestell and others, 1981; Stevens and others, 1987), recording incorporation of olistoliths shed from paleo-Pacific equatorial seamounts that were swept into the trench (Wright, 1982; Miller and Wright, 1987). An unusual occurrence of a single species of Tethyan-family coral in both the eastern Hayfork and eastern Klamath terranes contrasts with otherwise strongly differing faunas (Stevens and others, 1987). Together with stratigraphic and structural setting, biogeography perhaps supports the relationship of both terranes to a single convergent margin (Miller and Wright, 1987).

The arc and subduction complex formed in response to eastward subduction of the paleo-Pacific floor. The overriding plate was ensimatic in nature, thus, direct relations to coeval continental margin and arc assemblages have received little consideration. Nevertheless, indirect relations to western North America via marginal basin deposits of the Golconda allochthon and continent-derived sedimentary rocks in the arc basement have been considered likely by many workers. Possible along-strike variations in the convergent margin have also been little appreciated (Saleeby and others, in press).

DETRITAL ZIRCON POPULATIONS
AND ISOTOPIC RESULTS

A single sample of quartzose sandstone (subarkose of Folk, 1968) from the eastern Hayfork melange (Fig. 3) yielded zircon. Monocrystalline quartz dominates the sandstone; polycrystalline quartz aggregates and feldspar are sparsely present. Original grain rounding is obscured by metamorphic frittering of grain margins; some well rounded grains are well preserved and other grains show high sphericity (Fig. 4A) indicating a mature continental source (see criteria of Suttner and others, 1981). Argillaceous matrix recrystallized to biotite and muscovite indicate low greenschist facies metamorphism. A

Figure 4. A - Quartzose sandstone from the eastern Hayfork melange. Framework grains are mostly monocrystalline quartz with some alkali feldspar. Original grain shape is obscured by metamorphism. Picture width corresponds to ~2.9 mm. B - Detrital zircon from sandstone in A. Well rounded, dark grains predominate although other types are also present. Picture width corresponds to ~0.5 mm.

homogeneous separate of well rounded, pink to ruby-red zircon (Fig. 4B) was hand sorted for sample purification and enhancement of slight differences in color and morphology that were present in the sample. Nine fractions were dated by methods modified from those of Krogh (1973); some dispersion of ages was achieved (Fig. 5). $^{207}Pb/^{206}Pb$ ages cluster between 2.046 and 2.139 Ga, and generally correlate to slight changes in color (Table 1).

Proterozoic U-Pb and Pb-Pb average ages for the eastern Hayfork sample indicate that Precambrian rocks formed part of the ultimate source for recycled detritus. The oldest $^{207}Pb/^{206}Pb$ age obtained, 2.146 Ga, is a minimum age for the oldest component in the fraction; all $^{207}Pb/^{206}Pb$ ages cluster between 2.04 Ga and 2.15 Ga (Table 1). Best fit discordia lines were not calculated because relatively little dispersion of the data prevent calculation of meaningful intercept ages.

Detrital zircon suites are not comagmatic, hence they are commonly highly discordant, reflecting heterogeneous, multi-component source terranes. Homogenization of grain morphology by sedimentary recycling, however, reduces our ability to isolate multi-grain populations of possible age significance. The eastern Hayfork zircon fractions are remarkable for 1) yielding ages that are dramatically older than depositional age, 2) color and grain shape homogeneity, and 3) relatively little discordance (Fig. 5). All three characteristics point to a well-mixed, multi-cycle, ancient source. This contrasts strongly with results from other upper Paleozoic and lower Mesozoic clastic rocks within the Klamath Mountains (e.g., Miller and Saleeby, 1987, 1991; Miller and others, 1988; Fig. 6).

Clustering of the isotopic data do not imply ultimate sources of restricted age. A similar tightly clustered data set characterizes the Antelope Mountain Quartzite with $^{207}Pb/^{206}Pb$ ages from 2.03 Ga to 2.33 Ga, yet single crystal analyses yield $^{207}Pb/^{206}Pb$ ages as old as 2.95 Ga (Wallin, 1989). In light of this and the predominance of 2.0 to 2.1 Ga ages in reworked sedimentary rocks in western North America, $^{207}Pb/^{206}Pb$ ages for multi-grain detrital fractions are averages of younger and older components. Difficulty in characterizing end members partly results from homogenization of grain shape by multi-cycle sediment reworking.

Early Proterozoic average ages for detrital zircon in Permo-Triassic rocks have important geologic implications. The relatively short mean residence time of

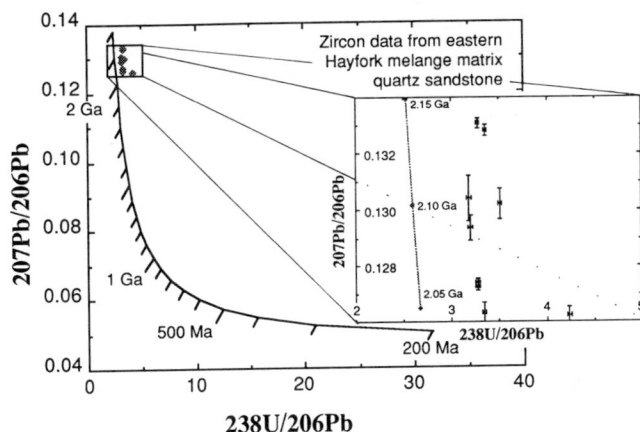

Figure 5. Concordia diagram (after Tera and Wasserburg, 1972) showing isotopic characteristics of detrital zircon from quartzose sandstone in the eastern Hayfork melange. Analytical uncertainties are shown on inset.

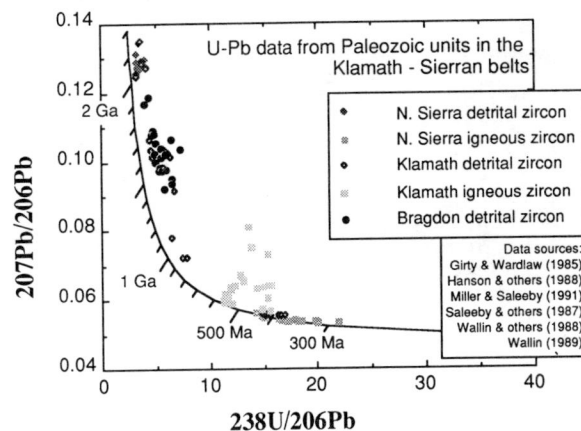

Figure 6. Concordia diagram (after Tera and Wasserburg, 1972) showing available isotopic data from older terranes in the Klamath Mountains and Sierra Nevada.

oceanic crust on the earth's surface (55 m.y. for modern crust) and young age of the oldest crust (180 Ma) preserved on the modern ocean floor (Howell and Murray, 1986) make detrital zircon ages a powerful tool for discriminating recycled ensialic sediment from younger oceanic sources. Isotopic data from detrital zircon that establish source components >500 m.y. older than the depositional age of the host clastic rocks clearly indicate an ultimate continental source. In the case of the eastern Hayfork rocks, isotopic characteristics are in accord with the

multi-cycle, quartz-rich provenance of the sandstone in which the zircon occur (Fig. 4A).

DISCUSSION

The eastern Hayfork terrane represents a Late Triassic and older sedimentary accretionary wedge that accumulated along an eastward-dipping subduction zone resulting from convergence of two oceanic plates (Wright, 1982). Eastward Permo-Triassic volcanic and volcaniclastic strata in the eastern Klamath terrane record arc magmatism

TABLE 1: ZIRCON ISOTOPIC AGE DATA

Fraction size(μm)	Fraction character[†]	Amount Analyzed (mg) [@]	Concentrations (ppm)			Atomic Ratios			Isotopic ages (Ma) [§]		
			^{238}U	$^{206}Pb^{*}$	$\frac{^{206}Pb}{^{204}Pb}$	$\frac{^{206}Pb^{*}}{^{238}U}$	$\frac{^{207}Pb^{*}}{^{235}U}$	$\frac{^{207}Pb^{*}}{^{206}Pb^{*}}$	$\frac{^{206}Pb^{*}}{^{238}U}$	$\frac{^{207}Pb^{*}}{^{235}U}$	$\frac{^{207}Pb^{*}}{^{206}Pb^{*}}$
62-80	C	0.3	797	217	5096	0.31417(226)	5.6466	0.13041(080)	1761	1923	2103
>120	C	0.4	448	118	4335	0.30498(157)	5.3547	0.12740(015)	1716	1878	2062
62-80	1 xo	0.5	69	14	1855	0.23602(122)	4.1064	0.12624(027)	1366	1656	2046
62-80	2 xo	0.5	141	36	5542	0.29929(152)	5.2110	0.12634(035)	1688	1855	2047
62-80	2 o	0.1	354	87	4605	0.28536(151)	5.1213	0.13022(054)	1619	1840	2101
62-80	3 x	0.1	1237	327	3569	0.30516(157)	5.3529	0.12728(015)	1717	1877	2061
62-80	3 o	0.1	431	111	2947	0.29865(153)	5.4672	0.13283(020)	1685	1896	2136
62-80	4 x	0.8	423	114	4227	0.31233(213)	5.5684	0.12936(045)	1752	1911	2089
62-80	4 o	0.9	159	42	2275	0.30574(159)	5.6072	0.13307(017)	1720	1917	2139

* Radiogenic:nonradiogenic Pb correction based on 40 picogram blank Pb (1:18.78:15.61:38.50) and initial Pb approximations: 206/204 = 15.66; 207/204 = 15.3; 208/204 = 35.3 (Proterozoic, Stacey and Kramers, 1975).

† Fractions are non-magnetic at 1.7 amps and side/front slopes of 2/20 on Franz Isodynamic Separator. Samples hand-picked to 99.9% purity prior to dissolution. Color and morphology codes: C=composite of size fraction, only sorted for purity, 0=lightest pink, 1=light pink, 2=medium pink, 3=dark pink, 4=rose and ruby red. Where colored fractions were subdivided by grain shape, o=rounded or well rounded, x=subangular to subrounded. Dissolution and chemical extraction technique modified from Krogh (1973).

@ Due to uncertainties in weight of sample, U and Pb concentration may be in error, but use of mixed ^{205}Pb-^{230}Th-^{235}U spike insures correct age.

§ Decay constants used in age calculation: $\lambda^{238}U$ =1.55125 x 10^{-10}, $\lambda^{235}U$=9.8485 x 10^{-10} (Jaffey and others, 1971); $^{238}U/^{235}U$ atom=137.88. Uncertainties in $^{206}Pb^{*}/^{238}U$ and $^{207}Pb^{*}/^{206}Pb^{*}$ are given as (±) in the last three figures. Uncertainties calculated by quadratic sum of total derivatives of ^{238}U and $^{206}Pb^{*}$ concentration and $^{207}Pb^{*}/^{206}Pb^{*}$ equations with error differentials defined as: 1. Isotope ratio determinations from standard errors ($\sigma/\sqrt{n}$) of mass spectrometer runs plus uncertainties in fractionation corrections based on multiple runs of NBS 981, 983, and U500 standards; 2. Spike concentrations from range of deviations in multiple calibrations with normal solutions; 3. Spike compositions from external precisions of multiple isotope ratio determinations; 4. Uncertainty in natural $^{238}U/^{235}U$ from Chen and Wasserburg (1981); and 5. Nonradiogenic Pb isotopic compositions from uncertainties in isotope ratio determinations of blank Pb and uncertainties in composition of initial Pb from Stacey and Kramers (1975).

(Fig. 7). Apparently conflicting pieces of evidence bear on the paleogeography of that plate margin. On the one hand, limestone blocks contain exotic Tethyan faunas, implying to some an origin far removed from the eastern Klamath arc and western North America (Stevens and others, 1990). The faunas occur in olistoliths, however, and were rafted into the trench on allochthonous seamounts on paleo-Pacific ocean floor (Miller and Wright, 1987), and therefore do not constrain the location of the trench along which they were accreted. More problematical differences between Permian McCloud (eastern Klamath)-type fauna and contemporaneous faunas on autochthonous North America have led Stevens and others (1990) to suggest longitudinal separation of at least 5000 km between the faunal provinces.

On the other hand, geologic and paleotectonic ties suggest an early relationship to western North America: subduction polarity is constrained by the paired arc and subduction complex and no younger suture lies east of the Klamath Mountains (Fig. 7). Wright (1982) first

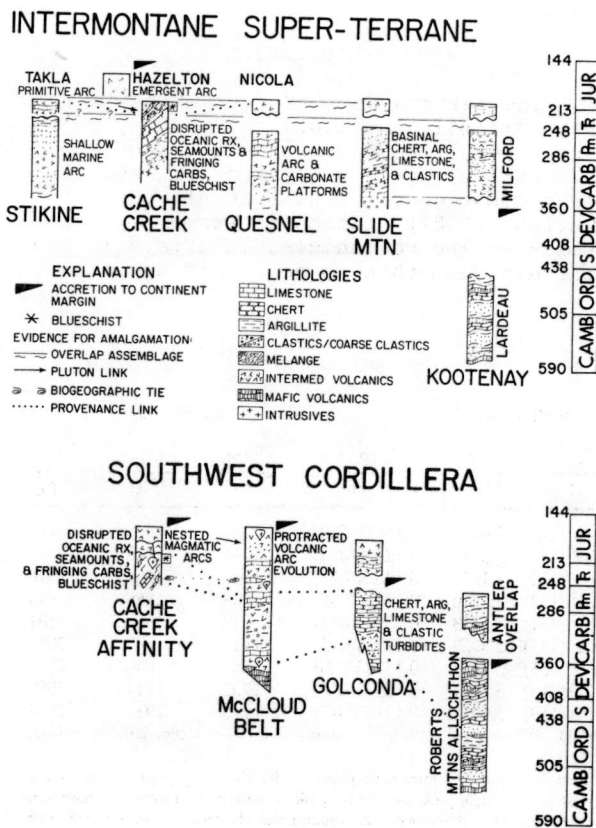

Figure 7. Petrotectonic evolution and paleogeographic relationships of Paleozoic and early Mesozoic belts in the southwestern Cordillera, near latitude 41°N and analogous relations in British Columbia (51°N). From Miller and Saleeby (1989).

noted the unusual occurrence of quartz-rich sandstone of probable continental provenance in an oceanic subduction complex. Quartzose sandstone contained in the eastern Hayfork melange is dominantly composed of monocrystalline quartz and conspicuously lacks lithic fragments with the exception of some quartz aggregates, implying derivation from contemporaneous, mature continental sheet sands rather than from an intermediate, predominantly oceanic, sedimentary source. Our new isotopic data confirm an ultimate continental origin for sand accreted along the trench. Multi-cycle reworking of zircon homogenized grain morphology, which hampered our efforts to enhance different-age components by hand sorting. The predominance of well rounded zircon, however, confirms a mature continental source.

Without exception, modern sedimentary accretionary wedges occur along convergent margins that intersect continental sediment sources somewhere along strike (Scholl and others, 1990). Little opportunity exists for sediment transport across a marine arc. Continent-derived detritus undergoes trench-parallel sediment transport within the bathymetric confines of the trench (Dickinson, 1982; Velbel, 1985; Underwood, 1986). Transport distances may exceed thousands of kilometers along strike, introducing continental detritus into oceanic settings. Post-accretion tectonic transport along the forearc or inner trench wall may account for further lateral translation parallel to the plate margin (Dickinson, 1982). These relations provide a basic paleogeographic context for Permo-Triassic subduction recorded in Klamath Mountains terranes.

By analogy, we infer that the oceanic convergent margin represented by the eastern Hayfork subduction complex and eastern Klamath arc intersected a continental source along strike. The zircon and quartz provenance data are non-unique, however, and are insufficient to discriminate between continents. Nonetheless, 2.1 Ga average ages from detrital zircon are common in western North America (summarized by Rubin and others, 1990; Miller and Saleeby, 1991), and invite (but do not require) consideration of a North American source.

Some pre-Permian recycled continental sedimentary rocks are present within the Klamath Mountains (e.g., Wallin, 1989; Miller and Saleeby, 1991) and in similar units in the Sierra Nevada (Bond and DeVay, 1980; Girty and Wardlaw, 1985), yet few warrant serious consideration as possible sources (Fig. 6). For instance, zircon data from Carboniferous turbidites in the eastern Klamath terrane show mixing of recycled continental and early Paleozoic magmatic components (Bragdon detrital zircon in Fig. 6; Miller and Saleeby, 1991). Isotopic heterogeneity rules it out as a possible source terrane. The only isotopically plausible local source units, multi-cycle quartzose sandstone of the Antelope Mountain

PERMO-TRIASSIC
PALEOGEOGRAPHY

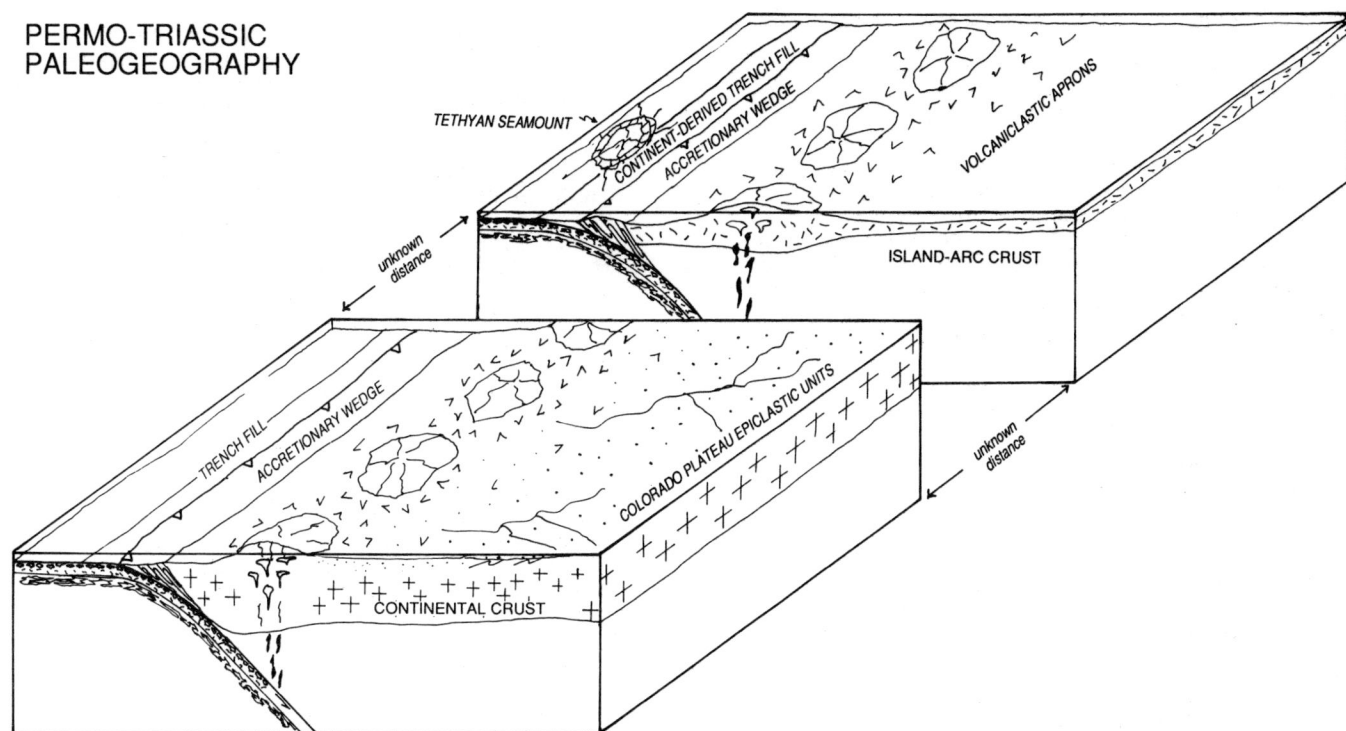

Figure 8. Possible along-strike relationship between the Klamath Permo-Triassic convergent margin and a continental source. Contemporaneous continent-derived quartzose sands were present on the Colorado Plateau and the Mojave Desert region, and may have been funnelled into the trench, transported along strike within the bathymetric confines of the trench, and accreted along the plate margin in a sedimentary accretionary wedge.

Quartzite and the Shoo Fly Complex (compare Fig. 5 and Fig. 6), are so restricted that volumetric insignificance preclude them as sources; if reworked into younger rocks, they would have been diluted by younger sediment sources (as in the case of the Carboniferous turbidites). Furthermore, such possible source rocks were buried beneath the Permo-Triassic arc edifice during accretion of the eastern Hayfork terrane.

A Permo-Triassic magmatic arc was constructed across the truncated North American continental margin in the Mojave Desert (Walker, 1988), and in a basinal setting in southwestern Nevada (Speed, 1977; Speed and others, 1977), but is unknown on the miogeocline or other continent-tied rocks further to the north, implying a possible location for the transition from continental to oceanic subduction. Late Triassic and older quartzose sandstone occurs in the Lower and Middle(?) Triassic Moenkopi Formation and the Upper Triassic Chinle Formations on the Colorado Plateau (Stewart and others, 1972a, 1972b) and their equivalents in the Mojave Desert (Walker, 1988). These formations are compositionally variable, yet they contain zircon-bearing quartzose sandstone that is petrographically similar to that studied from the eastern Hayfork terrane (see photomicrographs of

Cadigan, 1971; detrital zircon descriptions of Stewart and others, 1972a, 1972b). Although we know of no isotopic data for these units, they are interesting potential candidate sources and/or terrestrial equivalents for quartzose sandstone in eastern Hayfork melange.

Recent interpretations of biogeographic relations emphasize the large distances implied by differences between the North American, McCloud (eastern Klamath), and Tethyan Permian faunal provinces (Stevens and others, 1990). Many geologic data are not addressed by these evaluations, however, and we know of no way to reconcile known geologic relations with such biogeographic interpretations. That the Permo-Triassic Klamath convergent margin intersected a continental source seems certain. Whether the continental source was part of North America has not been uniquely determined from the zircon data, but has widespread support from regional geologic relations (e.g., Davis and others, 1978; Burchfiel and Davis, 1972, 1975; Miller, 1987, 1989; Miller and Saleeby, 1991). Analogous relationships between major Cordilleran terranes are also well documented in British Columbia (Monger, 1984; Fig. 7). The interpretation in Figure 8 is consistent with and motivated by regional geologic relations in the southwestern U.S. The stratigraphic and structural constraints presented here provide a context for new evaluations of biogeographic relations.

CONCLUSIONS

1. In the context of modern convergent margins, Permo-Triassic sedimentary accretionary wedge development in the Klamath Mountains implies intersection of an oceanic convergent margin with a continental sediment source. The unusual occurrence of quartzose sandstone in the eastern Hayfork terrane confirms a continental source.

2. Isotopic data from detrital zircon in such quartzose sandstone yield a narrow range of $^{207}Pb/^{206}Pb$ ages that cluster between 2.046 and 2.139 Ga, further characterizing this source as a mature, homogeneous continental source. Together with sandstone framework composition, these data imply a contemporaneous source of well-worked quartzose sand. Regional geologic relations and 2.0-2.1 Ga ages invite consideration of possible North American sources.

3. Permo-Triassic arc deposits developed on continental margin facies in the Mojave Desert suggest a possible location for the intersection of an oceanic convergent margin with the continent. Epiclastic rocks of appropriate age and provenance are represented in the Moenkopi and Chinle Formations on the Colorado Plateau, and their equivalents in the Mojave Desert. This speculation is testable by isotopic studies of detrital zircon from these possible equivalents.

ACKNOWLEDGEMENTS

J. E. Wright's detailed field studies in the eastern Hayfork terrane set the stage for isotopic studies reported here, and he first recognized the importance of mature quartzose sediment in an oceanic subduction complex. In addition, Wright generously provided Miller with sampling localities. Laboratory studies were supported by NSF Grant EAR 87-08266 awarded to J. B. Saleeby. D.S. Harwood, C.M. Rubin, and J.E. Wright provided helpful reviews.

REFERENCES CITED

Bond, G.C., and DeVay, J.C., 1980, Pre-Upper Devonian quartzose sandstones in the Shoo Fly Formation, northern California-Petrology, provenance and implications for regional tectonics: Journal of Geology, v. 88, p. 285-288.

Burchfiel, B.C., and Davis, G.A., 1972, Structural framework and evolution of the southern part of the Cordilleran orogen, western United States: American Journal of Science, v. 272, p. 97-118.

————, 1975, Nature and controls on Cordilleran orogenesis, western United States: Extensions of an earlier synthesis: American Journal of Science, v. 275A, p. 363-396.

Cadigan, R.A., 1971, Petrology of the Triassic Moenkopi Formation and related strata in the Colorado Plateau region: U.S. Geological Survey Prof. Paper 692, 70 p.

Chen, J.H., and Wasserburg, G.J., 1981, The isotopic composition of uranium and lead in Allende inclusions and meteoritic phosphates: Earth and Planetary Science Letters, v. 52, p. 1-15.

Davis, G.A., Monger, J.W.H., and Burchfiel, B.C., 1978, Mesozoic construction of the Cordilleran 'collage', central British Columbia to California, in Howell, D.G., and McDougall, K.A., eds., Mesozoic Paleogeography of the Western United States: Los Angeles, Pacific Section, Society of Economic Paleontologists and Mineralogists, Pacific Coast Paleogeographic Symposium 2, p. 1-32.

Dickinson, W.R., 1982, Compositions of sandstones in circum-Pacific subduction complexes and fore-arc basins: American Association of Petroleum Geologists Bulletin, v. 66, p. 121-137.

Folk, R.L., 1968, Petrology of sedimentary rocks: Austin, Texas, Hemphill's Book Store, 170 p.

Girty, G.H., and Wardlaw, M.S., 1985, Petrology and provenance of pre-Late Devonian sandstones, Shoo Fly Complex, northern Sierra Nevada, California: Geological Society of America Bulletin, v. 96, p. 516-521.

Hanson, R.E., Saleeby, J.B., and Schweickert, R.A., 1988, Composite Devonian island-arc batholith in the northern Sierra Nevada, California: Geological Society of America Bulletin, v. 100, p. 446-457.

Harper, G.D., and Wright, J.E., 1984, Middle to Late Jurassic tectonic evolution of the Klamath Mountains, California-Oregon: Tectonics, v. 3, p. 759-772.

Howell, D.G., and Murray, R.W., 1986, A budget for continental growth and denudation: Science, v. 233, p. 446-449.

Irwin, W.P., 1966, Geology of the Klamath Mountains province, in Bailey, E.H., ed., Geology of northern California: California Division of Mines and Geology Bulletin 190, p. 19-38.

————, 1972, Terranes of the western Paleozoic and Triassic belt in the southern Klamath Mountains, California: U.S. Geological Survey Professional Paper 800-C, p. C103-C111.

Irwin, W.P., and Galanis, S.P., Jr., 1976, Map showing limestone and selected fossil localities in the Klamath Mountains, California and Oregon: U.S. Geological Survey Miscellaneous Field Studies Map MF-749, scale 1:500,000.

Irwin, W.P., Jones, D.L., and Kaplan, T.A., 1978, Radiolarians from pre-Nevadan rocks of the Klamath Mountains, California and Oregon, in Howell, D.G., and McDougall, K.A., eds., Mesozoic Paleogeography of the Western United States: Los Angeles, Pacific Section, Society of Economic Paleontologists and Mineralogists, Pacific Coast Paleogeographic Symposium 2, p. 303-310.

Irwin, W.P., Wardlaw, B.R., and Kaplan, T.A., 1983, Conodonts of the western Paleozoic and Triassic belt, Klamath Mountains, California and Oregon: Journal of Paleontology, v. 57, p. 1030-1039.

Jaffey, A.H., Flynn, K.F., Glendenin, L.E., Bentley, W.C., and Essling, A.M., 1971, Precision measurement of half-lives and specific activities of ^{235}U and ^{238}U: Physical Reviews C, v. 4, 1889-1906.

Krogh, T.E., 1973, A low-contamination method for hydrothermal decomposition of zircon and extraction of U and Pb for isotopic age determinations: Geochimica Cosmochimica Acta, v. 46, p. 485-494.

Miller, M.M., 1987, Dispersed remnants of a northeast Pacific fringing arc - Upper Paleozoic island arc terranes of Permian McCloud faunal affinity, western U.S.: Tectonics, v. 6, p. 807-830.

————, 1989, Intra-arc sedimentation and tectonism: late Paleozoic evolution, eastern Klamath terrane, California: Geological Society of America Bulletin, v. 101, p. 170-187.

————, 1990, Episodic compressional tectonism in an intra-arc setting, eastern Klamath terrane, northern California, in Wiley, T.J., Howell, D.G., and Wong, F.L., eds., Terrane analysis of China and the Pacific rim: Houston, Circum-Pacific Council for Energy and Mineral Resources, Earth Science Series, v. 13, p. 133-146.

Miller, M.M., and Saleeby, J.B., 1987, Detrital zircon studies of the Galice Formation: Common provenance of strata overlying the Josephine ophiolite and Rogue island arc - western Klamath Mountains terrane: Geological Society of America Abstracts with Programs, v. 19, p. 772-773.

————, 1989, Accretionary tectonics: Examples from the North American Cordillera, in James, D.E., ed., Encyclopedia of Geophysics, New York, Van Nostrand Reinhold Company, Inc., p. 9-20.

————, 1991, Continental detrital zircon in Carboniferous ensimatic arc rocks, Bragdon Formation, eastern Klamath terrane, northern California: Geological Society of America Bulletin, v. 103, p. 268-276.

Miller, M.M., Saleeby, J.B., Harper, G.D., and Wright, J.E., 1988, Jurassic orogenesis within the Klamath Mountains: Episodic compressional tectonism and arc superposition in an evolving convergent margin: Geological Society of America Abstracts with Programs, v. 20, p. A274.

Miller, M.M., and Wright, J.E., 1987, Paleogeographic implications of Permian Tethyan corals from the Klamath Mountains, California: Geology, v. 15, p. 266-269.

Monger, J.W.H., Cordilleran tectonics: A Canadian perspective: Bulletin Societé Géologique de France, v. 7, p. 255-278.

Nestell, M.K., Irwin, W.P., and Albers, J.P., 1981, Late Permian (early Djulfian) Tethyan foraminifera from the southern Klamath Mountains, California:

Geological Society of America Abstracts with Programs, v. 13, p. 519.

Rubin, C.M., Miller, M.M., and Smith, G.M., 1990, Cordilleran mid-Paleozoic arc and basinal terranes and their basement: Evidence for tectonism in a convergent margin setting, in Harwood, D.S., and Miller, M.M., eds., Paleozoic and early Mesozoic paleogeographic relations between the Klamath Mountains, Sierra Nevada, and related terranes: Geological Society of America, Special Paper 255, p. 1-16.

Saleeby, J.B., Hannah, J.L., and Varga, R.J., 1987, Isotopic age constraints on middle Paleozoic deformation in the northern Sierra Nevada, California: Geology, v. 15, p. 757-760.

Saleeby, J.B., Busby-Spera, C.J., and five contributors, in press, Early Mesozoic tectonic evolution of the western U.S. Cordillera, in Burchfiel, B.C., Lipman, P., and Zoback, M.L., eds., The Cordilleran Orogen: Conterminous United States, Geological Society of America, Decade of North American Geology, The Geology of North America, v. G-3.

Scholl, D.W., von Huene, R., and Dieffenbach, H.L., 1990, Rates of sediment subduction and subduction erosion - Implications for growth of terrestrial crust: Eos, Transactions, American Geophysical Union, v. 71, p. 1576.

Speed, R.C., 1977, Island-arc and other paleogeographic terranes of late Paleozoic age in the western Great Basin, in Stewart, J.H., Stevens, C.H., and Fritsche, A.E., eds., Paleozoic Paleogeography of the Western United States, Los Angeles, Pacific Section, Society of Economic Paleontologists and Mineralogists, Pacific Coast Paleogeographic Symposium 1, p. 349-362.

Speed, R.C., MacMillan, J.R., Poole, F.G., and Kleinhampl, F.J., 1977, Diablo Formation, central western Nevada; composite of deep and shallow water Upper Paleozoic rocks, in Stewart, J.H., Stevens, C.H., and Fritsche, A.E., eds., Paleozoic Paleogeography of the Western United States, Los Angeles, Pacific Section, Society of Economic Paleontologists and Mineralogists, Pacific Coast Paleogeographic Symposium 1, p. 301-314.

Stacey, J.S., and Kramers, J.D., 1975, Approximation of terrestrial lead isotope evolution by a two-stage model: Earth and Planetary Science Letters, v. 26, p. 207-211.

Stevens, C.H., 1985, Reconstruction of Permian paleogeography based upon distribution of Tethyan faunal elements, in Dutro, J.T., Jr., and Pfefferkorn, H.W., eds., Compte Rendu/Neuvième Congrés international de stratigraphie et de géologie du carbonifère, 5: Paleontology, paleoecology, paleogeography: Carbondale, Southern Illinois University Press, p. 383-393.

Stevens, C.H., Miller, M.M., and Nestell, M.K., 1987, A new Permian Tethyan coral

652

from the Klamath Mountains, California: Journal of Paleontology, v. 61, p. 690-699.

Stevens, C.H., Yancey, T.E., and Hanger, R.A., 1990, Significance of the provincial signature of Early Permian faunas of the eastern Klamath terrane, *in* Harwood, D.S., and Miller, M.M., eds., Paleozoic and early Mesozoic paleogeographic relations between the Klamath Mountains, Sierra Nevada, and related terranes: Geological Society of America, Special Paper 255, p. 201-218.

Stewart, J.H., Poole, F.G., and Wilson, R.F., 1972a, Stratigraphy and origin of the Chinle Formation and related Upper Triassic strata in the Colorado Plateau Region: U.S. Geological Survey Professional Paper 690, 336 p.

————, 1972b, Stratigraphy and origin of the Triassic Moenkopi Formation and related strata in the Colorado Plateau Region: U.S. Geological Survey Professional Paper 691, 195 p.

Suttner, L.J., Basu, A., and Mack, G.H., 1981, Climate and the origin of quartz arenites: Journal of Sedimentary Petrology, v. 51, p. 1235-1246.

Tera, F., and Wasserburg, J.G., 1972, U-Th-Pb systematics in three Apollo 14 basalts and the problem of initial Pb in lunar rocks: Earth and Planetary Science Letters, v. 14, p. 281-304.

Tomlinson, A.J., and Wright, J.E., 1986, The Klamath-Sierran arc: A possible source region for Permian volcaniclastic and conglomeratic units within the Golconda allochthon: Geological Society of America Abstracts with Programs, v. 18, p. 193.

Underwood, M.B., 1986, Sediment provenance within subduction complexes — an example from the Aleutian forearc: Sedimentary Geology, v. 51, p. 57-73.

Velbel, M.A., 1985, Mineralogically mature sandstones in accretionary prisms: Journal of Sedimentary Petrology, v. 55, p. 685-690.

Walker, J.D., 1988, Permian and Triassic rocks of the Mojave Desert and their implications for timing and mechanisms of continental truncation: Tectonics, v. 7, p. 685-709.

Wallin, E.T., 1989, Provenance of lower Paleozoic sandstones in the eastern Klamath Mountains and the Roberts Mountains allochthon, California and Nevada: [Ph.D. thesis]: Lawrence, University of Kansas, Kansas, 152 p.

Wallin, E.T., Mattinson, J.M., and Potter, A.W., 1988, Early Paleozoic magmatic events in the eastern Klamath Mountains, northern California: Geology, v. 16, p. 144-148.

————, 1982, Permo-Triassic accretionary subduction complex, southwestern Klamath Mountains, northern California: Journal of Geophysical Research, v. 87, p. 3805-3818.

Wright, J.E., and Fahan, M.D., 1988, An expanded view of Jurassic orogenesis in the western United States Cordillera: Middle Jurassic (pre-Nevadan) regional metamorphism and thrust faulting within an active arc environment, Klamath Mountains, California: Geological Society of America Bulletin, v. 100, p. 859-876.

AGE AND TECTONIC SIGNIFICANCE OF METAMORPHIC ROCKS ALONG THE AXIS OF THE SIERRA NEVADA BATHOLITH: A CRITICAL REAPPRAISAL

Richard A. Schweickert and Mary M. Lahren
Department of Geological Sciences
University of Nevada, Reno
Reno, Nevada 89557

ABSTRACT

A continental crustal sliver referred to as the "Snow Lake block", with the approximate dimensions of the Salinian block, and represented by metamorphosed Proterozoic and Cambrian miogeoclinal rocks, lies within the axial parts of the Sierra Nevada batholith (Lahren, 1989; Lahren and Schweickert, 1989; Lahren and others, 1990). Previously, metamorphic rocks in roof pendants along the axis of the Sierra Nevada batholith between 36° and 38° N latitude have been considered part of the "Kings sequence," of probable Mesozoic age. However, most of the pendants are in fact undated by fossils and probably contain Paleozoic or Precambrian strata.

Our new studies, and a reappraisal of new and existing data on these pendants have led to the following hypotheses: The Snow Lake block extends 120 mi (200 km) southeastward from northern Yosemite National Park to the Kaweah River drainage and includes rocks in the following pendants: Snow Lake, Piute Mountain, Glen Aulin, May Lake, Iron Mountain, Quartz Mountain, Dinkey Creek, Patterson Mountain, Boyden Cave, and Sequoia Park. Probable stratigraphic units in this block include the Middle Cambrian Bonanza King Formation in addition to the Proterozoic Stirling Quartzite, Lower Cambrian Wood Canyon Formation and Zabriskie Quartzite, Lower to Middle Cambrian Carrara Formation, and Lower Triassic Fairview Valley Formation proposed by Lahren and others (1990). This sliver was displaced about 400 km northward along an intrabatholithic dextral strike-slip fault during Early Cretaceous time (Lahren, 1989; Lahren and Schweickert, 1989; Lahren and others, 1990; Schweickert and Lahren, 1990).

In addition, we propose that tectonostratigraphic units of the Western Sierra Nevada Metamorphic belt continue southeastward along the western edge of the Snow Lake block to latitude 36° N and possibly farther south. The lower Paleozoic Shoo Fly Complex, which lies adjacent to the Snow Lake block, includes rocks in parts of the following pendants: Lower Kings River, Lower Kaweah River, Tule River, and Kern Canyon.

The apparent juxtaposition of the Shoo Fly Complex with the Snow Lake block requires the presence of a major tectonic break between them. This structure, now largely obliterated by Cretaceous plutons of the Sierra Nevada batholith, may be the equivalent of the Golconda thrust in north-central Nevada.

The Paleozoic-Triassic Calaveras Complex lies west of the Shoo Fly Complex in parts of the Oakhurst, Lower Kings River, Lower Kaweah River, Yokohl Valley, and Tule River pendants, and the two terranes are inferred to be separated by the Calaveras-Shoo Fly thrust.

Paleozoic to Jurassic rocks of the Don Pedro terrane, including the Kings-Kaweah ophiolite belt, occur west of the Calaveras Complex, and are inferred to be separated from the Calaveras by the Sonora fault.

The new subdivisions of the pendants presented here are consistent with available stratigraphic, structural, petrologic, geochemical, and isotopic data. These subdivisions have important implications for batholithic evolution and for the regional tectonic framework of eastern and southern California.

INTRODUCTION

Metasedimentary and metavolcanic rocks in small roof pendants within the Sierra Nevada batholith between 36° and 38°N latitude (Fig. 1) are structurally complex, generally unfossiliferous, and highly metamorphosed. Their stratigraphic and structural affinities are uncertain and have long been controversial. Ironically, the few fossil localities in the region have led many geologists to the conclusion that all the rocks are Mesozoic in age, whereas in this paper we argue that most of the rocks are, in fact, Paleozoic in age. Lead and strontium isotopic data suggest that the axial part of the batholith in this region is underlain by a promontory of crust of continental character (Kistler and Peterman, 1973; Kistler, 1978; Chen and Tilton, 1978; Saleeby and Chen, 1978; Kistler, 1990; Saleeby and others, 1986), and this data has led to suggestions that major strike-slip faults or sutures lie concealed within the batholith (Saleeby and Chen, 1978; Kistler and others, 1980; Nokleberg, 1983; Robinson and Kistler, 1986; Saleeby and others, 1986; Kistler 1990). Major crustal boundaries also have been proposed to lie within this region based on studies of Sm and Nd isotopes, oxygen isotopes, and the mineral chemistry of plutonic rocks and their wallrocks (DePaolo, 1981; Ague and Brimhall, 1987, 1988; Kistler, 1990).

In Cooper, J.D., and Stevens, C.H., eds., 1991, **Paleozoic Paleogeography of the Western United States-II:** Pacific Section SEPM, Vol. 67, p. 653-676.

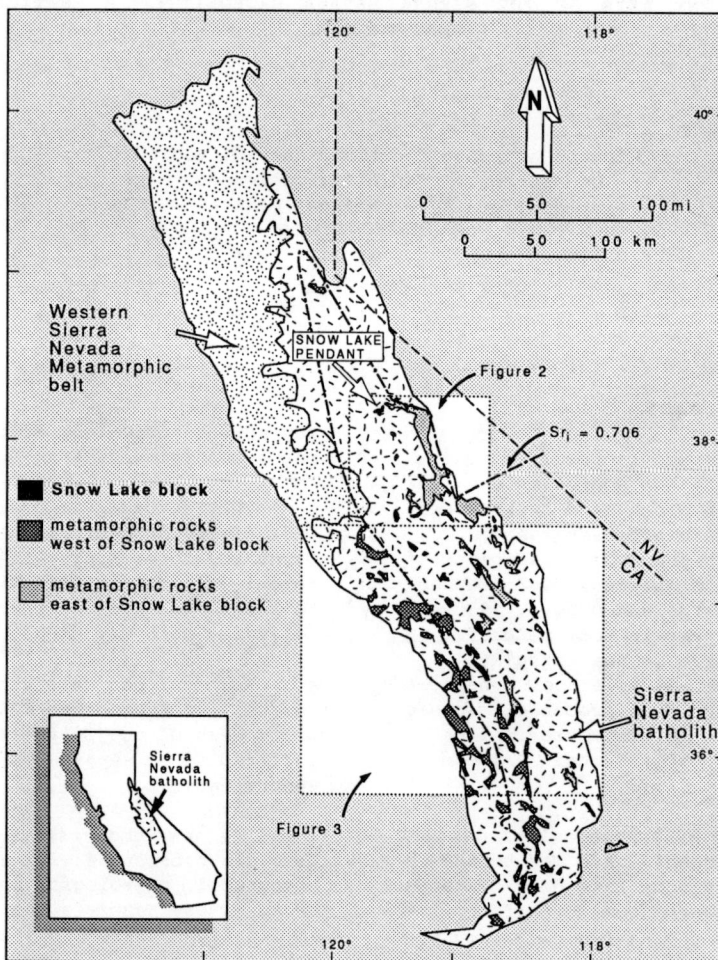

Figure 1. Simplified geologic map of the Sierra Nevada region, showing pendants of the Snow Lake block (black) and other metamorphic rocks within and west of the Sierra Nevada batholith.

Recent stratigraphic, structural, and geochronologic studies of metasedimentary rocks at Snow Lake pendant, near latitude 38° N (Fig. 1), have provided several lines of evidence that upper Proterozoic and Cambrian rocks of miogeoclinal affinity lie within the axial part of the batholith at that latitude (Lahren, 1989; Lahren and Schweickert, 1989; Lahren and others, 1990), and that these rocks have been translated about 400 km northward during the Early Cretaceous from the western Mojave region along the cryptic Mojave-Snow Lake fault. Correlation of rocks of Snow Lake pendant with rocks in the western Mojave Desert is supported by strong similarities in stratigraphy, structural development of metamorphic rocks, and age(s) and types of igneous intrusions with units in the western Mojave region, together with the absence of viable candidates for correlation in the region between Snow Lake pendant and the Mojave. Several predictions about the age and affinity of igneous rocks at Snow Lake based upon this correlation recently have been confirmed (Lahren and others, 1990).

We have conducted detailed and reconnaissance studies of a number of other pendants of metamorphic rocks south of latitude 38° N, and have re-evaluated previous interpretations of the age

and stratigraphic affinities reported previously for some of the pendants in this region. In this paper, we describe some of our detailed observations in Piute Mountain, Glen Aulin, and May Lake pendants (Fig. 2), and in addition, we present new interpretations of other pendants to the south based upon our reconnaissance and on existing maps and descriptions.

We suggest that a slice of miogeoclinal rocks, averaging 20 to 25 km wide (called the Snow Lake block), extends at least 200 km SSE from the Snow Lake pendant and includes rocks in Piute Mountain, Glen Aulin, Tuolumne Peak, May Lake, Iron Mountain pendants, Shuteye Peak, Dinkey Creek, Patterson Mountain, Boyden Cave, and Sequoia Park pendants, together with isolated exposures of metasedimentary rocks in Shuteye Peak, Kaiser Peak, and Shaver Lake quadrangles. The extent and geometry of the Snow Lake block north of Snow Lake pendant will not be addressed in this paper; preliminary evidence suggests it may trend abruptly eastward north of Snow Lake pendant (Schweickert and Lahren, 1991).

The Snow Lake block as defined here is bounded to the east by lower Paleozoic eugeoclinal rocks and by Paleozoic miogeoclinal units of the Inyo facies. Eugeoclinal rocks occur east of both the northern and southern ends of the Snow Lake block. East of the southern end of the Snow Lake block, eugeoclinal rocks occur in the Rattlesnake Creek, Hockett Peak, Bald Mountain, and Kennedy Meadows pendants (Dunne and others, 1988; Dunne and Suczek, this volume) and at Mineral King pendant (this paper; Fig. 3). East of the northern part of the Snow Lake block, eugeoclinal rocks of the Antler and Sonoma orogenic belts are exposed in the eastern parts of the Saddlebag Lake and Ritter Range pendants (Fig. 2). Outer shelf miogeoclinal rocks of the Inyo facies lie east of the central part of the Snow Lake block in the southern Ritter Range, Mount Morrison, Pine Creek, Bishop Creek, and Big Pine Creek pendants (Fig. 2,3).

Along the west side of the Snow Lake block (Fig. 2), we present new evidence that terranes of the Western Metamorphic belt extend the full length of the Sierra Nevada, possibly as far south as the Garlock fault. Lower Paleozoic rocks of the Shoo Fly Complex are exposed in pendants near Bass Lake, and the Lower Kings River, Lower Kaweah River, Tule River, and Kern Canyon pendants. Rocks related to the Paleozoic-Triassic Calaveras Complex form a narrow belt west of the Shoo Fly Complex in many of the same pendants.

PREVIOUS STUDIES

Confused nomenclature

The informal nomenclature applied to metamorphic rocks in pendants south and east of the Western Metamorphic belt has a long, complicated, and confusing history. A short summary follows.

Miller and Webb (1940) used the informal term "Kernville series" for metasedimentary rocks in some pendants in the Kern Canyon area south of latitude 36° N. At the same time, Durrell (1940) used the terms Lemon Cove schist, Homer quartzite, and Three Rivers schist for metamorphic rocks along the Lower Kaweah River. Based upon many years of mapping in the central Sierra Nevada, Bateman and Clark (1974) designated quartzite-rich strata in Iron Mountain, Strawberry Mine, Shuteye Peak, Dinkey Creek, Boyden Cave, and Mineral King pendants as "Kings sequence," named for fossiliferous strata in Boyden Cave pendant along the Kings River. In 1977, Schweickert and others, based on reconnaissance mapping by Saleeby, extended the name Calaveras Complex southward from the Western Sierra Nevada Metamorphic belt (henceforth Western Metamorphic belt) to include metasedimentary rocks in Oakhurst, Dinkey Creek, Boyden Cave, Lower Kings River, Lower Kaweah River, Sequoia Park, Mineral King, and Tule River pendants. In effect, the Kings sequence of Bateman and Clark (1974) was included in the Calaveras. The rationale was that quartzite-rich pendants (of the Kings sequence) were similar to an upper, quartzite-rich unit of the Calaveras Complex in the Western Metamorphic belt.

Subsequently, Saleeby and others (1978) reverted to the name Kings sequence for all pendants from Dinkey Creek south to the Tule River, and, in addition, extended the name to pendants farther south, including Kern Canyon, Isabella, and Tehachapi pendants. Saleeby (1978a,b, 1979) referred to mafic and ultramafic rocks along the

lower Kings and Kaweah Rivers as part of the Kings-Kaweah ophiolite belt, which he interpreted as the basement to part of the Kings sequence. Saleeby and others (1978) suggested that the western part of the Kings sequence, which contains some chert-argillite rocks, is equivalent to the Calaveras Complex of the Western Metamorphic belt, and that the eastern, quartzite-rich part of the Kings sequence was equivalent to the Kernville series of Miller and Webb (1940). Thus, all metamorphic rocks in the southern part of the Sierra south of Dinkey Creek pendant and east of the Kings-Kaweah ophiolite belt were included as Kings sequence. Saleeby (1981) and Saleeby and others (1987) continued this interpretation.

Schweickert (1981) narrowed the definition of the Calaveras Complex in the Western Metamorphic belt to exclude the quartzite-rich unit, which he recognized as being structurally distinct. The latter unit was designated the southern part of the Shoo Fly Complex. He argued that the Calaveras Complex does not extend south of about latitude 37° N. Saleeby (1981) used the term Calaveras sequence for rocks in the Western Metamorphic belt.

Terrane nomenclature began to supplant lithostratigraphic names in the 1980's. In 1982, Blake and others compiled a terrane map of California and assigned the Shoo Fly Complex to the Northern Sierra terrane, the Calaveras Complex to the Merced River terrane, and showed the Kings-Kaweah ophiolite belt as the Kaweah terrane. Blake and others (1982) and Nokleberg (1983) designated the Kings sequence of Saleeby and others (1978) as the Kings terrane. However, Nokleberg (1983) interpreted Dinkey Creek pendant as a slice of the Owens terrane to the east, and Snow Lake pendant as part of his High Sierra terrane. He included metavolcanic rocks of the Kings-Kaweah ophiolite belt, together with the Calaveras Complex, as parts of the Merced River terrane.

Blake and others (1982), Schweickert and Bogen (1983), Schweickert and others (1988), and Sharp (1988) recognized that a separate terrane of phyllitic and schistose Jurassic metasedimentary and metavolcanic rocks lies west of the Calaveras Complex and east of the Melones fault in the southern part of the Western Metamorphic belt; Blake and others (1982) designated this terrane the Don Pedro terrane. Schweickert and others (1988) used Blake and others' (1982) terminology, but Sharp (1988) referred to this same belt as the Sullivan Creek terrane. The southeastward extent of this terrane has not been discussed.

Rocks west of the Melones fault in the Western Metamorphic belt have been called the Foothills terrane by many workers. However, Edelman and Sharp (1989) recently grouped these rocks together with rocks of the Don Pedro terrane as the Tuolumne River terrane.

Based upon new studies at Snow Lake pendant and reconnaissance of other pendants, Lahren (1989), Lahren and Schweickert (1989), and Lahren and others (1990) argued that miogeoclinal strata of probable Proterozoic and Early Cambrian age occur in Snow Lake and Dinkey Creek pendants, and in the "type area" of the Kings sequence, Boyden Cave pendant. They removed these pendants from the

Kings sequence (terrane) and referred them to the Snow Lake block. In addition, they avoided use of the term Kings sequence (terrane), and referred to pendants near Lake Isabella (just south of Figure 3) as comprising the informal Isabella sequence, of Jurassic and possibly older age.

The nomenclature we use for metamorphic rocks in the region between latitudes 36° and 38° N (Fig. 1-3) is as follows:

1. Miogeoclinal strata of axial roof pendants we refer to as part of the Snow Lake block (after Lahren, 1989; Lahren and others, 1990). A strong case can be made that these rocks are of Proterozoic and Cambrian age.

2. Very strongly deformed pelitic and impure psammitic rocks that lie west of the Snow Lake block in pendants such as the Lower Kings River, Kaweah River, Tule River, and Kern Canyon, we provisionally assign to the lower Paleozoic Shoo Fly Complex.

3. Chert-argillite rocks in the eastern part of the Oakhurst pendant, the western part of the Lower Kings River pendant, the Lower Kaweah River pendant, the eastern part of the Yokohl Valley pendant, and part of the Tule River pendants we include within the Paleozoic to Triassic Calaveras Complex of the Western Metamorphic belt.

4. Schistose and phyllitic metasedimentary and metavolcanic strata in Oakhurst pendant and in some pendants to the south, together with Paleozoic ophiolitic rocks of the Kings-Kaweah ophiolite belt, we include within the Don Pedro terrane.

5. Stratigraphic units west of the Melones fault are included within the Foothills terrane.

6. We recommend that the terms "Kings sequence" and "Kings terrane" be abandoned because they have become a catchall for any and all metamorphic rocks between the Western Metamorphic belt and the Garlock fault, and include rocks of a vast spectrum of ages and tectonic and depositional environments.

Previous interpretations of ages of rocks in Sierran pendants

Paleozoic strata

Lithologic, stratigraphic, and structural features have led many workers to the conclusion that rocks in the pendants are mainly of Paleozoic age. Rose (1957 a,b, 1958) interpreted metasedimentary rocks in May Lake pendant to be of late Paleozoic age. Kistler and Bateman (1966) and Kistler and Peterman (1973) concluded that the metamorphic rocks at Dinkey Creek are early Paleozoic in age based on presumed structural similarities with lower Paleozoic rocks in the Mount Morrison pendant. Lockwood and Bateman (1976) suggested metasedimentary rocks in Shaver Lake quadrangle are of late Paleozoic age. Girty (1977 a,b, 1985) presented evidence that the rocks in the western part of Boyden Cave pendant are pre-Jurassic and probably Paleozoic in age.

Schweickert and others (1977) considered the rocks in most of the pendants to be of late Paleozoic age, and interpreted them to be overlain on the east by fossiliferous Triassic and Jurassic units. Walker (1980) suggested that both May Lake and Tuolumne Peak pendants contain Paleozoic rocks. Schweickert (1981) interpreted the metasedimentary rocks in these pendants as representing the southwesterly continuation of Proterozoic and lower Paleozoic strata of the Inyo Mountains region, appearing on the western limb of the Nevadan synclinorium. He argued that Mesozoic fossils at Boyden Cave and at Mineral King date strata in the eastern parts of these pendants, and that highly deformed rocks in the western parts could be much older.

Nokleberg (1983) adopted Kistler and Bateman's (1966) interpretation of the age of rocks in Dinkey Creek pendant, but considered them to be a slice of lower Paleozoic rocks enclosed within Mesozoic metasedimentary rocks of the Kings terrane. Saleeby and others (1986) noted that metasedimentary rocks in the region may have Paleozoic protolith ages, and that some rocks could be related to the lower Paleozoic Shoo Fly Complex.

Lahren (1989), Lahren and Schweickert (1989), and Lahren and others (1990) proposed that lower Paleozoic and Proterozoic miogeoclinal rocks of the Death Valley facies occur in Piute Mountain, Glen Aulin, May Lake, Dinkey Creek, Boyden Cave, and other smaller pendants.

Mesozoic strata

In contrast, several fossil discoveries in the region have led to the idea that the pendants contain mainly Mesozoic strata. Durrell (1940) and Christensen (1963) reported Triassic fossils from Mineral King pendant. Moore and Dodge (1962) and Jones and Moore (1973) reported Lower Jurassic fossils from a metasiltstone unit in the central part of Boyden Cave pendant. Because this unit was interpreted to underlie marble and quartzite to the west, Jones and Moore (1973) interpreted all rocks in pendants in this region to be of Mesozoic age. Bateman and Clark (1974) and Bateman (1981) inferred that rocks of the Kings sequence are of Triassic to Jurassic age based upon the fossil localities at Boyden Cave and Mineral King. Bateman (1981) also inferred that the Kings sequence either overlies, or partly correlates with, the upper quartzitic unit of the Calaveras Complex (now assigned to the Shoo Fly Complex).

The idea that all pendants south of latitude 37° N contain mainly Mesozoic rocks was adopted by Saleeby and others (1978) and Saleeby (1981), who reported additional Jurassic fossils at Boyden Cave, Mineral King, and near Lake Isabella. They interpreted the rocks in all the pendants in this region as a deep marine sequence transitional between silicic volcanic rocks to the east and mafic volcanic rocks to the west. They presented a model in which the Kings sequence was interpreted to consist of sedimentary and volcanic rocks of a single magmatic arc that developed across a cryptic boundary between continental crust in the eastern Sierra Nevada and oceanic crust in the western foothills.

Nokleberg (1981) reported a Lower Jurassic fossil from rocks in Strawberry Mine pendant.

Blake and others (1982) and Nokleberg (1983) designated the Kings terrane as Late Triassic to Early Jurassic in age. Triassic U-Pb ages on metavolcanic rocks in Mineral King pendant were reported by Busby-Spera (1984a, b, 1985, 1986). Saleeby and others (1986) interpreted the rocks of Dinkey Creek and Lower Kings River pendants as Triassic or Jurassic in age.

Discussion

The only fossils to have come from any of the pendants between 36° and 38° N latitude are of Triassic and Jurassic age, and based upon this fact, all rocks formerly included in the Kings sequence have been assumed by many authors to be Triassic and Jurassic in age. Unfortunately, these fossil occurrences have apparently resulted in confusion about the ages of rocks in the region. As discussed below, all the Mesozoic fossils now appear to have been collected from strata that lie above major unconformities or structural breaks between Mesozoic and older rocks. In addition, the Mesozoic strata appear to have been deposited in entirely different depositional settings and are generally less deformed than the probable Paleozoic rocks that lie beneath them to the west.

For example, at the Boyden Cave pendant, the Mesozoic fossil localities are located east of the contact between quartzite-rich rocks of probable shallow-water origin in the western part of the pendant and rocks of probable deep-water origin in the eastern part of the pendant (Jones and Moore, 1973; Saleeby and others, 1978; Moore and others, 1979; Girty, 1977b, 1985). The contact between these two sequences of rocks has been interpreted to be an unconformity or a fault (Girty, 1977b, 1985; Nokleberg, 1983; Moore and others, 1979). In addition, we present new evidence below that Mesozoic rocks at Mineral King lie unconformably upon highly deformed lower Paleozoic eugeoclinal rocks. Therefore, it probably is not valid to assume that isolated fossil occurrences determine the ages of all the metasedimentary rocks within the widely separated pendants within the Sierra Nevada.

A major point of this paper is that, despite the presence of Jurassic and Triassic fossils in some areas, Mesozoic rocks are probably lacking in most of the pendants in the axial parts of the batholith or, if present, probably represent overlap sequences on older rocks.

LITHOLOGIC, STRATIGRAPHIC, AND STRUCTURAL DESCRIPTIONS OF PENDANTS

In the following sections we summarize new and published data on the lithology, protoliths, structural features, and probable stratigraphy of pendants in four main groupings: (1) pendants containing shallow-water, inner miogeoclinal rocks of the Death Valley facies, which we consider to form remnants of the Snow Lake block; (2) pendants west of the Snow Lake block containing highly deformed quartzite, schist, amphibolite and other rocks that we suggest may be related to the lower Paleozoic Shoo Fly Complex; (3) pendants west of the Shoo Fly Complex; and (4) pendants east of the Snow Lake block that are east of the Mojave-Snow Lake fault, and that contain lower Paleozoic eugeoclinal rocks of the Antler orogenic belt or Paleozoic outer miogeoclinal strata of the Inyo facies. All of the pendants once may have had Mesozoic overlap sequences. The pendants of the Snow Lake block and pendants to the west are enclosed mainly within and between Early Cretaceous plutons of the Sierra Nevada batholith, while pendants to the east of the Snow Lake block occur both within Late Cretaceous plutons and as septa between Late Cretaceous plutons and Jurassic-Triassic plutons to the east.

Pendants of the Snow Lake block

Known or inferred pendants of the Snow Lake block extend southward from Snow Lake pendant near latitude 38° N to Sequoia Park pendants and possibly include some metasedimentary rocks in the southern part of Mineral King pendant near latitude 36° N (Fig. 1-3).

Snow Lake pendant

The Snow Lake pendant (Fig. 1,2) contains abundant quartzite, micaceous, feldspathic quartzite, quartz-mica schist, calc-silicate schist, and marble that are highly folded and thrust faulted. Despite the complex structure, quartzites and schists reveal abundant primary sedimentary structures that allow unequivocal determination of tops and facing directions of folds. We describe structural and stratigraphic relations (Lahren ,1989; Lahren and others, 1990) for comparison with other pendants to be described later.

Stirling Quartzite:
Quartzite, feldspathic quartzite, and minor schist, calc-silicate schist, and marble believed to be part of the Proterozoic Stirling Quartzite occur within the highest thrust sheet (Bigelow Peak thrust sheet) and the underlying thrust sheet (Buckskin thrust sheet). Small and large xenoliths of Stirling Quartzite also occur around the periphery of Sachse Monument pendant, 3 km southwest of Snow Lake. Based upon consideration of fold geometry, preserved stratigraphic thickness of the Stirling is estimated to be about 710 m.

Wood Canyon Formation:
Micaceous quartzite, quartz-biotite schist, marble, and calc-silicate schist that are correlated with the Proterozoic to Lower Cambrian Wood Canyon Formation overlie the Stirling Quartzite in the Buckskin thrust sheet and also occur within the lowest structural unit in the pendant. Individual lithologic subunits of the Wood Canyon here match closely members of the Wood Canyon in areas such as Death Valley and the Striped Hills in southern Nevada. Vertical burrows, *Skolithos*, and other signs of bioturbation occur widely in thin quartzite beds in the uppermost preserved part of the Wood Canyon. Preserved thickness of the Wood Canyon is about 375 m. A very small remnant of micaceous quartzite, quartz-mica schist, calc-silicate schist, and marble that resembles parts of the Wood Canyon occurs 10 km west of Snow Lake pendant, in Emigrant Basin (Fig. 2).

Zabriskie Quartzite:
The Zabriskie Quartzite is a distinctive,

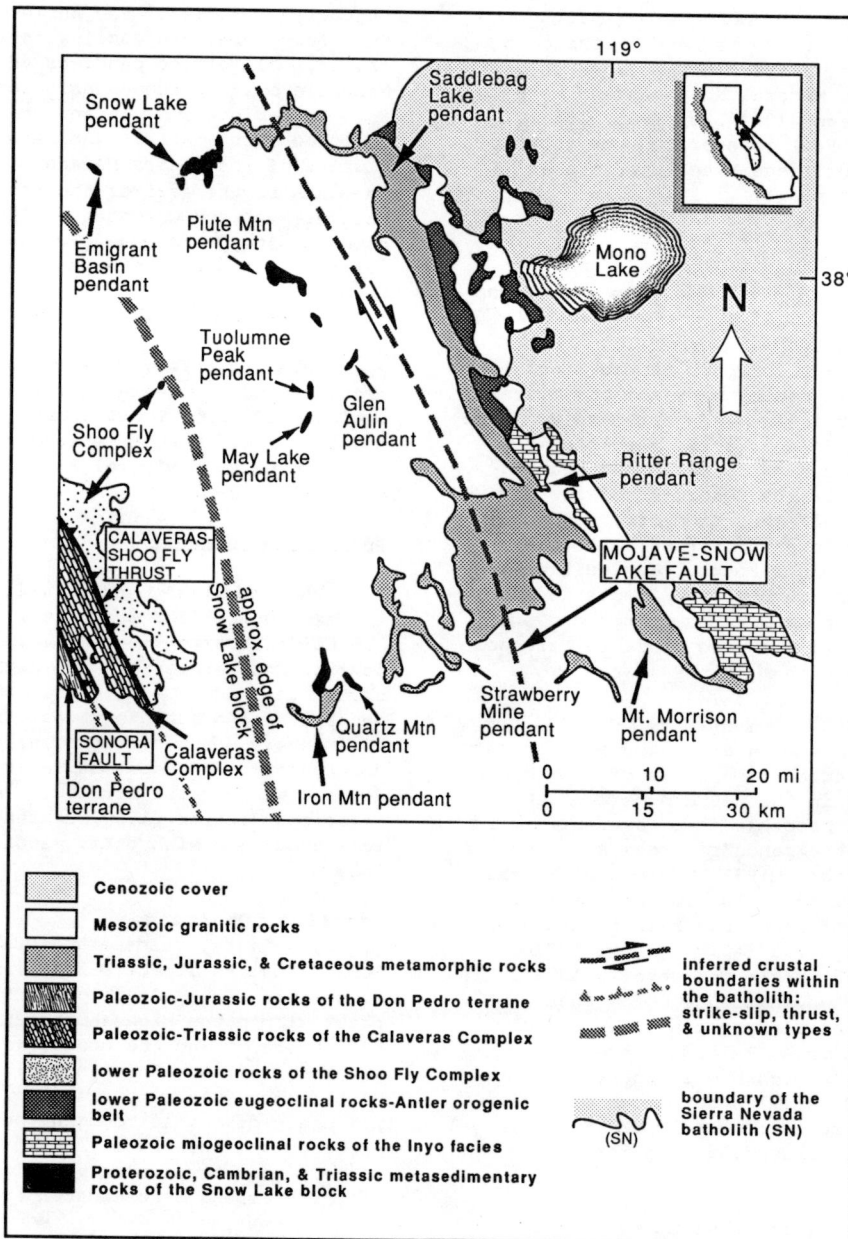

Figure 2. Sketch map of pendants in the vicinity of Yosemite National Park, between 37°30' and 38° 15' N latitude.

white, gray, and pink quartzite unit with well-preserved planar-tabular cross bedding, that occurs in the Quartzite Peak thrust sheet and structurally overlies *Skolithos*-bearing Wood Canyon Formation. Preserved thickness is about 180 m.

Carrara Formation:

Conformably overlying the Zabriskie is a transitional unit consisting of three cycles of pelite, quartzite, and thin carbonate, that we correlate with the Lower Cambrian Carrara Formation. *Skolithos* occurs sparsely within thin quartzite beds in the middle part of the Carrara. Preserved thickness is about 260 m. The uppermost part of the Carrara apparently is cut out by the Buckskin thrust.

Fairview Valley Formation:

The Stirling Quartzite in the Bigelow Peak

thrust sheet is unconformably overlain by a distinctive unit of red- to green-weathering, laminated, calcareous metamudstone and metasiltstone with minor conglomerate and marble that we correlate with the Lower Triassic Fairview Valley Formation of the Mojave region. This unit has a local basal conglomerate containing clasts of the underling Stirling Quartzite together with marble cobbles, and elsewhere has conglomerate dikes with a siltstone matrix that contain abundant cobbles and boulders of augite monzonite and monzonite porphyry. A prominent white marble unit 30 m thick occurs about 30 m above the base of the unit. Preserved thickness is estimated at about 500 m. North of the pendant, a large xenolith of the Fairview Valley appears to grade upward into gray lithic sandstone that in turn is overlain by silicic volcanic breccia.

Igneous rocks:

All stratified units are cut by mafic and felsic dikes of the Independence dike swarm, one of which has been dated at 150 Ma (Lahren and others, 1990). In addition, 150-Ma intrusions of biotite granite (Mattinson and Lahren, unpub. data) and gabbro-diorite (Lahren and others, 1990), probably related to the dikes, cut the stratigraphic units on the southern and southwestern ends of the pendant. One hundred eighteen-Ma granites (Mattinson and Lahren, unpub. data) intrude the pendant along its northeastern margin.

Structure:

A cryptic, early deformational event which produced large structural relief in the Stirling Quartzite and resulted in removal of overlying units, predated the deposition of the Fairview Valley Formation. Major east-vergent folding and thrusting of the Proterozoic to Cambrian and Lower Triassic rocks predated the 150-Ma intrusions, and may reflect an important Jurassic deformational event. Later episodes involving northwest and east-trending folds postdate the 150-Ma intrusions and predate 118-Ma granites. We suspect these Early Cretaceous deformational events are related to translation of the Snow Lake block to its present position. Lahren and others (1990) discussed numerous lines of evidence for correlation of the rocks of Snow Lake pendant with rocks of the western Mojave Desert.

Piute Mountain pendant

Piute Mountain pendant contains the first significant exposures of metamorphic rocks south of Snow Lake pendant (Fig. 2). Metasedimentary rocks at Piute Mountain pendant are predominantly clean quartzite and micaceous, feldspathic quartzite that locally contain thin interbeds of quartz-mica schist. The pendant also contains subordinate units of quartz-mica schist and calc-silicate schist and marble (Wahrhaftig, unpub. mapping; Lahren and Schweickert, unpub data). The combination of rock types and their abundance strongly suggest that these rocks are correlative with the Stirling Quartzite at Snow Lake pendant. Weakly deformed mafic dikes with the same general trend as mafic dikes of the Independence dike swarm at Snow Lake pendant (Lahren, 1989) are also present.

The megascopic structure of the pendant comprises a large, open to close, northwest-trending and plunging antiform that folds a foliation. This structure appears to be the result of the third deformational event that affected the pendant because small-scale intrafolial isoclinal folds (F_1) and close to tight asymmetric folds (F_2) which deform S_0 and S_1 are deformed by the large antiformal fold. Minor F_3 folds that are geometrically compatible with the antiform also occur at outcrop scale. Although the rocks are highly deformed, recrystallized, and locally transposed, top directions in preserved bedding indicate that the main northwest-trending antiform may be upright. The form and trend of the megascopic fold and its parasitic folds suggest that these third generation folds are correlative with similarly oriented folds at Snow Lake pendant where they are Early Cretaceous in age. East-west to east-northeast-trending spaced cleavage and late

minor folds, together with curvature of the hinge surface of the F_3 antiform, appear to represent a fourth generation of folds that is similar in orientation to Early Cretaceous F_4 folds at Snow Lake.

Thus, the structural development of metamorphic rocks at Piute Mountain is similar to that at Snow Lake, and we suggest that the structures in both pendants are broadly correlative.

Glen Aulin pendant

Glen Aulin pendant is a very small, northeast-trending remnant (Fig. 2), 0.6 km long by about 100 m wide, of metasedimentary rocks between two Cretaceous plutons west of the Tuolumne Intrusive Suite (Bateman and others, 1983). Several other smaller remnants occur to the west. The rocks at Glen Aulin are very highly recrystallized quartzite and micaceous quartzite or granofels, with minor calc-silicate gneiss. Layering commonly is obscure, but we observed evidence suggestive of at least three phases of folding. Small bosses of metamorphosed diorite and plagioclase porphyry may be parts of disrupted and deformed dikes cutting the metamorphic rocks. Although the intense recrystallization and small outcrop areas of these pendants give the rocks few distinctive features, they may consist of Stirling Quartzite.

Tuolumne Peak pendant

Tuolumne Peak pendant (Fig. 2) is a narrow septum of steeply west-dipping metamorphic rocks that is about 2 km long and about 1 km wide at its southern end. The pendant is entirely enclosed within the El Capitan Granite (Huber and others, 1989). According to Walker (1980), metasedimentary rocks of the pendant consist of interbedded quartzite, micaceous, feldspathic quartzite, spotted schist, calc-silicate schist and gneiss, skarn, tactite, and marble. Walker (1980) interpreted a small amphibolite body to be a metatuff or metasedimentary unit, although an igneous protolith also is possible. The rocks in the pendant are highly deformed with regional deformation predating pluton emplacement. The metasedimentary rocks were interpreted to be Paleozoic sediments deposited in a shallow marine environment (Walker, 1980).

May Lake pendant

May Lake pendant (Fig. 2) is an elongate, northeast-trending remnant about 2 km in length by 0.3 km wide. Like Glen Aulin pendant, it forms a thin septum between two Cretaceous plutons. Rose (1957a, b, 1958) reported that the pendant consists of upper Paleozoic miogeoclinal rocks, principally quartzite with an interlayered, thin, mappable tongue of pelitic rocks, and with marble and metaigneous rocks at its southwestern end. Rose (1957a, b; 1958) also mapped several small metamorphosed mafic dikes. Our study indicates that the quartzite consists of thin to thickly layered, vitreous quartzite with thin, micaceous partings. Original sedimentary bedding has been transposed into a composite layering, but planar-tabular cross-bedding is preserved locally. Minor marble and calc-silicate gneiss locally is interlayered with the quartzite. The pelitic unit consists of quartz-biotite-muscovite schist and micaceous quartzite. Rose (1957a, b, 1958) considered the metaigneous rocks at the southwest end of the pendant to be intermediate to silicic metavolcanic rocks, but our observations suggest these rocks consist of ortho- and paragneisses of uncertain parentage.

Two generations of northeast-trending isoclinal folds occur within the layering of the quartzite and schist, and these are refolded by two generations of west-northwest and northwest-trending late folds. Schistose mafic dikes cut across the structural layering, and are apparently deformed by the west-northwest-trending folds.

The quartzite and pelite units are identical with units assigned to the Stirling Quartzite to the north at Piute Mountain and Snow Lake. We do not know how (or if) the structures at May Lake correlate with those in pendants to the north.

Quartz Mountain pendant

Quartz Mountain pendant (Fig. 2), in the southwestern part of the Merced Peak quadrangle (Peck, 1980), is about 4 km long and 1 km wide. This pendant exposes mainly thinly bedded, micaceous quartzite and mica schist. These metasedimentary rocks appear to be highly deformed with mesoscopic folds deforming a bedding-parallel foliation or compositional layering. These rocks are similar lithologically to Wood Canyon-type rocks elsewhere in the Snow Lake block.

Rocks near Cold Springs Meadow (northern part of Iron Mountain pendant)

Metasedimentary rocks northwest and west of Cold Springs Meadow in the Iron Mountain pendant (Fig. 2) were mapped by Peck (1980) and by Huber (1982). Our reconnaissance studies indicate that the northernmost exposures of metasedimentary rocks are composed of dark gray schist and greenish calc-silicate schist. These rocks are highly deformed with no preserved bedding. The rocks are similar in lithology to parts of either the Wood Canyon Formation or the Stirling Quartzite at Snow Lake pendant.

West of these highly deformed rocks is a sequence of metasedimentary rocks with well-preserved bedding and sedimentary structures. Part of these rocks was mapped as quartz-pebble conglomerate by both Peck (1980) and Huber (1982). The conglomerate contains clasts of quartzite, marble, and monzonite. Monzonitic clasts are a distinctive feature of conglomerate in the Fairview Valley Formation in the western Mojave area (Miller, 1978). Green and brown, rusty-weathering, calcareous siltstone and mudstone is interbedded with the conglomerate. These metaconglomerates and metasiltstone-mudstone rocks are very similar to and probably correlative with the Fairview Valley Formation(?) at Snow Lake pendant and in the western Mojave. The highly deformed rocks east of these little deformed rocks are similar to Wood Canyon-type rocks at Snow Lake pendant, and although the contact between the two sequences of rocks is not exposed, we infer it to be a major unconformity between probable Fairview Valley rocks and highly deformed miogeoclinal rocks.

The metasedimentary rocks near Cold Springs Meadow are in fault contact with metavolcanic rocks of unknown age to the south (Peck, 1980; Huber, 1982).

Shuteye Peak quadrangle

Several pendants of metasedimentary rocks occur in the Shuteye Peak quadrangle (Huber, 1968)(Fig. 3). The rocks consist of quartzite, quartz-mica schist, calc-silicate hornfels, and tactite (Huber, 1968). Quartzite on Kaiser Ridge contains well developed cross bedding (Huber, 1968). We tentatively include these rocks in the Snow Lake block because of their lithologic similarities and because they lie directly between other exposures of rocks included in the Snow Lake block in the Merced Peak quadrangle (Peck, 1980) to the north and at Dinkey Creek pendant to the south. However, the possibility exists that some of these pendants, especially the more westerly ones, may contain rocks that may be correlative with Shoo Fly-type rocks.

Kaiser Peak quadrangle

Metamorphic rocks are exposed in the Kaiser Peak quadrangle (Fig. 3) within the Kaiser Ridge Wilderness (Hamilton, 1956; Bateman and others, 1971; duBray and Dellinger, 1980), the western part of which extends into the southeastern part of the Shuteye Peak quadrangle (Huber, 1968). The rocks consist of quartzite, biotite schist, pyroxene hornfels, marble, and calc-silicate hornfels (Hamilton, 1956; Bateman and others, 1971; duBray and Dellinger, 1980). The quartzite in Kaiser Peak quadrangle has well-preserved cross bedding (Bateman and others, 1971), as does the quartzite exposed on the western extension of Kaiser Ridge in the Shuteye Peak quadrangle (Huber, 1968). Bateman and others (1971) interpreted the cross bedding as having been formed by near-shore, ocean currents. We include these rocks within the Snow Lake block because of the presence of well-preserved cross bedding and rock types which are typical of the miogeoclinal rocks at Snow Lake pendant.

Shaver Lake quadrangle

Several pendants of metasedimentary rocks are present in the northeast part of the Shaver Lake quadrangle (Lockwood and Bateman, 1976)(Fig. 3). The largest two pendants occur within and along the western border of the granodiorite of Dinkey Creek. These pendants consist of shallow-water, metasedimentary rocks, including biotite schist, quartzite, calc hornfels, and marble suggested to be of late Paleozoic age by Lockwood and Bateman (1976). We tentatively include these pendants in the Snow Lake block on the basis of lithologic similarities. Two other sizable pendants in the northwest and southwest parts of the Shaver Lake quadrangle may be correlative with Shoo Fly-type rocks (described later).

Dinkey Creek pendant

Dinkey Creek pendant (Fig. 3) has been mapped by Krauskopf (1953), Kistler and Bateman (1966), Bateman and Wones (1972), and Merritt (1985, in

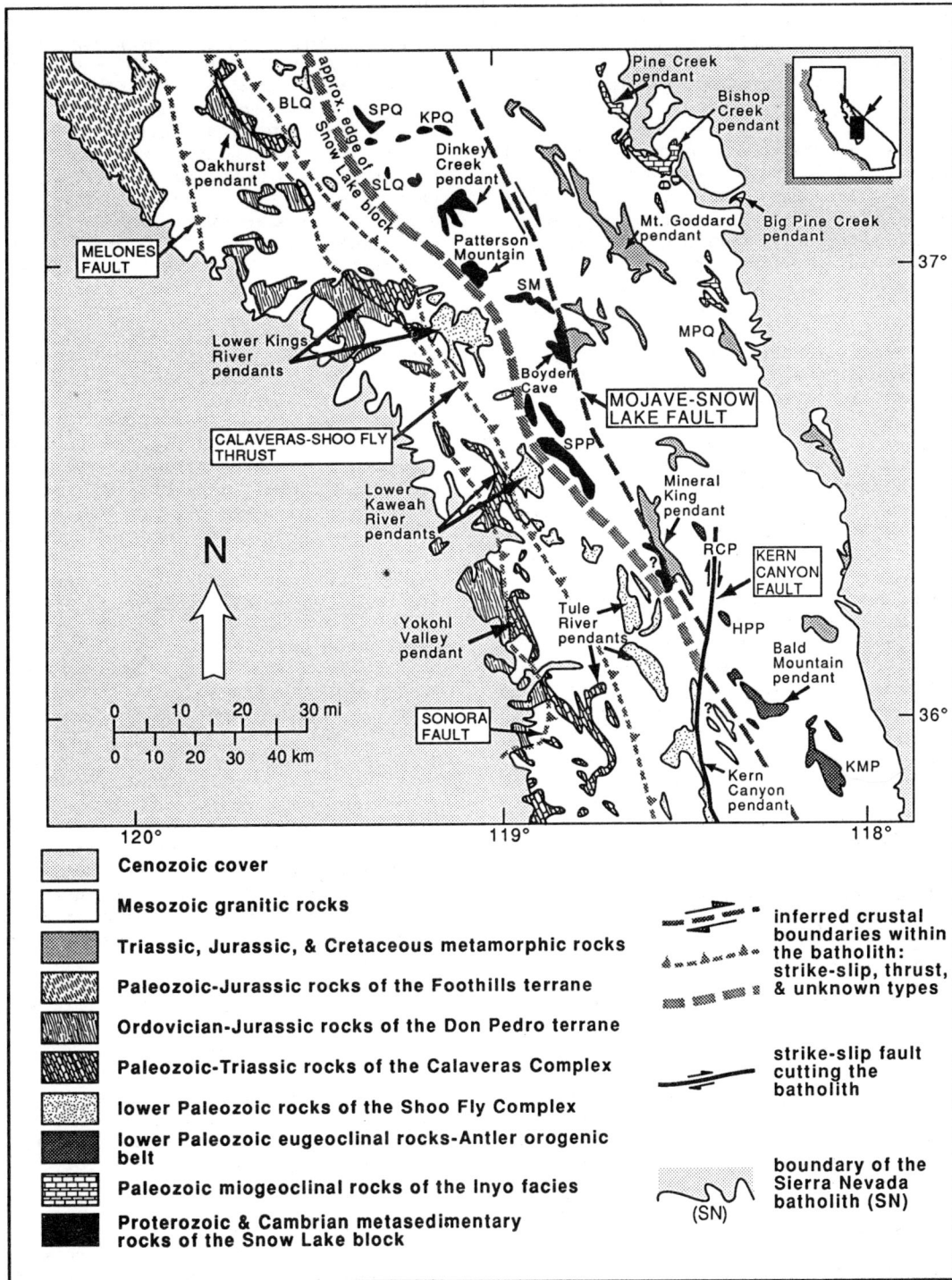

Figure 3. Sketch map of pendants within the batholith between 37°30'N and 35°45'N latitude. Abbreviations: BLQ, Bass Lake quadrangle; HPP, Hockett Peak pendant; KMP, Kennedy Meadows pendant; KPQ, Kaiser Peak quadrangle; MPQ, Mt. Pinchot quadrangle; RCP, Rattlesnake Creek pendant; SLQ, Shaver Lake quadrangle; SM, Spanish Mountain; SPP, Sequoia Park pendants; SPQ, Shuteye Peak quadrangle.

prep.). All of these authors have considered these rocks to be of miogeoclinal aspect, and Kistler and Bateman (1966) regarded the rocks as lithologically and structurally similar to lower Paleozoic rocks of the Mount Morrison pendant. According to Kistler and Bateman (1966), the stratigraphic sequence at Dinkey Creek, from oldest to youngest, is 1) marble, 2) calc-silicate rock, 3) biotite-andalusite hornfels, 4) white quartzite, and 5) schist. They estimated that the biotite-andalusite hornfels is about 240 to 300 meters thick, the white quartzite is about 450 to 600 meters thick, and the calc-silicate unit is about 60 meters thick. Merritt (1985, in prep.) noted that the structure is more complex than portrayed by Kistler and Bateman (1966), and that several ductile shear zones imbricate the rocks of the pendant. In particular, the biotite-andalusite hornfels unit structurally overlies the quartzite unit along a shear zone, and therefore stratigraphic relations are uncertain.

The quartzite is the most widespread unit, and in many outcrops it shows well-preserved planar, tangentially based crossbeds. Locally, it contains interbeds of pelitic rocks. The quartzite appears to underlie the schist unit, termed paragneiss by Merritt (1985). The schist unit consists of interlayered dark schist and micaceous quartzite, locally showing cross bedding, and it contains several layers of buff marble and calc-silicate hornfels. The biotite-andalusite hornfels of Kistler and Bateman (1966) is a schist lacking significant interstratified micaceous quartzite. The marble and calc-silicate rocks probably represent shallow marine limestone and limy siltstone and shale. Overall, the entire sequence resembles a shallow-water, miogeoclinal sequence. Merritt (in prep.) observed evidence of four phases of ductile folds, with shear zones and probable thrusts related to D_2 deformation.

Based upon comparison of the metamorphic rocks with those at Snow Lake pendant and rocks in the Death Valley region, Merritt (in prep.) has suggested that these rocks correlate with Proterozoic Johnnie Formation (biotite-andalusite schist), Proterozoic Stirling Quartzite (white quartzite unit), and Cambrian Wood Canyon Formation (paragneiss unit). Based upon the predominance of quartzite and micaceous quartzite, we believe that the rocks of Dinkey Creek pendant are distinctly unlike rocks of the Inyo facies, which lack thick, clean quartzite units.

Patterson Mountain pendant

Patterson Mountain pendant (Fig. 3), located within the southern part of the Huntington Lake quadrangle (Bateman and Wones, 1972) and the north part of the Patterson Mountain quadrangle, is about 7 km long and about 5 km wide. This remnant of metasedimentary rock, originally mapped by Krauskopf (1953), is composed of quartzite and schist. The rocks are highly recrystallized and appear to be paragneiss locally. These rocks, which are located about 8 km southeast of the southernmost exposures of Dinkey Creek pendant, are lithologically similar to other rocks in the Snow Lake block, and thus we provisionally include them in the Snow Lake block.

Boyden Cave pendant

Boyden Cave pendant (Fig. 3,4) has been mapped by Moore and Marks (1972) and Girty (1977a, 1985), and the geology has been discussed by Jones and Moore (1973), Moore and others (1979), Nokleberg and Kistler (1980), Schweickert (1981), Nokleberg (1983), and Saleeby and others (1978, 1990). It apparently has a northwestern extension on Spanish Mountain (Fig. 3). Boyden Cave has been one of the more controversial pendants, partly because of discoveries of Jurassic fossils in the east-central part of the pendant. Based upon the Jurassic fossils, many authors have interpreted all the rocks of the pendant as being of Jurassic age (Moore and Marks, 1972; Jones and Moore, 1973; Bateman and Clark, 1974; Saleeby and others, 1978;

Moore and others, 1979; Nokleberg, 1983). However, Girty and Nokleberg (in Moore and others, 1979; Nokleberg, 1983; Girty, 1985) argued that a major structural or stratigraphic break separates rocks in the western part of the pendant from the fossiliferous unit, and Girty (in Moore and others, 1979; Girty, 1985) and Schweickert (1981) proposed that rocks in the western part of the pendant form part of a Paleozoic miogeoclinal wedge. We believe strong evidence exists in support of the latter view, and that possible Proterozoic and Lower Cambrian shallow-water, miogeoclinal strata occur in the western part of the pendant.

Lithologic units in the western part of the pendant (Fig. 4) consist, from west to east, of a thick quartzite unit, a unit of biotite-andalusite schist, a thick marble unit, and a chaotic metasedimentary unit containing numerous blocks of the other units to the west. The chaotic unit is overlain to the east by a coherent unit of thinly bedded tuffaceous siltstone. The eastern part of the pendant is underlain by a thick sequence of mid-Cretaceous silicic and intermediate volcanic rocks and hypabyssal intrusions (Moore and Marks, 1972; Saleeby and others, 1990). The lithologic units are described below from west to east.

Quartzite unit:

The quartzite unit is the most extensive unit of the pendant, and is about 2.7 km wide (Fig. 4). Stratigraphic thickness is unknown, because the unit has been isoclinally folded. The unit consists of thin to thick-bedded quartzite with thin interbeds of pelitic schist. Thin interbeds of calc-silicate schist and marble occur locally. The quartzite commonly shows well-preserved cross bedding, including festoon, planar, and herringbone crossbeds (Girty, 1985). Some beds are pebbly, and most beds are medium- to coarse-grained quartzite. Sandstones are typically feldspathic quartzite, with up to 17 percent K-feldspar in some samples (Jones and Moore, 1973; Girty, 1977b; 1985). Chen (in Moore and others, 1979) reported Proterozoic ages for detrital zircons from feldspathic quartzite, suggesting the sedimentary rocks were derived from a Precambrian craton.

Moore and Marks (1972) regarded cross bedding in the quartzite unit as indicating tops are generally to the west. However, Girty (1985) noted that tops vary markedly throughout the unit because of isoclinal folding.

Pelitic intervals within the quartzite unit consist of thinly layered mica schist and dark, micaceous quartzite. Rare calc-silicate interbeds consist of thinly interlayered marble, calc-silicate gneiss, and calcareous quartzite, with layering averaging 1-2 cm. These thinly layered units reveal complex structure, including at least two phases of isoclinal folds, and all folds appear to deform a metamorphic layering. A 50-m-thick white marble unit occurs at the eastern contact of the quartzite unit with the biotite-andalusite schist unit.

Biotite-andalusite schist unit:

This unit consists of up to 0.5 km structural thickness of dark, micaceous quartzite and mica schist (Fig. 4). Most outcrops of this unit are thinly layered, with layers averaging about 5 to 10 cm, and occasional interbeds of buff-green,

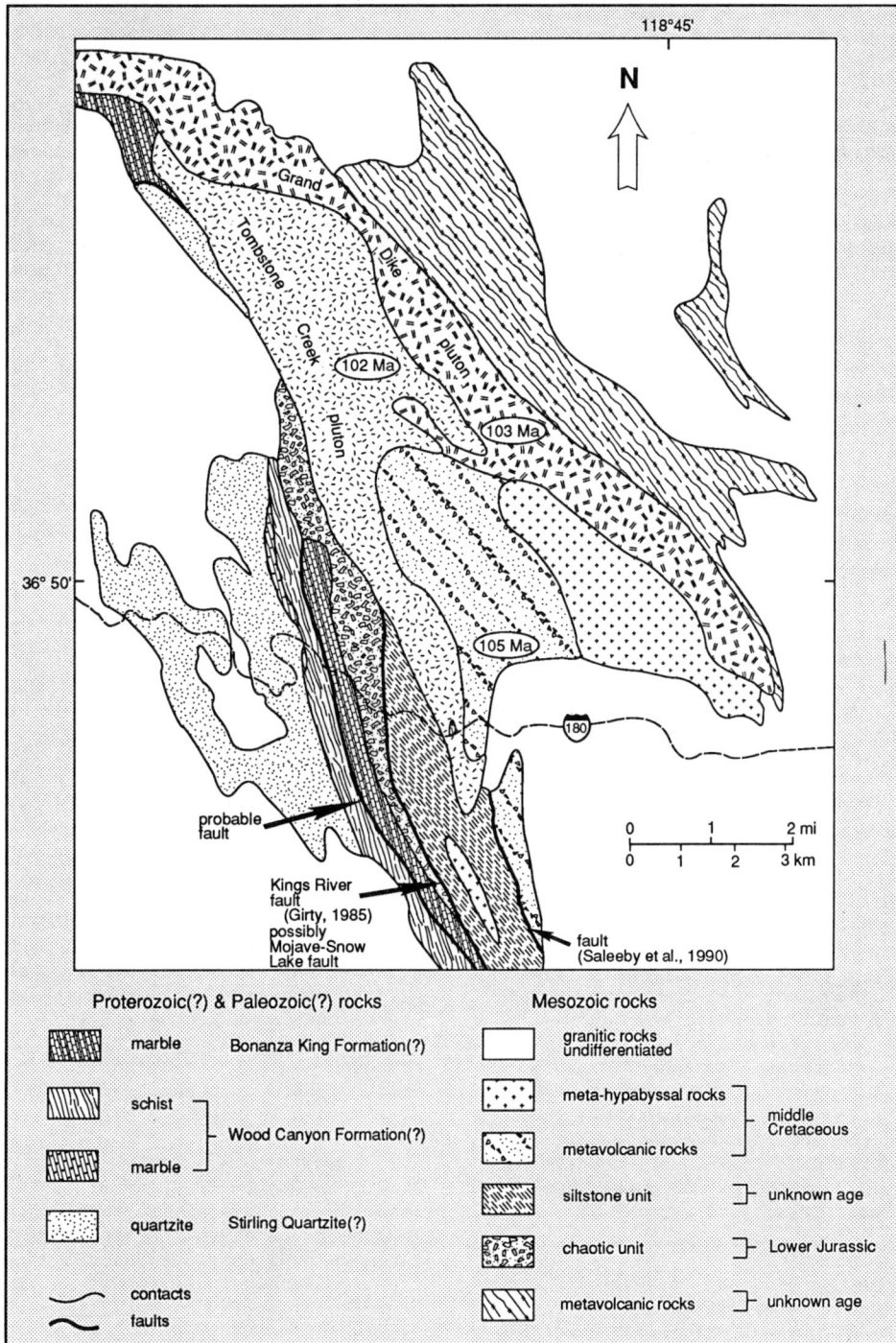

Figure 4. Geologic map of Boyden Cave pendant, after Moore and Marks (1972), Girty (1985), and Saleeby and others (1990). U-Pb zircon ages of plutonic and volcanic rocks from Saleeby and others (1990). Tentative correlations of Precambrian-Cambrian units shown in legend.

dolomitic marble occur within the schist. A prominent ledge of buff-white, dolomitic(?) marble about 20 m thick occurs about 150 m east of the contact with the quartzite unit. Suggestions of cross bedding occur in micaceous quartzites of this unit, but the presence of a strong fabric makes their identification uncertain.

Marble unit:

A massive unit of marble about 500 m thick

lies along the eastern edge of the schist unit (Fig. 4). It forms a very bold, prominent ridge in the central part of the pendant. Moore and Marks (1972) noted that resistant layers, ranging from less than 1 cm to about 30 cm in thickness, define bedding within the marble, and that the marble consists of calcite and dolomite. We observed that the resistant material forms ribs and irregular patches that give the marble a banded appearance in some outcrops and a mottled appearance in others.

According to Moore and Marks (1972) and Jones and Moore (1973), bedding at the north end of the marble unit defines a large fold closure, and they suggested it may indicate drag along a major fault on the west side of the marble unit. Such a fault, if present, would be consistent with the stratigraphic relations we propose below.

Chaotic unit:

A chaotic unit of argillite, containing small and large blocks of rocks from the other described units, evidently overlies the marble unit on its eastern contact (Moore and others, 1979; Girty, 1985), and rests directly upon the biotite-andalusite schist unit 2 km north of the highway (Fig. 4). According to Girty (1977b, 1985), this contact could be a fault or an unconformity. Lower Jurassic fossils collected from Boyden Cave pendant evidently came from this unit (Saleeby and others, 1978; Moore and others, 1979; Girty, 1985), and in view of the structural or stratigraphic break represented by the western contact of this unit, these fossils cannot be used to date the marble, schist, and quartzite units to the west (Girty, 1985). The matrix of the chaotic unit consists mainly of argillite and quartzose siltstone. Clasts within this matrix include calc-silicate schist, quartzite, marble, and rare volcanic rocks. Clasts range from centimeter-scale to several tens of meters across. A spectacularly folded block of calc-silicate schist that closely resembles beds within the quartzite unit evidently is a large block within the chaotic unit (Girty, pers. comm., 1989). The chaotic unit probably represents an olistostromal deposit shed off an area underlain by rocks like those to the west (Moore and others, 1979; Girty, 1985; Saleeby and others, 1978). Girty (pers. comm., 1989) suggested this unit was shed off a fault scarp.

Girty (1977a,b, 1985) summarized structural and textural data that indicates the units to the west of the chaotic unit were deformed and metamorphosed one or more times prior to the deposition of the chaotic unit. According to Nokleberg (1983) and Girty (1985), a major dextral fault (the Kings River fault) occurs within or along the eastern contact of the chaotic unit; it separates the chaotic unit from a well-bedded sequence of quartzose siltstone and tuffaceous siltstone to the east (described next).

Siltstone unit:

This unit (Fig. 4) consists of several hundreds of meters of slate, siltstone, and fine-grained quartzite, with minor calcareous layers, and with interbedded silicic tuffs near its top (Moore and Marks, 1972; Moore and others, 1979; Saleeby and others, 1978; Girty, 1985). This unit consistently tops to the east, and contains abundant sedimentary structures suggestive of turbidite deposition (Saleeby, in Moore and others, 1979; Saleeby and others, 1978). Metamorphic recrystallization and deformational structures are far less prominent in this unit than those to the west, and Girty (1977a,b, 1985; Girty, in Moore and others, 1979) argued that this unit has experienced only the latest of the deformations shown by the quartzite, schist, and marble units.

Fossils:

The exact origin of the much-discussed Jurassic fossils in Boyden Cave pendant is unclear. Jones and Moore (1973) described the fossils as having come from loose blocks derived from the metasiltstone unit (although they did not separate the chaotic unit from the siltstone unit as described here). Saleeby and others (1978) reported an Early Jurassic ammonite from the matrix of the chaotic unit, but Saleeby (in Moore and others, 1979) noted that this fossil was from an out-of-place block near the Kings River. Girty (1985) stated that the Lower Jurassic ammonites of Jones and Moore (1973) were derived from the chaotic unit, west of the Kings River fault. These interpretations indicate that the chaotic unit contains matrix and possibly some blocks of Early Jurassic age, and, in addition, suggest that the siltstone unit (as used here) has not been dated by fossils.

Eastern parts of the pendant:

The well-bedded siltstone unit is separated from metavolcanic rocks in eastern parts of the pendant by an elongate, sill-like pluton called the Tombstone Creek pluton (Fig. 4), along and north of the Kings River (Moore and Marks, 1972; Saleeby and others, 1990). South of the Kings River, Saleeby and others (1990) showed a fault between metavolcanic rocks and the siltstone unit. Indirect evidence that the Tombstone Creek pluton occupies a significant fault is provided by the presence of a 100-m-wide septum of metasedimentary rocks within the granite along the highway (Fig. 4). This septum consists of polyphase deformed micaceous quartzite, pelitic schist, and quartzite, that contrasts markedly with the little-deformed rocks of the siltstone unit to the west. Rocks of the septum are very similar in lithology and structure with the biotite-andalusite schist unit in the western part of the pendant. The septum may represent a fault slice of the schist unit or, less likely, may be a slide block. Saleeby and others (1990) reported U-Pb zircon ages of 102 Ma for the Tombstone Creek pluton and 105 Ma for the metavolcanic unit east of the pluton. These data suggest that the fault intruded by the southern part of the Tombstone Creek pluton is of mid-Cretaceous age.

Interpretation:

The Boyden Cave pendant is structurally complex and several faults and/or unconformities may separate different lithologic units. It now appears that only the chaotic unit is dated by Lower Jurassic fossils, and in view of possible structural breaks on both sides of the unit, these fossils provide little or no constraint on ages of units to the west or east. Nevertheless, it seems clear that units west of the chaotic unit are in general more complexly deformed and recrystallized than the chaotic unit and units to the east (except for the small septum within the Tombstone Creek pluton).

We believe that units west of the chaotic unit are best interpreted as pre-Mesozoic miogeoclinal strata, as suggested by Girty (1977a,b, 1985). There are no known sedimentary structures that would indicate turbidity current deposition as suggested by Saleeby and others (1978). Although stratigraphic relations among these units are unclear because of locally intense deformation and possible faults, the quartzite, schist, and marble units at Boyden Cave pendant have numerous lithologic, sedimentary, and structural features that are consistent with well-known stratigraphic units of the Death Valley facies and with units we have studied in detail at Snow Lake pendant. We suggest, on the basis of these similarities, that the quartzite, schist, and marble units could represent parts of the Stirling Quartzite, Wood Canyon Formation, and Bonanza King Formation, respectively.

The quartzite unit is a very good match for the Stirling Quartzite, both in the Death Valley region and at Snow Lake pendant, based upon its large apparent thickness, composition (feldspathic quartzite and quartzite), coarse overall grainsize, and sedimentary structures. We agree with Girty (1985) that this is a shallow-water unit.

The biotite-andalusite schist contains rocks that are identical to those within the Wood Canyon Formation, although the unit appears anomalously thin. Especially reminiscent of the lower part of the Wood Canyon Formation are dolomitic limestone marker beds in the lower part of the unit that are interbedded with thin, micaceous quartzites and pelites. These units are distinctive in the lower part of the Wood Canyon (Stewart, 1970). The upper part of the Wood Canyon would be missing in this interpretation.

The marble unit could represent some part of the Middle Cambrian Bonanza King Formation, based upon its great thickness, its dolomitic composition, and the banded and mottled textures, although other possibilities exist. If these interpretations are correct, a significant fault would be required along the western contact of the marble (as suggested by Moore and Marks (1972) and Jones and Moore (1973)) to explain the omission of the upper Wood Canyon, Zabriskie Quartzite, and Carrara Formation (Fig. 4). In the northern tip of the pendant, the marble unit apparently lies in contact with the quartzite unit, which may be consistent with a fault in this position.

The dextral Kings River fault of Nokleberg (1983) and Girty (1985) may be an exposure of the Mojave-Snow Lake fault, as suggested by Lahren and others (1990). This fault is older than 103 Ma, because it has been intruded by the Tombstone Creek and Grand Dike plutons (Fig. 4).

Sequoia Park pendants

The Sequoia Park pendants consist of three narrow, elongate, northwest-trending pendants ranging from 8 to 20 km long and 1.5 to 3 km across (Fig. 3). Several very small remnants of metamorphic rocks lie southwest of the largest body. These lie within the Kaweah River drainage and within Sequoia National Park. Ross (1958) mapped these pendants and provided brief descriptions of the various lithologic units within them. In addition, Saleeby and others (1978) provided descriptions of rocks along the main fork of the Kaweah River. The brief descriptions below are from Ross (1958), Saleeby and others (1978), and our own observations.

As in pendants to the north, the Sequoia Park pendants are dominated by thinly interlayered quartzite and schist, and contain local lenses of marble. We observed interlayered meter-scale calc-silicate and micaceous quartzite beds, and one 15-m-thick micaceous quartzite bed, showing possible relict cross bedding. The rocks in the largest pendant and the pendant along Redwood Creek show evidence of at least two generations of isoclinal folds, and therefore thicknesses are unknown.

The next most common lithology is quartzite, which is most abundant in the southeastern part of the largest pendant. Ross (1958) noted that the quartzite contains subordinate feldspar and mica. Saleeby and others (1978) reported east-facing graded bedding from interlayered quartzite and schist from the eastern part of the largest pendant, but we could not confirm this observation.

Several large, semicontinuous lenses of marble occur in the southern and central parts of the largest pendant, and in the central part of the northwestern pendant. Some lenses are up to 500 m across and over 2 km in length. Ross (1958) noted that the marble is white to dark gray, and is commonly banded with white and gray layers. The marble contains abundant interlayered calc-silicate schist.

Saleeby and others (1978) interpreted the rocks of the Sequoia Park pendants to represent flysch with intercalated olistostromes, and the lenses of quartzite and marble to be large olistoliths. We suggest, instead, based on lithologic and structural similarities with pendants to the north and poorly preserved planar cross stratification, that these rocks represent a strongly deformed shallow marine miogeoclinal sequence like that represented at Dinkey Creek and Boyden Cave. Rocks similar to the Stirling Quartzite, Wood Canyon Formation, Zabriskie Quartzite, and Carrara Formation occur here.

Western edge of Mineral King pendant

Christensen (1963) mapped lenses of marble along the western edge of Mineral King pendant (Fig. 3), where they lie in contact with Cretaceous plutonic rocks. Busby-Spera (1984b) interpreted this marble as the basal part of the Triassic section of the pendant. As discussed later, possible lower Paleozoic eugeoclinal rocks lie between the marble and the Triassic section. We examined this marble and found it to be a massive white to light bluish-gray, coarsely crystalline marble with an extremely strong blastomylonitic foliation. It locally contains convolute, rootless mushroom folds of the foliation, and contains tectonic lenses of micaceous quartzite and chert-argillite. Based upon the strong contrast in deformation with rocks to the east, we believe this marble possibly marks a major shear zone, and that the marble itself could be a small slice of miogeoclinal strata of the Snow Lake block. More detailed mapping is needed in this area to work out the structural relations, but this marble could provide a glimpse of the Mojave-Snow Lake fault.

Knopf and Thelen (1905) described quartzite as the main lithology in the southern tail of the Mineral King pendant. We are unaware of any modern descriptions of these rocks, but note that they lie approximately on trend with metamorphic rocks of

666

Sequoia Park and could represent the southeastern end of the Snow Lake block (Fig. 3). We do not know whether rocks of the Snow Lake block extend farther southeast, beyond the Kern Canyon fault.

Summary

Based upon published descriptions and on our reconnaissance, we believe a strong case can be made for the existence of highly deformed, but still recognizable, Proterozoic and Cambrian miogeoclinal rocks in pendants from latitude 38° N to about 36°30' N, a distance of about 200 km along the axis of the Cretaceous batholith. We infer that these small pendants together make up the original basement of the Snow Lake block, and that Mesozoic rocks within some of these pendants are either in tectonic contact with or unconformably overlie the miogeoclinal rocks. This view differs greatly from that of Bateman and Clark (1974), Saleeby and others (1978), and Nokleberg (1983), and other authors who interpreted rocks in all these pendants as Triassic-Jurassic in age. Instead, our interpretation is similar to that of Kistler and Bateman (1966), Girty (1977a,b, 1985), and Schweickert (1981), in which miogeoclinal rocks were inferred to occur in these pendants. An important difference between our current interpretation and that of the latter authors is that these pendants are regarded as containing rocks of the Death Valley facies, an inner facies of the miogeocline, rather than the quartzite-poor Inyo facies, an outer miogeoclinal facies.

Pendants west of the Snow Lake block

The lower Paleozoic Shoo Fly Complex occurs in the southeastern part of the Western Metamorphic belt of the Sierra Nevada, about 35 km west of Snow Lake, Piute Mountain, and May Lake pendants, and 15 km west of Iron Mountain-Quartz Mountain pendants (Fig. 2). It consists of a highly deformed assemblage of quartzitic rocks, schists, and amphibolites, together with Paleozoic orthogneisses. Although these rocks in some cases resemble rocks of the Snow Lake block, we believe they can be distinguished with some confidence on the basis of their lithology, structure, and associated intrusions.

Shoo Fly Complex

The Shoo Fly Complex in the southern part of the Western Metamorphic belt (Fig. 2) has been mapped by Merguerian (1985), Bhattacharyya (1986), Merguerian and Schweickert (1987), Bateman and Krauskopf (1987), Dodge and Calk (1987) and Schweickert and others (1988). The Shoo Fly is characterized by strongly deformed quartzite, quartzofeldspathic gneiss, mica-garnet schist, and rare marble, calc-silicate rocks, and amphibolite. Protoliths of the Shoo Fly Complex included quartzose turbidites, pebbly sandstone, siltstone and shale, limestone and calcareous siltstone, radiolarian chert, argillite, and minor mafic metavolcanic rocks. Highly deformed orthogneiss of Devonian to Permian age (Merguerian and Schweickert, 1987; Sharp, 1988) forms a distinctive part of the Shoo Fly, and is important in establishing a structural chronology. The orthogneiss originally had gabbroic, granitic, and syenitic protoliths.

In the southern part of the Western Metamorphic belt, intense deformation has essentially obliterated most primary structures in the metasedimentary rocks, and the rocks consist of blastomylonitic quartzite, paragneiss, and schist. The earliest known deformation predated intrusion of Devonian granitoids, and both were strongly deformed by an amphibolite-grade second deformation, which mylonitically deformed most rocks. A third intense deformation produced blastomylonitic shear zones and other zones of intense deformation, but some areas were not strongly deformed by this event. At least four later deformational events affected the Shoo Fly, some involving tight folds, but most produced minor crenulations and/or spaced cleavages. Several episodes of intrusion of plutons and dike swarms were interspersed between deformational events.

Important differences between the Shoo Fly Complex and pendants of the Snow Lake block are that the Shoo Fly is much more intensely deformed, protoliths where recognizable include turbidites, chert-argillite sequences, and minor volcanic rocks, and granitic orthogneisses are common. However, small, isolated exposures of Shoo Fly or of Snow Lake block could be difficult to classify. Paleogeographically, the Shoo Fly had a much different origin than rocks of the Snow Lake block, because the Shoo Fly represents deep marine, continentally-derived deposits that form the basement of a Devonian to Permian island arc sequence, whereas the Snow Lake block represents shallow marine, miogeoclinal strata.

Pendants containing rocks possibly equivalent to the Shoo Fly Complex

Pendants in Bass Lake quadrangle:

Pendants in Bass Lake quadrangle (Fig. 3) were mapped as quartzite and phyllite by Bateman (1989), who noted that they are lithologically similar to rocks of the Shoo Fly Complex along the Merced River to the north. Masses in the northeastern part of the quadrangle consist of rusty-weathering, mylonitic quartzite with lenses of light quartzite enclosed in a darker quartzite, and a highly irregular mylonitic fabric. Interlayered with the quartzite are lenses of dark quartz schist or paragneiss, similar to Shoo Fly rocks.

Lower Kings River pendants:

Metasedimentary rocks in the Lower Kings River pendants (Fig. 3) were mapped by MacDonald (1941) and Krauskopf (1953). Schweickert and others (1977) and Saleeby and others (1986) showed these pendants as containing chert-argillite rocks in their western parts, north of Pine Flat reservoir, and quartzite and schist in the eastern parts. The western parts of the pendants (Fig. 3) have rocks and structures that closely resemble those of chert-argillite rocks of the Calaveras to the north, and are so designated. The southeastern part of this group of pendants, east of Pine Flat reservoir, contains highly deformed quartzite and schist that is here tentatively assigned to the Shoo Fly Complex.

Lower Kaweah River pendants:

Durrell's (1940) Three Rivers schist, which occurs in the Lower Kaweah River pendants (Fig. 3) along the north and south forks of the Kaweah

River, consists of polyphase deformed black schist and marble with large lenses and boudins of black quartzite. The rocks are commonly mylonitic, and original bedding has been obliterated. The rocks are identical to highly deformed Shoo Fly rocks in areas to the north, and they contrast markedly with units to the west and with rocks in the Sequoia Park pendants to the east. We therefore assign them to the Shoo Fly Complex.

Tule River pendants:

The Tule River pendants (Fig. 3) were shown by Saleeby and others (1978) as containing interbedded quartzite and schist. They lie east of chaotic chert-slate-limestone-volcanic rocks in the Yokohl Valley pendant described by Saleeby (1978a, 1979). Our observations of the rocks on Slate Mountain (the large southern pendant) showed that these rocks consist of highly stained, locally black, impure quartzite and schist, with minor marble, and with locally strong fabrics. All primary structures appear to have been transposed into a mylonitic foliation, and a second foliation is locally strongly developed. These rocks bear a strong resemblance to the Shoo Fly Complex, and are unlike any of the rocks we have observed in the Snow Lake block.

Kern Canyon pendant

The Kern Canyon pendant (Fig. 3) was mapped by Miller and Webb (1940), Saleeby and others (1978), and Saleeby and Busby-Spera (1986). Ross (1987) described these rocks as characterized by dark, poorly sorted, poorly bedded quartzite with common pebbly intervals. Along the Kern River, siliceous to micaceous schist and phyllite occur locally, and large, lenticular bodies of marble occur in the northern lobe of the pendant. Our observations show that the dark quartzite and schist have strong tectonite fabrics that have obliterated primary structures. Locally, mylonitic foliation shows highly convolute folds. Saleeby and Busby-Spera (1986) and Busby-Spera and Saleeby (1990) showed that these rocks are unconformably overlain by mid-Cretaceous metavolcanic rocks. The metasedimentary rocks closely resemble the Shoo Fly Complex in both lithology and structural style, and we tentatively include them with the Shoo Fly. According to Ross (1987), these rocks are fairly distinctive, and extend about 30 km farther south.

Pendants west of the Shoo Fly Complex

Oakhurst pendant:

The Oakhurst (or Coarsegold) pendant is the first large pendant to the southeast of the Western Metamorphic belt (Fig. 1,3). It was mapped by Bateman and Busacca (1982), Bateman and others (1983), Krauskopf (1985), Bateman (1989), and was discussed by Russell and Cebull (1977). The northeastern edge of the pendant reportedly consists of highly deformed phyllite, quartzite, and metachert, all units that occur within the Calaveras Complex, 10 km to the north (Schweickert and others, 1988). Krauskopf (1985) noted these rocks lithologically resemble the Calaveras, and we therefore assign rocks in the eastern part of the pendant to the Calaveras Complex. Schistose and mylonitic metavolcanic rocks form a strip along the axis of the pendant and also occur within several smaller pendants south of the Oakhurst pendant (Bateman and others, 1982, 1983), together with

phyllitic metasedimentary rocks. We tentatively assign these latter rocks to the Don Pedro terrane (Fig. 3).

Lower Kings River pendants (western part):

Chert-argillite rocks in the central part of the Lower Kings River pendant (Fig. 3) have previously been assigned to the Calaveras Complex and we follow that usage here. These rocks have both lithology and structural style identical to that of the Calaveras Complex. The chert-argillite rocks are in fault contact with the Ordovician to Pennsylvanian Kings River ophiolite (Saleeby, 1978b; Shaw and others, 1987) to the west, which we assign to the Don Pedro terrane.

Lower Kaweah River pendants:

The Lemon Cove schist and Homer quartzite of Durrell (1940), in the eastern part of the Lower Kaweah River pendants (Fig. 3), consist of chert, argillite, and diamictite, with discontinuous marble lenses. Saleeby and others (1978) described these rocks as consisting of chert-argillite overlain by flysch, argillaceous units, and diamictites. We believe the flysch described by Saleeby and others (1978) is instead thinly layered chert and siltstone, highly recrystallized by surrounding plutons. These rocks are structurally and lithologically identical to the Calaveras Complex of the Western Metamorphic belt.

Yokohl Valley pendant:

Saleeby and others (1978) and Saleeby (1979) mapped and described the eastern part of the Yokohl Valley pendant as consisting of a chert-argillite melange with scattered blocks of bedded chert, limestone, and volcanic rocks in a matrix of chert, argillite, and slate. These rocks are in fault contact with the Kaweah serpentinite melange to the west (Saleeby, 1979). Some of the limestone blocks contain Permian Tethyan fusulinids (Saleeby and others, 1978). Lithologically, the parts of this unit with augite-porphyry andesitic-basaltic breccia and volcaniclastic slate and sandstone resemble units of the Don Pedro terrane, while other parts consisting mainly of chaotic chert-argillite resemble units in the Calaveras Complex. Nokleberg (1983) postulated that a fault separates the melange containing limestone blocks on the west from an eastern unit of chert-argillite olistostromes. We tentatively adopt Nokleberg's interpretation, which leads to an assignment of the Tethyan limestone blocks to the Don Pedro terrane, and chert-argillite rocks to the east to the Calaveras Complex. The alternative, which is not adopted here, is that the entire chaotic unit could be part of the Calaveras Complex. Rocks in the Kaweah serpentinite melange are also assigned to the Don Pedro terrane.

Summary

We suggest that many of the pendants that lie to the west of the Snow Lake block contain highly deformed quartzites, schists, and marble of the Shoo Fly Complex (Fig. 3). Although this interpretation is far from proven and other interpretations are possible, the rocks and structural styles of these pendants are markedly similar to those within the Shoo Fly Complex in the Western Metamorphic belt. We believe this is an important alternative to considering most of these pendants to consist of Jurassic or Triassic strata as proposed by Saleeby and others (1978) and Nokleberg (1983). In addition, these pendants are bordered to the west by occurrences of chert-argillite rocks that have strong lithologic and structural affinities to the Calaveras Complex (Fig. 3).

We therefore suggest that geologic relations and tectonostratigraphic units typical of the Western Metamorphic belt may extend at least as far south as latitude 36° N, along the western edge of the Snow Lake block. The nature of the boundary between the Snow Lake block and rocks of the Shoo Fly Complex is unknown; it could be exposed in some of the pendants between the Kings and Tule Rivers.

Pendants east of the Snow Lake block

East of the cryptic Mojave-Snow Lake fault, many pendants, including Saddlebag Lake, Ritter Range, Merced Peak, Mount Morrison, Mount Goddard, the eastern part of Boyden Cave, and Mineral King pendants (Fig. 1-3), expose thick sequences of Paleozoic metasedimentary and Mesozoic metavolcanic rocks. The Mesozoic rocks can be viewed as overlap sequences that perhaps tie together some of the basement terranes. However, these Mesozoic rocks provide little direct information as to the nature of the basement beneath them, and therefore will not be discussed here.

Pendants containing lower Paleozoic eugeoclinal rocks of the Antler orogenic belt

Saddlebag Lake pendant and the northern part of the Ritter Range pendant (Fig. 2), contain thick sequences of highly deformed chert, and argillite, with minor limestone, quartzite, and metavolcanic rocks in their eastern parts (Kistler, 1966; Schweickert and Lahren, 1987). Schweickert and Lahren (1987), and Greene and others (1989) showed that these rocks are structurally and lithologically identical to rocks of the Antler orogenic belt in the Candelaria Hills, 60 km to the east. In addition, these eugeoclinal rocks are unconformably overlain in Saddlebag Lake pendant by Permian and Triassic clastic rocks correlated with the Candelaria and Diablo Formations of west-central Nevada (Schweickert and Lahren, 1987). In the southeastern part of the Ritter Range pendant, the eugeoclinal rocks structurally overlie strata assigned to the outer miogeoclinal facies (Strobel, 1986; Greene and others, 1989).

Pendants containing lower to middle Paleozoic strata of probable miogeoclinal facies

Pendants containing miogeoclinal strata of early to late Paleozoic age include the southeastern part of the Ritter Range pendant (mentioned above), Mount Morrison, Pine Creek, Bishop Creek, Big Pine Creek pendants, and possibly pendants in the Mount Pinchot quadrangle (Fig. 2, 3). These pendants have been mapped and described by Kistler (1966), Strobel (1987), Rinehart and Ross (1964), Bateman (1965), Moore (1963), Moore and Foster (1980), Frazier and others (1986), and Willahan (1990). In general, these pendants contain thick sequences of shale and limestone with minor quartzite, and some bedded chert. Cambrian fossils have been reported from Big Pine Creek, Ordovician fossils have come from Bishop Creek and Mt. Morrison pendants, and Mississippian, Pennsylvanian and Permian fossils have been reported from Mt. Morrison pendant. Although many important stratigraphic relations are still unknown, it seems likely that these pendants can be regarded as part of the Inyo facies.

Pendants containing eugeoclinal rocks that probably have been displaced from the Antler orogenic belt

Metasedimentary rocks of probable early Paleozoic age have been identified in the Rattlesnake Creek, Hockett Peak, Bald Mountain, and Kennedy Meadows pendants by Dunne and others (1988; Dunne and Suczek, this volume)(Fig. 3). According to Dunne and others (1988), these rocks have close lithologic affinities with lower Paleozoic eugeoclinal rocks in the El Paso Mountains to the south. In addition, we have identified similar rocks along the western edge of Mineral King pendant, north of Vandever Mountain. We therefore infer that eugeoclinal strata similar to those of Saddlebag Lake and Ritter Range pendants form a crustal block east of the Snow Lake block from the Kaweah River to the southeast.

Mineral King pendant

Mineral King pendant (Fig. 3) contains a well-known sequence of Triassic volcanic and sedimentary rocks that have been mapped and described by Knopf and Thelen (1905), Christensen (1963), and Busby-Spera (1984a,b, 1985, 1986). The lowest stratigraphic unit described by Busby-Spera (1984b, 1986) consists of thinly bedded, fine-grained, siliceous, calcareous, and pelitic rocks of the White Chief unit, inferred to be a deep marine unit of Triassic age. This unit underlies a Triassic ash flow tuff. In White Chief Basin, north of Vandever Mountain, this unit consists of dark brown slate, slaty, pyritic argillite, sandstone, and pebbly conglomerate, locally with graded bedding indicating tops to the northeast. Beneath this unit to the west is a more strongly deformed unit of rusty-weathering chert-argillite, locally with rhythmic bedding, and locally chaotic in appearance. White to dark gray chert layers commonly show phosphatic streaks and nodules. Lenses of fine-grained quartzite, quartzose siltstone, and gray-black marble locally occur within this unit. The structural style, lithology, and especially phosphatic chert are identical to features in lower Paleozoic eugeoclinal rocks in areas like Saddlebag Lake and Ritter Range pendants, and are similar to features described in pendants to the southeast by Dunne and others (1988).

The contact between the highly deformed chert-argillite sequence and the overlying Triassic(?) unit is apparently an angular unconformity marked by lenses of essentially undeformed volcanic sandstone and volcanic conglomerate resting on highly deformed chert-argillite, and containing clasts of dark chert. We interpret this older unit to be part of the eugeoclinal sequence described by Dunne and others (1988). West of the eugeoclinal rocks is a large mass of marble mapped by Knopf and Thelen (1905) and Christensen (1963). This highly deformed marble was mentioned in an earlier section as possibly representing a small slice of the Snow Lake block in a shear zone along the western edge of the Mineral King pendant. The existence of eugeoclinal strata in Mineral King pendant supports the interpretation that the Mojave-Snow Lake fault lies near or along the western and southern parts of the pendant (Schweickert and Lahren, 1990).

DISCUSSION: INFERRED FRAMEWORK OF THE SIERRA NEVADA BATHOLITH

Figure 5 shows the inferred distribution of basement rock types within the Sierra Nevada batholith, together with tectonostratigraphic units of the Western Metamorphic belt. It also shows our interpretation of the position of the Mojave-Snow Lake fault and several other tectonic boundaries of note. Isotopic boundaries inferred by Saleeby and others (1986), Saleeby (1990), and Kistler (1990) are shown for reference on Figure 1.

We suggest that six principal basement types occur within the central and southern parts of the Sierra Nevada batholith. Proterozoic and Cambrian miogeoclinal strata (and presumably Proterozoic crystalline basement) of the Snow Lake block lie along the axis of the batholith at least as far south as latitude 36° 15' N. West of the Snow Lake block, tectonostratigraphic units of the Western Metamorphic belt may continue to the southern end of the Sierra as thin, easterly-dipping thrust sheets.

Lower Paleozoic strata related to the Shoo Fly Complex form a continuous belt as far south as the Kern Canyon area, and are tentatively shown extending to the Garlock fault. If we are correct in assigning these rocks to the Shoo Fly Complex, the corollary is that they form part of the basement of a Devonian to Permian island arc, strata of which are preserved in the northern Sierra Nevada (Fig. 5).

The western boundary of Shoo Fly Complex is inferred to be a continuation of the Calaveras-Shoo Fly thrust as far south as the Kern Canyon area. The Paleozoic to Triassic Calaveras Complex (Bateman and others, 1985) can be identified as far south as the Yokohl Valley area, and may extend farther south, as shown.

The Jurassic Sonora fault, which forms the structural boundary between the Don Pedro terrane and the Calaveras Complex, is inferred to extend as far south as the Tule River area. Ordovician to Pennsylvanian ophiolitic rocks of the Kings-Kaweah ophiolite belt are interpreted to be part of the basement of the Don Pedro terrane and to lie west of the southern continuation of the Sonora fault.

East of the Snow Lake block, rocks of the Antler and Sonoma orogenic belts can be traced into the batholith in the Saddlebag Lake and Ritter Range pendants, west of which they appear to have been truncated (at least in part) by the Mojave-Snow Lake fault. Miogeoclinal strata of the Inyo facies form the basement of the eastern part of the batholith between the Ritter Range pendant and the eastern part of the Boyden Cave pendant, and probably in areas to the southeast.

Displaced eugeoclinal rocks of inferred early Paleozoic age (Geraci and others, 1987; Dunne and others, 1988; Dunne and Suczek, this volume) form the basement to the batholith from the vicinity of Mineral King pendant southeast to the El Paso Mountains. These rocks are grouped with the El Paso terrane (Dunne and Suczek, this volume). Interpretations of Stone and Stevens (1988) and Walker (1988), as discussed below, necessitate a sinistral late Paleozoic fault along the northeastern edge of the El Paso terrane, as shown on Figure 5. This structure may actually have been responsible for the truncation of the Antler orogenic belt at Saddlebag and Ritter Range pendants, and parts of it (the northern parts) may have been subsequently reactivated as a dextral fault during displacement on the Mojave-Snow Lake fault.

There have been several previous interpretations of the existence of major breaks within the southern half of the Sierra Nevada batholith. In particular, much discussion has centered about a structure called the "Foothills suture," which forms the boundary between oceanic crustal rocks to the west and rocks of continental affinity to the east. Saleeby and others (1978, 1986), and Saleeby (1981) showed the "Foothills suture" in a position that coincides in part with the Calaveras-Shoo Fly thrust of this paper. In contrast, the "Foothills suture" of Schweickert (1981), and Nokleberg (1983), south of latitude 37° N, coincides approximately with the inferred southern extension of the Sonora fault, which separates the Calaveras Complex and the Don Pedro terrane. Nokleberg (1983), however, extended the Foothills suture northward along the eastern side of the Western Metamorphic belt, and postulated that the "Kings River fault" extends along the axis of the batholith from north of Mineral King to the western side of Snow Lake pendant. He also proposed a "San Joaquin fault" that parallels the Kings River fault, and lies west of the Saddlebag, Ritter Range, and Snow Lake pendants. Saleeby and others (1986) also inferred that "intrabatholithic breaks" occur east of the Foothills suture, one coincident with the Shoo Fly-Snow Lake contact of this paper and one close to the position of the Mojave-Snow Lake fault.

Burchfiel and Davis (1981) postulated that eugeoclinal rocks of the El Paso terrane represented displaced slices of the Antler orogenic belt, and, together with their views of truncation events along the cordilleran margin, this interpretation necessitated cryptic left-lateral faults of early Mesozoic age in the southern Sierra, one coincident with the $Sr_1=0.706$ line. Stone and Stevens (1988) and Walker (1988) modified this interpretation, and have postulated a cryptic late Paleozoic left-lateral fault east of the El Paso terrane is responsible for truncation of the Antler orogenic belt. Based on new work in the southeastern Sierra, Dunne (1989) recognized a major early (?) Mesozoic shear zone near the western edge of the El Paso terrane, but still east of our proposed trace of the Mojave-Snow Lake fault. Kistler (1990) inferred this early Mesozoic(?) shear zone is a major tectonic boundary extending northwestward from the El Paso Mountains through Mineral King and Boyden Cave pendants, and then curving more westerly to join the Melones fault zone in the Western Metamorphic belt.

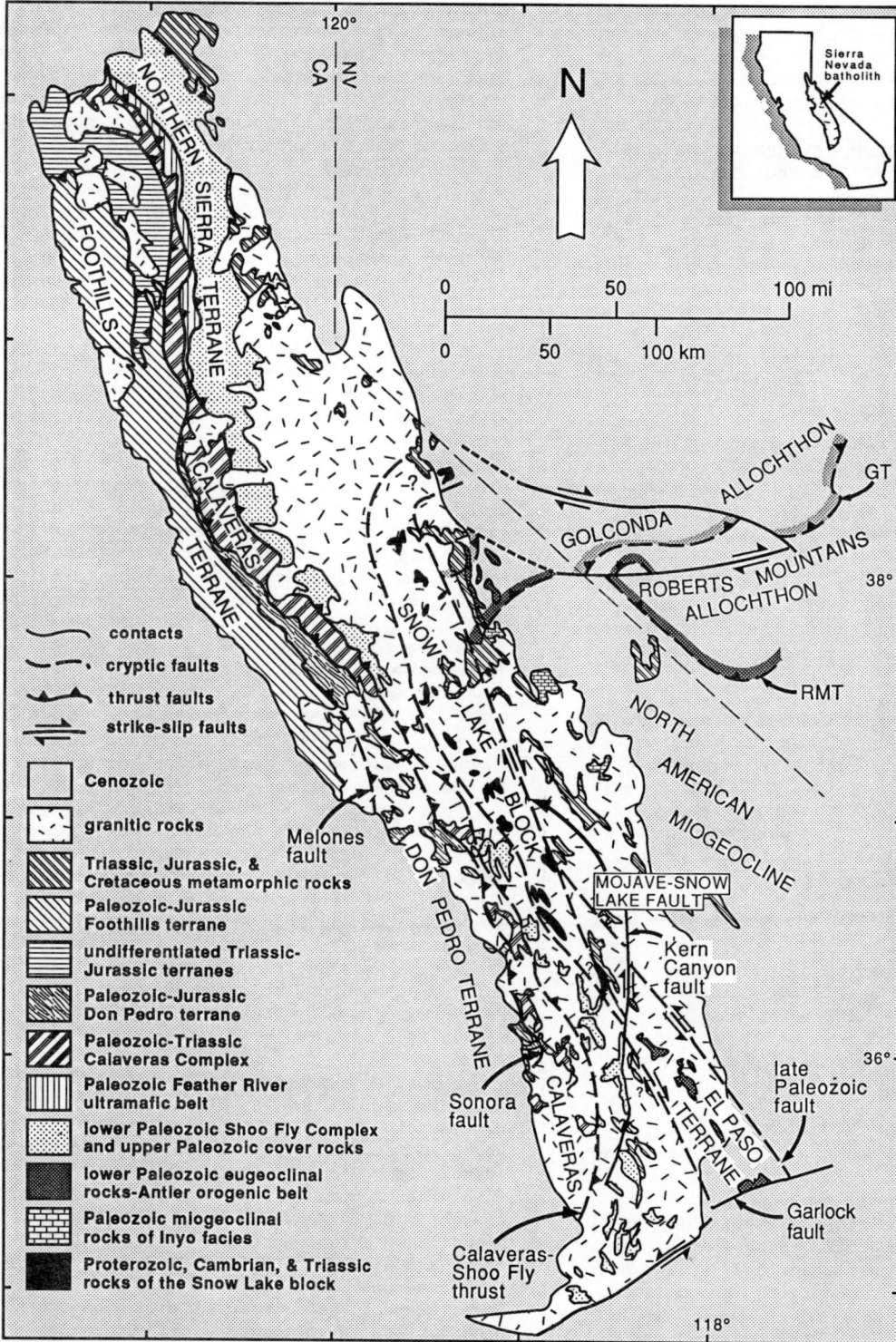

Figure 5. Tectonic sketch map of the Sierra Nevada region, showing inferences
about the nature of major crustal boundaries and terranes within the Sierra
Nevada batholith. See text for discussion, and see Schweickert and others
(1988) and Edelman and Sharp (1989) for discussions and nomenclature of
terranes in Western Metamorphic belt. Undifferentiated Triassic-Jurassic
terranes--Jurassic, Triassic, and older sedimentary, volcanic, and ophiolitic
rocks in melanges, including parts of Fiddle Creek and Slate Creek units of
Edelman and Sharp (1989); Northern Sierra terrane includes Shoo Fly Complex and
upper Paleozoic cover rocks in the northern Sierra Nevada; GT--Golconda thrust;
RMT--Roberts Mountains thrust.

While we agree in general that important structures probably occur at some of these locations, the structural breaks depicted on Figure 5 have been deduced independently from stratigraphic and structural data reviewed here. Furthermore, our interpretations of their geometry and significance differ significantly in many respects from those of earlier workers.

The southern extension of the Sonora fault is viewed here as a Middle to Late Jurassic suture between rocks accreted earlier to North America and a variety of Paleozoic ophiolitic units and Jurassic island arc sequences to the west (see Schweickert and others (1988), Sharp (1988), and Edelman and Sharp (1989) for discussions of the Sonora fault and other terranes in the Western Metamorphic belt). The Calaveras-Shoo Fly thrust in the southern Sierra Nevada is viewed as a Triassic boundary against which the Calaveras accretionary prism developed. The Shoo Fly-Snow Lake boundary, as noted by Schweickert and Lahren (1990), could be an expression of the Sonoma suture belt. The Mojave-Snow Lake fault is likely a strike-slip fault (Lahren and others, 1990). Although we regard the boundaries west of the Snow Lake block as most likely to have been thrust faults, the possibility that they have in addition experienced strike-slip motions, as suggested by Saleeby (1981), Nokleberg (1983), Saleeby and others (1986), and Kistler (1990) cannot be excluded.

TECTONIC IMPLICATIONS

The geometry of basement terranes in Figure 5 has wide-ranging implications both for evolution of the Sierra Nevada batholith and for the regional tectonic framework of the southwestern Cordillera. Some of these implications are discussed below.

Implications for batholithic evolution

Figure 5 provides a new framework for interpretation of geochemical, isotopic, and petrological data from the Sierra Nevada batholith, and for understanding mechanisms of emplacement of the batholith.

1. The combined areas of the Shoo Fly Complex and the Snow Lake block show a nearly perfect correspondence with the distribution of "I-SCR" plutons of Ague and Brimhall (1988). According to these authors, these strongly contaminated and reduced, I-type plutons formed as a result of extensive assimilation of pelite-rich metasedimentary rocks that are characteristic of the Kings terrane. However, the Kings terrane, as depicted by Nokleberg (1983), included many pendants east and west of the area of I-SCR plutons. Our proposed divisions of the pendants in the southern Sierra Nevada are more consistent with the distribution of these distinctive plutons, arguing that magmatic assimilation of metasedimentary rocks of the Snow Lake block and the Shoo Fly Complex contributed to the distinctive petrological character of these plutons.

2. Although the boundaries drawn on Figure 5 have been based entirely upon lithologic and structural features in metamorphic rocks in the various pendants, we are encouraged by the near coincidence of the $Sr_i=0.706$ line of Kistler (1990) and Saleeby (1990) with the inferred western boundary of the Shoo Fly Complex (compare Figures 1 and 5). This line has been inferred by Kistler (1990, and other references) to delineate the approximate western edge of Precambrian sialic crust in the western United States. We suggest that Shoo Fly metasedimentary rocks in the region between latitude 36° and 39° N contain sufficient detritus derived from Proterozoic cratonic rocks that they have imparted the same isotopic signature to magmas as Proterozoic crust itself. In contrast, chert-argillite rocks of the Calaveras Complex west of the $Sr_i=0.706$ line apparently contain little continentally derived detritus.

3. Many of the plutons of the Sierra Nevada batholith probably were emplaced along shear zones or thrusts, and some may have been emplaced during continued or renewed movement along these crustal boundaries.

Implications for regional tectonic framework

1. Figure 5 is the first terrane map of the Sierran region that is fully consistent with lithologic, stratigraphic, structural, isotopic, geochemical, and regional data. It extends the Snow Lake block to the Tule River drainage and the geology of the Western Metamorphic belt as far south as the Garlock fault. It also delineates several fundamental structural breaks within the batholith in addition to the Mojave-Snow Lake fault.

2. Based upon a reconstruction of the Mojave-Snow Lake fault by Schweickert and Lahren (1990), the cryptic Shoo Fly/Snow Lake contact may represent a southern continuation of the Sonoma suture separating the late Paleozoic island arc and its basement (Shoo Fly Complex) from inner miogeoclinal rocks of North America (Snow Lake block). This further suggests that along the southern continuation of the Sonoma belt, the Inyo facies of the Cordilleran miogeocline is missing. A similar relation exists in the Mojave block, south of the Garlock fault. This may have important ramifications for evolution and/or truncation of the southwestern part of the Cordilleran miogeocline.

3. Fossiliferous Triassic-Jurassic rocks occur in a narrow belt near the eastern edge of the Snow Lake block in Boyden Cave, Mineral King, and Isabella pendants, but they lie structurally or unconformably upon probable Cambrian and older strata. Those at Mineral King occur east of the Mojave-Snow Lake fault, where they rest unconformably upon eugeoclinal strata of probable early Paleozoic age. The relation of the Jurassic strata in the Isabella pendants to the Snow Lake block is unknown because they lie east of the proto-Kern Canyon fault (Busby-Spera and Saleeby, 1990). Jurassic strata in Strawberry Mine pendant (Nokleberg, 1981) occur within the region of the Snow Lake block, but their relation to rocks of the Snow Lake block is unknown. We infer that they are part of an overlap sequence.

4. The southern parts of the Shoo Fly Complex on Figure 5 would restore to a position 200 to 300 km south of the San Bernardino Mountains in the reconstruction of Schweickert and Lahren (1990).

Therefore, these rocks may once have extended south through the present Peninsular Ranges into northern Baja California. Interestingly, Gastil (1986) described eugeoclinal lower Paleozoic rocks in northern Baja California that could be similar to Shoo Fly rocks.

5. The Paleozoic-Triassic Calaveras Complex, which has been interpreted by many authors as part of an early Mesozoic subduction complex, forms a continuous, narrow belt in the southern Sierra Nevada west of the Shoo Fly Complex. Its distribution suggests that a Triassic or younger convergent plate boundary extended the full length of the present Sierra Nevada (and, if restored, far to the south of the Mojave region).

6. 480 Ma and 300 Ma ophiolitic elements of the Kings-Kaweah belt (Saleeby, 1978b,1979; Shaw and others, 1988), together with overlying chaotic deposits bearing Permian Tethyan fusulinids, are here interpreted to occur within the Don Pedro terrane. Since they lie outboard of the Calaveras and Shoo Fly Complexes, they probably were accreted to the North American margin in Jurassic time.

7. The "paleo-basement reconstruction" of Saleeby and others (1978) regarded rocks in all the pendants in the axial parts of the batholith between 35° and 37° N latitude as forming a Triassic-Jurassic facies transition from a continental arc in the east to an oceanic island arc in the west. Although the interpretation that several major intrabatholithic breaks occur within the batholith (Saleeby and others, 1986) conflicts with this reconstruction, it has nevertheless been cited as important evidence that ties the island arc rocks to the west (in the Foothills and Don Pedro terranes) to the continental arc rocks in the east by Late Triassic time. Based on evidence presented here, the rocks in these pendants include Proterozoic-Cambrian, lower Paleozoic, and Triassic-Jurassic strata each having different depositional, structural, and metamorphic histories, that have been juxtaposed across fundamental crustal boundaries. Our analysis provides an important new alternative explanation for the diverse rocks in Sierran roof pendants and the complex relationships between them.

These implications underscore the importance of new and detailed studies in all the pendants of the Sierra Nevada, and in addition, they suggest important new avenues for study, such as relating emplacement mechanisms of the Sierra Nevada western batholith to important crustal boundaries, evaluating the northern continuation of the Mojave-Snow Lake fault, and investigating possibilities that cryptic tectonic boundaries occur within other cordilleran batholiths.

ACKNOWLEDGEMENTS

Our research in the central and southern Sierra Nevada has been supported by grant no. NSF-EAR-89-03963. We are grateful to George Dunne, Gary Girty, and Nancy Merritt for helpful discussions, and to Brian Lahren for help with fieldwork. We thank Rich Schultz for valuable suggestions on illustration preparation. We also thank Cal Stevens for his helpful review of this paper, but hasten to add that we alone are responsible for errors of fact or interpretation.

REFERENCES CITED

Ague, J.J., and Brimhall, G.H., 1987, Granites of the batholiths of California: Products of local assimilation and regional-scale crustal contamination: Geology, v. 15, p. 63-66.

_____ 1988, Regional variations in bulk chemistry, mineralogy, and the compositions of mafic and accessory minerals in the batholiths of California: Geol. Soc. America. Bull., v. 100, p. 891-911.

Bateman, P.C., 1965, Geology and tungsten mineralization of the Bishop District, California: U.S. Geol. Survey Prof. Paper 470, 208p.

_____ 1981, Geologic and geophysical constraints on models for the origin of the Sierra Nevada batholith, California, in Ernst, W.G., ed., The Geotectonic Development of California: Englewood Cliffs, N.J., Prentice-Hall, p. 71-86.

_____ 1989, Geologic map of the Bass Lake quadrangle, west-central Sierra Nevada, California: U. S. Geol. Survey Geol. Quadrangle Map GQ-1656, scale 1:62,500.

Bateman, P.C. and Wones, D., 1972, Geologic map of Huntington Lake quadrangle, central Sierra Nevada, California: U.S.Geol. Survey Geol. Quadrangle Map GQ-987, scale 1:62,500.

Bateman, P.C., and Clark, L.D., 1974, Stratigraphic and structural setting of the Sierra Nevada batholith, California: Pacific Geol., v. 8, p. 79-89.

Bateman, P.C., and Busacca, A.J., 1982, Geologic map of the Millerton Lake quadrangle, west-central Sierra Nevada, California: U. S. Geol. Survey Geol. Quadrangle Map GQ-1548, scale 1:62,500.

Bateman, P.C., and Krauskopf, K.B., 1987, Geologic map of the El Portal quadrangle, west-central Sierra Nevada, California: U. S. Geol. Survey Misc. Field Studies Map MF-1998, scale 1:62,500.

Bateman, P.C., Lockwood, J.P., and Lydon, P.A., 1971, Geologic map of the Kaiser Peak quadrangle, central Sierra Nevada, California: U.S. Geol. Survey Geol. Quadrangle Map GQ-894, scale 1:62,500.

Bateman, P.C., Busacca, A.J., Marchand, D.E., and Sawka, W.N., 1982, Geologic map of the Raymond quadrangle, Madera and Mariposa Counties, California: U. S. Geol. Survey Geol. Quadrangle Map GQ-1555, scale 1:62,500.

Bateman, P.C., Busacca, A.J., and Sawka, W.N., 1983, Cretaceous deformation in the western foothills of the Sierra Nevada, California: Geol. Soc. America Bull., v. 94, p. 30-42.

Bateman, P.C., Kistler, R.W., and Peck, D.L., 1983, Geologic map of the Tuolumne Meadows quadrangle, Yosemite National Park, California: U. S. Geol. Survey Geol. Quadrangle Map GQ-1570, scale 1:62,500.

Bateman, P.C., Harris, A.G., Kistler, R.W., and Krauskopf, K.B., 1985, Calaveras reversed: Westward younging is indicated: Geology, v. 13, p. 338-341.

Bhattacharyya, T., 1986, Tectonic evolution of the Shoo Fly formation and the Calaveras complex, central Sierra Nevada, California (Ph.D. thesis): Santa Cruz, Univ. of California.

Blake, M.C., Howell, D.G., and Jones, D.L., 1982, Preliminary tectonostratigraphic terrane map of California: U.S. Geol. Survey Open-file

Report OF-82-593.

Burchfiel, B.C., and Davis, G.A., 1981, Triassic and Jurassic tectonic evolution of the Klamath Mountains-Sierra Nevada geologic terrane, *in* Ernst, W.G., ed., The Geotectonic Development of California: Englewood Cliffs, N. J., Prentice-Hall, p. 50-70.

Busby-Spera, C.J., 1984a, The lower Mesozoic continental margin and marine intra-arc sedimentation at Mineral King, California, *in* Crouch, J.K., and Bachman, S.B., eds., Tectonics and sedimentation along the California margin: Los Angeles, Pacific Section, Society of Economic Paleontologists and Mineralogists, v. 38, p. 135-156.

_____ 1984b, Large volume rhyolite ash flow eruptions and submarine caldera collapse in the lower Mesozoic Sierra Nevada, California: Jour. Geophys. Res., v. 89, p. 8417-8428.

_____ 1985, A sand-rich submarine fan in the Lower Mesozoic Mineral King caldera complex, Sierra Nevada, California: Jour. Sed. Pet., v. 55, p. 376-391.

_____ 1986, Depositional features of rhyolitic and andesitic volcaniclastic rocks of the Mineral King submarine caldera complex, Sierra Nevada, California: Jour. Volcanol. and Geothermal Res., v. 27, p. 43-76.

Busby-Spera, C.J., and Saleeby, J.B., 1990, Intra-arc strike-slip fault exposed at batholithic levels in the southern Sierra Nevada, California: Geology, v. 18, p. 255-259.

Chen, J.H., and Tilton, G.R., 1978, Lead and strontium isotopic studies of the southern Sierra Nevada batholith, California (abs.): Geol. Soc. America Abs. with Programs, v. 10, p. 99-100.

Christensen, M.N. 1963, Structure of metamorphic rocks at Mineral King, California: Calif. Univ. Pubs. Geol. Sci., v. 42, p. 159-198.

DePaolo, D.J., 1981, A neodymium and strontium isotopic study of the Mesozoic calc-alkaline granitic batholiths of the Sierra Nevada and Peninsular Ranges, California: Jour. Geophys. Res.,v. 86, p. 10470-10488.

Dodge, F.C.W., and Calk, L.C., 1987, Geologic map of the Lake Eleanor quadrangle, central Sierra Nevada, California: U.S. Geol. Survey Geol. Quadrangle Map GQ-1639, scale 1:62,500.

DuBray, E.A., and Dellinger, D.A., 1980, Geologic, aeromagnetic, and geochemical anomaly maps of the Kaiser Ridge Wilderness, central Sierra Nevada, California: U.S. Geol. Survey Misc. Field Studies Map MF-1181, scale 1:62,500.

Dunne, G.C., 1989, The Kern plateau synbatholithic shear zone, southern Sierra Nevada, California (abs.): EOS, v. 70, p. 1308.

Dunne, G.C., Suczek, C.A., Dybel, B.E., Godwin, M., Kofoed, J.C., Lindquist, J.C., and Swanson, B., 1988, Newly recognized Paleozoic eugeoclinal strata in the southern Sierra Nevada, California (abs.): Geol. Soc. America Abs. with Programs, v. 20, p. A149.

Durrell, C., 1940, Metamorphism in the southern Sierra Nevada northeast of Visalia, California: Calif. Univ. Pubs. Geol. Sci. Bull., v. 24, p. 1-118.

Edelman, S.H., and Sharp, W.D., 1989, Terranes, early faults, and pre-Late Jurassic amalgamation of the western Sierra Nevada metamorphic belt, California: Geol. Soc. America Bull., v. 101, p. 1420-1433.

Frazier, M., Stevens, C.H., Berry, W., Smith, B.M.,

and Varga, R., 1986, Relationship of the Sierran Coyote Creek pendant to the adjacent Inyo Mountains, east-central California (abs.): Geol. Soc. America Abs. with Programs, v. 16, p. 106.

Gastil, G., 1986, Terranes of peninsular California and adjacent Sonora, *in* Howell, D.G., ed., Tectonostratigraphic terranes of the circum-Pacific region: Houston, Circum-Pacific Council for Energy and Mineral Resources, p. 273-284.

Geraci, J., Fischer, P., and Dunne, G., 1987, Indian Wells Canyon roof pendant, SE Sierra Nevada, California: Another fragment of the Paleozoic cordilleran eugeocline? (abs.): Geol. Soc. America Abs. with Programs, v. 19, p. 381.

Girty, G.H., 1977a, Multiple regional deformation and metamorphism of the Boyden Cave roof pendant, central Sierra Nevada, California (M. A. thesis): Fresno, California State University, 82p.

_____ 1977b, Cataclastic rocks in the Boyden Cave roof pendant, central Sierra Nevada, California (abs.): Geol. Soc. America Abs. with Programs, v. 9, p. 423.

_____ 1985, Shallow marine deposits in Boyden Cave roof pendant, west-central Sierra Nevada: Calif. Geol., v. 38, p. 51-55.

Greene, D.C., Schweickert, R.A., and Strobel, R.J., 1989, Possible westward continuation of the Roberts Mountains thrust in the northern Ritter Range pendant (NRP), eastern Sierra Nevada, California (abs.): Geol. Soc. America Abs. with Programs, v. 21, p. 86-87.

Hamilton, W., 1956, Geology of the Huntington Lake area, Fresno County, California: Calif. Div. Mines Special Report 46.

Huber, N.K., 1968, Geologic map of the Shuteye Peak quadrangle, Sierra Nevada, California: U.S. Geol. Survey Geol. Quadrangle Map GQ-728, scale 1:62,500.

_____ 1982, Geologic map of the Mount Raymond roadless area, central Sierra Nevada, California: U.S. Geol. Survey Misc. Field Studies Map MF-1417-A, scale, 1:62,500.

Huber, N.K., Bateman, P.C., and Wahrhaftig, C., 1989, Geologic map of Yosemite National Park and vicinity, California: U.S. Geological Survey Misc. Investigations Map I-1874, scale 1:125,000.

Jones, D.L., and Moore, J.G., 1973, Lower Jurassic ammonite from the south-central Sierra Nevada, California: Jour. Research U. S. Geol. Surv., v. 1, p. 453-458.

Kistler, R.W., 1966, Geologic map of the Mono Craters quadrangle, Mono and Tuolumne Counties, California: U.S. Geol. Survey Geol. Quadrangle Map GQ-462, scale 1:62,500.

_____ 1978, Mesozoic paleogeography of California: A viewpoint from isotope geology, *in* Howell, D.G., and McDougall, K.A., eds., Mesozoic Paleogeography of the Western United States: Los Angeles, Pacific Section, Society of Economic Paleontologists and Mineralogists, Pacific Coast Paleogeography Symposium 2, p. 277-282.

_____ 1990, Two different lithosphere types in the Sierra Nevada, California: Geol. Soc. America Mem. 174, p. 271-281.

Kistler, R.W., and Bateman, P.C., 1966, Stratigraphy and structure of the Dinkey Creek roof pendant in the central Sierra Nevada,

674

California: U.S. Geol. Surv. Prof. Paper 524-B, p. B1-B14

Kistler, R.W., and Peterman, Z.E., 1973, Variations in Sr, Rb, K, Na, and initial $^{87}Sr/^{86}Sr$ in Mesozoic granitic rocks and intruded wall rocks in central California: Geol. Soc. America Bull., v. 84, p. 3489-3512.

Kistler, R.W., Robinson, A.C., and Fleck, R.J., 1980, Mesozoic right-lateral fault in eastern California (abs.): Geol. Soc. America Abs. with Programs, v. 12, p. 115.

Knopf, A., and Thelen, P., 1905, Sketch of the geology of Mineral King, California: Calif. Univ. Pubs. Geol. Sci. Bull., v. 4, p. 227-262.

Krauskopf, K.B., 1953, Tungsten deposits of Madera, Fresno, and Tulare Counties, California: Calif. Div. Mines and Geol. Special Report 35, 83p.

_____ 1985, Geologic map of the Mariposa quadrangle, Mariposa and Madera Counties, California: U.S. Geol. Survey Geol. Quadrangle Map GQ-1586, scale 1:62,500.

Lahren, M.M., 1989, Tectonic studies of the Sierra Nevada: Structure and stratigraphy of miogeoclinal rocks in Snow Lake pendant, Yosemite-Emigrant wilderness; and TIMS analysis of the Northern Sierra terrane (Ph.D. thesis): Reno, University of Nevada, 260p.

Lahren, M.M., and Schweickert, R.A., 1989, Proterozoic and Lower Cambrian miogeoclinal rocks of Snow Lake pendant, Yosemite-Emigrant Wilderness, Sierra Nevada, California: Evidence for major Early Cretaceous dextral translation: Geology, v. 17, p. 156-160.

Lahren, M.M., Schweickert, R.A., Mattinson, J.M., and Walker, J.D., 1990, Evidence of uppermost Proterozoic to Lower Cambrian miogeoclinal rocks and the Mojave-Snow Lake fault: Snow Lake pendant, central Sierra Nevada, California: Tectonics, v. 9, p. 1585-1608.

Lockwood, J.P. and Bateman, P.C., 1976, Geologic map of the Shaver Lake quadrangle, central Sierra Nevada, California: U.S. Geol. Survey Geol. Quadrangle Map GQ-1271.

MacDonald, G.A., 1941, Geology of the western Sierra Nevada between the Kings and San Joaquin Rivers, California: Calif. Univ. Pubs. Geol. Sci. Bull., v. 26, p. 215-285.

Merguerian, C., 1985, Stratigraphy, structural geology, and tectonic implications of the Shoo Fly Complex and the Calaveras-Shoo Fly thrust, central Sierra Nevada, California (Ph. D. thesis): New York, Columbia University, 255p.

Merguerian, C., and Schweickert, R.A., 1987, Paleozoic gneissic granitoids in the Shoo Fly Complex, central Sierra Nevada, California: Geol. Soc. America Bull., v. 99, p. 699-717.

Merritt, N.J., 1985, The Dinkey Creek pendant, central Sierra Nevada, California: A pre-Nevadan(?) deep crustal shear zone (abs.): Geol. Soc. America Abs. with Programs, v. 17, p. 369.

_____ in prep., Polyphase deformation of possible Proterozoic to Lower Cambrian miogeoclinal rocks of the Dinkey Creek pendant, central Sierra Nevada, California (M. S. thesis): Reno, University of Nevada.

Miller, E.L., 1978, The Fairview Valley Formation: A Mesozoic intraorogenic deposit in the southwestern Mojave Desert, *in* Howell, D.G., and McDougall, K.A., eds., Mesozoic Paleogeography of the Western United States:

Los Angeles, Pacific Section, Society of Economic Paleontologists and Mineralogists, Pacific Coast Paleogeography Symposium 2, p. 277-282.

Miller, W.J., and Webb, R.W., 1940, Descriptive geology of the Kernville quadrangle, California: Calif. Div. Mines Jour. Mines and Geol., v. 36, p. 343-378.

Moore, J.G.,1963, Geology of the Mount Pinchot quadrangle, southern Sierra Nevada, California: U.S.Geol. Survey Bull. 1130, 152p.

Moore, J.G., and Dodge, F.C.W., 1962, Mesozoic age of metamorphic rocks in the Kings River area, southern Sierra Nevada, California: U. S. Geol Survey Prof. Paper 450B, p. B19-B21.

Moore, J.G., and Marks, L.Y., 1972, Mineral resources of the High Sierra primitive area, California: U.S. Geol. Survey Bull. 1371-A, 40p.

Moore, J.G., Nokleberg, W.J., Chen, J.H., Girty, G.H., and Saleeby, J., 1979, Geologic guide to the Kings Canyon highway, central Sierra Nevada, California: Cordilleran Section, Geol. Soc. America, 75th Ann. Mtg., 33p.

Moore, J.N., and Foster, C.T., Jr., 1980, Lower Paleozoic metasedimentary rocks in the east-central Sierra, California: Geol. Soc. America Bull., v. 91, p. 37-43.

Nokleberg, W.J., 1981, Stratigraphy and structure of the Strawberry Mine roof pendant, central Sierra Nevada, California: U.S. Geol. Survey Prof. Paper 1154, 18p.

_____ 1983, Wallrocks of the central Sierra Nevada batholith, California: A collage of accreted tectono-stratigraphic terranes: U.S. Geol. Survey Prof. Paper 1255, 28p.

Nokleberg, W.J., and Kistler, R.W., 1980, Paleozoic and Mesozoic deformations in the central Sierra Nevada, California: U. S. Geol. Survey Prof. Paper 1145, 24p.

Peck, D.L., 1980, Geologic map of the Merced Peak quadrangle,central Sierra Nevada, California: U.S. Geol. Survey Geol. Quadrangle Map GQ-1531, scale 1:62,500.

Rinehart, C.D., and Ross, D.C., 1964, Geology and mineral deposits of the Mount Morrison quadrangle, Sierra Nevada, California, U.S. Geol. Survey Prof. Paper 385, 106p.

Robinson, A.C., and Kistler, R.W., 1986, Maps showing isotopic dating in the Walker Lake 1° by 2° quadrangle, California and Nevada: U.S. Geol. Survey Misc. Field Studies Map MF-1382N, scale 1:250,000.

Rose, R.L., 1957a, Geology of the May Lake area, Yosemite National Park (Ph.D. dissertation): Berkeley, Univ. of California, 215p.

_____ 1957b, Andalusite- and corundum-bearing pegmatites in Yosemite National Park, California: American Mineralogist, v. 42, p. 635-647.

_____ 1958, Metamorphic rocks of the May Lake area, Yosemite Park, and a metamorphic facies problem (abs.): Geol. Soc. America Bull., v. 69, p. 1703.

Ross, D.C., 1958, Igneous and metamorphic rocks of parts of Sequoia and Kings Canyon National Parks, California: Calif. Div. Mines Spec. Report 53, 24p.

_____ 1987, Metamorphic framework rocks of the southern Sierra Nevada, California: U. S. Geol. Survey Open File Report OF-87-81, 74p.

Russell, L.R., and Cebull, S.E., 1977, Structural-metamorphic chronology in a roof pendant near Oakhurst, California: Implications for tectonics of the western Sierra Nevada: Geol. Soc. America Bull., v. 88, p. 1530-1534.

Saleeby, J., 1978a, Fieldguide for the Kings-Kaweah ophiolitic melange, southwest Sierra Nevada, California: Field trip guide, Geological Society of America Penrose Conference on Melanges, April, 22p..

———— 1978b, Kings River ophiolite, southwest Sierra Nevada foothills, California: Geol. Soc. America Bull., v. 89, p. 617-636.

———— 1979, Kaweah serpentinite melange, southwest Sierra Nevada foothills, California: Geol. Soc. America Bull., v. 90, p. 29-46.

———— 1981, Ocean floor accretion and volcanoplutonic arc evolution of the Mesozoic Sierra Nevada, in Ernst, W.G., ed., The Geotectonic Development of California: Englewood Cliffs, N.J., Prentice-Hall, p. 132-182.

———— 1990, Progress in tectonic and petrogenetic studies in an exposed cross-section of young (100 Ma) continental crust, southern Sierra Nevada, California, in Salisbury, M.H., and Fountain, D.M., eds., Exposed cross-sections of the continental crust: The Netherlands, Kluwer Academic Publishers, p. 137-158.

Saleeby, J. and Chen, J.H., 1978, Preliminary report on initial lead and strontium isotopes from ophiolitic and batholithic rocks, southwestern foothills, Sierra Nevada, California: U.S. Geol. Survey Open-file Report OF-78-701.

Saleeby, J.B., and Busby-Spera, C., 1986, Stratigraphy and structure of metamorphic framework rocks Lake Isabella area, southern Sierra Nevada, in Dunne. G.C., ed., Mesozoic and Cenozoic structural evolution of selected areas, east-central California: Los Angeles, Geol. Soc. America, Cordilleran Section, meeting guidebook, trip 14, p. 81-94.

Saleeby, J.B., Sams, D.B., and Kistler, R.W., 1987, U/Pb zircon, strontium, and oxygen isotopic and geochronological study of the southernmost Sierra Nevada batholith, California: Jour. Geophys. Res.,v. 92, p. 10,443-10,466.

Saleeby, J.B., Goodin, S.E., Sharp, W.D., and Busby, C.J., 1978, Early Mesozoic paleotectonic-paleogeographic reconstruction of the southern Sierra Nevada region, in Howell, D.G., and McDougall, K.A., eds., Mesozoic Paleogeography of the Western United States: Los Angeles, Pacific Section, Society of Economic Paleontologists and Mineralogists, Pacific Coast Paleogeography Symposium 2, p. 311-336.

Saleeby, J., Kistler, R.W., Longiaru, S., Moore, J.G., and Nokleberg, W.J., 1990, Middle Cretaceous silicic metavolcanic rocks in the Kings Canyon area, central Sierra Nevada, California: Geol. Soc. America Mem. 174, p. 251- 270.

Saleeby, J.B., Speed, R.C., Blake, M.C., Allmendinger, R.W., Gans, P.B., Kistler, R.W., Ross, D.C., Stauber, D.A., Zoback, M.L., Griscom, A., and McCulloch, D.S., 1986, Centennial continent/ocean transect #10, C-2 central California offshore to Colorado Plateau: Geol. Soc. America, 63p.

Schweickert, R.A., 1981, Tectonic evolution of the Sierra Nevada Range, in Ernst, W.G., ed., The Geotectonic Development of California: Englewood Cliffs, N.J., Prentice-Hall, p. 87-131.

Schweickert, R.A., and Bogen, N.L., 1983, Tectonic transect of Sierran Paleozoic through Jurassic accreted belts: Pacific Section, Soc. Econ. Paleontologists and Mineralogists, 22p.

Schweickert, R.A., and Lahren, M.M., 1987, Continuation of Antler and Sonoma orogenic belts to the eastern Sierra Nevada, California, and Late Triassic thrusting in a compressional arc: Geology, v. 15, p. 270-273.

———— 1990, Speculative reconstruction of a major Early Cretaceous(?) dextral fault zone in the Sierra Nevada: Implications for Paleozoic and Mesozoic orogenesis in the western United States: Tectonics, v. 9, p. 1609-1630.

———— 1991, Preliminary interpretation of metamorphic rocks between Sonora Pass, California, and the Wassuk Range, Nevada: Implications for oroclinal bending (abs.): Geol. Soc. America Abs. with Programs, v. 23.

Schweickert, R.A., Saleeby, J., Tobisch, O.T., and Wright, W.H., III, 1977, Paleotectonic and paleogeographic significance of the Calaveras Complex, western Sierra Nevada, California, in Stewart, J.H., Stevens, C.H., and Fritsche, A.E., eds., Paleozoic paleogeography of the United States: Los Angeles, Pacific Section, Society of Economic Paleontologists and Mineralogists, Pacific Coast Paleogeography Symposium 1, p. 381-394.

Schweickert, R.A., Merguerian, C., and Bogen, N.L., 1988, Deformational and metamorphic history of Paleozoic and Mesozoic basement terranes in the western Sierra Nevada metamorphic belt, in Ernst, W.G., ed., Metamorphism and Crustal Evolution of the Western United States: Englewood Cliffs, N.J., Prentice-Hall, p. 789-822.

Sharp, W.D., 1988, Pre-Cretaceous crustal evolution in the Sierra Nevada region, California, in Ernst, W.G., ed., Metamorphism and Crustal Evolution of the Western United States: Englewood Cliffs, N.J., Prentice-Hall, p. 823-864.

Shaw, H.F., Chen, J.H., Saleeby, J.B., and Wasserburg, G.J., 1987, Nd-Sr-Pb systematics and age of the Kings River ophiolite, California: Implications for depleted mantle evolution: Contr. Mineral. Petrol., v. 96, p. 281-290.

Stewart, J.H., 1970, Upper Precambrian and Lower Cambrian strata in the southern Great Basin, California and Nevada: U.S. Geol. Survey Prof. Paper 620, 206p.

Stone, P., and Stevens, C.H., 1988, Pennsylvanian and Early Permian paleogeography of east-central California: Implications for the shape of the continental margin and the timing of continental truncation: Geology, v. 16, p. 330-333.

Strobel. R.J., 1986, Stratigraphy and structure of the Paleozoic rocks in the Rush Creek drainage, Northern Ritter Range pendant, California (M.S. thesis): Reno, University of Nevada, 123p.

Walker, J.D., 1988, Permian and Triassic rocks of the Mojave Desert and their implications for timing and mechanisms of continental truncation: Tectonics, v. 7, p. 685-709.

676

Walker, M.G., 1980, Petrogenesis of the Tuolumne
    Peak pendant and its border igneous intrusives
    central Sierra Nevada, California (B.S.
    thesis): Arcata, California, Humboldt State
    University, 86p.
Willahan, D.E., 1990, Biostratigraphy of upper
    Paleozoic rocks in the Mount Morrison roof
    pendant, Sierra Nevada: Evidence for its
    original paleogeographic position (abs.):
    Geol. Soc. America Abs. with Programs, v. 22,
    p. 94.

EARLY PALEOZOIC EUGEOCLINAL STRATA IN THE KERN PLATEAU PENDANTS, SOUTHERN SIERRA NEVADA, CALIFORNIA

George C. Dunne
Dept. of Geological Sciences
California State University
Northridge, California 91326

Christopher A. Suczek
Department of Geology
Western Washington University
Bellingham, Washington 98225

ABSTRACT

Exposed across the Kern Plateau of the southern Sierra Nevada is a northwest-trending belt of roof pendants at least 100 km long. Although no age-diagnostic fossils have been recovered from these greenschist- to amphibolite-grade rocks, most previous mapping efforts in this region inferred pendant strata to be of Mesozoic age. Our study of these pendants shows that all of them are composed of a distinctive assemblage of lithosomes that is unlike Mesozoic assemblages in the region. The predominant lithosome forming the pendants, which we refer to as the 'host' lithosome, is composed of thin-bedded, very fine-grained, carbonaceous argillite, sandy argillite, and quartz sandstone. Interstratified with host lithosome rocks are less abundant 'marker' lithosomes including very pure quartz arenite, limestone and arenaceous limestone, basalt, conglomerate, amphibolite of uncertain protolith, and sulfide-rich stratiform barite. Resting unconformably(?) on the foregoing strata is an equally metamorphosed 'overlap' lithosome consisting of conglomerate, quartz arenite, calc-silicate rock and felsic tuff. Pendant strata below this overlapping conglomerate are interpreted to have been deposited mostly as turbidites and other gravity flows in oxygen-poor water on the continental rise or slope and to be lithostratigraphically correlative with continental rise and slope strata of Ordovician, Devonian and Mississippian age exposed both in the El Paso Mountains to the south and in the Roberts Mountains allochthon of Nevada. The overlapping conglomerate unit was deposited in shallower water near or in a volcanic arc; we provisionally correlate this lithosome with the Early Permian Bond Buyer sequence in the El Paso Mountains. Recognition of early Paleozoic continental slope or rise strata in Kern Plateau pendants greatly expands the region within which such strata are recognized and suggests that other poorly studied pendants in the southern Sierra may contain pre-Mesozoic strata as well.

INTRODUCTION

The original location and shape of the Paleozoic continental margin of the southwestern U.S. Cordillera have been the subjects of considerable speculation during the past 25 years. By the mid-1960's (e.g.,

Ross, 1966), it was well established that structurally juxtaposed miogeoclinal[1] and eugeoclinal Paleozoic facies belts that are coincident with the continental margin trend southwestward across Nevada and into the eastern fringe of the Sierra Nevada magmatic arc in eastern California. Hamilton and Myers (1966) were the first to note that these Paleozoic facies belts did not reappear on the west side of the arc. Subsequently, there occurred the discovery in Sonora, Mexico by Poole and Hayes (1971) of Paleozoic miogeoclinal strata similar to those in eastern California, as well as the recognition in the Mojave Desert by Smith and Ketner (1970) and Burchfiel and Davis (1972) of Paleozoic eugeoclinal strata exposed much farther south than expected. These discoveries led to proposals by Burchfiel and Davis (1972), Silver and Anderson (1974), and Davis and others (1978) that the continental margin and its bounding facies belts had been left-laterally truncated in Permo-Triassic and(or) mid-Mesozoic time, resulting in the southeastward transport of large blocks of continental margin strata. Eugeoclinal strata in and south of the El Paso Mountains (fig. 1, inset map) were interpreted to lie within a relatively small, parallochthonous sliver of continental margin rocks (Davis and others, 1978). Dickinson (1977, 1981), noting common shape characteristics of modern continental margins, presented an alternative interpretation. He speculated that the trend of the margin in eastern California had a *primary* abrupt jog from southwest to southeast trend, which reflected the intersection of a southwest-trending rift margin with a southeast-trending transform fault. Subsequent studies in northwestern Mexico and southeastern California (summarized in Stewart and others, 1990, p.188-189) have been used by various workers to both support and refute the left-lateral truncation/transport hypothesis.

Resolution of the controversy regarding

---

[1] For the sake of brevity and euphony we use *miogeocline* to refer to that portion of the continental margin consisting of the continental shelf, and we use *eugeocline* to refer to that portion of the continental margin consisting of the continental slope, rise and adjacent ocean floor, without bias as to the width of the ocean basin or as to how far oceanward beneath the slope and rise thinned continental crust might extend.

In Cooper, J.D., and Stevens, C.H., eds., 1991, **Paleozoic Paleogeography of the Western United States-II:** Pacific Section SEPM, Vol. 67, p. 677-692.

the truncation hypothesis has been stymied in part by the paucity of exposures of paleo-geographically distinctive Paleozoic strata in the vicinity of the proposed truncation or bend structure. We have recognized such strata of likely Paleozoic age preserved in several pendants in the southern Sierra Nevada. In this paper we describe these pendant strata, characterize their likely depositional environment(s), suggest likely correlations for them, and speculate upon their implications for the truncation vs. bend argument, as well as for other tectonic phenomena in this portion of California.

Local Geologic and Geographic Setting

The pendants containing strata of probable Paleozoic age are located in the Kern Plateau region (fig. 1), a remnant of a broad, undulating upland of the southern Sierra that has been deeply incised by the Kern River drainage system. These pendants, which we refer to collectively as the Kern Plateau pendants, comprise a colinear, 100-km-long, northwest-trending belt across the Plateau. Three major pendants or clusters of pendants (Indian Wells, Kennedy[2], and Bald Mountain) and two smaller pendants (Hockett and Rattlesnake) compose this suite (fig. 1). The names we use for these pendants are derived from prominent nearby geographic features.

The pendants were intruded by scattered lensoidal to sill-like bodies of hornblende diorite and gabbro, hornblende quartz diorite and hornblende-biotite tonalite before being engulfed by granitoid plutons that have known or inferred ages ranging from Early Triassic to Late Cretaceous (Ross, 1987; Dunne and others, *in press*). Plutons that are predominantly of Jurassic age, and which locally contain northwest-trending mafic dikes reminiscent of the Independence dike swarm of eastern California, crop out northeast of the pendants, whereas plutons that are predominantly of Late Cretaceous age crop out southwest of the pendants. Intruded by the latter group, and structurally juxtaposed against the southwest margins of the three largest pendants, are variably mylonitized remnants of Triassic and Middle Jurassic plutons. These mylonites comprise a portion of a major early Mesozoic, synbatholithic, ductile shear zone (Dunne, 1989; Kistler, 1990; Dunne and others, *in press*).

Access to and exposure of most of these pendants are good. Numerous graded dirt roads maintained by the U.S. Forest Service and the Bureau of Land Management as well as one county-maintained paved road to Kennedy Meadows (Tulare County Road J41) provide ready access to the Indian Wells, Kennedy and Bald Mountain pendants. The smaller Hockett Peak and Rattlesnake pendants are more remote and require day-long hikes to reach.

---

[2] Kennedy pendant has also been referred to (e.g., Kistler, 1990) as Ninemile pendant, after a canyon that is more than 5 km east of the nearest point in the pendant. Because of this remoteness, we have chosen not to employ this older, little-used name.

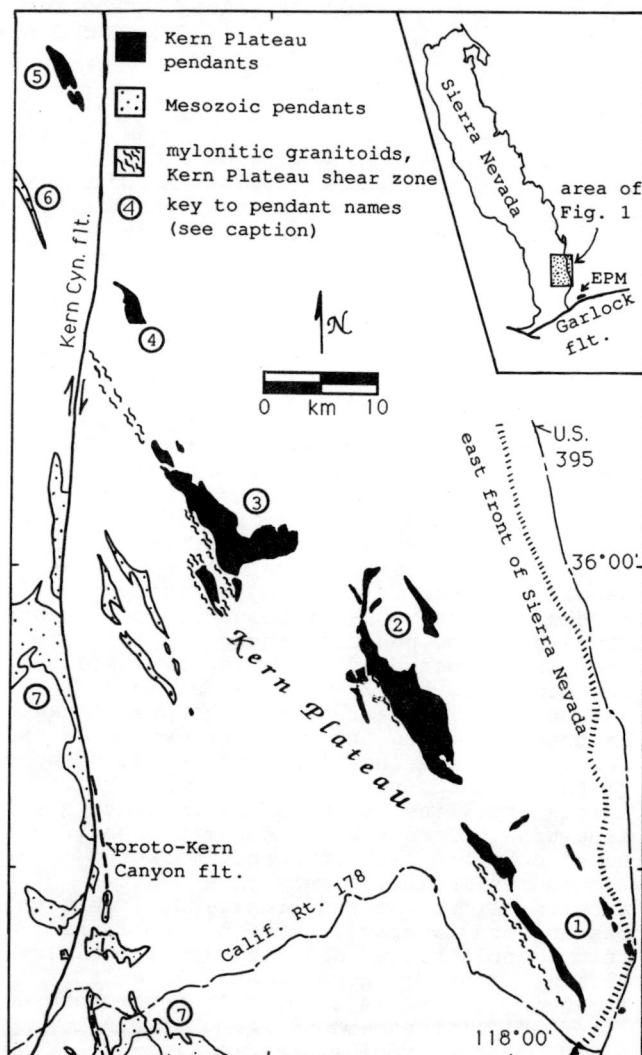

Figure 1. Location map for Kern Plateau pendants and some other geologic features of the southern Sierra Nevada. Pendants are identified by number as follows: 1=Indian Wells; 2=Kennedy; 3=Bald Mountain; 4=Hockett; 5=Rattlesnake; 6=Mineral King; 7=Lake Isabella. EPM of inset regional map = El Paso Mountains.

Previous Studies

Miller and Webb (1940) presented the first published geologic map and report that refer to some of these pendants. They assigned strata of the Kennedy and Bald Mountain pendants to their 'Kernville series', and they speculated that these strata were of Carboniferous age, even though they noted the presence of rare, recryst-allized forms reminiscent of graptolites. In his mapping of the Indian Wells pendant, Dibblee (1954) speculated that these strata might be of Paleozoic age. In regional geologic syntheses of pendants of the central and southern Sierra, Saleeby and others (1978) and Nokleberg (1983) both speculat-ively assigned pendant strata to the Kings sequence of Mesozoic age. Moore and Sisson (1985) similarly inferred that strata in the Rattlesnake pendant are of Mesozoic age. In reexaminations of all of the Kern Plateau pendants south of Rattlesnake during the 1980's, du Bray and Dellinger (1981), Bergquist and Nitkiewicz (1982), and Taylor and others (1986) assigned a Mesozoic or undivided Mesozoic/Paleozoic age to pendant strata, whereas Diggles and others (1987) assigned a Paleozoic(?) age to Indian Wells pendant strata following the earlier reasoning of Miller and Webb (1940). In his regional compilation map of the southern Sierra, Ross (1987) speculated that these pendant strata might be of Paleozoic age based on the reported occurrence of stratiform barite within them. Dunne and others (1988) and Suczek and Dunne (1989) outlined initial results of the first extended study that focused on these pendants and marshalled several lines of evidence favoring a Paleozoic age. No age-diagnostic fossils have yet been recovered from the pendants.

Metamorphic And Deformational Overprints

Strata of the Kern Plateau pendant have been thoroughly recrystallized during emplacement of surrounding plutons, and they have been multiply deformed as well. We provide a brief summary of these metamorphic and structural overprints so that the reader might more fully appreciate the limitations and uncertainties inherent in interpreting the stratigraphy and depositional environments of these rocks.

Based on metamorphic mineral assemblages in both pelitic and basic (basalt + amphibolite) rocks, we have identified the metamorphic facies of pendant strata to be low-pressure variants of the upper greenschist (biotite zone) to lower amphibolite facies (cf. Turner, 1981). Nearly all pelitic rocks contain the present mineral assemblage quartz + biotite ± muscovite ± chlorite(retrograde in origin) + graphite. Locally present within this main assemblage are albite and andalusite, and less commonly garnet and(or) cordierite. In basic rocks, the most common present mineral assemblages include actinolite(tremolite) ± albite + chlorite + epidote ± biotite ± sphene ± calcite + opaque minerals. Locally, hornblende and plagioclase are present in lieu of actinolite(tremolite) and albite respectively. Rare garnet porphyroblasts were also noted. Saleeby and others (1978) and Elan (1985) reported similar metamorphic mineral assemblages in most other pendants in the southern Sierra. Barton and Hanson (1989) provide insights as to why such facies are widespread within Sierran pendants.

A complex, superposed array of ductile and brittle structures is present in pendant strata. Bedding almost everywhere strikes northwest and dips steeply northeast, with local dip reversals near pluton contacts. The most obvious structures within the pendants are mesoscopic and rare macroscopic tight, planar-limbed, reclined folds. A pervasive northwest-trending, steeply dipping, penetrative foliation that seems to be axial planar to these folds commonly cuts bedding at angles of less than 10°. Hinge lines of most folds plunge moderately to steeply to the northwest or northeast and are commonly parallel or subparallel to a steeply plunging, penetrative linear fabric that consists of diffuse, streaky mineral lineations, stretched clasts, and stretched vesicles in basaltic rocks. Two younger sets of more open, rounded mesoscopic folds lacking axial plane cleavage are locally superimposed on this older fold set. These younger sets, the relative ages of which remain undetermined, trend northeast and northwest. Finally, a northeast-trending crenulation cleavage is locally developed in some clay-rich intervals.

We have recognized one possible thrust fault in the northeast portion of the Kennedy pendant and numerous strike-slip faults in both the Kennedy and Bald pendants. Strike-slip faults, which cause lateral separations of a few meters to a few hundred meters, comprise two -- possibly conjugate -- fault sets; faults striking ~N78°E display left separations, whereas faults striking ~N32°E display right separations. If truly conjugate, these faults may represent an episode of late Mesozoic, northeast-southwest-directed contraction of the pendants and their host granitoids that is correlative with a similar episode inferred elsewhere in the Sierra (Segall and others, 1990) and in eastern wallrocks of the Sierra (Dunne, 1986, p. 16)

In terms of macroscopic-scale structure, the pendant sections have the semblance of homoclines. Modest numbers of depositional top features suggest relatively consistent facing directions of northeast in the Indian Wells and Kennedy pendants and southwest in the Bald Mountain pendant. However, we consider this 'simple-structure' interpretation to be provisional in light of the abundance of mesoscopic structures and fabrics and the lack of paleontologic age control. It is likely that further study of these rocks will reveal greater structural complexity than we presently can document.

PENDANT STRATIGRAPHY

Introduction

We interpret strata in all of the pendants of the Kern Plateau as being generally equivalent because all are composed of similar assemblages of distinctive lithosomes, the relative abundances of which are broadly similar from pendant to pendant. Stratigraphic and sedimentologic features of strata are best preserved in the Kennedy and Bald Mountain pendants, and it is from these pendants that we have derived the bulk of our insights regarding protolith characteristics. Figure 2 presents simplified geologic maps of these two pendants and the locations of especially good exposures of most lithosomes.

We adhere to the following nomenclatural and descriptive conventions. (1) We use the term lithosome to designate a recognizable package of one or more interbedded lithologies, the sum properties of which indicate a specific or narrow range of depositional conditions. (2) We have interpreted and

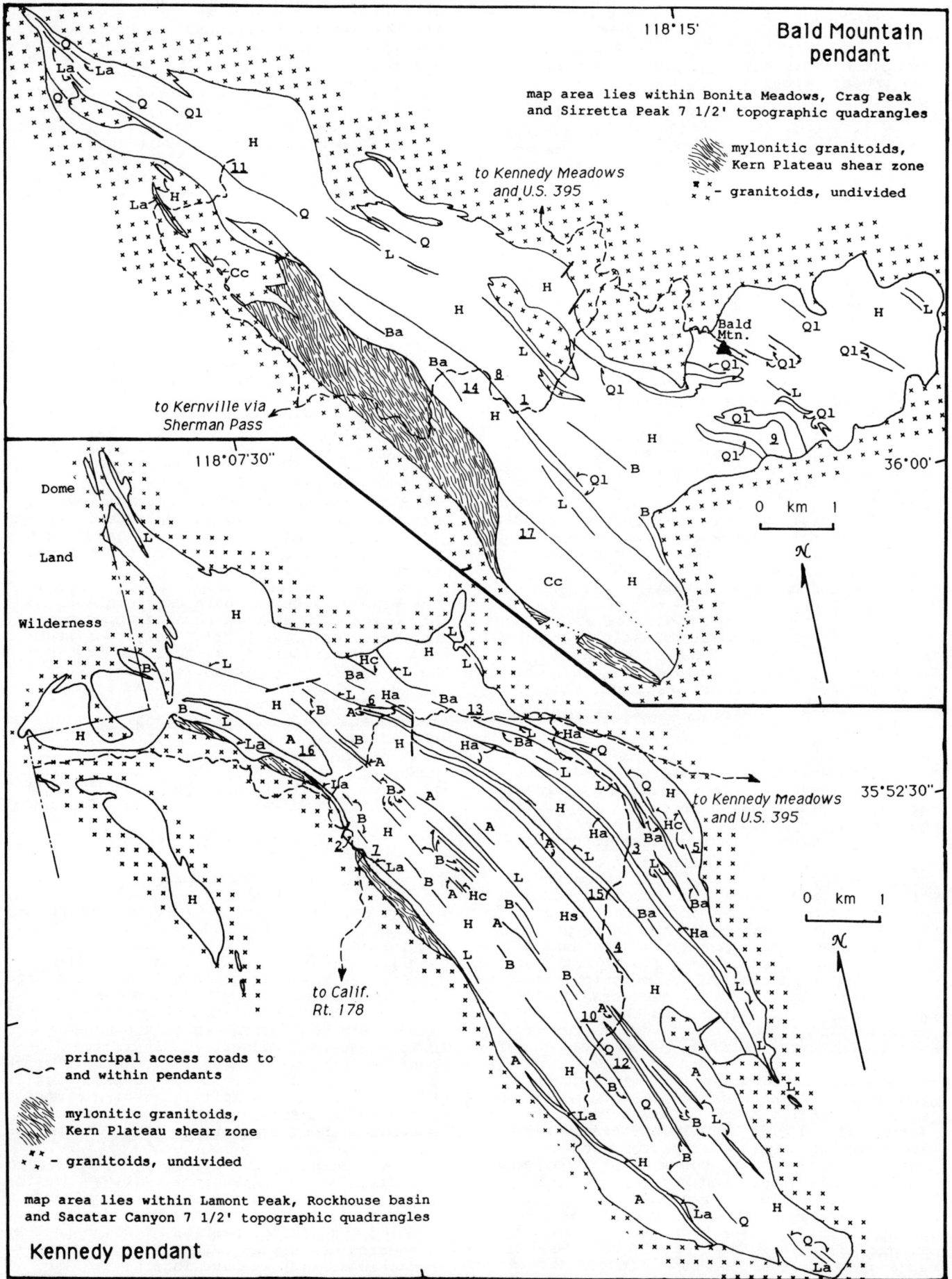

118°15'

Bald Mountain pendant

map area lies within Bonita Meadows, Crag Peak and Sirretta Peak 7 1/2' topographic quadrangles

mylonitic granitoids, Kern Plateau shear zone

x - granitoids, undivided

to Kennedy Meadows and U.S. 395

118°07'30"

to Kernville via Sherman Pass

Bald Mtn.

36°00'

0 km 1

N

Dome

Land

Wilderness

to Kennedy Meadows and U.S. 395

35°52'30"

0 km 1

N

to Calif. Rt. 178

principal access roads to and within pendants

mylonitic granitoids, Kern Plateau shear zone

x - granitoids, undivided

map area lies within Lamont Peak, Rockhouse basin and Sacatar Canyon 7 1/2' topographic quadrangles

Kennedy pendant

Figure 2. Geologic sketch maps of the main bodies of the Kennedy and Bald Mountain pendants, showing distribution of selected sublithosomes and marker lithosomes. Lateral continuity of marker lithosomes commonly inferential owing to middling quality of exposure. Underlined numbers refer to specific locations noted in the text. Lithosome codes from Table 1. Simplified from field mapping of scale 1:24000.

TABLE 1 - KERN PLATEAU PENDANT LITHOSOMES

Lithosome**, (MapCode)*	~% of Section	Protolith(s)
Host(H)	85	argillite and fine-grnd qtz. arenite; carbonaceous and locally phosphatic
cherty argillite (Ha)	1	carbonaceous cherty argillite
shale (Hs)	2	carbonaceous shale
conglomerate (Hc)	<1	mostly intraclast-pebble & rare extrabasinal quartzite conglomerate
Limestone (L)	2	limestone with variably abundant thin, siliceous interbeds
quartz arenite/ limestone (La)	1	interbedded limestone, qtz. arenite, with some 'hybrid' limy arenite
Limy quartz arenite (Ql)	1	limy quartz arenite to arenaceous limestone
Quartz arenite (Q)	1	blue-gray, massive-appearing, very pure quartz arenite
Basalt (B)	1	basaltic conglom./ breccia, with rare pillowed flows; limestone, mudstone beds;
Baritic (Ba)	1	stratiform barite, sericitized(?) clayey sandstone, all with Fe-sulfide, Mn mineralization
Amphibolite (A)	3	may derive from multiple protoliths
Argillite-pebble conglomerate (Ca)	1	argillite-pebble conglom. with rare outsized clasts of limy quartz arenite & amphibolite
Capping conglomerate (Cc)	1	subrounded to subangular quartzite- & amphibolite-pebble to cobble conglom., quartzite, and interbedded felsic tuff

* Map code refers to lithosome identifiers used on geologic maps of Fig. 2.
** Indented, lower case rock unit names refer to sublithosomes of the Host and Limestone lithosomes.

described all but one lithosome in terms of their protolith lithologies. The exception is amphibole-rich rocks of uncertain and possibly diverse origins to which we apply the name amphibolite. (3) We apply the name argillite to rocks composed of mixtures of clay and silt in which neither constituent is less abundant than 10%. (4) We use two terms for rocks composed mainly of sand: arenite for those with less than 10% matrix and sandstone for those with 10% or more matrix

We have recognized nine principal lithosomes within the Kern Plateau pendants, together with a few additional, locally mapped variants -- sublithosomes -- of some of these (Table 1). We use the term 'host' lithosome to describe the most abundant group of strata, which we interpret to represent ongoing 'background' deposition in a broadly stable, quiet, relatively unchanging depositional environment. Periodic, relatively brief and abrupt changes in this background environment led to the deposition of numerous relatively thin, laterally extensive 'marker' lithosomes that are interstratified within the host lithosome. We describe pendant lithosomes in the order in which they are listed in Table 1.

Host Lithosome

The host lithosome, composed of thin-bedded to laminated argillite, and including some sand-, chert-, shale- and conglomerate-rich variants, forms about 89% of the rocks in the pendants and is found throughout the stratigraphic section. Argillite and sandy argillite in parallel, laterally continuous laminae, very thin beds, and thin beds make up the bulk of the host lithosome. Representative exposures of such strata are present in many roadcuts (fig. 2, locs. 1, 2). Strata vary from medium to dark gray, with the latter being carbonaceous. The main components of the rocks are quartz silt and clay minerals. Phosphate blebs within beds are common. The argillite incorporates rare laminae of quartz siltstone and limy argillite and thin, graded beds of quartz sandstone. Although laterally persistent beds are characteristic, there occur low in the Bald Mountain section thin laminae of phosphatic argillite that are discontinuous and pinch out over distances of a few to tens of centimeters. Small-scale ripple cross-strata are fairly common, usually in Bouma bcd and cde sequences. Some beds contain sparse small intraclasts of carbonaceous shale and cherty argillite having maximum diameters of as much as a centimeter or two. On the east flank of Bald Mountain, a horizon of carbonaceous argillite contains burrows

oriented perpendicular to bedding planes. These, together with rare, recrystallized features reminiscent of graptolites observed in some shale beds of the host lithosome at Kennedy pendant, are the only fossils so far recognized in pendant strata.

Three distinctive variants of the host that we refer to as sublithosomes have in some areas been mapped separately. These are cherty argillite, carbonaceous shale, and conglomerate. The cherty argillite sublithosome (fig. 2, loc. 3), which makes up about 1% of pendant strata, is dark gray to black and occurs in beds 0.5 to 2 cm thick which locally contain pale-gray phosphatic blebs. Bounding surfaces of beds are locally somewhat wavy, and some have discontinuous laminae. The shale sublithosome (fig. 2, loc. 4), which makes up about 2% of pendant strata, is composed of variably carbonaceous black to greenish-gray shale, which is laminated and which less commonly contains small-scale ripples. In the southeastern part of the Bald Mountain pendant, one outcrop contains soft-sediment slump folds. The conglomerate sublithosome, which makes up less than 1% of pendant strata, has two distinct lithologic variants: (1) intraclast conglomerate composed of elongate rip-ups in an argillite or sandy argillite matrix that occurs in 0.1- to 0.5-m-thick, laterally discontinuous beds throughout the section; (2) conglomerate composed of subrounded to rounded granule-, pebble-, and rare cobble-size clasts of quartz arenite and rare argillite that occurs in 0.1- to 5-m-thick, laterally more extensive intervals that have been observed only in the upper few hundred meters of the Kennedy and Bald Mountain pendants (fig. 2, loc. 5).

Limestone Lithosome

The limestone lithosome and its quartz arenite/limestone sublithosome together compose about 3% of pendant sections. They are present at all stratigraphic levels, and they form intervals 1 to 18 m thick that are intercalated with host lithosome strata along sharp contacts. The principal lithology of the lithosome (~65% of total) is medium- to dark-gray limestone in thin to very thin beds and laminae that locally display low-angle cross bedding. Although these rocks are intensely deformed and recrystallized to marble, apparent bedding thicknesses and their association with fine-grained siliciclastic rocks suggest that they originated as fine-grained limestones, probably lime mudstone or wackestone. No variations in crystal size suggesting the presence of allochems were noted. Variable amounts of argillite and cherty argillite in laminae and thin beds are interstratified with limestone in almost all exposures (fig. 2, loc. 6). In one interval that is stratigraphically low in the Kennedy pendant, several 4- to 6-cm-thick packets of limestone and argillite display soft-sediment slump structures.

In the quartz arenite/limestone sublithosome, which composes about 35% of the limestone lithosome, very thin to medium beds of limestone may have originally been lime wackestone or packstone, judging by their thicker bedding and the coarser grained siliciclastic rocks associated with them. They are interbedded with thin to very thin beds of quartz sandstone, quartz arenite and quartz siltstone, and less commonly with limy quartz arenite (fig. 2, loc. 7). Strata of this sublithosome commonly grade laterally into limestone or arenaceous limestone.

Limy Quartz Arenite Lithosome

This lithosome is composed of about 45% limy quartz arenite which is interbedded with quartz-arenaceous limestone(18%), muddy quartz sandstone(17%), argillite(14%) and minor amounts of shale, limestone, and carbonaceous limestone. It is present in northwesternmost exposures of the Kennedy pendant and is more widespread and abundant in the Bald Mountain pendant where excellent exposures are present in roadcuts (fig. 2, loc. 8).

The medium-gray to locally rusty weathering limy quartz arenite is composed of moderately well sorted, well rounded, medium to fine quartz sand, with some very fine sand and silt and rare coarse sand. Strata are not always discernible, but in places they can be distinguished based on differential weathering, scattered shale chips along bedding planes, and rare graded beds. Most are in thin and very thin beds and laminae. Thin beds are commonly wavy and change thickness along strike; some fill small channels. Rare shale intraclasts occur near the bases of some beds. Laminae are usually planar. Low-angle cross strata in sets 5 to 10 cm thick are common, and locally climbing ripples were observed. Rare graded beds topped by laminae are Bouma ab intervals, but most Bouma sequences begin with b or c. Limy quartz arenite grades into thin laminae of muddy quartz sandstone composed of fine to very fine quartz sand that contains variable amounts of clay. This facies is fissile and nonresistant. Also associated, but commonly poorly exposed, are laminated argillite and shale in packets up to 1 m thick.

The light- to medium-gray quartz-arenaceous limestone lithology commonly occurs in gradational contact with limy quartz arenite as well as in separate, laterally extensive intervals 1 to 10 m thick. A few much thicker, broadly lensoidal intervals that may represent major channel fillings are present at the south margin of the Bald Mountain pendant (fig.2, loc.9). This lithology commonly is massive appearing to faintly bedded. Where especially well preserved, it displays flat to slightly wavy thin beds to thin laminae in packets mostly 0.5 to 3 cm thick that are separated by laminae of argillite or, uncommonly, limestone. The quartz sand is moderately well sorted and generally medium to fine grained, but in some thin to medium beds, the quartz sand is coarse. Cross-stratified sets as much as 10 cm thick are moderately common. In the well exposed section in the Bald Mountain pendant, discontinuous laminae, small-scale scours, and soft-sediment slump folds occur in one area. Some bedding surfaces are lined with shale chips. Thin beds of limestone and carbonaceous limestone are interbedded, and one locality in the Bald Mountain pendant has small chert nodules.

Quartz Arenite Lithosome

Massive-appearing, blue-gray quartz arenite is the dominant lithology in this lithosome, which forms the most striking marker horizon in the pendants. It is present throughout all the pendants, but especially thick and prominent examples are present high in the Kennedy and Bald Mountain pendants (fig. 2, locs. 10, 11). Outcrops are nearly continuous over lateral distances of several kilometers. Contacts commonly are covered by large blocks of float, but where visible are sharp against the host lithosome.

Rusty-weathering, massive-appearing, thick beds of well sorted fine- to medium-grained and less common coarse-grained quartz sand in quartz cement compose about 95% of this lithosome. Rare grading and clusters of intraclasts indicate bed boundaries. Composition overall averages 99% quartz, with clay making up the remainder. Beds are amalgamated and have sharp, locally scoured bases. Most beds are even and continuous, but some pinch and swell. Near the north end of the Bald Mountain pendant, amalgamated beds of coarse and very coarse sand and granules, in sequences 2 to 4 m thick, are abruptly cut out along strike and appear to be filling channels. Commonly associated with the thick beds of quartz arenite are very thin beds, laminae, and thin laminae of fine-grained quartz sand with a few percent scattered clay and clay partings. The common association of a set of laminae with a thin set of cross stratification is suggestive of Bouma b,c intervals of turbidites. In some outcrops, a base of structureless, ungraded quartz arenite probably represents Bouma a. Bouma d laminae are rare in this lithosome, and Bouma e intervals have not been recognized

Basalt Lithosome

Strata of the basalt lithosome are present in the three largest pendants, but they are most abundant and best preserved in the Kennedy pendant. The basalt lithosome occurs in intervals that range in thickness from 2 to 30 m, and these occur both as single horizons and as clusters of basalt horizons interstratified with host lithosome strata. Many individual basalt sequences can be traced laterally for less than 0.5 km before they pinch out, but commonly another, overlapping basalt sequence positioned slightly higher or lower stratigraphically continues on.

Basalt takes the form both of basalt-clast conglomerate and breccia (90% of lithosome) as well as pillowed to rubbly lava flows. Basalt-clast conglomerate and breccia is most abundant in the Kennedy pendant where it occurs as laterally extensive individual beds and as clusters of beds separated by thin sequences of limestone or argillite (fig. 2, loc. 12). Clasts, which range in size up to 0.75 m but average 3 cm in diameter, are typically subrounded to subangular. Cross sections of some larger clasts reveal concentric texture zones suggestive of pillows. Most beds are composed mostly or entirely of basalt clasts having a variety of vesicular and massive textures, but with rare clasts of argillite locally mixed in. A few beds composed mostly of argillite with rare limestone clasts were also recognized. Clast sorting ranges from good in some beds to very poor in others, the later state giving rise to the term breccia where clasts are also less rounded. Almost all the conglomerates and breccias are clast supported, with interstices filled most commonly with fine basaltic material, but locally with argillite or limestone.

Basalt lava flows are a few to several meters thick. Basalt is variably vesicular; vesicles are most commonly 2 to 5 mm in diameter and mostly calcite-filled. The rubbly portions of these flows are commonly set in a limestone matrix; they may represent flow-margin rubble and(or) discrete lenses of separately derived breccia. Thin, discontinuous limestone veneers commonly rest upon the rare pillowed sequences. Thin sections of basalt reveal that little of the original mineralogy or texture is preserved. Rare phenocrysts of plagioclase up to 2 mm are present, and mafic clots that might be remnants of hornblende and/or pyroxene phenocrysts are present in a few samples.

Eight samples of basalt from the Kennedy and Indian Wells pendants were chemically analyzed (Table 2). With the exception of some relatively low silica values, major element abundances are broadly similar to ocean island basalts (cf. Engel and others, 1965).

Baritic Lithosome

We use the name baritic lithosome for an assemblage of related lithologies and a spatially related alteration overprint. This lithosome is present at two stratigraphic levels in the Kennedy pendant, but at only one level in the Bald Mountain pendant (fig. 2, locs. 13, 14). The two most prominent baritic intervals can be traced semi-continuously for distances of 4.5 and 2.5 km respectively. Along these reaches, the thickness of the lithosome varies in gentle pinch-and-swell fashion, but it most commonly has a thickness of 5 to 40 m. This lithosome is composed of variably clay-rich fine quartz sandstone(65%), stratiform barite(30%), limestone(4%) that is locally barite bearing, and rare, thin amphibolite beds(1%).

Variably clay-rich (now mostly muscovite) fine- to medium-grained quartz sandstone occurs in diffusely bounded beds 1 to 6 cm thick that contain faint to prominent internal lamination with typical spacing of a few to several millimeters. Laminations are defined by varying abundances of muscovite and by discontinuous laminae of pyrite. Intervals of this rock are in sharp contact with barite beds and provide the dominant and distinctive host lithology within which most barite beds are interstratified. Colors for this rock are commonly yellowish gray to very pale orange; less pyritized samples are very pale gray. Stratiform barite typically occurs as sharply bounded beds which vary in thickness from 10 to 120 cm in gentle pinch-and-swell fashion, locally pinching out entirely over lateral distances of several tens of meters. Colors for freshly broken barite are essentially the same as for the enclosing clayey quartz sandstone. Many beds contain diffuse, parallel, darker color bands that are 0.2 to 2 cm thick and marked by concentrations of pyrite and (or) manganese oxides. The number of barite beds at a given exposure ranges from zero to six, with three or four beds being most common. Quartz sand grains are present in all barite samples we examined. In the purest barite samples these grains constitute only a few percent of the rock and are scattered irregularly through it, but such beds commonly grade vertically into quartz-dominated mixtures in which quartz and barite are arranged in diffuse layers parallel to bedding. Limestone beds

TABLE 2 - GEOCHEMICAL DATA FOR VOLCANIC ROCKS, KERN PLATEAU PENDANTS

| | Basaltic Lithosome | | | | | | | | Tuff in Capping Conglomerate | Lithosome |
| | Kennedy Pendant | | | | Indian Wells Pendant | | | | | |
Sample	D11a-7	D49-7	Dx-7	39-5	G3-6	G59c-6	G115-7	G132-7	Da-7	D16-7
SiO_2	51.0	46.6	41.3	43.8	40.2	47.9	38.6	50.7	65.6	73.8
TiO_2	3.14	2.70	3.76	3.44	3.20	2.18	3.76	2.41	0.30	0.15
Al_2O_3	15.7	14.6	15.9	16.1	10.6	13.1	13.6	13.7	17.8	13.2
Fe_2O_3	9.65	11.8	14.2	9.52	12.1	11.5	10.7	9.96	2.67	1.81
MnO	0.16	0.17	0.13	0.11	0.18	0.17	0.19	0.19	0.08	0.03
MgO	5.87	8.10	5.55	3.90	10.3	7.45	6.04	7.14	0.63	0.14
CaO	6.47	9.00	5.80	9.17	15.6	11.9	17.5	8.79	2.37	1.04
Na_2O	4.51	2.39	2.00	2.67	0.56	3.22	2.03	3.80	5.24	3.81
K_2O	1.57	1.91	5.62	4.06	1.46	0.48	0.60	0.39	5.52	5.35
P_2O_5	0.74	0.39	0.92	0.52	0.53	0.24	0.78	0.23	0.06	0.05
LOI	0.62	0.70	3.47	4.77	1.00	0.70	3.85	0.77	1.15	0.39
Sum:	99.7	99.5	99.2	98.3	96.3	98.9	97.8	98.2	99.2	99.7
Rb (ppm)	31	33	70	48	37	19	24	21	158	154
Sr	570	370	290	279	453	306	737	354	262	63
Ba	830	8170	4330	1780	3870	334	447	661	437	190
Cs	3	4	4	1	3	1	1	1	2	<1
Y	29	25	35	30	27	23	38	26	55	40
Zr	370	170	450	239	245	104	354	162	309	235
Nb	80	60	110	42	57	17	85	23	21	15
Hf	1	1	1	1	6	3	8	1	2	1
Ta	6	3	9	4	5	2	6	2	2	2
Th	6.3	3.4	9.5	3.8	6.8	1.8	8.8	2.9	22.3	15.6
U	0.8	0.7	0.9	0.7	1.5	0.7	1.9	0.5	3.1	1.8
La	69.3	30.3	80.7	39.5	57.0	18.5	69.7	30.2	60.3	47.7
Ce	131.0	59.1	153.0	81.5	105.0	38.8	133.0	58.6	126.0	92.6
Pr	17.0	8.1	19.0	10.1	12.9	5.3	16.5	7.5	14.9	11.2
Nd	58.0	33.2	67.8	43.7	52.5	24.2	65.6	31.3	60.0	44.9
Sm	10.0	6.7	11.3	8.2	8.6	5.4	11.7	6.9	11.3	8.3
Eu	4.01	9.65	7.02	3.08	3.40	2.26	3.97	2.53	1.32	0.63
Gd	9.8	7.6	11.3	8.3	8.2	5.7	10.3	6.7	11.1	8.1
Tb	1.5	1.2	1.9	1.1	1.0	0.8	1.4	0.8	1.6	1.1
Dy	6.5	5.6	8.0	6.2	6.1	4.7	7.7	5.2	10.4	7.4
Ho	1.18	0.93	1.38	1.14	1.06	1.95	1.46	1.00	2.18	1.60
Er	2.4	1.9	2.9	2.4	2.7	2.4	3.5	2.8	6.1	4.4
Tm	0.3	0.2	0.3	0.2	0.3	0.3	0.4	0.3	0.8	0.5
Yb	2.0	1.9	2.1	1.7	2.0	1.8	2.8	2.1	6.0	4.1
Lu	0.38	0.44	0.41	0.25	0.36	0.34	0.42	0.38	0.86	0.56

All analyses by X-Ray Assay Labs, Inc. All samples agate milled.

are present within the baritic lithosome in the Kennedy pendant, and they commonly mark the approximate top and(or) bottom of the baritic lithosome. At the southeasternmost exposure of the baritic lithosome at Kennedy, baritic limestone is the most abundant lithology present. Where least recrystallized and mineralized, limestone consists of 0.5- to 3-m-thick, light- to medium-gray beds with variably developed color and(or) grain-size banding. Samples examined in thin section contain 3 to 10% sand grains, and most also contain a few to almost 50% barite grains.

Imposed upon the baritic lithosome in all outcrops is a diagnostic overprint of mineralization comprised of abundant pyrite and of coatings of manganese oxides, both commonly concentrated along bedding planes.

We suspect that the muscovite-rich quartz sandstone lithology of the baritic lithosome -- found only rarely elsewhere in the pendants -- may also be a product of mineralization, perhaps having originated from intense sericitization of clayey quartz sandstone. We did not see any obvious feeder veins or any asymmetry of mineralization development that would indicate that mineralizing fluids had moved upward into the baritic lithosome exposures through substrate strata. Taylor and others (1986) found no significant concentrations of sulfides other than pyrite in or around exposures of this lithosome.

Amphibolite Lithosome

We have mapped a number of intervals of

amphibole-rich rocks. Metamorphic reconstitution of these rocks has been so thorough that protoliths cannot readily be determined; thus we refer to this lithosome in terms of its present metamorphic lithology. Amphibolites are most abundant in the Kennedy pendant where as many as 12 different intervals, ranging in thickness from 2 to 80 m, are present in some cross-pendant traverses. Some amphibolite exposures are distinctly stratiform, have faint remnants of what may be bedding, can be traced laterally for as much as 2 km, and locally contain rounded pebbles of other lithologies (fig. 2, loc. 15). At least some of these horizons almost certainly had a sedimentary protolith. Less common, thicker, in part lensoidal, amphibolite bodies (fig. 2, loc. 16) may represent mafic to ultramafic intrusions or fault-bounded masses.

Amphibolite samples examined in thin section are composed predominantly of magnesium-rich minerals including, in order of decreasing abundance, tremolite, chlorite, and phlogopite. A few samples contain serpentine, talc, and rare diopside and olivine. We are presently evaluating two possible origins for the amphibolites: 1) They may have originated as argillite that was subjected both to extensive Mg-metasomatism by seafloor or sub-seafloor hydrothermal fluids originating from thermal springs (Koski and others, 1990) and to additional metasomatic alteration during regional metamorphism; 2) Some -- especially those containing olivine -- may have originated as ultramafic rocks that may have been emplaced as sedimentary detritus, as sills, or, in the case of laterally discontinuous bodies, perhaps as tectonic lenses. We are presently obtaining geochemical data for selected amphibolites in order to test these hypotheses.

Argillite-Pebble Conglomerate Lithosome

The argillite-pebble conglomerate lithosome crops out only in the Indian Wells pendant where it forms just one interval that is as much as 60 m thick and that extends laterally for approximately 2 km. It commonly rests upon typical host lithosome and locally upon limy arenite strata with a sharp, slightly discordant contact. Its upper contact is less well exposed, but it may be gradational with host strata.

Rock types composing the clasts in this medium-gray conglomerate are a close match for rock types composing the underlying strata in the Indian Wells pendant. In decreasing order of abundance, clasts are argillite and less abundant dolomitic argillite(~55% of clasts), quartz arenite(~30%), limy arenite(~10%) and amphibolite(~5%). Most clasts have maximum diameters ranging from 0.5 to 2 cm and are subrounded to rounded. Scattered through the conglomerate are rare, exceptionally large clasts that have maximum diameters of up to 35 cm and are composed of arenaceous limestone or of amphibolite. The matrix for conglomerate clasts is slightly argillaceous, slightly dolomitic to limy quartz siltstone.

Locally suspended in the matrix are scattered grains of quartz sand and very rare feldspar up to 1 mm in diameter.

Capping Conglomerate Lithosome

What we provisionally interpret as the youngest lithosome exposed in the Kennedy and Bald Mountain pendants is a distinctive conglomerate unit unlike those conglomerates that are interpreted to be stratigraphically intercalated with host strata of the pendant sequences. We refer to this as the capping conglomerate lithosome. The largest exposure of this lithosome is along the southwest margin of the Bald Mountain pendant where the conglomerate can be observed to rest with sharp apparent conformity upon strata of the host lithosome (fig. 2, loc. 17). Exposures of this same conglomerate lithosome, engulfed and thoroughly disrupted by granitoid plutons, are present near the northwestern corner of the Kennedy pendant. We think it likely that these latter conglomerate exposures once capped the Kennedy pendant section as they do the Bald Mountain section.

Conglomerate(~65% of lithosome), quartz arenite(~16%), dolomitic quartz arenite and dolomitic argillite(~16%), and felsic tuff (~3%) comprise the capping conglomerate lithosome. Conglomerate occurs in two end-member lithologies as well as gradations between these end members. One end-member lithology, which we refer to informally as the quartzitic type, is comprised predominantly of quartz arenite clasts in a quartz arenite matrix, whereas the other member, which we refer to informally as the calcareous type, is comprised of amphibolite clasts in a dolomitic quartz arenite matrix. At Bald Mountain, the quartzitic conglomerate rests directly upon strata of the host lithosome, and exposures of the calcareous conglomerate become gradually more abundant toward the southwest ("stratigraphically higher" in the section). Quartzitic conglomerate consists of faintly bedded light-gray, poorly sorted, variably rounded quartz arenite and rare silt- to clay-rich clasts set in a quartz arenite matrix. Clasts as large as 30 cm were noted, but more typically clasts are 1 to 4 cm in greatest dimension. Among poorly expressed bedding 10 cm to 1 m thick, we observed a few scoured contacts, rare crude grading and one set of cross strata.

Calcareous conglomerate is typically gray green, and clasts are composed of various proportions of quartzite and amphibolite. They are well rounded, poorly sorted, and range in maximum dimension up to 6 cm, with 2 cm being a common size. Clasts are set in a matrix of dolomitic quartz arenite which comprises as much as 70% of the rock. Interbedded with conglomerate are clast-free intervals composed of gray-green, faintly bedded, dolomitic quartz arenite that grades to reddish-brown dolomitic argillite or to pinkish-gray quartz arenite.

Also interbedded with conglomerate at both the Bald Mountain and Kennedy exposures is light-gray to medium-gray felsic tuff.

Tuff occurs as one or more sharply bounded interbeds 10 to >25 m thick. At the Bald Mountain exposure, only one tuff bed was noted, lying within 1 m of the base of the lithosome. The most distinctive outcrop feature of the tuff is abundant dark-gray, 1-cm-long, flattened, biotite-rich clots which we interpret to be recrystallized pumice lapilli. Subequant, 1-mm-long phenocrysts of plagioclase and K-feldspar comprise several percent of most specimens, and they are set in a matrix composed of subequal amounts of quartz, K-feldspar, and plagioclase. All samples contain a few percent biotite, and one sample additionally contains sparse blue-green hornblende. Two samples of tuff, representing the greatest mineralogic diversity of Kennedy pendant exposures, were chemically analyzed (Table 2). In standard classification schemes based upon both major and trace element chemistry, these rocks are high-K dacite and rhyolite.

INFERRED DEPOSITIONAL SETTINGS

Although metamorphic and deformational overprints have obscured many primary depositional features of pendant strata, we have extracted sufficient lithostratigraphic information from most of these rocks to assign them to likely depositional environments with reasonable confidence. Below, we address each lithosome in turn.

Host Lithosome

The even, continuous laminae and thin beds of argillite that make up most of the host lithosome must have been deposited in quiet water, and therefore below storm wave base. Base-cut-out Bouma sequences demonstrate that much deposition was by turbidity currents, although reworking by contour-currents of some of the host cannot be ruled out. Host lithosome strata fit Mutti and Ricci Lucchi (1972) turbidite facies D and G and so are interpreted as outer fan or basin-plain deposits[3] (Suczek and Dunne, 1989). The conglomeratic sublithosome exposures that are present high in the Kennedy pendant section fits Mutti and Ricci-Lucchi facies A and is interpreted as a suprafan deposit. The presence of abundant carbonaceous material and common phosphate blebs throughout the lithosome and the absence, except in one locality, of burrows suggest that deposition was in an environment low in oxygen, either on the slope or rise within the oxygen minimum zone or within a restricted basin. Paucity of soft-sediment deformation features argues against open slope deposition.

Limestone Lithosome

The original depositional components of the limestones, the proportions of allochems

and lime mud, and even stratal thicknesses cannot be reconstructed owing to the superimposed deformation and metamorphism; hence their depositional environment remains uncertain. Judging by the associated strata, they probably originated as carbonate turbidites or grain flows onto a submarine fan. The carbonate in this and in the limy quartz arenite lithosome suggests a source in warm, carbonate-productive water on the continental shelf.

Limy Quartz Arenite Lithosome

The mixed siliciclastic-carbonate sediment of this lithosome forms Bouma sequences beginning with the laminated b or rippled c intervals, base-cut-out turbidites of Mutti and Ricci Lucchi (1972) facies D and E. Less commonly, coarser, graded or massive intervals with rip-up clasts form the lower interval of Bouma ab and abcd sequences 10 to 30 cm thick that fit Mutti and Ricci Lucchi turbidite facies C. Rare climbing ripples suggest locally developed overbank deposits. Carbonaceous material and lack of burrows suggest oxygen-poor conditions.

Quartz Arenite Lithosome

We assign the amalgamated quartz-sand beds characteristic of this lithosome to Mutti and Ricci Lucchi turbidite facies B, and we assign quartz-granule conglomerates near the top of the section to their facies A. Although many beds have scoured bases, we observed no large-scale channeling, and prominent thickening-up and thinning-up sequences are lacking. Therefore, we interpret these sediments as having been deposited on the distal part of a suprafan as it prograded across the outer submarine fan deposits below it.

Basalt Lithosome

The chemical composition of the basalt, in combination with some distinctive lithostratigraphic characteristics, together provide information concerning the depositional environment of the basalt lithosome. As noted above, the chemistry of these rocks shows substantial similarity with alkalic ocean island and seamount basalts. Additionally, plots of less mobile trace elements on the discrimination diagrams of Wood and others (1979) and Pearce and Cann (1973) reveal that Kern Plateau basalts plot in or close to the "within-plate" compositional field (fig 3), as would be expected for ocean island basalts. As a basis for comparison, compositional fields defined by 28 least altered samples of alkalic basalt in Paleozoic continental rise and slope rocks of the Roberts Mountains allochthon in Nevada, as reported by Madrid (1987) and Watkins and Browne (1989), are outlined in figure 3. With the exception of slightly depleted Hafnium values for Kern Plateau basalts, the two basalt suites are similar.

Some lithostratigraphic characteristics of the basalt are also informative. Most obviously, the basalt lithosome is interstratified with host lithosome strata for

---

[3] As do many workers, we utilize the facies concepts of Mutti and Ricci-Lucchi (1972) to help interpret the likely depositional settings of these deep-water marine rocks, but it should be kept in mind that these concepts have not been fully tested on modern submarine fans (cf. Shanmugan and others, 1985).

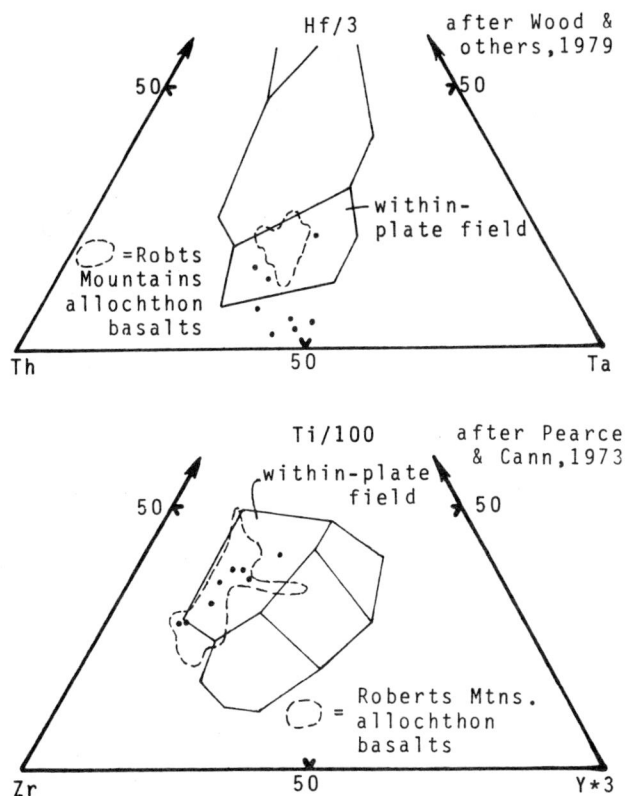

Figure 3. Plots of geochemical data from eight Kern Plateau basalt samples (bold dots) on descrimination diagrams of Wood and others, 1979 and of Pearce and Cann, 1973. See text for discussion.

which there is abundant independent evidence of an off-shelf marine environment, and the pillowed nature of some of the basalt is consistent with such a submarine environment. The coarse, fragmental texture of most exposures together with the local intermixing of host lithosome clasts probably originated in either of two ways, or as a hybrid of these two: 1) Faulting on the seafloor created steep scarp exposures of basalt and some host strata that shed debris aprons across portions of the slope or rise; 2) Erosion of a seafloor volcanic edifice generated debris aprons that spread across portions of the slope or rise, and during transit this debris picked up clasts of host lithosome rock from the substrate. Holcomb and Searle (1990) have shown that, in addition to normal erosional sloughing, slopes of seamounts and islands commonly shed large submarine slides, the distal portions of which might be composed of debris like that which composes much of the basalt lithosome.

Baritic Lithosome

We use two lines of evidence to deduce the likely depositional environment for this lithosome: (1) similarities between the baritic lithosome and stratiform barite exposed elsewhere in the North American Cordillera; (2) similarities between the baritic lithosome and presently forming barite/sulfide deposits. First, virtually

all of the lithostratigraphic characteristics of the baritic lithosome are common in stratiform barite deposits of the Antler allochthon in Nevada (Papke, 1984, his Table 6), and all of these Nevada barite deposits are interpreted to have been deposited on the continental rise or lower slope (Poole and others, in press). Second, active hot spring systems on the deep ocean floors are seen to be emitting fluids that precipitate barite, metal sulfides and manganese and that cause associated alteration of seafloor sediments (e.g., Morton and others, 1990). The many similarities between these modern sulfide/barite depositional systems and stratiform barite/sulfide deposits of early Paleozoic age in the Cordillera have led most recent workers (e.g., Poole, 1988; Murchey and others, 1987; Winn and Bailes, 1987) to conclude that Cordilleran stratiform barite/sulfide deposits probably originated via such a mechanism as well.

Argillite-Pebble Conglomerate Lithosome

This lithosome, which best fits Mutti and Ricci-Lucchi (1972) facies A, is interpreted to have been deposited in the same environment as the host lithosome strata that envelope it, that is to say below wave base and probably on the continental slope or rise. This interpretation is based on the observation that both the clasts and matrix are similar to -- and thus have potential sources within -- the immediate substrate of the conglomerate and probably were derived from some portion of that substrate that was subjected to an unusual and possibly widespread episode of erosion. Such erosion may have occurred on the rise in response to faulting or folding that caused uplift there, or it may have occurred on higher portions of the slope in response to changes in sea level (Carr and others, in press).

Capping Conglomerate Lithosome

Two characteristics of the capping conglomerate lithosome suggest that it was deposited in an environment substantially different from the quiet, deep water setting of underlying strata. First, the abundance of relatively coarse clasts and generally mature compositions and textures indicates a substantial change in the source for sediment, and the bedding style is more characteristic of debris flows than of the turbidites and grain flows that predominate in the underlying strata. An exact depositional environment cannot be determined, but it must have been in shallower water, and could even have been subaerial. Second, the volcanic rocks in this lithosome have compositions that differ from the basalts in the underlying strata in that they are much more evolved (rhyolite and dacite), and some characteristics of their minor and trace element compositions are similar to volcanic rocks formed in arc settings rather than within-plate settings. For example, on spidergram plots the felsic tuff data are observed to have decided troughs at Nd and Ta, a feature common to all subduction-generated arc rocks (e.g. Thompson and others, 1984). Also, on a

TiO_2 vs. Zr diagram (Pearce, 1982, his fig. 11), the tuff data plot within or close to compositional fields of arc volcanic rocks rather than close to fields of within-plate volcanic rocks.

Summary of Depositional Environments

We believe that pendant strata older than the capping conglomerate were deposited in the distal portions of a submarine fan in the oxygen-minimum zone of the continental rise or lower slope, and perhaps within a fault-controlled basin. They are predominantly basin-plain and outer fan sediments, with a sandy depositional lobe (suprafan) high in the section. The restricted range in composition and grain sizes of sediment supplied to the depositional site requires proximity to sources of both highly mature siliciclastic sediment and carbonate sediment. We envision this source to be the outer shelf along a passive continental margin, where occasional slowing of deposition of siliciclastic sediment allowed carbonate production. Clastic sediments were transported to the depositional site primarily as turbidity flows and less commonly as grain flows and debris flows.

Both the basalt and the hydrothermal fluids giving rise to the baritic lithosome needed passageways to the seafloor surface, and faults may have served as such passageways as well as given rise to scarps that shed basaltic debris down the submarine fan. Evidence for extensional faulting coeval with basaltic seafloor volcanism and with barite mineralization is nearly universal in eugeoclinal strata elsewhere in the Cordillera (Poole and others, 1991; Murchey and others, 1987). Lonsdale (1979) described active hydrothermal barite and manganese mineralization on a deep-water submarine fan near a fault, a setting similar to that which we envision for Kern Plateau pendant barite.

We believe that the capping conglomerate accumulated in relatively shallow water in or near a subduction-related volcanic arc, and thus reflects a substantial change in depositional environment.

REGIONAL CORRELATIONS

To the south in the Mojave Desert (El Paso Mountains-Pilot Knob-Goldstone-Lane Mountain belt) and to the north in some central Sierran pendants near Tioga Pass, as well as more widely in central Nevada, there are preserved pre-batholithic rock assemblages that we interpret to be lithostratigraphically correlative with those in the Kern Plateau pendants. We briefly review the characteristics of the Mojave Desert and central Nevada exposures.

El Paso Mountains-Pilot Knob-Goldstone-Lane Mountain Belt

Dark, siliciclastic-dominated, early Paleozoic eugeoclinal strata in this group of exposures have been the subject of recent studies by Carr and others (1984), Carr and others (in press), and Miller and Sutter (1982). We focus on the El Paso Mountains section because it contains the thickest and best preserved sections and because it is located just 20 km southeast of the Indian Wells pendant (fig. 1, inset map). Carr and others (1984) have documented the presence in the El Paso Mountains (abbreviated hereafter as EPM) of several partially coeval sequences of eugeoclinal strata deposited over much of Paleozoic time. These include an Upper Cambrian through Devonian eugeoclinal sequence, an Upper Cambrian, Ordovician and Silurian(?) transitional[4] (outer shelf/upper slope) sequence, a Lower Mississippian 'Antler' conglomerate and argillite sequence, upper Lower(?) Mississippian 'Antler' foreland basin deposits, and complex sequences of Pennsylvanian to Upper Permian overlap strata that range in composition from deep-water tubidites to andesitic volcanic rocks and shallow-water conglomerate.

There exist substantial similarities as well as some distinctive differences between some of these EPM sequences and certain intervals of strata in the Kern Plateau pendants. Portions of the Kennedy pendant strata are similar to the EPM transitional sequence in terms of the relative abundances and general nature of dark argillite strata, in the presence in both areas of distinctive interbedded blue-gray quartz arenites and very dark-gray shale or argillite with shale partings, and in the presence of thin limestone intervals that have characteristics of sediment gravity flows. The most notable difference between these Kennedy pendant strata and the EPM transitional sequence is the presence of the baritic lithosome in the former. Another lithostratigraphic similarity exists between the Antler conglomerate and the argillite-pebble conglomerate lithosome of the Indian Wells pendant. These two units, consisting mostly of argillite pebbles and less common, cobbles and boulders of arenaceous limestone and greenstone, are uncanny lithologic matches for one another. The Bald Mountain pendant sequence has broad similarities to the Devonian eugeoclinal strata of EPM in that both sequences are volumetrically dominated by carbonaceous argillite, interstratified within which are relatively abundant quartz arenite, limestone, and limy arenite hybrids. Basalt intervals are present in both sequences as well. However, the EPM section lacks the baritic lithosome. Finally, the capping conglomerate lithosome in the pendants is similar to one of the overlap sequences in the EPM (Bond Buyer sequence), which is composed of conglomerate, quartzite, calc-silicate rock and interlayered andesite, the latter having yielded an Early Permian U/Pb zircon date of 280 Ma (Martin and Walker, 1990). The capping conglomerate lithosome in the pendants contains a strikingly similar array of rock types, with the exception that the volcanic component is more felsic.

---

4 We include this environment within our broader definition of the eugeocline.

Antler Allochthon, Central Nevada

The Antler allochthon is comprised of outer shelf to upper slope (transitional) and lower slope to rise (eugeoclinal) strata that were thrust over coeval shelfal (miogeoclinal) strata during the Antler orogeny in Late Devonian through Early Mississippian time (Poole and others, 1991). Madrid (1987) recognized several distinctive regional characteristics of these early Paleozoic strata: 1) The sections are volumetrically dominated by dark, carbonaceous argillite, shale, chert and quartz siltstone. 2) Quartz arenite and much less abundant feldspathic arenite, mostly emplaced as grainflows, are the second most abundant constituents. Distinctive interbedded arenite/shale intervals are widespread in Ordovician sections. 3) Relatively thin limestone intervals, deposited mostly as grainflows and turbidites, comprise a minor component of all sections. 4) Alkalic basalt is a widespread minor component of all sections. 5) Arenaceous limestone that grades to limy quartz arenite, and less abundant limy argillite, are distinctive components of many Devonian sections. 6) Stratiform barite deposits, commonly associated with sulfide-enriched horizons, are widespread in Upper Ordovician and Upper Devonian sequences. 7) Many sections are unconformably capped by lithostratigraphically variable overlap sequences of quartzite, limestone and conglomerate (Diablo 'units' of Speed and others, 1977) deposited in varied deep to shallow water environments during Middle Pennsylvanian to Late Permian time. All seven of these widespread lithostratigraphic characteristics of Roberts Mountains allochthon strata have counterparts in the Kern Plateau pendant sections, and for this reason we believe the pendant strata are correlative with those in the allochthon.

In summary, the many lithostratigraphic similarities of Kern Plateau pendant strata with those in the El Paso Mountains and the Antler allochthon lead us to believe that pendant strata are broadly coeval with the strata in these other two regions and that strata in all three regions accumulated in similar depositional settings, namely the continental slope and(or) rise of western North America. The most convincing specific lithostratigraphic matches between dated sections in these other two regions and the pendant sections suggest that part of the Indian Wells pendant section is of Early Mississippian age, part of the Kennedy pendant section is of Ordovician age, part of the Bald Mountain pendant section is of Devonian age, and that early Paleozoic strata in the latter two pendants are overlapped by a conglomerate/volcanic overlap unit which we provisionally correlate with the Early Permian Bond Buyer sequence in the El Paso Mountains. Viable alternative correlations for the conglomerate/volcanic overlap unit would be with the mid-Paleozoic Grizzly Formation of the Northern Sierra Nevada terrane (Girty, this volume) or with Triassic conglomerate that marks the birth of the Sierran volcanic arc (Dunne, 1986).

IMPLICATIONS

Recognition of probable early Paleozoic eugeoclinal strata and late Paleozoic overlap rocks in the Kern Plateau pendants has implications for several aspects of regional geology:

1) Recognition of the correct paleogeographic affinity and probable age of these pendant strata roughly doubles the known extent of eugeoclinal exposures in the Mojave Desert and southern Sierra region. Upon removing left slip on the Garlock fault, these exposures are seen to form a northwest-trending belt at least 200 km long, for which we utilize the name El Paso terrane, after Blake and others (1982). This terrane not only preserves eugeoclinal strata in its uppermost portions, but it is also intruded by Mesozoic plutons whose initial Strontium isotopic signature indicates that they rose through deep lithosphere having oceanic composition (Kistler and Ross, 1990; Kistler, 1990). Thus, this terrane may represent a complete lithospheric section of oceanic composition.

2) Other little-studied pendants in the southern Sierra northwest of the Kern Plateau also may contain Paleozoic strata of eugeoclinal affinities, and we believe that a renewed round of careful examination of pendants in this region is needed. Questions that need to be addressed include: 1) How far north does the El Paso terrane extend within the Sierra?; 2) Might there be preserved in some of these pendants portions of other Paleozoic eugeoclinal terranes of the southwest Cordillera located outboard of the Roberts Mountains allochthon, such as the Northern Sierra Nevada/Havallah sequence (Harwood and Murchey, 1990) or the Shoo Fly Complex (Schweickert and Lahren, 1990, and this volume)?

3) The substantial change in depositional environment indicated by the capping conglomerate lithosome marks a profound paleogeographic change, namely regional uplift and the onset of subduction-related arc magmatism. If we are correct in correlating these rocks with similar Early Permian rocks in the El Paso Mountains, then this lithosome supplements previously available evidence from the Mojave Desert and east-central California indicating that Permian uplift, tectonism and magmatism may have been widespread (cf. Walker, 1988).

4) Although Paleozoic eugeoclinal sections in the El Paso Mountains, Kern Plateau pendants and Antler allochthon of Nevada share many lithostratigraphic similarities, thus suggesting deposition in broadly similar continental slope and rise environments, these sections nonetheless have enough subtle differences between them to leave open the possibility that they accumulated in different locations along the continental margin. If one emphasizes these differences, and at the same time is persuaded by Dickinson's (1977) vision of an irregularly shaped continental margin which may have trended southeastward down the

length of the southern Sierra, then one might reasonably conclude --as do we-- that the Mojave Desert/southern Sierra eugeoclinal terrane has not *necessarily* experienced large latitudinal tectonic transport.

5) Kistler (1990), noting that granitoids on opposite sides of the pendant belt have distinctly different isotopic properties, proposed that the belt marks the approximate position of a profound tectonic boundary of Jurassic age across which different lithospheric terranes have been juxtaposed. Mylonitic granitoids comprising the Kern Plateau shear zone that abuts the southwest margins of the pendants (fig. 2) are among the few surviving outcrops of this mostly cryptic boundary. The El Paso terrane may lie just northeast of the main tectonic boundary and thus be most closely allied with -- and not necessarily greatly displaced relative to -- the miogeocline to the east, or it may lie just southwest of the main tectonic boundary and thus be substantially out of place with respect to the miogeocline, or it may lie within a fault-bounded lozenge that is substantially allochthonous with respect to *both* sides of the tectonic boundary. Determining which of these three possibilities is correct is one major focus of our ongoing study of the Kern Plateau pendants.

ACKNOWLEDGMENTS

Principal financial support for this research was provided by the Petroleum Research Fund, administered by the American Chemical Society, with abundant additional support provided by the Department of Geological Sciences, California State University.,Northridge. Mike Carr provided much advice and encouragement throughout this project and generously shared his unpublished data concerning eugeoclinal strata in the Mojave Desert. Tom Anderson, Dave Harwood, Raul Madrid, Meagan Miller, Scott Paterson, Barney Poole, and Jason Saleeby have visited us in the field and provided helpful insights as our research progressed. Larry Collins and Sorena Sorenson provided much help in evaluating the metamorphism. Mike Diggles, Ron Kistler, Glenn Roquemore, Rich Schweickert, Cal Stevens and Gary Taylor have generously shared information regarding Sierran pendant rocks and(or) surrounding granitoids. Undergraduate students from CSUN, including Mark Davis, Brad Dyble, Josh Feffer, Pat Fischer, Jeff Geraci, Mike Godwin, Jeff Kofoed, John Lindquist, Andy Modugno, Frank Tepley, and Brian Swanson, assisted with field and laboratory work. We are grateful to all of these organizations and individuals for their assistance. Finally, we thank Mike Carr, Gene Fritsche, Rachel Gulliver and Dave Liggett for providing critiques of various portions of this manuscript that led to its considerable improvement.

REFERENCES CITED

Barton, M. and Hanson, B., 1989, Magmatism and the development of low-pressure metamorphic belts: Implications from the western United States and thermal modeling: Geological Society of America Bulletin, v. 101, p. 1051-1065.

Bergquist, J. and Nitkiewicz, A., 1982, Geologic map of the Domeland Wilderness and contiguous roadless areas, Kern and Tulare Counties, California: U.S. Geological Survey Map MF-1395-A, scale 1:48,000.

Blake, M. Jr., Howell, D. and Jones, D., 1982, Preliminary tectonostratigraphic terrane map of California: U.S. Geological Survey Open File Report 82-593, scale 1:750,000.

Burchfiel, B. and Davis, G., 1972, Structural framework and evolution of the southern part of the Cordilleran orogen, western United States: American Journal of Science, v. 272, p. 97-118.

Carr, M., Christiansen, R., and Poole, F., 1984, Pre-Cenozoic geology of the El Paso Mountains, southwestern Great Basin, California--A summary, in Lintz, J., ed., Western geological excursions: Geological Society of America, a guidebook prepared for the 1984 Annual Meeting, Reno, Nevada, v. 4, p.84-93.

Carr, M., Harris, A., Poole, F. and Fleck, R., in press, Stratigraphy and structure of Paleozoic outer continental-margin rocks in Pilot Knob Valley, north-central Mojave Desert, California: U.S. Geological Survey Bulletin.

Davis, G., Monger, J., and Burchfiel, C., 1978, Mesozoic construction of the Cordilleran "collage", central British Columbia to central California, in Howell, D, and McDougall, K., eds., Mesozoic paleogeography of the western United States: Pacific Section Society of Economic Paleontologists and Mineralogists, Pacific Coast Paleogeography Symposium 2, p. 1-32.

Dickinson, W., 1977, Paleozoic plate tectonics and the evolution of the Cordilleran continental margin, in Stewart, J., Stevens, C. and Fritsche, E., eds., Paleozoic paleogeography of the western United States: Pacific Section Society of Economic Paleontologists and Mineralogists, Pacific Coast Paleogeography Symposium 1, p. 137-155.

Dickinson, W, 1981, Plate tectonics and the continental margin of California, in Ernst, W.G., ed., The geotectonic development of California; Englewood Cliffs, New Jersey, Prentice-Hall, Inc., p. 1-28.

Diggles, M., Dellinger, D. and Conrad, J., 1987, Geologic map of the Owens Peak and Little Lake Canyon Wilderness Study areas, Inyo and Kern Counties, California: U.S. Geological Survey Map MF-1927-A, scale 1:48,000.

Dibblee, T., 1954, Geology of the Inyokern 15-minute quadrangle, Kern County, California: U.S. Geological Survey Open File Map 59-31, scale 1:62,500.

du Bray, E. and Dellinger, D., 1981, Geologic map of the Golden Trout Wilderness, southern Sierra Nevada, California: U.S. Geological Survey Map MF-1231-A, scale 1:48,000.

Dunne, G., 1986, Mesozoic evolution of the southern Inyo Mountains, Darwin Plateau, and Argus and Slate Ranges: an overview, *in* Dunne, G., compiler, Mesozoic and Cenozoic structural evolution of selected areas, east-central California: Geological Society of America, Cordilleran Section Fieldtrip Guidebook and Volume, Trips 2 and 14, 94 p.

Dunne, G., 1989, The Kern Plateau synbatholithic shear zone, southern Sierra Nevada, California [abs.]: EOS, Transactions, American Geophysical Union, v. 70, no. 43, p. 1308.

Dunne, G., Suczek, C., Dybel, B., Godwin, M., Kofoed, J., Lindquist, J., and Swanson, B., 1988, Newly recognized Paleozoic eugeoclinal strata in the southern Sierra Nevada, California [abs.]: Geological Society of America Abstracts with Programs, v.20, no. 7, p. 149.

Dunne, G., Saleeby, J. and Farber, D., *in press*, Early synbatholithic ductile faulting in the southern Sierra Nevada: new U/Pb age and geobarometric constraints for the Kern Plateau fault zone [abs.]: Geological Society of America Abstracts with Programs, v. 23, no. 3.

Elan, R., 1985, High-grade contact metamorphism at the Lake Isabella north shore roof pendant, southern Sierra Nevada, California: Los Angeles, University of Southern California, M.S. dissertation,, 201 p.

Engel, A., Engel, C., and Havens, R., 1965, Chemical characteristics of oceanic basalts and the upper mantle: Geological Society of America Bulletin, v. 76, p. 719-734.

Hamilton, W., and Myers, W., 1966, Cenozoic tectonics of the western United States: Reviews of Geophysics, v. 4, p. 509-549.

Harwood, D. and Murchey, B., 1990, Biostratigraphic, tectonic, and paleogeographic ties between upper Paleozoic rocks in the Northern Sierra terrane, California and the Havallah sequence, Nevada, *in* Harwood, D. and Miller, M., eds., Paleozoic and early Mesozoic paleogeographic relations; Sierra Nevada, Klamath Mountains, and related terranes: Geological Society of America Special Paper.255, p.157-173.

Holcomb, R., and Searle, R., 1990, Abundance of giant landslides from oceanic volcanoes [abs.]: EOS, Transactions, American Geophysical Union, v. 71, p. 1578.

Kistler, R.,1990, Two different lithospheric types in the Sierra Nevada, California, *in* Anderson, J. L., ed., The nature and origin of Cordilleran magmatism: Geological Society of America Memoir 174, p. 271-281.

Kistler, R., and Ross, D., 1990, A strontium isotopic study of plutons and associated rocks of the southern Sierra Nevada and vicinity, California: U.S. Geological Survey Bulletin 1920, 20 p.

Koski, R., Zierenberg, R., Shanks, W. and Campbell, A., 1990, Chemical characteristics of sulfide deposits, hydrothermal fluids, and sediment alteration at Escanaba Trough, southern Gorda Ridge [abs.]: EOS, Transactions, American Geophysical Union, v. 71, p. 1565.

Lonsdale, P., 1979, A deep-sea hydrothermal site on a strike-slip fault: Nature, v. 281, p. 531-534.

Madrid, R., 1987, Stratigraphy of the Roberts Mountains allochthon in north-central Nevada: Palo Alto, Stanford University Ph. D. dissertation, 341 p.

Martin, M. and Walker, D., 1990, New stratigraphic relationships from the shadow Mountains, western Mojave Desert: implications for late Paleozoic paleogeography [abs.]: Geological Society of America Abstracts with Programs, v.22, no. 3, p.64.

Miller, J. and Webb, R., 1940, Descriptive geology of the Kernville [30'] quadrangle, California: California Journal of Mines and Geology, v. 36, p. 343-378.

Miller, E., and Sutter, J., 1982, Structural geology and ^{40}Ar-^{39}Ar geochronology of the Goldstone-Lane Mountain area, Mojave Desert, California: Geological Society of America Bulletin, v. 93, p. 1191-1207.

Moore, J., and Sisson, T., 1985, Geologic map of the Kern Peak quadrangle, Tulare County, California: U.S. Geological Survey Map GQ-1584, scale 1:62,500.

Morton, J., Zierenberg, R. and Groschel-Becker, H., 1990, Geologic setting of massive sulfide deposits in the sediment-covered Escanaba Trough, southern Gorda Ridge; An overview [abs.]: EOS, Transactions, American Geophysical Union, v.71, p.1565.

Murchey, B., Madrid, R. and Poole, F., 1987, Paleozoic bedded barite associated with chert in western North America, *in* Hein, J., ed., Siliceous sedimentary rock-hosted ores and petroleum: New York, Van Nostrand Reinhold Co., p. 269-283.

Mutti, E., and Ricci Lucchi, F., 1972, Le torbiditi dell'Appennino settentrionale: introduzione all'analisi di facies: Memorie della Societa Geologica Italiana, v. 11, p. 161-199.

Nokleberg, W., 1983, Wallrocks of the central Sierra Nevada batholith, California: a collage of accreted tectono-stratigraphic terranes: U.S. Geological Survey Professional Paper, 1255, 28 p.

Papke, K., 1984, Barite in Nevada: Nevada Bureau of Mines and Geology Bulletin 98, 123 p.

Paterson, S., 1989, A reinterpretation of conjugate folds in the central Sierra Nevada, California: Geological Society of America Bulletin, v. 101, p. 248-259.

Pearce, J., 1982, Trace element characteristics of lavas from destructive plate boundaries, *in* Thorpe, R.,ed., Andesites: New York, John Wiley and Sons, p. 525-548.

Pearce, J., and Cann,J. 1973, Tectonic setting of basic volcanic rocks determined using trace element analysis: Earth and Planetary Science Letters, v. 19, p. 290-300.

Poole, F., 1974, Flysch deposits of the Antler foreland basin, western United States, *in* Dickinson, W., ed., Tectonics and sedimentation: SEPM Special Publication 22, p. 58-82.

Poole, F., 1988, Stratiform barite in Paleozoic rocks of the western United States, *in* Zachrisson, E., ed., Proceedings of the Seventh Quadrennial IAGOD Symposium, Lulea, Sweden: E.

692

Schweizerbartsche Verlagsbuchhandlung, Stuttgart, p. 309-319.

Poole, F., and Hayes, P., 1971, Depositional framework of some Paleozoic strata in northwestern Mexico and southwestern United States [abs]: Geological Society of America Abstracts with Programs, v. 3, no. 2, p. 179.

Poole, F., Stewart, J., Palmer, A., Sandberg, C., Madrid, R., Ross, R., Hintze, L., Miller, M. and Wrucke, C., in press, Latest Precambrian to latest Devonian time: development of a continental margin, in Burchfiel, B., Lipman, P. and Zobeck, M., eds.,, The Cordilleran orogen; conterminous U.S.: Boulder, Colorado, Geological Society of America. Geology of North America, v. G3.

Ross, D., 1966, Stratigraphy of some Paleozoic formations in the Independence quadrangle, Inyo County, California: U.S. Geological Survey Professional Paper 396, 62 p.

Ross, D., 1987, Generalized geologic map of the basement rocks of the southern Sierra Nevada, California: U.S. Geological Survey Open File Report 87-276, 28 p.

Saleeby, J., Goodin, S., Sharp, W. and Busby, C., 1978, Early Mesozoic paleotectonic-paleogeographic reconstruction of the southern Sierra Nevada region, in Howell, D. and McDougall, K., eds., Mesozoic paleogeography of the western United States: Pacific Section Society of Economic Paleontologists and Mineralogists, Pacific Coast Paleogeography Symposium 2, p. 311-336.

Schweickert, R. and Lahren, M., 1990, Speculative reconstruction of the Mojave-Snow Lake fault; Implications for Paleozoic and Mesozoic orogenesis in the western United States: Tectonics, v. 9, p. 1609-1630.

Segall, P., McKee, E., Martel, S. and Turrin, B., 1990, Late Cretaceous age of fractures in the Sierra Nevada batholith, California: Geology, v. 18, p. 1248-1251.

Smith, G., and Ketner, K., 1970, Lateral displacement on the Garlock fault, southeast California, suggested by offset sections of similar metasedimentary rocks: U,S, Geological Survey Professional Paper 700-D, p. D1-D9.

Suczek, C., and Dunne, G., 1989, Depositional facies and composition of Paleozoic eugeoclinal strata in the southern Sierra Nevada, California [abs.]: Geological Society of America Abstracts with Programs, v. 21, no. 5, p. 149.

Silver, L., and Anderson, T., 1974, Possible left-lateral early to middle Mesozoic disruption of the southwestern North American craton margin [abs.]: Geological Society of America Abstracts with Programs, v. 6, p. 955.

Speed, R., MacMillan, J., Poole, F. and Kleinhampl, F., 1977, Diablo Formation, central western Nevada; composite of deep and shallow water Upper Paleozoic rocks, in Stewart, J., Stevens, C. and Fritsche, E, eds., Paleozoic paleogeography of the western United States: Pacific Section Society of Economic Paleontologists and Mineralogists, Pacific Coast Paleogeography Symposium 1, p. 301-314.

Stewart, J., Poole, F., Ketner, K., Madrid, R., Roldan-Quintana, J. and Amaya-Martinez, R., 1990, Tectonics and stratigraphy of the Paleozoic and Triassic southern margin of North America, Sonora, Mexico, in Gehrels, G. and Spencer, J.,eds., Geologic excursions through the Sonoran Desert Region, Arizona and Sonora: Geological Society of America Meeting, Cordilleran Section 86th, Fieldtrip Guidebook, p. 183-202.

Shanmugan, G., Damuth, J. and Moiola, R., 1985, Is the turbidite facies association scheme valid for interpreting ancient submarine fan environments?: Geology, v. 13, p. 234-237.

Taylor, G., Lloyd, R., Alfors, J., Burnett, J., Stinsonm, M., Chapman, R., Silva, M., Bacon, C. and Anderson, T., 1986, Mineral resource potential of the Rockhouse Basin Wilderness Study Area, Kern and Tulare Counties, California: California Division of Mines and Geology Special Report 157, 31 p.

Thompson, R., Morrison, M., Henry, G. and Parry, S., 1984, An assessment of the relative roles of crust and mantle in magma genesis: an elemental approach: Philosophical Transactions, Royal Society of London, v. A310, p. 549-590.

Turner, F., 1981, Metamorphic petrology: New York, McGraw-Hill, 403 p.

Walker, D., 1988, Permian and Triassic rocks of the Mojave Desert and their implications for timing and mechanism of continental truncation: Tectonics, v. 7, p. 685-709.

Watkins, R. and Browne, Q., 1989, An Ordovician continental-margin sequence of turbidites and seamount deposits in the Roberts Mountains allochthon, Independence Range, Nevada: Geological Society of America Bulletin, v. 101, p. 731-741.

Winn, R. and Bailes, R., 1987, Stratiform lead-zinc sulfides, mudflows, and turbidites: Devonian sedimentation along a submarine fault scarp of extensional origin, Jason deposit, Yukon Territory, Canada: Geological Society of America Bulletin, v. 98, p. 528-539.

Wood, D., Joron, J.-L. and Treuil, M., 1979, A reappraisal of the use of trace elements to classify and descriminate between magma series erupted in different tectonic settings: Earth and Planetary Science Letters, v. 45, p. 326-336.

THE SOUTHERNMOST LENS OF THE UPPER DEVONIAN GRIZZLY FORMATION, NORTHERN SIERRA NEVADA, CALIFORNIA: EVIDENCE FOR A TRENCH-SLOPE DEPOSITIONAL SETTING

Gary H. Girty, Alysa M. Keller, Kristi R. Franklin
and Robert C. Stroh

Department of Geological Sciences
San Diego State University
San Diego, California 92040

ABSTRACT

The Upper Devonian Grizzly Formation forms five laterally discontinuous lenses that lie unconformably above the remnants of a lower to middle Paleozic subduction complex (i.e., the Shoo Fly Complex). Each lens grades upward into the Upper Devonian to Lower Mississippian (?) deep-marine volcanogenic deposits of an island arc. The results of stratigraphic and sedimentologic studies of the southernmost lens of the Grizzly Formation are reported here. The results of these studies suggest that the southernmost lens of the Grizzly Formation accumulated at the base of a submarine canyon that in cross sectional view was approximately 7 km wide and 0.6 km deep. Bedded sandstone breccias and laminated argillites make up the upward fining fill of the submarine canyon. The former deposits are inferred to be the results of mass transport down a relatively steep submarine slope, whereas laminated argillites are probably pelagic or hemipelagic deposits.

Given the stratigraphic setting of the southernmost lens of the Grizzly Formation above the deposits of a subduction complex (i.e., the Shoo Fly Complex), we infer that it accumulated in a trench-slope setting. In trench-slope settings, tectonic processes associated with continuing accretion of trench wedges probably produce oversteepened canyon walls, and, together with periodic seismic shocks, generate the kinds of mass flow deposits observed in the Grizzly Formation.

INTRODUCTION

In the northern Sierra Nevada, California, the Upper Devonian to Lower Mississippian (?) Sierra Buttes and Taylor Formations represent the remnants of an island arc that either fringed the western North American continental margin or formed in some currently unknown location in the vast paleo-Pacific Ocean, Panthalassa (D'Allura and others, 1977; Dickinson, 1977; Schweickert and Snyder, 1981; Hanson, 1983; Harwood, 1988; Speed, 1979; Hannah and Moores, 1986; Miller and others, 1984; Hanson and Schweickert, 1986) (Fig. 1). Volcanoclastic rocks in the Sierra Buttes and Taylor Formations were deposited in a deep marine environment (Hanson and Schweickert, 1986), and unconformably overlie the pre-Upper Devonian Shoo Fly Complex. Work completed in the Shoo Fly Complex suggests that it formed either: (1) as a continental-rise submarine-fan system that subsequently was imbricated and accreted to a subduction complex (Schweickert and Snyder, 1981), or (2) as successively deposited and accreted sand-rich trench wedges (Girty and others, 1990).

Mapping by Girty (1983), Hanson (1983), Hanson and Schweickert (1986), and Stroh (1990) shows that the Grizzly Formation consists of five laterally discontinuous lenses lying stratigraphically between the Shoo Fly Complex and the middle Paleozoic volcanogenic rocks of the Sierra Buttes and Taylor Formations. Hanson and Schweickert (1986) interpreted the discontinuous nature of the Grizzly Formation to be due to its deposition in a series of "channels" cut into the Shoo Fly Complex. Although the Grizzly Formation marks the transition from subduction related activity in the Shoo Fly Complex to island arc volcanism in the Sierra Buttes and Taylor Formations, it has not been studied in detail. Hence, we undertook a detailed stratigraphic and sedimentologic study of the southernmost lens of the Grizzly Formation (Fig. 1). The results of previous work in the Grizzly Formation are described in Hanson (1983) and Hanson and Schweickert (1986). These papers also summarize limited biostratigraphic data which indicate that the upper part of the Grizzly Formation is Late Devonian in age.

THE SOUTHERNMOST LENS OF THE GRIZZLY FORMATION

Overview

Between Culbertson Lake and Blue Lake, the Grizzly Formation crops out as an irregularly shaped lens that is approximately 7 km long and 0.6 km thick just south of Culbertson Lake (Fig. 1). It is composed of sandstone breccia, sandstone, and argillite, and is metamorphosed to the greenschist facies or albite-epidote-hornfels facies of contact metamorphism. Primary sedimentary features are well

In Cooper, J.D., and Stevens, C.H., eds., 1991, **Paleozoic Paleogeography of the Western United States-II**: Pacific Section SEPM, Vol. 67, p. 693-701.

693

694

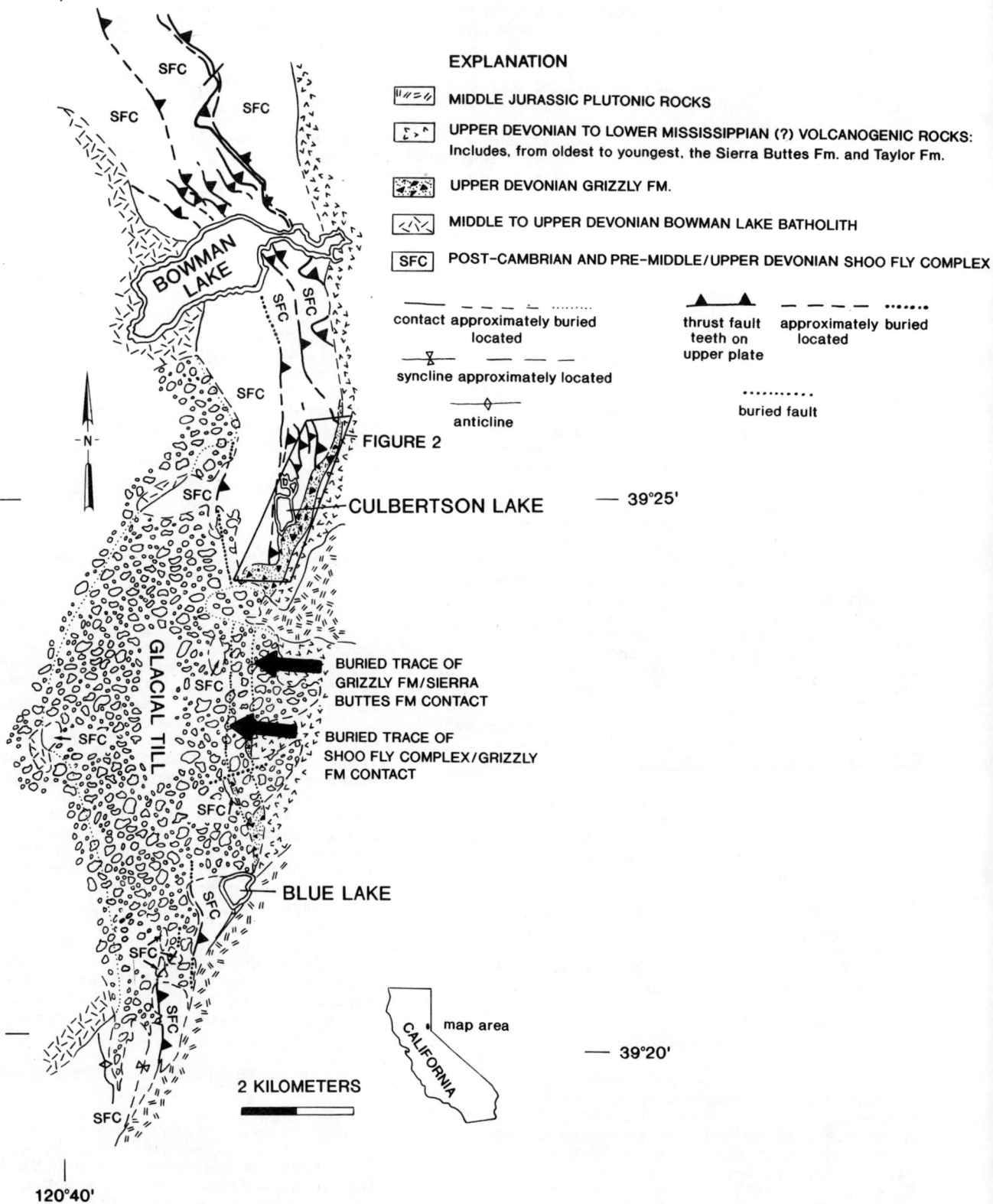

Figure 1. Simplified geologic map showing the lense-shaped outcrop pattern of the Grizzly Formation between Culbertson and Blue Lakes, and the location of Figure 2.

preserved, and as a result the prefix "meta" is not used in this report.

Bedding in the Grizzly Formation near Culbertson Lake strikes northwestward and dips uniformly eastward (Figs. 1 and 2). In contrast, bedding in the Grizzly and Sierra Buttes Formations, near Blue Lake, strikes northwestward and dips steeply to the northwest (Fig. 1). Observations of graded beds consistently indicate that the Grizzly and Sierra Buttes Formations are overturned in this area. The area between the overturned section at Blue Lake and the stratigraphically upright section near Culbertson Lake is covered by glacial till (Fig. 1). As a result, we concentrated our studies in the well exposed outcrops of the Grizzly Formation east of Culbertson Lake (Figs. 1 and 2).

Our mapping in the Culbertson Lake area indicates that the Grizzly Formation contains large blocks of chert (approximately 10 - 20 m long and 8 - 10 m thick), limestone (about 10 m long and 5 m thick), and basalt (approximately 35 m long and 10 m thick) in its lower part, and abundant finer grained pebble to cobble breccias in its upper part (Fig. 2). The large boulder-sized blocks occur in the thickest part of the Grizzly Formation, which crops out in the steep rugged walls of a cirque. Exposures in the cirque walls are poor, and as a result of steep topography, abundant colluvium, and brush, we were unable to measure a section of the lower boulder breccias of the Grizzly Formation. However, at several locations we observed that the boulder-sized blocks are separated by bedded breccias like those exposed in the upper part of the Grizzly Formation. Thus, the coarsest breccias, as judged from the largest observable clast size, occur in the thickest part of the Grizzly Formation and only in its lower half. The boulder breccias grade upward into pebble and cobble breccias that make up the upper half of the Grizzly Formation.

In order to document the stratigraphic and sedimentologic characteristics of the upper parts of the Grizzly Formation, we measured four sections, bed-by-bed, in the well exposed outcrops east of Culbertson Lake. Detailed descriptions and locations of each of the four sections are presented in Franklin (1990) and Keller (1990). Two of the four sections are illustrated in figures 3 and 4. The locations of these two representative sections are shown in figure 2.

The results of our mapping, stratigraphic, and sedimentologic studies indicate that the upper part of the Grizzly Formation is composed of seven distinctive facies which we designate (1) matrix-supported breccias, (2) clast-supported breccias, (3) thick-bedded sandstones, (4) thin-bedded sandstones, (5) turbidites, (6) tuffaceous argillites, and (7) laminated argillites. Our observations between Culbertson Lake and Blue Lake suggest that these seven facies are characteristic of the

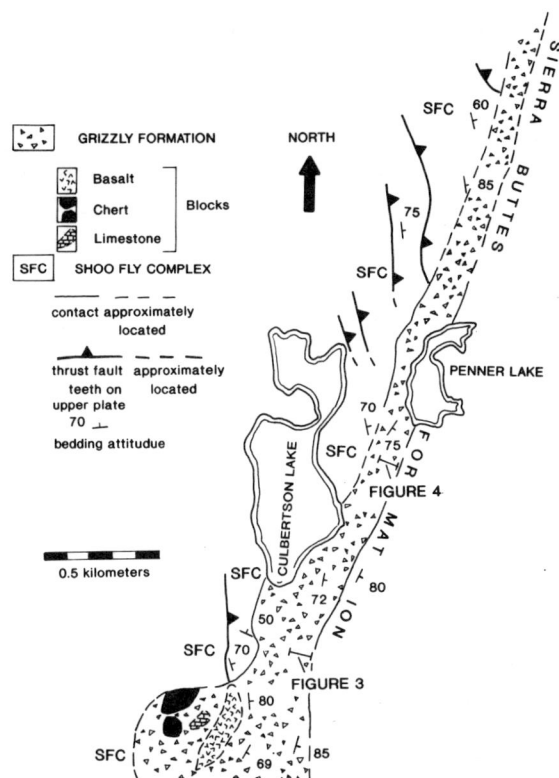

Figure 2. Map of the Grizzly Formation near Culbertson Lake. Note the locations of Figures 3 and 4.

Grizzly Formation as a whole, with the one exception that breccias are coaser grained in the lower half of the formation south of Culbertson Lake as described earlier in this paper. The general characteristics of each of the seven facies are described briefly in the following sections and are illustrated in Figure 5.

Matrix-Supported Breccia

Breccia beds make up 31% to 56% of the four measured sections (e.g., Figs. 3 and 4) (Franklin, 1990; Keller, 1990). Eighty-four percent of these beds are matrix-supported, and 16% are clast-supported. Approximately 19% of matrix-supported breccia beds display reverse grading in their lower parts, whereas 31% exhibit normal coarse-tail grading. The remaining proportion of matrix-supported breccia beds are disorganized, displaying no primary internal sedimentary structures (Figs. 3, 4, and 5). Some matrix-supported breccia beds contain clasts whose long axes are aligned parallel, and a few beds exhibit silty laminae draped over and around pebble-sized clasts. Clast populations are poorly sorted, ranging from fine pebble to boulder; coarse pebbles, however, predominate (Figs. 3, 4, and 5). Bed thicknesses range from 0.35 m to 27 m, with an average thickness of 4.1 m. Clast types are mostly sandstone, but also include, in order of decreasing abundance, argillite and basalt (Figs. 3 and 4).

696

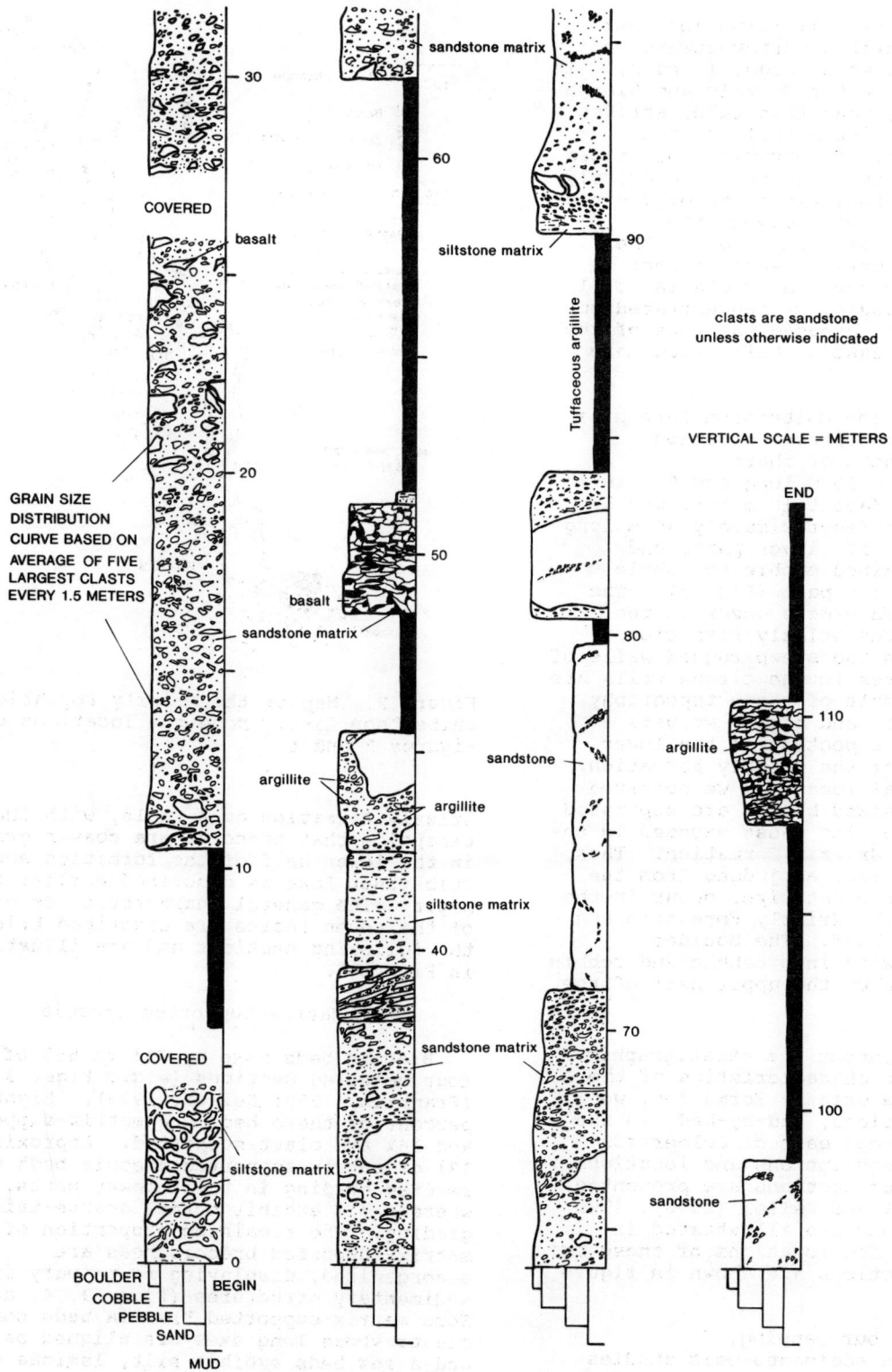

Figure 3. Stratigraphic column of the upper part of the
Grizzly Formation. Location of Figure 3 is shown in Figure
2. See Figure 5 for explanation of ornamentation pattern.

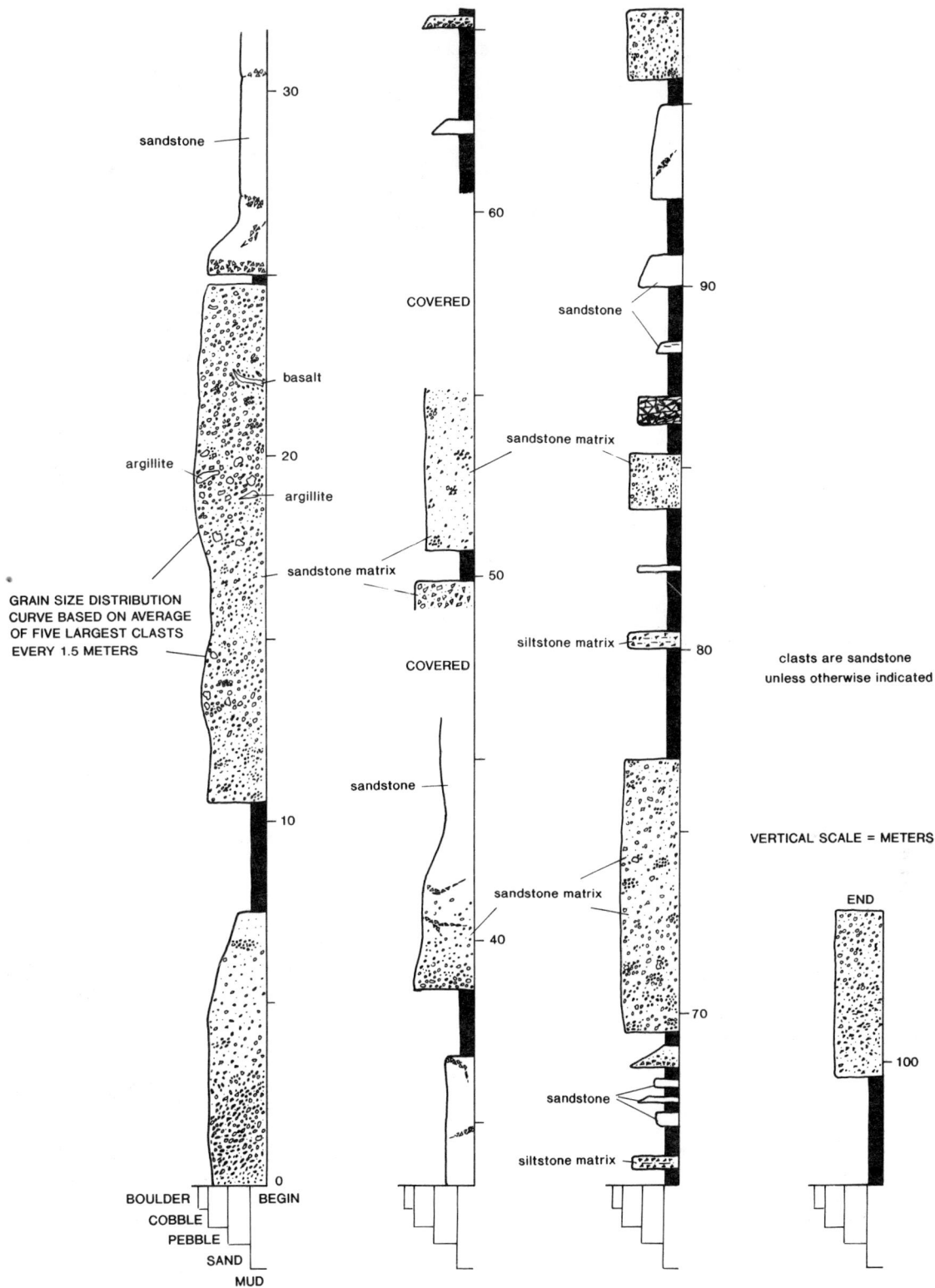

Figure 4. Stratigraphic column of the upper part of the
Grizzly Formation. Location of Figure 4 is shown in Figure
2. See Figure 5 for explanation of ornamentation pattern.

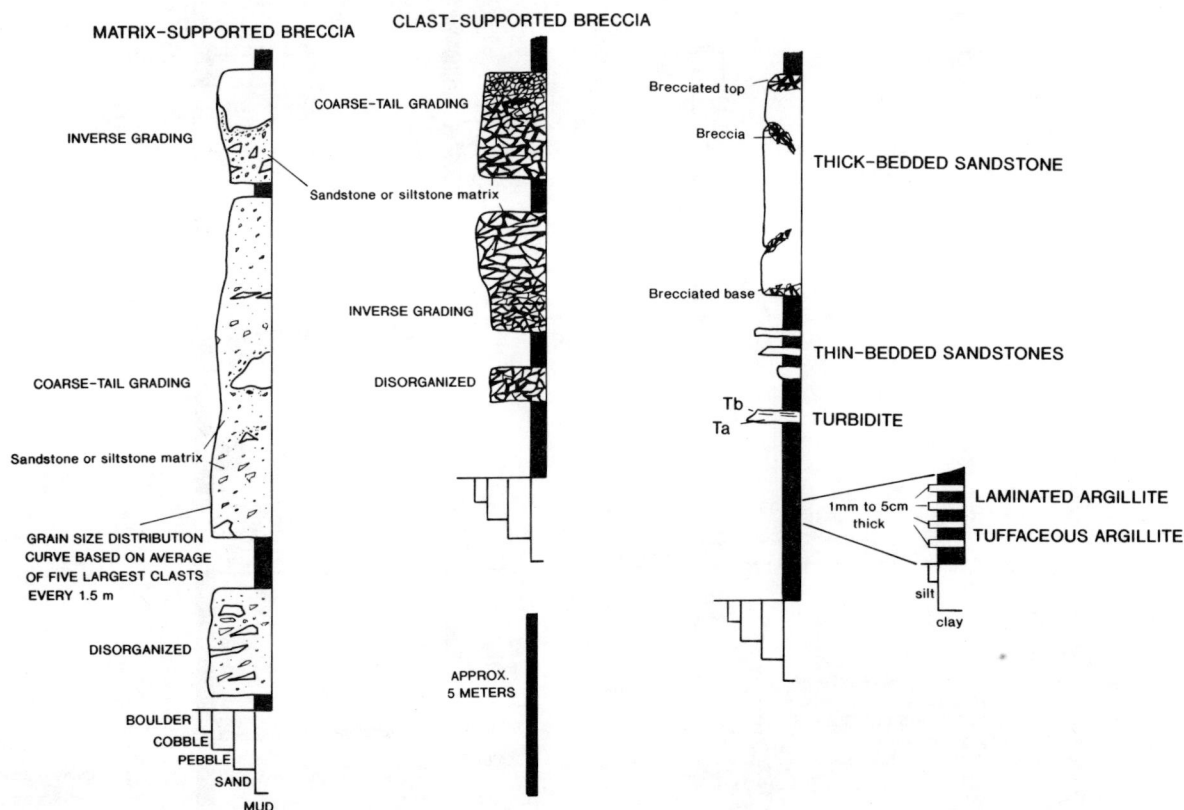

Figure 5. Schematic illustration of the sedimentological characteristics of the seven facies that make up the Grizzly Formation. Ornamentation pattern is the same as that used in Figures 3 and 4.

Clasts are generally subangular.

The matrix in matrix-supported breccias varies from quartz-rich siltstone to coarse-grained quartz sandstone (Figs. 3, 4, and 5). Grains comprising the matrix are moderately to poorly sorted.

Clast-Supported Breccia

Clast-supported breccias make up less than 1% and up to 26% of the four well studied sections (e.g., Figs. 3, 4) (Franklin, 1990; Keller, 1990). Of the total number of observed breccia beds, 16% are clast-supported. These beds range from graded to disorganized (Figs. 3, 4, and 5). Clast size ranges from fine pebble to boulder; however, fine cobble is most common. Clast-supported breccia beds range in thickness from 0.35 m to 10.5 m, but on average are 3.5 m thick. Clasts are predominantly subangular, although they range from subrounded to angular. Clasts are mostly sandstone, but also include, in order of decreasing abundance, argillite and basalt (Figs. 3 and 4).

The matrix of the clast-supported breccias ranges from fine-to coarse-grained sandstone, but medium-grained sandstone predominates. Most matrix grains are subangular and moderately sorted.

Thick-bedded Sandstone Facies

Sandstones make up 2% to 21% of the four measured sections (e.g., Figs. 3, 4) (Franklin, 1990; Keller, 1990). Thirty-seven percent of these beds are assigned to the thick-bedded sandstone facies. In this facies, bed thicknesses range from 2.0 to 10.8 m, but the average thick-bedded sandstone is 6.2 m thick. Ungraded beds composed of detritus ranging from fine-grained sand to coarse-grained pebble are characteristic of the thick-bedded sandstone facies; coarse-grained sandstone, however, is most common.

Most thick-bedded sandstones can not be traced laterally for distances greater than 20 meters. They commonly are brecciated along their upper and lower contacts, and along internal fractures (Figs. 3, 4, and

5). Clasts in breccia developed within thick-bedded sandstones are commonly pebble size (Fig. 5).

Thin-bedded Sandstone Facies

Thirteen percent of the four measured columns is composed of thin-bedded sandstones (e.g., Figs. 3, 4) (Franklin, 1990; Keller, 1990). These beds, on average, are 33 cm thick, but range in thickness from 15 cm to 88 cm. Most of the thin-bedded sandstone beds are normally graded, although ungraded and inversely graded beds are also present (Fig. 5). Detritus making up thin-bedded sandstones is poorly sorted, subangular to subrounded, and ranges from fine sand to coarse pebble; coarse sand, however, predominates.

Turbidite Facies

Turbidites make up less than 1% of the four well studied sections (Franklin, 1990; Keller, 1990). Beds assigned to this facies contain both the Ta and Tb Bouma intervals (Bouma, 1962) (Figs. 3, 4, and 5). Grain size in such beds ranges from silt to fine pebble. Ta intervals display either coarse-tailed grading or are ungraded. Tb intervals are composed of fine- to very fine-grained sand and silt arranged in plane parallel laminations that are 1 mm to 5 mm thick. Turbidite beds are normally 28 cm to 47 cm thick, with an average bed thickness of 35 cm.

Laminated Argillite Facies

Laminated argillites make up approximately 25% to 33% of the sections that we studied (Franklin, 1990; Keller, 1990). They commonly display 1 mm to 5 cm thick plane parallel laminations (Figs. 3, 4, and 5). In the four measured sections, intervals composed wholly of argillite range in thickness from 7 cm to 25 m.

Tuffaceous Siltstone

In one section, a single discontinuous 5.5 m thick tuffaceous argillite bed was measured (Fig. 3). This bed exhibits 2 mm to 2 cm thick plane parallel laminations, and contains abundant silt-sized grains of feldspar.

DISCUSSION

The southernmost lens of the Grizzly Formation was deposited in a marine environment unconformably on the Shoo Fly Complex, which represents the sandstone-rich remnants of a lower to middle Paleozoic subduction complex (Schweickert and Snyder, 1981; Hanson and Schweickert, 1986; Girty and others, 1990). The lense-shaped character of the Grizzly Formation betweem Culbertson Lake and Blue Lake suggests that it is the infill of a large canyon, approximately 7 km wide and 0.6 km deep (Fig. 1) (Stroh, 1990). The coarse-grained and angular to subangular character of the breccia clasts in the inferred canyon fill

indicate minimal transportation distances from a nearby source. The compositions and great abundance of quartz-rich sandstone breccia clasts clearly point to the underlying Shoo Fly Complex as their source (Figs. 3 and 4). The above observations imply that the southernmost lens of the Grizzly Formation accumulated in a setting where submarine mass wasting and erosional processes were active, presumably at or near the base of a submarine canyon. Given the fact that the Grizzly Formation was deposited on the Shoo Fly Complex, it therefore follows that the Grizzly Formation accumulated in a trench-slope setting.

Mass transport processes common in slope settings along modern convergent margins include rock falls, slides, and sediment gravity flows (Cook and others, 1982; Field, 1987). These processes (i.e., rock falls, slides, and sediment gravity flows) can produce clast-supported breccias, glide or slump blocks; and matrix-supported breccias, graded and ungraded sandstones, and turbidites, respectively. In convergent-margin settings, such deposits probably accumulate along the bases of slope segments bounded by thrust faults (Fig. 6).

Given the varieties of sedimentological processes that probably operated in the inferred trench-slope setting for the southernmost lens of the Grizzly Formation, we interpret the clast-supported breccias to be the deposits of rock falls, and the matrix-supported breccias to be the products of debris flows. Thin-bedded sandstones are most likely the deposits of grain, fluidized, or liquified flows, whereas turbidites were derived from turbidity currents. Laminated and tuffaceous argillites are the products of pelagic and/or hemipelagic deposition. Thick-bedded sandstones and the large boulder-sized blocks in the lower part of the Grizzly Formation probably represent large pieces of the underlying Shoo Fly Complex that were undercut by submarine erosion and fell or slid into the lower end of the submarine canyon in which the Grizzly Formation accumulated. A schematic diagram illustrating our paleogeographic/paleotectonic model for the Grizzly Formation is shown in Figure 6.

CONCLUSIONS

The Grizzly Formation between Culbertson Lake and Blue Lake represents the upward-fining fill of the lower part of a submarine canyon that was cut into a lower to middle Paleozoic subduction complex. The fill of the submarine canyon consists of clast-supported breccias, matrix-supported breccias, thick-bedded sandstones, thin-bedded sandstones, turbidites, tuffaceous argillites, and laminated argillites. Given the stratigraphic setting of the Grizzly Formation above the Shoo Fly Complex in the Culbertson Lake/Blue Lake area, we infer that the Grizzly Formation represents a submarine trench-slope accumulation. In

700

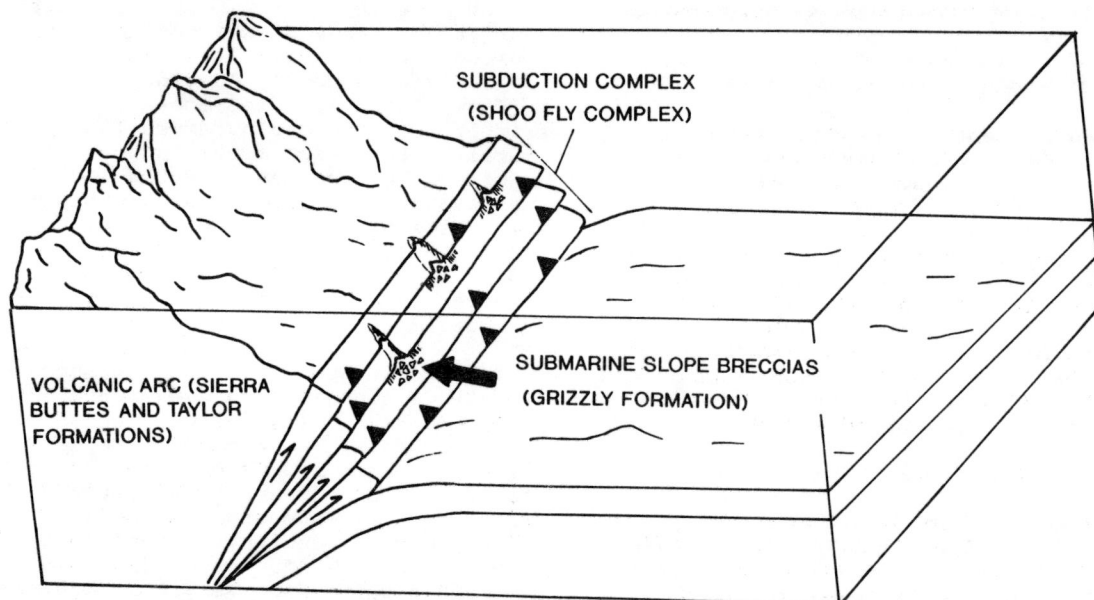

Figure 6. Schematic, not-to-scale, paleogeographic and paleotectonic model for the Grizzly Formation.

trench-slope settings continuing tectonic activity associated with the offscraping of trench wedges can produce oversteepened submarine canyon walls, and along with seismic activity can generate the mass flow deposits observed in the Grizzly Formation (Fig. 6).

ACKNOWLEDGEMENTS

Support for work reported here was provided by NSF grant EAR 8902382 awarded to G. H. Girty. Professors John D. Cooper and Richard Miller provided constructive reviews of this paper.

REFERENCES CITED

Bouma, A. H., 1962, Sedimentology of Some Flysch Deposits: A Graphic Approach to Facies Interpretation: Amsterdam, Elsevier, 163 p.

Cook, H. E., Field, M. E., and Gardner, J. V., 1982, Characteristics of sediments on modern and ancient continental slopes, in Scholle, P. A., and Spearing, D., eds., Sandstone Depositional Environments, American Association of Petroleum Geologists, Tulsa, Oklahoma, p. 329-364.

D'Allura, J. A., Moores, E. M., and Robinson, L., 1977, Paleozoic rocks of the northern Sierra Nevada: Their structural and paleogeographic implications, in Stewart, J. H., Stevens, C. H., and Fritsche, A. E., eds., Paleozoic Paleogeography of the Western United States: Society of Economic Paleontologists and Mineralogists, Pacific Section, Pacific Coast Paleogeography Symposium 1, p. 395-408.

Dickinson, W. R., 1977, Paleozoic plate tectonics and the evolution of the Cordilleran continental margin, in Stewart, J. H., Stevens, C. H., and Fritsche, A. E., eds., Paleozoic Paleogeography of the Western United States: Society of Economic Paleontologists and Mineralogists, Pacific Section, Pacific Coast Paleogeography Symposium 1, p. 137-155.

Field, M. E., 1987, Sediment mass-transport in basins: Controls and patterns, in Gorsline, D. S., ed., Depositional Systems in Active Margin Basins, Society of Economic Paleontologists and Mineralogists, Pacific Section, volume 54, p. 1-24.

Franklin, K. R., 1990, Sedimentology and stratigraphy of the Penner Lake facies of the Grizzly Formation, northern Sierra Nevada, California: Part I [unpublished Senior thesis]: San Diego, California, San Diego State University, 21 p.

Girty, G. H., 1983, The Culberston Lake allochthon, a newly identified structural unit within the Shoo Fly Complex: Sedimentologic, stratigraphic, and structural evidence for extension of the Antler orogenic belt to the northern Sierra Nevada, California [unpublished Ph.D. thesis]: New York, New York, Columbia University, 155 p.

Girty, G. H., Gester, K. C., and Turner, J. B., 1990, Pre-Late Devonian geochemical, stratigraphic, sedimentologic, and structural patterns, Shoo Fly Complex, northern Sierra Nevada, California, in Harwood, D. S., and Miller, M. M., eds., Paleozoic and Early Mesozoic Paleogeographic Relations; Sierra Nevada, Klamath Mountains, and Related Terranes: Geological Society of America Special Paper 255 (in press).

Hannah, J. L., and Moores, E. M., 1986, Age relationships and depositional environments of Paleozoic strata, northern Sierra Nevada, California: Geological Society of America Bulletin, v. 97, p. 787-797.

Hanson, R. E., 1983, Volcanism, plutonism, and sedimentation in Late Devonian island arc setting, northern Sierra Nevada, California [unpublished Ph.D. thesis]: New York, New York, Columbia University, 345 p.

Hanson, R. E., and Schweickert, R. A., 1986, Stratigraphy of mid-Paleozoic island-arc rocks in part of the northern Sierra Nevada, Sierra and Nevada Counties, California: Geological Society of America Bulletin, v. 97, p. 986-998.

Harwood, D. H., 1988, Tectonism and metamorphism in the northern Sierra terrane, northern California, in Ernst, W. G., ed., Metamorphism and Crustal Evolution of the Western United States (Rubey volume VIII): Englewood, Cliffs, New Jersey, Prentice-Hall, p. 765-788.

Keller, A. M., 1990, Sedimentology and stratigraphy of the Penner Lake facies of the Grizzly Formation, northern Sierra Nevada, California: Part II [unpublished Senior thesis]: San Diego, California, San Diego State University, 22 p.

Miller, E. L., Holdsworth, B. K., Whiteford, W. B., and Rodgers, D. 1984, Stratigraphy and structure of the Schoonover sequence, northeastern Nevada: Implications for Paleozoic plate-margin tectonics: Geological Society of America Bulletin, v. 95, p. 1063-1076.

Schweickert, R. A., and Snyder, W. S., 1981, Paleozoic tectonics of the Sierra Nevada and adjacent regions, in Ernst, W. G., ed., The Geotectonic Development of California (Rubey volume 1): Englewood Cliffs, New Jersey, Prentice-Hall, p. 182-201.

Speed, R. C., 1979, Collided Paleozoic platelet in the western United States: Journal of Geology, v. 87, p. 279-292.

Stroh, R. C., 1990, Evidence for an Ordovician to Late Devonian angular unconformity between the Shoo Fly Complex and the Sierra Buttes Formation, northwestern Sierra Nevada, California [unpublished Senior thesis]: San Diego, California, San Diego State University, 24 p.

THE PRE-UPPER DEVONIAN LANG AND BLACK OAK SPRINGS SEQUENCES, SHOOFLY COMPLEX, NORTHERN SIERRA NEVADA, CALIFORNIA: TRENCH DEPOSITS COMPOSED OF CONTINENTAL DETRITUS

Gary H. Girty, Larry D. Gurrola,
Greg W. Taylor, Marci J. Richards, and Melissa S. Girty

Department of Geological Sciences
San Diego State University
San Diego, California 92040

ABSTRACT

The post-Cambrian and pre-Upper Devonian Shoo Fly Complex is composed of, in descending structural order, the Sierra City melange, Culbertson Lake allochthon, Duncan Peak allochthon, and Lang sequence. Recently completed mapping in the Lang sequence suggests that it can be subdivided into, from oldest to youngest, the Lang and Black Oak Spring sequences. Stratigraphic and sedimentologic data indicate that large portions of the Lang and Black Oak Spring sequences are composed of massive sandstones, turbidites, and mudstones/argillites.

Massive sandstones are on average 0.6 m thick, and are mostly graded beds. Turbidities include Ta-present and Ta-absent varieties that display two or more Bouma intervals. The average Ta-present bed is 0.5 m thick, whereas the average Ta-absent bed is about 0.2 m thick. Mudstones/argillites occur as intercalations between sandstone beds or as rather homogeneous sections as thick as 12 meters. The sedimentological characteristics of sandstone beds observed and measured in the Lang and Black Oak Spring sequences are indicative of sediment deposited by subaqueous sediment gravity flows such as high density and dilute turbidity currents.

Twenty sections measured in the Lang and Black Oak Spring sequences contain both upward thickening and coarsening and upward thinning and fining cycles composed of four or more beds; however, some intervals show no vertical trends. Upward thickening and coarsening cycles are on average 3.2 m thick, and primarily are composed of massive sandstones and lesser amounts of Ta-present turbidites. Upward thinning and fining cycles are on average 4.2 m thick. They are composed of massive sandstones and Ta-present turbidites in their lower parts, and Ta-present and Ta-absent turbidites and mudstones/argillites in their upper parts. These observations suggest that upward thickening and coarsening cycles are progradational lobes probably produced by channel migration. Upward thinning and fining cycles are most likely channel-fill deposits.

Our data for the Lang and Black Oak Spring sequences consistently indicate a middle fan facies association and a general absence of an outer fan facies association. These characteristics are suggestive of deposition in a trench setting where axial channels and possibly unconfined axial flow played an important role in distributing sediment.

Petrographic studies indicate that sandstones in the Lang and Black Oak Spring sequences are mostly metamorphosed quartz arenites and quartz wackes. U-Pb analyses of detrital zircons suggest a Proterozoic or possibly Archean source, or a source composed of clastic material derived from Proterozoic or Archean rocks. Thus, petrographic and U-Pb detrital zircon data suggest that sand in the Lang and Black Oak Spring sequences was derived from a continental source.

Some of our colleagues have suggested that sandstones in the Shoo Fly Complex were deposited as part of an extensive continental-rise submarine-fan/slope system adjacent to western North America prior to the Late Devonian-Early Mississippian Antler orogeny. However, the results of work summarized here and in the literature consistently indicate that sandstone/argillite sequences in the Shoo Fly Complex were deposited in a trench setting where deformation and sedimentation were intimately linked processes. Based on the generally quartz-rich compositions of sandstones, and the results of U-Pb studies on detrital zircons, the trench in which the Shoo Fly Complex developed must have been located near enough to a continental landmass to have received detritus derived from it. Published paleontological data, and published geochemical data derived from phosphatic lenses in the Shoo Fly Complex, imply that the trench developed adjacent to the western North American margin perhaps within $\pm$ 50° of the paleoequator.

INTRODUCTION

The Shoo Fly Complex, northern Sierra Nevada, California, is a metamorphosed pre-Upper Devonian eugeoclinal assemblage of sandstone, mudstone, chert, limestone, basalt, gabbro, and ultramafic rock that is

In Cooper, J.D., and Stevens, C.H., eds., 1991, **Paleozoic Paleogeography of the Western United States-II:** Pacific Section SEPM, Vol. 67, p. 703-716.

unconformably overlain by the Upper Devonian Grizzly and Sierra Buttes Formations (Hanson and Schweickert, 1986; Girty and others, 1990). It has been subdivided into four generally northwestwardly striking and northeastwardly dipping lithotectonic units by Schweickert and others (1984) and Harwood (1988). These fault-bounded units are, in descending structural order, the Sierra City melange, Culbertson Lake allochthon, Duncan Peak allochthon, and Lang sequence (Fig. 1). Detailed mapping by Taylor (1986) and Richards (1990) indicates that the Lang sequence can be subdivided into the Lang and Black Oak Spring sequences (Fig. 1).

This paper summarizes the results of recently completed sedimentological and U-Pb detrital zircon studies in the Lang and Black Oak Spring sequences of Taylor (1986) and Richards (1990). These new data, in conjunction with the results of previous studies, suggest that sandstone/argillite sequences in the Lang and Black Oak Spring sequences were deposited in a submarine trench that had developed close enough to a continental landmass to have received sand-sized detritus derived from it. Before presenting our new data we briefly review paleogeographic and paleotectonic constraints derived from each of the major lithotectonic units in the Shoo Fly Complex.

SIERRA CITY MELANGE

The Sierra City melange contains blocks of sandstone, bedded chert, limestone, gabbro, and basalt embedded in a variably sheared mudstone or serpentinite matrix (Schweickert and others, 1984; Girty and Pardini, 1987). The portion of the melange exposed near Bowman Lake has been studied in some detail, and is composed of blocks of sandstone, bedded chert, limestone, and volcanic-clast bearing debris flows enclosed in a variably foliated matrix of mudstone. Girty and Pardini (1987) interpreted the melange near Bowman Lake to have formed as a result of mass wasting in a trench-slope environment (Fig. 1). Other parts of the melange have not been studied in detail, and as a result little is known about the sedimentological or tectonic processes which may have led to their development.

Feldspar-rich sandstones in the Sierra City melange near Bowman Lake contain detritus derived from a dissected magmatic arc, whereas feldspar-poor sandstones are composed of debris derived from a more siliceous continental-like source (Fig. 1) (Girty and Pardini, 1987). U-Pb data derived from detrital zircon extracted from a single block of feldspar-rich sandstone suggest a Late Cambrian or Early Ordovician (506 $\pm$ 22 Ma) source-rock age, whereas other U-Pb detrital zircon data from feldspar-poor sandstones in the melange indicate a Precambrian source (Girty and Wardlaw, 1984; Pardini, 1986; Girty and Pardini, 1987).

The Sierra City melange north of the area shown in Figure 1 has yielded several important pieces of information regarding its origin. The Montgomery Limestone, a series of lenses or blocks within the melange near the crest of Grizzly Ridge, has yielded Late Ordovician megafossils and conodonts that have a North American affinity (Hannah and Moores, 1986). Phosphatic nodules in argillaceous rocks in the Lakes Basin area yielded Ordovician to Silurian radiolarians (Varga and Moores, 1981), and Saleeby and others (1987) report a Late Silurian (423 +5/-15 Ma) U-Pb zircon age for a siliceous tuff. The rocks sampled by Varga and Moores (1981) and Saleeby and others (1987) occur in or near Lakes Basin in the Sierra City melange as defined by Bond and Schweickert (1981).

Hsu (1968) concluded that melange blocks must be older than the matrix in which they occur. Thus, the above data imply that the Sierra City melange is younger than about Late Silurian and older than Late Devonian, the age of the rocks which unconformably overlie it.

CULBERTSON LAKE ALLOCHTHON

The Culbertson Lake allochthon includes; (1) alkalic basalts interstratified with bedded radiolarian cherts (i.e., the lower part of the Bullpen Lake Sequence in Figure 1); (2) probable allodapic limestones interstratified with bedded cherts (i.e., the upper part of the Bullpen Lake sequence in Figure 1); (3) bedded cherts composed of variable proportions of radiolaria and sponge spicules that are probably younger than Cambrian and older than Late Devonian (i.e., cau-1 and cau-2 in Figure 1) (Girty and others, 1990; D. L. Jones, 1985, personal communication; R. Barber, 1990, personal communication); and (4) abundant quartz arenite and quartz wacke composed of continentally derived detritus (i.e., the Poison Canyon and Red Hill units in Figure 1) (Girty and Wardlaw, 1985; Girty and others, 1990).

Sandstone beds in the allochthon are commonly graded and/or exhibit two or more of the Bouma intervals. Thus, Girty and others (1990) interpreted sandstone beds to have been derived from sediment gravity flows.

Sandstone-dominated sequences in the Culbertson Lake allochthon are bounded by generally westward verging pre-Late Devonian thrust faults and are arranged in upward thickening and coarsening sequences (Fig. 1) (Girty and others, 1990). The upper portions of the upward thickening and coarsening sequences contain variably thick, upward thinning and fining channel-fill sequences, whereas the lower portions contain variably thick and laterally discontinuous upward thickening and coarsening lobe deposits that overlie pelagic cherts and hemipelagic mudstones. Pre-Late Devonian sedimentological and structural patterns in the Culbertson Lake allochthon suggest that it is the record of deposition and deformation in a trench setting (Girty and others, 1990). Detailed

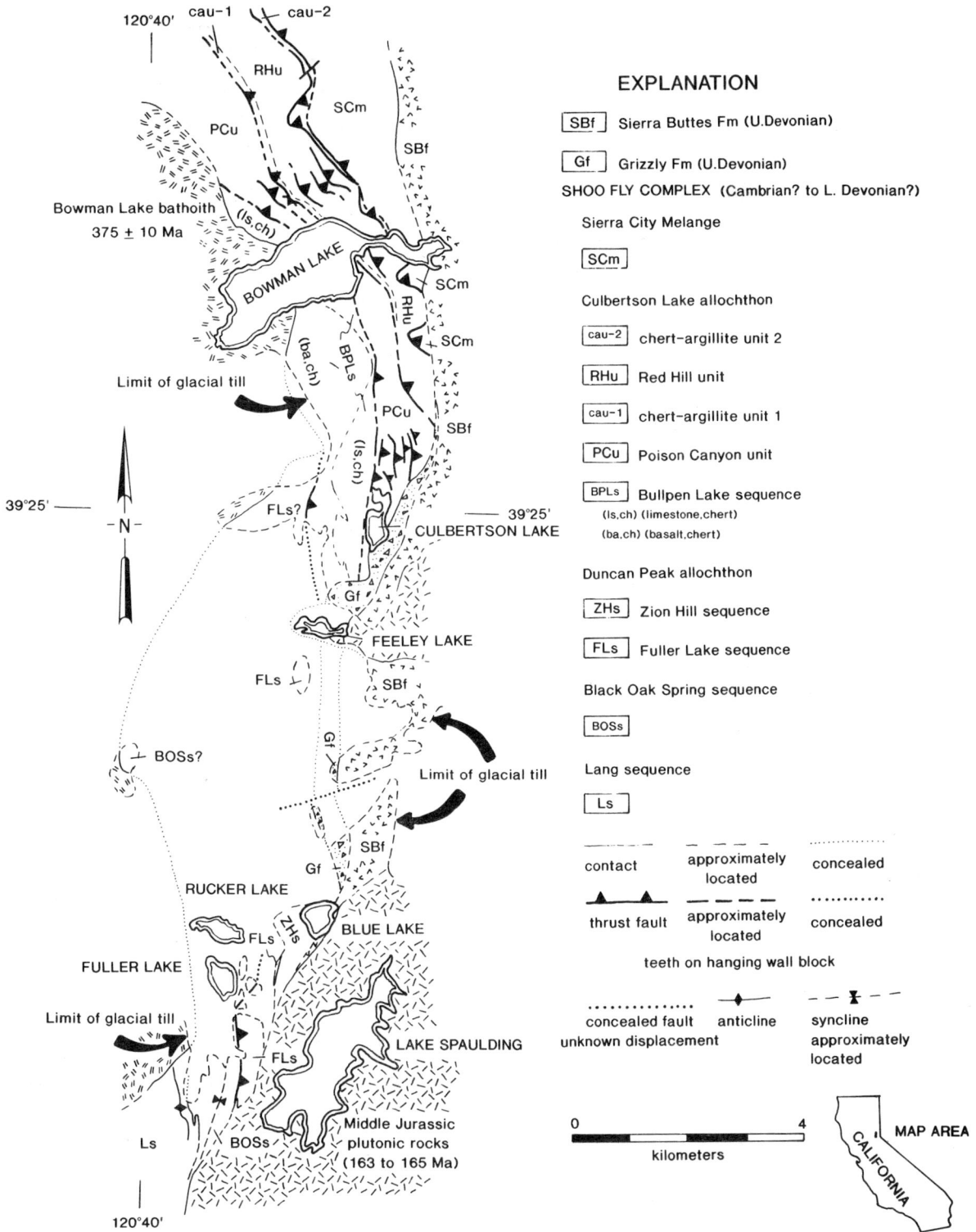

EXPLANATION

| SBf | Sierra Buttes Fm (U.Devonian)

| Gf | Grizzly Fm (U.Devonian)

SHOO FLY COMPLEX (Cambrian? to L. Devonian?)

Sierra City Melange

| SCm |

Culbertson Lake allochthon

| cau-2 | chert-argillite unit 2

| RHu | Red Hill unit

| cau-1 | chert-argillite unit 1

| PCu | Poison Canyon unit

| BPLs | Bullpen Lake sequence
(ls,ch) (limestone,chert)
(ba,ch) (basalt,chert)

Duncan Peak allochthon

| ZHs | Zion Hill sequence

| FLs | Fuller Lake sequence

Black Oak Spring sequence

| BOSs |

Lang sequence

| Ls |

Figure 1. Simplified geologic map of the Shoo Fly Complex between Bowman and Spaulding Lakes.

petrologic and U-Pb detrital zircon studies indicate that the trench developed close enough to some continental landmass to have received detritus derived from it (Girty and Wardlaw, 1985).

DUNCAN PEAK ALLOCHTHON

Recent mapping between Blue Lake and Lang Crossing indicates that the Duncan Peak allochthon of Schweickert and others (1984) and Harwood (1988) consists of, in descending structural order, the Zion Hill

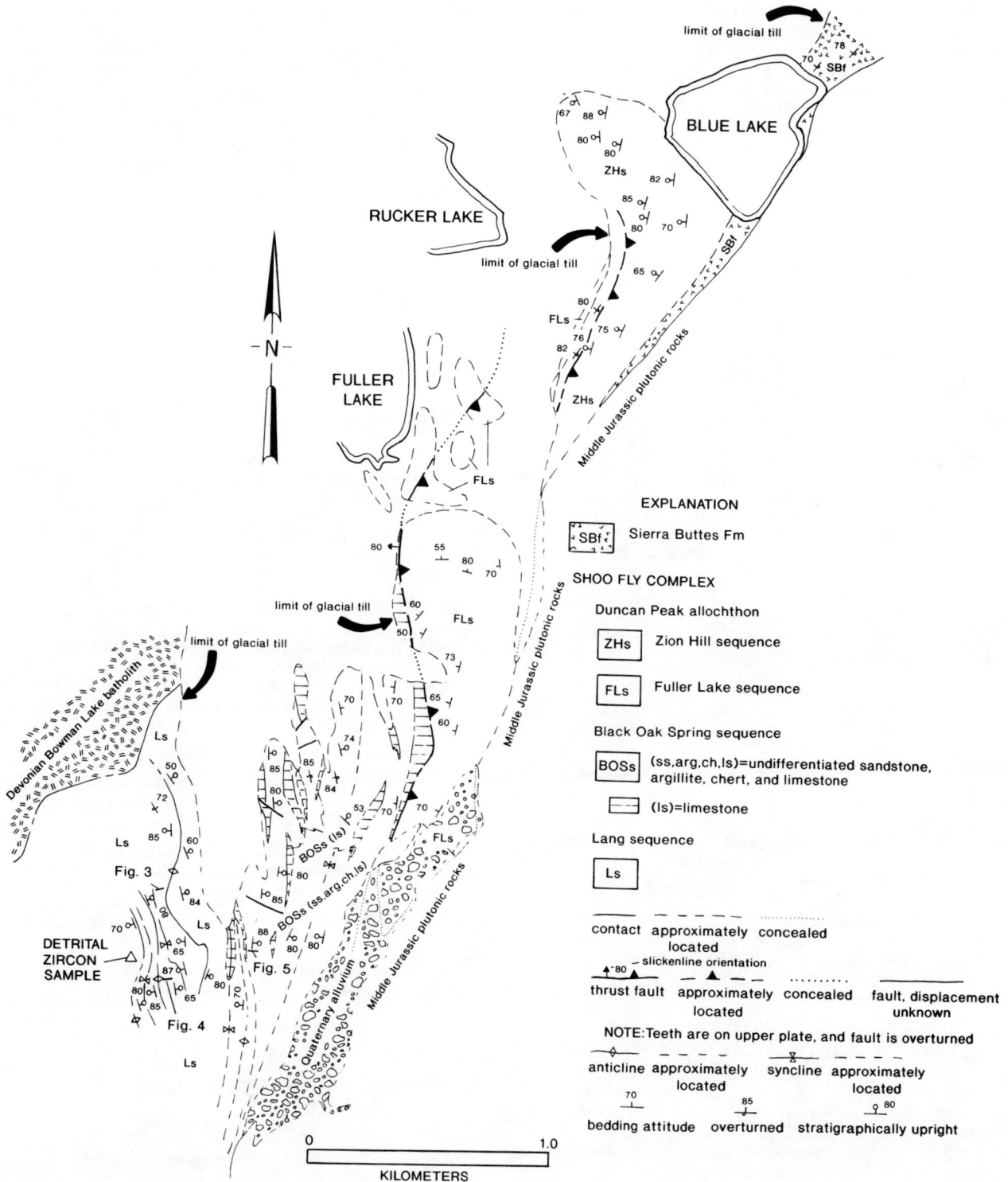

Figure 2. Detailed geologic map of the southern part of Figure 1 showing the location of the detrital zircon sample and figures 3, 4, and 5.

and Fuller Lake sequences (Figs. 1 and 2) (Taylor, 1986; Richards, 1990). Sedimentological aspects of the Zion Hill and Fuller Lake sequences have not been studied, but detailed mapping suggests that the former unit consists of about 90 m of sandstone and argillite, whereas the latter unit is composed of about 90 m of bedded chert, argillite, and minor sandstone (Richards, 1990). A detailed U-Pb study of detrital zircons extracted from a single bed of quartz-rich sandstone in the Zion Hill sequence indicates a Precambrian continental source (Richards, 1990).

LANG AND BLACK OAK SPRING SEQUENCES

The Black Oak Spring sequence is a newly defined unit that structurally underlies the Fuller Lake sequence (Figs. 1 and 2) (Taylor, 1986; Richards, 1990). It was previously included in the Lang-Halsted of Schweickert and others (1984), and in the Lang sequence of Harwood (1988). The Black Oak Spring sequence is composed of about 300 m of bedded chert, argillite, limestone, and sandstone (Taylor, 1986; Richards, 1990).

Limestones in the Black Oak Spring sequence are variable in thickness (2 m to 46 m), and are laterally discontinuous (Fig. 2). They are recrystallized and highly deformed, and as a result, little sedimentological data have been extracted from these rocks. Bedded chert lenses, 1 m to 5 m thick, also are extensively recrystallized and locally highly deformed. Attempts to extract conodonts and radiolarians from cherts and limestones in the Black Oak Spring sequence have been unsuccessful.

The contact between the Black Oak Spring sequence and the underlying Lang sequence is not well exposed (Figs. 1 and 2). Lacking evidence that it is structurally controlled, we assume that the contact is depositional. Abundant graded beds near the inferred depositional contact consistently indicate that the Lang sequence is older than the Black Oak Spring sequence (Fig. 2). The Lang sequence is composed of at least 300 m of highly folded and faulted sandstone and argillite.

In the following sections we summarize the results of our sedimentological and stratigraphical studies of relatively undeformed sandstone/argillite sequences in the Lang and Black Oak Springs sequences. Although the Lang and Black Oak Springs sequences have been metamorphosed to the hornblende hornfels facies by nearby Middle Jurassic plutons, and are multiply deformed (Fig. 1), protoliths and sedimentary structures are moderately to well preserved and easily identifiable in most places (Taylor, 1986; Richards, 1990; McNulty, 1990). For ease of discussion, the prefix "meta" is not used in this paper.

SANDSTONE/ARGILLITE IN THE LANG AND BLACK OAK SPRINGS SEQUENCES

Although the Lang and Black Oak Springs sequences are folded and faulted, mapping at a scale of 1:12,000 revealed many locations where exposure was of good quality, and graded beds, festoon ripple cross-laminations, flame structures, and sole marks indicated a uniform stratigraphic facing direction for a considerable thickness of section (Fig. 2) (Taylor, 1986; McNulty, 1990; Richards, 1990). These areas normally occur on the limbs of tight folds that formed prior to the emplacment of the 375 ± 10 m.y. old Bowman Lake batholith (Fig. 2) (Taylor, 1986; Hanson and others, 1988; McNulty, 1990; Richards, 1990). In order to document sedimentologically significant characteristics and vertical stratigraphic trends, we measured, bed-by-bed, 19 sections in the Lang sequence, and 1 section in the Black Oak Spring sequence. Descriptions and locations of each measured section can be found in Taylor (1986) and Gurrola (1990). Representative sections are shown in Figures 3, 4, 5. Our analysis of these 20 sections and our observations during mapping suggest that sandstone/argillite successions in the Lang and Black Oak Spring sequences are composed of three major lithofacies which we designate massive sandstone, turbidite, and mudstone/argillite. The general sedimentological characteristics of each of these lithofacies are described below, and are shown schematically in Figure 6, which also serves as an explanation of the symbols used in Figures 3, 4, and 5.

Because of the ubiquitous presence of folds, faults, glacial till, and colluvium, we were unable to trace laterally any given bed in our measured sections beyond about 90 meters. In addition, the rocks in our study area have undergone multiple phases of folding and cleavage development (Taylor, 1986; Richards, 1990; McNulty, 1990). A paleocurrent analysis would require undeforming paleoflow directions through at least three phases of folding and accompanying plastic distortions. Unfortunately, the magnitudes and directions of the principal strains associated with each phase of deformation are not known. Thus, a paleocurrent analysis was not undertaken during this study.

Massive Sandstones

Massive sandstone beds account for 49% to 66% of the sandstone beds in the 20 measured sections (e.g., Figs. 3, 4, and 5). Sediment ranges from very fine sand to pebble, but very coarse sand to pebble are the most commonly observed grain sizes. Bedding thickness ranges from 5.0 cm to 3.0 m. The average massive sandstone bed is 0.6 m thick.

Most (84%) massive sandstone beds are graded; non-graded beds are less common (16%) (e.g., Figs. 3, 4, and 5). Graded beds display distribution, coarse-tail, or inverse grading.

Amalgamated sequences composed of massive sandstone commonly include two to nine sandstone beds, and range in thickness from 1.0 m to over 11.3 m (e.g., Figs. 3, 4, and 5). Planar and scoured sandstone bases occur in these intervals.

Most (>50%) massive sandstone beds have flat lower contacts, but scoured basal contacts also were observed. Locally, scours range from 1 cm to 6 cm deep, and are typically filled with coarse sand to pebble conglomerate. Load casts, flame structures, and flute casts are uncommon (e.g., Figs. 3, 4, and 5).

708

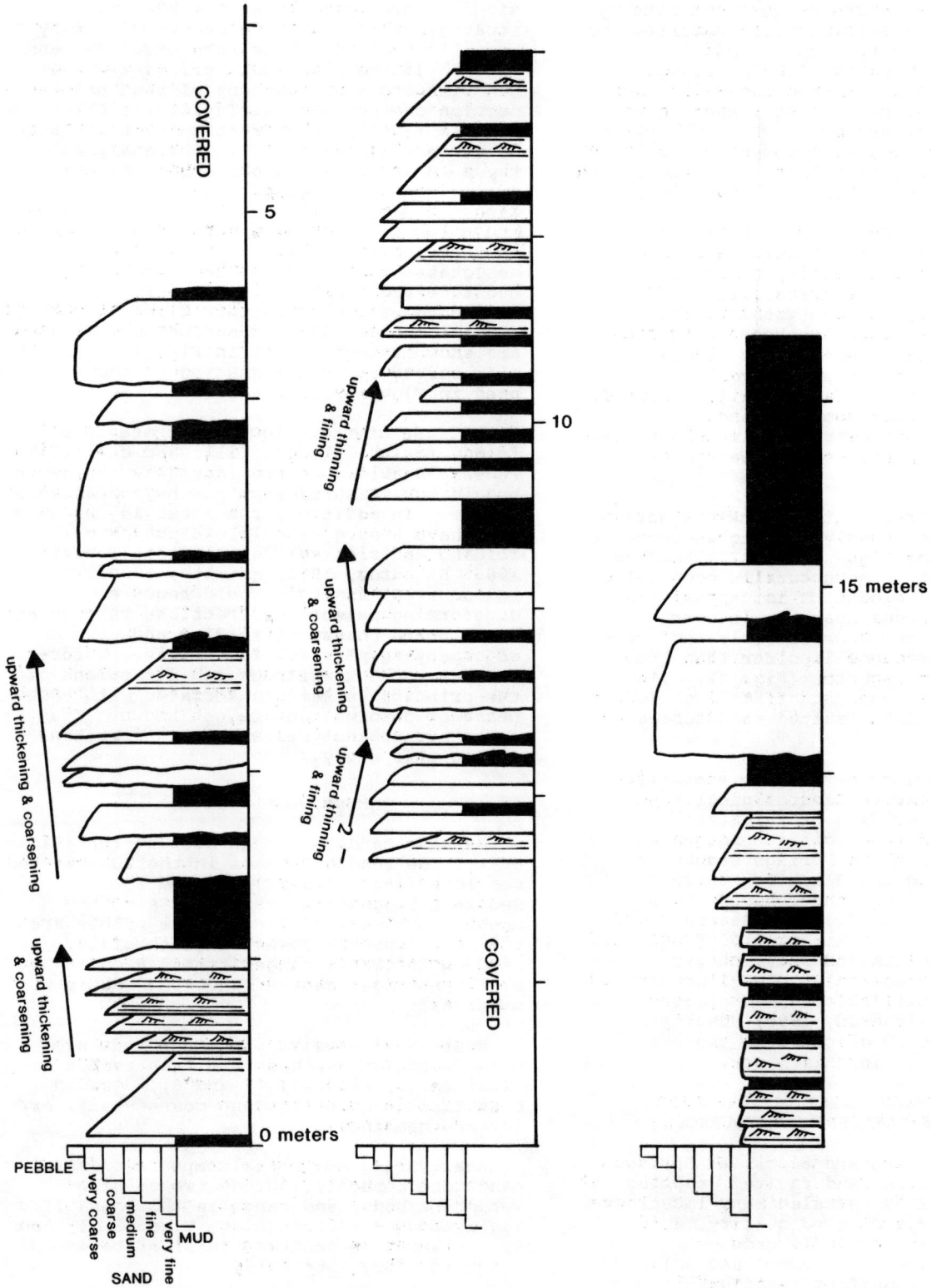

Figure 3. Detailed stratigraphic column of a small portion of the Lang sequence. See Figure 6 for an explanation of the symbols used in this illustration, and Figure 2 for the location of the measured section.

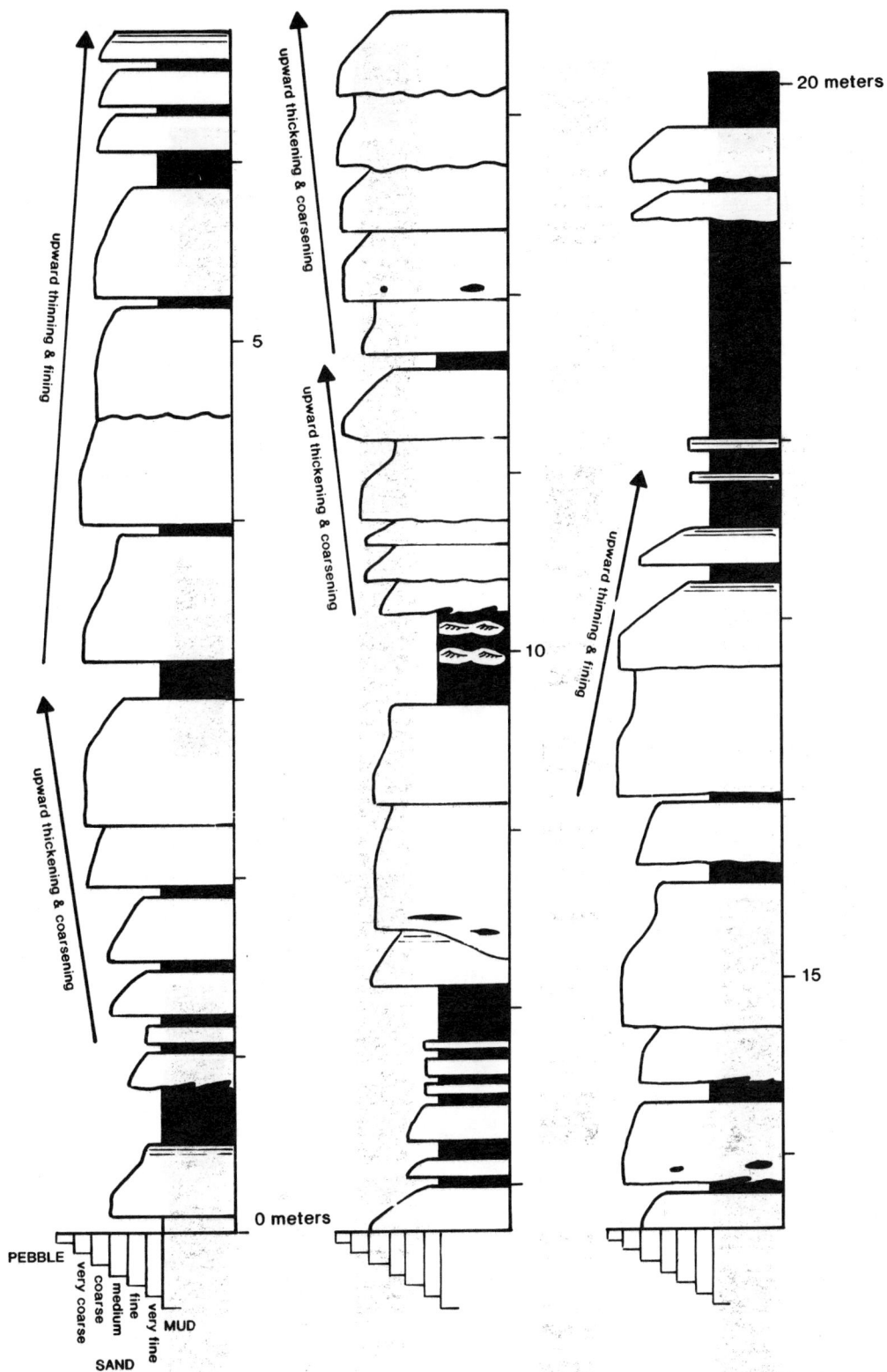

Figure 4. Detailed stratigraphic column of a small portion of the Lang sequence. See Figure 6 for an explanation of the symbols used in this illustration, and Figure 2 for the location of the measured section.

710

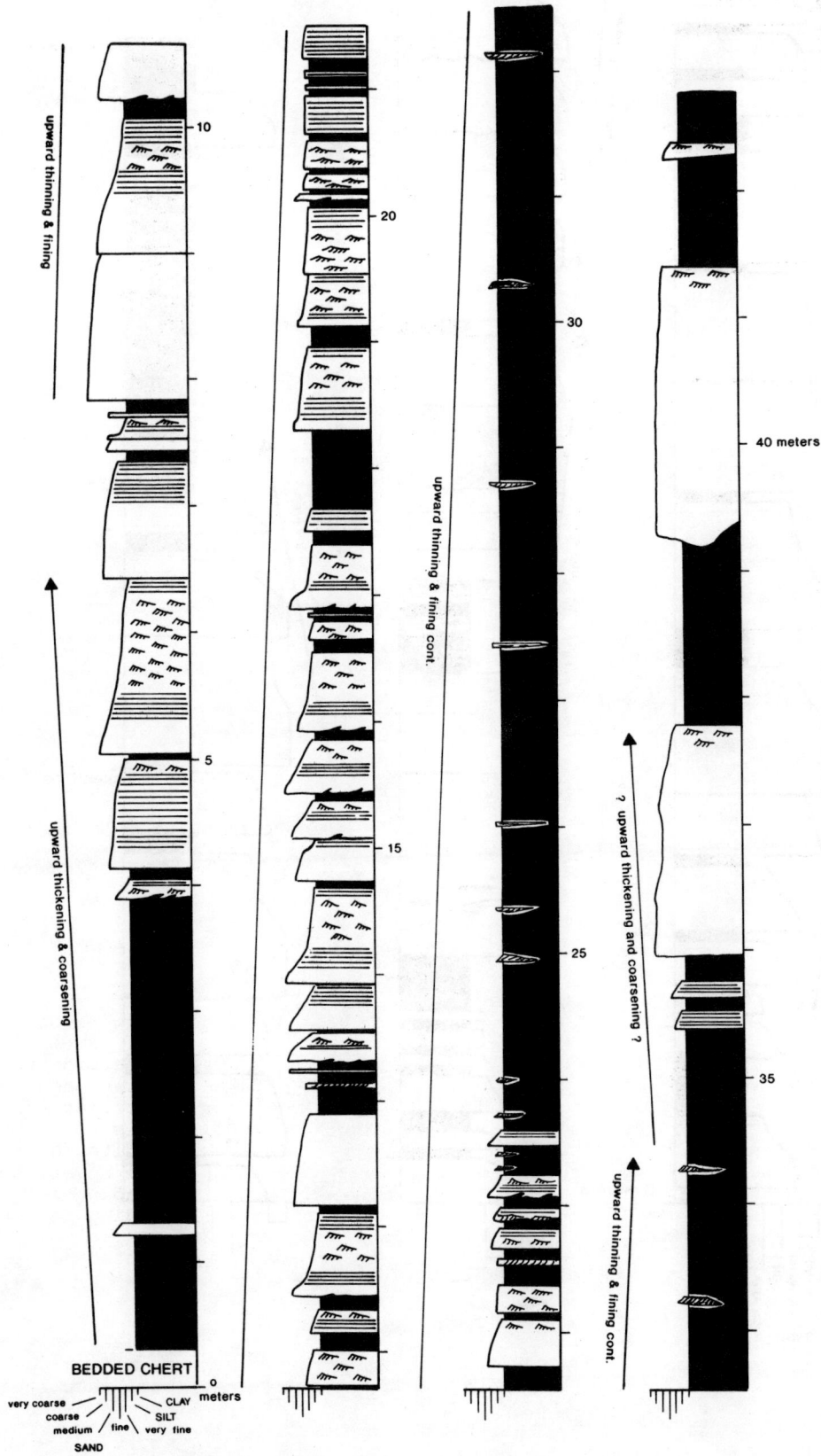

Figure 5. Detailed stratigraphic column of a small portion of the Black Oak Spring sequence. See Figure 6 for an explanation of the symbols used in this illustration, and Figure 2 for the location of the measured section.

MASSIVE SANDSTONES

TURBIDITES

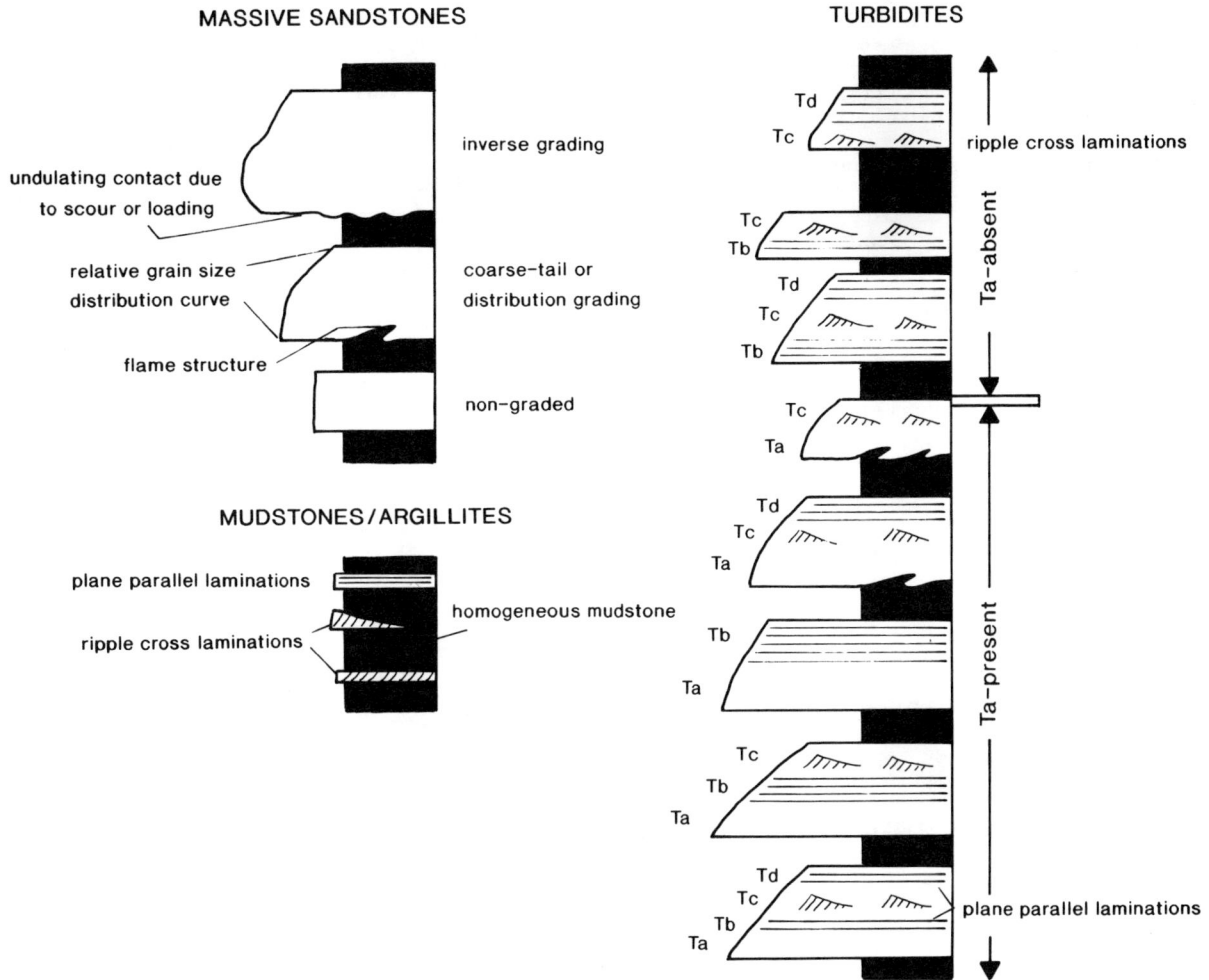

inverse grading

undulating contact due
to scour or loading

relative grain size
distribution curve

flame structure

coarse-tail or
distribution grading

non-graded

MUDSTONES/ARGILLITES

plane parallel laminations

ripple cross laminations

homogeneous mudstone

Td
Tc   ripple cross laminations

Ta-absent

Tc
Tb

Td
Tc
Tb

Tc
Ta

Td
Tc
Ta

Tb
Ta

Ta-present

Tc
Tb
Ta

Td
Tc
Tb   plane parallel laminations
Ta

Figure 6. General characteristics of major lithofacies recognized
in the Lang and Black Oak Spring sequences. This illustration also
serves as an explanation for the various sedimentological symbols
used in Figures 3, 4, and 5.

Turbidites

Ta-present Turbidites

The ideal and complete turbidite includes
all the Bouma intervals which are, from
bottom to top; (1) a graded Ta interval; (2)
a Tb interval consisting of plane parallel
laminae; (3) a Tc interval containing
convolute or ripple cross laminae; (4) a Td
interval consisting of plane parallel
laminae; and a Te interval composed of
homogeneous mudstone (Bouma, 1962). During
our studies we included the Te interval as
part of the mudstone/argillite lithofacies.
Between 16% and 18% of the sandstone beds in
our 20 sections are turbidites (e.g., Figs.
3, 4, and 5). Of the turbidites in our
sections, Ta-present beds account for 76%
and Ta-absent types account for 24%.
Thicknesses of Ta-present turbidites range
from 10 cm to 2.6 m. The average Ta-present
bed is 0.5 m thick.

Ta-present turbidite beds include Tab,
Tabc, or Tabcd types (e.g., Figs., 3, 4, 5).
More than 80% of the Ta-present turbidites
are composed of Tab intervals. Tabc and
Tabcd types are less common. Detritus in
Ta-present beds ranges from pebble to very
fine sand. Most Ta-present turbidites have
very coarse sand bases that grade upward to
fine sand tops. Distribution grading is
most common, and coarse-tail grading is less
frequent. Basal contacts are commonly
planar, whereas scoured lower contacts, or
flame structures are less common.

Ta-absent Turbidites

Ta-absent turbidites make up 24% of the
turbidites present in our measured sections
(e.g., Figs. 3, 4, and 5). They range in
thickness from 3.0 cm to 27.0 cm, and are
composed of medium-grained sand to very-
fine-grained sand or silt.

Ta-absent turbidites commonly contain the Tbc or Tcd Bouma intervals (Figs. 3, 4, and 5). Tb intervals are on average 8 cm thick, and are composed of plane parallel laminated fine-grained sandstone. Laminations are commonly 1 cm to 4 cm thick. Tc intervals are most commonly ripple cross-laminated, very-fine-grained sandstone and siltstone. They typically have planar or scoured lower contacts, and are 4 cm to 10 cm thick. Convolute laminations and starved or climbing ripples are uncommon forms of Tc intervals. Td intervals are on average 2.5 cm to 5 cm thick. They are composed mostly of laminated very-fine-grained sandstone and siltstone.

Mudstone/Argillite

Mudstone/argillite beds occur throughout the sections measured in the Lang and Black Oak Springs sequences (e.g., Figs. 3, 4, and 5). They vary from less than 2.5 cm to over 12.0 m in thickness, and occur as intercalations between massive sandstone beds and/or turbidites, or as sequences capping upward thinning and fining cycles. Mudstone/argillite beds are homogeneous, locally display very thin laminae (3 mm to 5 mm thick), or exhibit ripple cross laminations (e.g., Figs. 3, 4, 5).

INTERPRETIVE DEPOSITIONAL SETTING

The sedimentological characteristics of sandstone beds observed and measured in the Lang and Black Oak Spring sequences are indicative of sediment deposited by subaqueous sediment gravity flows such as high density and dilute turbidity currents (Lowe, 1982, Nelson and Nilsen, 1984). In the modern world, and in the ancient rock record, thick accumulations of sandstone derived from sediment gravity flows commonly occur, or are inferred to have occurred, at the mouths of submarine canyons that transect slopes along passive continental margins, trenches, and borderland basins such as those in the coastal region of California (Howell and Normark, 1982; Underwood and Bachman, 1982; Nelson and Nilsen, 1984; Lash, 1985, 1986; Thornburg and Kulm, 1987). In these settings, sand or sandstones derived from sediment gravity flows may form extensive submarine fan systems composed of upper, middle, and lower fan sub-environments (Howell and Normark, 1982; Nelson and Nilsen, 1984). The sedimentological characteristics of sediment deposited in each of these sub-environments are largely based upon their recognition in well documented ancient submarine fans (Nelson and Nilsen, 1984). The upper fan is composed of channelized massive sandstones and conglomerates. In contrast, the middle fan is comprised of massive sandstones and turbidites that form both upward thickening and coarsening and upward thinning and fining sequences. The lower fan consists of thin-bedded mudstones and laterally continuous turbidites that lack the Ta interval. Submarine fan progradation produces a vertical sequence that begins with basin plain sediments followed by the

deposits of the lower, middle, and upper fan (Howell and Normark, 1982).

Sandstone-dominated depositional systems in trenches may contain not only the deposits of submarine fans, but also sediment deposited in trench-parallel axial channels that function as a distributary network along the trench between submarine canyons which act as points of sediment supply (Schweller and Kulm, 1978; Underwood and Bachman, 1982; Lash, 1985, 1986; Thornburg and Kulm, 1987). Sediment gravity flows moving along the trench floor also can be unconfined (Schweller and Kulm, 1978; Underwood and Bachman, 1982). Such flows spread sediment along and parallel to the trench floor in sheet-like form. Well developed vertical trends are unlikely in trench sediments deposited by unconfined axial flow (Underwood and Bachman, 1982).

Ancient examples of deposition within an axial channel are dominated by thick successions of channelized massive sandstones and Ta-present turbidites characteristic of the inner and middle fan facies associations, and lack a well developed outer fan facies association (Underwood and Bachman, 1982). In some ancient trench deposits sheet-like depositional units that exhibit no vertical trends and are composed of massive sandstones and Ta-present turbidites may be the result of unconfined axial flow (Underwood and Bachman, 1982).

Our sections from the Lang and Black Oak Spring sequences contain both upward thickening and coarsening and upward thinning and fining cycles composed of four or more beds; however, some intervals show no vertical trends (e.g., Figs. 3, 4, and 5) (Taylor, 1986; Gurolla, 1990). Upward thickening and coarsening cycles are on average 3.2 m thick, and are composed mostly of massive sandstones and lesser amounts of Ta-present turbidites. The average sandstone-to-mudstone ratio in upward thickening and coarsening cycles is 8:1. Upward thinning and fining cycles are on average 4.2 m thick, and generally are composed of massive sandstones and Ta-present turbidites in their lower parts, and Ta-present and Ta-absent turbidites and mudstones/argillites in their upper parts (e.g., Figs. 3, 4, and 5). The average sandstone-to-mudstone ratio in upward thinning and fining sequences is 5:1. We infer that the upward thickening and coarsening cycles are progradational lobes probably produced by channel migration, whereas the upward thinning and fining cycles are most likely channel-fill deposits (cf., Nelson and Nilsen, 1984).

Our data for the Lang and Black Oak Spring sequences consistently indicate a middle fan facies association and a general absence of an outer fan facies association. Hence, we infer that sandstone/argillite sequences in the Lang and Black Oak Spring sequences were deposited in a trench setting where axial channels and possibly unconfined axial flow played an important role in distributing sediment. Our paleogeographic and paleotectonic model for the Lang and Black Oak Spring sequences is illustrated in Figure 7.

Thin section petrography indicates that sandstones in the Lang and Black Oak Spring sequences largely are metamorphosed quartz arenites and quartz wackes (Taylor, 1986; Richards, 1990). These observations suggest that the sandstones contain detritus derived from a continental source. In order to test this idea, we collected approximately 50 kg of sandstone from a single bed in the Lang sequence for U-Pb detrital zircon study. The location of our detrital zircon sample is shown on Figure 2.

DETRITAL ZIRCONS

Separation Techniques and Characteristics

Utilizing conventional crushing and heavy liquid techniques approximately 60 mg of detrital zircon was extracted from the 50 kg sandstone sample (Richards, 1990). Zircons were then subdivided into several fractions based on their magnetic susceptibilites with a Frantz Magnetic Barrier Laboratory Separator. From the nonmagnetic fraction a zircon population coarser than 200 mesh was collected using standard sieve cloth. From this relatively coarse fraction, 0.15 mg of round pink, 0.17 mg of round clear, 0.7 mg of subhedral pink, and 0.1 mg of euhedral pink zircon were handpicked with surgical tweezers for U-Pb analyses.

Analytical Procedures and Results

The four handpicked zircon splits were analyzed for their U-Pb isotopic compositions on a 30.5-cm, AVCO mass spectrometer equipped with an electron multiplier and programmable magnetic field scanning and digital output at the Baylor Brooks Institute of Isotope Geology, San Diego State University. Analytical techniques were modified from Krough (1973). Laboratory blanks averaged 0.1 nG total Pb. Data were corrected for modern blank Pb and nonradiogenic model Pb (Stacey and Kramers, 1975). Assuming that the use of the Stacey and Kramers (1975) model Pb values for approximately 2.2-b.y.-old rocks is valid, analytical uncertainties are estimated to be better than ± 1.0% for U/Pb ratios, and better than ± 0.5% for radiogenic Pb/Pb ratios at the 95% confidence level. Data reduction followed procedures outlined in Ludwig (1989). The results of our analyses are given in Table 1. Ages were calculated using the decay constants of Jaffey and others (1971).

The four analyzed zircon fractions have variable Precambrian U/Pb and Pb/Pb ages (Table 1). In general, such behavior is expected of mixtures of zircon of different ages and different lead-loss histories. Note, however, that the Pb/Pb ages for the four fractions vary from about 1.8 to 2.4 billion years. For detrital zircons, the Pb/Pb ages are minimum estimates of the ages of the oldest possible zircon components. Thus, our U-Pb data suggest that the source for sandstones in the Lang sequence included Proterozoic or possible Archean rocks, or older clastic rocks composed of Proterozoic or Archean debris. This result, in combination with our limited petrographic data, suggests that sandstones in the Lang sequence, and presumably sandstones in the Black Oak Spring sequence, are composed of continentally-derived detritus.

DISCUSSION/CONCLUSIONS

Schweickert and Snyder (1981) suggested that the Shoo Fly Complex is a part of an accretionary system that includes the Roberts Mountains allochthon in north-central Nevada. The Roberts Mountains allochthon was emplaced onto the edge of western North America during the Late Devonian-Early Mississippian Antler orogeny as the result of a collison between western North America and an east-facing island arc (Schweickert and Snyder, 1981; Speed and Sleep, 1982). Schweickert and Snyder (1981) interpreted Cambrian sandstones in the Roberts Mountains allochthon (e.g., the Harmony Formation), and sandstones in the Poison Canyon unit of the Culbertson Lake allochthon as the deposits of an extensive submarine fan system that developed in a rise setting adjacent to the western North American margin prior to the Antler orogeny. In contrast, Girty and others (1990) suggested that all sandstones in the Culbertson Lake allochthon were deposited in a trench setting where deformation and sedimentation were intimately linked processes, and that rocks in the Shoo Fly Complex may not be related directly to rocks in the Roberts Mountains allochthon.

Data presented in this paper allow the paleogeographic and paleotectonic model suggested by Girty and others (1990) to be extended to the Lang and Black Oak Spring sequences. For example, the lack of a well defined outer fan association and the relative abundance of thick bedded, channelized and non-channelized massive sandstones and Ta-present turbidites in the Black Oak Spring and Lang sequences are traits more consistent with the trench-wedge model than the continental-rise model of Schweickert and Snyder (1981). Thus, we interpret the Shoo Fly Complex to be a series of trench-wedges that were accreted to a subduction complex (Fig. 7).

Based on the generally quartz-rich compositions of sandstones, and the results of our U-Pb studies on detrital zircons, the trench in which the Shoo Fly Complex developed must have been located near enough to a continental landmass to have received sand-sized detritus derived from it. Limited paleontological data, when coupled with geochemical data derived from phosphatic lenses from the Shoo Fly Complex, imply that the trench developed adjacent to the western North American margin perhaps within ± 50° of the paleoequator (Hannah and Moores, 1986; Varga, 1982). Unfortunately,

714

Table 1. U-Pb detrital zircon data from the Lang Sequence.

Fraction	wt.(mg)	Pb (ppm)	U (ppm)	208/206	207/206	204/206	206/238	207/235	207/206	7/6 err.±
round pink	0.15	71.7	178	0.16021	0.14037	0.000089	0.3604	6.918	0.1392	
							1984	2101	2217	2
round clear	0.17	17.7	47	0.19669	0.16034	0.000698	0.3286	6.858	0.1514	
							1831	2093	2362	2
subeuhedral pink	0.7	62.7	174	0.16321	0.13582	0.000089	0.3231	5.999	0.1347	
							1805	1976	2160	2
euhedral	0.1	5.1	20	0.16545	0.13023	0.001536	0.2292	3.459	0.1084	
							1331	1518	1790	5

Column span header: Pb isotopic composition — Unspiked, multiplier corrected (208/206, 207/206, 204/206); radiogenic ratios* age (m.y.) (206/238, 207/235, 207/206)

*Radiogenic Pb corrected for blank (206Pb/204Pb=18.87; 207Pb/204Pb=15.66; 208Pb/204Pb=38.53) and initial Pb (Stacey and Kramers (1975) 2200 m.y. old Pb: 206Pb/204Pb=14.742; 207Pb/204Pb=15.083; 208Pb/204Pb=34.394).

Lead ratios were corrected for mass fractionation of 0.1±0.03% per mass unit and the uranium ratio was corrected using 0.12±0.06% per mass unit based on replicate analyses of NBS 981, 983 and U-050.

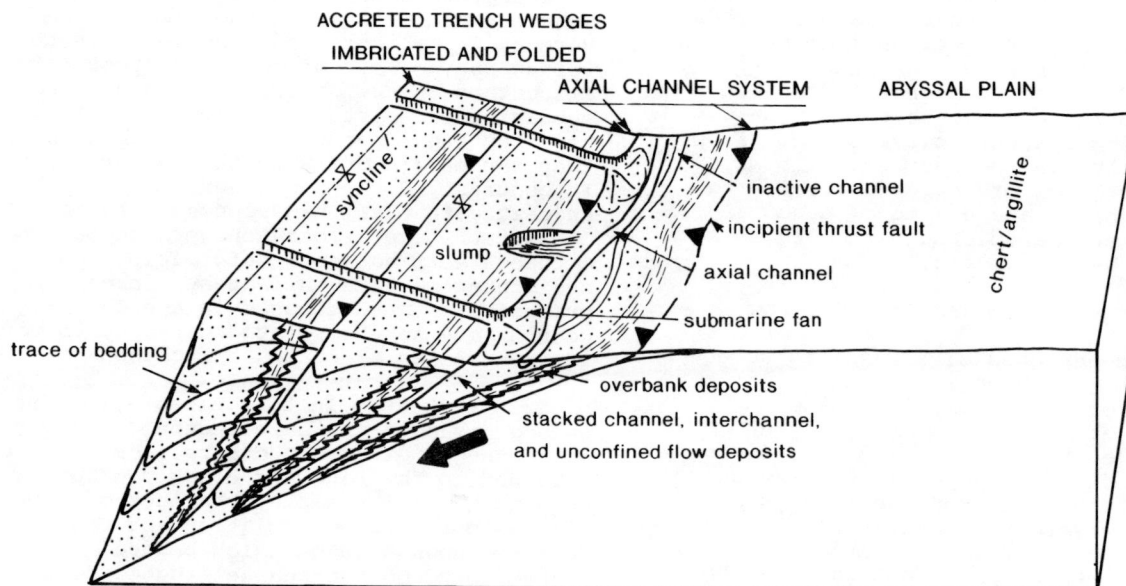

Figure 7. Schematic, not-to-scale, block diagram illustrating the paleogeographic and paleotectonic setting for the Shoo Fly Complex.

the exact location of source rocks, and the positions of the trench and the North American continental margin within the ± 50° of latitudinal constraint remain problematical. Such a conclusion may be unsatisfying, but it is an inherent aspect of our current level of understanding of the Cordilleran accretionary collage (Coney and others, 1980). Continuing detailed field studies such as the ones summarized here, are our only hope of unraveling the complex history of early to middle Paleozoic deposition and accretion along the western Cordilleran margin.

ACKNOWLEDGEMENTS

Work reported here was funded by a grant from the National Science Foundation (EAR 8902382) awarded to G. H. Girty. Professors R. Miller and J. D. Cooper provided reviews of this paper.

REFERENCES CITED

Bond, G. C., and Schweickert, R. A., 1981, Significance of pre-Devonian melange in the Shoo Fly Complex, northern Sierra Nevada, California: Geological Society of America Abstracts with Programs, v. 13, p. 46.
Bouma, A. H., 1962, Sedimentology of Some Flysch Deposits: A Graphic Approach to Facies Interpretation: Amsterdam, Elsevier, 163 p.
Coney, P. J., Jones, D. L., and Monger, J. W. H., 1980, Cordilleran suspect terranes: Nature, v. 288, p. 329-333.

Girty, G. H., and Pardini, 1987, Provenance of sandstone inclusions in the Paleozoic Sierra City melange, Sierra Nevada, California: Geological Society of America Bulletin, v. 98, p. 176-181.

Girty, G. H., and Wardlaw, 1984, Was the Alexander terrane a source of feldspathic sandstones in the Shoo Fly Complex, Sierra Nevada, California?: Geology, v. 12, p. 339-342.

Girty, G. H. and Wardlaw, 1985, Petrology and provenance of pre-Late Devonian sandstones, Shoo Fly Complex, northern Sierra Nevada, California: Geological Society of America Bulletin, v. 96, p. 516-521.

Girty, G. H., Gester, K. C., and Turner, J. B., 1990, Pre-Late Devonian geochemical, stratigraphic, sedimentologic, and structural patterns, Shoo Fly Complex, northern Sierra Nevada, California, in Harwood, D. S., and Miller, M. M., eds., Paleozoic and Early Mesozoic Paleogeographic Relations; Sierra Nevada, Klamath Mountains, and Related Terranes: Geological Society of America Special Paper 255, p. 43-56.

Gurrola, L. D., 1990, Sedimentological characteristics of metamorphosed sandstones in the Lang sequence, northern Sierra Nevada, California [unpublished Senior thesis]: San Diego, California, San Diego State University, 38 p.

Hannah, J. L., and Moores, E. M., 1986, Age relationships and depositional environments of Paleozoic strata, northern Sierra Nevada, California: Geological Society of America Bulletin, v. 97, p. 787-797.

Hanson, R. E., and Schweickert, R. A., 1986, Stratigraphy of mid-Paleozoic island-arc rocks in part of the northern Sierra Nevada, Sierra and Nevada Counties, California: Geological Society of America Bulletin, v. 97, p. 986-998.

Hanson, R. E., Saleeby, J. B., and Schweickert, R. A., 1988, Composite Devonian island-arc batholith in the northern Sierra Nevada, California: Geological Society of America Bulletin, v. 100, p. 446-457.

Harwood, D. S., 1988, Tectonism and metamorphism in the northern Sierra terrane, northern California, in Ernst, W. G., ed., Metamorphism and Crustal Evolution of the Western United States (Rubey volume VII): Englewood Cliffs, New Jersey, Prentice-Hall, p. 765-788.

Howell, D. G., and Normark, W. R., 1982, Sedimentology of submarine fans, in Scholle, P. A., and Spearing, D., eds., Sandstone Depositional Environments: American Association of Petroleum Geologists, Memoir No. 31, Tulsa, Oklahoma, p. 365-404.

Hsu, K. J., 1968, Principles of melanges and their bearing on the Franciscan-Knoxville paradox: Geological Society of America Bulletin, v. 79, p. 1063-1074.

Jaffey, A., Flynne, K. F., Glendenin, L. E., Bentley, W. C., and Essling, A. M., 1971, Precision measurements of half-lives and specific activities of ^{235}U and ^{238}U: Physical Review, C4, p. 1889-1906.

Krogh, T. E., 1973, A low-contamination method for hydrothermal decomposition of zircon and extraction of U and Pb for isotopic age determination: Geochimica et Cosmochimica Acta, v. 37, p. 485-494.

Lash, G. G., 1985, Recognition of trench fill in orogenic flysch sequences: Geology, v. 13, p. 867-870.

Lash, G. G., 1986, Sedimentology of channelized turbidite deposits in an ancient (early Paleozoic) subduction complex, central Appalachians: Geological Society of America Bulletin, v. 97, p. 703-710.

Lowe, D. R., 1982, Sediment gravity flows: II: Depositional models with special references to the deposits of high-density turbidity currents: Journal of Sedimentary Petrology, v. 52, p. 279-297.

Ludwig, K. L., 1989, PBDAT for MS-DOS, a computer program for IBM-PC compatibles for processing raw Pb-U-Th isotope data, verison 1.05 revised April 1989: U.S. Geological Survey Open File Report 88-542, 39 p.

McNulty, B. A., 1990, Polyphase deformation in the lower Paleozoic Lang sequence, northern Sierra Nevada, Calif-ornia [unpublished M.S. thesis]: San Diego, California, San Diego State University, 76 p.

Nelson, C. H. and Nilsen, T. H., 1984, Modern and Ancient Deep-sea Fan Sedimentation: Society of Economic Paleontol- ogists and Mineralogists, lecture notes for Short Course No. 14, Tulsa, Oklahoma, 404 p.

Pardini, C. H., 1986, Petrological and structural analysis of the Sierra City melange, northern Sierra Nevada, California [unpublished M.S. thesis]: San Diego, California, San Diego State University, 87 p.

Richards, M. J., 1990, Shoo Fly Complex, Lake Spaulding area, northern Sierra Nevada, California: An early Paleozoic accretionary complex? [unpublished M.S. thesis]: San Diego, California, San Diego State University, 78 p.

Saleeby, J., Hannah, J. L., and Varga, R. J., 1987, Isotopic age constraints on middle Paleozoic deformation in the northern Sierra Nevada, California: Geology, v. 15, p. 757-760.

Schweickert, R. A., and Snyder, W. S., 1981, Paleozoic plate tectonics of the Sierra Nevada and adjacent regions, in Ernst, W. G., ed., The Geotectonic Development of California (Rubey volume 1): Englewood Cliffs, New Jersey, Prentice-Hall, p. 182-201.

Schweickert, R. A., Harwood, D. S., Girty, G. H. and Hanson, R. E., 1984, Tectonic development of the northern Sierra terrane: An accreted late Paleozoic island arc and it basement (Guidebook): Geological Society of America National Meeting, Reno, Nevada, 50 p.

Schweller, W. J. and Kulm, L. D., 1978,
Depositional patterns and channelized
sedimentation in active eastern Pacific
trenches, in Stanley, D. J., and Kelling,
G., eds., Sedimentation in Submarine
Canyons, Fans, and Trenches: Stroudsburg,
Pennsylvania, Dowden, Hutchinson, and
Ross, Inc., p. 311-324.

Speed, R. C., and Sleep, N. H., 1982, Antler
orogeny and foreland basin: A model:
Geological Society of America Bulletin,
v. 93, p. 815-828.

Stacey, J. S., and Kramers, J. D., 1975,
Approximation of terrestrial lead isotope
evolution by a two-step model: Earth and
Planetary Science Letters, v. 26, p. 207-
221.

Taylor, G. W., 1986, Structural,
sedimentological, and petrological
setting of the Lang-Halsted sequence and
Duncan Peak chert, lower Shoo Fly
Complex, northern Sierra Nevada,
California [unpublished M.S. thesis]: San
Diego, California, San Diego State
University, 110 p.

Thornburg, T. M. and Kulm, L. D., 1987,
Sedimentation in the Chile trench:
Depositional morphologies, lithofacies,
and stratigraphy: Geological Society of
America Bulletin, v. 98, p. 33-52.

Underwood, M. B., and Bachman, S. B., 1982,
Sedimentary facies associations within
subduction complexes, in Leggett, J. K.,
ed., Trench-Forearc Geology: Geological
Society of London Special Publication No.
10, p. 537-550.

Varga, R. J., 1982, Implications of
Paleozoic phosphorites in the northern
Sierra Nevada range: Nature, v. 297, p.
217-220.

Varga, R. J. and Moores, E. M., 1981, Age,
origin, and significance of an
unconformity that predates island-arc
volcanism in the northern Sierra Nevada:
Geology, v. 9, p. 512-518.

UPPER DEVONIAN AND LOWER MISSISSIPPIAN ISLAND-ARC AND BACK-ARC
DEPOSITS IN THE NORTHERN SIERRA NEVADA, CALIFORNIA

David S. Harwood
U. S. Geological Survey
Menlo Park, California 94025

James C. Yount
U. S. Geological Survey
University of Nevada
Reno, Nevada 89557-0047

Victor M. Seiders
U. S. Geological Survey
Menlo Park, California 94025

ABSTRACT

Upper Devonian and Lower Mississippian island-arc volcanic rocks in the northern Sierra Nevada grade southward into chert- and quartz-rich conglomeratic turbidites of the Picayune Valley Formation. The quartzose debris was derived from the south. Sedimentary structures and bedding characteristics suggest that the quartzose clastic rocks were deposited in a channelized inner or middle deep-sea fan environment in an arc-marginal basin.

Sandstone petrology and pebble counts from the Picayune Valley Formation indicate a provenance rich in chert, quartz sandstone, and argillite and poor in plutonic and volcanic rocks. Conglomerate lenses interbedded with Lower Mississippian volcaniclastic rocks in the arc sequence were derived from part of the arc basement that was uplifted during active volcanism. These conglomerates, though richer in chert and poorer in quartz sandstone and monocrystalline quartz, are similar to quartzose rocks in the Picayune Valley Formation. Modal data and pebble counts of quartzose clastic rocks from various units derived from the Antler orogenic highland in Nevada also are similar to comparable data from the Picayune Valley Formation. The petrographic data strongly suggest that all of these quartzose clastic rocks form a single petrologic province with a common provenance. Quartzose detritus in the Picayune Valley Formation may have been derived from the Antler orogenic highland or from the Shoo Fly Complex exposed in uplifted segments of the arc basement. Similarities between quartzose clastic rocks in different parts of the Antler orogen and those in the Sierran arc argues sharply against an exotic origin for the Sierran arc terrane.

INTRODUCTION

In the northern Sierra Nevada, Upper Devonian and Lower Mississippian arc-volcanic rocks rest unconformably on deformed lower Paleozoic rocks of the Shoo Fly Complex (Fig. 1). Vent-proximal volcanic rocks in the northern part of the terrane grade southward into arc-apron deposits that contain chert-rich conglomerate lenses and interbeds of hemipelagic mudstone. East of the Talbot fault zone (Fig. 1), distal volcaniclastic rocks, derived from the arc, are interbedded with chert- and quartz-rich conglomeratic

turbidites derived from an uplifted area that lay generally south of the island arc.

The paleogeographic setting of Late Devonian and Early Mississippian arc volcanism is imperfectly understood but the transition from island arc to basinal epiclastic deposition, viewed in regional context, suggests that the arc evolved adjacent to North America (Harwood and Murchey, 1990). For example, this period of island-arc volcanism coincided with the emplacement of lower Paleozoic eugeoclinal rocks in the Roberts Mountains allochthon onto the continental margin in Nevada during the Antler orogeny (Roberts and others, 1958; Speed and Sleep, 1982). Thrusting of the allochthon produced the Antler orogenic highland that shed chert-quartz-rich debris eastward into a linear foreland basin that extended at least from southwestern Idaho to southeastern California (Poole; 1974; Poole and Sandberg, 1977; Nilsen, 1977). Orogenic clastic debris shed oceanward from the Antler orogenic highland has been identified in the Golconda allochthon where it is interbedded with Lower Mississippian volcaniclastic debris probably derived from the northern Sierra arc terrane (Miller and others, 1984; Harwood and Murchey, 1990; Whiteford, 1990). Harwood and Murchey (1990) viewed the orogenic clastic debris as an important lithologic link between the northern Sierran arc and the Havallah basin that tied the arc terrane to the Antler orogen and the North American continental margin during the Late Devonian and Early Mississippian.

This report presents data on the composition and depositional patterns in Upper Devonian and Lower Mississippian orogenic turbidites and the arc volcanic deposits that are transitional into them. The coarse chert-quartz-rich clastic rocks are compared to similar orogenic clastics in other parts of the western Cordillera.

GEOLOGY OF UPPER DEVONIAN AND LOWER MISSISSIPPIAN ROCKS

Arc Rocks

Arc basement

Upper Devonian epiclastic and volcanic rocks rest unconformably on the Shoo Fly Complex, which consists of four regionally extensive thrust blocks referred to informally as the (ascending); 1) Lang sequence, 2)

In Cooper, J.D., and Stevens, C.H., eds., 1991, **Paleozoic Paleogeography of the Western United States-II:** Pacific Section SEPM, Vol. 67, p. 717-733.

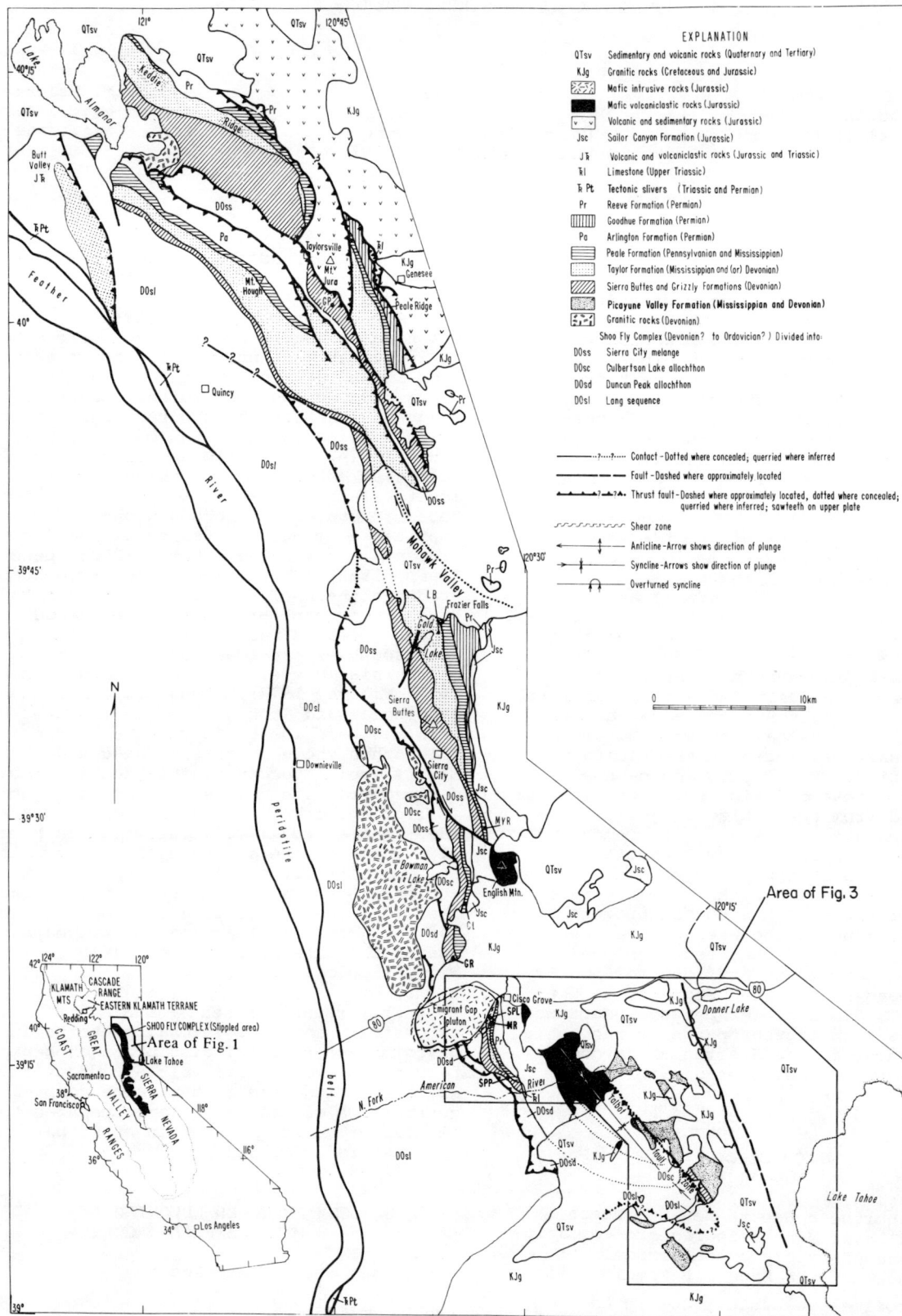

Figure 1. Geologic map of the northern Sierra terrane north of 39°N. latitude showing the distribution of Paleozoic and Mesozoic arc volcanic rocks and the location of Figures 2 and 3. Modified from Harwood and Murchey (1990). MYR, Middle Yuba River; CL, Culbertson Lake; MR, Monumental Ridge; GR, Grouse Ridge; SPL, SP Lake; SPP, Sugar Pine Point.

Duncan Peak allochthon, 3) Culbertson Lake allochthon and 4) the Sierra City melange (Schweickert and others, 1984). The Lang sequence consists primarily of quartzose turbidites deposited in a continental slope and rise environment (Bond and DeVay, 1980) and the Duncan Peak allochthon is composed of radiolarian chert and siliceous argillite. The lower part of the Culbertson Lake allochthon consists of alkali basalt, interpreted to be a remnant seamount, capped with chert and limestone (Girty, 1983), whereas the upper part of the allochthon consists of chert, hemipelagic mudstone, and an upward thinning and fining sequence of massive, amalgamated sandstones interpreted by Girty and others (1990) to be channelized trench-fill deposits. The Sierra City melange consists of blocks of serpentinite, gabbro, greenstone, chert, and limestone in a matrix of broken sandstone and pelite.

This structural assemblage of diverse deep-water deposits is generally regarded to be a subduction-related accretionary complex (Schweickert and others, 1984; Girty and others, 1990) but its time of formation and paleogeographic affinity to North America are not firmly established. Late Middle to Late Ordovician conodonts, extracted from a lens of limestone in the Lang sequence (Harwood and others, 1988), currently provide the oldest depositional age for the complex. Limestone blocks in the Sierra City melange have yielded Late Ordovician conodonts (Hannah and Moores, 1986) and megafossils (Potter and others, 1990) that have North American affinities. Zircons from a felsic tuff interbedded in the Sierra City melange give a concordant late Early Silurian U/Pb age (423 ± 10 Ma) (Saleeby and others, 1987), which is the youngest

depositional age obtained, thus far, in the Shoo Fly Complex. Discordant U/Pb ages on

zircon populations from the Bowman Lake batholith (Fig. 1), which intrudes thrust blocks in the Shoo Fly, range in age from late Middle to Late Devonian (Hanson and others, 1988). The isotopic ages suggest that the thrust blocks were assembled between the late Early Silurian and the Late Devonian. Hanson and others (1988) interpreted the Bowman Lake batholith to be a magmatic source for the overlying Upper Devonian arc-volcanic rocks.

Vent-proximal arc deposits

Hanson and Schweickert (1986) provide a detailed description of the Upper Devonian and Lower Missippian vent-proximal arc-volcanic rocks in the vicinity of Sierra Buttes (Fig. 1) that is summarized here so that comparisons can be made with distal deposits. The volcanic rocks locally overlie the Grizzly Formation (Fig. 2) that consists of polymict conglomerate, breccia, and quartzose sandstone interbedded with chert and pelite. Epiclastic debris in the Grizzly was derived from the Shoo Fly Complex and deposited in deep water as sandstone turbidites and channelized conglomeratic debris flows. Chert interbeds in the Grizzly have produced mid-Famennian conodonts and a large slump block of shallow-water sandstone and limestone, included in the deep-water deposits, produced Frasnian megafossils (Hanson and Schweickert, 1986).

The Sierra Buttes Formation conformably overlies the Grizzly Formation and rests unconformably on the Shoo Fly Complex where the Grizzly is missing. Hanson and Schweickert (1986) divided the submarine volcanic rocks of the Sierra Buttes Formation into four informal members, lettered A through D in ascending order. Member A consists of

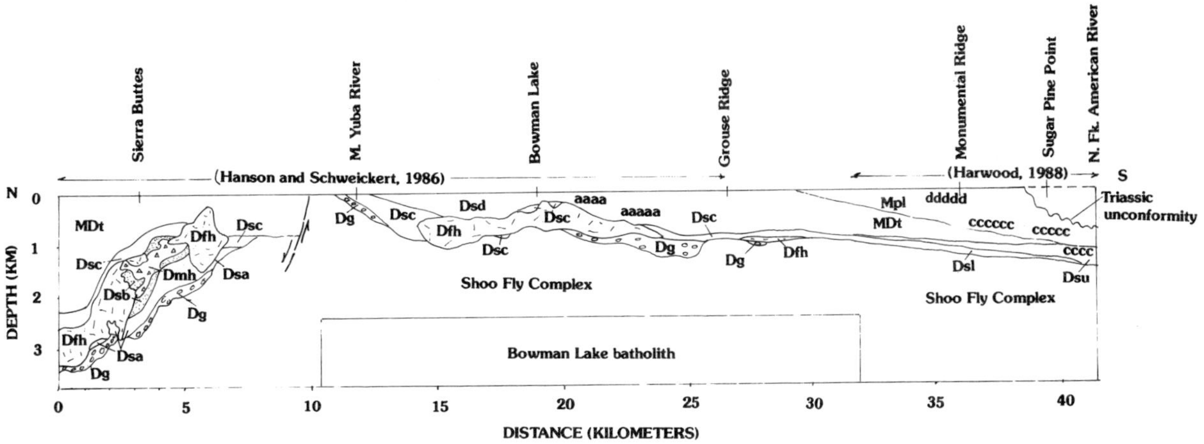

Figure 2. North-south cross section showing distribution of units in Upper Devonian and Lower Mississippian volcanic rocks between Sierra Buttes and North Fork of the American River (expanded from Hanson and Schweickert, 1986). Datum is chert member of the Peale Formation. Dg, Grizzly Formation; Dsa, member A of Sierra Buttes Formation; Dsb, member B of Sierra Buttes Formation; Dsc, member C of Sierra Buttes Formation; Dsd, member D of Sierra Buttes Formation of Hanson and Schweickert (1986); MDt, Taylor Formation; Mpl, lower member of Peale Formation; Dmh, mafic hypabyssal intrusive rocks; Dfh, felsic hypabyssal rocks, aaa, intrusive andesite; ddd, debris-flow deposits in lower member of Peale Formation; ccc, chert-granule conglomerate in upper member of Sierra Buttes Formation and lower member of Peale Formation.

720

270 m of coarse-grained felsic and andesitic lapilli tuff and tuff breccia with interbeds of subaqueous ash-fall tuff and phosphate-streaked black chert. Member B is 300 m thick and consists predominantly of massive, coarse-grained felsic breccia and tuff breccia with minor interbeds of fine-grained, felsic tuffaceous turbidites. Member C is like member A but andesitic debris is sparse and felsic pumice is more abundant. Member D consists of 800 m of fine-grained andesitic tuff and tuffaceous turbidites that contain blocky andesitic debris-flow deposits up to 24 m thick. Hanson and Schweickert (1986) mapped member D only south of the Middle Yuba River (Fig. 2) and assigned the 1000 m of blocky andesitic tuff breccia and flows that overlie member C north of Sierra Buttes (Fig. 2) to the Taylor Formation. They separated their member D from the Taylor by the presence of felsic hypabyssal rocks in member D, but their section (Fig. 2) shows felsic intrusive rocks cutting the Taylor at Sierra Buttes so the distinguishing criterion is questionable and their member D is considered here to be equivalent to the Taylor.

Hypabyssal felsic and mafic intrusive rocks are abundant in the vent-proximal volcanic rocks. The hypabyssal rocks contain locally well developed intrusive hyaloclastic breccia and peperite indicating intrusion into wet, unlithified deposits (Hanson, 1982; Brooks and others, 1982; Hanson and Schweickert, 1986). Hanson and others (1988) reported a concordant U/Pb age of 368 Ma on zircon from a felsic sill that intrudes the Sierra Buttes Formation. This age is consistent with mid-Famennian conodonts obtained from black, phosphate-streaked chert at several localities in the Sierra Buttes Formation (Hanson and Schweickert, 1986).

North of Sierra Buttes, Upper Devonian and Lower Mississippian vent-proximal deposits form a northward-thickening wedge more than 3 km thick (Durrell and D'Allura, 1977). The youngest volcanic deposits are alkalic tuff, tuff breccia, and minor flows of the lower member of the Peale Formation that contains lenses of Lower Mississippian (late Kinderhookian) limestone (Harwood, 1988; 1991). Harwood and Murchey (1990) proposed that the wedge of vent-proximal deposits collected in an intra-arc graben bounded on the south by a normal fault that developed over the north end of the Bowman Lake batholith (Fig. 2). The area over the batholith was uplifted, probably due to magmatic heating (Hanson and Schweickert, 1986) and Upper Devonian and Lower Mississippian volcanic rocks either were eroded from or not deposited over the uplifted area. Above the north end of the batholith north of the Middle Yuba River (Fig. 2), the chert member of the Peale Formation, which ranges in age from Early Mississippian to Middle Pennsylvanian (Harwood, 1988), rests locally on the Shoo Fly Complex. The uplifted area apparently formed a barrier of unknown dimensions that separated thick, blocky vent-proximal debris and lava flows in the north from thinner, finer grained arc-apron deposits to the south.

Volcanic-apron deposits

South of Cisco Grove (Fig. 1), the Upper Devonian and Lower Mississippian volcanic sequence consists of the Sierra Buttes, Taylor, and lower member of the Peale Formation. Some units in this section are lithologically similar to vent-proximal deposits and establish continuity between parts of the arc. However, hypabyssal intrusive rocks are absent and distal volcaniclastic units are present that do not occur to the north. The distribution of arc-apron units and their spatial relation to the Picayune Valley Formation, discussed in a following section of this report, are shown in Figure 3.

The lower member of the Sierra Buttes Formation (Harwood, 1983) consists of coarse-grained felsic tuff breccia that contains abundant intraclasts of phosphate-streaked black chert. Typically the unit is massive and the only lithologic variation is in the amount of chert intraclasts, which may mark different debris flows. Thin beds of fine-grained, parallel laminated felsic tuff occur locally near the top of the member. The lower member has a maximum thickness of 400 m but it pinches out beneath the upper member near the North Fork of the American River. The lower member is considered equivalent to member C of Hanson and Schweickert (1986).

The upper member of the Sierra Buttes Formation is 200 m thick and consists of thin tabular beds and paper-thin parallel laminae of very fine-grained volcaniclastic sandstone, siltstone and tuffaceous mudstone now metamorphosed to slate. Sandstone beds, 2 to 10 cm thick, are commonly graded and contain parallel- or ripple cross-lamination at the top. They appear to be Tad or Tade turbidites (Bouma, 1962). Thinner beds are commonly parallel laminated and locally ripple cross laminated. Lenses of fine- to medium-grained felsic tuff breccia and chert-granule conglomerate occur sparsely with volcaniclastic sandstone in the upper part but the member does not appear to be characterized by distinct bedding sequences.

The upper member of the Sierra Buttes Formation is overlain by brick-red weathering, dark-green andesitic crystal-lithic turbidites assigned to the Taylor Formation. Beds range in thickness from 2 to 20 cm and grade up from coarse andesitic sand to parallel laminated, ultra-fine, andesitic tuff at the top. Coarse pumiceous fragments, more felsic than the matrix, occur locally below the parallel laminated tops of some beds (Fig. 4A). Lenticular andesitic debris flow deposits, less than 2 m thick, occur sparsely in the turbidites. The andesitic turbidites have a maximum thickness of 200 m but they pinch out north of the North Fork of the American River.

The lower member of the Peale Formation overlies the Taylor and consists predominantly of thin-bedded tuffaceous siltstone and black mudstone. Lenses of felsic tuff, volcanic- and chert-fragment pebbly mudstone, volcaniclastic debris-flow deposits and chert-granule conglomerate are scattered in the tuffaceous

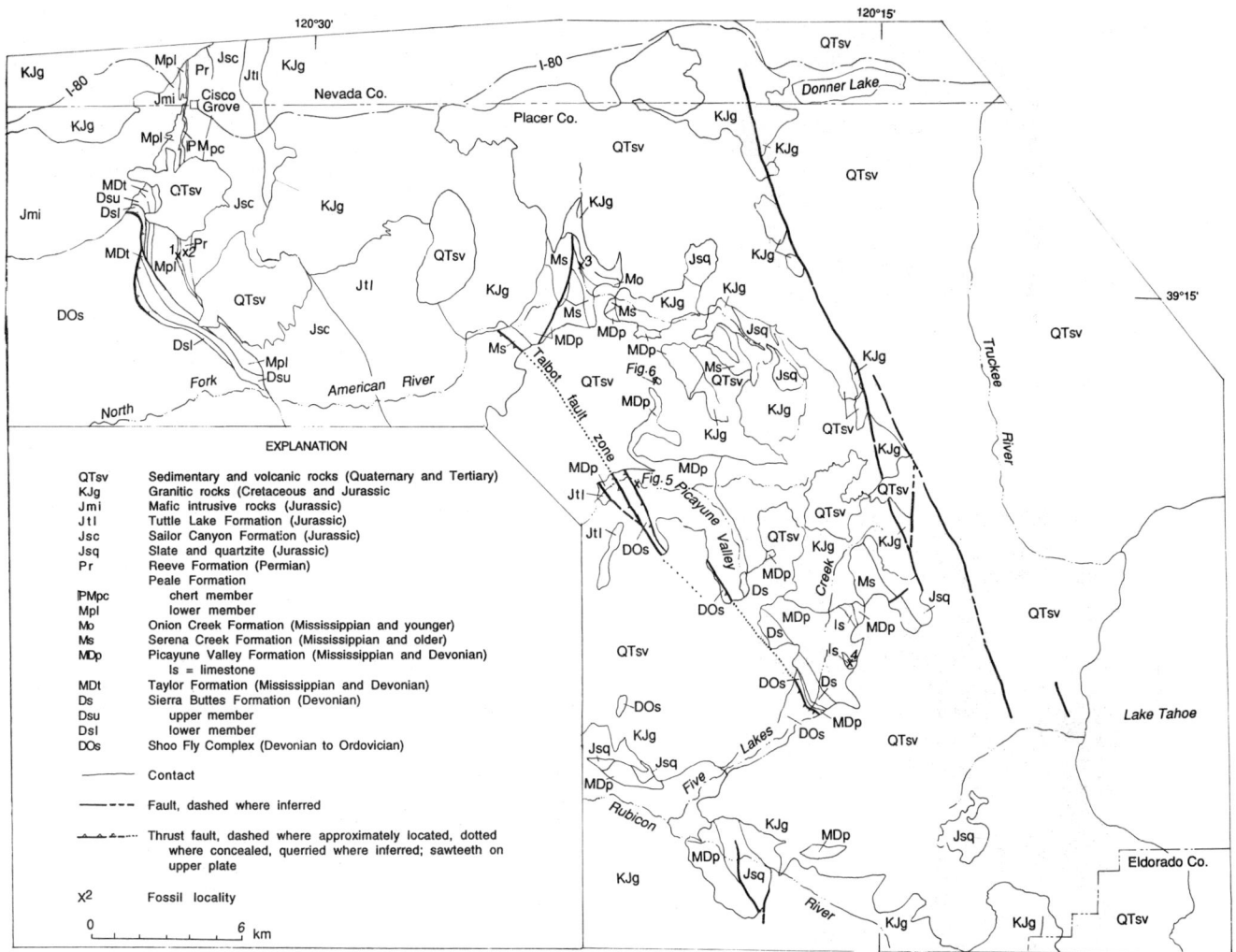

Figure 3. Simplified geologic map of part of eastern Placer County, California showing spatial relations of Upper Devonian and Lower Mississippian arc-apron deposits to Picayune Valley Formation.

siltstone. Chert-granule conglomerate lenses occur near the base of the member on the south slope of Monumental Ridge (Figs. 2, 3) and sandy debris flow deposits with boulders of felsic and mafic volcanic rocks and jasper occur near SP Lakes southeast of Cisco Grove. Thin-bedded tuffaceous siltstone and black mudstone locally contain intraformational slump folds and layer-parallel extension zones (Fig. 4B) that indicate southward, down-to-basin, movement.

Basin Deposits

Chert- and quartz-rich turbidites of the Picayune Valley Formation (Harwood, 1991) occur east of the Talbot fault zone and rest unconformably on the Shoo Fly Complex in Five Lakes Creek (Fig. 3). Lindgren (1897) originally mapped these rocks as part of the Jurassic Sailor Canyon Formation but lithologic and limited paleontologic data indicate they are mid-Paleozoic in age and correlative with the Upper Devonian and Lower Mississippian arc-volcanic rocks to the west. Limestone in the turbidite sequence produced

silicified echinoderm debris (Fig. 3, no. 4) indicative of a Silurian or Devonian age (J.T. Dutro, 1988, written commun.). Calc-silicate rocks in the overlying Serena Creek Formation (Harwood, 1991) contain coarse-ribbed bivalve fragments (Fig. 3, no. 3) that are definitely Paleozoic (post-Ordovician) in age and one deformed brachiopod fragment suggests a mid-Paleozoic, possibly Silurian to Mississippian age (J.H. Pojeta, 1984, written commun.). The turbidites contain interbeds and mappable units of felsic volcaniclastic rocks that are lithologically similar to those in the Sierra Buttes Formation. Andesitic volcaniclastic rocks interbedded with the turbidites and with the felsic volcaniclastic rocks correlate with andesitic rocks in the Sierra Buttes or Taylor Formations. In Five Lakes Creek (Fig. 3), the volcaniclastic rocks occur near the base of the turbidites and indicate that early basin deposition coincided with the Late Devonian and Early Mississippian arc volcanism.

Distal volcaniclastic rocks

Distal volcaniclastic rocks assigned to the Sierra Buttes Formation form a section

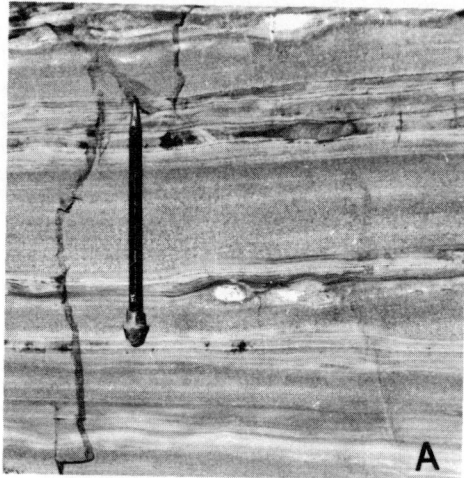

Figure 4. A. **Thin-bedded, graded andesitic** turbidities of the Taylor Formation on Monumental Ridge. Note felsic pumice clasts to right of eraser on pencil (pencil 12 cm long). B. Intrafolial layer-parallel extension of interbedded tuffaceous siltstone (white) and hemipelagic mudstone (black) in the lower member of the Peale Formation on south slope of Monumental Ridge. Pencil 14 cm long.

about 150 m thick in Five Lakes Creek. Most of the section consists of fine-grained tuff without phenocrysts or macroscopic framework grains. Variable amounts of microscopic plagioclase and monocrystalline quartz occur in a recrystallized matrix of calcic amphibole and pale brown biotite. Tuff beds, which range in thickness from 2 to 40 cm, are not visibly graded but commonly are parallel laminated in the upper part. Color variations from light- to dark-gray within thicker beds

probably reflect grain-size variations but this could not be confirmed in thin section. Lenses of white-weathering quartz-phyric tuff that contain scattered black, phosphate-streaked chert intraclasts and thin lenses of tuff with abundant black chert intraclasts occur near the base of the volcaniclastic section. Chert clasts are rounded and have a maximum diameter of 3 cm.

Massive felsic debris-flow deposits, up to 2 m thick, occur sporadically in the fine-grained tuff. Monocrystalline quartz and felsic lithic fragments, 5 to 10 cm in size, comprise most of the rock and some deposits also contain scattered black chert intraclasts up to 5 cm long. No primary sedimentary structures were recognized in the debris-flow beds. Mafic debris-flow deposits are less abundant but thicker than their felsic counterparts. These deposits are matrix supported and contain abundant rounded to subrounded, aphyric and porphyritic mafic clasts and sparse quartz-phyric felsite clasts.

Picayune Valley Formation

The Picayune Valley Formation is composed of chert- and quartz-rich conglomerate, quartzose sandstone, and black pelite. The rocks have been recrystallized by Mesozoic plutons of the Sierra Nevada batholith and pelitic beds are cordierite-andalusite horn-fels. Grain size, sedimentary structures, bedding style, and the vertical arrangement of beds in the Picayune Valley Formation show many of the features associated with channel-ized inner or middle deep-sea fan environments (Walker and Mutti, 1973; Howell and Normark, 1982). Specifically, these features indicate an abundance of Facies A, Facies B, and Facies D turbidites as described by Mutti and Ricci Lucchi (1972). Two measured sections (Figs. 5, 6), which are located on Figure 3, illustrate the major sedimentologic features found in the coarser parts of the formation.

Coarse-grained parts of the formation consist mainly of matrix-supported pebble conglomerate, pebbly sandstone, and coarse-grained sandstone. Cobble-size clasts occur in some conglomerates (Fig. 7A) and some pebbly mudstones contain folded pelite-sand-stone clasts that suggest slumping (Fig. 7B). Gritty sandstone and medium-grained sandstone are interbedded in some coarse intervals but are less common than pebble conglomerate, pebbly sandstone (Fig. 7C) and coarse-grained sandstone. Thin pelite intervals separate some, but not all, sandstone and conglomerate beds. Fine-grained intervals consist of pelite (hornfelsed mudstone) and thin interbeds of fine- to medium-grained sand-stone. Subtle color variations may indicate silt to clay gradations in the pelite but it is impossible to consistently separate original siltstone from clay-rich intervals in the metamorphosed pelite.

Despite the metamorphism, sedimentary structures are remarkably well preserved in some parts of the formation. Conglomerate and pebbly sandstone beds are rarely coarse-tail graded and commonly they are normally graded

Figure 5 (left column):

Top not exposed

56m
50
40
30
20
10
8
6
4
2
0

Base not exposed

Pelite — SS (fine, med., crs.) — Grit — Congl. and debris flow

Graded quartzose sandstone; coarse- to fine-grained, parallel laminated. Scattered pelite rip-up clasts; fines upward to fine sandstone and parallel-laminated pelite.

Pelite rip-up clasts

Pebble line

Pelite rip-up clasts

Pebble line

Pelite rip-up clasts

Reverse graded

Graded, matrix-supported pebble conglomerate; clasts mostly black chert with lesser amounts of quartzite and felsitic rocks.

Covered

Covered

Thin-bedded (1-2 cm) pelite-fine sandstone. Strong parallel laminations throughout; cross-laminations common; ripples sparse.

EXPLANATION

Cross-lamination — Ripple — Parallel lamination — Upward thinning-fining sequence

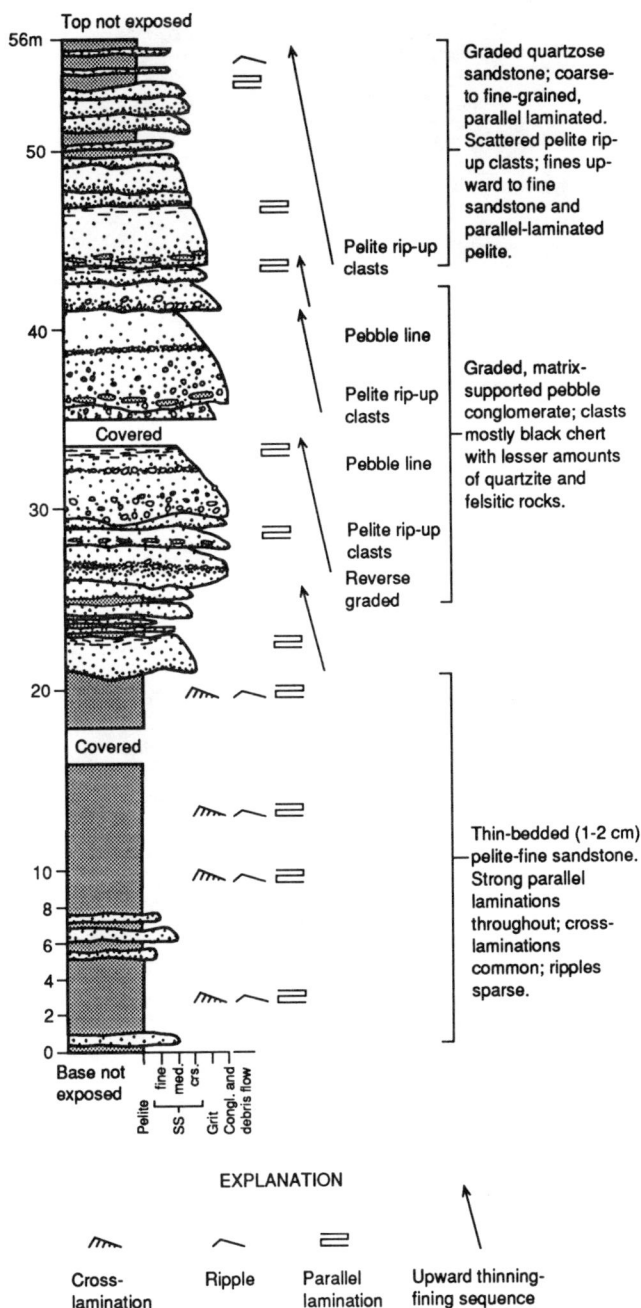

Figure 5. Measured section of conglomeratic channel-fill turbidites in Picayune Valley Formation. Location of section shown in Figure 3.

Figure 6 (right column):

Top not exposed

64m
60
50
40
30
20
10
8
6
4
2
0

Base not exposed

Pelite — SS (fine, med., crs.) — Grit — Congl. and debris flow

Thin bedded (1-4 cm) pelite, fine quartz sandstone

Pebbly mudstone debris flow

Upward thinning and fining, amalgamated coarse quartzite with scattered granule conglomerate lenses

Clastic dike

Pebbly mudstone debris flow; largest quartzite clast 30 cm in diameter

Upward thinning and fining chert-quartz-granule conglomerate; coarse- to medium-grained quartzite

Upward thinning and fining quartzite and quartz-granule conglomerate

Pebbly mudstone debris flow

Upward thinning and fining amalgamated quartzose sandstone and graded chert-quartz-granule conglomerate

Thin-bedded (1-4 cm) pelite, fine-grained quartz sandstone

Poorly sorted, poorly graded, clast-supported cobble to pebble conglomerate with abundant pelite rip-up clasts; overlain by poorly sorted, coarse sandstone with 2 to 10 cm thick lens of pebble conglomerate

Thin-bedded (1-3 cm) pelite and fine sandstone; parallel laminated throughout; cross laminations common; ripples sparse

Figure 6. Measured section of conglomeratic channel-fill turbidites in Picayune Valley Formation. Location of section shown on Fig. 3.

Body text (left column):

from pebble conglomerate at the base to medium- or coarse-grained sandstone, or locally, pelite at the top. Reverse grading is present in some pebble conglomerate beds. Bouma Ta and Tae (Bouma, 1962) sequences are the most common in conglomerate and pebbly sandstone, but Tab sequences occur locally. Coarse- and medium-grained sandstone beds are massive, weakly graded, and commonly contain parallel lamination near the top. Bouma Tab and Tb intervals are common in this size range. Fine-grained intervals contain well-developed parallel laminations in the pelite and common cross-lamination and parallel lami-

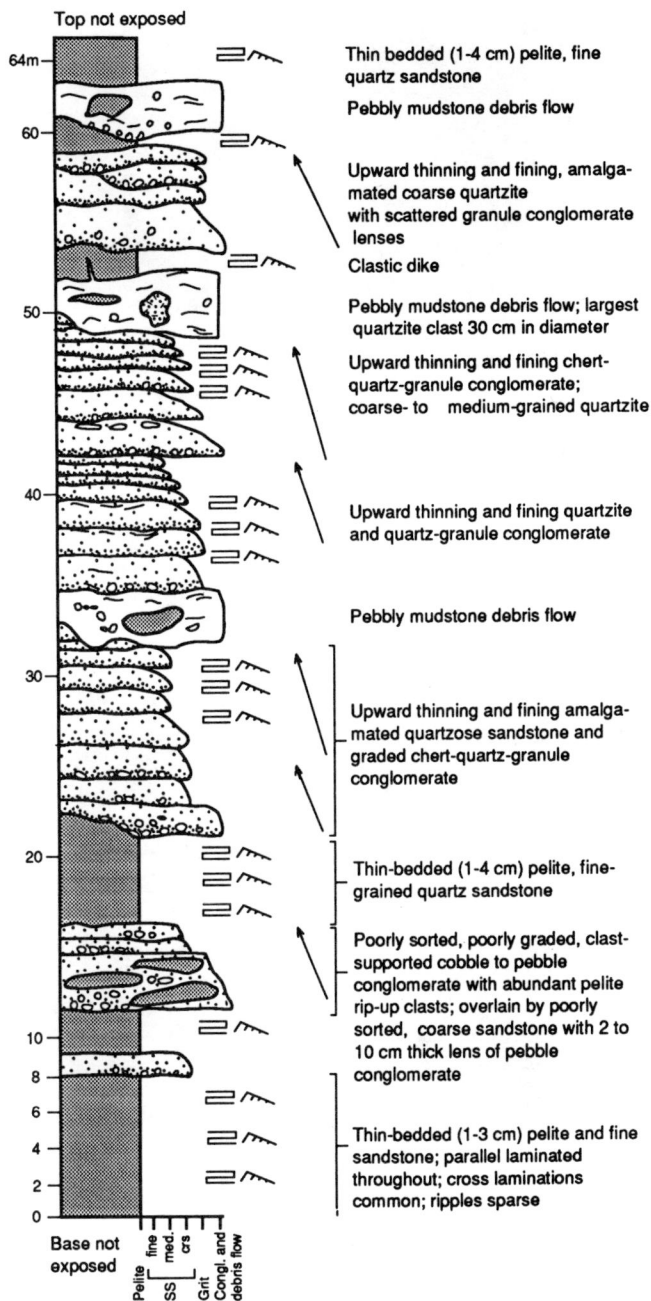

Body text (right column):

nation in the interbedded sandstone beds. Fine-grained sandstone beds show ripple tops with internal ripple cross-lamination (Fig. 7D). Multiple ripple crests have wavelengths of 10 to 20 cm. Bouma Tbc, Tbcd, and Tcde intervals are the most common sequences in the fine-grained rocks.

The coarse-grained intervals are organized into thinning- and fining-upward sequences that vary in thickness from 10 to 30 m (Fig. 5, 6). Individual conglomerate beds range in thickness from 40 cm to 5 m. Most of the thicker beds show evidence of scour at the

Figure 7. Lithologic and bedding characteristics of rocks in the Picayune Valley Formation. A. Chert pebble and cobble conglomerate with abundant rip-up clasts of pelite. Pen 13 cm long. B. Chert pebble layers in coarse-grained quartrose sandstones. Coin 19 mm in diameter. C. Folded intraclast of pelite and quartzite in matrix supported pebbly mudstone. D. Thin bedded pelite (dark) and medium-grained quartzite showing parallel and ripple cross-laminations. Dark pits in pelitic layers are weathered cordierite porphyroblasts.

base that locally reaches a depth of 20 cm. Some sandstone beds are amalgamated. Fine-grained intervals consist of 1 to 5 cm thick pelite beds and 3 to 20 cm thick quartz sandstone beds with sparse quartz sandstone beds as much as 70 cm thick. Bedding contacts are generally conformable and sharp but some sandstone beds grade up into pelite. Slumped intervals, consisting of contorted pelite and thin sandstone, 30 to 70 cm thick, occur locally. No pattern of vertical bedding organization or textural variations was seen in the fine-grained intervals.

Graded conglomerate and coarse-grained sandstone units contain many of the features of Mutti and Ricci Lucchi's (1972) turbidite Facies A, the arenaceous-conglomeratic facies. Thick graded beds organized into thinning-upward intervals, coarse-grained sandstone beds with parallel laminated tops, abundant pelite rip-up clasts, and rare sandstone injection or diking along some contacts support a Facies A interpretation. The thick, parallel-bedded and parallel-laminated sandstones that commonly overlie the conglomerate beds resemble Facies B, the arenaceous facies of Mutti and Ricci Lucchi (1972). Facies A and B deposits result from channelized grain-flow processes (Mutti and Ricci Lucchi, 1972; Walker and Mutti, 1973).

Most of the fine-grained sections fit descriptions of Facies D, the pelitic arenaceous facies of Mutti and Ricci Lucchi

(1972). The persistence of parallel bedding, the abundance of delicate parallel lamination, the predominance of pelite over sandstone, and the gradational contacts between many sandstones and the overlying pelite are properties that best correlate with Facies D turbidites. Minor slumped intervals within the fine-grained Facies D deposits represent Facies F, the chaotic facies, of Mutti and Ricci Lucchi (1972).

The association of Facies A and Facies B deposits overlain and underlain by Facies D and minor Facies F deposits strongly suggests a middle to possibly inner submarine fan environment (Mutti and Ricci Lucchi, 1972; Walker and Mutti, 1973; Howell and Normark, 1982). Facies A conglomerate and Facies B sandstones represent mid-fan channel deposits and Facies D pelite and sandstones represent interchannel overbank deposits. Slumped Facies F intervals could have originated on the outer flanks of channel levees or, possibly, could have been derived from steeper slopes above the fan apex in an inner-fan setting.

PETROGRAPHY

Modal compositions of fine- to medium-grained sandstones are used most commonly for provenance studies because they can be compared directly to detrital suites from different tectonic basins determined by Dickinson and Suczek (1979). In the northern Sierra Nevada, however, contact metamorphism east of the Talbot fault zone has obscured grain boundaries and produced variable amounts of cordierite, andalusite, biotite, calcic-amphibole, feldspar, and quartz from the matrix and pelitic laminae in fine- and medium-grained sandstones. By necessity, therefore, modal analyses presented in Table 1 were made on grits and pebble conglomerates with grains ranging in size from 1 to 10 mm. Similar sized material was point counted from regionally metamorphosed, but lower grade rocks, west of the Talbot fault zone and we believe the data are comparable but the difference in grain size must be considered if these modes are compared to those of finer grained, unmetamorphosed rocks.

Modes were determined primarily to characterize and compare volcaniclastic rocks and chert granule conglomerates in the volcanic-apron deposits with similar rocks in the Picayune Valley Formation. Unfortunately, volcaniclastic rocks east of the Talbot fault zone contain so much metamorphic amphibole that modes could not be determined on those rocks. Modes were made on pebbly sandstone and granule conglomerate in the Picayune Valley Formation, which are compared to modes of granule conglomerates in the lower member of the Peale Formation.

Modal data for volcaniclastic rocks and granule conglomerates in the lower member of the Peale and for conglomerates in the Picayune Valley Formation are given in Table 1 and are shown graphically in Figure 8. Granule conglomerates from the lower member of the Peale are not distinguishable from granule conglomerates in the Picayune Valley Formation in Figure 8A or B because these plots do not isolate monocrystalline quartz from other quartz-rich detritus. However, when monocrystalline quartz is used as one plotting parameter (Fig. 8C), the granule conglomerates of the Picayune Valley Formation define a separate field enriched in monocrystalline quartz relative to the chert granule conglomerates in the lower member of the Peale.

Modal differences shown in Figure 8C are clearly shown in the petrography of the respective granule conglomerates. Clasts in granule conglomerates from the Picayune Valley Formation are supported in a matrix of monocrystalline quartz (Fig. 9A). Granule conglomerates in the lower member of the Peale, on the other hand, are commonly clast supported (Fig. 9B) and only locally contain sparse matrix of very fine-grained tuffaceous silt.

Clasts in the lower member of the Peale are predominantly rounded to subangular fragments of light- to dark-gray chert with lesser amounts of quartz arenite and siltstone. Quartz arenite clasts commonly contain large

Table 1. Detrital modes of coarse sandstone and granule conglomerate: lower member Peale Formation and Picayune Valley Formation.

| Unit: lower member, Peale Formation / Picayune Valley Formation | | | | | | | | | | | | | | | | | | | | | | | | | | | | | |
| Lithology: Volcaniclastic Debris-flow Deposits / Chert Granule Conglomerate / Chert-Quartz Granule Conglomerate | | | | | | | | | | | | | | | | | | | | | | | | | | | | | |
Locality no.	1	2	3	4	5	6	7	8	mean of 8	std. dev.	9	10	11	12	13	14	15	mean of 6	std dev.	16	17	18	19	20	21	22	23	mean of 8	std. dev.
Chert	1.5	2.2	-	2.4	.5	-	-	1.8	1.0	1.0	34.9	54.2	75.7	73.2	74.2	38.9	54.3	61.7	14.9	23.9	20.7	37.5	25.7	27.8	68.3	62.6	58.3	40.6	19.4
Black slate/mudstone	-	-	-	-	-	-	-	-	-	-	-	-	-	-	-	-	-	-	-	7.9	7.6	7.5	10.3	3.8	4.1	2.4	28.6	9.0	8.3
Siltstone/argillite	.5	2.7	.8	5.2	4.2	-	.2	.5	1.8	2.0	-	30.9	15.5	3.4	4.4	26.2	7.0	14.6	11.7	-	-	22.6	12.4	13.0	10.5	8.4	-	8.4	8.1
Quartz sandstone	-	-	-	-	-	-	-	-	-	-	4.2	2.4	2.5	3.4	7.7	-	4.1	3.4	2.5	-	-	-	6.9	3.0	-	2.5	-	1.6	2.5
Polycrystalline quartz	.6	-	-	-	-	-	.8	-	.2	.3	-	-	-	1.9	-	-	-	.3	.8	-	-	-	-	-	-	-	-	-	-
Monocrystalline quartz	.8	1.1	4.6	1.5	4.3	4.7	1.8	.6	2.4	1.8	-	2.1	-	.9	1.0	2.8	7.4	2.4	2.6	18.5	44.7	19.7	29.0	29.7	13.7	19.4	6.2	22.6	11.7
Volcanic rocks	22.8	32.7	69.0	1.5	52.5	32.4	29.1	20.2	32.5	20.5	15.6	.2	2.5	1.6	1.4	6.0	-	1.9	2.2	9.9	1.5	-	-	-	-	-	-	1.4	3.5
Polycrystalline feldspar	10.6	4.9	5.9	10.1	18.0	-	-	-	6.2	6.4	8.4	-	-	-	-	-	-	-	-	8.1	5.3	-	-	.3	-	-	-	1.7	3.2
Monocrystalline feldspar	18.6	22.7	2.4	43.6	-	46.4	48.7	33.1	26.9	19.2	2.4	-	.5	.3	.6	-	.5	.3	.3	-	-	-	-	-	-	-	-	-	-
Devitrified glass	.9	-	-	12.5	.4	1.4	-	5.2	2.5	4.4	-	-	-	-	-	-	-	-	-	-	-	-	-	-	-	-	-	-	-
Matrix	43.8	33.8	17.3	23.3	19.9	14.3	19.9	38.0	26.3	10.8	31.3	10.1	3.4	17.1	8.9	26.2	26.7	15.4	9.6	31.7	19.6	12.7	15.7	22.4	3.4	4.6	6.9	14.6	9.8
Opaque oxides	-	-	-	-	-	-	.2	.6	.1	.2	-	-	-	-	-	-	-	-	-	-	-	-	.6	-	.2	-	-	.1	-
Other	-	-	-	.4	-	-	-	.1	.1	.1	-	-	-	-	-	-	-	-	-	-	-	-	-	-	-	-	-	.1	.2
Total (percent)	100.1	100.1	100.0	100.1	100.2	100.0	99.9	100.0	100.0		99.9	99.9	100.1	99.9	100.1	100.1	100.0	100.0		100.0	100.0	100.0	100.2	100.0	100.0	99.9	100.0	100.0	
Grains counted	671	551	629	553	579	364	437	668	557		450	614	530	615	519	501	633	569		719	608	584	612	630	533	417	564	583	
Q	5	5	5	6	5	6	2	4	5	1	57	65	81	94	93	56	90	80	16	62	82	65	73	78	85	89	69	76	10
F	52	42	10	70	22	54	61	54	46	20	16	0	0	0	1	0	1	0	1	12	7	0	0	0	0	0	0	2	5
L	43	53	84	25	72	40	37	42	49	19	27	35	19	6	6	44	9	20	16	26	11	35	27	22	15	11	31	22	9
C	6	6	0	11	1	0	0	6	4	4	60	62	79	90	85	55	83	76	14	57	69	56	47	58	82	82	67	65	13
Lv	92	87	99	65	92	100	99	92	91	11	27	0	2	2	1	8	0	2	3	24	5	0	0	0	0	0	0	3	8
Ls	2	7	1	24	7	0	1	2	5	8	13	38	19	8	14	37	17	22	12	19	26	44	53	42	18	18	33	32	14
C	6	6	0	10	1	0	0	7	4	4	60	61	79	89	84	53	75	74	14	40	28	43	31	36	71	66	62	47	17
Qm	3	3	6	7	7	12	6	2	6	3	0	2	0	1	1	4	10	3	4	31	60	23	34	38	14	20	7	28	16
Lt	91	91	94	83	92	88	94	91	90	4	40	37	21	10	15	43	15	23	14	29	12	34	35	26	15	14	31	25	9

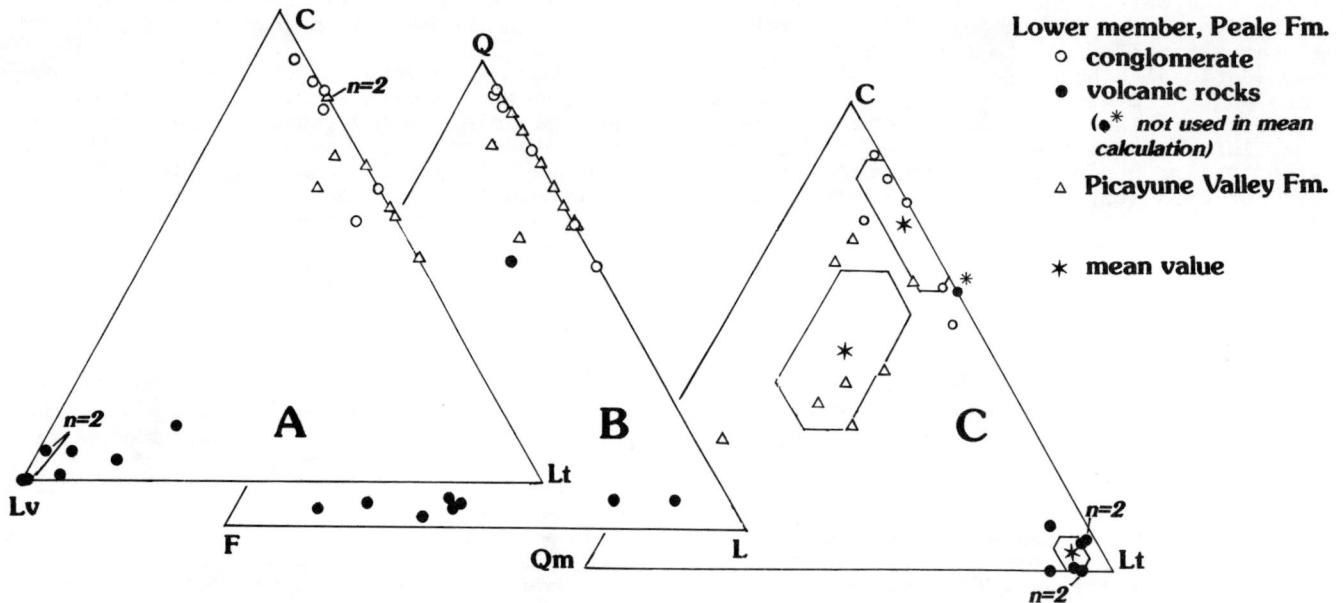

Figure 8. Modal composition of granule conglomerates in Picayune Valley Formation shown relative to provenance fields determined by Dickinson and Suczek (1979). A; Qp = chert, Ls = all sedimentary lithic grains except chert, Lv = volcanic lithic grains. B; Q = chert, monocrystalline quartz, polycrystalline quartz, and quartz sandstone. L = all sedimentary lithic grains except those in Q; F = feldspar. C; Qp = chert, Qm = monocrystalline quartz, Lt = total lithic grains minus chert.

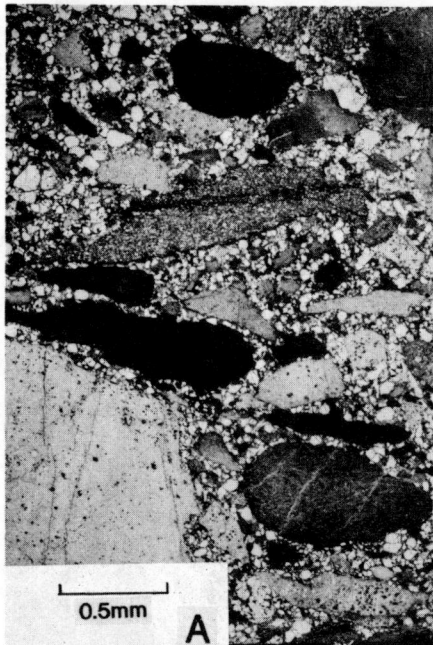

Figure 9. A. Granule conglomerate from Picayune Valley Formation showing abundance of monocrystalline quartz in matrix. B. Granule conglomerate from lower member of Peale Formation.

flakes of detrital muscovite, which is a common detrital mineral in quartz arenites in the Shoo Fly Complex. Detrital muscovite grains are commonly oriented parallel to a

weak metamorphic foliation defined by flattened grains in the quartz arenite clasts. Siltstone clasts and pelite clasts also show a weak metamorphic fabric defined by parallel oriented fine-grained muscovite and chlorite. Foliation in these clasts is randomly oriented with respect to a weakly penetrative, probably Jurassic, foliation in the matrix of the conglomerate lenses that is defined by oriented flakes of fine-grained muscovite. The quartz arenite and siltstone clasts were derived from a source area that was metamorphosed to low greenschist grade prior

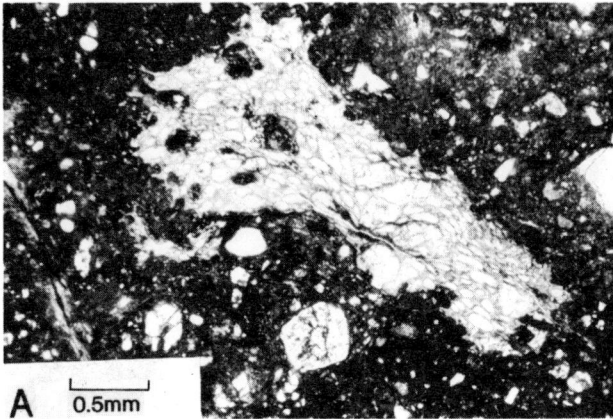

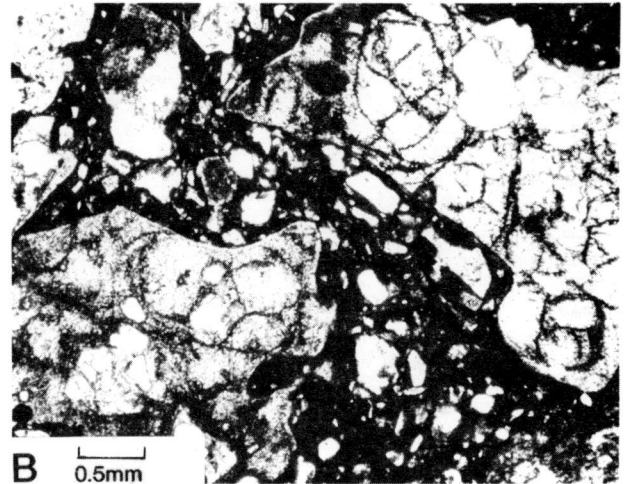

Figure 10. Pyroclastic fragments in volcani-clastic mudstones of lower member of Peale Formation. A. Pumice fragment showing delicate, flame-like boundaries and original out-line of long-tube gas pores; B. Blocky glass fragments showing embayed outlines and perlitic fractures.

to the Late Devonian.

Volcaniclastic rocks in the lower member of the Peale contain variable amounts of chert that occur as angular rip-up clasts. Radiolarians and sponge spicules are common in the chert clasts. Volcanic lithic fragments are volcaniclastic and pyroclastic. The most common volcaniclastic detritus is composed of coarse plagioclase laths, more or less oriented, in a matrix of plagioclase microlites. Polycrystalline plagioclase aggregates without matrix are also abundant. Alkali feldspar occurs as blotchy perthite and only rarely as microcline. Primary mafic minerals are remarkably scarse in the volcanic-lithic fragments and only a few augite grains were seen in the rock matrix. Pyroclastic material occurs as devitrified pumice fragments that either show delicate, flame-like grain boundaries and primary porous textures (Fig. 10A) or blocky grains with embayed boundaries and perlitic fractures (Fig. 10B). Under crossed nicols the pumice fragments are seen as mats of fine-grained chlorite.

Clasts in granule conglomerates of the Picayune Valley Formation are angular to subrounded. Although chert clasts are abundant in the Picayune Valley Formation, black mudstone and argillite comprise significant proportions of the conglomerate. Black mudstone clasts are generally fine-grained hornfelsed pelite fragments that contain sparse porphyroblasts of cordierite or andalusite set in a matrix of fine-grained biotite, muscovite, and black carbonaceous material. The microscopic texture of black mudstone clasts is like that of the pelitic interbeds in the Picayune Valley Formation and the mudstone is interpreted to be rip-up clasts. Argillite fragments contain microscopic laminae relatively rich in fine-

grained quartz that alternate with biotite-rich lamellae. Because of Mesozoic contact metamorphism, we were unable to determine if the argillite fragments were metamorphosed prior to incorporation in the conglomerate. Quartz arenite clasts are most commonly angular to subrounded fragments of fine-grained quartzite. Grains in most quartzite fragments are annealed and show 120° intersections. Sparse quartzite clasts are composed of intensely flattened and sutured grains. Quartz grains in the arenite clasts are significantly smaller than the monocrystalline quartz grains in the matrix indicating separate source rocks rather than a progressive breakdown of quartz arenite clasts to form the matrix monocrystalline quartz. Some quartz arenite clasts may be rip-up fragments from the Picayune Valley Formation but others were clearly derived from a low-grade metamorphic terrane. Grain boundaries in the monocrystalline quartz are locally serrated, commonly marked by lines of carbonaceous(?) inclusions, and annealed to 120° intersections. Sparse monocrystalline quartz aggregates contain detrital zircon and some quartz grains are rutilated. Lines of carbonaceous(?) debris and unidentified high relief material cross some homogeneous grains. No feldspar, biotite, or hornblende was found in the monocrystalline quartz aggregates suggesting that the quartz was not derived from a plutonic or gneissic source. The monocrystalline quartz may have been derived from a source composed of coarse-grained quartzite. Feldspar and volcanic lithic fragments were found in two samples from the Talbot fault zone. Plagioclase occurs as single grains and as crystal aggregates; microcline is exceedingly rare. Volcanic lithic fragments commonly contain plagioclase laths in a matrix of plagioclase microlites. Scaley mats of untwinned albite(?) occur commonly and are interpreted to be devitrified pumice fragments. Chert fragments are commonly black and foliated. Some contained ellipsoidal radiolarian tests now recrystallized to fine-grained polycrystalline quartz.

Because of the difference in grain size and analytical procedure, pebble counts of

TABLE 2 PEBBLE COUNTS FROM CONGLOMERATE IN LOWER MEMBER OF PEALE FORMATION AND PICAYUNE VALLEY FORMATION

UNIT	LOWER MEMBER PEALE FORMATION					PICAYUNE VALLEY FORMATION								
LOCALITY NO.	1	2	3	MEAN OF 3	STD. DEV.	4	5	6	7	8	9	10	MEAN OF 7	STD. DEV.
CHERT	89.2	89.4	85.0	87.9	2.5	61.9	54.3	48.7	55.8	63.8	58.7	64.7	58.3	5.8
MUDSTONE/SLATE/ARGILLITE	8.8	4.2	10.5	7.8	3.3	17.7	18.6	8.9	18.5	14.1	32.8	28.8	19.9	8.2
QUARTZ SANDSTONE	1.7	2.0	3.1	2.3	0.7	18.0	25.7	32.2	18.5	18.0	7.4	5.1	17.8	9.5
OTHER SANDSTONE	-	0.8	0.3	0.4	0.4	-	1.1	1.6	0.8	0.8	-	-	0.6	0.6
VEIN QUARTZ	-	-	-	-	-	0.3	-	-	0.3	0.3	-	-	0.1	0.2
FELSIC VOLCANIC ROCKS	-	2.0	-	0.7	1.1	0.3	0.3	0.3	-	-	-	0.3	0.2	0.2
OTHER VOLCANIC ROCKS	-	-	-	-	-	-	-	-	-	-	-	-	-	-
GRANITIC ROCKS	-	-	-	-	-	-	-	-	-	-	-	0.3	-	0.1
OTHER	0.3	1.7	1.1	1.0	0.7	1.7	-	8.4	6.2	3.0	1.1	0.8	3.0	3.1
TOTAL (PERCENT)	100.0	100.1	100.0	100.1		99.9	100.0	100.1	100.1	100.0	100.0	100.0	99.9	
PEBBLES COUNTED	352	357	354	354		344	350	382	373	362	363	354	361	
Q	2	2	4	3	1	23	32	40	25	22	11	7	23	11
I	0	2	0	0	1	0	0	0	0	0	0	1	0	0
C	98	96	96	97	1	77	68	60	75	78	89	92	77	11
Lss	91	94	89	91		82	81	90	80	85	67	71	80	
Lsu	9	4	11	8		18	19	10	20	15	33	29	20	
Lv	0	2	0	1		0	0	0	0	0	0	0	0	

coarse conglomerates (Table 2) differ considerably from the modes of granule conglomerates but, like those modes, they reveal sharp differences between the Picayune Valley Formation and the lower member of the Peale. Pebble counts were made on cut, etched slabs according to the techniques described by Seiders and Blome (1988). Conglomerates from the lower member of the Peale are distinctly chert rich and contain small minor amounts of argillite and minor quartz arenite. Volcanic fragments are sparse. Conglomerates from the Picayune Valley Formation are less rich in chert (mean 58 percent versus 88 percent) and contain significantly more pelite (mean 20 percent versus 8 percent) and quartzite (mean 18 percent versus 2 percent) clasts than conglomerates in the lower member of the Peale. Furthermore, the matrix of coarse conglomerates in the Picayune Valley Formation is composed of monocrystalline quartz whereas the matrix of conglomerates in the lower member of the Peale is volcanic siltstone.

PROVENANCE

Compositional differences in the conglomerates from the Picayune Valley and lower member of the Peale indicate different source areas. Chert formed part of both sources but it was the dominant lithology in the provenance of conglomerates in the lower member of the Peale. Quartzite and argillite, as well as chert, formed erodable bedrock in the provenance of the Picayune Valley Formation. The abundance of monocrystalline quartz in conglomerates of the Picayune Valley Formation and its relative paucity in the Peale indicates that chert-rich conglomerate lenses in the lower member of the Peale are not thin, distal deposits of the Picayune Valley Formation.

Source of epiclastic detritus - Peale Formation

Interstratification of chert-rich conglomerate lenses and volcaniclastic rocks in the lower member of the Peale indicates deposition in a common arc-apron environment. Rare mixing of volcaniclastic debris in the epiclastic conglomerates lenses, however, points to two separate sources within the arc terrane. The dominant volcaniclastic debris was eroded from the margin of an active island arc that supplied variable amounts of juvenile tephra (Fig. 10) to the arc-apron. Epiclastic chert-rich debris was derived from the arc basement possibly along the uplifted area over the Bowman Lake batholith or a similar area now removed from view. The chert-rich Duncan Peak allochthon is the most obvious intra-arc source for the epiclastic debris but parts of the Culbertson Lake allochthon also might have supplied abundant chert and little quartz arenite detritus.

Source of the Picayune Valley detritus

Data from the Picayune Valley Formation place several constraints on the source of detritus. Based on modal data and pebble counts, the provenance of the Picayune Valley Formation contained quartz sandstone, chert, and argillite and may have contained mudstone or slate but those components may be intra-clasts. Plutonic and volcanic rocks were not significant in the source terrane that supplied the chert- and quartz-rich debris. Monocrystalline quartz was derived from quartzite metamorphosed to low grade but there is no indication that high-grade schist and gneiss were present in the source area. Six paleocurrent determinations based on the orientation of trough axes of ripple cross laminations from the area of Figure 6 indicate sediment transport from the south (range 151°-230°; mean 194°). Rocks in the source area were no younger than Late Devonian.

On the basis of these criteria, the Shoo Fly Complex and the Roberts Mountains allochthon represent potential sources for detritus in the Picayune Valley Formation. Both the Shoo Fly and the Roberts Mountains allochthon contain large tracts of pre-Late Devonian quartzite, chert, and argillite. Both belts of rocks extend south of the Picayune Valley Formation (Fig. 11) and, thus one or both of the terranes could have shed debris northward into an arc-marginal basin. Although the Picayune Valley Formation rests unconformably on the Shoo Fly north of the 39th parallel, we cannot rule out the possibility that the Shoo Fly was uplifted and eroded in its southern exposures during the mid-Paleozoic. The Roberts Mountains allochthon, on the other hand, formed the Antler orogenic highland that clearly shed chert- and quartz-rich debris low in feldspar and volcanic detritus into adjacent parts of the Antler orogen (Dickinson and others, 1983).

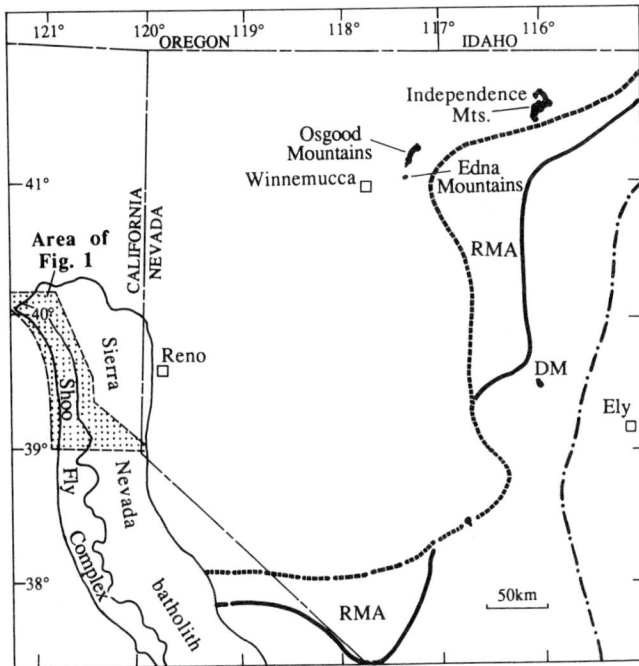

Figure 11. Regional map of eastern California and Nevada showing area of exposure of Roberts Mountains allochthon (RMA), inferred eastern extent of the Golconda thrust (short dashed line), eastern limit of Lower Mississippian deposits in Antler foreland basin (Poole and Sandberg, 1977) (dash-dot line), and areas mentioned in the text (DM, Diamond Mountains).

In figure 12 the sandstone petrology of the Picayune Valley Formation is compared with that of the Peale Formation and with several rock suites from the Antler orogenic belt of Nevada. For each suite the mean composition and a dispersion field defined by the standard deviations is presented.

In terms of QFL (Fig. 12A), the mean clast composition and dispersion field of the Picayune Formation rocks are similar to those of the quartzose rocks of the Peale Formation and to suites from Antler foreland basin

deposits in the Diamond Mountains (Fig. 11) (Harbaugh, 1980), the Antler overlap assemblage in the Osgood and Edna Mountains (Fig. 11) (Saller, 1980), and to Lower Mississippian rocks in the Schoonover sequence in the Independence Mountains (Fig. 11) (Whiteford, 1990). All of these suites plot near the Q-L side of the triangle, centered near 75 percent Q, reflecting their richness in quartzose detritus and their poverty in feldspar. A suite of Upper Mississippian-Pennsylvanian sandstone from the Schoonover sequence (Whiteford, 1990) is relatively richer in Q (mean 91 percent). A suite of volcaniclastic rocks from the Peale Formation plots near the F-L join. Volcaniclastic rocks from the Lower Mississippian part of the Schoonover sequence (Whiteford, 1990) overlap part of the field of volcaniclastic rocks of the Peale but are more diverse and are generally richer in Q and F. One rock from the Peale (Table 1, locality 9) is intermediate in composition between the quartzose and volcanic suites.

The quartzose suites show more diversity in Figure 12B, representing chert, lithic sedimentary (Ls), and lithic volcanic (Lv) grains. The Picayune Valley Formation falls near the chert-Ls side of the triangle centered near 65 percent chert. Closely similar are suites from the Antler foreland basin (Harbaugh, 1980) and from Upper Mississippian and Pennsylvanian rocks of the Schoonover sequence (Whiteford, 1990). The Peale Formation quartzose suite is similar but richer in chert (mean, 76 percent). Whiteford (1990) divided Lower Mississippian rocks of the Schoonover sequence into two suites based on grain size. The coarser suite (>1mm), which is the one most comparable in grain size to the Picayune Valley data, overlaps less than half of the Picayune Valley dispersion field and is generally richer in lithic volcanic grains (Lv). Whiteford's (1990) finer grained suite of Lower Mississippian rocks, as well as the Antler overlap assemblage of Saller (1980), are richer in both Ls and Lv relative to the Picayune Valley suite. Volcaniclastic suites from the Peale Formation and the Schoonover sequence (Whiteford, 1990) plot near the Lv corner of the diagram. One Peale rock is intermediate between the volcaniclastic and quartzose suites.

Pebble counts of conglomerate suites are shown in Figure 12C, representing the compositions in terms of chert, quartz sandstone, and all other clasts taken together. The mean of the Picayune Valley Formation suite lies at 58 percent chert, 18 percent quartz sandstone, and 24 percent other clasts. The pebble suite in Lower Mississippian rocks of the Schoonover sequence (Fagan, 1962; Whiteford, 1990) partly overlaps the Picayune Valley field but is slightly poorer in chert (mean, 51 percent). Conglomerates from the Antler overlap assemblage (Saller, 1980) lie adjacent to the Picayune Valley field, richer in quartz sandstone (mean 30 versus 18 percent) and more variable in chert content.

Conglomerate in the Peale Formation is

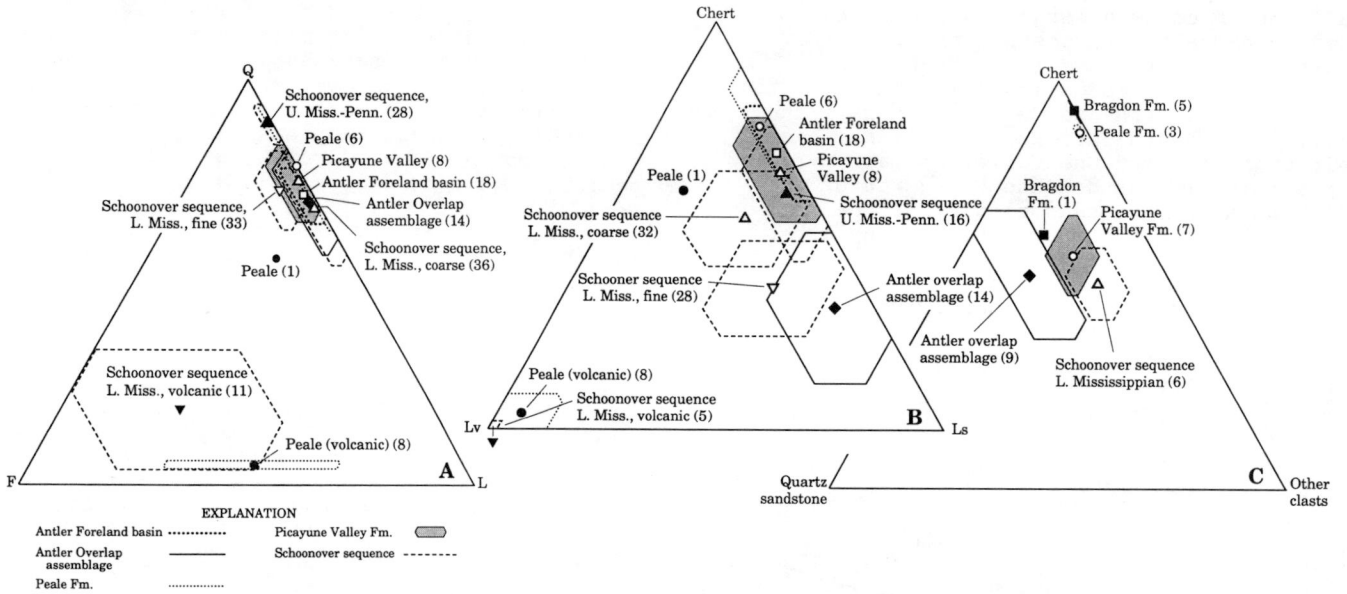

Figure 12. Sandstone petrology (A and B) and conglomerate clast compositions (C) of rocks in the lower member of the Peale Formation and the Picayune Valley Formation compared to rocks from the Antler orogenic belt in Nevada and the Bragdon Formation in the southeastern Klamath Mountains. Q, monocrystalline quartz, polycrystalline quartz, chert, and quartz sandstone; L, all lithic grains except those in Q; Ls, all sedimentary rock lithic grains except chert; Lv, all volcanic lithic grains, F, feldspar. For each suite, mean compositions are shown by a geometric symbols with an envelope defined by the standard deviations of the three components represented. Numbers in parentheses are the numbers of samples in the groups.

richer in chert and very low in quartz sandstone (mean, 85 percent chert, 2 percent quartz sandstone, and 10 percent other clasts). The Peale is similar to most conglomerate in the Bragdon Formation (Seiders, unpublished data), an Upper Devonian(?) and Mississippian stratigraphic unit in the southeastern Klamath Mountains (Miller and Cui, 1989). One Bragdon conglomerate, however, is relatively rich in quartz sandstone and other clasts, and is similar to conglomerate in the Picayune Valley Formation. Miller and Cui (1989) found greater variability in the amounts of chert and quartzite versus argillite, schist, and sandstone in conglomerate from the northern part of the Bragdon Formation.

The sandstone petrology and pebble counts show both similarities and contrasts between the Picayune Valley and Peale Formations and stratigraphic units in the Antler orogenic belt of Nevada. The similarities are most strongly seen in the Q-F-L diagram (Fig. 12A), but are also evident in the chert-Ls-Lv diagram (Fig. 12B). The dissimilarities in Figure 12B should not be overemphasized, because some of the differences may result from differences in the grain size of the material studied and in possible differences in methodology between workers. Taken together, the data in figure 12 strongly suggest that all of the quartzose clastic rocks originated within a single petrographic province. The present geographic distribution of the suites (Fig. 11) suggests that the province was large and some of the compositional diversity may result from the large geographic separation of the samples.

The similarity of the Picayune Valley Formation with that in the Schoonover sequence (Whiteford, 1990) is especially important because of its paleographic implications. The Schoonover sequence, like the Picayune Valley Formation, contains interbedded volcaniclastic and quartzose detrital rocks. Whiteford (1990) suggested that the Schoonover sequence represents a basin that was simultaneously receiving debris from a volcanic arc, the Antler highland, and the continental shelf. Similarly, the Picayune Valley Formation might have received sediment from the volcanic arc to the north and from the Antler orogenic highland to the south. We cannot rule out, however, the possibility that the quartzose detritus in the Picayune Valley Formation was derived from a different part of the Shoo Fly Complex in the basement of the volcanic arc, like the quartzose detritus in the Peale and Grizzly Formations. The sharp difference in pebble composition between the Peale and Picayune Valley Formations (Fig. 12C) implies a different but not necessarily a remote source area for detritus in the Picayune Valley Formation. A similar contrast in pebble composition occurs within the Bragdon Formation.

SUMMARY AND CONCLUSIONS

Upper Devonian and Lower Mississippian island-arc volcanic rocks in the northern Sierra Nevada grade southward from coarse,

blocky vent-proximal deposits to relatively
fine-grained volcaniclastic arc-apron
deposits. Distal volcaniclastic rocks
interfinger with coarse-grained, conglomeratic
quartzose turbidites of the Picayune Valley
Formation. Quartzose detritus in the Picayune
Valley Formation was derived from the south
and deposited in an inner or middle deep-sea
fan environment in an arc-marginal basin.

Lower Paleozoic rocks of the Shoo Fly
Complex, which form the arc basement, were
uplifted on an intra-arc horst that developed
over the north end of the Late Devonian Bowman
Lake batholith during active volcanism.
Chert-rich detritus, eroded from the Shoo Fly
Complex in the intra-arc horst, was transported southward and deposited on the arc
apron as chert-granule conglomerate lenses
interbedded with Lower Mississippian volcaniclastic rocks of the lower member of the Peale
Formation. Pebbly sandstone and conglomerate
in the Picayune Valley Formation are coarser
grained and richer in monocrystalline quartz,
quartz sandstone, and argillite than
conglomerate in the Peale Formation and
suggest that the Picayune Valley Formation was
derived from a different source area.

Comparative modal data and conglomerate
pebble counts (Fig. 12) from quartzose rocks
in the Picayune Valley Formation, the Peale
Formation, and various units in the Antler
orogenic belt, however, show remarkable
similarities as well as some contrasts. The
sandstone petrology and pebble counts strongly
suggest that all of these quartzose rocks
define a single petrologic province with a
common provenance of lower Paleozoic rocks in
the Shoo Fly Complex and the Roberts Mountains
allochthon. The samples analyzed in Figure 12
come from reginally widespread areas and the
differences in the modal data and pebble
counts may reflect, in large measure,
geographic variations in the stratigraphy or
structural stacking arrangement of the lower
Paleozoic rocks.

If the limited modal data and pebble
counts, shown in Figure 12, are representative
of the geographic region, they imply that
lower Paleozoic rocks of the Shoo Fly Complex
and the Roberts Mountains allochthon were
paleogeographic related, by deposition (Bond
and DeVay, 1980; Schweickert and others, 1984)
or structural juxtapositioning, prior to the
Late Devonian. The Picayune Valley Formation
may have been derived from the Antler orogenic
highland or an uplifted part of the Shoo Fly
Complex south of the area where it is overlain
unconformably by the Upper Devonian and Lower
Mississippian arc-volcanic rocks. This
ambiguity is frustrating, but it strongly
suggests that the Upper Devonian and Lower
Mississippian arc, and its basement, evolved
along the North American continental margin.

ACKNOWLEDGMENTS

The manuscript was thoughtfully reviewed
and greatly improved by M. D. Carr, D. M.
Miller and M. M. Miller.

REFERENCES CITED:

Bouma, A.H., 1962, Sedimentology of some
flysch deposits: Amsterdam, Elsevier,
168 p.

Bond, G.C., and DeVay, J.C., 1980, Pre-upper
Devonian quartzose sandstones in the Shoo
Fly Formation, northern California--
Petrology, provenance, and implications
for regional tectonics: Journal of
Geology, v. 88, p. 285-308.

Brooks, E.R., Wood, M.M., and Garbutt, P.L.,
1982, Origin and metamorphism of peperite
and associated rocks in the Devonian
Elwell Formation, northern Sierra Nevada,
California: Geological Society America
Bulletin v. 93, p. 1208-1231.

Dickinson, W.R., and Suczek, C.A., 1979, Plate
tectonics and sandstone compositions:
American Association of Petroleum
Geologists Bulletin, v. 63, p. 2164-2182.

Dickinson, W.R. Harbaugh, D., Saller, A.H.,
Heller, P.L., and Snyder, W.S., 1983,
Detrital modes of upper Paleozoic
sandstones derived from Antler Orogen in
Nevada: Implications for nature of
Antler orogeny: American Journal of
Science, v. 283, p. 481-509.

Durrell, C., and D'Allura, J.A., 1977, Upper
Paleozoic section in eastern Plumas and
Sierra counties, Sierra Nevada,
California: Geological Society of
America Bulletin, v. 88, p. 844-852.

Fagan, D.J., 1962, Carboniferous cherts,
turbidites, and volcanic rocks in the
northern Independence Range, Nevada:
Geological Society of America Bulletin,
v. 73, p. 595-612.

Girty, G.H., 1983, The Culbertson Lake
allochthon - A newly identified
structural unit in the Shoo Fly Complex -
Sedimentological, stratigraphic, and
structural evidence for extension of the
Antler orogenic belt to the northern
Sierra Nevada, California· Ph.D.
dissertation, Columbia Univ., New York,
N.Y., 155 p.

Girty, G.H., Gester, K.C., and Turner, J.B.,
1990, Pre-Late Devonian geochemical,
stratigraphic, sedimentologic, and
structural patterns, Shoo Fly Complex,
northern Sierra Nevada, California, in
Harwood, D.S., and Miller, M.M., eds.,
Paleozoic and early Mesozoic paleographic
relations: Sierra Nevada, Klamath
Mountains and related terranes: Boulder,
Colorado, Geological Society of America
Special Paper 255.

Hannah, J.L., and Moores, E..M., 1986, Age
relationships and depositional
environments of Paleozoic strata,
northern Sierra Nevada, California:
Geological Society of America Bulletin,
v. 97, p. 787-797.

732

Hanson, R.E., 1982, An intrusive andesitic hyaloclastite complex in an Upper Devonian island-arc sequence, Sierra Buttes Formation, northern California: Geological Society of America Abstracts with Programs, v. 14, p. 170.

Hanson, R.E., and Schweickert, R.A., 1986, Stratigraphy of mid-Paleozoic island-arc rocks in part of the northern Sierra Nevada, Sierra and Nevada counties, California: Geological Society of American Bulletin, v. 97, p. 986-998.

Hanson, R.E., Saleeby, J.B., and Schweickert, R.A., 1988, Composite Devonian island-arc batholith in the northern Sierra Nevada, California: Geological Society of America Bulletin, v. 100, p. 446-457.

Harbaugh, D.W., 1980, Depositional facies and provenance of the Mississippian Chainman Shale and Diamond Peak Formation, central Diamond Mountains, Nevada [M.S. thesis]: Stanford, California, Stanford University, 81 p.

Harwood, D.S., 1983, Stratigraphy of upper Paleozoic volcanic rocks and regional unconformities in part of the northern Sierra terrane, California: Geological Society of America Bulletin, v. 94, p. 413-422.
_____ 1988, Tectonism and metamorphism in the northern Sierra terrane, northern California, in Ernst, W.G., ed., Metamorphism and crustal evolution of the western United States: Rubey Volume 7: Englewood Cliffs, New Jersey, Prentice-Hall, p. 764-788.

_____, 1991, Stratigraphy of Paleozoic and lower Mesozoic rocks in the northern Sierra terrane, California: U. S. Geological Survey Bulletin, 1957, in press.

Harwood, D.S., Jayko, A.S., Harris, A. G., Silberling, N.J., and Stevens, C. H., 1988, Permian-Triassic rocks slivered between the Shoo Fly Complex and the Feather River periodite belt, northern Sierra Nevada, California: Geological Society of America Abstracts with Programs, v. 20, p. 167-168.

Harwood, D.S., and Murchey, B.L., 1990, Biostratigraphic, tectonic, and paleogeographic ties between upper Paleozoic volcanic and basinal rocks in the northern Sierra terrane, California, and the Havallah sequence, Nevada, in Harwood, D.S., and Miller, M.M., eds., Paleozoic and early Mesozoic paleogeographic relations; Sierra Nevada, Klamath Mountains, and related terranes: Boulder, Colorado, Geologic Society of America Special Paper 255.

Howell, D.G., and Normark, W.R., 1982, Sedimentology of Submarine Fan in Scholle, P.A., and Darwin Spearing, eds., Sandstone Depositional Environments, American Association of Petroleum Geologists Memoir 31, p. 365-404;

Lindgren, W., 1897, Description of the gold belt description of the Truckee quadrangle, California: U.S. Geological Survey Geol. Atlas, Folio 39, 8 p.

Miller, E. L., Holdsworth, B.K., Whiteford, W.B., and Rogers, D., 1984, Stratigraphy and structure of the Schoonover Sequence, northeastern Nevada: Implications for Paleozoic plate margin tectonics: Geological Society of America Bulletin, v. 95, p. 1063-1076.

Miller, M. M., 1989, Intra-arc sedimentation and tectonism; Late Paleozoic evolution, eastern Klamath terrane, California: Geological Society of America Bulletin, v. 101, p. 170-187.

Miller, M. M., and Cui, B., 1989, Carboniferous arc-related sedimentation, Klamath Mountains, California; Hybrid fan characteristics and dual sediment provenance: Canadian Journal of Earth Sciences, v. 26, p. 927-940.

Mutti, E. and Ricci Lucchi, F., 1972, Le torbiditi dell 'Apennino settentrionale: introduzione all 'analisi di facies: Memoir Society Geology Italy 11, p. 161-199. English translation in International Geology Review, 1978, v. 20, no. 2, p. 125-166,

Nilsen, T. H., 1977, Paleogeography of Mississippian turbidites in south-central Idaho, in Stewart, J.H., Stevens, C.H., and Fritsche, E.A., eds., Paleozoic paleogeography of the western United States, Pacific Coast Paleogeography Symposium 1 Pacific Section, Society of Economic Paleontologists and Mineralogists, p. 275-299.

Poole, F.G., 1974, Flysch deposits of the Antler foreland basin, western United States, in Dickinson, W.R., ed., Tectonics and sedimentation: Society of Economic Paleontologists and Mineralogists Special Publication 22, p. 58-82.

Poole, F.G., and Sandberg, C.A., 1977, Mississippian paleogeography and tectonics of the western United States, in Stewart, J.H., Stevens, C. H., and Fritsche, A.E., eds., Paleozoic paleogeography of the western United States, Pacific Coast Paleogeography Symposium 1 Pacific Section, Society of Economic Paleontologists and Mineralogists, p. 67-85.

Potter, A.W., Watkins, R., Boucot, A.J., Elias, R.J., Flory, R.A., and Rigby, J.K., 1990, Biogeography of the Upper Ordovician Montgomery Limestone, Shoo Fly Complex, northern Sierra Nevada, California, and comparisons of the Shoo Fly Complex with the Yreka terrane, in Harwood, D.S., and Miller, M. M., eds., Paleozoic and early Mesozoic paleogeographic relations; Sierra Nevada, Klamath Mountains, and related terranes;

Boulder, Colorado, Geological Society or America Special Paper 255.

Roberts, R. J., Hotz, P.E. Gilluly, J., and Ferguson, H.G., 1958, Paleozoic rocks of north-central Nevada: American Association of Petroleum Geologists Bulletin, v. 42, p. 2813-2857.

Saleeby, J.B., Hannah, J.L., and Varga, R.J., 1987, Isotopic age constrains on middle Paleozoic deformation in the northern Sierra Nevada, California: Geology, v, 15, p. 757-760.

Saller, A.H., 1980, Depositional setting of post-Antler Pennsylvanian strata in north-central Nevada [M.S. thesis]: Stanford, California, Stanford California, 118 p.

Schweickert, R.A., Harwood, D.S., Girty, G.H., and Hanson, R.E., 1984, Tectonic development of the Northern Sierra terrane: An accreted late Paleozoic island-arc and its basement, in Lintz, J., ed., Western geological excursions: Geological Society of America 1984 Annual Meeting Field Trip Guides, v. 4, p. 1-65.

Seiders, V.M., and Blome, C.D., 1988, Implications of upper Mesozoic conglomerate for suspect terrane in western California and adjacent areas: Geological Society of America Bulletin, v. 100, p. 374-391.

Speed, R.C., and Sleep, N.H., 1982, Antler orogeny and foreland basin: A model: Geological Society of America Bulletin, v. 93, p. 815-828.

Walker, R.G., and Mutti, E., 1973, Turbidite facies and facies associations, in G.V. Middleton and A.H. Bouma, eds., Turbidites and deep-water sedimentation, Society of Economic Paleontologists and Mineralogist, Pacific Section, Short Course Notes, p. 119-157.

Whiteford, W.B., 1990, Paleogeographic setting of the Schoonover sequence, Nevada, and implications for the late Paleozoic margin of western North America in Harwood, D.S., and Miller, M.M., eds., Paleozoic and early Mesozoic paleogeographic relations; Sierra Nevada, Klamath Mountains, and related terranes: Boulder, Colorado, Geological Society of America Special Paper 255, p. 115-136.

STRATIGRAPHY, SEDIMENTOLOGY, AND DEPOSITIONAL CONDITIONS OF LOWER PALEOZOIC WESTERN-FACIES ROCKS IN NORTHEASTERN NEVADA

Keith B. Ketner, U.S. Geological Survey, MS 939,
Federal Center, Denver CO 80225

ABSTRACT

Siliceous, basinal (western facies) rocks, comprising the Roberts Mountains allochthon in northeastern Nevada, form a consistent stratigraphic sequence of regional extent.

Lower to lower Middle Ordovician rocks are composed largely of quartz and chert sand, sand-size bioclastic carbonate, sand-size phosphate, limestone-clast conglomerate, graptolite-rich shale and two kinds of bedded chert. Locally, the lower Middle Ordovician beds contain strata-bound lead, zinc, and iron sulfides. Upper Middle to lower Upper Ordovician beds are typically composed of thick, black, graptolite-rich shale, and sporadic thin beds of chert and quartz sandstone. Uppermost Ordovician strata consist of black, thick-bedded chert and sporadic beds of dark micritic limestone. The Ordovician sequence includes very scarce, thin, pods of mafic igneous rock.

Lower Silurian beds are composed of black, white, and brightly-colored chert that contains significant deposits of lead, zinc, and iron sulfides or, in surface exposures, abundant cavities from which sulfides were dissolved. Middle to lower Upper Silurian rocks are typically muscovite-rich, K-feldspar-rich, limy quartz siltstone, and shale, locally weakly mineralized with lead and zinc. Uppermost Silurian beds have not been identified in the region and Lower to Middle Devonian beds are scarce.

Unconformably overlying the Ordovician-Silurian sequence are Upper Devonian beds composed of bedded chert, bedded barite, quartz sandstone, feldspathic, micaceous quartz siltstone, sandstone and conglomerate composed mainly of radiolarian-bearing chert clasts, and detrital limestone.

The sedimentary and stratigraphic features of the lower Paleozoic western facies rocks suggest the following: Ordovician and Silurian detrital sediment was derived mainly from provenance terranes external to the basin of deposition. Much of this sediment may have been derived from terranes to the west of the basin of deposition, in as much as potential sources to the east and channels of transport across the miogeocline in appropriate places and at appropriate stratigraphic positions have not been located. Upper Devonian detrital sediment, to a large extent, was derived from internal provenance terranes indicating tectonism in the basin of deposition.

Erosion within the basin of deposition cannot be explained by a drop in sea level for it occurred during a time of generally high, and rising, sea level. The western facies sequence includes a very wide range of rock types but one kind of rock is conspicuously absent: plagioclase-rich, chloritic sandstone of a type normally abundant near volcanic arcs. Most of the bedded chert, like the sulfide and sulfate deposits, is probably of submarine hydrothermal origin. The depositional setting of the western facies sequence in northeastern Nevada is compatible with certain previously described models: an extensional basin on the continental margin with provenance terranes on both sides, sporadically cut longitudinally by transpressional-transtensional faults.

INTRODUCTION

This report concerns well dated sequences of lower Paleozoic allochthonous basinal facies ("western-facies" in contrast with "eastern-facies" shelf carbonate) in a part of northeastern Nevada extending from the southern Tuscarora Mountains to the HD Range (Figs. 1,2, Table 1). Like earlier reports on the same subject (Ketner, 1977, 1980), it is a progress report of a long term investigation. These strata constitute part of the Roberts Mountains allochthon. Sedimentary rocks included in this study are the Ordovician Vinini Formation and associated unnamed Silurian, Devonian, and lowest Mississippian strata. The Ordovician Valmy Formation and associated strata of north-central Nevada are excluded. Stratigraphic and lithic data are based on detailed mapping in several mountain ranges by the author, examination of many hundreds of thin sections, and scores of paleontological age determinations, mainly by U.S. Geological Survey specialists. Descriptions of strata-bound sulfides are based on extensive surface observations of gossan and examination of drill cores from the Adobe Range and southern Independence Mountains.

In Cooper, J.D., and Stevens, C.H., eds., 1991, **Paleozoic Paleogeography of the Western United States-II:** Pacific Section SEPM, Vol. 67, p. 735-746.

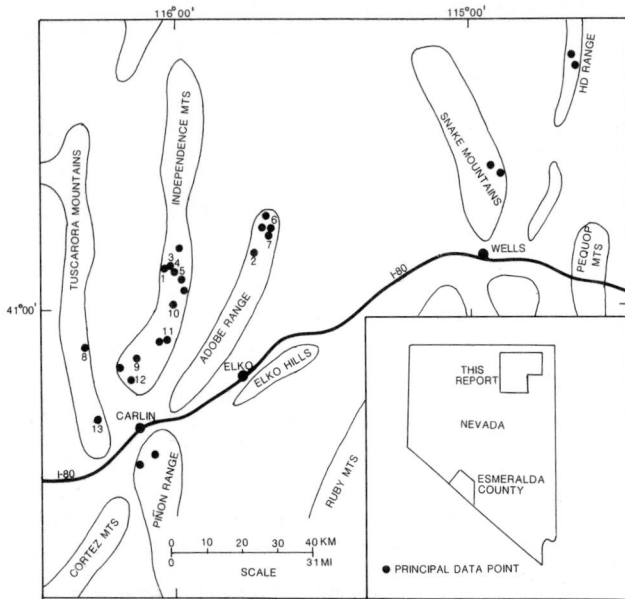

Figure 1. Index map of northeastern Nevada showing principal locations of sources of data. Data points listed in Table 1: 1-Basco Field; 2-Coal Mine Basin; 3-Lone Mountain Creek; 4-Cold Creek; 5-Sproul Cabin; 6-Badger Spring; 7-Cherry Spring; 8-Simon Creek; 9-Cottonwood Creek; 10-Blue Basin Creek; 11-Swales Mountain; 12-Hadleys Windmill; 13-Marys Mountain

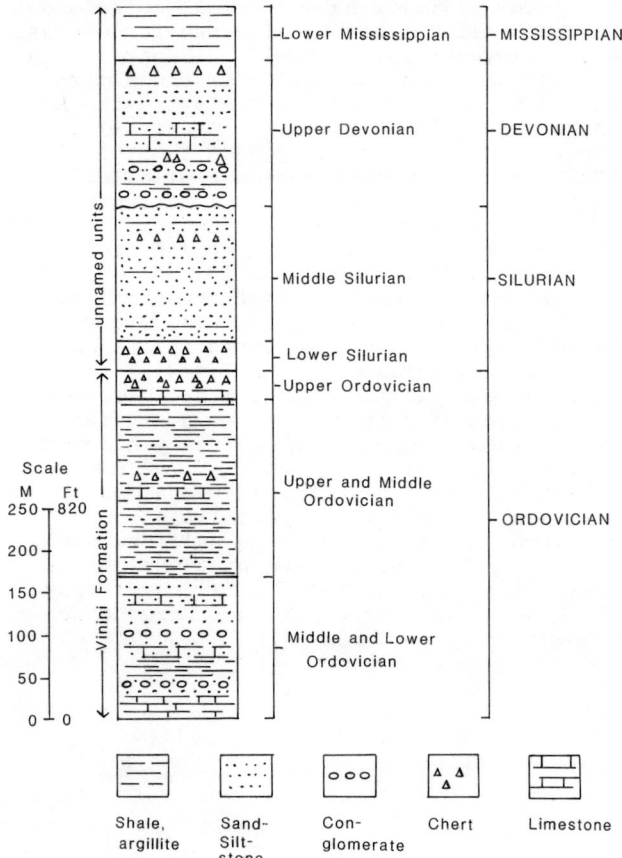

Figure 2. Diagrammatic column of the strata discussed in this report. Thicknesses and lithic composition shown are generalized from scattered localities.

Table 1--Exposures of well-dated typical lower Paleozoic allochthonous rocks in northeastern Nevada

Locality name	Quadrangle	UTM coordinates	
UPPER UPPER DEVONIAN TO LOWER LOWER MISSISSIPPIAN			
Basco Field	Blue Basin	81100E	50200N
Coal Mine Basin	Coal Mine Basin	07100E	54800N
LOWER UPPER DEVONIAN			
Lone Mountain Creek	Blue Basin	82500E	48600N
Cold Creek	Singletree Creek	85800E	46900N
Sproul Cabin	Singletree Creek	87425E	46620N
Coal Mine Basin	Coal Mine Basin	05050E	53500N
Coal Mine Basin	Coal Mine Basin	05550E	53500N
LOWER MIDDLE TO LOWER UPPER DEVONIAN			
Badger Spring	Coal Mine Basin	11050E	62100N
Badger Spring	Coal Mine Basin	11200E	61700N
Badger Spring	Coal Mine Basin	13200E	62100N
MIDDLE TO LOWER UPPER SILURIAN			
Lone Mountain Creek	Blue Basin	81350E	50550N
Badger Spring	Coal Mine Basin	10750F	62250N
Badger Spring	Coal Mine Basin	10150E	64100N
LOWER SILURIAN			
Badger Spring	Coal Mine Basin	10900E	62150N
Cherry Spring	Coal Mine Basin	10997E	60230N
UPPER UPPER ORDOVICIAN			
Badger Spring	Coal Mine Basin	11050E	61950N
Simon Creek	Rodeo Creek NE	59900E	29650N
Cottonwood Creek	Schroeder Mountain	72600E	24750N
UPPER MIDDLE TO LOWER UPPER ORDOVICIAN			
Lone Mountain Creek	Blue Basin	82500E	50500N
Blue Basin Creek	Blue Basin	83900E	41450N
Lone Mountain	Blue Basin	83850E	49500N
Badger Spring	Coal Mine Basin	10200E	64050N
Cherry Spring	Coal Mine Basin	11600E	60200N
Cottonwood Creek	Swales Mountain	74400E	26350N
Simon Creek	Rodeo Creek NE	60080E	29580N
LOWER TO LOWER MIDDLE ORDOVICIAN			
Badger Spring	Coal Mine Basin	11000E	62150N
Lone Mountain Creek	Blue Basin	82550E	50350N
Lone Mountain Creek	Blue Basin	81400E	49850N
Simon Creek	Rodeo Creek NE	60400E	29160N
Blue Basin Creek	Blue Basin	84050E	40050N
Swales Mountain	Swales Mountain	80800E	28200N
Simon Creek	Rodeo Creek NE	60300E	29100N
Hadley's Windmill	Schroeder Mountain	71450E	19000N
Marys Mountain	Carlin	63800E	10800N

STRATIGRAPHY AND PETROLOGY

Ordovician Vinini Formation

The Vinini Formation of central and northern Nevada and the Palmetto Formation in Esmeralda County, southern Nevada, are virtually identical in lithic composition, stratigraphic sequence, and age. In this report the term "Vinini" is applied to a distinctive, mappable stratigraphic sequence of Early, Middle, and Late Ordovician age. Although the term "Palmetto" was first applied to western-facies Ordovician rocks of southern Nevada in 1902 and has priority over the term "Vinini", the latter is preferable because it is used more widely and too firmly entrenched in the literature to be displaced.

Detailed stratigraphic studies of the Vinini Formation have not been made in the past because of structural complexity and poor exposures. However, a very general sequence was recognized by several observers in the "type" area of the Vinini, the Roberts Mountains of central Nevada. Merriam and Anderson (1942), who originally described the formation, Winterer (1968), Stanley and others (1977), and Finney and Perry (1989) reported that the lower part of the "type" Vinini section contains much more carbonate rock and quartz sand than the upper part, which is composed almost entirely of shale and chert. Among the published descriptions of the Vinini in northern Nevada, Lovejoy (1959), Riva (1970), Smith and Ketner (1975), and Ketner (1969, 1975) observed a similar lithic distinction between its lower and upper parts, and Ketner (1980, 1986) called attention to remarkable similarities between the Vinini and correlative units in Texas, Arkansas, and Mexico.

A large accumulation of precise paleontological age determinations from distinctive lithic units within the Vinini in northern Nevada now makes the stratigraphy of the Vinini much more clear and shows that its internal stratigraphic sequence is regional in extent.

Lower member of Vinini Formation

The lower member of the Vinini (Lower to lower Middle Ordovician) is distinguished by its extreme heterogeneity. Commonly it is composed of both siliceous and carbonate arenites, limestone conglomerate, shale, chert, siltstone, dolostone, and greenstone.

Characteristically, the lower member consists of rocks that are more coarse grained and more limy than other members of the Vinini. Sand-size components form the bulk of the unit and conglomerates are common. Although most arenites are heterogeneous in composition and difficult to classify, some have a relatively simple composition and can be classified as calcarenite, quartz arenite, etc. With the exception of certain quartz arenites, all of the detrital rocks in the lower member are composed of generally poorly sorted and poorly rounded clasts. Attributes of turbidity current deposits, such as graded bedding and sole marks, are displayed by some arenite beds, but other beds in the same sequence are entirely devoid of such features. Thick sequences of monotonously similar graded beds, such as those that constitute typical flysch deposits, have not been found. Instead, the sequence of beds within the lower member is characteristically randomly varied in thickness, composition, texture, and sedimentary structure, reflecting a constantly changing environment. The base of the lower member is nowhere clearly exposed in northeastern Nevada and its substrate is unknown.

Components of Arenites and Congomerates

The following types of sand-size and larger clasts are present in various combinations in strata of the lower member.

1. Quartz. These grains range in size from silt to coarse sand. The silt and very fine sand size grains tend to be angular, but the coarser sizes are commonly well rounded. Almost invariably quartz grains are single crystals and most are strained.

2. Fragments of calcareous organisms. Among these are parts of recognizable animals such as pelmatozoans, trilobites, gastropods, cephalopods, ostracodes, and brachiopods. Calcispheres are common also, but sand-size fragments of the alga *Nuia* (Ross and others, 1988b) are by far the most common calcareous organic constituent and form the bulk of calcarenites in the lower part of the lower member.

3. Fragments of phosphatic organisms. Among the fossil debris are fragments of graptolites, the crustacean *Caryocaris*, inarticulate brachiopod shells, and conodonts. Graptolite debris is abundant and ubiquitous in the finer-grained rocks; brachiopod shell fragments are common and sparsely disseminated in many kinds of rocks, but fragments of *Caryocaris* seem to be concentrated in only a few thin beds in which they are the principal constituent.

4. Fragments of siliceous organisms. Siliceous sponge spicules are abundant constituents of some beds. Together with various proportions of other sediments, they form thinly laminated, commonly cross-bedded, siliceous sandstones. Where the proportion of spicules is very high, these rocks have the gross appearance and composition of chert. However, where the proportion is small and other sand-size components predominate, they have the appearance of cherty sandstone, dolostone, or calcarenite. Axial canals of spicules commonly display a reticulate network of black opaque material.

5. Calcareous fecal(?) pellets. These are the most abundant sand-size particles in the upper part of the lower member. They generally are associated with clastic particles of similar size in well-bedded strata, suggesting that they have been reworked.

6. Dolomite rhombs and glauconite grains. Dolomite rhombs are abundant sand-size constituents of cherty, spicule-rich rocks. Their angularity suggests they were formed nearly in place rather than transported from distant source areas. Glauconite grains are crudely equant but not rounded and are sparsely distributed. Whether they formed in situ or were derived from a distant source is difficult to determine.

7. Sand-size and grit-size rock fragments. Angular, equant, or tabular clasts of dolostone, limestone, shale, and chert are common constituents. Clasts of mafic igneous rock and of lithic, aphanitic, untextured phosphorite, as opposed to organic phosphate, are much less common.

8. Pebbles, cobbles, and boulders of sedimentary rock. Clasts in this size range are generally tabular; less commonly they are equant. They are generally only corner-rounded, suggesting relatively little transport. Their composition is that of

associated strata including dolomitic chert and cherty dolostone, micrite, *Nuia*-rich calcarenite and pelletal calcarenite.

Calcarenite

Calcarenite, a very abundant rock type in the lower member of the Vinini, forms beds generally a few centimeters thick that are commonly internally massive and non-graded. However, at Blue Basin Creek, some calcarenite beds display sporadic evidence of deposition by turbidity currents, such as grading and sole marks. Although the calcarenites are commonly associated with black shale, their color in outcrop ranges from light gray and brown to black. Clasts of calcarenite incorporated in Ordovician conglomerate have the same range of colors, indicating that some of the lighter shades are original and are not due entirely to differential oxidation of the present outcrops. Calcarenites in the lower part of the lower member are composed primarily of sand-size fragments of the alga *Nuia*. Other constituents are silt- to sand-size quartz grains, glauconite, fragments of phosphate rock, and fragments of both phosphatic and calcareous brachiopods. Calcarenites of the upper part of the lower member are composed of pelletal carbonate rather than the *Nuia* fragments characteristic of the older beds.

Mature Quartz Arenite

Mature quartz sandstone, cemented with a small proportion of ferruginous dolomite, and quartzite cemented with silica constitute a small part of the Vinini in some areas. Stratigraphic units are commonly about two to three meters thick and internal bedding is absent or indistinct. These rocks are similar to the characteristic quartzites of the Ordovician Valmy Formation in central Nevada, but are thinner and less common. Grain size ranges from fine to coarse and the median is in the medium sand-size range. Grains are well rounded and moderately well sorted. These arenites do not contain fossils, but a quartzite bed at Swales Mountain is dated as Early Ordovician on the basis of associated fossiliferous shale beds. The age of another quartzite at Lone Mountain Creek is very well established as late Early to early Middle Ordovician by fossils in strata above and below. Quartz sandstone in the "type" Vinini in the Roberts Mountains is of early Middle to late Middle Ordovician age (Finney and Perry, 1989, and this volume).

Bedded Detrital Pseudo-chert

Although resembling radiolarian-bearing cherts in macroscopic appearance, most "chert" beds in the lower member are siliceous arenites composed primarily of silica-cemented sponge spicules, chert grains, sparse dolomite rhombs, quartz sand and silt, rock fragments, and calcareous bioclastic debris. This kind of chert is commonly intimately associated with detrital dolostone, and there is a complete gradation between chert nearly free of dolomite rhombs to dolostone with lenticular blebs of chert. Similar beds of pseudo-chert were observed in Upper Devonian strata. This kind of chert contrasts with that of the upper part of the Vinini, which is nearly pure, radiolarian-bearing chert associated with fine-grained pelagic limestone and shale.

Heterogeneous Arenite

These complex detrital rocks are intermediate in composition among distinct end members such as calcarenite, quartz arenite, and detrital chert. The heterogeneous arenites are composed of various proportions of quartz; calcareous, silicic, and phosphatic bioclastic debris; glauconite; and rock fragments. The latter include siltstone, chert, limestone, dolostone, phosphate rock, and, rarely, fine-grained mafic igneous rock. Only calcareous fecal(?) pellets and larger quartz grains are well rounded. All other constituents tend to be angular and poorly sorted. Beds a few centimeters thick may be either graded or ungraded. Parallel beds, cross-beds, and climbing ripples are fairly common, but examples of sole marks and convolute lamination are generally scarce. The extreme lateral and vertical heterogeneity of composition, textural detail, and bedding types suggest randomly varied depositional conditions that only rarely yielded mature sediments of simple mineralogical composition.

Conglomerate

Limestone conglomerate is an abundant constituent of rocks in the lower member of the Vinini Formation and chert-clast conglomerate is associated with limestone conglomerate at one location, Swales Mountain. The conglomerate forms tabular beds, not irregular or lenticular pods, generally a few centimeters to a few meters thick and are interbedded with calcarenite. Larger clasts, which range to 0.7 m in diameter, are tabular with rounded edges; smaller ones are tabular to equant. Most clasts are composed of calcarenite identical to that in associated calcarenite beds, but some are composed of chert. Almost all conglomerates are clast supported. The matrix, which forms a relatively small part of the rock, is commonly calcarenite, dolomite, or chert. The large size of the clasts and their poor rounding, close packing, and composition indicate that they constitute intraformational conglomerate whose clasts have been transported only a short distance. Conodonts in clasts of conglomerate and in calcarenite interbedded with the conglomerate are identical.

Other Strata

These include radiolarian-bearing (non-detrital) chert, shale, siltstone, and greenstone. Chert beds are shades of gray or black; bedding surfaces range from planar to knobby to crinkly. Shale is black, silty, calcareous, and rich in graptolites. Siltstone is quartzose, but commonly contains

abundant feldspar and mica. Greenstone of
basic to intermediate composition is a poorly
exposed and scarce component. It occurs as
small (several meters in largest dimension)
irregular and tabular bodies. Pillow
structures were observed in some exposures,
but in most occurrences it is not certain
whether the greenstone represents lavas or
intrusive dikes and sills of unknown age.

Limestone of Badger Spring

A unique limestone body of late Early
to early Middle Ordovician age (Ketner and
Ross, 1990; Ross and others, 1988a) is
exposed near Badger Spring in the northern
Adobe Range and on the north slope of Lone
Mountain in the southern Independence
Mountains. In the Adobe Range, the limestone
of Badger Spring crops out sporadically for a
distance of about 11 km and attains a
thickness of several tens of meters. Each of
the outcrops could be considered a separate
lens but here they are regarded as isolated
outcrops of a single originally continuous
sheet whose outcrop belt was segmented either
structurally or erosionally and was partially
buried by alluvium and colluvium. The
limestone of Badger Spring lies on Devonian
rocks with a thrust fault contact and its
original substrate is therefore unknown. It
is overlain unconformably, with a clearly
exposed sedimentary contact, by Lower
Silurian bedded black chert and limy
turbiditic siltstone. The age of all units
is well established. The limestone of Badger
Spring displays sedimentary features and a
shelly fauna remarkably similar to those of a
mudmound in the shelf carbonate sequence at
Meiklejohn Peak in southern Nevada (Ross and
others, 1975), but there the mudmound is
overlain conformably by Middle and Upper
Ordovician carbonate rocks. At Badger
Spring, the overlying Silurian unit is a
relatively deep-water deposit. Hypotheses of
origin of this enigmatic sequence are
discussed in a subsequent section.

Sulfides

Iron, zinc, and lead sulfides occur
sporadically in the upper (lower Middle
Ordovician) part of the lower member of the
Vinini Formation. The best-known occurrence
is in the southern Independence Mountains,
where the mineralized zone crops out and was
also encountered in drilling.

Fauna, Age, and Correlatives

The age of the lower member was
determined principally on the basis of
abundant graptolites from shale and abundant
conodonts from limy beds. Most calcareous
shelly fossils are too fragmented to supply
significant evidence of age, but in one area
redeposited trilobites were identified.
Although many of the conodonts probably occur
in clasts, and therefore were redeposited,
most collections are composed of compatible
assemblages and give ages that are consistent
with those determined from graptolites in
associated shale. Collections from 32
localities indicate an age range for the

lower member of early Early to early Middle
Ordovician.

Similar correlative rocks are fairly
well exposed in Esmeralda County, southern
Nevada, at Miller Mountain (Stewart, 1979,
F15 locality), Palmetto Mountain (McKee,
1968, localities 27 and 32), Oasis Divide,
and Railroad Pass. In the Marathon area of
Texas, the approximately correlative
allochthonous Marathon Limestone, Alsate
Shale, and Fort Pena Formation, collectively,
are lithically and stratigraphically similar
to strata in the lower member of the Vinini
(Ketner, 1980; King, 1937). In the Ouachita
Mountains of Arkansas, the partially
correlative allochthonous sequence that
consists of the Collier Shale, Crystal
Mountain Sandstone, Mazarn Shale, and Blakely
Sandstone is also similar, but the sequence
there contains a higher proportion of quartz
sandstone and shale than the lower member of
the Vinini Formation in northern Nevada and
more closely resembles the Valmy Formation
than the Vinini.

Environment of Deposition

The abundance of coarse debris derived
from a shallow-marine environment and the
near absence of volcanic detritus indicate
that the lower member was deposited adjacent
to an emergent or nearly emergent shelf
terrane and far from any emergent magmatic
arc terrane. Clasts derived from erosion
within the basin itself, other than those of
the "rip-up" conglomerates, are relatively
scarce.

Intraformational, rip-up, limestone
conglomerates similar to those in the lower
member of the Vinini are a characteristic
feature of contemporaneous carbonate
formations of the shelf environment in
western Utah (Nolan, 1935; Staatz and Carr,
1964; Morris and Lovering, 1961; Hintze,
1974). In the shelf environment, such
conglomerates, composed of poorly rounded and
sorted clasts lithically similar to
underlying undisturbed beds, are correctly
interpreted as very shallow-water deposits.
The Vinini conglomerates may have been
deposited in relatively deep water by swift
turbidity currents, but their close
resemblance to contemporaneous conglomerates
of the shelf, and the general absence of
convolute lamination, suggest the possibility
that the Vinini conglomerates may actually be
shallow-water deposits also

Middle Member of Vinini Formation

The middle member of the Vinini
Formation (upper Middle to lower Upper
Ordovician) is much more uniform laterally
and stratigraphically than the lower member.
Regionally, it is a fine-grained carbonaceous
shale that locally contains thin beds of
micritic limestone, mature quartz sandstone
or quartzite, and radiolarian-bearing chert.
Graptolites constitute the principal fauna.
The limestone beds, which are concentrated
near the top of the member, are black
micrites composed almost entirely of
calcareous spicules less than 0.5 mm long and

calcispheres about 0.1 mm in diameter. Many of the limestone beds are recrystallized to form large regular cone-in-cone structures elongated perpendicular to the bedding, whereas other beds contain randomly oriented plumose calcite crystals. A third variety consists of veins of sparry calcite subparallel to bedding. Graptolites and conodonts were obtained from limestone beds of the middle member, but graptolites are less abundant in limestone than in the associated shale.

Chert beds of the middle member contain radiolarians or other spheroidal organic skeletons rather than spicules, which commonly constitute a large part of chert beds in the lower member. However, the radiolarians are a minor constituent and most of the chert is free from any sign of organic origin.

The sparse sandstone or quartzite beds of the middle member are composed of rounded, moderately well-sorted quartz grains and sparse ferruginous carbonate cement.

The boundary between middle and lower members is gradational over a narrow interval. Graptolites and conodonts indicate that the middle member ranges from Middle to early Late Ordovician in age. Thickness of the middle member is typically several hundred meters.

Lithically similar, and approximately correlative, units of the middle member are: shale and argillite of the Phi Kappa Formation in central Idaho (Carter and Churkin, 1977); a thick shale unit in the Palmetto Formation in southwestern Nevada; shale and argillite of the Guayacan Group in northern Mexico (Ketner and Noll, 1987); the Woods Hollow Shale in the Marathon area of Texas (McBride, 1969); and the Womble Shale in Arkansas (Ross and others, 1982).

Environment of Deposition

The predominantly dark color, fine grain, and lithic uniformity of the middle member, in contrast with the lower member, suggest an environment of deeper-water, farther from sources of terrigenous sediment, and almost devoid of internal or marginal tectonism.

Upper Member of Vinini Formation

The upper member (Upper Ordovician) is composed mainly of thick-bedded, radiolarian-bearing, black chert. Chert concretions as much as 0.5 m in diameter are common. Thin micritic limestone strata, similar to those in the upper part of the middle member, are present among the chert beds in some exposures. The upper member is commonly only a few meters thick and rarely a few tens of meters. In combination with the overlying Lower Silurian chert, it crops out fairly well and tends to form low ridges.

Graptolites in argillaceous beds between the chert beds and conodonts in micritic limestone interbeds indicate a Late Ordovician age for the upper member. Radiolarians are present in the chert, but generally are not abundant.

The uppermost part of the upper member locally contains pyrite, galena, and sphalerite, but sulphide mineralization is more prevalent in the directly overlying Silurian rocks. These occurrences are described in a subsequent section.

Lithically similar stratigraphic units approximately correlative with the upper member are: black chert beds in the upper part of the Palmetto Formation of southwestern Nevada; siliceous shale at the top of the Phi Kappa Formation in central Idaho (Carter and Churkin, 1977); black chert beds in the Guayacan Group of northern Mexico (Ketner and Noll, 1987); the Maravillas Chert of the Marathon area of Texas (McBride, 1970); and the Bigfork Chert and Polk Creek Shale of Arkansas (Miser and Purdue, 1929; Ross and others, 1982).

Environment of Deposition

The upper member was deposited under conditions similar to those of the middle member in relatively deep water, far from sources of terrigenous sediment and under tectonically quiescent conditions.

Lower Silurian strata

The Lower Silurian is represented by a sequence of thin-bedded black and strongly pigmented chert, and unusually thick-bedded white chert somewhat similar to the lower Paleozoic novaculite beds of Arkansas and Texas. This unit, which is generally a few meters thick, invariably lies concordantly on graptolite-bearing Upper Ordovician thick-bedded, black chert and beneath graptolite-bearing Middle Silurian siltstone. A sparse graptolite fauna of Early Silurian age occurs within the chert unit near Cherry Spring in the northern Adobe Range. Radiolarians and spicules are notably scarce and some beds are entirely free from evidence of organic origin.

The Lower Silurian chert is extensively mineralized with silver-bearing iron, lead, and zinc sulfides that weather to form a conspicuous gossan described in a subsequent section. Although this mineralized zone occurs almost entirely within the Lower Silurian chert unit, in some areas the uppermost part of the Upper Ordovician black chert unit may also be mineralized.

The lower, novaculite part of the Caballos Formation in the Marathon area, Texas is a probable correlative of the Lower Silurian chert of northeastern Nevada. Fossils are scarce in both of these units, but both consist partly of thick-bedded white chert that concordantly overlies well-dated Upper Ordovician black chert units.

Environment of Deposition

The presence of thick-bedded white

chert suggests an environment somewhat different from that of the directly underlying, uniformly black chert of the Upper Ordovician and from some of the Silurian beds. Lack of terrigenous sediments in both the Upper Ordovician and Lower Silurian strata suggests great distance from sources of sediments but the white color of some beds suggests the possibility that the water was occasionally oxygenated.

Middle to lower Upper Silurian Strata

Middle Silurian beds (and sporadic occurrences of lower Upper Silurian strata) are composed of black to mainly pale green, tan-weathering siltstone or fine-grained sandstone. Graded bedding and abundant sole marks indicate a turbiditic origin. Quartz grains constitute most of the rock, but feldspar and muscovite are abundant and the unit can be identified with the aid of a hand lens by their presence. Drill cores from this unit indicate that, unlike most other constituents of the western facies assemblage, much of the micaceous siltstone unit is light colored, even below the weathered zone. Other strata are black to commonly light-colored bedded chert and shale. Bedding surfaces are marked by abundant meandering trails and other trace fossils. Graptolites, preserved on the few surfaces free from bioturbation, indicate Middle to early Late Silurian age. The thickness of this unit is a few hundred meters.

Correlatives are the Elder Sandstone of the Shoshone Range, which is similar in mineral composition, but tends to be more coarse-grained (Gilluly and Gates, 1965; Girty and others, 1985), and the Blaylock Sandstone of Arkansas and Oklahoma (Miser and Purdue, 1929). The resemblance of the Silurian siltstone unit to the Blaylock is especially remarkable considering the distance that separates them; both units are distal turbidites consisting mainly of graptolite-bearing micaceous, feldspathic siltstone marked by abundant trace fossils and sole marks.

Environment of Deposition

The light color, even in subsurface samples, of much of the siltstone and chert in these strata suggests the possibility of oxygenated depositional conditions for at least part of the time, but the uniformly fine grain, graded beds, and trace fossils indicate considerable depth and distance from the basin margin.

Lower and Middle Devonian Strata

Lower and Middle Devonian beds are sparse, thin, and very poorly exposed in the study area. The scarcity of uppermost Silurian to Middle Devonian beds in the area of this report, as well as generally in the region, is part of the evidence of a widespread unconformity near the Middle-Upper Devonian boundary. Other evidence is discussed in a subsequent section.

Upper Devonian and Lowermost Mississippian Strata

Description

Lower Upper Devonian western-facies rocks are exposed extensively in the southern Independence Mountains, Adobe Range, and southern Snake Mountains. Upper Upper Devonian units are exposed in the southern Independence Mountains and Adobe Range and uppermost Devonian to lowermost Mississippian units are exposed in the northern Adobe Range and southern Snake Mountains. Upper Devonian exposures are more extensive than those of Early to Middle Devonian age and are generally unconformable on older units.

Like the lower member of the Vinini Formation, the Upper Devonian strata are characterized by abundance of clastic sediment and extreme lateral and vertical variations in lithic composition. Common lithic types are coarse, clast-supported chert conglomerate, chert grit nearly free from quartz grains, heterogeneous carbonate-cemented sandstone, limy siltstone, brown-weathering silty shale, dark cherty limestone, argillite, abundant dark lustreless bedded chert, and bedded barite. Small bodies of greenstone are a scarce component. Lowermost Mississippian beds are commonly dark siliceous silty argillite with sporadic sandy beds and scarce limestone and bedded barite.

Components of Arenites and Conglomerate

Quartz grains in sandstone and sandy limestone are rounded and as much as 1.0 mm in diameter. Silt-size grains are composed mainly of quartz and K-feldspar. Chert clasts are mainly of sand to pebble size and are composed of radiolarian-bearing chert apparently derived from within the basin of deposition. Calcite grains, commonly of sand size, are composed largely of *Nuia*, probably derived from erosion of Lower Ordovician limestone within the basin. *Nuia* is indigenous only to the Cambrian and Ordovician. Some of the sand-size clasts are composed of black, vesicular, microlitic basalt and devitrified glass. In a few thin beds the basalt clasts may constitute as much as three percent of the clasts, but they are totally absent from most stratigraphic units. Considering the circumstances in which these igneous clasts occur, their most likely source is not contemporaneous volcanism, but rather, erosion of lava flows in unconformably underlying Ordovician rocks.

Conglomerate and Grit

Some conglomerate and grit units of Late Devonian age consist of limestone clasts, but more commonly they are composed of poorly rounded to angular clasts of bedded chert. Limestone clasts may have been derived from shelf rocks, although that is uncertain, but the chert clasts are composed of radiolarian-bearing chert that certainly was derived from western-facies basinal deposits. Some of the limestone-clast conglomerates are similar to rip up-clast conglomerate of the lower member of the Vinini and could represent a shallow-water environment. The conglomerate and grit are clast-supported and associated with beds that only sporadically exhibit turbidite features.

Quartz Sandstone

Sandstone is commonly carbonate cemented and grades into sandy limestone. Clasts are mainly well-rounded quartz, but carbonate and rock fragments are common. Some exhibit features of turbidites such as Bouma sequences, and sole marks, but these features are not universal, and nowhere do recognizable turbiditic sequences form a major part of the Upper Devonian suite.

Bedded Chert

Bedded chert is a common constituent and is normally not easily distinguished from chert of other ages. However, much of it is black, commonly lacking surface lustre, and generally contains white pods or concretions a few centimeters in longest dimension. Most chert in Devonian rocks contains widely scattered radiolarians and some is composed largely of spicules, but the bulk of the chert is free from skeletal debris.

Bedded Barite

Bedded barite in the southern Independence Mountains and Adobe Range is enclosed in a variety of rock types of latest Devonian age. Barite in the southern Snake Mountains is in black argillite of earliest Mississippian age. The barite deposits include thin, laterally continuous beds of pure barite that are probably primary sediments. In one deposit, however, abundant, unequivocal textural evidence indicates that barite replaces greenstone extensively (Ketner, 1975), and in others the microscopic evidence indicates barite sequentially replaces limestone, chert, including chert pebbles, and finally quartz sand.

Pre-Upper Devonian Unconformity

Upper Devonian beds commonly lie discordantly across Ordovician and Silurian units. This discordance is proved to be an unconformity rather than a fault by unusually clear relations in the southern Independence Mountains. There, limy sandstones in the Devonian unit contain clasts composed of the Cambrian to Ordovician alga *Nuia*, angular basaltic clasts, and reworked Ordovician conodonts. The *Nuia*-bearing clasts and the basaltic clasts are lithically identical to beds in the directly underlying Ordovician unit.

Regionally, in Nevada, fossil localities in western-facies rocks of late Late Silurian age are virtually unknown (Berry and Boucot, 1970) and in northern Nevada those of Late Devonian age greatly outnumber those of Early and Middle Devonian age. These facts and the coarsely clastic nature of large tracts of Upper Devonian western-facies rocks, the presence of radiolarian-bearing chert clasts and other clasts derived from underlying strata indicate sporadic emergence and erosional unconformities of Devonian age in the western-facies assemblage.

Whereas the detrital components of the Ordovician and Silurian units seem to have been derived mainly from outside the basin of deposition, those of the Upper Devonian unit apparently were derived largely from provenance terranes within the basin. Because sea level was unusually high during the Late Devonian (Johnson and Sandberg, 1989), such erosion must have been caused by differential uplift within the basin rather than by a drop in sea-level.

SULFIDE DEPOSITS

Two strata-bound gossans were identified in surface exposures of the western facies sequence: one of extremely widespread distribution at the base of the Silurian and the other of relatively restricted distribution in the lower Middle Ordovician. The basal Silurian gossan is well developed and well exposed in the northern Adobe Range (Ketner and Ross, 1990), especially near the southern boundaries of secs. 17 and 18, T. 38 N., R. 56 E. In the Blue Basin quadrangle of the southern Independence Mountains, the gossan is exposed just below the Silurian siltstone in a unit (unit Oc of Ketner, 1974) that includes the uppermost Ordovician black chert unit and the lowermost Silurian chert unit. In the HD Range, it occurs in the Agort Chert of Riva (1970) which also includes both the uppermost Ordovician and lowermost Silurian chert. The Middle Ordovician gossan is known to crop out in the southern Independence Mountains and in the southern Snake Mountains. A third, very weakly mineralized zone in Middle Silurian siltstone was observed in drill core from one locality in the Adobe Range, but a distinct gossan in surface exposures at that horizon has not been observed. A possible fourth mineralized zone within strata of the middle member of the Vinini was reported by K.B. Riedell (written commun., 1989).

The most extensive, and most conspicuous, gossan occurs almost entirely within the Lower Silurian chert unit. In some areas the uppermost part of the Upper Ordovician black chert unit also may be mineralized. The gossan consists of discolored chert, chert containing scattered angular to rounded cavities, boxwork, iron oxide accumulations, and iron oxide- or quartz-cemented collapse breccia. Colors typically include bright blue-green, bright red, and caramel brown. Some exposures exhibit unusual patterns such as black and white zebra stripes. Various oxidation states of iron account for the green, red, and brown colors. Spectrographic analyses of the gossan indicate anomalous values of lead, zinc, and silver. Average lead content in 100 samples of gossan from widely scattered outcrops in the northern Adobe Range is more than 2000 ppm; zinc content more than 1300 ppm; silver content of forty percent of the samples is anomalously high, ranging from 1 to 15 ppm. Barium content ranges from 50 to 5000 ppm and is, therefore, about normal for unmineralized western facies rocks (100-3000 ppm). Subsurface samples from the southern Independence Mountains and northern Adobe Range show the primary iron mineral to be pyrite and the primary lead and zinc minerals to be galena and sphalerite.

TECTONIC SETTING, PROVENANCE, AND DEPOSITIONAL CONDITIONS

Owing to pervasive low-angle faulting, mostly of Mesozoic and Tertiary ages, the tectonic setting of the basin in which the Vinini Formation and directly overlying strata were deposited is difficult to determine. However, certain data regarding the substrate of the basin, the depth of water, and the provenance of clastic sediments can contribute to a solution of the problem.

The depositional substrate of the Vinini is unknown from presently available data in northeastern Nevada, but relations elsewhere supply pertinent information. In northeastern Washington, the Ledbetter Slate, the lithic and age equivalent of the Vinini, lies with sedimentary contact on shallow-water Cambrian limestone and dolostone (Schuster, 1976a, b). In southwestern Nevada, the Palmetto Formation, the twin of the Vinini, lies with sedimentary contact on Cambrian rocks, and the Cambrian sequence includes shallow-water shelf units. By interpolation, the original substrate of the Vinini Formation in northeastern Nevada also may have been Cambrian shelf units.

The provenance of clastic sediments in the western-facies sequence of northeastern Nevada is partly outside the depositional basin and partly within it. The detrital components of Ordovician and Silurian strata such as quartz, K-feldspar, muscovite, shelfal bioclastic debris and the argillaceous components of shale and argillite were most likely derived from marginal subaerial or shallow-marine environments. Similar clasts in the Upper Devonian to lowermost Mississippian strata also may have been derived from peripheral terranes, but rock fragments, constituting much of the strata of this age, such as radiolarian-bearing chert, *Nuia*-bearing limestone, quartzite, and basalt are lithically identical to strata in the underlying Ordovician and Silurian units and must have been derived from intrabasinal source areas.

Most of the siliceous debris, including quartz sand, feldspar, and mica, comprising the Ordovician and Silurian rocks must have come from a terrane other than the eastern shelf area (Ketner, 1977; Madrid, 1987). The stratigraphy of the carbonate shelf sequence to the east is well known and neither channels of appropriate ages nor siliceous provenance terranes of sufficient size, and exposed at appropriate times and places, have been identified.

Volcanic ash of any composition and feldspathic debris eroded from mafic to intermediate volcanic terranes are conspicuously absent from the western-facies rocks. Almost all of the detrital feldspar is orthoclase, microcline, or sodic plagioclase ultimately derived from granitic terranes. The sand-size quartz grains are typically strained which also indicates a plutonic rather than volcanic source. Evidently the western facies rocks of northeastern Nevada were deposited far from any volcanic arc and there is therefore no compelling reason to relate the depositional basin to arc-forming processes.

Most of the Ordovician to lowermost Mississippian basinal sediments of northeastern Nevada were deposited in water considerably deeper than that on the shelf. The evidence is in the presence of unoxidized carbon compounds, the presence of turbidites and other strata composed of shelf debris, and the absence of indigenous shallow-water fauna. However, extensive, thin, tabular beds of clast-supported conglomerate, especially abundant in the Lower to lower Middle Ordovician and Upper Devonian, may reflect deposition locally above wave base.

Although the radiolarian-bearing cherts of the western facies in northeastern Nevada (as opposed to the less abundant detrital pseudo-chert and spicule cherts) commonly contain radiolarian skeletons, these are an incidental and very minor component and many beds are entirely free from any such evidence of organic origin. In northeastern Nevada such cherts tend to be associated stratigraphically with sulfide and sulfate deposits of hydrothernal exhalative origin. Most of the sulfides are in the uppermost Ordovician to lowermost Silurian chert interval and most barite is in the generally chert-rich Upper Devonian. These associations indicate the possibility of a hydrothermal origin for most of the chert also.

At Badger Spring, in the northern Adobe Range, the early Middle Ordovician shallow-water limestone buildup or mudmound, described in a previous section of this report, lies on Devonian rocks with a thrust fault contact of Mesozoic age. It is overlain unconformably by Lower Silurian, graptolite-bearing bedded black chert and Middle Silurian micaceous limy siltstone. Three hypotheses of origin are considered: (1) The mudmound is an olistolith that slid as a giant intact slab, or sheet, into the Vinini basin from the adjacent shelf; (2) it plunged downslope from the adjacent shelf, not as an intact slab, but, as a fragmentary debris flow; (3) it formed in shallow water on a tectonically elevated portion of the Vinini basin. The fact that the mudmound is overlain unconformably by lower Silurian deep-water sediments rather than by Upper Ordovician strata of the Vinini Formation indicates that it was immersed in deep water not immediately after deposition in early Middle Ordovician, but, at the end of Ordovician or the very beginning of Silurian. If it had slid into the basin intact from the shelf edge at that time it would be overlain by Upper Ordovician shelf carbonate. If it had slid into the basin as a debris flow it would display the fragmental texture typical of consolidated debris flows generally and would likely include clasts derived from other shelf strata. The lack of any overlying Ordovician strata, lack of pervasive fragmental textures, and lack of heterogeneous composition suggest the third alternative: that it was deposited in shallow water on a tectonically elevated portion of the basin, was exposed to erosion in Late Ordovician, and was then lowered tectonically at the end of the Ordovician.

In general, the heterogeneity of rocks of Ordovician to earliest Mississippian age, the widespread presence of conglomerate, especially in Devonian, the apparent variety of water depths, and the presence of unconformities suggest intermittent tectonism in and adjacent to the basin during deposition. Extreme fluctuations of sea level could not be the cause of intrabasinal erosion, because the adjacent Ordovician to Devonian shelf sequence was deposited under shallow-water conditions almost continuously. Moreover, sea level was high and rising between Middle and early Late Devonian--the time of a regional unconformity in the western-facies basin (Johnson and Sandberg, 1989).

Exhalative lead and zinc sulfides and barite deposits similar to those of northeastern Nevada are generally considered to have been deposited adjacent to deeply penetrating, high-angle faults in an extensional or transtensional-transpressional basin or trough (Large, 1981; Gordey and others, 1987; Winn and Bales, 1987). Although direct evidence of basin-bounding faults was not observed in northeastern Nevada, their presence would be masked by the structural complexity and obscure exposures typical of the region, and they are assumed to be associated with the sulfide and sulfate deposits there as elsewhere. The depositional circumstances in northeastern Nevada probably were analogous to those of the the Selwyn Basin of northwestern Canada as described by Carne (1979) and Carne and Cathro (1982), and enlarged upon by others, including Gordey and others (1987), Winn and Bales (1987), Lydon and others (1985), Goodfellow and Jonnason, (1984), and Turner and others (1989) and to exhalative sulfide-barite basins generally (Large, 1981). For northeastern Nevada, the analogy with the Selwyn basin is strikingly precise as previously indicated (Ketner, 1983). In both areas Ordovician and Silurian siliceous sediments were deposited in a deep basin floored by continental-shelf or slope strata and were overlain unconformably by Devonian to Lower Mississippian rocks composed partly of internally derived siliceous clastic sediments. In both areas the Middle to Upper Silurian consists of bioturbated, light-colored siltstone indicative of a period of sea-water ventilation and oxygenation. In both areas, strata-bound lead and zinc deposits are concentrated in basal Silurian beds and barite deposits are concentrated in Upper Devonian strata.

The picture that emerges for the depositional conditions of the western facies of northeastern Nevada is compatible with that depicted by Large (1981) for submarine exhalative lead-zinc basins generally: extensional epicontinental basins bounded by, and internally segmented by, high-angle extensional or strike-slip faults with provenance terranes on both sides and within the basin. In northeastern Nevada, in accordance with this model, Ordovician and Silurian sediment entered the basin from both east and west. Movement on internal or bounding high-angle faults during and at the end of the Ordovician accounted for rip-up-clast conglomerates and the anomalous stratigraphic position of the limestone of Badger Spring. Movement on internal or bounding high-angle faults accounted for the local accumulation of Devonian conglomerates. Widespread uplift within the basin resulted in the unconformity beneath the Upper Devonian rocks. Hydrothermal solutions rising along the high-angle faults produced the lead-zinc, barite, and chert deposits. In this model, subduction processes and island arcs play no direct part, which accounts for the absence of arc-derived volcanic detritus in the western facies of northeastern Nevada.

ACKNOWLEDGMENTS

Knowledge of the stratigraphy of the dismembered western facies assemblage depends entirely on an abundance of precise paleontological dating. I am grateful to a large number of paleontologists for such dating. Because of the nature and age of the rocks, graptolites, conodonts, and radiolarians are the most valuable taxa for dating. Graptolite determinations were supplied by Reuben J. Ross, Jr., William B. N. Berry, and Claire Carter. Conodont determinations were made by Anita G. Harris, Charles A. Sandberg, John E. Repetski, Anna Dombrowski, John W. Huddle (deceased) and Chauncey G. Tillman (deceased). Radiolaria determinations were by Benita L. Murchey and David L. Jones. In addition to supplying a large number of graptolite determinations, "Rube" Ross accompanied me in the field on many occasions and shared his extensive knowledge of the stratigraphy and sedimentation of Ordovician rocks. Suggestions by Forrest G. Poole, Charles H. Thorman, Charles A. Sandberg, and K. Brock Riedell greatly improved the manuscript.

REFERENCES CITED

Berry, W.B.N., and Boucot, A.J., 1970, Correlation of North American Silurian rocks: Geological Society of America Special Paper 102, 289 p.

Carne, R.C., 1979, Geological setting and stratiform lead-zinc-barite mineralization, Tom Claims, Macmillan Pass, Yukon Territory: (Canadian) Department of Indian and Northern Affairs, Report no. 1979-4, 30 p.

Carne, R.C., and Cathro, R.J., 1982, Sedimentary exhalative (sedex) zinc-lead-silver deposits, northern Canadian Cordillera: Canadian Institute of Mining and Metallurgy Bulletin, v. 75, no. 840, p. 66-78.

Carter, Claire, and Churkin, Michael, Jr., 1977, Ordovician and Silurian graptolite succession in the Trail Creek area, central Idaho--a graptolite zone reference section: U.S. Geological Survey Prof. Paper 1020, 37 p., 7 plates.

Finney, S.C., and Perry, B.D., 1989, Depositional history of the Vinini Formation; results from the type area, Roberts Mountains, Nevada: Geological Society of America, Abstracts with Programs, v. 21, no. 5, p. 78.

Gilluly, James, and Gates, Olcott, 1965, Tectonic and igneous geology of the northern Shoshone Range, Nevada: U.S. Geological Survey Prof. Paper 465.

Girty, G.H., Reiland, D.N., and Wardlaw, M.S., 1985, Provenance of the Silurian Elder Sandstone, north-central Nevada: Geological Society of America Bulletin, v. 96, p. 925-930.

Goodfellow, W.D., and Jonasson, I.R. 1984, Ocean stagnation and ventilation defined by $\delta^{34}S$ secular trends in pyrite and barite, Selwyn Basin, Yukon: Geology, v. 12, p. 583-586.

Gordey, S.P., Abbott, J.G., Tempelman-Kluit, D.J., and Gabrielse, H., 1987, "Antler" clastics in the Canadian Cordillera: Geology, v. 15, p. 103-107.

Hintze, L.F., 1974, Preliminary geologic map of the Notch Peak quadrangle, Millard County, Utah: U.S. Geological Survey Miscellaneous Field Studies Map MF-636.

Johnson, J.G. and Sandberg, C.A., 1989, Devonian eustatic events in the western United States and their biostratigraphic responses, in McMillan, N.J., Embry, A.F., and Glass, D.J., eds., Devonian of the World: Calgary, Canadian Society of Petroleum Geologists, Memoir 14, vol. 3, p. 171-178, 2 figs. (date of imprint, 1988).

Ketner, K.B., 1969, Ordovician bedded chert, argillite, and shale of the Cordilleran eugeosyncline in Nevada and Idaho: U.S. Geological Survey Professional Paper 650-B, p. B23-B34.

_____1974, Preliminary geologic map of the Blue Basin quadrangle, Elko County, Nevada: U.S. Geological Survey Miscellaneous Field Studies Map, MF-559.

_____1975, Replacement barite deposit, southern Independence Mountains, Nevada: U.S. Geological Survey Journal of Research, v. 3, p. 547-551.

_____1977, Deposition and deformation of lower Paleozoic western facies rocks, northern Nevada, in Stewart, J.H., Stevens, C.H., and Fritsche, A.E., eds., Paleozoic Paleogeography of the Western United States: Pacific Section SEPM Pacific Coast Paleogeography Symposium no. 1, p. 251-258.

_____1980, Stratigraphic and tectonic parallels between Paleozoic geosynclinal siliceous sequences in northern Nevada and those of the Marathon uplift, Texas, and Ouachita Mountains, Arkansas and Oklahoma, in Fouch, T.D. and Magathan, E.R., eds., Paleozoic Paleogeography of west-central United States: Rocky Mountain Section SEPM, Paleogeography Symposium v. 1, p. 363-369.

_____1983, Strata-bound, silver-bearing iron, lead, and zinc sulfide deposits in Silurian and Ordovician rocks of allochthonous terranes, Nevada and northern Mexico: U.S. Geological Survey, Open-File report 83-792, 7 p.

_____1986, Eureka Quartzite in Mexico?--tectonic implications: Geology, v. 14, p. 1027-1030.

Ketner, K.B., and Noll, J.H., Jr., 1987, Preliminary geologic map of the Cerro Cobachi area, Sonora, Mexico: U.S. Geological Survey Miscellaneous Field Studies Map, MF-1980, 1:50,000.

Ketner, K.B., and Ross, R.J., Jr., 1990, Geologic map of the northern Adobe Range, Elko County, Nevada: U.S. Geological Survey Miscellaneous Investigations Map I-2081, Scale 1:24,000.

King, P.B.,1937, Geology of the Marathon region, Texas: U.S. Geological Survey Professional Paper 187, 148 p.

Large, D.E., 1981, Sediment-hosted submarine exhalative lead-zinc deposits--A review of their geological characteristics and genesis, in Wolf, K.H., ed., Handbook of strata-bound and stratiform ore deposits, Part III, v. 9: Amsterdam, Elsevier, p. 469-507

Lovejoy, D.W., 1959, Overthrust Ordovician and Nannies Peak intrusive, Lone Mountain, Elko County, Nevada: Geological Society of America Bulletin v. 70, p. 539-564.

Lydon, J.W., Goodfellow, W.D., and Jonasson, I.R., 1985, A general genetic model for stratiform baritic deposits of the Selwyn Basin, Yukon Territory and District of Mackenzie: Geological Survey of Canada, Paper 85-1A, p. 651-660.

Madrid, R.J.J., 1987, Stratigraphy of the Roberts Mountains allochthon in north-central Nevada: Stanford University PhD dissertation (unpublished), 340 p.

McBride, E.F., 1969, Stratigraphy and sedimentology of the Woods Hollow Formation (Middle Ordovician), Trans-Pecos Texas: Geological Society of America Bulletin, v. 80, p. 2287-2302.

_____, 1970, Stratigraphy and origin of Maravillas Formation (Upper Ordovician), West Texas: American Association of Petroleum Geologists Bulletin, v. 54, no. 9, p. 1719-1745.

McKee, E.H., 1968, Geology of the Magruder Mountain area, Nevada-California: U.S. Geological Survey Bulletin 1251-H, p. H1-H40.

Miser, H.D. and Purdue, A.H., 1929, Geology of the De Queen and Caddo Gap quadrangles, Arkansas: U.S. Geological Survey Bulletin 808, 195 p.

746

Morris, H.T., and Lovering, T.S., 1961, Stratigraphy of the east Tintic Mountains, Utah: U.S. Geological Survey Professional Paper 361, 145 p.

Merriam, C.W., and Anderson, C.A., 1942, Reconnaissance survey of the Roberts Mountains, Nevada: Geological Society of America Bulletin, v. 53, p. 1675-1727.

Nolan, T.B., 1935, The Gold Hill mining district, Utah: U.S. Geological Survey Professional Paper 177, 172 p.

Riva, John, 1970, Thrusted Paleozoic rocks in the northern and central HD Range, northeastern Nevada: Geological Society of America Bulletin, v. 11, no. 9, p. 2689-2716.

Ross, R.J., Jr., Jaanusson, Valdar, and Friedman, Irving, 1975, Lithology and origin of Middle Ordovician calcareous mudmound at Meiklejohn Peak, southern Nevada: U.S. Geological Survey Professional Paper 871, 48 p.

Ross, R.J., Jr., and others, 1982, The Ordovician System in the United States--correlation chart and explanatory notes: International Union of Geological Sciences Publication no. 12, 73 p.

Ross, R.J., Jr., James, N.P., Hintze, L.F., and Ketner, K.B., 1988a, Early mid-Ordovician paleogeography of Basin ranges: Geological Society of America, Cordilleran Section, Abstracts with programs, p. 226.

Ross, R.J.,Jr., Valusek, J.E., and James, N.P, 1988b, *Nuia* and its environmental significance: New Mexico Bureau of Mines and Mineral Resources Memoir 44, p. 115-121.

Schuster, J.E., 1976a, Geology of the Clugston Creek area, Stevens County, Washington: Washington Department of Natural Resources, Division of Geology and Energy Resources, Open-File Report, 26 p.

Schuster, J.E., 1976b, Geology of the contact between the Metaline Limestone and Ledbetter Slate in the Clugston Creek area, Stevens County, Washington: Geological Society of America, Abstracts with Programs, v. , p. 408.

Smith, J. F., Jr., and Ketner, K.B., 1975, Stratigraphy of Paleozoic rocks in the Carlin-Pinon Range area, Nevada: U.S. Geological Survey Professional Paper 867-A, 87 p.

Staatz, M.H., and Carr, W.J., 1964, Geology and mineral deposits of the Thomas and Dugway Ranges, Juab and Tooele Counties, Utah: U.S. Geological Survey Professional Paper 415, 189 p.

Stanley, K.O., Chamberlain, C.K., and Stewart, J.H., 1977, Depositional setting of some eugeosynclinal Ordovician rocks and structurally interleaved Devonian rocks in the Cordilleran mobile belt, Nevada, *in* Stewart, J.H., Stevens, C.H., and Fritsche, A.E., eds., Paleozoic paleogeography of the Western United States: Pacific Section SEPM Pacific Coast Paleogeography Symposium, v. 1, p. 259-274.

Stewart, J.H., 1979, Geologic map of the Miller Mountain area, Esmeralda County, Nevada: U.S. Geological Survey Open File Map 79-1145.

Turner, R.J., Madrid, R. J., and Miller E. L., Roberts Mountains allochthon: Stratigraphic comparison with lower Paleozoic outer continental margin strata of the northern Canadian Cordillera: Geology, v. 17, p. 341-344.

Winn, R.D., Jr., and Bailes, R.J., 1987, Stratiform lead-zinc sulfides, mudflows, turbidites: Devonian sedimentation along a submarine fault scarp of extensional origin, Jason deposit, Yukon Territory, Canada: Geological Society of America Bulletin, v. 98, p. 528-539.

Winterer, E.L., 1968, Tectonic erosion in the Roberts Mountains, Nevada: Journal of Geology, v. 76, p. 347-357.

DEPOSITIONAL SETTING AND PALEOGEOGRAPHY OF ORDOVICIAN VININI FORMATION, CENTRAL NEVADA

Stanley C. Finney and Bruce D. Perry
Department of Geological Sciences
California State University, Long Beach 90840

ABSTRACT

The Vinini Formation (Ordovician, graptolite zones 4-15) was deposited on a deep-marine basin plain and continental rise located adjacent to the western margin of North America. Its stratigraphy records pelagic sedimentation and also progradation and retrogradation of a submarine fan, the spread of contourites, and submarine volcanism.

Graptolite and conodont biostratigraphy permit the depositional history of the Vinini to be compared to Ordovician eugeoclinal strata in other parts of the Roberts Mountains allochthon and in the autochthonous miogeoclinal shelf sequence. Such a comparison reveals that patterns of sedimentation were similar throughout the continental rise and basin plain and that major events affecting sedimentation on the shelf also affected sedimentation on the rise and basin plain. In particular, two major episodes of quartz sand deposition are recognized.

The first occurred in the late Ibexian to early Whiterockian and was initiated by a major drop in relative sea level. Submarine channels directed huge volumes of quartz sands, eroded from the craton, across the carbonate shelf and down onto the rise and basin plain, leading to the progradation of submarine fans. By the middle Whiterockian, a subsequent rise in relative sea level shut off the supply of quartz sands, and led to the retrogradation of the submarine fans and a return of pelagic sedimentation of organic-rich graptolite shales on the continental rise and basin plain.

The second episode occurred in the late Whiterockian to early Mohawkian. Although sea level was at a high stand, the volume of quartz sands was so large that it prograded across the shelf and spilled off onto the continental rise and basin plain. By the middle Mohawkian, the sand supply to the basin was again shut off, perhaps by an even higher rise in relative sea level. Except for the two periods of quartz sand deposition and the sporadic eruption of submarine volcanics in the Ibexian and Whiterockian, Ordovician sedimentation on the continental rise and basin plain was largely pelagic.

The fact that the deposition of eugeoclinal strata of the Roberts Mountains Allochthon is closely tied to that of autochthonous miogeoclinal strata of North America is evidence that the RMA is not an exotic terrane.

INTRODUCTION

Detailed study of the Vinini Formation in its type area has proved to be critical for unraveling the depositional history and paleogeography of Ordovician strata in the Roberts Mountains allochthon (RMA). From a historical perspective, Merriam and Anderson's (1942) recognition of the nature and age of the Vinini Formation, when they first defined it in the Roberts Mountains, is a milestone in the history of research on the western Cordillera. This work, together with subsequent work in other ranges in Nevada by Roberts (1949; Roberts and others, 1958), Gilluly and Gates (1965) and other USGS geologists, resulted in the discovery of the major tectonic event affecting the western margin of North America in the Paleozoic, i.e. the emplacement of the Roberts Mountains allochthon along the Roberts Mountains thrust during the Antler orogeny.

The Roberts Mountains allochthon (RMA) is a complexly deformed basinal assemblage of Ordovician to Devonian quartzite, sandstone, siltstone, shale, argillite, limestone, chert, and greenstone. These rocks have been called the "western" assemblage or facies (Merriam and Anderson, 1942; Roberts and others, 1958), and have been referred to as eugeoclinal by Stewart and Poole (1974) who concluded that they represent continental rise deposits. The rocks are perdominately siliciclastic, but are sufficiently variable that they have been subdivided vertically and laterally into numerous stratigraphic units (e.g. Vinini Formation, Valmy Formation, Palmetto Formation, Petes Summit Formation, Basco Formation, Elder Sandstone, Slaven Chert, Snow Canyon Formation, McAfee Quartzite, etc.). The RMA was thrust at least 150 km east to southeastward over a lower Paleozoic carbonate platform sequence, the so-called "eastern" assemblage, continental shelf, or miogeoclinal suite, that is the lower autochthonous plate of the Roberts Mountains thrust.

The rocks of the RMA are critical for unraveling the geologic history of western North America during the early and middle Paleozoic. Although they have been mapped extensively and studied throughout Nevada,

In Cooper, J.D., and Stevens, C.H., eds., 1991, **Paleozoic Paleogeography of the Western United States-II:** Pacific Section SEPM, Vol. 67, p. 747-766.

748

interpretation of their depositional history and reconstruction of their paleogeographic setting have been severely limited by poor biostratigraphic control. Structural complexities and the lack of continuous exposures have hindered most attempts to measure and describe the stratigraphy. And even when relatively complete and structurally coherent stratigraphic sections were described (Miller and Larue, 1983; Watkins and Browne, 1989) or when widely separated stratigraphic sequences were correlated (Ketner, 1977; Watkins and Browne, 1989; Turner and others, 1989), correlations were at the system or series level, never at the zonal level. Graptolites and conodonts are the only common fossils in the strata of the RMA that are useful for biostratigraphic control. Although both suffer limitations (e.g. graptolite shales are poorly exposed, and limestones that yield conodonts are not common), graptolite zones can be identified at many stratigraphic levels in the eugeoclinal strata, and with the less common conodonts, allow for correlation with the miogeoclinal strata.

Graptolite-rich shales are a distinctive lithology in the eugeoclinal sequence. Graptolites were first reported from western assemblage strata in Nevada more than 100 years ago (White, 1874; Gilbert, 1875), but they were not collected in earnest until the 1940s, 50s and 60s, when many of the outcrops of western assemblage rocks were initially mapped. Faunal lists and age determinations were published by, among others, Merriam and Anderson (1942), Roberts (1951), Lovejoy (1959), Ross and Berry (1963), Kay and Crawford (1964), Gilluly and Gates (1965) Churkin and Kay (1967), Riva (1970), and McKee (1976). Most of the collections were of a reconnaissance nature with too few specimens to make precise zonal correlations and usually not tied into measured stratigraphic sections that would have allowed for inter-regional correlation of lithofacies. And even when graptolites were collected so that zones could be recognized (Kay, 1962; Kay and Crawford, 1964; McKee, 1976; Churkin and Kay, 1967; Watkins and Browne, 1989; Riva, 1970), correlations were limited to a few areas or were done at the series level.

We have collected graptolites from the Vinini Formation in the Roberts Mountains for the last four years. These collections allow us to reconstruct the entire stratigraphy of the Vinini exposed in the Roberts Mountains, to correlate its lithofacies to those of allochthonous eugeoclinal sequences in other ranges in Nevada, and, with the aid of conodonts, to correlate the stratigraphy of the Vinini with that of the autochthonous miogeoclinal sequence. These correlations enable us to make several definitive interpretations of the depositional history and paleogeographic setting of the Ordovician eugeoclinal strata in the RMA. We have focused our interest on the quartz sands that compose abundant and distinctive quartzites and sandstones. We find that these sands were widely deposited during two

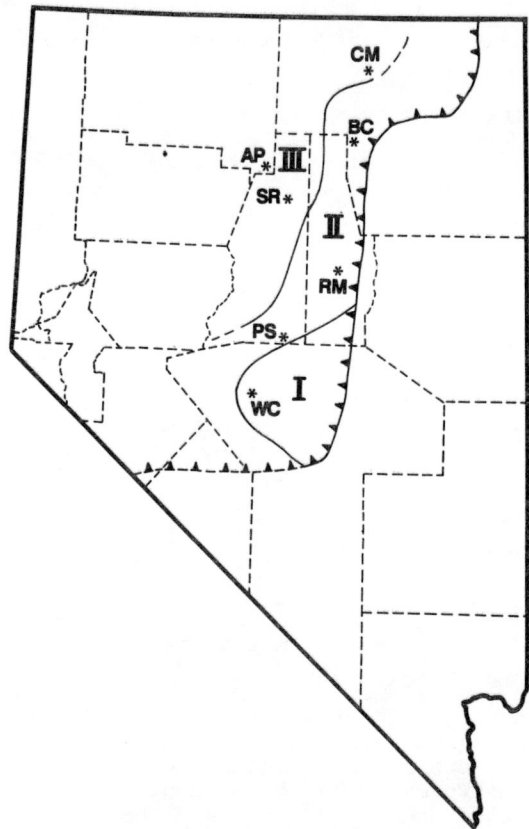

Figure 1. Index map of Nevada showing localities mentioned in text, depositional provinces of Stewart and Poole (1974) within RMA, and eastern extent of Roberts Mountains thrust (arrowed line). Area I is shale-limestone depositional province (transitional assemblage at outer edge of miogeoclinal suite); Area II is shale-chert depositional province (inner belt of eugeoclinal suite); Area III is chert-shale-quartzite-greenstone depositional province (outer belt of eugeoclinal suite). Localities are Wall Canyon (WC) in southern Toiyabe Range, Petes Summit (PS) in northern Toquima Range; Roberts Mountains (RM), Basco Creek (BC) in southern Independence Range, California Mountain quadrangle (CM) in northern Independence Range, northern Shoshone Range (SR), and Antler Peak area (AP) at Battle Mountain.

episodes, one in the late Ibexian to early Whiterockian and one in the latest Whiterockian to early Mohawkian. The sands of the younger episode are coeval with, and lithologically similar to, those of the Eureka Quartzite of the miogeoclinal sequence. In fact, we interpret them as Eureka sands that spilled off the shelf and into the basin, and present them as definitive evidence that the eugeoclinal strata of the RMA were deposited immediately adjacent to the North American continent. We conclude that the sands of the older episode were also derived from the North American craton, and interpret them as being associated with major relative sea-level changes

that are recorded in the succession of
lithofacies in the eugeoclinal sequence.

DEFINITION OF VININI FORMATION AND ITS
RELATIONSHIPS TO COEVAL STRATIGRAPHIC
UNITS

The Vinini Formation composes much of
the RMA in central Nevada (Fig. 1). In the
Roberts Mountains, Merriam and Anderson
(1942) mapped all the rocks of the upper
plate of the Roberts Mountains thrust as the
Vinini Formation, naming it after Vinini
Creek where it is relatively well exposed
(Fig. 2). They recognized two informal
units: the lower Vinini, readily distinguished by abundant quartzite and sandstone,
and the upper Vinini, composed of shale and
chert. They reported lower Ordovician and
middle Ordovician graptolites from the
members, respectively. Subsequently, Ross
and Berry (1963) reported on graptolites in
the U.S. Geological Survey collections from
the Vinini in central Nevada. All were
recognized as Ordovician, which became the
generally accepted age of the formation. As
exposures of the RMA were mapped and studied
in other ranges in Nevada, it was soon
discovered that the RMA included Cambrian,
Silurian, and Devonian strata (e.g. Roberts,
1964; Gilluly and Gates, 1965). In 1968,
Murphy reported that some of the rocks
mapped as Vinini by Merriam and Anderson
(1942) in the Roberts Mountains contained
Devonian fossils. Subsequently, Murphy and
others (1978, 1984) described the nature of,
and mapped, the Devonian strata, as well as
Ordovician strata, of the RMA in the
northern Roberts Mountains, and Minnick
(1975) reported Devonian limestone (based on
conodonts identified by Dr. R.L. Ethington)
in the Vinini Formation in the Tyrone Gap
area, southeast of the Roberts Mountains.
Limestone samples gathered recently from
exposures of the Vinini on Henderson Creek
Road south of the Roberts Mountains have
yielded Devonian conodonts (identified by
R.L. Ethington). All of these areas were
mapped as the Ordovician Vinini Formation by
Merriam and Anderson (1942) and by Roberts
and others (1967).

The Devonian strata in the Vinini
Formation in the Roberts Mountains consist
of shale, chert, and limestone that are
easily distinguished from similar lithologies in the Ordovician strata (Murphy and
others, 1984). For example, Devonian shales
are siliceous and have pink and orange
weathered surfaces in contrast to Ordovician
shales that are often organic-rich and
weather to white, bluish-gray, and brown.
The Devonian strata occur in thrust slices
above the Roberts Mountains thrust and
structurally beneath thrust slices with
Ordovician strata (Murphy and others, 1978,
1984). We have observed this same relationship on the east and south sides of the
Roberts Mountains.

Because the Vinini Formation was
considered to be Ordovician for so long and
because Silurian and Devonian strata can
often be readily distinguished lithologically from Ordovician strata not only in the
Roberts Mountains but also in other ranges
(e.g. Ketner, 1974, 1977), we revise the
Vinini Formation to include only Ordovician
strata. No formal name is given to the
Devonian strata. We retain a two-fold
division of the Vinini into formal Lower and
Upper members, the stratigraphy and biostratigraphy of which are described below and in
Figure 2. Perhaps the members should be
elevated to formational rank with the Vinini
raised to the rank of group. The members
are distinctive, and can be mapped throughout the Roberts Mountains even in different
thrust plates. We postpone making such a
decision, however, until we can study the
Vinini in other mountain ranges in much
greater detail.

The Vinini Formation is the most widely
recognized Ordovician stratigraphic unit in
the shale and chert depositional province of
Stewart (1980), which is regarded as the
inner belt of the eugeocline by Stewart and
Poole (1974). Some of the other stratigraphic units recognized within this
province should also be referred to the
Vinini Formation. For example, the Basco
Formation defined by Lovejoy (1959) in the
southern Independence Range (Fig. 1) is so
similar to the Vinini of the Roberts
Mountains that Ketner (1974) mapped it as
Vinini, and the Clipper Canyon sequence in
the northern Toquima Range, which was named
and divided into four formal formations by
Kay and Crawford (1964), was later recognized and mapped as the Vinini Formation by
McKee (1976).

Besides the Vinini Formation of the
shale and chert depositional province,
Ordovician strata of two other depositional
provinces compose significant parts of the
RMA. The chert-shale-quartzite-greenstone
province, referred to as the outer belt of
the eugeocline by Stewart and Poole (1974),
is represented by extensive outcrops of the
Valmy Formation in north-central Nevada that
are generally to the north and west of
outcrops of the Vinini (Fig. 1). Although
the Valmy is composed of the same lithologies as those found in the Vinini Formation,
it has long been distinguished from the
Vinini by its greater amount of quartzite
and greenstone and lesser amount of shale
(Roberts, 1964; Gilluly and Gates, 1965;
Churkin and Kay, 1967). Our work in the
Roberts Mountains and the Toquima Range and
earlier studies (Merriam and Anderson, 1942;
Murphy and others, 1978; Kay and Crawford,
1964) clearly demonstrate, however, that the
Vinini Formation does include significant
amounts of quartzite and greenstone.
Although the varying proportions of shale,
quartzite, and greenstone can still be used
to distinguish the Vinini and Valmy Formations, e.g., quartzites commonly occur in
much thicker beds in the Valmy, the differences between them are not as great as
generally accepted. They are here considered to be very similar facies. Another
distinct facies of Ordovician strata in the
RMA, the shale-limestone province (Fig. 1),
is represented by outcrops in the southern

750

Toquima and Toiyabe ranges and includes the
Perkins Canyon Formation (Kay and Crawford,
1964) and the Zanzibar Limestone and Toquima
Formation (Ferguson, 1924). Stewart and
Poole (1974) refer to this facies as the
transitional assemblage at the outer edge of
the miogeocline and interpret it as continental slope deposits. This facies is
distinct from the Vinini and Valmy.
Quartzites are restricted to one thin
stratigraphic interval. Greenstones are
absent. Shaley to platy limestone and
calcareous shale are the predominant and
distinguishing lithologies of the facies.

Widespread outcrops of Ordovician strata
in Esmeralda County and parts of Mineral and
Nye Counties are mapped as the Palmetto
Formation. Various outcrops have been
referred to either the chert-shale-quartzite-greenstone province or the shale-chert
province (Stewart, 1980). Although the
Palmetto is coeval with the Vinini, its
depositional relationships are not considered here because its stratigraphy and
biostratigraphy are not known in sufficient
detail.

LITHOSTRATIGRAPHY, BIOSTRATIGRAPHY AND
SEDIMENTOLOGY OF VININI FORMATION
IN ROBERTS MOUNTAINS

Areas of Study

The Vinini Formation crops out extensively on both the eastern and western
slopes of the Roberts Mountains. On the
east side, a 3000 m section was measured
from the upper part of the Dry Creek
drainage across to, and down, Vinini Creek
(Fig. 2, locality 1). The section is within
a very thick, widespread thrust plate that
includes most of the Vinini, both Lower and
Upper members in sequence, and extends along
most of the eastern side of the Roberts
Mountains. The base of the section is low
in the sandstone unit of the Lower Member,
and at a thrust fault that is underlain by
Middle Ordovician shales and chert. The
Middle Ordovician shales and chert compose a
thin thrust slice that, in turn, overlies
thrust slices containing Devonian strata
(Murphy and others, 1978, 1984). A fault
terminates the section at its top, high in
the Upper Ordovician. Structurally isolated
outcrops of Vinini along the valley floor of
Vinini Creek are of latest Ordovician age
(i.e. younger than the highest beds in the
measured section) and supplement the
stratigraphy of the measured section.

Measurements in the section (Fig. 2)
reflect some structural thickening, yet they
are probably only small overestimations of
true stratigraphic thickness. Numerous
tight folds and normal faults repeating
section are common and were taken into
account when discovered in the measured
section. Nonetheless, substantial well-
exposed stratigraphic intervals display
little deformation. The fact that strikes
are consistently north-south, dips are consistently to the east in the vicinity of the
section, and age determinations from

graptolite collections are consistently
younger upsection also indicate that
measurements in the section are reasonable.

Two short sections were measured high on
the western side of the Roberts Mountains
(Fig. 2). Although extending through only a
short interval of the Lower Member, the
section at locality 2 is important because
it includes the base of the sandstone
interval that composes the upper and
thickest part of the Lower Member. Beneath
the sandstones is an interval of black
shales, the base of which is not exposed
because the shales are cut out partly by a
fault. Another short section was measured
at locality 3, directly across the fault
from locality 2. The section is within and
near the base of the sandstone interval of
the Lower Member. It includes a channel
filled with submarine volcanic flow breccia.

Locality 4 (Fig. 2) on the south side of
Meadows Canyon is within the thesis area of
Mr. Paul Emsbo (Colorado School of Mines).
It is important because there a thick pile
of submarine volcanic flows, flow breccias,
and volcaniclastics is underlain and
overlain by sandstones of the Lower Member
of the Vinini.

Lithostratigraphy

Introduction

As revised here, the Vinini Formation
consists of a Lower Member with a black
shale interval overlain by a sandstone
interval and an Upper Member of alternating
intervals of shale and chert and a distinctive siltstone interval (Fig. 2).

Lower Member

The lower black shale interval is 140
meters thick and consists entirely of black
siliceous shale. Graptolites are common
where they have been found restricted to a
few thin intervals. The base of the shales
is cut out by faulting.

The upper sandstone interval, with a
true stratigraphic thickness that may
approach the 1832 m measured at locality 1,
consists of quartzite, quartz arenite,
quartz wacke, calcareous sandstone, siltstone, shale, limestone, and rare chert and
conglomerate. On the southwest side of the
Roberts Mountains, greenstones representing

Figure 2. Stratigraphic columns for
portions of Vinini Formation exposed at
localities 1-4 in the Roberts Mountains
(inset map). Berry's (1960) graptolite
zonation as recently revised. Biostratigraphic positions within Lower Member of
sections exposed at Localities 3 and 4 is
uncertain because of lack of biostratigraphic control. Inset map shows location
of long measured section (1) that extends
southeasterly from upper Dry Creek drainage
to Vinini Creek. Contact between the Lower
and Upper Members (LV and UV, respectively)
is mapped across the measured section.

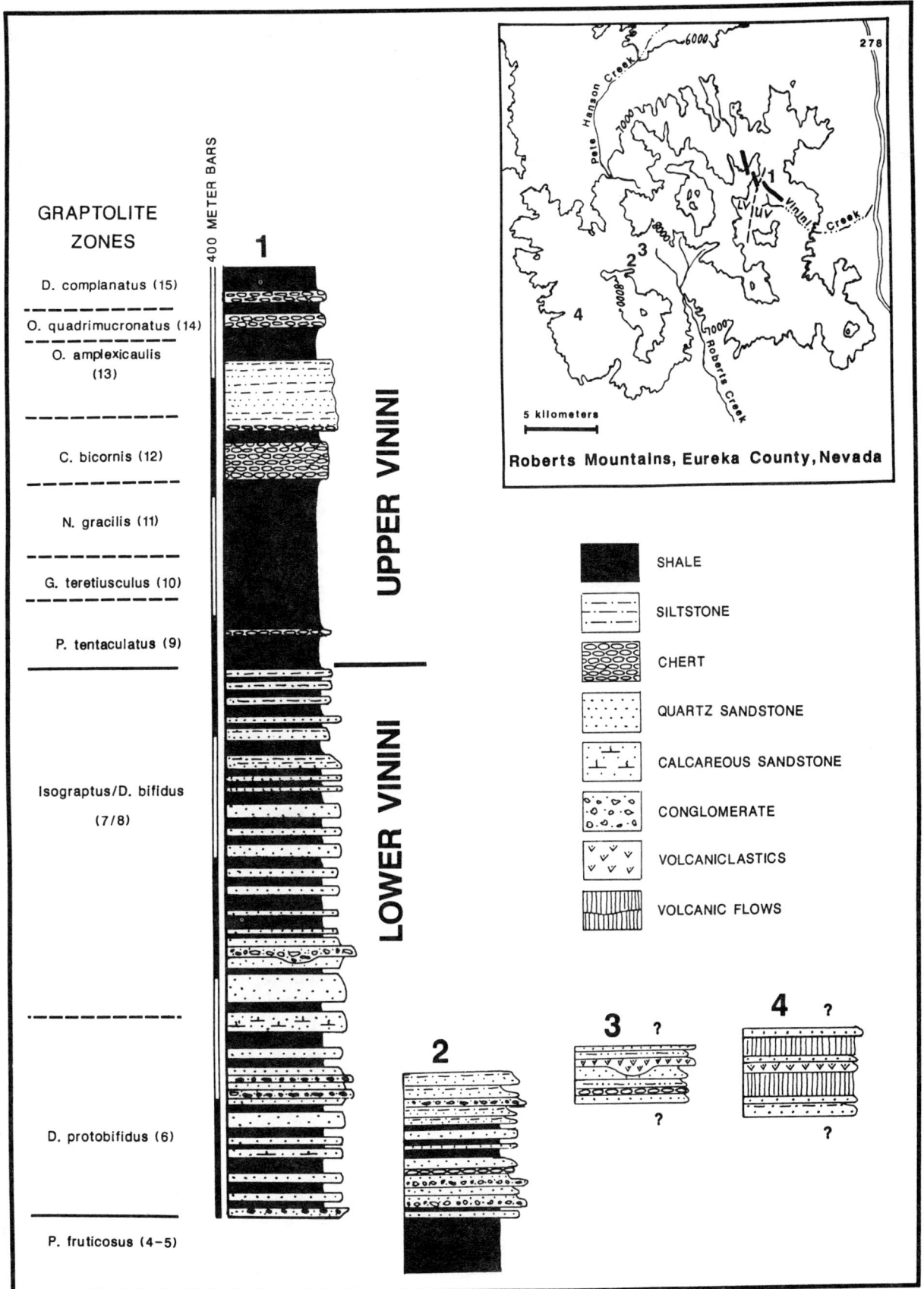

Roberts Mountains, Eureka County, Nevada

submarine flows, flow breccias, and volcaniclastics occur within the sandstone interval at a number of localities. They are especially well exposed in a 250+ meter-thick pile at locality 4 and in a 10 meter-deep channel at locality 3.

Sandstones, siltstones, and shale occur in approximately equal proportions in, and comprise most of, the upper sandstone interval. The float that covers the land surface is, however, dominated by sandstone and siltstone. Shale outcrops are sparse, but typically occur as thin interbeds between much thicker sandstone beds, and are exposed only after considerable excavation. The sandstones range from nearly pure quartzites and quartz arenites with little matrix to matrix-supported quartz wackes with clay matrix to calcareous sandstones with quartz grains supported in a micrite or sparite matrix. Quartz grains are medium to coarse, very well rounded, and well sorted, though sorting is locally bimodal. Bedding thickness ranges from 5 cm to 3 m, but is generally in the range of 10-30 cm. Thick beds (1-3 m) form ledges that can be traced from a few tens to several hundred meters along strike. These massive beds generally pinch out abruptly. Sedimentary structures are common and include cross-bedding, graded beds, parallel lamination, and flute casts. Trace fossils are common on many bedding planes. However, many sandstone beds are massive and lack sedimentary structures and trace fossils. Except for a few light colored quartzites, all sandstones are dark, generally brown to gray. Graptolites and conodonts are the only fossils found in the sandstones. Graptolites were found in only a few beds of calcareous sandstone, but where they occur they are abundant and well preserved. Conodonts were extracted from all beds with graptolites, as well as several without them.

Quartz siltstones are light brown and finely laminated to cross laminated. They occur in uniform beds 0.3 to 5 cm thick. Flute casts and a variety of trace fossils are common.

Although shale composes up to one-third of the sandstone interval, it is rarely exposed as either outcrop or float. It is generally found as thin interbeds (2-10 cm thick) between much thicker beds of sandstone. The shale is generally black, but light brown when silty. Graptolites are common.

Limestones are generally gray micrites, occur in beds 5-30 cm thick, and commonly include appreciable rounded quartz silt to sand-sized grains. They display cross-bedding, parallel laminations, and flute casts. A few contain graptolites.

Chert is rare in the sandstone interval. It occurs in beds 5-15 cm thick and is gray to black.

Conglomerates are rare but prominent. Seven individual beds were found in the measured section. Each is thick (0.3-3.0 m), but can be traced only a few tens of meters laterally before pinching out. Clast size ranges from 3 to 12 cm, and include chert, limestone, siltstone, and sandstone. Clasts are generally matrix supported, and most are aligned parallel to bedding.

No greenstone was found on the east side of the Roberts Mountains, but it is common on the west side. The greenstone occupies a channel 10 m deep and 20 m wide at locality 3. The channel is within the sandstone interval and filled with angular vesicular blocks, 10-50 cm in diameter, that float in a matrix of volcanogenic sandstone. The sandstone is arranged in three crude fining upward sequences with irregular bottoms. Paul Emsbo (personal communication, 1990) has measured a 250 m thick succession of greenstones at locality 4. The greenstones are within the sandstone interval and include flows, flow breccias, and volcaniclastics. Angular and rounded volcanic clasts are as large as 1 m in diameter; the greenstones also contain rare limestone and chert clasts.

The uppermost 75 m of the sandstone interval in the measured section is fine grained and grades upward into the overlying Upper Member (Figs. 2 and 3). The highest sandstone bed of typical thickness (20-30 cm) is 100 m below the top of the Lower Member. Above it, sandstone beds rapidly decrease in thickness and abundance; siltstone and shale increase in abundance. But because the shale is highly weathered, siltstone dominates the float. Siltstone beds are uniform in thickness, finely laminated, and display abundant trace fossils. They decrease in thickness upsection in the uppermost part of the Lower Member, and the shale intervals between individual siltstone beds thicken. The base of the Upper Member (and the top of the Lower Member) is placed at the top of the highest siltstone bed, which occurs at 1832 m in the measured section.

Upper Member

Shale and chert dominate the Upper Member (Fig. 2). This fact, together with the general absence of sandstone, readily distinguishes the Upper Member from the Lower Member. Argillite is common in the Upper Member; limestone is rare. Siltstone composes much of a distinctive 170 m thick interval in the upper part of the Upper Member, and the little sandstone that occurs in the Upper Member is restricted to this interval.

Most of the shale is black on fresh surfaces, but some is brown. Weathered surfaces display a variety of colors that include black, bluish gray, white, olive, green, and brown. Black shales are organic rich and contain abundant graptolites. Other shales are highly siliceous and regarded as argillites. Besides graptolites, the crustacean Caryocaris is common in the black shales. Trilobites and inarticulate brachiopods are rare.

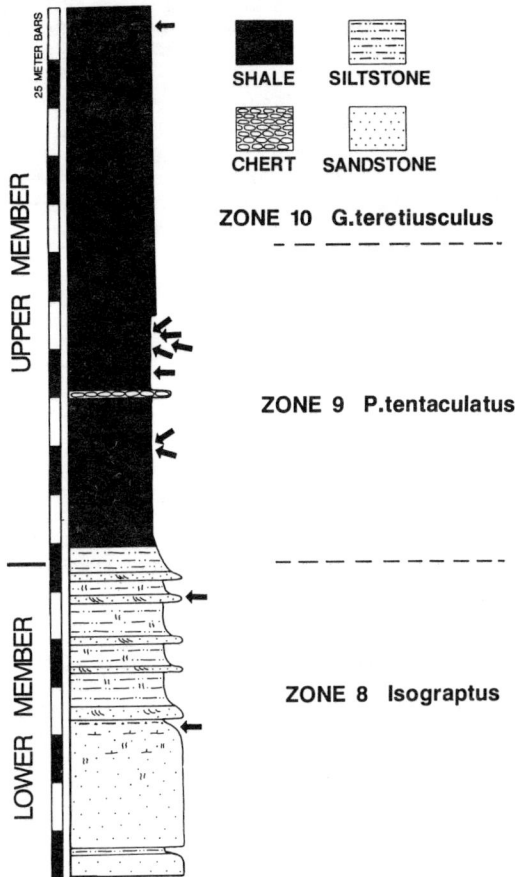

Figure 3. Detailed stratigraphic column for 450 m of measured section at locality 1 (Fig. 2) that illustrates the nature of the boundary between the Upper and Lower Members. Arrows indicate levels of grapto-lite collections.

Chert is distinctive in the Upper Member. It occurs in intervals up to 100 m thick, but typically 5-10 m thick, and forms prominent ledges. Rarely, it occurs as single beds. The chert is everywhere bedded with bed thickness averaging 10-15 cm. Chert beds are commonly separated by thin interbeds of siliceous shale. Bedding surfaces are undulatory, and trace fossils occur in some chert beds. Colors of both fresh and weathered chert range from green to white, cream, black, yellow, and rusty brown.

The distinctive interval of siltstone begins at 2675 m in the measured section, immediately above a prominent 12 m thick interval of bedded chert. It includes abundant shale, occasional sandstone, and rare chert beds. The siltstone is light gray to brown on fresh surfaces and red to tan on weathered surfaces. Individual beds have a consistently uniform thickness, with bed thickness ranging from 2-5 cm and averaging 3 cm. Sedimentary structures and trace fossils are absent. The sandstone consists of medium to coarse grains of quartz and chert and occur in beds 2-5 cm

thick.

The base of the Upper Member is distinct and easy to locate in the field (Fig. 3). It is placed at the top of the highest siltstone bed of the Lower Member. In the measured section, the lowest 30 m of the Upper Member are composed of tan to olive green weathering shale and are followed by more than 300 m of black shale that weathers bluish gray and is rich in graptolites.

The top of the Upper Member is cut out by faulting. The top of the measured section is within the lower part of grapto-lite Zone 15. A small, structurally isolated outcrop of shale and chert in the valley floor of Vinini Creek contains graptolites from the upper part of Zone 15.

Biostratigraphy

Introduction

Graptolites were the key to unraveling the lithostratigraphy of the Vinini Forma-tion and conodonts also proved valuable. Because graptolites were identified, and approximate zonal determinations made, as they were collected in the field, the lithofacies could be placed in a temporal sequence and correlated between exposures as they were examined in the field and as the measured section (locality 1) was construc-ted. This was possible in spite of problems presented by poor exposures, monotonous lithologies, and complex structure.

Graptolite abundance varies considerably in the Vinini. There are black shale intervals where specimens cover nearly every bedding surface through stratigraphic intervals 100 meters thick, and there are black shale intervals where specimens are found in only a few thin beds separated by barren intervals several meters thick. There are thick intervals of argillite and chert with only rare "scrappy" specimens found in a few thin beds. And there are thick stratigraphic intervals where grapto-lites occur in only a few beds, but are extremely abundant.

The graptolite fauna of the Vinini is clearly part of the widespread, probably tropical, Pacific Province. Many of its species also occur in Idaho, the Marathon region of Texas, the Ouachita Mountains, British Columbia, the Yukon, Australia, and South China. Berry's (1960) zonation, developed in Texas and previously employed in the Basin Ranges by Ross and Berry (1963) and revised over the years (e.g., Dover and others, 1980; Ross and others, 1982), is used for the Vinini (Fig. 2).

Conodonts collected from calcareous sandstones were important for confirming graptolite-based age determinations, especially for the base and top of the sandstone interval, and for correlating the lithofacies of the Vinini into those of the miogeoclinal sequence.

754

Lower Member

Graptolites were collected from several beds in the relatively thin lower black shale interval. Species identified include Pendeograptus fruticosus, Tetragraptus akzharensis, T. quadribrachiatus, Didymograptus (Expansograptus) nitidus, D. (E.) similis, Goniograptus thureaui, Holmograptus sp., and Sigmagraptus sp. and indicate a correlation with Zones 4-5 (Fig. 2), which Williams and Stevens (1988) refer to as a single zone of P. fruticosus.

No graptolites have yet been discovered in the lowest beds of the upper sandstone interval. Conodonts collected from a sandstone bed 3 m above the base of the interval indicate a correlation with the uppermost Fillmore to lowest Wah Wah limestones of the Ibex region of Utah (R.L. Ethington, personal communication), which in turn indicates a correlation with the Ninemile Formation of central Nevada, the upper Ibexian Series (Ross and others, 1982), and Zone 6 of Berry (Ross and others, 1982). Accordingly, the base of the sandstone interval is correlated with the base of graptolite Zone 6 (Fig. 4) on the basis of conodonts.

Many of the shale interbeds in the sandstone interval contain graptolites, but only a few species, such as Xiphograptus svalbardensis, X. declinatus, and Didymograptus (E.) extensus, which makes a precise zonal determination difficult. However, a few calcareous sandstone beds in the middle and upper parts of the sandstone interval have yielded large, diverse species assemblages that include Tetragraptus bigsbyi, T. quadribrachiatus, T. serra serra, T. taraxacum, Didymograptus (Didymograptellus) bifidus, D. (D.) paraindentus, Xiphograptus svalbardensis, X. declinatus, Isograptus cf. primulus, I. victoriae victoriae, I. v. divergens, I. v. cf. divergens, I. subtilis, I. caduceus australis, Pseudisograptus dumosus, P. manubriatus koi, P. gracilis, Oncograptus upsilon biangulatus, Cardiograptus morsus, Undulograptus austrodentatus, Pseudotrigonograptus ensiformis, Holmograptus sp., and Dichograptus sp. These species indicate a correlation with Berry's Zone 7/8 (although Zone 8 follows Zone 7 in Texas, they have been demonstrated to be laterally equivalent in other parts of North America). Conodonts are abundant in the same sandstone beds and support a correlation with the base of the Whiterockian Series (base of Antelope Valley Limestone) at its type section in the Monitor Range (Fig. 4). A similar graptolite fauna occurs with lowest Whiterockian shelly fossils in the Cow Head Group of Newfoundland (Williams and Stevens, 1988; R.J. Ross, Jr., personal communication). The highest collection from the sandstone interval is from 20 m below the top of the Lower Member (Fig. 3). It too indicates a correlation with Zone 7/8.

Upper Member

The lowest graptolite collection from the Upper Member is from 60 m above its base (Figs. 2 and 3). The species assemblage is large and occurs in several more collections over the next 60 m. The assemblage includes Brachiograptus sp., Pseudobryograptus incertus, Tetragraptus quadribrachiatus, Phyllograptus nobilis, Didymograptus cognatus, D. compressus, D. dubitatus, D. nodosus, Isograptus forcipiformis, Bergstroemograptus crawfordi, Glossograptus acanthus, G. hincksii, Cryptograptus tricornis, Paraglossograptus tentaculatus, Amplexograptus confertus, A. differtus, Climacograptus riddellensis, and Diplograptus? decoratus. It clearly indicates a correlation with Zone 9 of Berry. It is followed immediately at 1952 m in the section by beds with a species assemblage indicative of Zone 10 (Fig. 2). The assemblage includes Pterograptus lyricus, G. hincksii, C. tricornis, A. differtus, and Pseudoclimacograptus angulatus. Species assemblages correlative with Zone 10 occur as high as 2288 m in the measured section. The highest collections in Zone 10 include Didymograptus sp., Reteograptus geinitzianus, Nemagraptus sp., Glossograptus ciliatus, Cryptograptus tricornis, Glyptograptus euglyphus, G. teretiusculus, and Pseudoclimacograptus angulatus. Two small collections at 2364 m and 2440 m in the section include poor specimens of Nemagraptus gracilis, Dicellograptus sextans, D. gurleyi, Glossograptus sp., C. tricornis, Climacograptus meridionalis, Glyptograptus euglyphus, and Pseudocliamcograptus modestus. This interval is correlated with Zone 11. Zone 12 is also represented by only two small collections at 2485 m and 2500 m that include Reteograptus geinitzianus, Dicellograptus sp., Climacograptus bicornis, Climacograptus sp., Glyptograptus euglyphus, Orthograptus calcaratus acutus, and Pseudoclimacograptus sp.

The next higher graptolite collection is at 2860 m from a shale immediately above the 170 m thick siltstone interval (Fig. 2). It includes Dicranograptus nicholsoni, Climacograptus spiniferus, and Orthograptus amplexicaulis, and is correlated with Zone 13. Collections from higher in this zone include C. typicalis and Corynoides sp. Zone 14 is represented by only one collection from 3032 m. It includes Dicellograptus flexuosus, Pleurograptus sp., Climacograptus tubuliferus, Orthograptus amplexicaulis, Glyptograptus daviesi, Neurograptus sp., and Plegmatograptus sp. Zone 14 is immediately followed by Zone 15, which extends upward to the top of the section at 3170 m. Zone 15 is recognized by the presence of Dicellograptus minor, Climacograptus miserabilis, and C. tubuliferus.

A small, structurally isolated outcrop of shale and chert in the valley floor of Vinini Creek contains the youngest graptolites in the Vinini. These include Dicellograptus minor, D. ornatus, Tangyagraptus typicus, Climacograptus hastatus, and C. longispinus, which indicate a correlation with a level high in Berry's Zone 15. This is the first record of T. typicus outside of China where it occurs below, but close to, the Ordovician-Silurian boundary.

Sedimentology

Introduction

The lithologic character and fossil content of the Vinini Formation clearly indicate that it was deposited in a deep marine setting. Questions remain, however, as to the exact depositional setting, in particular its paleogeographic relationship to North America, and its depositional history. The only previous sedimentological study of the Vinini is that of Stanley and others (1977), but it included very little of the Lower Member, and because of the lack of biostratigraphic control, questions of depositional history and paleogeographic setting could not be readily addressed. Sedimentologic studies of the coeval Valmy (Miller and Larue, 1983; Watkins and Browne, 1989) are similarly limited.

On the basis of provenance and process of deposition, the lithologies of the Vinini can be divided into four lithosomes: quartz sandstones-calcareous sandstones-limestone-conglomerate; siltstone; shale-chert; and greenstones.

Quartz Sandstones, calcareous Sandstones, Limestone, and Conglomerate

Much of the Lower Member is composed of quartz and calcite grains that had, respectively, emergent cratonic and shallow-water shelf origins (Fig. 2). Quartzites and quartz arenites are composed of greater than 98% medium- to coarse-grained, well-sorted, well- rounded quartz grains, commonly with abraded overgrowth rims and undulose extinction, indicating a reworked sand with an ultimate plutonic or metamorphic origin. In quartz wackes the quartz grains are supported in a matrix of illitic clay that composes up to 35% of the rock. The calcareous sandstones are variable mixtures of fine to medium, well rounded quartz grains, skeletal debris, ooliths, and peloids with a matrix of micrite and illite clay. Skeletal grains include ostracods, trilobites, brachiopods, echinoderms, and sponge spicules. Such fossils are common in coeval shelfal strata of the miogeoclinal suite. Quartz grains also compose much of the matrix of conglomerates, which are matrix supported pebble to cobble conglomerates with parallel-aligned clasts composed predominately of chert and limestone. Limestone lithologies represented in the clasts include pelsparite, biopelmicrosparite, peloidal-skeletal biomicrite, and oosparite.

In the Lower Member, these lithologies are interbedded with each other and with siltstone and shale (Fig. 2). The arrangement of these lithologies and their sedimentary structures suggest the sediment was deposited by turbidity flows in a prograding submarine fan. Quartzites and quartz arenites are the dominant lithologies in much of the Lower Member, but in some parts of the section quartz wackes and calcareous sandstones are also common. The quartzites and quartz arenites are massive to horizontally stratified and rarely cross stratified. Beds of quartz wacke and calcareous sandstone display both types of stratification with horizontal stratification in the lower part of a bed replaced by cross stratification in the upper part of the bed. Bedding of sandstones thickens gradually upsection to as much as 3 m. Several outcrops show within them a gradational change upward from massive, coarse quartzite to cross-bedded, finer grained calcareous sandstones capped by thinly bedded hemipelagic mudstone and represent complete a-d Bouma turbidites. Most turbidites, however, are ab, abc, or bc sequences with sharp lower and upper contacts, separated by 2-30 cm graptolite shale beds of pelagic origin. Flute and groove casts are common on the sole of each turbidite sandstone bed. The rare beds of somewhat pure limestone (micrite and fossiliferous micrite) may represent overbank deposits from the calcareous turbidites. The conglomerates coarsen and thicken (from 0.3 to 3 m) upsection in the interval of approximately 400 to 900 m in the measured section (Fig. 2). All conglomerates have sharp, irregular bottom surfaces. Some grade up into massive quartzites; others have sharp upper contacts and are overlain by shale. The conglomerates are interpreted as representing sedimentation within submarine fan channels. The fact that they and the turbidites coarsen and thicken upsection and are interbedded with pelagic shales suggests that the stratigraphic interval in which they occur was deposited during fan progradation, from a lower fan to an upper mid fan or upper fan environment (Walker, 1984). Above 900 m, bed thickness decreases and becomes more uniform. Thick massive beds are few, and most of these can be traced long distances laterally indicating that the sands were not confined to channels. Instead, channels were filled and abandoned, and the sands were spreading across the surface of the fan.

Siltstones

Siltstones occur with the sandstones of the Lower Member as local thin interbeds; they are abundant and the coarsest lithology in the uppermost part of the Lower Member; and they compose much of a distinctive 170 m-thick interval in the Upper Member. They are everywhere thin and of uniform thickness and display horizontal laminations and climbing ripple cross laminations. Tool marks are ubiquitous on bottom surfaces, and bioturbation, when present, occurs as well-preserved inclined burrows and surface traces. Graptolites found on bedding surfaces are highly fragmented. These observations are consistent with those of Stanley and others (1977) who concluded that these siltstones are representative of contour current deposits as described by Bouma and Hollister (1973).

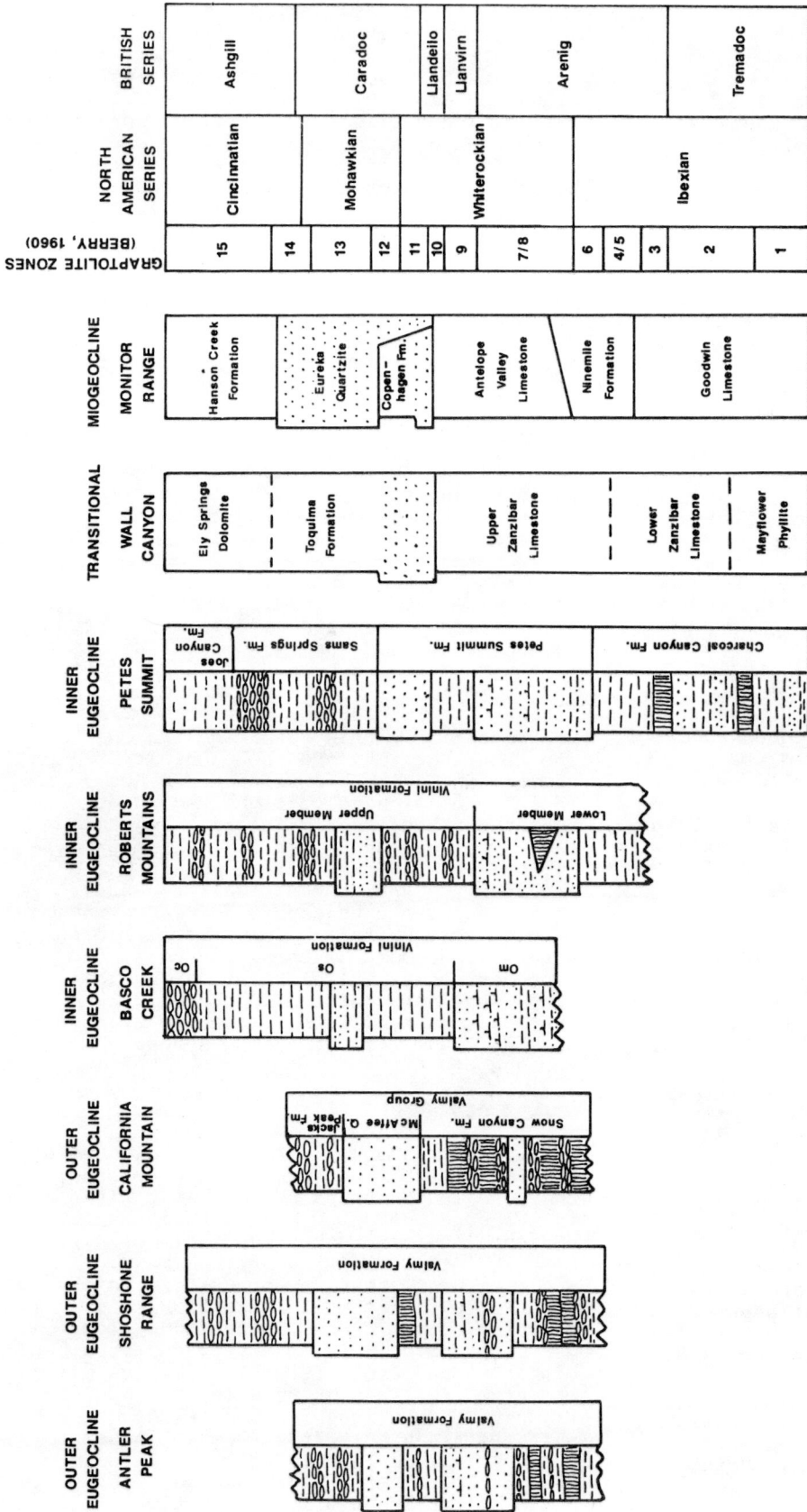

Figure 4. Biostratigraphic correlation of sections of Ordovician strata representing the depositional provinces of Stewart and Poole (1974). Lithologic legend same as Figure 3 except calcareous stratigraphic intervals at the Wall Canyon and Monitor Range sections are shown without patterns. Sources of information: Monitor Range (Ross, 1977; Ross and others, 1982); Wall Canyon (Ross, R. J., Jr. and Poole, F. G., unpublished data and field reconnaissance); Petes Summit (Kay, 1962; Kay and Crawford, 1964; McKee, 1976; field reconnaissance); Roberts Mountains (this report); Basco Creek (Lovejoy, 1959; Ketner, 1974; field reconnaissance); California Mountain (Churkin and Kay, 1967; Miller and Larue, 1983; Watkins and Browne, 1989); Shoshone Range (Gilluly and Gates, 1965; Madrid, 1987); Antler Peak (Roberts, 1964; Madrid, 1987).

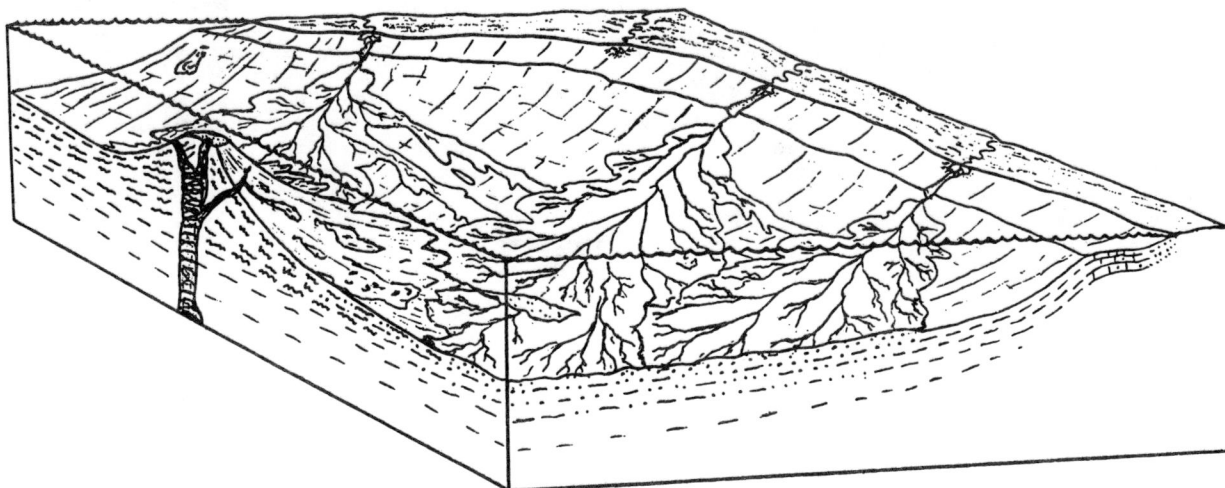

Figure 5. Generalized paleogeographic reconstruction of western margin of North America showing carbonate shelf, submarine canyons feeding submarine fans on continental rise, pelagic sedimentation on basin plain, and submarine volcanics.

Shale and Chert

Pelagic sedimentation is represented by shale and chert. Clay settling through the water column and dead planktonic and pelagic organisms such as graptolites, radiolarians, inarticulate brachiopods, and Caryocaris served as the source for these sediments. A relatively deep marine setting is interpreted for the shales. They are generally organic rich and completely lack fossils of benthic organisms. No evidence of bioturbation was discovered in contrast to observations reported by Stanley and others (1977). The fact that the chert is interbedded with the shales supports the interpretation of deposition in a deep marine setting for the chert as well.

Greenstones

The fact that the greenstones occur within the lower part of sandstone interval of the Lower Member indicates that they erupted on the seafloor. The volcaniclastics in the channel at locality 3 and the volcaniclastics and flow breccias at locality 4 are considered to be slope deposits related to seamounts, as interpreted by Watkins and Browne (1989) for similar greenstones in the Valmy Formation. Large (10-30 cm) clasts of limestone occur in the flow breccias at Locality 4. These might represent shallow-water carbonates on the sides of seamounts that were eroded by, and incorporated within, the flows as they were erupted and moved down the sides of the seamount. The occurrence of greenstones within the sandstones is evidence that the flow deposits of the seamounts encroached upon submarine fans, prograding into the basin (Fig. 5).

Depositional Setting and History

The entire Vinini Formation appears to have been deposited in a continental rise to basin plain setting. Pelagic deposition of shale was continuous through the Ordovician, but it was overwhelmed periodically by the construction of a submarine fan, the spread of contourites, periodic pelagic sedimentation of biogenic silica, and the eruption of submarine volcanics.

Black shales of Zones 4-5 represent the earliest record of eugeoclinal sedimentation in the Roberts Mountains. Clay and graptolite skeletons rained down on what was probably a basin plain situated close to a continental slope. In the late Ibexian (Zone 6), this slow, quiet pattern of sedimentation was overwhelmed by the sudden influx of turbidites as a submarine fan prograded out onto the basin plain. Early in the buildup of the fan, submarine volcanism erupted in close proximity to the fan, and flows, flow breccias, and volcaniclastics spread across the surface of the fan (Fig. 5). Progradation of the fan continued well into the time represented by Zone 7/8, when the upper portion of the fan occupied the area. But subsequently, channels in the upper fan were filled and abandoned, turbidite deposition gradually ceased, and bottom currents began reworking finer grained terrigenous clastics, as represented by the siltstones in the upper part of Zone 7/8 (Fig. 2). Finally, a change to entirely pelagic sedimentation is recorded at the top of Zone 9 (the contact between the Lower and Upper members) where the last siltstone bed of the Lower Member occurs.

The gradual change in environment from

prograding fan to coutourite deposition and finally to pelagic sedimentation may have been caused by a number of factors, but a rise in relative sea level that began in the early Whiterockian (base of Zone 7/8) is considered the most likely.

When the flow of sand into the basin ceased, sediment accumulation continued, but only because of the pelagic sedimentation of clay. However, by the Mohawkian (Zone 12), pelagic sedimentation of biogenic silica was periodically overwhelming the supply of clay, and bottom currents reintroduced silt and sand into the basin for a short time, only to be replaced again by pelagic sedimentation until the end of the Ordovician.

REGIONAL SYNTHESIS

Determination of the depositional history and paleogeographic setting of the Vinini Formation in the Roberts Mountains provides only a partial solution to a much larger problem, namely the depositional history and paleogeographic setting of all the eugeoclinal strata in the RMA. By means of graptolite and conodont biostratigraphy, it is now possible, with some limitations where biostratigraphic data are poor or absent, to compare the depositional history of Ordovician strata throughout the RMA and to correlate it with that of the autochthonous miogeoclinal sequence (Fig. 4). This provides regional insights critical for determining the paleogeographic setting of the eugeoclinal strata.

Provenance of Siliclastics

When we first worked out the stratigraphy of the Vinini in the Roberts Mountains (Finney and others, 1989), the significance of the siltstone interval in the Upper Member was not readily apparent; instead we were intrigued by the prominent sandstones of the Lower Member first described by Merriam and Anderson (1942) and considered characteristic of Ordovician eugeoclinal strata (Ketner, 1966). Because these sandstones proved to be decidedly older than the widespread Eureka Quartzite of the miogeoclinal sequence and largely correlative with the lowest Antelope Valley Limestone (Fig. 4), it was difficult to conclude that their provenance was the North American craton to the east. How could such tremendous quantities of sand have been transported from the craton to the outboard Vinini basin without leaving a trace of their passage in the stratigraphy of the inboard shelf, which was represented by a shallow marine carbonate platform? This question begged for an answer, especially in light of the fact that when tremendous quantities of sand, as represented by the Eureka Quartzite, were eventually accumulating on the shelf, none appeared to be spilling off the shelf into that part of the basin represented by the Vinini in the Roberts Mountains.

We (Finney and others, 1989) entertained the hypothesis that the provenance of the sands of the Lower Member was continental crust located to the west of the Vinini basin. This hypothesis has long been championed for eugeoclinal sands by Ketner (1966, 1977) and Gilluly and Gates (1965), and Madrid (1987) recently proposed a western source for feldspathic sands in the Valmy Formation. In contrast, Churkin (1974), Miller and Larue (1983), and Watkins and Browne (1989) proposed an eastern, North American source for sands in the Valmy, which are considerably younger than those of the Lower Member of the Vinini. Madrid also concluded that quartzose sands in the Valmy, in contrast to feldspathic sands, were derived from the east. In spite of their detailed sedimentologic analysis of the Vinini sands in the Roberts Mountains, Stanley and others (1977) could not determine the direction of origin of the sand, and neither could Stewart and Poole (1974) in their regional study of the eugeoclinal suite. Determination of the direction of the origin of the sands is, of course, critical for interpreting the geologic history of western North America in terms of plate tectonics.

Correlation of sections at Wall Canyon and Petes Summit with that in the Roberts Mountains (Figs. 1 and 4) presents a solution to the problem. Although the section at Wall Canyon is allochthonous and composes a thrust plate of the RMA, it is composed of strata that are transitional between those of the eugeoclinal and miogeoclinal sequences and that were deposited at the outboard edge of the miogeoclinal suite (Stewart and Poole, 1974). The Ordovician is composed of bedded limestone and dolomite, platy to shaly weathering limestone, silty limestone, mudstone, and thick massive beds of quartzite. The quartzite composes the lower 35 m of the Toquima Formation. Graptolites are common in thin shale interbeds and include Nemagraptus gracilis and Climacograptus bicornis, on which basis the quartzites can be confidently correlated with Zones 11 and 12 of Berry (R.J. Ross, Jr. and F.G. Poole, personal communciation). The quartzites are thus correlative with the lower part of the Eureka Quartzite of the miogeoclinal sequence (Figure 4). The lower part of the Eureka Quartzite correlates with the Pygodus anserinus and Amorphognathus tvaerensis conodont zones (Harris and others, 1979), which correlate, in turn, with graptolite Zones 11 and 12 (Ross and others, 1982). The fact that there are no sands at Wall Canyon equivalent to the sandstones of the Lower Member of the Vinini may be due to the transitional position of the Wall Canyon sequence.

Prominent, massive quartzite beds are also well known at Petes Summit, and graptolite age determinations clearly indicate that these likewise are correlative with Zones 11 and 12 of Berry (Kay, 1962; Kay and Crawford, 1964; McKee, 1976) and, thus, with the lower part of the Eureka Quartzite (Fig. 4). However, the quartzite

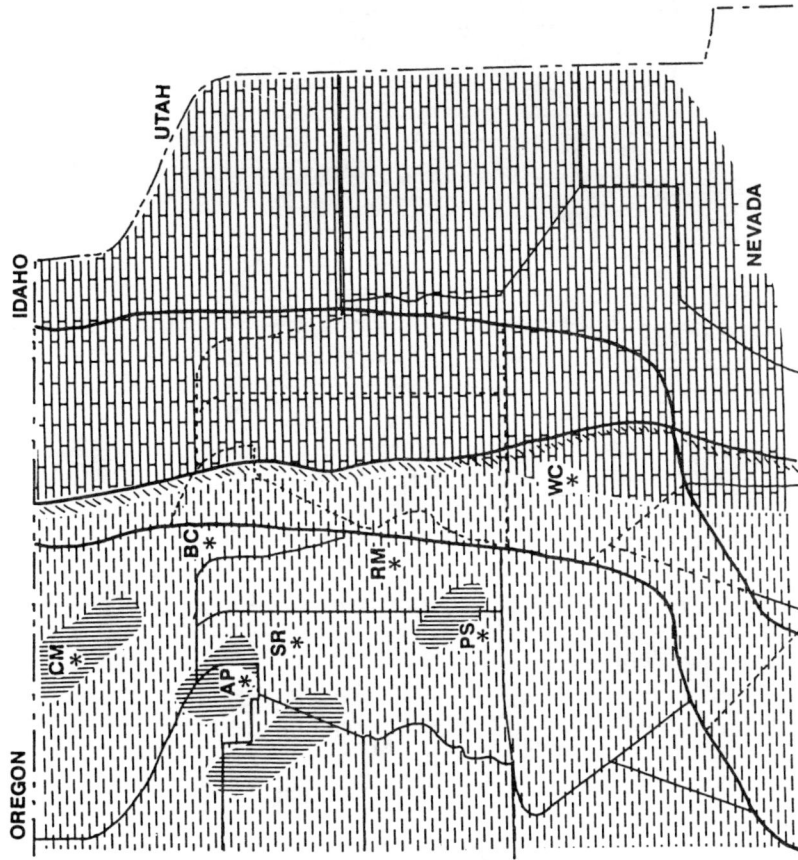

Figure 6. Paleogeologic maps of five depositional phases during the Ordovician, plotted on palinspastic base of Stewart and Poole (1974). Position of continental shelf edge and slope represented by heavy line with inclined hachures. Reconstruction of basin from information in this paper. Reconstruction of shelf modified from Ross (1974, 1977) and Ross and others (1989). Legend for Figs. 6A-E. Figs. 6B-E on following pages.

A. Paleogeologic map for upper Ibexian showing pelagic sedimentation and submarine volcanism in basin and carbonate deposition on shelf.

submarine volcanics

pelagic sediments

contourites

turbidites/shelfal sands

slope carbonates

shelf carbonates

land

Upper Ibexian (zone 4/5)

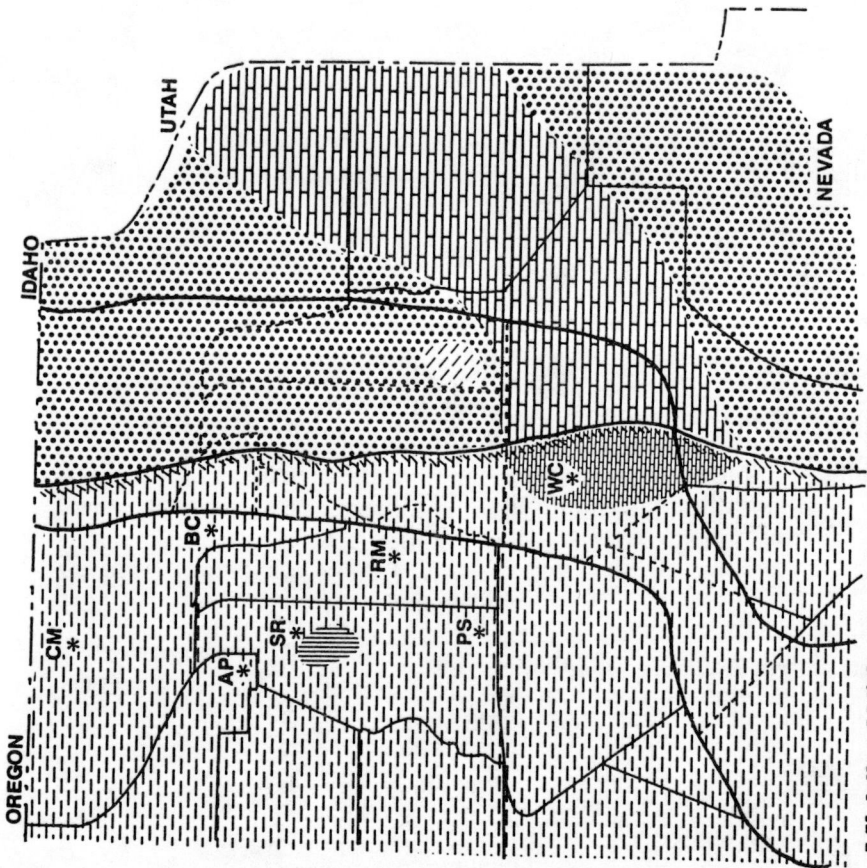

Middle Whiterockian (zone 9-10)

Figure 6C. Paleogeologic map for middle Whiterockian showing quartz sands stranded on shelf, local uplift of outer shelf, and pelagic sedimentation in basin.

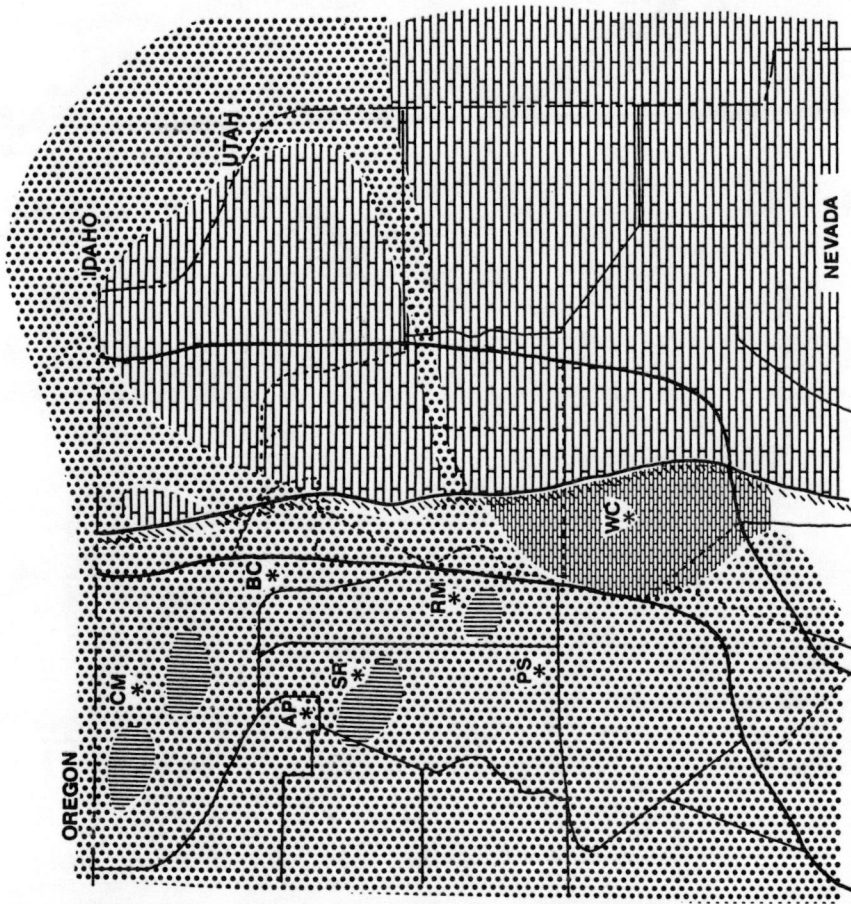

Lower Whiterockian (zone 7/8)

Figure 6B. Paleogeologic map for lower Whiterockian showing quartz sands on shelf flowing by way of hypothetical shelf channels and submarine canyons into basin and also showing submarine volcanism is basin.

761

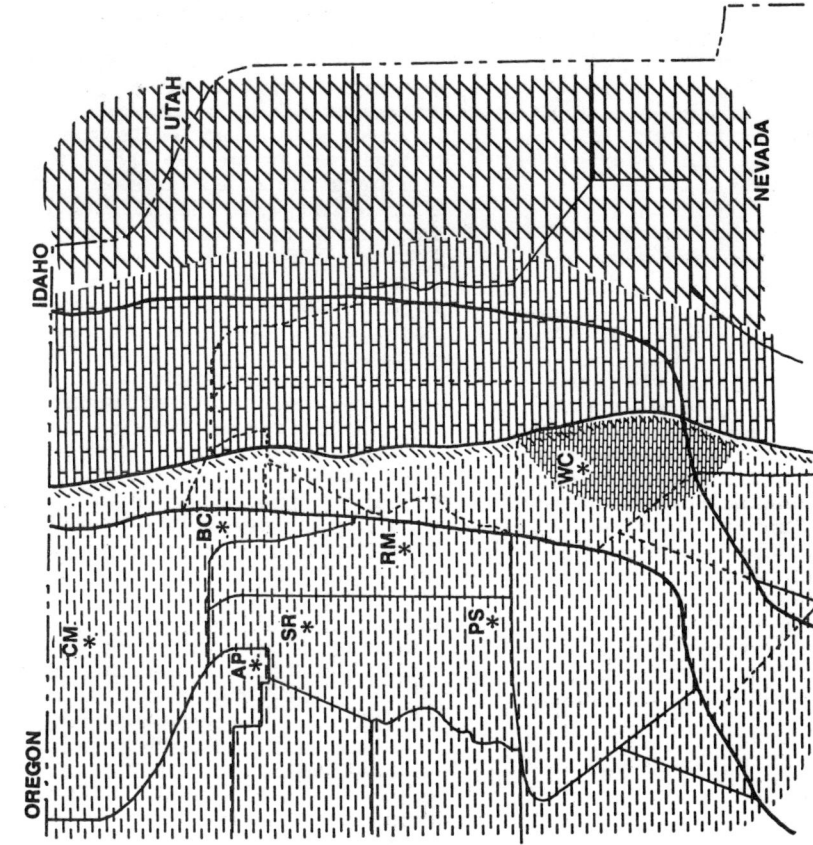

Lower Cincinnatian (zone 14)

Upper Whiterockian – Lower Mohawkian (zone 11-12)

Figure 6E. Paleogeologic map for lower Cincinnatian showing return of carbonate deposition on shelf and pelagic sedimentation in basin.

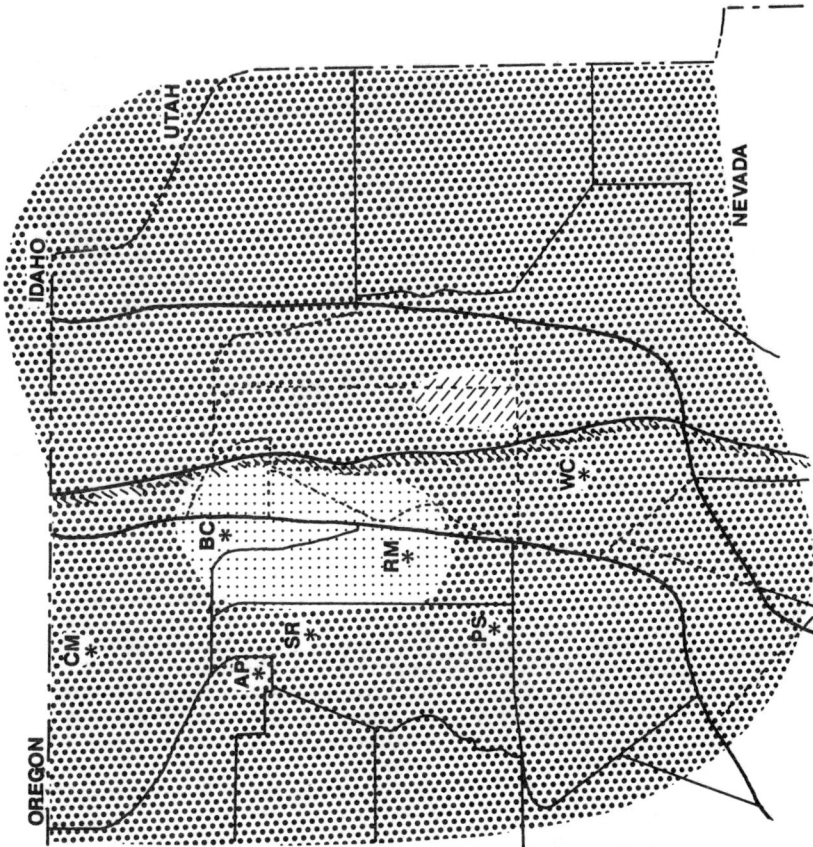

Figure 6D. Paleogeologic map for upper Whiterockian to lower Mohawkian showing quartz sands completely inundating shelf and spilling out into basin, local uplift of outer shelf, and deposition of contourites in Roberts Mountains and Basco Creek areas.

beds at Petes Summit are not the only sandstones within the Vinini Formation in the Petes Summit area. To the east of Petes Summit, the prominent quartzite beds are underlain by graptolitic shales below which there is an appreciable thickness of sandstones referred to the lower part of the Petes Summit Formation of Kay and Crawford (1964). Whereas the Eureka Quartzite and the coeval quartzites at Wall Canyon and Petes Summit are pure massive orthoquartzites, the sandstones in the lower part of the Petes Summit Formation are similar to those of the Lower Member of the Vinini in that they are highly varied, include quartz arenite, quartz wacke, and calcareous sandstone, and occur generally as beds of medium thickness interbedded with shale and siltstone. Graptolite age determinations (Kay, 1962; Kay and Crawford, 1964; McKee, 1976) indicate that the sandstones in the lower part of the Petes Summit Formation and the Lower Member of the Vinini are closely correlative (Fig. 4). On the basis of lithologic similarity and temporal equivalence, we conclude that these two sandstones are the same. We also conclude, for the same reasons, that the quartzites at Petes Summit are the same as the quartzites at Wall Canyon and the Eureka Quartzite. These younger quartzites represent a single stratigraphic unit continuous from the miogeoclinal sequence to the eugeoclinal sequence (Fig. 4), and are the conclusive evidence that the eugeoclinal sequence was deposited in close proximity to, and is a deep-water basinal facies of, the miogeoclinal sequence.

It is also evident that the quartz sands of the Lower Member of the Vinini, being in the same depositional sequence with the younger quartzites, were deposited close to the western margin of North America. The siliciclastic and calcareous sediments that compose submarine fans and contourite deposits of the Lower Member must have been derived from an emergent craton and shallow shelf and transported down a slope. We prefer the interpretation that this slope was the western margin of North America. The alternative interpretation of a major cratonic source to the west of the basin is unreasonable. It would require that the sands flowed completely across the basin before accumulating on its eastern, North American, side in a submarine fan.

The shelf in Nevada was the site of shallow-water deposition of the Antelope Valley Limestone (Ross and others, 1989) and thus the likely source of the calcareous sediment of the fan (Fig. 6B). It may have been emergent locally and subject to erosion and the production of siliciclastic sediment. Significant unconformaties occur in the Lower Ordovician in shelf sequences in the Cortez and Toquima ranges and the Roberts Mountains. However, much greater quantities of quartz sand could have been derived from a more inboard region of the shelf. Quartz sands of the Swan Peak and Kinnickinic quartzites were being deposited contemporaneously on the shelf in Idado and Utah (Ross, 1976). These sands, derived from an emergent cratonic source, may have crossed the carbonate shelf of Nevada and flowed down the slope to the rise by way of submarine canyons (Fig. 6B). In their interpretation of younger quartz sands (Eureka equivalents) in the Valmy Formation, Miller and Larue (1983) suggest that shelf channels directed coarser quartz grain sizes across the outer shelf and into the heads of submarine channels. To accumulate as submarine fans, the sands must have flowed into the basin from point sources. Submarine channels feed by shelf channels were most likely these point sources.

The significance of the siltstone interval in the Upper Member of the Vinini in the Roberts Mountains is now obvious. It is a distal, basinal facies of the Eureka Quartzite, with which it correlates (Fig. 4), and represents another episode of siliciclastic sediment spilling into the basin from the North American craton (Fig. 6D). In this episode, however, quartz sands of the Eureka Quartzite prograded across the entire shelf, and they spilled off into the basin in the vicinity of Wall Canyon and Petes Summit to the south where massive beds of pure orthoquartzite accumulated and also to the north as discussed below. However, this was not the case in the area of deposition of the Vinini of the Roberts Mountains. The reason might be related to unconformities in shelf sequences in the Cortez and Toquima ranges (McKee and Ross, 1969; Ross and others, 1989). They indicate that the outer margin of the shelf was tectonically active and locally emergent (Fig. 6D). The part of the continental rise represented by Vinini of the Roberts Mountains may have been downslope from an emergent margin of the shelf so that the direct flow of quartz sands to it was blocked. The finer grained fraction of siliciclastic sediment that reached the area, i.e. the silt, could have been transported laterally into the area by contour currents from other areas of the rise that were receiving the direct flow of siliciclastics off the shelf (Fig. 6D). A similar siltstone interval occurs in the upper Vinini Formation at Basco Creek (Fig. 4). It overlies a predominately shale interval that in turn overlies highly variable sandstones similar in lithology and age to those of the Lower Member of the Vinini.

Although graptolite age determinations of lithofacies in the Valmy are very uncertain, it appears that there are two bodies of sandstone equivalent to those in the Vinini (Fig. 4). Thin gray sandstones interbedded with mudstones occur with Arenig graptolites in the Snow Canyon Formation in the California Mountain quadrangle (Churkin and Kay, 1967; Watkins and Browne, 1989). These are separated by overlying shale, chert, and greenstone from younger, massive quartzites (the McAfee Quartzite) that are well dated by graptolites and are correlative with the Eureka Quartzite (Churkin and Kay, 1967; Miller and Larue, 1983; Watkins and Browne, 1989). The Valmy stratigraphy is very disrupted structurally in the Shoshone Range and at Antler Peak. Graptolites are sparse. Nevertheless, Madrid (1987) has recognized two sandstone intervals separated by an interval of shale, argillite, chert, and greenstone. Although graptolite age determinations are poor, they indicate that the shales between the sandstones may be correlative with shales between the two sandstone intervals in the California Mountain quadrangle and the Vinini Formation.

The younger sandstone interval in the Vinni and Valmy, correlative with the Eureka Quartzite, varies considerably. It consists of pure, massive orthoquartzites a few tens of meters thick at Wall Canyon and Petes Summit. It is represented by siltstones deposited by contour currents at Roberts Mountains and Basco Creek, and at California Mountain it is composed of exceptionally thick (up to 100 m) turbiditic orthoquartzite units. Several different and widely separated submarine canyons must have carried the flow of siliclastics down from the shelf and into the basin to produce these differences, and a similar situation probably controlled deposition of the lower sandstone interval. No doubt, the flow of sand down widely separated canyons was not precisely synchronous, although it was probably controlled to some degree by relative sea level changes, as discussed below. Variation in the quartz sand supply from different submarine canyons and the possible delivery of feldspathic sands from the west (Madrid, 1987) is probably the cause of much of the lithologic variation recognized between the Vinini and Valmy formations.

Sea-Level Controls on Sedimentation

It is easy to interpret the stratigraphic succession of eugeoclinal strata as a record of eustatic sea-level changes. A major drop and subsequent rise in sea level is an obvious explanation for the progradation and subsequent retrogradation of the submarine fan recorded in the Lower Member of the Vinini Formation, especially in light of the fact that much of the North American craton was subaerially exposed when submarine fan deposition commenced. Yet other controlling factors cannot be ruled out, and, in fact, regional subsidence and local tectonism contemporaneously affected the shelf during deposition of the Antelope Valley Limestone (Ross and others, 1989). Because the Ordovician record of eustasy is poorly understood, we consider sea-level changes in a relative sense and regard them as the result of eustatic sea-level changes, sedimentation, and tectonism.

Sea level was relatively high during the Ibexian; most of the craton was flooded and was the site of shallow-water carbonate deposition (Ross, 1976; Ross and others, 1982). In the earliest Whiterockian, however, most of the craton was exposed and sedimentation continued only on its margins.

But by the latest Whiterockian, shallow marine sedimentation again advanced onto the craton. This pattern indicates a high stand of relative sea level through much of the Ibexian, a rapid drop in relative sea level at the end of the Ibexian, and a subsequent rise in relative sea level through the Whiterockian and into the earliest Mohawkian.

The rapid drop in relative sea level in the latest Ibexian is approximately coincident with the base of the lower sandstone interval in the eugeoclinal sequence, especially as recorded in the Vinini Formation of the Roberts Mountains. This drop in sea level and its subsequent rise may have controlled the progradation and subsequent retrogradation of the submarine fans represented by the lower sandstone interval. Lowering of sea level is only one of several possible causes of fan progradation, but it is favored for a number of reasons. A sea-level low stand certainly existed at the beginning of the Whiterockian, given that much of the craton was subaerially exposed and not experiencing orogenesis. Tectonism did affect the western shelf, but it primarily involved subsidence that accompanied progradation of carbonate sediment of the Antelope Valley Limestone (Ross and others, 1989). Local uplift on the shelf was not extensive enough to account for the tremendous volume of submarine fan deposits in the eugeoclinal suite. Sedimentation on the shelf was dominated by shallow-water carbonates (Fig. 6B); major influxes of quartz sands onto the Nevada shelf did not occur until the latest Whiterockian. As summarized by Fraser (1989), fan progradation is accelerated during low stands because turbidity flows, capable of long-distance transport onto lower fan and basin areas, occur more frequently.

The Whiterockian transgression did not extend far onto the craton until well into graptolite Zone 9 (Ross and others, 1982). Its effects, however, could have been recorded earlier on the margins of the craton, and that appears to have been the case on the continental rise as recorded in the Vinini Formation. The upper part of the Lower Member represents a gradational change from channel abandonment to the replacement of turbidity flows by contour currents, and finally to the dominance of pelagic sedimentation. This return to pelagic sedimentation is recorded throughout the eugeoclinal suite as a prominent interval of organic-rich graptolite shale (Fig. 6C). In light of the fact that quartz sands were prograding southward along the shelf, a rise in sea level is the most probable mechanism for stranding the quartz sands on the shelf, deactivating the fan, and reestablishing pelagic sedimentation on the continental rise. The rise in sea level may have started at the beginning of Zone 7/8, which correlates with the base of the Whiterockian.

Sea level did not have a strong in-

764

fluence on the deposition of the upper sandstone interval because of a tremendous influx of quartz sands. During the latest Whiterockian to early Mohawkian, most of the craton was again inundated and flooded. Relative sea level was at a high stand. Yet the western shelf was covered by a prograding blanket of quartz sands (the Eureka Quartzite), and these sands spilled off the shelf and down onto the continental rise and basin plain in great volumes (Fig. 6D). Finney (1986) recognized a eustatic drop in sea level in the middle Mohawkian at the base of Zone 13 followed rapidly by a rise. Although the sea level drop was presented as the cause of turbidity flows on the southern margin of North America, it does not appear to have had a visible effect on sedimentation on the western shelf and rise. In fact, deposition of turbidites and contourites was replaced by pelagic sedimentation at approxmately that time (Fig. 6E). The renewed rise in sea level may have again stranded siliciclastics on the shelf; perhaps, it even inundated the source area, shutting off erosion and covering the source area with younger sediments.

Relationship of RMA to North America

Stratigraphic ties between eugeoclinal and miogeoclinal suites indicate that the rocks of the RMA were deposited adjacent to the western margin of North America and do not compose an exotic terrane. Stratigraphic ties also indicate that, during eastward Antler thrusting of 145 Km, rocks of the RMA did not experience significant translational displacement. There is, for example, the alignment of contourites deposits in the eugeoclinal suite in the Roberts Mountains and at Basco Creek with emergent shelf features in the miogeoclinal suite in the Toquima and Cortez ranges (Fig. 6D).

SUMMARY

The paleogeographic setting and depositional history of eugeoclinal and transitional strata of the RMA are illustrated in Figures 5 and 6.

The strata of the eugeocline were deposited on the continental rise and basin plain immediately adjacent to the western margin of North America. The transitional strata were deposited on the slope leading from the shelf to the rise.

During the Ibexian, sea level was high. Carbonates accumulated on the shallow shelf and on the slope; pelagic sediments accumulated on the basin plain where submarine volcanics were periodically erupting.

By the late Ibexian, quartz sands were prograding southwards along inboard regions of the shelf, and with an abrupt drop in sea level, the sands were directed into submarine canyons and flowed out onto the continental rise and basin plain to build up prograding submarine fans. Submarine volcanics continued to erupt and flowed

across the fan surfaces. Terrigenous clay mixed with carbonate sediment accumulated on the slope.

With a rise in relative sea level through the Whiterockian, the supply of sand to the submarine canyons was cut off even though the sands were still accumulating across much of the shelf and clay/carbonate mixtures were being deposited on the slope. Except for isolated volcanics, the rise and basin plain experienced only pelagic sedimentation.

Although there may have been a minor lowering of sea level in the early Mohawkian, sea level in general continued rising through the late Whiterockian and Mohawkian. However, the influx of quartz sands on the shelf was so great that in the late Whiterockian and earliest Mohawkian they spilled off the shelf, probably by way of submarine canyons, and flowed out across the slope and onto the rise and basin plain. Those areas not in the path of the flows, still received finer siliciclastics carried in by contour currents.

By the middle Mohawkian, the rise in sea level shut off the supply of sand to the slope, rise and basin. Pelagic sedimentation was reestablished, and by the Cincinnatian carbonates were again accumulating on the shelf.

ACKNOWLEDGMENTS

Acknowledgment is made to the donors of the Petroleum Research Fund, administered by the American Chemical Society, for support of this research. The Scholarly and Creative Activities Committee at CSULB is also thanked for release time to carry out this research. Several colleagues provided help and information that were critical to the success of the project. R.L. Ethington graciously processed all our conodont samples, identified the specimens, and provided age determinations. R.J. Ross, Jr. and F.G. Poole provided us with their unpublished data from the measured section they described and collected at Wall Canyon, and F.G. Poole kindly gave of his time to walk S. Finney through the section. In the initial stages of this project, K. Ketner spent considerable time in the field leading S. Finney through the stratigraphy at Basco Creek, Antler Peak, the Shoshone Range, and other areas in northern Nevada. Paul Emsbo showed us much of the field area for his Masters' thesis, including the greenstones at Locality 4. M.A. Murphy demonstrated to us the sequence of thrust slices in the RMA in the northern Roberts Mountains, and J.D. Cooper gave valuable advice on sedimentologic aspects of the research. We gratefully received critical comments on the manuscript from J.D. Cooper, R.L. Ethington, M.A. Murphy, F.G. Poole, and R.J. Ross, Jr., although we have not taken all their advice.

REFERENCES CITED

Berry, W. B. N., 1960, Graptolite faunas of the Marathon Region, West Texas: Texas Univiversity, Bureau of Economic Geology, Publication 6005, 179 p.

Bouma, A. H., and Hollister, D. D., 1973, Deep ocean basin sedimentation, in Turbidites and Deep-water Sedimentation: Pacific Section SEPM Short Course Lecture Notes, p. 79-118.

Churkin, M., Jr., 1974, Paleozoic marginal ocean basin-volcanic arc systems in the Cordilleran foldbelt, in Dott, R. J., and Shaver, R. H., eds., Modern and Ancient Geosynclinal Sedimentation: Society of Economic Paleontologists and Mineralogists Special Publication No. 19, p. 174-192.

Churkin, M., Jr., and Kay, M., 1967, Graptolite-bearing Ordovician siliceous and volcanic rocks, Northern Independence Range, Nevada: Geological Society of America Bulletin, v. 78, p. 651-668.

Dover, J. H., Berry, W. B. N., and Ross, R. J., Jr., 1980, Ordovician and Silurian Phi Kappa and Trail Creek Formations, Pioneer Mountains, Central Idaho - Stratigraphic and Structural Revisions, and New Data on Graptolite Faunas: U.S. Geological Survey Professional Paper 1090, p. 1-54.

Ferguson, H. G., 1924, Geology of ore deposits of the Manhattan District, Nevada: U. S. Geological Survey Bulletin 723, p. 1-163.

Finney, S. C., 1986, Graptolite biofacies and correlation of eustatic, subsidence, and tectonic events in the Middle to Upper Ordovician of North America: Palaios, v. 1, p. 435-461.

Finney, S. C., Perry, B. D., and Cooper, J. D., 1989, Depositional history of the Vinini Formation: results from the type area, Roberts Mountains, Nevada: Geological Society of America Abstracts with Programs, v. 21, no. 5, p. 78.

Fraser, G. S., 1989, Clastic Depositional Sequences: Processes of Evolution and Principles of Interpretation: Englewood Cliffs, New Jersey, Prentice Hall, 459 p.

Gilbert, G. K., 1875, Report upon the geology of portions of Nevada, Utah, California, and Arizona examined in the years 1871 and 1872: U.S. Geographical and Geological Surveys W. 100th Meridian Report, v. 3, p. 21-187.

Gilluly, J., and Gates, O., 1965, Tectonic and igneous geology of the northern Shoshone Range, Nevada: U.S. Geological Survey Professional Paper 465, 153 p.

Harris, A. G., Bergstrom, S. M., Ethington, R.L., and Ross, R. J., Jr., 1979, Aspects of Middle and Upper Ordovician Conodont Biostratigraphy of Carbonate Facies in Nevada and Southeast California and Comparison with Some Appalachian Successions: Brigham Young University Geological Studies, v. 26, no. 3, p. 7-43.

Kay, M., 1962, Classification of Ordovician Chazyan shelly and graptolite sequences from central Nevada: Geological Society of America Bulletin, v. 73, no. 11, p. 1421-1430.

Kay, M. and Crawford, J. P., 1964, Paleozoic facies from the miogeosynclinal to the eugeosynclinal belt in thrust slices, central Nevada: Geological Society of America Bulletin, v. 75, no. 5, p. 425-454.

Ketner, K. B., 1966, Comparison of Ordovician eugeosynclinal and miogeosynclinal quartzites of the Cordilleran geosyncline: U.S. Geological Survey Professional Paper 550-C, p. C54-C60.

Ketner, K. B., 1974, Preliminary geologic map of the Blue Basin Quadrangle, Elko County, Nevada: U.S. Geological Survey Miscellaneous Field Studies Map MF-559.

Ketner, K. B., 1977, Deposition and deformation of lower Paleozoic western facies rocks, northern Nevada: in Stewart, J. H., Stevens, C. H., and Fritsche, A. E., eds., Paleozoic Paleogeography of the Western United States, Pacific Coast Paleogeography Symposium 1, Pacific section, Society of Economic Paleontology and Mineralogy, p. 251-258.

Lovejoy, D. W., 1959, Overthrust Ordovician and the Nannie's Peak intrusive, Lone Mountain, Elko County, Nevada: Geological Society of America Bulletin, v. 70, no.5, p. 539-564.

Madrid, R. J., 1987, Stratigraphy of the Roberts Mountains allochthon in north-central Nevada [unpublished dissertation]: Stanford, California, Stanford University, 336 p.

McKee, E. H., 1976, Geology of the northern part of the Toquima Range, Lander, Eureka, and Nye Counties, Nevada: U.S. Geological Survey Professional Paper 931, 49 p.

McKee, E. H., and Ross, R. J., Jr., 1969, Stratigraphy of Eastern Assemblage Rocks in a Window in Roberts Mountains Thrust, Northern Toquima Range, Central Nevada: American Association of Petroleum Geologists Bulletin, v. 53, no. 2, p. 421-429.

Merriam, C. W. and Anderson, C. A., 1942, Reconnaissance survey of the Roberts Mountains, Nevada: Geological Society of America Bulletin, v. 53, p. 16751728.

Miller, E. L., and Larue, D. K., 1983, Ordovician quartzite in the Roberts Mountains allochthon, Nevada: Deep sea fan deposits derived from cratonal North America, in Stevens, C. H., ed., Pre-Jurassic Rocks in Western North American Suspect Terranes: Los Angeles, Society of Economic Paleontologists and Mineralogists, Pacific Section, p. 91-102.

Minnick, E. P., 1975, Structure and stratigraphy of the Ordovician Vinini Formation, Tyrone Creek area, Eureka County, Nevada [unpublished M.S. thesis]: Athens, Ohio, Ohio University, 55 p.

Murphy, M. A., 1968, Sequence of units in the upper plate of the Roberts Mountains thrust fault, northwestern Roberts Mountains, Nevada: Geological Society of America Abstracts for 1968, Special Paper 121, p. 212.

Murphy, M. A., McKee, E. H., Winterer, E. L., Matti, J. C., and Dunham, J. B.,

766

1978, Preliminary geologic map of the Roberts Creek Mountains quadrangle, Nevada: U.S. Geological Survey Open-File Report 78-376, 2 sheets.

Murphy, M. A., Power, J. D., and Johnson, J. G., 1984, Evidence for Late Devonian movement within the Roberts Mountains allochthon, Roberts Mountains, Nevada: Geology, v. 12,p. 20-23.

Riva, J., 1970, Thrusted Paleozoic rocks in the northern and central HD Range, northeastern Nevada: Geological Society of America Bulletin, v. 81, p. 2689-2716.

Roberts, R. J., 1949, Structure and stratigraphy of the Antler Peak quadrangle, North-Central Nevada: Geological Society of America Bulletin, v. 60, p. 1917.

Roberts, R. J., 1951, Geology of the Antler Peak Quadrangle, Nevada: U.S. Geological Survey Quadrangle Map Series, GQ-10.

Roberts, R. J., 1964, Stratigraphy and structure of the Antler Peak quadrangle Humboldt and Lander counties Nevada: U.S. Geological Survey Professional Paper 459-A, 93 p.

Roberts, R. J., Hotz, P. E., Gilluly, J., and Ferguson, H. G., 1958, Paleozoic rocks of north-central Nevada: American Association of Petroleum Geologists Bulletin, v. 42, p. 28132857.

Roberts, R. J., Montgomery, K. M., and Lehner, R. E., 1967, Geology and mineral resources of Eureka County, Nevada: Nevada Bureau of Mines and Geology Bulletin 64, 152 p., 12 pls.

Ross, R. J., Jr., 1976, Ordovician sedimentation in the western United States, in Bassett, M. G., ed., The Ordovician System: Proceedings of a Palaeontological Association Symposium, Birmingham, September 1974: Cardiff, University of Wales Press and National Museum of Wales, p. 73-105.

Ross, R. J. Jr., 1977, Ordovician paleogeography of the western United States, in Stewart, J. H., Stevens, C. H., and Fritsche, A. E., eds., Paleozoic Paleogeography of the Western United States, Pacific Coast Paleogeography Symposium I, Pacific Section, Society of Economic Paleontologists and Mineralogists, p. 19-38.

Ross, R. J., Jr and Berry, W. B. N., 1963, Ordovician graptolites of the Basin Ranges in California, Nevada, Utah, andIdaho: U.S. Geological Survey Bulletin 1134, 177 p.

Ross, R. J., Jr., Amsden, T. W., Bergstrom, D., Bergstrom, S. M., Carter, C., and 22 others, 1982, The Ordovician System in the United States, Correlation chart and explanatory notes: International Union of Geological Sciences Publication No. 12, 73 p.

Ross, R. J., Jr., James, N. P., Hintze, L. F., and Poole, F. G., 1989, Architecture and evolution of a Whiterockian (early middle Ordovician) carbonate platform, Basin Ranges of western U.S.A.: Controls on Carbonate Platform and Basin Development, Society of Economic Paleontologists and Mineralogists Special Publication No. 44, p. 167-185.

Stanley, K. O., Chamberlain, C. K., and Stewart, J. H., 1977, Depositional setting of some eugeosynclinal Ordovician rocks and structurally interleaved Devonian rocks in the Cordilleran mobile belt, Nevada: in Stewart, J. H., Stevens, C. H., and Fritsche, A. E., eds., Paleozoic Paleogeography of the Western United States, Pacific Coast Paleogeography Symposium 1, Pacific Section, Society of Economic Paleontologists and Mineralogists, p. 259-274.

Stewart, J. H., 1980, Geology of Nevada: Nevada Bureau of Mines and Geology Special Publication 4, 136 p.

Stewart, J. H., and Poole, F. G., 1974, Lower Paleozoic and upper-most Precambrian Cordilleran miogeocline, Great Basin, Western United States, in Dickinson, W. R., ed., Tectonics and Sedimentation: Society of Economic Paleontologists and Mineralogists Special Publication 22, p. 28-57.

Turner, R. J. W., Madrid, R. J., and Miller, E. L., 1989, Roberts Mountains allochthon: Stratigraphic comparison with lower Paleozoic outer continental margin strata of the northern Canadian Cordillera: Geology, v. 17, p. 341-344.

Walker, R. G., 1984, Turbidites and associated coarse clastic deposits, in R. G. Walker, ed., Facies Models, 2nd edition: Geoscience Canada Reprint Series 1.

Watkins, R. and Browne, Q. J., 1989, An Ordovician continental-margin sequence of turbidite and seamount deposits in the Roberts Mountains allochthon, Independence Range, Nevada: Geological Society of America Bulletin, v. 101, p. 731-741.

White, C. A., 1874, Preliminary report upon invertebrate fossils collected by the expeditions of 1871, 1872, and 1873, with descriptions of new species, in Wheeler, G.M.: U.S. Geographical and Geological Explorations and Surveys W. 100th Meridian Report, p. 5-27.

Williams, S. H., and Stevens, R. K., 1988, Early Ordovician (Arenig) graptolites of the Cow Head Group, western Newfoundland, Canada: Palaeontographica Canadiana No. 5, 167 p.

DEPOSITIONAL PROVINCES IN NEVADA DURING THE FAMENNIAN

Kenneth S. Coles
Dept. of Earth and Atmospheric Sciences
Purdue University
West Lafayette, IN 47907

ABSTRACT

The depositional setting of deformed upper Famennian and lowermost Kinderhookian strata near the base of the Roberts Mountains allochthon in Nevada constrains the paleogeography of the region at the time it was undergoing a transition from the shelf-slope setting of the early Paleozoic to the foreland basin and highland of the Antler orogeny.

The Pinecone sequence and correlative rocks of latest Devonian age in central and northeastern Nevada consist of black chert and argillite, commonly with nodular phosphate. Deposition took place in a detritus-starved, oxygen-poor slope or foredeep setting east of the advancing, but still submerged, Roberts Mountains allochthon. The short time interval between deposition of the Pinecone and the end of thrust emplacement of the allochthon indicates that the Pinecone is less far-travelled than much of the allochthon.

During late Famennian time, a number of contrasting provinces of deposition existed in the vicinity of Nevada. First, the lower black shale unit in the Leatham Member of the Pilot Shale, in eastern Nevada and western Utah, formed in the dysaerobic, deep subtidal belt described by Sandberg and coworkers. Second, a bathyal(?) province in central Nevada, to the west of the Pilot, contained black chert and phosphate, including the Pinecone sequence, in a zone of strong surface water productivity. Also present, but rare, were beds of carbonate detritus with a probable provenance to the east, and olistoliths(?) of quartz sandstone resembling sandstone known in the Roberts Mountains allochthon, which lay to the west. Third, a continental borderland, including the Roberts Mountains allochthon, was undergoing deformation. Fourth, pelagic oceanic rocks and greenstones formed in a predominantly oxygenated environment somewhere beyond the borderland. These pelagic rocks were later rearranged among several terranes, including the Golconda allochthon.

INTRODUCTION

The stratigraphic/rock record of the Great Basin region of the western U.S. indicates that Paleozoic time included episodes of relative stability or gradual change in paleogeography, as well as times of rapid change. The Devonian-Mississippian Antler orogeny (Roberts and others, 1958) was one of the most significant times of rapid change. A primary feature of this orogeny in Nevada was the Roberts Mountains allochthon, a deformed wedge of Cambrian to Devonian, deep marine sedimentary rocks, which was thrust eastward over coeval shelf strata. The allochthon formed a subaerial highland, the Antler orogenic belt, which shed detritus eastward during the Mississippian into a deep, narrow foreland basin. Here I use the term *Antler orogenic belt* for the belt of rocks deformed during the Antler orogeny, whereas the term *Antler highland* denotes the positive source of sediment within the orogenic belt.

The quoted age span of Antler deformation (*sensu stricto*; as defined by P. B. King in Nilsen and Stewart, 1980), Late Devonian-Early Mississippian, is based upon structural and stratigraphic evidence (Roberts and others, 1958; Smith and Ketner, 1968; Nilsen and Stewart, 1980; see discussion below). Internal imbrication of the associated allochthon began in the Late Devonian (Murphy and others, 1984a) and motion of the allochthon ceased by the Kinderhookian or Osagean (Early Mississippian; Johnson and Pendergast, 1981; Speed and Sleep, 1982; 1983). The contemporary paleogeography is poorly resolved, however, in the region involved in deformation and transport. Outstanding questions (see Nilsen and Stewart, 1980) concerning the Late Devonian to Kinderhookian time interval include: 1) the succession of sedimentary environments as the shelf became a foreland basin, 2) the location and timing of uplifts and downwarps during the transition, and 3) the sources and depocenters of detritus.

Upper Famennian to lowermost Kinderhookian strata near the base of the Roberts Mountains allochthon in Nevada provide critical information about the paleogeography of the region at the time of transition from the early Paleozoic shelf-slope setting to the foreland basin and highland configuration of the Mississippian. In this paper, following a review of the regional geologic setting, I summarize the results of field and laboratory studies that will be documented in detail elsewhere (see also Coles, 1988; 1989). These results are the basis for a discussion of latest Devonian paleogeography.

Reexamination of stratigraphic and structural evidence indicates that several depositional provinces lay west of the

In Cooper, J.D., and Stevens, C.H., eds., 1991, **Paleozoic Paleogeography of the Western United States-II:** Pacific Section SEPM, Vol. 67, p. 767-782.

767

768

shallow to deep subtidal environments of Utah and eastern Nevada late in the Devonian. The first province was a hemipelagic to starved foredeep. The Roberts Mountains allochthon in late Famennian time was part of a continental borderland west of the foredeep. Coarse-grained detritus was lacking at or near the leading edge of the allochthon at this time, and the Antler highland was not yet subaerially exposed in central Nevada. Beyond the borderland lay a pelagic oceanic realm. Phosphate deposition took place in parts of the foredeep, borderland, and pelagic provinces whenever upwelling and decreased detrital input provided favorable conditions.

Terminology

Several features discussed herein have more than one name in the literature. I use the name Antler orogeny, as narrowly defined (see King, quoted in Nilsen and Stewart, 1980) for the events of Late Devonian-Early Mississippian age noted above. Some workers in western North America have termed any tectonic activity or sedimentation of Devonian or Mississippian age "Antler." The term *orogeny* also has had different genetic connotations over the years; the suggestions of Trexler and others (1990) regarding new terminology may resolve this ambiguity.

The mid-Paleozoic stratigraphic unit in the Toquima Range of Nevada described herein has been correlated to the Vinini Formation, the Slaven Chert, and/or the Pinecone Formation. I here prefer the local name, Pinecone, originally used by Kay and Crawford (1964; Crawford, 1958), as the stratigraphic range involved does not correspond to that of the other named units where they occur in the region. The Pinecone is a disrupted to broken formation (terminology of Raymond, 1984); hence it has in some cases been given the lithodemic names *sequence* (after Kay and Crawford, 1964) or *assemblage* (Murchey and others, 1987), rather than formation. These terms are not entirely satisfactory, however, as they have other meanings. For purposes of the present discussion I shall employ the name Pinecone sequence (informal).

REGIONAL SETTING

Initiation of the Early Paleozoic Cordilleran Margin

The western margin of the North American craton lay in the vicinity of Utah in the early Paleozoic (near the Wasatch line; Stewart, 1980; Fig. 1). The western limit of Precambrian continental crust has been assumed to lie farther west, near the isopleth marking an initial strontium isotopic ratio of 0.7060 for Mesozoic plutons (I_{Sr} = 0.706 line on Figure 1, from Kistler and Peterman, 1978; Elison and others, 1990; but see also Farmer and DePaolo, 1983, who placed the crustal limit as much as 100 km farther east). During the latest Precambrian and Early Cambrian, widespread deposition of clastics occurred within the Cordilleran miogeocline, a term used by Stewart and Poole

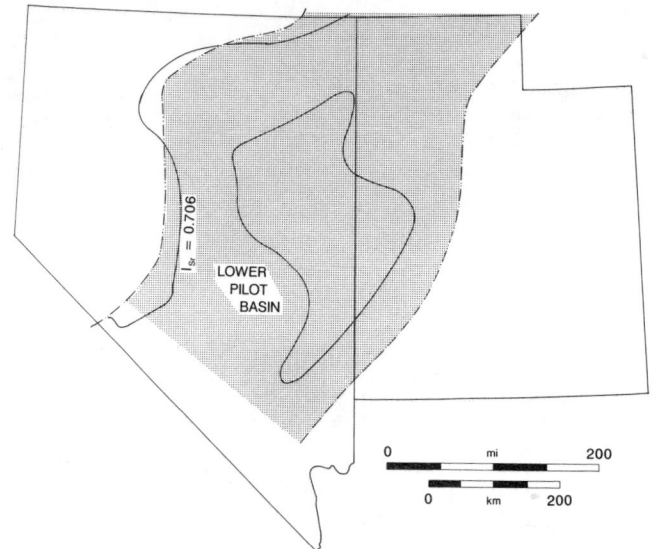

Figure 1: Location of the lower Paleozoic Cordilleran miogeocline (shaded) in Utah and Nevada. The approximate limits of the miogeocline are from Stewart and Poole (1974) and Stewart (1980). The line of initial 87Sr/86Sr = 0.706 is from Kistler and Ross (1990) and Elison and others (1990). Maximum extent of Lower Pilot (protoflysch) basin of Late Devonian age is from Sandberg and others (1988).

(1974) for shelf and slope strata of the continental margin west of the North American craton. The initiation of the miogeocline appears to have been the result of a latest Precambrian rifting event (Stewart, 1972; Stewart and Suczek, 1977; Armin and Mayer, 1983; Bond and others, 1983; 1984; 1985; Bond and Kominz, 1984).

From the Middle Cambrian to the Devonian, miogeoclinal carbonate and sandstone were deposited at gradually decreasing rates in a shallow-water shelf environment (Stewart and Poole, 1974; Stewart and Suczek, 1977). The Paleozoic succession generally thickens westward, away from the North American craton, although second-order variations in thickness of units younger than Cambrian are locally present in Nevada and Utah (Armstrong, 1968; Stewart and Poole, 1974, their figures 5-8) West of the miogeocline, in deeper water, lay chert, shale, quartzite and greenstone lavas (Stewart and Poole, 1974; for more on Paleozoic stratigraphy of the continental margin in the western U.S. see Kay, 1951; Roberts and others, 1958; Silberling and Roberts, 1962; Kay and Crawford, 1964; Armstrong, 1968; Stewart and Poole, 1974; Stewart, 1980; Dickinson and others, 1983; Stevens, 1986; and papers in Stewart, Stevens, and Fritsche, 1977; Fouch and Magathan, 1980; and this volume).

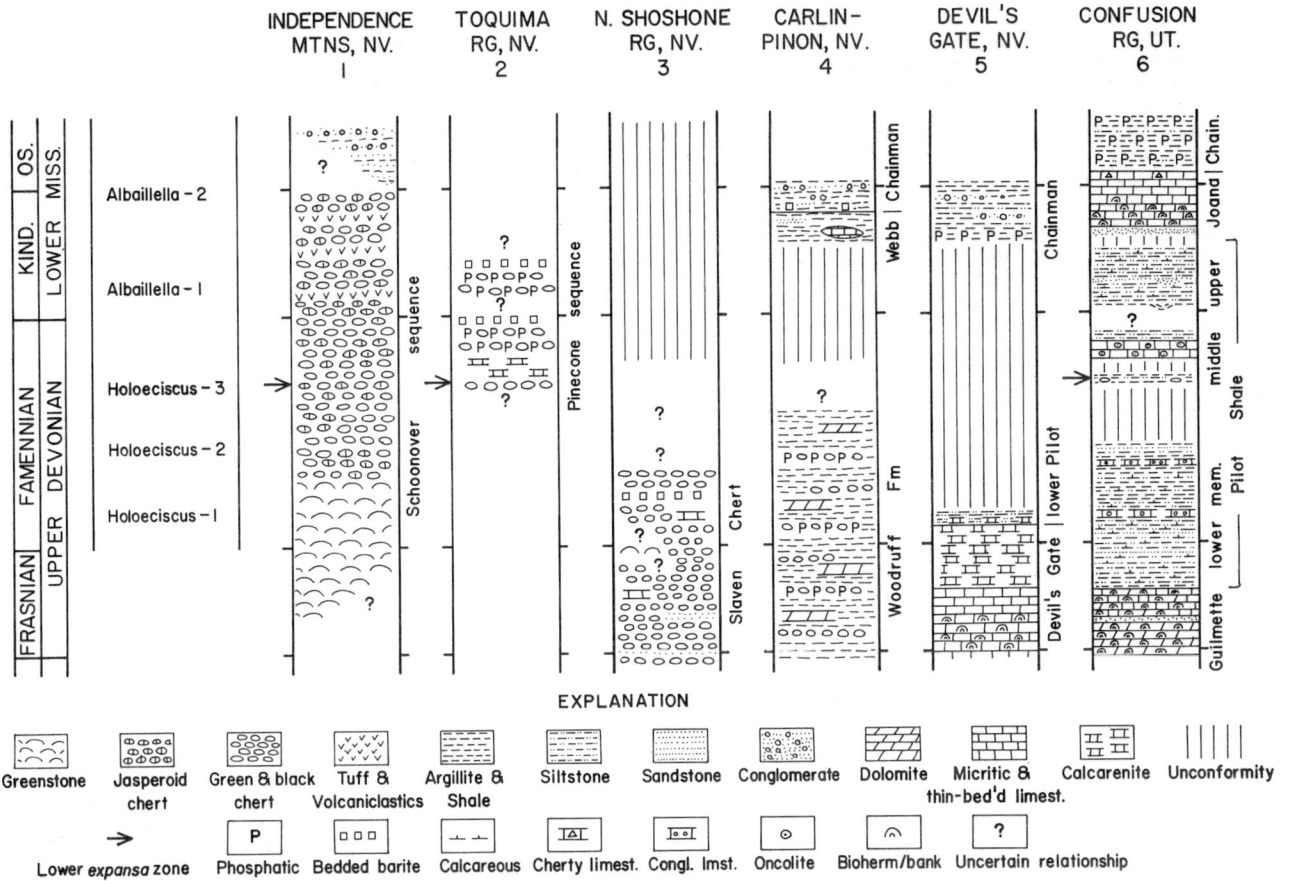

Figure 2: Correlation of several successions of Upper Devonian and Lower Mississippian strata in Nevada and Utah. Location of the sections is given on Figure 4. Section 1 lies in the Golconda allochthon and sections 2, 3 and at least the Devonian part of 4 are within the Roberts Mountains allochthon. Radiolarian and conodont zones have not yet been correlated with certainty (see Holdsworth and Jones, 1980). The Lower *expansa* zone (Ziegler and Sandberg, 1984; discussed in text) is marked by by the arrows (position in section 1 is estimated). Conodonts of this zone occur in the Woodruff Formation in the southern Fish Creek Range (Figure 4; Poole and others, 1983). The span of deposition of the Pinecone sequence is uncertain, but includes faunal zones discussed in text. Sources of data: 1) Miller and others, 1984; 2) Coles, 1988; 3) Gilluly and Gates, 1965; Poole and others, 1977; Wrucke and Jones, 1978; Jones and others, 1979; 4) Smith and Ketner, 1975; Poole and others, 1977; Johnson and Pendergast, 1981; 5) Merriam, 1963; Sandberg and Poole, 1977; Johnson and Pendergast, 1981; 6) Hose, 1966; Gutschick and Rodriguez, 1979; Gutschick and others, 1980; Sandberg and Gutschick, 1980; Sandberg and others, 1980. Radiolarian zones from Holdsworth and Jones, 1980.

Late Devonian to Early Mississippian Tectonics and Sedimentation

The Late Devonian to Early Mississippian Antler orogeny was a major post-rift tectonic event in Nevada (Roberts and others, 1958; Smith and Ketner, 1968; Burchfiel and Davis, 1972; 1975; Dickinson, 1977; 1981; Stewart, 1980; Nilsen and Stewart, 1980; Schweickert and Snyder, 1981; and others). The following summary is based on the geologic studies cited, as well as biostratigraphic studies (Sandberg and Poole, 1977; Sandberg and others, 1982; 1988), to which the reader is referred for a more detailed description of events.

During the Frasnian to early Famennian (termed herein Phase A), deposition of shallow-water carbonates on the western portion of the shelf, in eastern Nevada, gave way to basinal deposition of clastic sediments of the lower member of the Pilot Shale (Fig. 2). Gravity flows and other clastic sediments, deposited at rates of 32-160 m/m.y., were interspersed between areas of very slow deposition (4.5-6 m/m.y., rates given as compacted rock, Sandberg and Poole, 1977; Poole and others, 1977). The lower member of the Pilot Shale, which consists primarily of siltstone and mudstone, has been termed the Antler protoflysch by Poole and others (1977) and Sandberg and Poole (1977;

Figs. 1, 2), who proposed that the source of at least some clastic detritus was to the west. West of the Pilot Shale basin, in central Nevada, bedded (stratiform-stratabound) barite, alkalic volcanic rocks, and lamprophyre dikes are present in Upper Devonian pelagic strata. The igneous activity may be related to faulting that formed intraformational conglomerates in the associated barite deposits (Papke, 1984; Madrid, 1987; Dubé, 1988)

A hiatus of mid- to late Famennian age (Phase B) developed over much of eastern Nevada following deposition of the protoflysch. Erosion and non-deposition were widespread, but deposition of the lower black shale of the Leatham Member (also known as the middle member) of the Pilot Shale upon this erosion surface occurred locally. The lower black shale consists primarily of cherty shale, mudstone, and siltstone, with a lag sandstone at the base (Fig. 2; Poole, 1974; Sandberg and Poole, 1977; Gutschick and Rodriguez, 1979; for further discussion of the Upper Devonian stratigraphy of eastern Nevada see Poole and others, 1977; Sandberg and Dreesen, 1984).

In Early Mississippian time (Phase C) a deformed thrust stack of lower Paleozoic siliciclastic and volcanic rocks, the Roberts Mountains allochthon, formed a highland in central Nevada (Poole, 1974; Poole and Sandberg, 1977; Fig. 3). The outer, western portion of the shelf, which lay to the east, deepened rapidly into a narrow trough (Antler flysch trough, or foreland basin) and filled with clastic debris eroded from this highland (Roberts and others, 1958; Kay and Crawford, 1964; Burchfiel and Davis, 1972; 1975; Nilsen and Stewart, 1980; Stewart, 1980; Johnson and Pendergast, 1981; Speed and Sleep, 1982; Dickinson and others, 1983). By the Pennsylvanian, the flysch trough in Nevada had been filled, and deposition of shallow-water carbonate and siliciclastic sediment, punctuated by erosional events, resumed (see summaries in Rich, 1977; Dickinson and others, 1983).

The appearance of the Pilot Shale basin, as well as igneous activity, probable faulting, and mineralization during Phase A, imply a tectonic cause. Explanations of this tectonism invoke diverse causes. Gutschick and Rodriguez (1979, p. 53) attributed the downwarping of the shelf to compressional flexure folding. Johnson and Pendergast (1981, p. 650) also considered the Frasnian events a result of crustal shortening. In contrast, extension has been proposed by Turner (1985), Madrid (1987, p. 311 et seq.), and Dubé (1988, p. 242) to account for Frasnian to mid-Famennian volcanism and tectonics.

Most models of the Antler orogeny in Nevada focus on Phase C and attribute the rapid Mississippian subsidence of the flysch trough to flexural loading of the edge of the continental lithosphere by the thrust sheets of the Roberts Mountains allochthon (Johnson and Pendergast, 1981; Speed and Sleep, 1982;

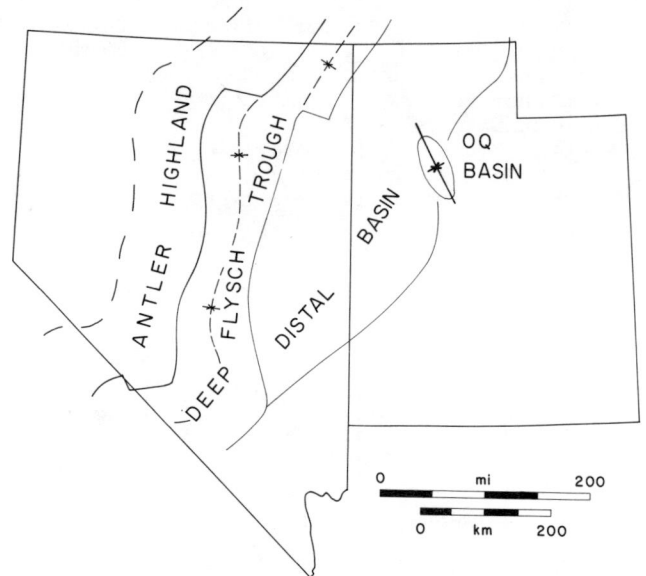

Figure 3: Location of tectonic provinces of Late Paleozoic age in Utah and Nevada (Poole, 1974; Poole and Sandberg, 1977; Jordan and Douglass, 1980; Kluth and Coney, 1981; Kluth, 1986). The Antler highland, flysch trough, and distal basin formed in the Mississippian. The Oquirrh Basin did not form until the Pennsylvanian.

Dickinson and others, 1983). Calculations by Speed and Sleep (1982) suggested that the trough was less than 200 km wide and adjacent to the thrust wedge. A flexural upwarp of a few hundred meters in amplitude would be expected 200 or more km east of the trough (Speed and Sleep, 1982). Thus, a given location on the former shelf in eastern Nevada first would have experienced uplift, followed by subsidence, as the trough-upwarp pair migrated eastward ahead of the thrust load. The starved Pinecone basin (discussed below) of latest Devonian age (Phase B), west of the eastern Nevada shelf (see Coles, 1988; Coles and Snyder, 1985), and the mid- to late Famennian unconformity (also Phase B) represent the earliest manifestation of the trough and upwarp, respectively (Johnson and Pendergast, 1981; Speed and Sleep, 1982; Dickinson and others, 1983; Goebel, 1990). Goebel (1990) used the same model to postulate existence of a second downwarp east of the upwarp and ascribed the subsidence of the shelf (Phase A) to passage of this downwarp. By the Mississippian (Phase C), the Antler highland had appeared in central Nevada and the trough had migrated eastward to become the Antler foreland basin.

STARVED DEPOSITS IN CENTRAL NEVADA

Strata deposited late in the history of the early Paleozoic continental margin provide several constraints on Famennian and Kinderhookian paleogeography of Nevada.

Rocks named the Pinecone Formation by Kay and Crawford (1964) and correlative units are associated with the Roberts Mountains allochthon, but their character and history contrast with most other allochthonous strata. The results summarized below are based upon stratigraphic and structural studies (Coles, 1988; 1989). My observations are primarily from the Toquima Range, Nevada; inferences regarding correlative units are based mainly on descriptions by other workers.

Depositional Setting

The Pinecone sequence consists, in order of decreasing abundance, of chert, phosphatic chert and argillite, and lesser argillite, bedded barite, and detrital limestone. The chert and argillite are commonly dark-gray to black. Siliciclastic detritus, other than clay-size material, was lacking in the environment of deposition. Several blocks of quartz sandstone tens of meters in size, otherwise absent in the Pinecone, may be synsedimentary slide blocks. Alternatively, they may be fault-bounded slices of other units, such as the Ordovician Vinini Formation (McKee, 1976). Some of the chert forms a distinct stratigraphic unit; the balance of the Pinecone contains a mix of all rock types owing both to original association (e.g. phosphatic chert and bedded barite) and subsequent tectonic mixing. Much of the Pinecone is now a broken formation.

The Pinecone sequence reflects pelagic, hemipelagic, and mass-gravity flow sedimentation. The phosphatic chert and argillite is best explained by deposition on the sea floor beneath a zone of surface upwelling of nutrient-rich water. The high biologic productivity in such surface waters commonly leads to deposition of sediment rich in silica, phosphate, and organic material. Comparison of the phosphatic rocks with Neogene to Recent examples suggests that the depositional setting for the Pinecone was probably bathyal and adjacent to the continent (Coles and Snyder, 1985; Coles and Varga, 1988). The argillite and detrital limestone represent hemipelagic and rare gravity flow input. No major land areas were contributing eroded material, although the possible slide blocks of quartz sandstone could be olistoliths(?) from nearby submarine paleohighs.

Age

The Pinecone sequence comprises some of the youngest rocks in the Roberts Mountains allochthon, most of which is Cambrian to Devonian strata (Roberts and others, 1958; Stewart and Poole, 1974; Madrid, 1987). Radiolarians recovered from phosphate nodules in the Pinecone are latest Devonian (late Famennian) to earliest Mississippian age (*Holoeciscus*-3 and *Albaiella*-1 faunas; B. Murchey, pers. comm., 1986). A conodont fauna from a detrital limestone outcrop within phosphatic chert is of late Famennian age (Lower *expansa* through Middle *praesulcata* zones, K. Denkler and K. Schindler, pers.

comm., 1986; equivalent to old Upper *styriacus* to *costatus* zone, Sandberg 1979; Ziegler and Sandberg, 1984). Poole and Sandberg (1975) found conodonts of the Lower *expansa* (equivalent to old Upper *styriacus*) conodont zone in a cherty limestone layer associated with phosphatic nodules in the Pinecone.

Structural history

Deformation of the Pinecone is pre-Late Mississippian in age, and the unit was incorporated into the Roberts Mountains allochthon during the Antler orogeny. The deformation in the Pinecone and the upper few tens of meters of the underlying autochthon is represented by 1) open to isoclinal folds having north-trending hingelines (main phase folds), and by 2) gently to moderately dipping faults, some of which show reverse offsets. In at least one locality in the Toquima Range, phosphatic chert, identical to dated uppermost Devonian-lowermost Mississippian chert several km distant, shows main-phase folds that are erosionally truncated and overlapped by gently-dipping, unfolded strata of the Wildcat Peak Formation (Coles, 1988). The Wildcat Peak ranges in age from Late Mississippian to Permian (Verville and others, 1985). The folds are thus Mississippian in age, and I infer that main-phase folds in the remainder of the Pinecone are also Mississippian. The faults, some imbricating Pinecone strata, are not directly dated but are consistent with the same E-W or ESE-WNW shortening as the main-phase folds. I interpret the faults and folds as representing an early Mississippian, Antler deformation.

The Pinecone sequence could have been deposited either in "piggyback" basins that developed on top of the growing Roberts Mountains allochthon, or within a foredeep in front of the allochthon. Older rocks of the Roberts Mountains allochthon have not been found structurally or depositionally beneath the Pinecone. In addition, coarse-grained detritus, expected in a piggyback setting, is rare in the Pinecone, and detritus that is present indicates provenance other than the rocks of the allochthon (Coles, 1988). These observations suggest that the Pinecone was deposited in a foredeep setting and was a late addition to the base of the allochthon as the allochthon moved eastward across the foredeep/continental slope toward the shelf. Therefore, the Pinecone was transported a lesser distance than most of the Roberts Mountains allochthon, accounting for the difference in depositional setting and structural position of the two successions, a scenario also presented by Murphy and others (1984a) and Jansma and Speed (1985) for Devonian to Mississippian rocks elsewhere in Nevada (see also Kay and Crawford, 1964; McKee, 1976; Johnson and Pendergast, 1981; Speed and Sleep, 1982; and Oldow, 1984).

Regional distribution

The Pinecone sequence formed part of a more widespread Famennian to Kinderhookian unit in Nevada (Fig. 2). The Upper Devonian portion of the Woodruff Formation of northeastern Nevada (Smith and Ketner, 1968; 1975) shows striking similarities to the Pinecone. The Woodruff consists predominantly of gray to black shale, mudstone, dolomite, and chert and contains white spheres that may be phosphate nodules. Although much of the Woodruff is older than the Pinecone, the uncertainties in age, especially for the latter unit, permit an overlap between the units (Fig. 2).

The Webb Formation is Early Mississippian (Kinderhookian, Smith and Ketner, 1968; 1975) and originally was considered to postdate thrusting and deformation of the Roberts Mountains allochthon (Smith and Ketner, 1968); more recently the Webb has been reported as also involved in the deformation (Johnson and Pendergast, 1981; Murphy and others, 1984a; 1984b; Jansma and Speed, 1985). The Webb consists chiefly of mudstone and claystone with quartz and chert silt grains, and some sandstone and limestone (Smith and Ketner, 1975). Smith and Ketner noted (1975, p. A38) that the Woodruff and Webb appear similar but "the Woodruff contains much chert, the Webb almost none," and "no units in the Woodruff are as coarse grained as some of the sandy beds in the Webb." Hereafter, I shall treat the Woodruff, Pinecone, and Webb together as representing the record of starved to hemipelagic settings that existed in Nevada prior to the formation of the Antler foreland basin.

A possible objection to the significant role of the Pinecone sequence in paleogeographic reconstruction is that deposition of the unit was not widespread or long-lived. I argue, however, that the aforementioned correlative units indicate that the detritus-poor or starved deposit accumulated at one time or another across much of central Nevada. Although deposition was not long-lived in most places, in total it did span considerable time. Hemipelagic input was significant locally; elsewhere it was absent. Chert was less abundant in the latter part of the interval as clastic detrital input increased. The Pinecone, Woodruff, and Webb units reflect an environment that contrasts markedly with those that preceded and followed.

FAMENNIAN TO EARLY KINDERHOOKIAN
PALEOGEOGRAPHY

Available biostratigraphic data for the Upper Devonian of the western U.S. provide a useful temporal framework for analysis of sedimentary rocks of the region. Many publications have documented conodont zones in Utah and Nevada, as well as areas to the north and south, and used these data to reconstruct paleogeography (Sandberg and Poole, 1977; Poole and others, 1977; Sandberg, 1976; 1979; Sandberg and Gutschick, 1979; Gutschick and Rodriguez, 1979; Sandberg and others, 1980; 1982; 1988; Sandberg and Dreesen, 1984). Although Mesozoic and Cenozoic tectonism has altered the

distribution of Upper Devonian rocks in eastern Nevada and Utah, the general paleogeographic patterns are preserved. The objective herein is to extend the paleogeographic reconstruction westward into a region where relative displacements of strata, and thus the original relationships of the depositional provinces, are less well-known, owing to additional tectonism of Paleozoic age. This inherent uncertainty in interpretation notwithstanding, it is possible, using structural and stratigraphic data, to clarify what depositional settings lay west of the rocks now in eastern Nevada and to place those settings in proper relative positions.

Depositional environments

Strata of Famennian and early Kinderhookian age record a number of environments from tidal to deep marine. Correlation of stratigraphic sections across Utah and Nevada is limited by uncertainties in ranges of different fossil groups (conodonts and radiolarians) and variation in sampling density. Nevertheless, a story emerges. Devonian strata in Nevada and Utah have been grouped into several belts (Stewart and Poole, 1974; Stewart, 1980). From east to west these belts are: 1) a carbonate and quartzite province; 2) a limestone and shale province; 3) a shale and chert province; and 4) a chert province (Stewart, 1980, p. 32). By Famennian time, however, at least one of these belts had shifted position and several new units define additional provinces.

Figure 4 is adapted from Sandberg and Dreesen (1984) and shows the distribution of the Lower *expansa* conodont zone, which falls high in the Famennian stage. The names of selected stratigraphic units that contain this zone are indicated (Fig. 4), and the units also are shown on the correlation chart (Fig. 2). An erosional event in eastern Nevada that removed parts of the Pilot Shale (unconformity on Fig. 4) is of later origin, according to Sandberg and others (1982; 1988). Sandberg and Dreesen interpreted shallow subtidal to moderately deep subtidal settings for carbonate and quartzite strata in Utah and southern Nevada (limestone pattern on Fig. 4, corresponds to carbonate and quartzite province of Stewart and Poole, 1974). Deep subtidal deposits in a dysaerobic (oxygen-poor) setting lay in eastern Nevada, as recorded in basal beds (lower black shale) of the Leatham Member (middle member) of the Pilot Shale (Fig. 4; Sandberg and Dreesen, 1984; Gutschick and Rodriguez, 1979), which lay east of the earlier position of the limestone and shale province. The Woodruff Formation and the Pinecone sequence correspond to the shale and chert province (cross hatch, Fig. 4) and resemble the poorly oxygenated strata of the Pilot, with the addition of nodules and lenses of phosphate.

The Devonian Slaven Chert (Gilluly and Gates, 1965) of the chert province is part of the Roberts Mountains allochthon. Madrid (1987) gave the age span as Early Devonian to

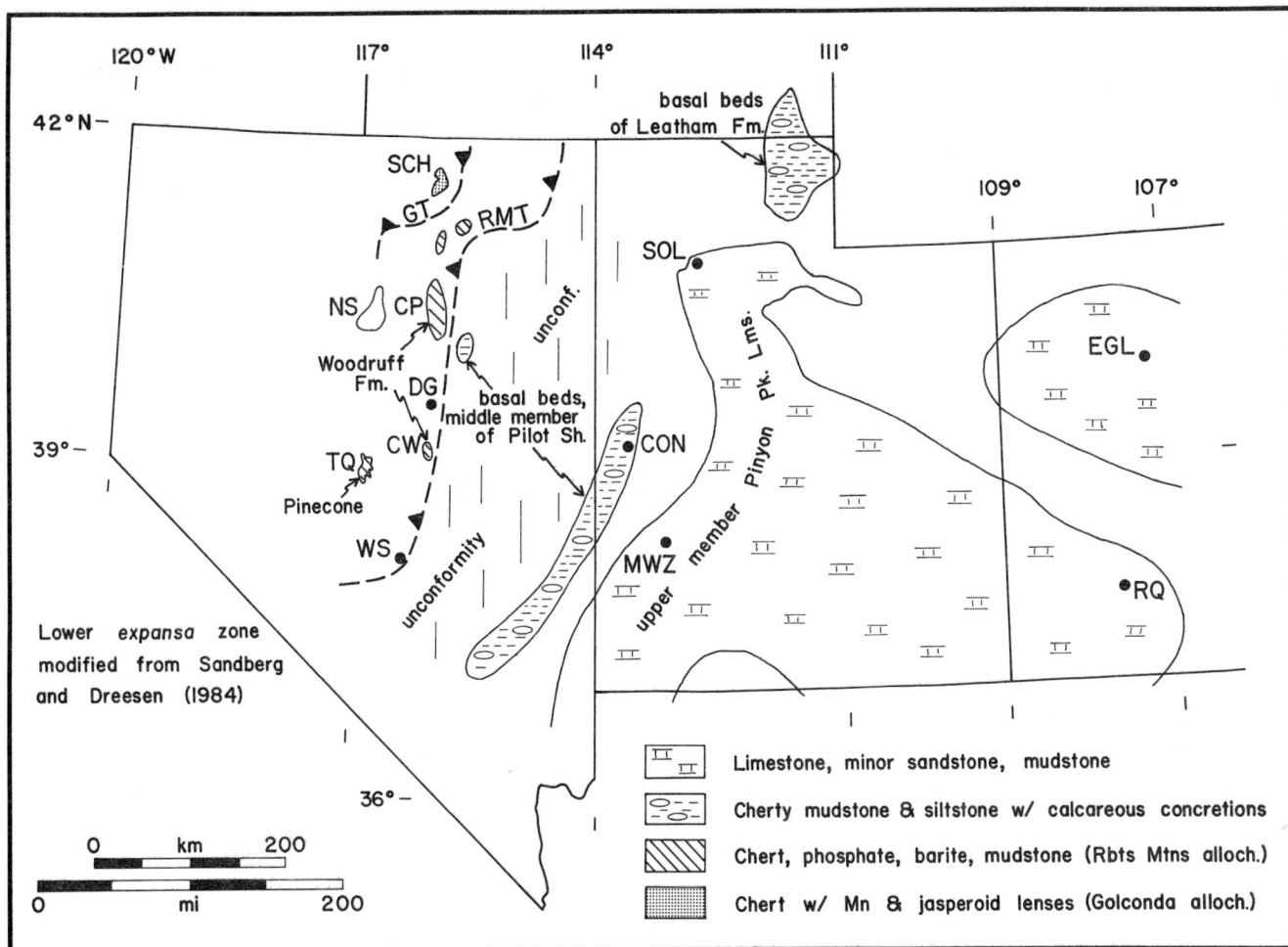

Figure 4: Lithofacies distribution of the Lower *expansa* zone (upper Famennian) in Nevada and Utah, modified from Sandberg and Dreesen (1984). The record in much of Nevada is missing owing to a younger unconformity. CON, Confusion Range; CP, Carlin-Pinon Range area; CW, Cockalorum wash, southern Fish Creek Range; DG, Devil's Gate; EGL, East Glenwood Canyon; GT, Golconda thrust; MWZ, Mowitza Mine, Star Range; NS, North Shoshone Range; RMT, Roberts Mountains thrust; RQ, Rockwood Quarry, San Juan Mountains; SCH, Schoonover sequence in Independence Mountains; SOL, South Lakeside Mountains; TQ, Toquima Range; WS, Warm Springs.

Late Devonian and reported sandstone, bedded barite, and volcanic units, as well as black chert and argillite, as young as Famennian or probable Famennian. At least part of the Scott Canyon Formation (Roberts, 1964) is also Devonian, and lies in the chert province (Stewart, 1980; Madrid 1987).

The Schoonover sequence in northeast Nevada includes Famennian cherts, which locally are manganiferous or jasperoid (Miller and others, 1984; Fig. 2). Below, I treat the Schoonover as representing oceanic sedimentation west of the Roberts Mountains allochthon (interpretation of Miller and others, 1984). Miller and others (1984) discussed possible tectonic settings for the Schoonover and the distance between it and the Roberts Mountains allochthon (see also Speed, 1979; Snyder and Brueckner, 1983; Madrid, 1987; Dubé, 1988).

Detrital Sources and Sinks and Timing of Antler Highland Development

Detrital rocks of Famennian age in Nevada and Utah indicate the Antler orogenic belt had not yet formed a subaerial highland during Phase B. The lower black shale of the Leatham Member of the Pilot Shale (within the Lower *expansa* zone, shown on Fig. 4; Sandberg and Dreesen, 1984) is part of the late Famennian depositional cycle (Sandberg and Poole, 1977) and accumulated at rates of 20 m/m.y. or less in Nevada and Utah (Gutschick and Rodriguez, 1979). A problem noted by Gutschick and Rodriguez (1979, p. 54-55) concerns the source of detritus in this unit. They thought the source was the newly-formed Antler orogenic belt to the west, and pointed out that a flysch counterpart of the Pilot should exist in areas west of the Pilot Shale and east of the orogenic belt (Fig. 5A).

A: After Figure 12 of Gutschick and Rodriguez (1979).

B: Modified from Sandberg and Dreesen (1984).

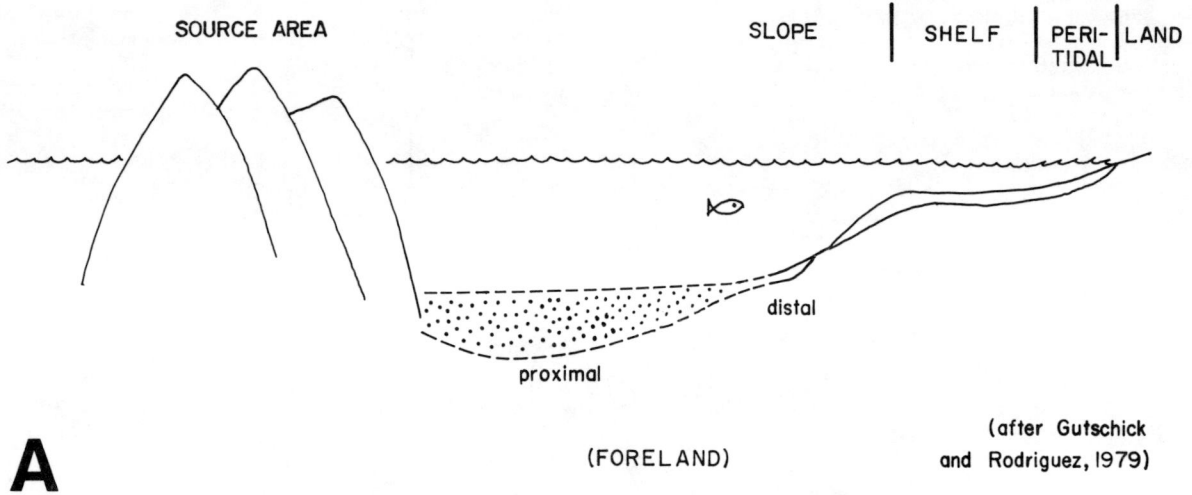

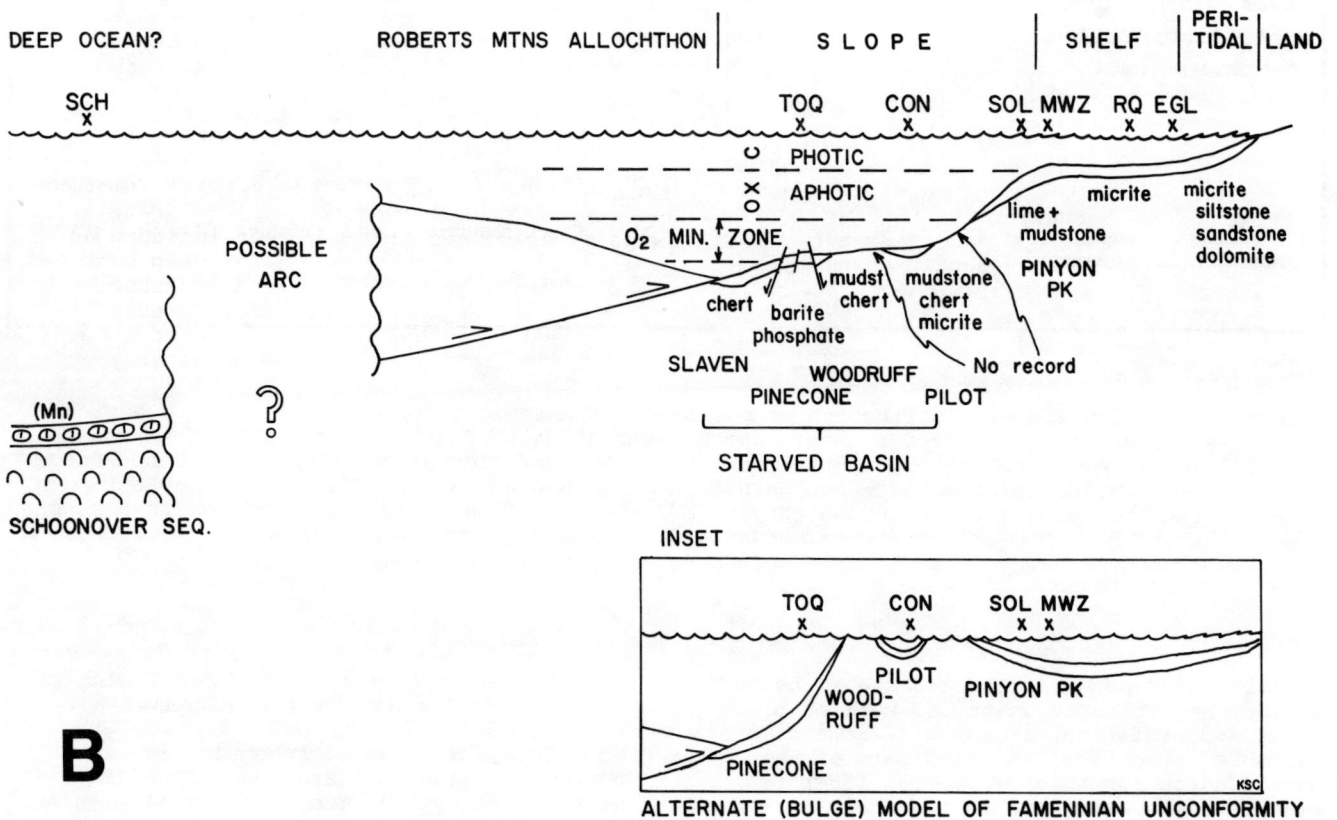

Figure 5: East-west cross sections showing alternative interpretations of the Lower *expansa* conodont zone.
A: After Figure 12 of Gutschick and Rodriguez (1979).
B: Modified from Sandberg and Dreesen (1984). Inset shows alternative interpretation of tectonic setting of the Pinecone sequence and Pilot Shale discussed in text.

Such a flysch counterpart is, however, lacking. On the contrary, the lack of coarse-grained clastics in the Pinecone sequence, which was deposited west of the Pilot Shale and near the future site of the Antler orogenic belt, suggests that fine-grained siliciclastic input into the Pilot was not accompanied by flysch sedimentation to the west (Fig. 2). The detrital component in Lower *expansa* zone strata decreases westward across Utah and Nevada (Figs. 4, 5B), rather than increasing as predicted by Gutschick and Rodriguez (1979).

A critical element of the late Famennian reconstruction of Gutschick and Rodriguez (1979) is an emergent Antler orogenic belt. The Antler highland was apparently not yet the source of large amounts of flysch (cf. phase 1 of Johnson and Pendergast, 1981). Either the Antler highland was not yet subaerially exposed, or it did not have relief sufficient to shed large volumes of detritus eastward at this time. The source of the shale in the Leatham Member of the Pilot Shale must therefore have lain along the north-south strike of the basin, to the east on the continent, and/or to the west on the shelf between Pilot and Woodruff. The apparent lack of a highland in the late Famennian suggests it also was not the westerly source, inferred by Poole and others (1977), and Sandberg and Poole (1977), of the older Frasnian protoflysch (cf. Sandberg and others, 1988). Provenance and paleocurrent studies of the lower and middle members of the Pilot Shale and correlative units may indicate the source and transport direction of the clastic sediment.

Two contrasting models (Fig. 5B) are possible for the paleogeography in eastern Nevada at the time of the upper Famennian Lower *expansa* (old Upper *P. styriacus*, Sandberg, 1979; Ziegler and Sandberg, 1984) conodont zone. Sandberg and Dreesen (1984) showed a nondepositional (or subsequently eroded) slope in their reconstruction of the Lower *expansa* zone (shown in modified form in Figure 5B). These authors analyzed the biofacies without implying a specific tectonic cause for the geography (Sandberg, 1976; Sandberg and Dreesen, 1984). Others (Sandberg and Poole, 1977; Gutschick and Rodriguez, 1979) have postulated a subaerially emergent area just prior to Lower *expansa* time. (Inset in Figure 5B). Did the emergent area persist, or did it become a submerged slope?

The depositional and tectonic history of strata, such as the Pinecone sequence and the Woodruff Formation, that lay west of the Pilot basin are critical to understanding the subsidence and uplift history of the continental margin. If a subaerial erosion surface existed on the outer shelf to the east during deposition of the Pinecone sequence, the only effect on the Pinecone seems to have been deposition of rare detrital limestone and/or slumping of larger slide blocks of carbonate into the sequence (see Coles, 1988). Shallow-water carbonates, which might be expected on the western margin

of a subaerial shoal on the outer shelf in eastern Nevada (Fig. 4), have not been reported. The submerged slope model for eastern Nevada appears more likely at present. Study of sedimentary facies and stratigraphy of the Woodruff Formation would clarify whether a shelf (emergent) or slope (submerged) environment lay between the Woodruff and the Leatham Member of the Pilot Shale.

Relative Positions and Spacing of Facies.

Significant questions concerning Famennian paleogeography are: 1) at the time of deposition where were Famennian strata now in the Roberts Mountains allochthon and the Golconda allochthon? and 2) how far were they from the shallower-water rocks now in eastern Nevada?

Two lines of evidence indicate that the Pinecone is less far-travelled than the bulk of the Roberts Mountains allochthon. First, some rocks now in the allochthon were already undergoing deformation by the late Famennian (Murphy and others, 1984a), whereas the Pinecone, the Mississippian Webb Formation, and possibly the youngest strata of the Slaven Chert and the Woodruff Formation, postdated the onset of deformation and later were incorporated into the allochthon (Johnson and Pendergast, 1981; Speed and Sleep, 1982; 1983; Poole and others, 1983; Jansma and Speed, 1985). Late Famennian paleogeographic elements thus included an incipient allochthon and a foredeep destined to be incorporated into the allochthon by early Mississippian time. The foredeep corresponds to the starved basin province predicted in stage A of the model of Speed and Sleep (1982; see also Jansma and Speed, 1985). I infer that the shorter time span of deformation implies lesser transport of the starved foredeep sediments than for the older strata in the allochthon.

Could the Roberts Mountains allochthon still be very far-travelled, with some parts simply more remote than others? The second line of evidence involves phosphatic cherts in the Roberts Mountains allochthon and other terranes. The settings of comparable Neogene to Recent phosphatic cherts imply that the Devonian phosphatic cherts in the terranes most likely formed along the continental margin of western North America (Varga, 1982; Coles and Snyder, 1985; Coles and Varga, 1988). The province extended along the length of western margin of North America; the east-west width of the province would depend upon the width of the zone of upwelling and high surface productivity. I speculate, based upon younger analogues, that width was the order of a few hundred km or less.

Evidence collected by Madrid (1987) also supports a local origin for the Roberts Mountains allochthon. Madrid examined in detail the allochthon in north-central Nevada and argued, based upon detrital provenance and basalt chemistry, that the strata of the allochthon accumulated upon thinned continental crust of North America, rather than oceanic crust beyond the edge of the continent. Turner (1985) described a Late Devonian extensional event that controlled barite deposition along the western margin of North America, and considered the Roberts Mountains allochthon a displaced fragment of the same barite province that is preserved in autochthonous parts of the Canadian Cordillera.

Phosphatic Borderland

I have argued that the allochthon was present west of the starved province during the late Famennian (Fig. 5B). Nevertheless, phosphate deposition does not require a continuous barrier or a restricted basin. Phosphatic chert of Holocene age has formed along the upper continental slope of Peru-Chile facing the open ocean. This is an area of strong wind-driven upwelling, and the deposits have formed in association with the zone of minimum dissolved O_2 (Burnett and others, 1983). I envision a similar setting for the Famennian of the western U.S., with the addition of a submerged borderland of highs and básins (Fig. 6A) analogous to the Neogene setting of coastal California, where upwelling and phosphate deposition occurred (e.g. Pisciotto and Garrison, 1981; Garrison and others, 1987). This reconstruction is consistent with the lack of an Antler highland at this time.

Phosphate in the upper Famennian to Kinderhookian Pinecone sequence, the upper Kinderhookian basal Webb Formation at Devil's Gate, Nevada, and the Osagean to Meramecian phosphatic member of the Chainman Shale in the Confusion Range, Utah (Fig. 2, sections 2, 4, 5, and 6, see references listed in caption; Fig. 4), suggests that water depths and conditions favorable for phosphate deposition migrated eastward across Nevada ahead of the advancing thrust load. Erosion occurred east of phosphatic chert deposition in the Famennian, and shoaling was present east of phosphatic deposition in the late Kinderhook as well. This shoal, the Joana Limestone bank (Fig. 2), is in eastern Nevada and western Utah (Gutschick and others, 1980; Johnson and Pendergast, 1981). Eastward migration of phosphate deposition during the Antler orogeny parallels the migration of other environments, including clastic sedimentation (Fig. 6B). Bradley and Kusky (1986) used cratonward migration of the features of a foreland environment, including erosion, faulting of the carbonate platform, onset of shale deposition, and onset of turbidite deposition, to track the motion of the Ordovician Taconic allochthon westward into New York state. A similar analysis would be of interest in the case of the Antler orogeny; phosphate deposition, however, also is influenced by oceanic circulation patterns, amount of clastic input, and vertical crustal movements. Speed and Sleep (1982, p. 825) noted that other migrating features of the Antler foreland were not dated precisely enough to estimate rate of motion the Roberts Mountains allochthon.

Paleogeographic Reconstruction

Key elements in a paleogeographic reconstruction for the late Famennian and earliest Kinderhookian are:

• the shallow to deep subtidal environments documented by Sandberg and coworkers;

• detritus-starved sedimentation west of, and in places overlapping, the dysaerobic zone;

• source(s) of the carbonate debris and rare detritus transported into the otherwise detritus-starved province;

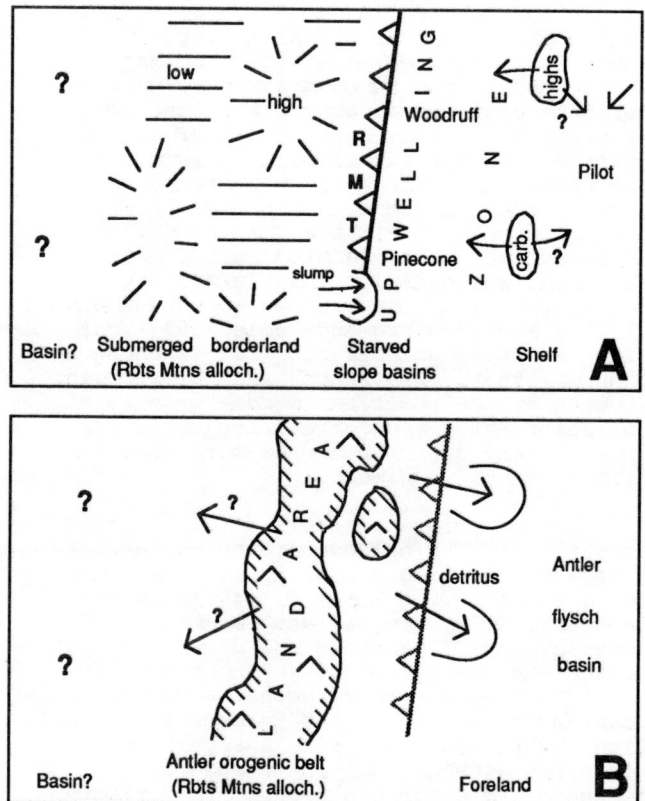

Figure 6: Paleogeographic sketch maps of central Nevada.
A (top) shows approximately late Famennian time. The Roberts Mountains allochthon is still submerged; uplifted carbonate highs provided detritus to the Woodruff-Pinecone basin. Transport of detritus eastward from these highs is speculative. RMT is Roberts Mountains thrust.
B (bottom) shows late Kinderhook time. The Roberts Mountains thrust has ceased moving (some workers, however, considered it active until the Osagean). It is uncertain whether detritus was shed westward from the subaerial Antler orogenic belt, and whether it was transported into the Schoonover-Havallah basin or some other basin (see text).

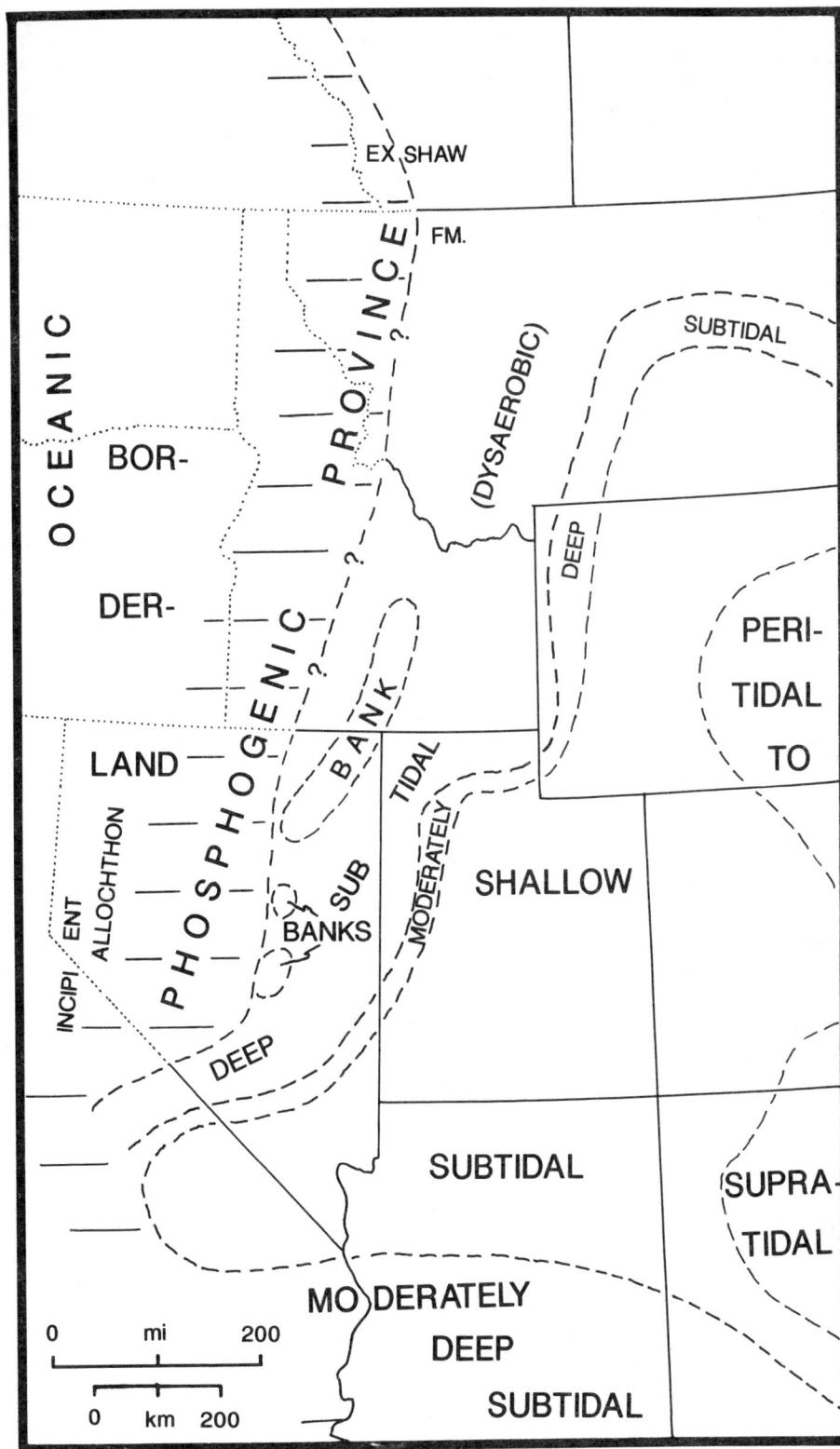

Figure 7: Paleogeography of the Lower *expansa* conodont zone (late Famennian), modified from Sandberg and Dreesen (1984) and Sandberg and others (1988).

• the onset of deformation and transport of the Roberts Mountains allochthon, but the lack of significant derived detritus, other than slump blocks(?). The Roberts Mountains allochthon probably did not form a continuous barrier to ocean circulation. Some rocks destined to be accreted to the allochthon still were being deposited to the east;

• upwelling, high productivity, and phosphate deposition scattered over parts of the continental margin and borderland;

• an oceanic province that included volcanism and pelagic sedimentation.

Parts of the borderland and pelagic realm were later fragmented and now are found in one or more terranes.

The reconstruction presented here (Figs. 5B, 6A, 7) is a modification of the Lower *expansa* zone model of Sandberg and others (1988). This zone illustrates the setting during the late Famennian and earliest Kinderhookian, given the change expected in the position of the Roberts Mountains allochthon as it migrated eastward (cf. Johnson and Pendergast, 1981; Speed and Sleep, 1982).

Uncertainties limit the resolution of paleogeography in Figure 7. Several stratigraphic units are not precisely dated by fossils but were taken into consideration because they appear likely to overlap the Lower *expansa* zone. The base map has not been corrected for post-Mississippian deformation, and the location of depositional provinces in the western part of the map is schematic, as paleolatitude data are lacking. A volcanic arc was active adjacent to North America at this time, according to some previous workers, though this remains controversial (Moores, 1970; Burchfiel and Davis, 1972; 1975; Churkin, 1974; Dickinson, 1977; 1981; Stewart, 1980; Schweickert and Snyder, 1981; Speed and Sleep, 1982; Dickinson and others, 1983; Miller and others, 1984; Dubé, 1988; Burchfiel and Royden, 1991). As noted by Madrid (1987), the Roberts Mountains allochthon and areas to the east do not record significant arc volcanism, although alkalic volcanic rocks appear throughout much of the succession in the Roberts Mountains allochthon. Any arc present in the region depicted on Figure 7 presumably lay somewhere in the oceanic province. I regard (Coles, 1988) the Schoonover sequence also as part of the oceanic province, following the conclusions of Miller and others (1984); the Schoonover basin was distinct from the part of the borderland that included the incipient Roberts Mountains allochthon (Fig. 5B).

The paleogeographic reconstruction yields several important constraints on the Antler orogeny. The Roberts Mountains allochthon was still west of the Toquima Range area in the late Famennian, and the Antler highland was not yet shedding siliciclastic material. Erosion of highs (shown as submarine on Figures 6A and 7) on the outer shelf supplied at least some carbonate debris and fine-grained siliciclastic material to the Pinecone/Woodruff in the foredeep between the allochthon and the shelf.

The paleogeographic reconstruction could be refined by documentation of the facies relations, sources and transport directions of detritus, and structural history of units such as the Pilot Shale, the Woodruff Formation, and the Pinecone sequence. A worthwhile, though ambitious, goal for future work in Famennian paleogeography is the compilation of palinspastic maps. The difficulties involved in estimating Cenozoic extension and the crustal shortening of several earlier episodes are great, yet the understanding of paleogeography cannot be considered complete without such effort. Armstrong (1968) and Stewart and Poole (1974) made palinspastic base maps for discussions of the Paleozoic; refinement of these maps could be the next step. The work of Marzolf (1988) on the Triassic and Jurassic of the southern Great Basin and adjacent regions is an example of how effective the restoration of stratigraphic units to original relative positions can be in answering outstanding questions and directing further research.

CONCLUSIONS

Four major depositional provinces (a deep subtidal province of cherty shale and mudstone, a bathyal(?) phosphatic chert province, a continental borderland of submerged highs and lows, and a pelagic oceanic realm) lay in Nevada late in the Famennian. Stratigraphic, sedimentologic, structural, and paleontologic evidence indicates that east-directed folding and thrusting were underway in Nevada by the late Famennian but were still confined to the continental borderland. The Antler orogenic belt still was submerged and a foredeep to the east was dominated by chert and phosphate rather than carbonate or siliciclastic deposition. The foredeep strata were later overridden by and accreted to the Roberts Mountains allochthon during the climactic, early Mississippian phase of the Antler orogeny.

ACKNOWLEDGMENTS

I thank Hannes Brueckner, Nick Christie-Blick, Gerard Bond, Ian Dalziel, Tom Dubé, Richard Hanson, Raul Madrid, Steve Marshak, François Megard, Barney Poole, and Doug Rhett for discussions and comments on these ideas in the office and field. Walt Snyder and Rich Schweickert have generously shared many of their ideas and insights. Lyn Karwaki and Kaj Hoernle ably and enthusiastically assisted me in the field. Chris Coles gave me a great view of the Monterey Formation in his Cessna 180, N2991K. Mary Ann Luckman and Connie Sancetta helped me extract microfossils, and Bonnie Murchey, Anita Harris, Kirk Denkler and Kate Schindler of the U.S. Geological Survey provided crucial fossil identifications. John Cooper, Bill Melhorn, and Walt Snyder read the manuscript. Portions of this study were supported by the Department of Geological Sciences at Columbia University, a Shell Companies Foundation Graduate Research Fellowship, the Geological Society of America, Sigma Xi (the Scientific Research Society), Phillips Petroleum Company, and the American Association of Petroleum Geologists.

REFERENCES CITED

Armstrong, R. L., 1968, The Cordilleran miogeosyncline in Nevada and Utah: Utah Geological and Mineralogical Survey Bulletin 78, 58 p.

Armin, R. A., and Mayer, L., 1983, Subsidence analysis of the Cordilleran miogeocline: Implications for timing of late Proterozoic rifting and amount of extension: Geology, v. 11, p. 702-705.

Bond, G. C., and Kominz, M. A., 1984, Construction of tectonic subsidence curves for the early Paleozoic miogeocline, southern Canadian Rocky Mountains: Implications for subsidence mechanisms, age of breakup, and crustal thinning: Geological Society of America Bulletin, v. 95, p. 155-173.

Bond, G. C., Kominz, M. A., and Devlin, W. J., 1983, Thermal subsidence and eustasy in the lower Paleozoic miogeocline in western North America: Nature, v. 306, p. 775-779.

Bond, G. C., Nickeson, P. A., and Kominz, M. A., 1984, Breakup of a supercontinent between 625 Ma and 550 Ma: New evidence and implications for continental histories: Earth and Planetary Science Letters, v. 70, p. 325-346.

Bond, G. C., Christie-Blick, N., Kominz, M. A., and Devlin, W. J., 1985, An early Cambrian rift to post-rift transition in the Cordillera of western North America: Nature, v. 316, p. 742-745.

Bradley, D. C., and Kusky, T. M., 1986, Geologic evidence for rate of plate convergence during the Taconic arc-continent collision: Journal of Geology, v. 94, p. 667-681.

Burchfiel, B. C., and Davis, G. A., 1972, Structural framework and evolution of the southern part of the Cordilleran orogen, western United States: American, Journal of Science, v. 272, p. 97-118.

Burchfiel, B. C., and Davis, G. A., 1975, Nature and controls of Cordilleran orogenesis, western United States: Extensions of an earlier synthesis: American Journal of Science, v. 275-A, p. 363-396.

Burchfiel, B. C., and Royden, L. H., 1991, Antler orogeny: A Mediterranean-type orogeny: Geology, v. 19, p. 66-69.

Burnett, W. C., Roe, K. K., and Piper, D. Z., 1983, Upwelling and phosphorite formation in the ocean, *in* Suess, E., and Thiede, J., eds., Coastal Upwelling, Part A: New York, Plenum Press, p. 377-397.

Churkin, M., Jr., 1974, Paleozoic marginal ocean basin-volcanic arc systems in the Cordilleran foldbelt, *in* Dott, R. H., Jr., and Shaver, R. H., eds., Modern and Ancient Geosynclinal Sedimentation: Tulsa, Oklahoma, Society of Economic Paleontologists and Mineralogists Special Publication 19, p. 174-192.

Coles, K. S., 1988, Stratigraphy and structure of the Pinecone sequence, Roberts Mountains allochthon, Nevada, and aspects of mid-Paleozoic sedimentation and tectonics in the Cordilleran geosyncline: [Ph.D. thesis] New York, Columbia University, 246 p.

Coles, K. S., compiler, 1989, Geologic map of Northumberland Pass area, Toquima Range, Nevada [scale 1:24,000]: Reno, Nevada, Nevada Bureau of Mines and Geology Open-File Report 89-2.

Coles, K. S., and Snyder, W. S., 1985, Significance of lower and middle Paleozoic phosphatic chert in the Toquima Range, central Nevada: Geology, v. 13, p. 573-576.

Coles, K. S., and Varga, R. J., 1988, Early to middle Paleozoic phosphogenic province in terranes of the southern Cordillera, western United States: American Journal of Science, v. 288, pp. 891-924.

Crawford, J. P., 1958, Structure and stratigraphy of the Northumberland-Mill Canyon area, northern Toquima Range, Nevada: [Ph.D. thesis] New York, Columbia University, 59 p.

Dickinson, W. R., 1977, Paleozoic plate tectonics and the evolution of the Cordilleran continental margin, *in* Stewart, J. H., Stevens, C. H., and Fritsche, A. E., eds., Paleozoic Paleogeography of the Western U.S.: Los Angeles, Pacific Section Society of Economic Paleontologists and Mineralogists, Pacific Coast Paleogeography Symposium 1, p. 137-155.

Dickinson, W. R., 1981, Plate tectonics and the continental margin of California, *in* Ernst, W. G., ed., The Geotectonic Development of California: Rubey Volume 1: Englewood Cliffs, New Jersey, Prentice-Hall, p. 1-28.

Dickinson, W. R., Harbaugh, D. W., Saller, A. H., Heller, P. L., and Snyder, W. S., 1983, Detrital modes of upper Paleozoic sandstones derived from Antler orogen in Nevada: Implications for nature of Antler orogeny: American Journal of Science, v. 283, p. 481-509.

Dubé, T. E., 1988, Tectonic significance of Upper Devonian igneous rocks and bedded barite, Roberts Mountains allochthon, Nevada, U.S.A., *in* McMillan, N. J., Embry, A. F., and Glass, D. J., eds., Devonian of the World, Proceedings of the Second International Symposium on the Devonian System, Volume II, Sedimentation: Calgary, Alberta, Canada, Canadian Society of Petroleum Geologists Memoir 14, v. II, p. 235-249.

Elison, M. W., Speed, R. C., and Kistler, R. W., 1990, Geologic and isotopic constraints on crustal structure of the northern Great Basin: Geological Society of America Bulletin, v. 102, p. 1077-1092.

Farmer, G. L., and DePaolo, D. J., 1983, Origin of Mesozoic and Tertiary granite in the western United States and implications for pre-Mesozoic crustal structure, 1. Nd and Sr isotopic studies in the geocline of the northern Great Basin: Journal of Geophysical Research, v. 88, p. 3379-3401.

Fouch, T. D., and Magathan, E. R., eds., 1980, Paleozoic Paleogeography of the West-Central U.S.: Denver, Colorado, Rocky Mountain Section Society of Economic Paleontologists and Mineralogists, Rocky Mountain Paleogeography Symposium 1, 431 p.

Garrison, R. E., Kastner, M., and Kolodny, Y., 1987, Phosphorite and phosphatic rocks in the Monterey Formation and related Miocene units, coastal California in Ingersoll, R. V., and Ernst, W. G., eds., Cenozoic Basin Development of Coastal California: Rubey Volume 6: Englewood Cliffs, New Jersey, Prentice-Hall, p. 348-381.

Gilluly, J., and Gates, O., 1965, Tectonic and igneous geology of the northern Shoshone Range, Nevada: U.S. Geological Survey Professional Paper 465, 153 p.

Goebel, K. A., 1990, Stratigraphic relationships associated with forebulge migration, late Devonian to early Mississippian Antler foreland, eastern Nevada: Geological Society of America Abstracts with Programs, v. 22, no. 7, p. A321.

Gutschick, R. C., and Rodriguez, J., 1979, Biostratigraphy of the Pilot Shale (Devonian-Mississippian) and contemporaneous strata in Utah, Nevada, and Montana: Brigham Young University Geology Studies, v. 26, part 1, p. 37-63.

Gutschick, R. C., Sandberg, C. A., and Sando, W. J., 1980, Mississippian shelf margin and carbonate platform from Montana to Nevada, in Fouch, T. D., and Magathan, E. R., eds., Paleozoic Paleogeography of the West-Central U.S.: Denver, Colorado, Rocky Mountain Section Society of Economic Paleontologists and Mineralogists, Rocky Mountain Paleogeography Symposium 1, p. 111-128.

Holdsworth, B. K., and Jones, D. L., 1980, Preliminary radiolarian zonation for Late Devonian through Permian time: Geology, v. 8, p. 281-285.

Hose, R. K., 1966, Devonian stratigraphy of the Confusion Range, west-central Utah: U.S. Geological Survey Professional Paper 550B, B36-B41.

Jansma, P. E., and Speed, R. C., 1985, Antler foreland basin tectonics: New data: Geological Society of America Abstracts with Programs, v. 17, p. 363.

Johnson, J. G., and Pendergast, A. G., 1981, Timing and mode of emplacement of the Roberts Mountains allochthon, Antler orogeny: Geological Society of America Bulletin, v. 92, p. 648-658.

Jones, D. L., Wrucke, C. T., Holdsworth, B., and Suczek, C. A., 1979, Greenstone in Devonian Slaven Chert in north-central Nevada: U.S. Geological Survey Professional Paper 1150, p. 81.

Jordan, T. E., and Douglass, R. C., 1980, Paleogeography and structural development of the Late Pennsylvanian to Early Permian Oquirrh Basin, northwestern Utah, in Fouch, T. D., and Magathan, E. R., eds., Paleozoic Paleogeography of the West-Central U.S.: Denver, Colorado, Rocky Mountain Section Society of Economic Paleontologists and Mineralogists, Rocky Mountain Paleogeography Symposium 1, p. 217-238.

Kay, M., 1951, North American geosynclines: Geological Society of America Memoir 48, 143 p.

Kay, M., and Crawford, J. P., 1964, Paleozoic facies from the miogeosynclinal to the eugeosynclinal belts in thrust slices, central Nevada: Geological Society of America Bulletin, v. 75, p. 425-454.

Kistler, R. W., and Peterman, Z. E., 1978, Reconstruction of crustal blocks of California on the basis of initial strontium isotopic compositions of Mesozoic granitic rocks: U.S. Geological Survey Professional Paper 1071, 17 p.

Kistler, R. W., and Ross, D. C., 1990, A strontium isotopic study of plutons and associated rocks of the southern Sierra Nevada and vicinity, California: U.S. Geological Survey Bulletin 1920, 20 p.

Kluth, C. F., 1986, Plate tectonics of the ancestral Rocky Mountains, in Peterson, James A., ed., Paleotectonics and Sedimentation in the Rocky Mountains: Tulsa, Oklahoma, American Association of Petroleum Geologists Memoir.41, p. 353-369.

Kluth, C. F., and Coney, P. J., 1981, Plate tectonics of the ancestral Rocky Mountains: Geology, v. 9, p. 10-15.

Madrid, R. J., 1987, Stratigraphy of the Roberts Mountains allochthon in north-central Nevada: [Ph.D. thesis] Stanford, California, Stanford University, 342 p.

Marzolf, J. E., 1988, Reconstruction of extensionally dismembered lower Mesozoic sedimentary basins: Geological Society of America Abstracts with Programs, v. 20, no. 3, p. 178 and 211.

McKee, E. H., 1976, Geology of the northern part of the Toquima Range, Lander, Eureka, and Nye Counties, Nevada: U.S. Geological Survey Professional Paper 931, 49 p.

Merriam, C. W., 1963, Paleozoic rocks of Antelope Valley, Eureka and Nye Counties, Nevada: U.S. Geological Survey Professional Paper 423, 67 p.

Miller, E. L., Holdsworth, B. K., Whiteford, W. B., and Rodgers, D., 1984, Stratigraphy and structure of the Schoonover sequence, northeastern Nevada: Implications for Paleozoic plate-margin tectonics: Geological Society of America Bulletin, v. 95, p. 1063-1076.

Moores, E., 1970, Ultramafics and orogeny, with models of the U.S. Cordillera and Tethys: Nature, v. 228, p. 837-842.

Murchey, B. L., Madrid, R. J., and Poole, F. G., 1987, Paleozoic bedded barite associated with chert in western North America, in Hein, J. R., ed., Siliceous Sedimentary Rock-Hosted Ores and

Petroleum: New York, Van Nostrand Reinhold, p. 269-283.

Murphy, M. A., Power, J. D., and Johnson, J. G., 1984a, Evidence for Late Devonian movement within the Roberts Mountains allochthon, Roberts Mountains, Nevada: Geology, v. 12, p. 20-23.

Murphy, M. A., Power, J. D., and Johnson, J. G., 1984b, Reply on Evidence for Late Devonian movement within the Roberts Mountains allochthon, Roberts Mountains, Nevada: Geology, v. 12, p. 445-446.

Nilsen, T. H., and Stewart, J. H., 1980, The Antler orogeny -- Mid-Paleozoic tectonism in western North America: Geology, v. 8, p. 298-302.

Oldow, J. S., 1984, Spatial variability in the structure of the Roberts Mountains allochthon, western Nevada: Geological Society of America Bulletin, v. 95, p. 174-185.

Papke, K. G., 1984, Barite in Nevada: Nevada Bureau of Mines and Geology Bulletin 98, 125 p.

Pisciotto, K. A., and Garrison, R. E., 1981, Lithofacies and depositional environments of the Monterey Formation, California, in Garrison, R. E., and Douglas, R. G., eds., The Monterey Formation and Related Siliceous Rocks of California: Los Angeles, Pacific Section Society of Economic Paleontologists and Mineralogists, Publication 15, p. 97-122.

Poole, F. G., 1974, Flysch deposits of Antler Foreland basin, western United States, in Dickinson, W. R., ed., Tectonics and Sedimentation: Tulsa, Oklahoma, Society of Economic Paleontologists and Mineralogists Special Publication 22, p. 58-82.

Poole, F. G., and Sandberg, C. A., 1975, Allochthonous Devonian eugeosynclinal rocks in Toquima Range, central Nevada, Geological Society of America Abstracts with Programs, v. 7, p. 361.

Poole, F. G., and Sandberg, C. A., 1977, Mississippian paleogeography of the western United States, in Stewart, J. H., Stevens, C. H., and Fritsche, A. E., eds., Paleozoic Paleogeography of the Western U.S.: Los Angeles, Pacific Section Society of Economic Paleontologists and Mineralogists, Pacific Coast Paleogeography Symposium 1, p. 67-85.

Poole, F. G., Sandberg, C. A., and Boucot, A. J., 1977, Silurian and Devonian paleogeography of the western United States, in Stewart, J. H., Stevens, C. H., and Fritsche, A. E., eds., Paleozoic Paleogeography of the Western U.S.: Los Angeles, Pacific Section Society of Economic Paleontologists and Mineralogists, Pacific Coast Paleogeography Symposium 1, p. 39-65.

Poole, F. G., Sandberg, C. A., and Green, G. N., 1983, Allochthonous Devonian eugeosynclinal rocks in southern Fish Creek Range of central Nevada: Geological Society of America Abstracts with Programs, v. 15, p. 304.

Raymond, L. A., 1984, Classification of melanges, in Raymond, L. A., ed., Melanges: Their Nature Origin, and Significance: Geological Society of America Special Paper 198, p. 7-20.

Rich, M., 1977, Pennsylvanian paleogeographic patterns in the western U. S., in Stewart, J. H., Stevens, C. H., and Fritsche, A. E., eds., Paleozoic Paleogeography of the Western U.S.: Los Angeles, Pacific Section Society of Economic Paleontologists and Mineralogists, Pacific Coast Paleogeography Symposium 1, p. 87-111.

Roberts, R. J., 1964, Stratigraphy and structure of the Antler Peak quadrangle, Humboldt and Lander Counties, Nevada: U.S. Geological Survey Professional Paper 459-A, 93 p.

Roberts, R. J., Hotz, P. E., Gilluly, J., and Ferguson, H. G., 1958, Paleozoic rocks of north central Nevada: American Association of Petroleum Geologists Bulletin, v. 42, p. 2813-2857.

Sandberg, C. A., 1976, Conodont biofacies of Late Devonian Polygnathus styriacus zone in western United States, in Barnes, C. R., ed., Conodont Paleoecology: Geological Association of Canada Special Paper 15, p. 171-186.

Sandberg, C. A., 1979, Devonian and Lower Mississippian conodont zonation of the Great Basin and Rocky Mountains: Brigham Young University Geology Studies, v. 26, part 3, p. 87-105.

Sandberg, C. A., and Dreesen, R., 1984, Late Devonian icriodontid biofacies models and alternate shallow-water conodont zonation, in Clark, D. L., ed., Conodont Biofacies and Provincialism: Geological Society of America Special Paper 196, p. 143-169.

Sandberg, and Gutschick, R. C., 1979, Guide to conodont biostratigraphy of Upper Devonian and Mississippian rocks along the Wasatch Front and Cordilleran hingeline, Utah: Brigham Young University Geology Studies, v. 26, part 3, p. 107-133.

Sandberg, C. A., and Gutschick, R. C., 1980, Sedimentation and biostratigraphy of Osagean and Meramecian starved basin and foreslope, in Fouch, T. D., and Magathan, E. R., eds., Paleozoic Paleogeography of the West-Central U.S.: Denver, Colorado, Rocky Mountain Section Society of Economic Paleontologists and Mineralogists, Rocky Mountain Paleogeography Symposium 1, p. 129-147.

Sandberg, C. A., and Poole, F. G., 1977, Conodont biostratigraphy and depositional complexes of Upper Devonian cratonic-platform and continental-shelf rocks in the western United States, in Murphy, M. A., Berry, W. B. N., and Sandberg, C. A., eds., Western North America: Devonian: Riverside, California, Campus Museum Association, Department of Earth Sciences, University of California, Campus Museum Contribution 4, p. 144-182.

Sandberg, C. A., Poole, F. G., and Gutschick, R. C., 1980, Devonian and Mississippian stratigraphy and conodont zonation of Pilot and Chainman shales, Confusion Range, Utah, in Fouch, T. D., and

782

Magathan, E. R., eds., Paleozoic Paleogeography of the West-Central U.S.: Denver, Colorado, Rocky Mountain Section Society of Economic Paleontologists and Mineralogists, Rocky Mountain Paleogeography Symposium 1, p. 71-79.

Sandberg, C. A., Gutschick, R. C., Johnson, J. G., Poole, F. G., and Sando, W. J., 1982, Middle Devonian to Late Mississippian history of the overthrust belt region, western United States, in Powers, R. B., ed., Geologic Studies of the Cordilleran Thrust Belt: from Alaska to Mexico, Volume II: Denver, Colorado, Rocky Mountain Association of Geologists, p. 691-719.

Sandberg, C. A., Poole, F. G., and Johnson, J. G., 1988, Upper Devonian of the western United States, in McMillan, N. J., Embry, A. F., and Glass, D. J., Devonian of the World, Proceedings of the Second International Symposium on the Devonian System, Volume I, Regional Syntheses: Calgary, Alberta, Canada, Canadian Society of Petroleum Geologists Memoir 14, v. I, p. 183-220.

Schweickert, R. A., and Snyder, W. S., 1981, Paleozoic plate tectonics of the Sierra Nevada and adjacent regions, in Ernst, W. G., ed., The Geotectonic Development of California: Rubey Volume 1: Englewood Cliffs, New Jersey, Prentice-Hall, p. 182-202.

Silberling, N. J., and Roberts, R. J., 1962, Pre-Tertiary stratigraphy and structure of northwestern Nevada: Geological Society of America Special Paper 72, 58 p.

Smith, J. F., Jr., and Ketner, K. B., 1968, Devonian and Mississippian rocks and the date of the Roberts Mountains thrust in the Carlin-Pinon Range area, Nevada: U.S. Geological Survey Bulletin 1251-I, 18 p.

Smith, J. F., Jr., and Ketner, K. B., 1975, Stratigraphy of Paleozoic rocks in the Carlin-Pinon Range area, Nevada: U.S. Geological Survey Professional Paper 867-A, 87 p.

Snyder, W. S., and Brueckner, H. K., 1983, Tectonic evolution of the Golconda allochthon, Nevada: Problems and perspectives, in Stevens, C. H., ed., Pre-Jurassic Rocks in Western North American Suspect Terranes: Los Angeles, Pacific Section Society of Economic Paleontologists and Mineralogists, Publication 32, p. 103-123.

Speed, R. C., 1979, Collided Paleozoic microplate in the western United States: Journal of Geology, v. 87, p. 279-292.

Speed, R. C., and Sleep, N. H., 1982, Antler orogeny and foreland basin: A model: Geological Society of America Bulletin, v. 93, p. 815-828.

Speed, R. C., and Sleep, N. H., 1983, Reply on Antler orogeny and foreland basin: A model: Geological Society of America Bulletin, v. 94, p. 685-686.

Stevens, C. H., 1986, Evolution of the Ordovician through Middle Pennsylvanian carbonate shelf in east-central California: Geological Society of America Bulletin, v. 97, p. 11-25.

Stewart, J. H., 1972, Initial deposits in the Cordilleran geosyncline: Evidence of a late Precambrian (<850 m.y.) continental separation: Geological Society of America Bulletin, v. 83, p. 1345-1360.

Stewart, J. H., 1980, Geology of Nevada: Nevada Bureau of Mines and Geology Special Publication 4, 136 p.

Stewart, J. H., and Poole, F. G., 1974, Lower Paleozoic and uppermost Precambrian Cordilleran miogeocline, Great Basin, western United States, in Dickinson, W. R., ed., Tectonics and Sedimentation: Tulsa, Oklahoma, Society of Economic Paleontologists and Mineralogists Special Publication 22, p. 28-57.

Stewart, J. H., and Suczek, C. A., 1977, Cambrian and latest Precambrian paleogeography and tectonics in the western United States, in Stewart, J. H., Stevens, C. H., and Fritsche, A. E., eds., Paleozoic Paleogeography of the Western U.S.: Los Angeles, Pacific Section Society of Economic Paleontologists and Mineralogists, Pacific Coast Paleogeography Symposium 1, p. 1-17.

Stewart, J. H., Stevens, C. H., and Fritsche, A. E., eds., 1977, Paleozoic Paleogeography of the Western U.S.: Los Angeles, Pacific Section Society of Economic Paleontologists and Mineralogists, Pacific Coast Paleogeography Symposium 1, 502 p.

Trexler, J. H., Jr., Snyder, W. S., Cashman, P. H., Gallegos, D. M., and Spinosa, C. M., 1990, An orogenic hierarchy: Examples from western North America: Geological Society of America Abstracts with Programs, v. 22, no. 7, p. A328.

Turner, R. J., 1985, A Late Devonian stratiform Pb-Zn and stratiform barite metallogenic event in the North American Cordillera: Geological Society of America Abstracts with Programs, v. 17, p. 414.

Varga, R. J., 1982, Implications of Palaeozoic phosphorites in the northern Sierra Nevada Range: Nature, v. 297, p. 217-220.

Verville, G. J., Drowley, D. D., Baesemann, J. F., and James, S. L., 1985, Age, correlation, and tectonic significance of Wildcat Peak Formation, northern Toquima Range, Nevada [abs.]: American Association of Petroleum Geologists Bulletin, v. 69, p. 869.

Wrucke, C. T., Jr., and Jones, D. L., 1978, Allochthonous Devonian chert in northern Shoshone Range in north-central Nevada: U.S. Geological Survey Professional Paper 1100, p. 70-71.

Ziegler, W., and Sandberg, C. A., 1984, Palmatolepis-based revision of upper part of standard Late Devonian conodont zonation, in Clark, D. L., ed., Conodont Biofacies and Provincialism: Geological Society of America Special Paper 196, p. 179-194.

SEDIMENTARY ROCKS OF THE GOLCONDA TERRANE:
PROVENANCE AND PALEOGEOGRAPHIC IMPLICATIONS

A. Elizabeth Jones
Department of Geology and Geophysics
University of California
Berkeley, California 94720

ABSTRACT

Lithologic and biostratigraphic data from previously little studied regions of the Golconda terrane in north-central Nevada provide new constraints on paleogeographic interpretations of accreted Paleozoic terranes of western North America. The Golconda terrane is an upper Paleozoic assemblage of clastic, volcanic, carbonate and deep ocean pelagic rocks.

Three types of sedimentary rocks characterize the Golconda terrane: 1) bedded chert and argillite, 2) intrabasinal and 3) extrabasinal clastic rocks. Thick sequences of bedded chert and argillite that are not interbedded with other coarse clastic sedimentary rocks represent abyssal plain sediments deposited across large expanses of the Paleozoic ocean floor. Three intrabasinal clastic subfacies reflect the vigorous sedimentary, volcanic, and tectonic processes active within the Golconda basin. The carbonate-black chert-basalt subfacies represents a typical Mississippian seamount assemblage characterized by extensive relief and environmental variability in a restricted area. The red chert-basalt subfacies is a Mississippian to Permian volcanically active, open ocean setting such as a basal seamount or a spreading ridge. The melange subfacies, containing large blocks of the other facies, formed as a result of tectonic and sedimentary processes related to accretion of the Golconda terrane. Three extrabasinal subfacies have lithologic and provenance characteristics that link them to adjacent accreted Paleozoic terranes and to the continental margin. The lithic subfacies was derived from a reworked continental margin, possibly the Osgood Mountains terrane or the Harmony Formation. The calcareous and quartzose siliclastic subfacies was derived from redeposited oceanic sedimentary rocks, probably the Roberts Mountains terrane. The calcarenite subfacies was derived from a mature continental margin source region, probably the Antler overlap sequence. The different facies of the Golconda terrane represent deposition in a basin with many different paleoenvironmental settings. Some of these settings were near the continental margin, whereas some show no evidence of nearby cratonal debris, suggesting that they formed in regions of the basin far removed from the craton.

The sedimentologic records in the Golconda terrane and the Roberts Mountains terrane do not support the interpretation

that the Antler orogeny was the result of an arc-continent collision. The Roberts Mountains basin and the Golconda basin did not evolve sequentially by the closing and re-opening of the same basin. A long-lived, episodic, transpressive, transform regime is proposed as the mechanism responsible for the accretion of numerous Paleozoic terranes to the margin of western North America throughout the Paleozoic.

INTRODUCTION

Complex Mesozoic tectonic events in the western United States have greatly obscured evidence that reveals the Paleozoic tectonic history of the region. Accordingly, Paleozoic plate tectonic models of this area still have room for broad interpretation. This analysis of the sedimentary rocks in the Golconda terrane provides new constraints for interpretations of paleogeographic and paleoenvironmental settings within the Golconda basin. The Golconda terrane is an upper Paleozoic assemblage of clastic, volcanic, carbonate and deep-ocean pelagic rocks, which are fragments of a paleo-Pacific ocean basin. Today, the remnants of this ocean basin are scattered across the mountain ranges of northern and central Nevada (Fig. 1).

Historical data has shaped interpretations of the Golconda terrane. The paucity of fossil data for the ages of these "eugeosynclinal" rocks was the greatest difficulty that restricted early workers' interpretations (Roberts and others, 1958). Few fusulinid and conodont collections from clastic units provided broad Pennsylvanian-Permian age constraints for the Havallah and Pumpernickle Formations, type formations of the Golconda terrane. This age data lead to an interpretation that restricted early all of the rocks of the Golconda allochthon postdated the rocks of the structurally underlying Roberts Mountains allochthon (Burchfiel and Davis, 1972; Burchfiel and Davis, 1975; Speed, 1979).

The age framework for both the Golconda and Roberts Mountains terranes has only been established in the last decade with the development of Paleozoic radiolarian biostratigraphy and extraction techniques (Pessagno and Newport, 1972; Holdsworth and Jones, 1980a, 1980b). This has revealed an expanded age range for the bedded chert and clastic rocks in the Golconda terrane dating back to the early Mississippian and even the Late Devonian (Miller and others, 1984;

In Cooper, J.D., and Stevens, C.H., eds., 1991, **Paleozoic Paleogeography of the Western United States-II:** Pacific Section SEPM, Vol. 67, p. 783-800.

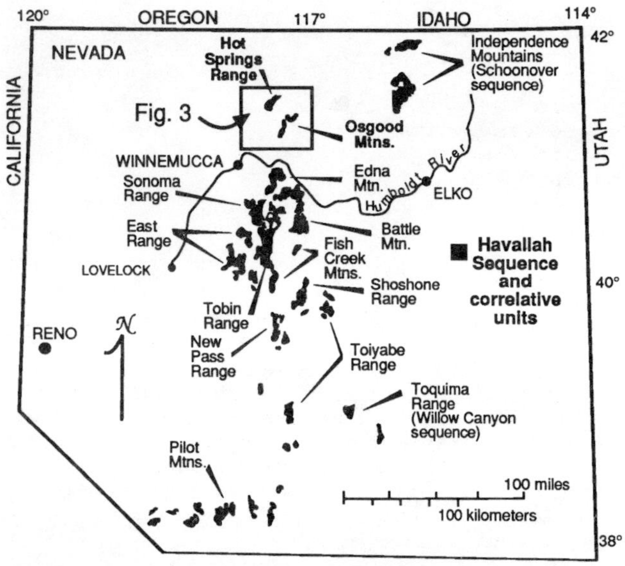

Figure 1. Exposures of the Golconda terrane in northern and central Nevada.

Stewart and others, 1986). These rocks overlap in age with the youngest rocks of the Roberts Mountains terrane (Murchey, 1989). New tectonic models have not been developed which incorporate this expanded collection of age information.

New paleogeographic and paleoenvironmental interpretations of the Golconda terrane require integration, on a regional scale, of age, structural, and lithologic data from the varied sedimentary and volcanic rocks which comprise the terrane. This study combines new biostratigraphic and lithostratigraphic data from previously little studied regions of the Golconda terrane with information collected elsewhere by others. The goals are to establish a regional framework within which the lithologic assemblages of the Golconda terrane can be characterized, to demonstrate important inconsistencies between presently accepted plate-tectonic models of the Paleozoic in the western United States and new biostratigraphic and lithologic information, and to offer ideas which will help to resolve these inconsistencies.

SEDIMENTARY ROCKS OF THE GOLCONDA TERRANE

Three major facies of sedimentary rocks are preserved in the Golconda terrane: bedded chert and argillite, intrabasinal, and extrabasinal clastic rocks. The intrabasinal facies has three subfacies: carbonate-black chert-basalt, red chert-basalt, and melange. The extrabasinal facies also has three subfacies: lithic, calcareous and quartzose siliclastic, and calcarenite.

Chert-Argillite facies

During the Paleozoic, extensive siliceous pelagic sediment generated by radiolarians, sponges, and other silica producing organisms covered much of the ocean floors. Fine laminae and graded beds in thick sequences of Late Devonian through

Mississippian, and middle Pennsylvanian through Permian gray and green radiolarian chert and siliceous argillite in the Independence Mountains (Units DMca and PPca of Miller and others, 1984) and at Battle Mountain (parts of LU-1 and LU-2 of Murchey, 1989) reflect a pelagic, quiet-water environment (Fig. 2). The thickness of these sequences, the absence of interbedded clastic material, and the delicate sedimentary structures they preserve, all suggest that they represent abyssal plain sediment deposited on the Paleozoic ocean floor.

Intrabasinal clastic facies

The sedimentological characteristics of clastic, volcaniclastic and bioclastic rocks in the Osgood Mountains and the Hot Springs Range (Fig. 1) indicate that they formed in regions within the Golconda basin experiencing active basaltic volcanism and tectonism. This volcanic and tectonic activity generated marked topographic relief, turbidity currents, grain flows, debris flows, and melange. Three different intrabasinal subfacies: carbonate-black chert-basalt, red chert-basalt, and melange are recognized.

Carbonate-black chert-basalt subfacies

Massive, spiculitic, dark gray and black chert; fossiliferous Mississippian limestone; basalt tuffs, flows, pillows, and breccias; and carbonate and basalt conglomerates characterize the Home Ranch subterrane of the Golconda terrane in the Osgood Mountains and Hot Springs Range of northern Nevada (Fig. 3,4). Conodonts, corals, and brachiopods from this assemblage are late Kinderhookian through Chesterian (Mississippian) in age (Anita Harris; Cal Stevens; and Rex Hanger, written communs., 1990). Crystal-rich basalt sandstone layers 5 to 15 cm thick are interbedded with laminated, cherty, black tuff (Fig. 4A,B). Graded beds containing broken plagioclase and perthite phenocrysts, and altered volcanic (basalt?) clasts show soft-sediment slumping and bed offsets. Clast-supported volcaniclastic sandstones and breccias have clasts of chert, argillite, basalt, pumice, recrystallized glass shards, and rare quartz siltstone (Fig. 4C,D). Conglomeratic layers have rounded to subrounded clasts of basalt, argillite and chert 10 to 20 cm in diameter. Bioclastic hash in beds tens of meters thick is interbedded with basalt flows (Fig. 4E). Matrix-supported, graded carbonate conglomerates have fragments of pillow basalt and vesicular flows in a limestone matrix (Fig. 4F). Olistostromal debris-flow deposits containing large blocks of fossiliferous limestone and basalt are the most chaotic unit within the Home Ranch subterrane (Fig. 4G).

A seamount setting with extensive relief and environmental variability in a restricted area can explain the close stratigraphic juxtaposition of so many different units in the carbonate-black chert-basalt subfacies of the Home Ranch subterrane. Facies with similar characteristics, also interpreted to

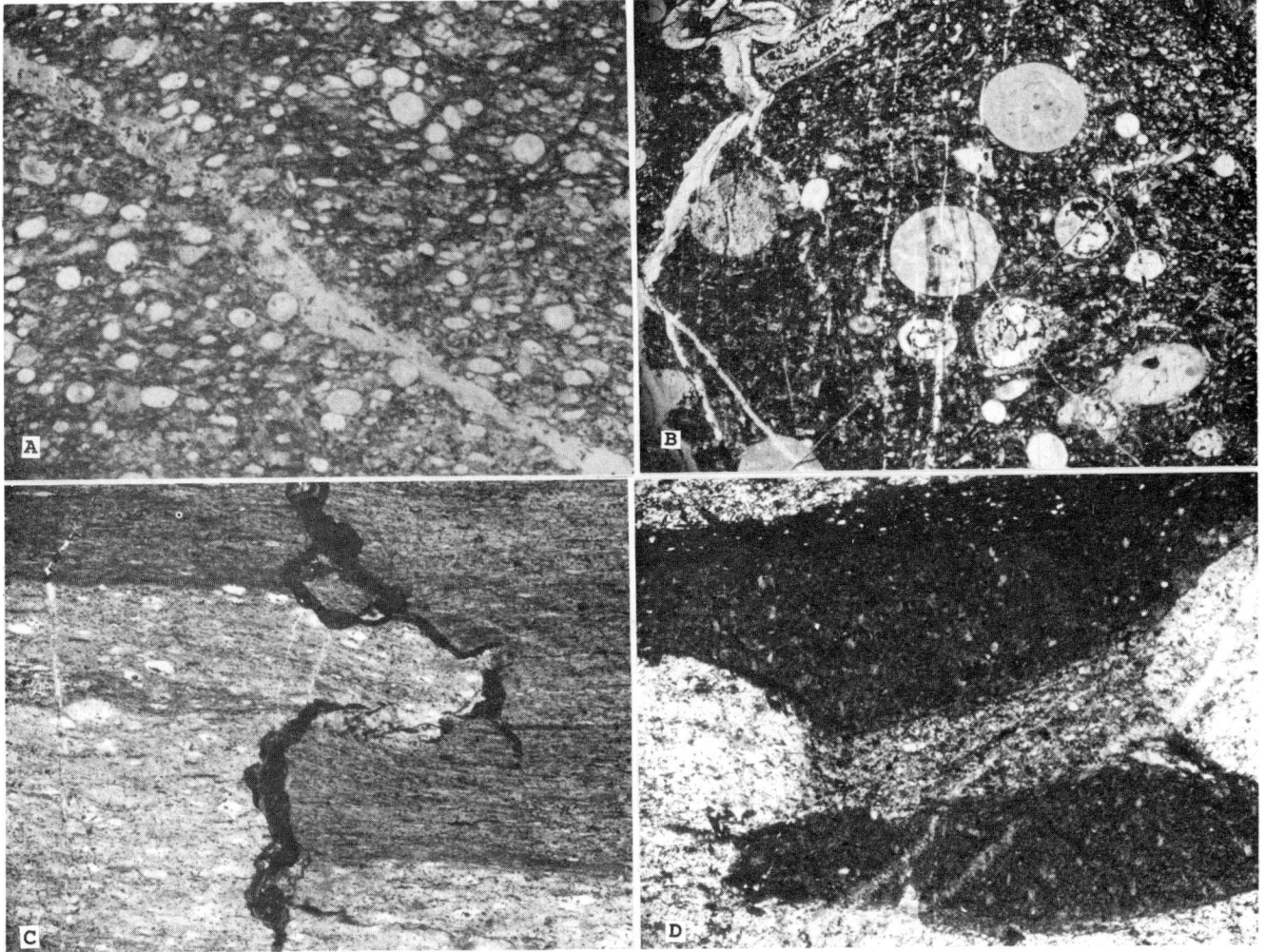

Figure 2. Biogenic and sedimentary features of bedded cherts from the Golconda terrane. **A:** Radiolarian-rich chert. **B:** Spiculitic chert. **C:** Chert and argillite laminae offset by pressure solution zone. **D:** Worm burrows in microspiculite. Field of view is about 1.8 cm.

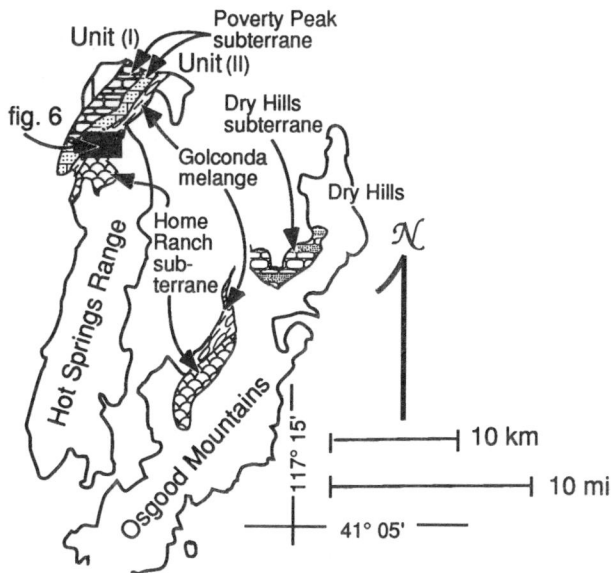

Figure 3. Golconda subterranes in the Osgood Mountains and the Hot Springs Range discussed in text. See figure 1 for location.

represent a Paleozoic seamount, have been described from the Ordovician Snow Canyon Formation in the Independence Mountains (Watkins and Browne, 1989), and from the Carboniferous and Permian Akiyoshi terrane of southwest Japan (Sano and Kanmera, 1988).

Red chert-basalt subfacies

Interbedded bright red and green radiolarian chert and argillite, pillow basalts, volcanic breccias, and basalt tuff (Fig. 5A,B), which typify the red chert-basalt subfacies of the Golconda terrane, are found in Unit (I) of the Poverty Peak subterrane in the northern Hot Springs Range (Jones, 1988)(Fig. 3). Radiolarians and conodonts extracted from the red ribbon cherts and siliceous argillites in the Poverty Peak Unit (I) range in age from middle Osagean (*typicus* through *anchoralis-latus* Conodont zone (Anita Harris, written commun., 1990) through early Leonardian (*Neogondollella intermedia*, Bruce Wardlaw, written commun., 1990). A continuous stratigraphic section of red chert and argillite is not preserved in the Poverty Peak subterrane, but a composite section

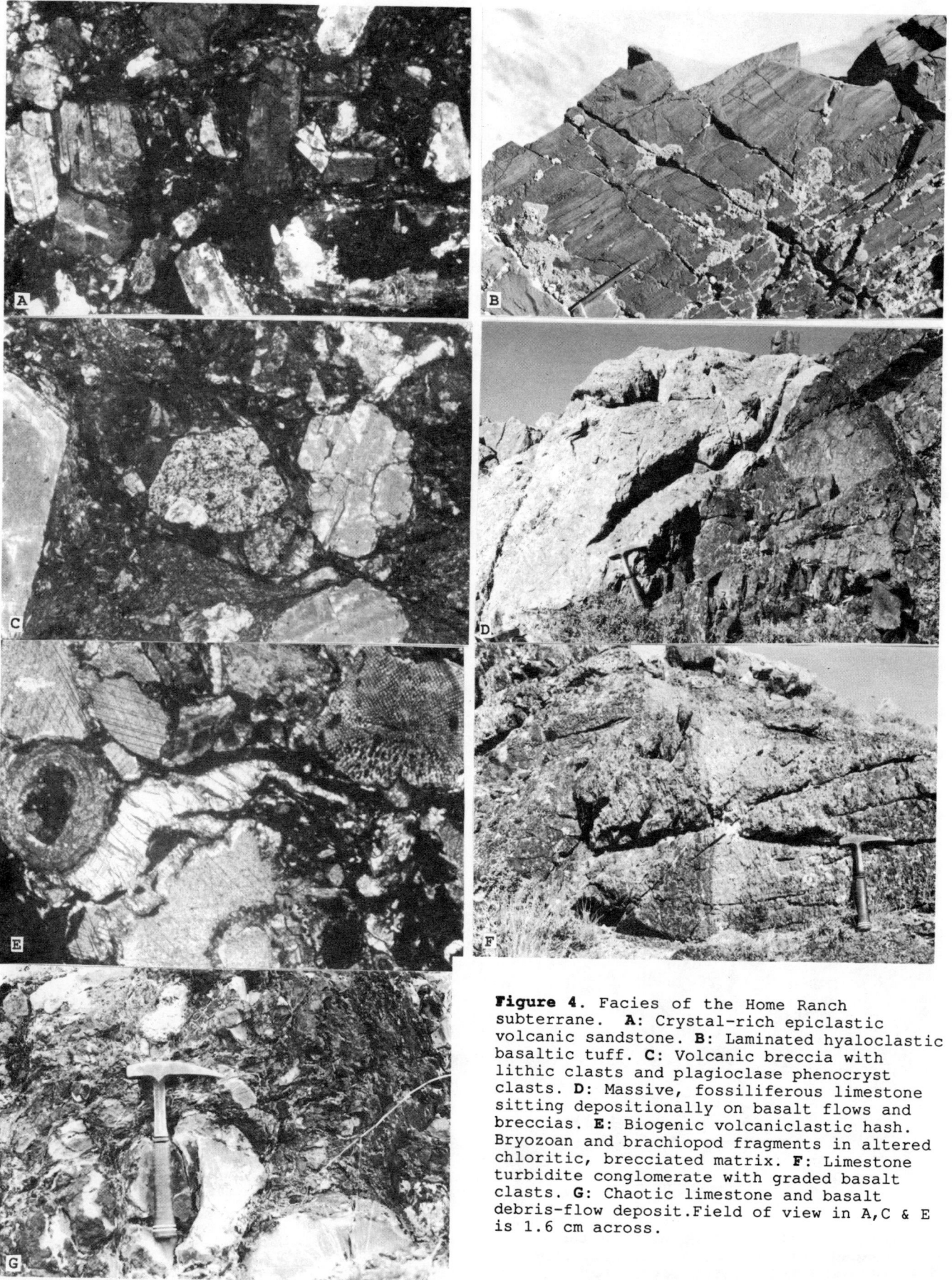

Figure 4. Facies of the Home Ranch subterrane. **A:** Crystal-rich epiclastic volcanic sandstone. **B:** Laminated hyaloclastic basaltic tuff. **C:** Volcanic breccia with lithic clasts and plagioclase phenocryst clasts. **D:** Massive, fossiliferous limestone sitting depositionally on basalt flows and breccias. **E:** Biogenic volcaniclastic hash. Bryozoan and brachiopod fragments in altered chloritic, brecciated matrix. **F:** Limestone turbidite conglomerate with graded basalt clasts. **G:** Chaotic limestone and basalt debris-flow deposit. Field of view in A, C & E is 1.6 cm across.

Figure 5. Poverty Peak Unit (I) subterrane. **A:** Vertically dipping pillow basalts. **B:** Depositional contact between basalt tuff (on right) and radiolarian chert (on left) with pale zone of silicic alteration in between. Field of view approx. 1.8 cm **C:** Plagioclase rich volcaniclastic sandstone. Field of view about 1 cm. **D:** Red chert breccia from Schoonover Sequence in the Independence Mountains. Angular radiolarian chert fragments are embedded in an argillaceous matrix. Field of view approx. 1.6 cm.

appears to represent most of this time (data in prep.). Hyaloclastic, graded tuff layers in between pillows serve as geopetal structures showing original orientations. Well-graded, volcaniclastic, basaltic tuff and sandstone interlayered with red and green radiolarian chert and argillite is composed of angular, broken and reworked plagioclase phenocrysts, basalt clasts with fine feldspar laths, magnetite, biotite, rare quartz grains, altered pyroxene phenocrysts, chlorite, sphene, and rutile (Fig. 5C). Massive basalt breccia contains clasts of plagioclase phenocryst-bearing basalt, red chert, green chert, massive basalt, altered hornblende, biotite, and rare zircons. In the Schoonover sequence in the Independence Range, the red chert-basalt subfacies includes a red chert breccia (Fig. 5D) that has angular radiolarian chert clasts in an argillite matrix.

Bright red and green iron- and manganese-rich sediment in today's oceans is closely associated with fast-spreading ridges (Kennett, 1982). The red chert-basalt subfacies is interpreted to have formed in a deep-marine setting with exhalative oxidizing vents (Snyder and Brueckner, 1983) near a

spreading ridge or seamount. Interbedded red and green radiolarian chert, volcanic, and volcaniclastic rocks in the Poverty Peak subterrane indicate ongoing basalt volcanism from Mississippian to Permian time in a deep, open-ocean setting. This volcanism was far enough removed from the boundaries of the basin to be unaffected by the influx of extrabasinal clastic material throughout much of its history.

The red chert-basalt subfacies also is exposed in the Big Mike Mine area of the Tobin Range (Snyder and Brueckner, 1983), at the Black Diablo Mine in the Sonoma Range, at Edna Mountain as unit PPc of Erickson and Marsh (1974a), in the Willow Canyon Formation in the Toquima Range (Laule and others, 1981), at Battle Mountain (Roberts, 1964) as basal unit LU-2 of Murchey (1982), and in the Independence Mountains as the Cap Winn member of Fagan (1962).

Melange subfacies

A large belt of melange in the Hot Springs Range is structurally sandwiched between the Poverty Peak subterrane and the Home Ranch subterrane (Fig. 3,6). The

788

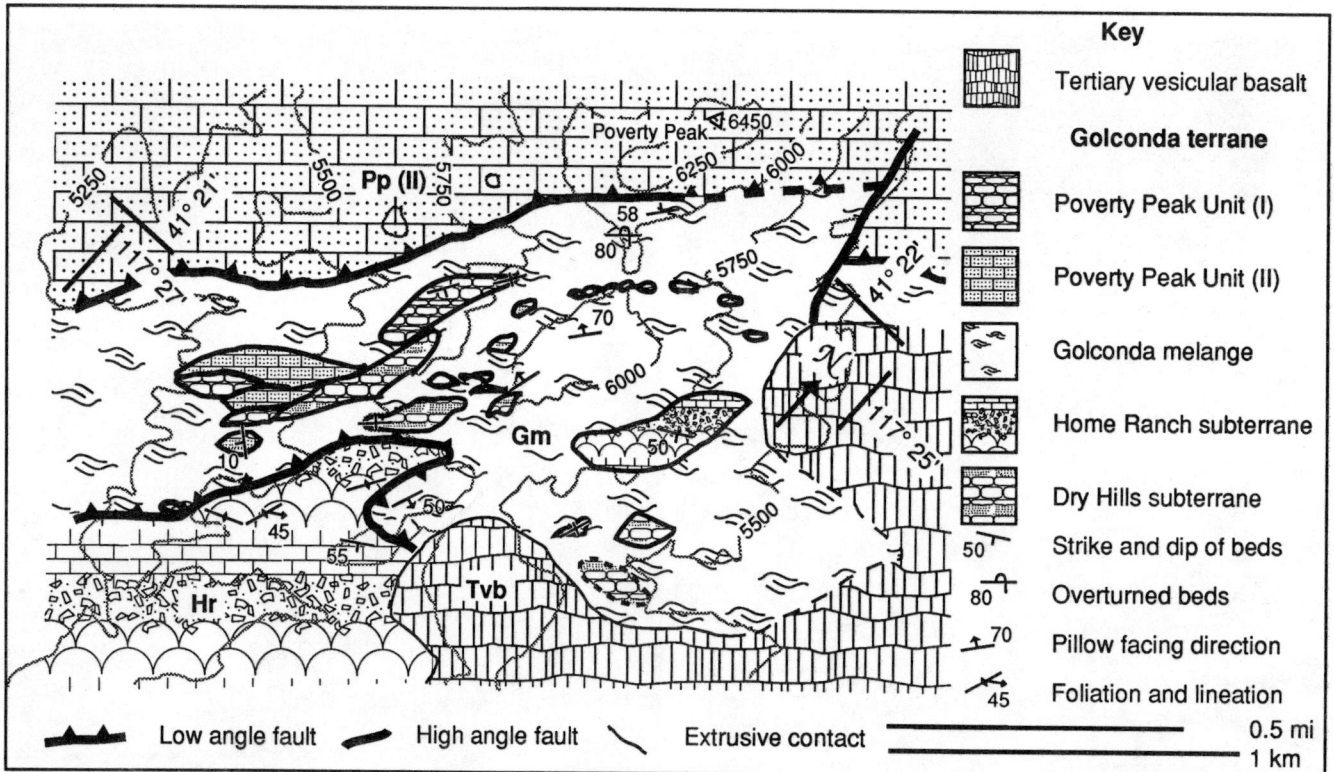

Figure 6. Geologic map of Golconda melange in the northern Hot Springs Range. See Figure 3 for location. Large blocks of different subfacies are embedded in the Golconda melange.

melange contains slabs, tens of meters long, of bedded red chert characteristic of Poverty Peak Unit (I), interbedded limey siltstone and quartzite typical of Poverty Peak Unit (II) (discussed below), massive limestone and volcaniclastic rocks typical of the seamount facies of the Home Ranch subterrane, and blocks of dark gray and green chert interbedded with lithic sandstone similar to lithic facies found in the Dry Hills subterrane in the Osgood Mountains (discussed below). The ages of the blocks in the melange include almost the entire spectrum of ages in the Golconda terrane from the earliest Mississippian to the late Early Permian. These blocks, and large, massive (unsheared) inclusions of pillow basalt and porphyritic diabase, are surrounded by a sheared basaltic tuff and argillite matrix (Fig. 7A). In the Osgood Mountains, the same melange subfacies was originally mapped as parts of the Farrel Canyon Formation and the Goughs Canyon Formation by Hotz and Willden (1964). Although the character of the melange is basically sedimentary (Fig. 7B), it has aspects suggesting both syn- and post-depositional tectonic activity. The Golconda melange most closely resembles Type I melange described by Cowan (1985). Pinch and swell boudins of chert and sandstone clasts are evidence of extreme layer-parallel extension of the beds (Fig. 7C,D). Locally, the boudins have been strongly folded (Fig. 7E) within a ductile, soft-sediment matrix (Fig. 7F).

The clast orientation in the Golconda melange is subparallel to the foliation that defines the north- to northeast-trending fold axes of the principal Sonoman deformation, which resulted in isoclinal folding of units throughout the Golconda terrane. If pre-Sonoman sedimentary slumping caused the stratal disruption that created the melange, the boudin-layering caused by the slumping would form subparallel to the more coherent sub- and suprajacent beds. These boudin-layers would then be isoclinally folded along the same fold axes as the other layers during the Sonoman deformation. This scenario would generate the observed orientations of foliations, bedding planes, and boudin layers in the melange and the adjacent units, and argues for pre-Sonoman sedimentary melange formation, but fails to explain the cause of the initial sediment disruption that formed the melange. Clasts in the melange are derived from many different depositional settings, originally separated from one another by unknown distances. The juxtaposition of clasts from the different subterranes into a common setting, therefore, must have occurred during the accretion which brought the different subterranes together, and hence argues for syn-tectonic melange formation. Melange formation that included both soft-sediment and syn-tectonic mechanisms would account for the characteristics in the Golconda melange. It is not clear whether all of the melange formed at the same time, or if it developed over an extended period of time.

Extrabasinal clastic facies

Clastic debris derived from outside the Golconda basin is interbedded with radiolarian and spiculitic chert in many

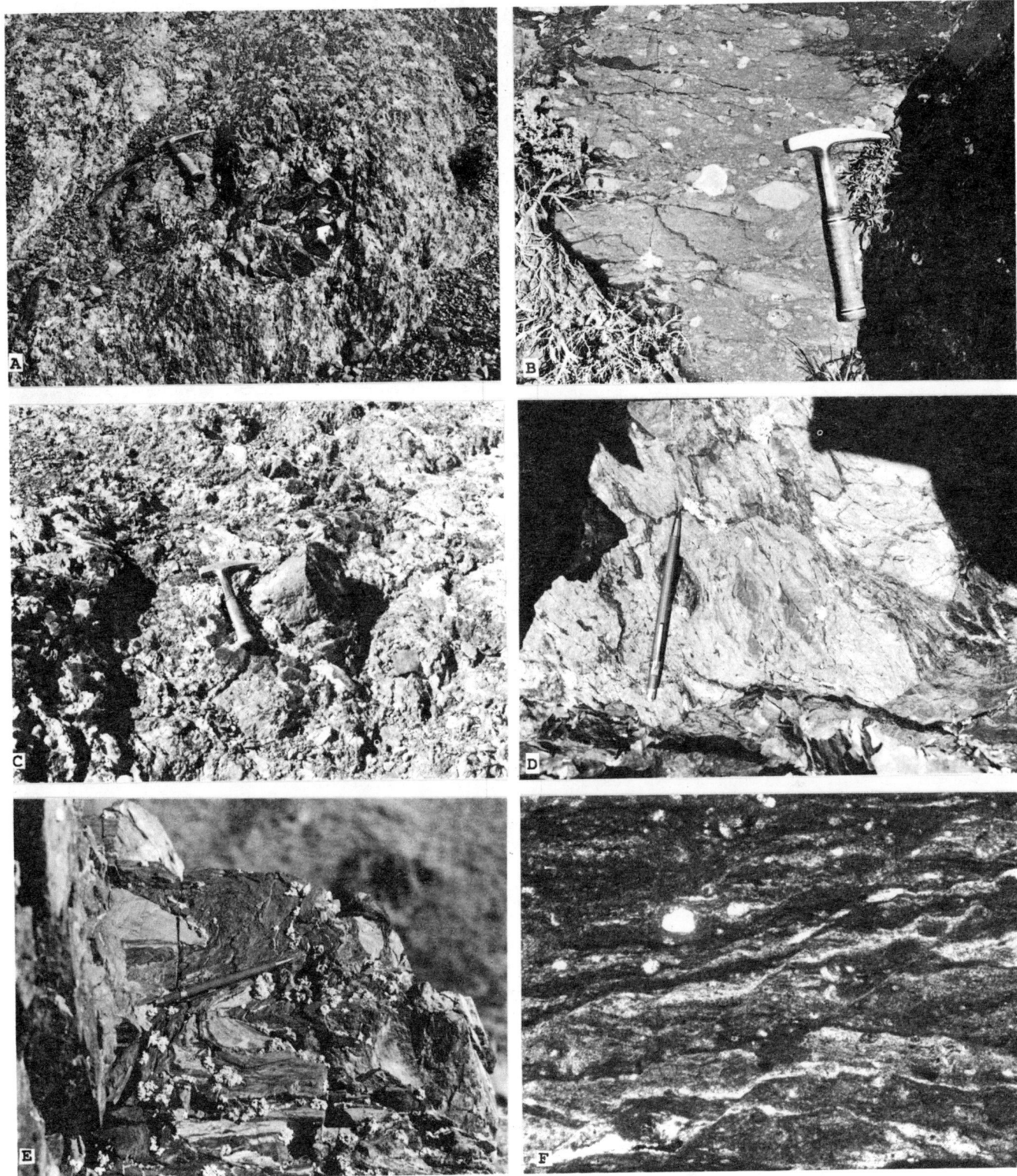

Figure 7. Golconda melange. **A:** Pillow basalt in sheared basaltic tuff matrix. **B:** Unsheared soft-sediment mudflow with clasts of chert and sandstone. **C:** Siltstone block in black argillite melange matrix. **D:** Soft-sediment shearing and boudinage of chert clasts in silty mudstone matrix. **E:** Folded and stretched chert and argillite boudins in mudstone matrix. **F:** Soft-sediment shearing and attenuation in melange matrix. Field of view about 1 cm.

regions of the Golconda terrane. Extrabasinal clastic rocks are defined by lithologic characteristics typical of cratonally derived sediments, or by ages incompatible with an origin from within the Golconda basin. These rocks have lithologic characteristics that link them to adjacent accreted Paleozoic terranes and the continental margin. The three extrabasinal clastic subfacies are: lithic, calcareous and quartzose siliclastic, and calcarenite.

Lithic subfacies

The petrography of Mississippian lithic and volcanogenic sandstones in the Schoonover sequence in the Independence Mountains has been described in detail by Whiteford (1984, in press). Volcaniclastic sandstone (arkose to feldspathic litharenite) is intercalated with more quartzose, massive, feldspathic litharenite and quartz arenite (Mss and Msv units of Miller (1984) and Whiteford (in press)). Whereas the feldspathic litharenite and quartz arenite are clearly extrabasinally derived rocks, the source of the volcanogenic sandstones in the Schoonover sequence is unclear. Although Whiteford (in press) interprets them to be arc-derived, insufficient evidence is presented to preclude them from being locally derived intermediate to mafic sea-floor volcaniclastic rocks. The lithic sandstone interbedded with the volcanogenic sandstone, however, is coeval with and compositionally very similar to the lithic sandstone in the Farrel Canyon Formation of the Dry Hills subterrane in the Osgood Mountains (Fig. 3). The Dry Hills subterrane is dated by radiolarians and conodonts in the Farrel Canyon Formation as Mississippian and early Pennsylvanian (McCollum and McCollum, 1989; Anita Harris, written commun., 1990; data in prep.). The greatest influx of clastic sediment occurred in middle Osage through Meramec (late early through early late Mississippian) time, but the age range of the interbedded chert appears to extend from the Kinderhookian (early Mississippian) through the Atokan (middle Pennsylvanian). Roberts (1964) described a similar facies in the Pumpernickle Formation at Battle Mountain, Silberling (1975) described a comparable clastic unit in the Havallah sequence in the Sonoma Range, and Erickson and Marsh (1974a) recognized this facies as the PPq unit of the Pumpernickle Formation in the Golconda quadrangle at Edna Mountain. Important distinctions between these lithic sandstones and other clastic facies of the Golconda terrane often have been overlooked.

Modal data (Fig. 8; Whiteford, 1984, 1990) from the sandstone of the Dry Hills subterrane in the Osgood Mountains reflect a provenance of a recycled orogen (Dickinson and others, 1983a) and suggest that it was derived from a reworked continental margin, rather than a reworked oceanic terrane. This sandstone, a lithic graywacke, is interbedded with radiolarian- and conodont-rich, green to dark gray ribbon chert and argillite, and rare limestone and tuffaceous lenses (Fig. 9A). Graded beds, partial Bouma turbidite sequences, cross laminae, and channels in the cherts and argillites interbedded with the coarser sandstone are evidence of extensive current activity in this part of the Golconda basin (Fig. 9B). The lithic sandstone is matrix-supported with few sedimentary features other than coarsely graded beds. The grain size is fine to medium (0.5-2 mm) except for rare pebble beds. Clast components include chert, quartz, feldspar, mica, lithics, and trace heavy minerals, in a silicified clay matrix that constitutes 25% to 75% of the rock by volume.

The sandstone contains 2-8% black or dark grey-green chert clasts. A few angular chert pebbles contain radiolarians (Fig. 9C), although no diagnostic faunas have yet been recovered. Quartz clasts (excluding chert) compose 40% to 60% of the Dry Hills sandstone. Monocrystalline and polycrystalline quartz (Fig. 9D) comprises the bulk of the grains. The monocrystalline quartz grains apparently are derived from fragments of similar polycrystalline clasts. The faint to strong undulating extinction of these quartz clasts is evidence of their pervasive strain history. The polycrystalline grains have gently sutured, irregular boundaries typical of moderately stressed vein quartz or recrystallized quartzite. In general, the boundaries are not highly-sutured metamorphic types. The quartz clasts are angular and poorly sorted (Fig 9C) with one important exception. In about 25% of the samples, 0.5% of well-rounded rutilated quartz grains (Fig. 9E) stand out in contrast to the angular grains. The freshest samples of sandstone contain 1 to 5% feldspar, mostly potassium feldspar and perthite, and less plagioclase (Fig. 9F). Up to 2% detrital muscovite was visible in about 25% of the samples counted. The higher percentages of both detrital muscovite and feldspar were counted in the fresher samples, suggesting that low grade metamorphism in some cases has obscured the original mineralogy. About 0.5% heavy minerals, mostly rounded pink zircons, and rarely sphene and tourmaline, were counted in half of the samples. Argillite rip-up and quartz arenite clasts were the only identifiable lithic fragments in the Dry Hills sandstone.

Calcareous and quartzose siliclastic subfacies

Calcareous and quartzose siliclastic rocks in the Golconda terrane originally were described in detail from Battle Mountain (Roberts, 1964) as the Jory and Mill Canyon members of the Havallah Formation. With detailed biostratigraphic and structural analyses, Murchey (1989) has recognized this "Jory facies" as a middle Mississippian through Permian stratigraphic sequence (Lithologic Unit 2 of Murchey, 1989). Chert- and quartzite-pebble clastic horizons are interbedded with Mississippian through middle Pennsylvanian radiolarian chert, whereas younger calcareous siliclastic rocks, which contain reworked Permian microfossils, do not appear to have chert interbeds (Murchey, 1989). Similar quartzose siliclastic rocks are described in detail from the Schoonover sequence in the Independence Mountains as the

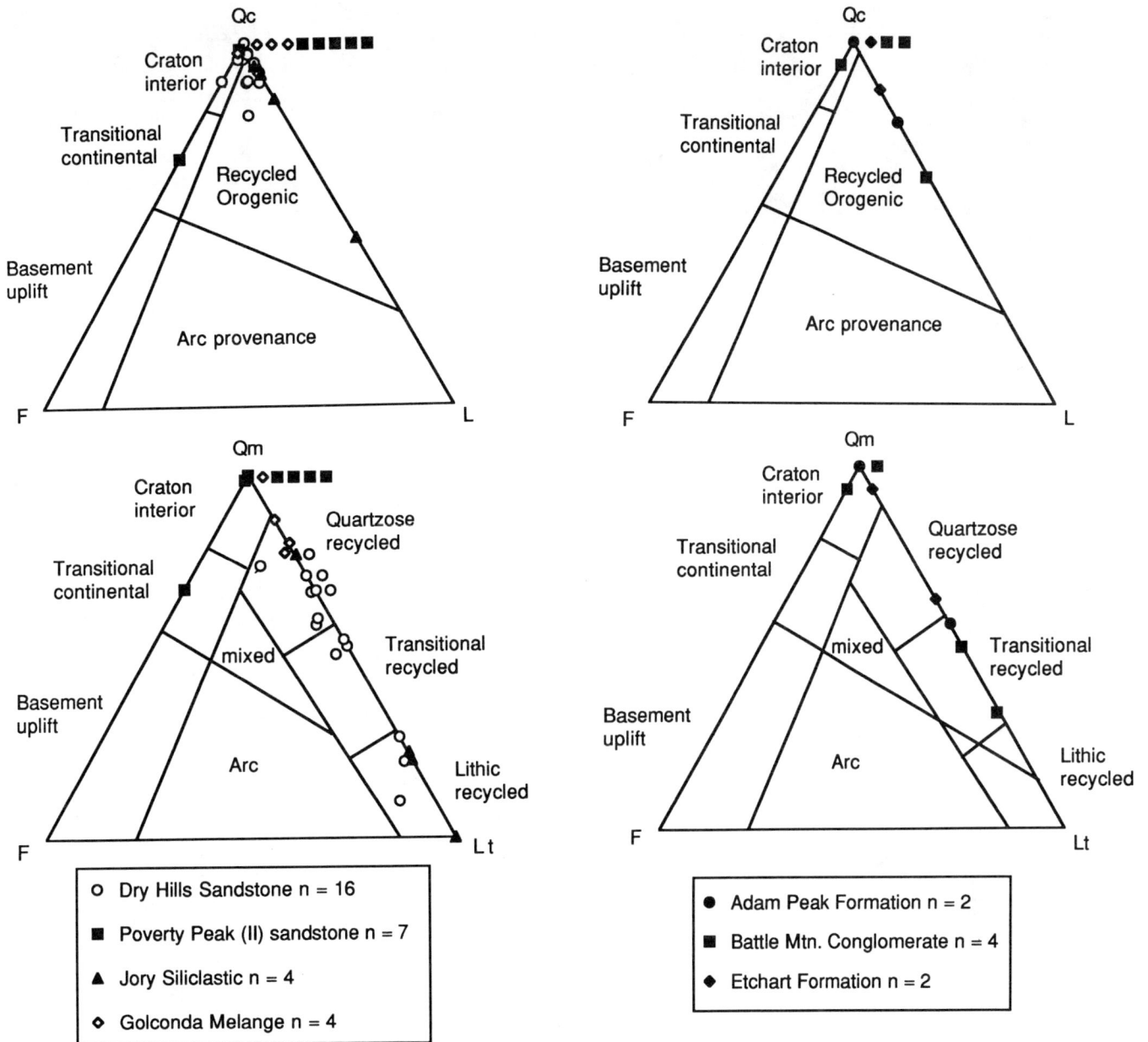

Figure 8. Modal data for sandstones of the Golconda terrane (left) and the Antler overlap sequence (right). See Whiteford (1984; in press) for detailed modal data on Schoonover sandstones similar to both the Dry Hills sandstone and the Jory siliclastic sandstone, and Dickinson and others (1983b) for more comprehensive data on the overlap sequence. Qc includes polycrystalline and monocrystalline quartz and chert clasts, F includes plagioclase, potassium feldspar, and perthite. Qm is monocrystalline quartz only. L is sedimentary and volcanic clasts excluding bedded chert and polycrystalline quartz. Lt includes bedded chert, polycrystalline quartz, sedimentary and volcanic clasts.

upper Mississippian and Pennsylvanian unit Css of Miller and others (1984) and Whiteford (in press).

Modal and biostratigraphic data from these sandstones (Whiteford, 1984, in press; Murchey, 1989; this paper, Fig. 8) support the interpretation that they were derived from reworked oceanic rocks of the Antler orogenic belt, primarily the Roberts Mountains terrane (Dickinson and others, 1983b; Snyder and Brueckner, 1983; Miller and others, 1984). These calcareous and quartzose sandstones have either a calcite

cement or quartz clastic matrix with a great diversity of clasts that includes rounded green and red radiolarian chert pebbles, black and dark gray chert pebbles, vein quartz, gray limestone, rare intermediate volcanics, potassium feldspar, quartz arenite siltstone and sandstone, shale, argillite, sutured metamorphic polycrystalline quartzite, bioclasts, and large phosphate nodules with small angular quartz fragments (Fig. 10)(see Murchey, 1989 for stratigraphic descriptions). Murchey (1989) recovered Middle Devonian, Late Devonian, and early and middle Mississippian radiolarians from chert

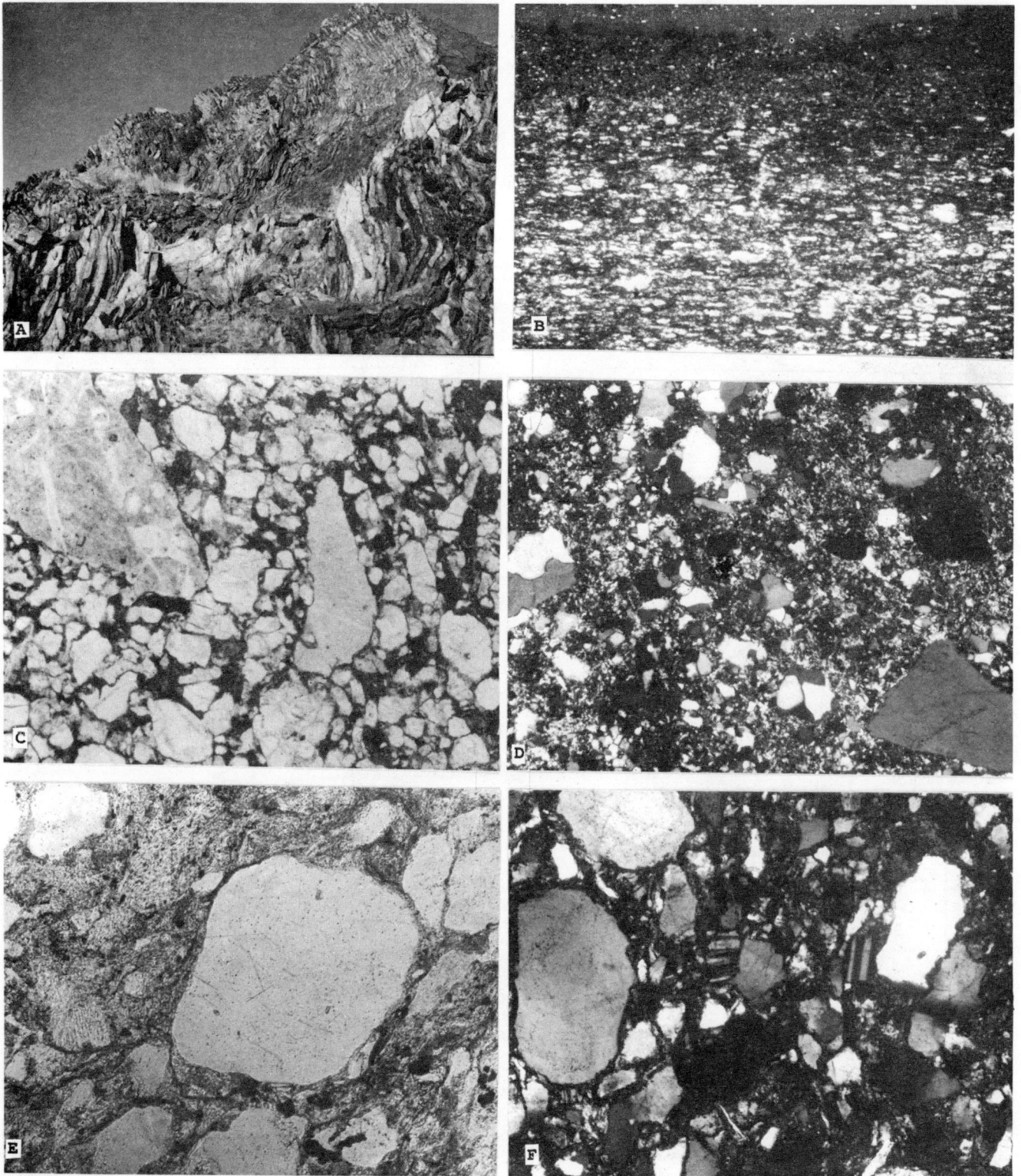

Figure 9 Dry Hills subterrane. **A:** Folded radiolarian cherts. **B:** Graded radiolarian and spiculitic chert. **C:** Angular radiolarian chert and quartz clasts in sandstone, plane light. **D:** Angular mono- and polycrystalline quartz fragments in matrix-supported sandstone, crossed nicols. **E:** Rutilated, rounded, quartz grain surrounded by angular quartz grains in sandstone. **F:** Plagioclase, detrital muscovite and angular quartz clasts in sandstone, crossed nicols. Field of view in B & C is 1.6 cm, D is 0.8 cm, E is 0.2 cm and F is 0.8 cm.

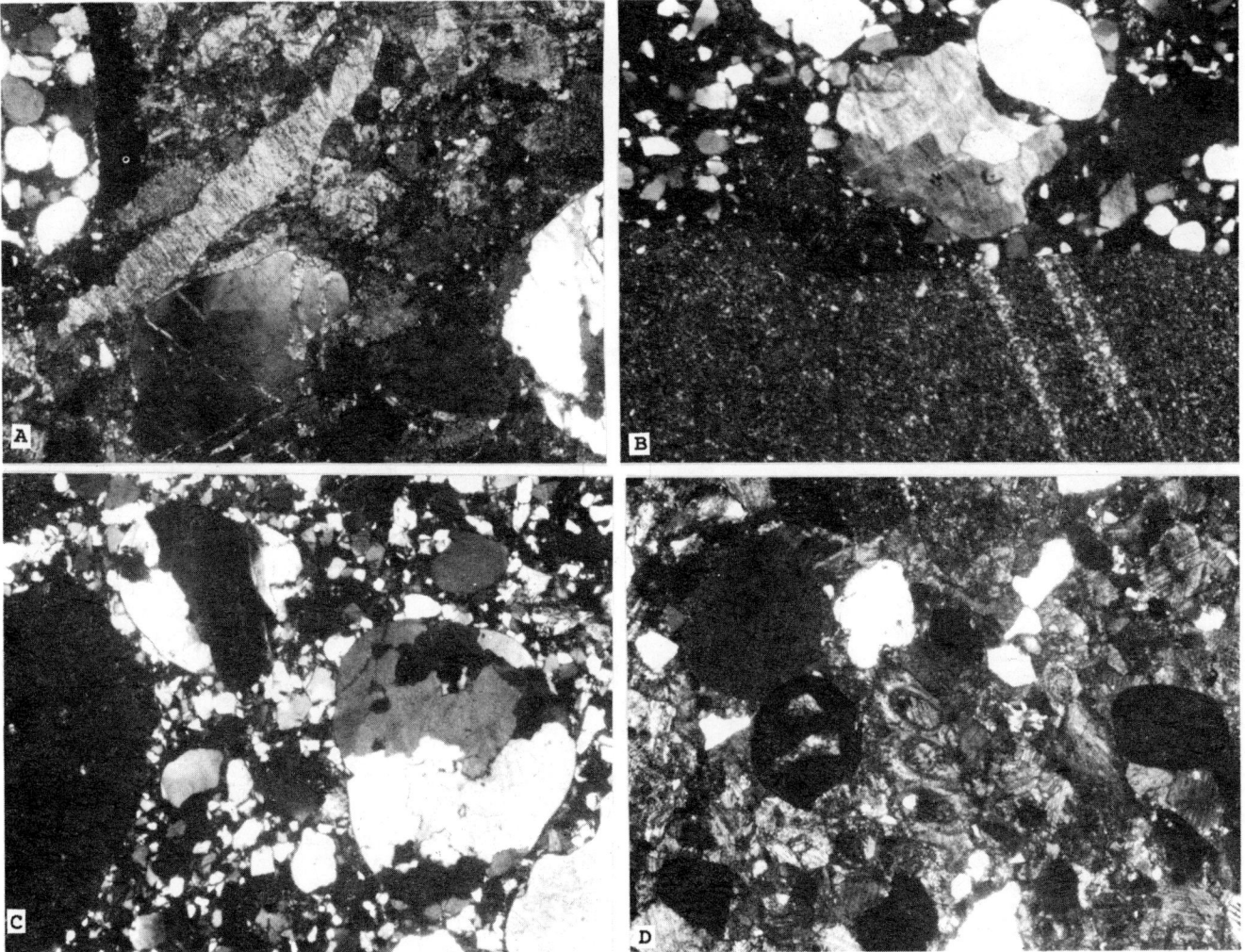

Figure 10. Calcareous and siliclastic "Jory" facies. See Murchey (1989; in press) for detailed lithologic descriptions. **A**: Well-rounded quartz siltstone, undulating monocrystalline quartz, and sutured metamorphic quartz clasts in calcareous sandstone. **B**: Coarse sandstone dike intrusion into fine siltstone. **C**: Well-rounded radiolarian chert and polycrystalline quartz clasts in fine grained quartz sandstone matrix. **D**: Quartz fragments, bioclasts and phosphate nodules in calcareous sandstone. Field of view in all photos is 1.6 cm.

clasts, and Silurian conodonts (Anita Harris, written commun., 1990) were recovered for this study from a conglomerate bed of the Jory facies.

Calcarenite subfacies

A Leonardian (late Early Permian) calcarenite (Anita Harris, written commun., 1990) in Unit (II) of the Poverty Peak subterrane (Fig. 3) is a striking contrast to the lithic sandstones in the Dry Hills subterrane and the calcareous and quartzose siliclastic sandstone of the Jory facies. At the type locality of the Havallah Formation at Hoffman Canyon in the Tobin Range silty limestone, limey siltstone, and argillite interbeds contain Early Permian conodonts (Stewart and others, 1986). Erickson and Marsh (1974b) recognized a similar facies in the Iron Point quadrangle as the Pennsylvanian to Early Permian PPpq unit of the Pumpernickle Formation. Pennsylvanian and Permian limestone turbidites also have

been reported from the Independence Range (Miller and others, 1984).

Modal data from the calcarenite subfacies (Fig. 8) reflect a more mature source region than the other extrabasinal clastic subfacies, and suggest that it was not directly derived from reworked orogenic material. Poverty Peak Unit (II) is tightly folded together with the Poverty Peak Unit (I) of red and green chert and basalt. The beds of both subfacies have the same structural trend suggesting they were a continuous stratigraphic sequence before they were folded and faulted together. Limited biostratigraphic data also support this interpretation. Poverty Peak Unit (II) is characterized by rhythmic interlayers of limey quartz siltstone and sandy limestone with common soft-sediment deformation (Fig. 11), current ripples, and *Nerites*-type ichnofossils. The limey and quartzose siltstones have angular to subrounded undulating extinction quartz clasts, rare

Figure 11 Calcarenite subfacies of Poverty Peak Unit (II). Soft sediment folding in silty limestone bed interlayered with quartz sandstone.

detrital mica, feldspar and heavy minerals, and scattered dolomite rhombs. They contain less than 10% total clasts in either a quartzose or micritic limey matrix. No chert clasts are present in the Poverty Peak Unit (II) sandstone and siltstone except at a narrow horizon that is interpreted to represent the depositional transition from the underlying Poverty Peak Unit (I) chert and basalt assemblage. A 5-cm layer of volcanic sandstone with quartz, broken plagioclase, volcanic, and chert clasts characterizes this horizon.

Provenance of Extrabasinal clastic facies

The three extrabasinal clastic subfacies in the Golconda terrane represent three distinct sources of clastic debris. These sources can be linked to a complex series of tectonic events that affected both the rocks of terranes presently adjacent to the Golconda terrane, and rocks farther east within the upper Paleozoic continental margin. Both the lithic and calcareous and quartzose siliclastic facies reflect recycled orogen source terranes but the latter has a more siliceous composition that includes lower Paleozoic oceanic debris and chert clasts. The calcarenite facies reflects a mature calcareous and quartz-rich source region and lacks evidence of extensive orogenic-derived debris.

Regional tectonic setting

Late Devonian and Mississippian tectonic events along the western North American margin were remarkably complex and are still incompletely understood. The Antler orogenic belt of Roberts and others (1958) includes Ordovician to early Mississippian oceanic rocks of the Roberts Mountains terrane, deformed in the Antler orogeny. The Mississippian Antler foreland basin in eastern Nevada contains a 2000-m-thick section of siliclastic sedimentary rocks that record tectonic events in the adjacent Antler

orogenic belt to the west. Sedimentologic and lithologic ties indicate that the sedimentary rocks of the Antler foreland basin were derived from rocks of the Roberts Mountains terrane (Roberts and others, 1958; Brew, 1971; Poole, 1974; Harbaugh and Dickinson, 1981; Dickinson and others, 1983b; Trexler and Nitchman, 1990). Siliclastic sedimentation in the Antler foreland basin, principally chert and quartzite debris, began in the Osagean and continued into the Pennsylvanian (Trexler and Nitchman, 1990) indicating that accretion of the Roberts Mountains terrane to the continental margin had commenced by early Mississippian time.

Structurally juxtaposed with the Roberts Mountains terrane in north-central Nevada is the arkosic sandstone of the Harmony Formation, and the Precambrian and lower Paleozoic Osgood Mountains terrane consisting of the Osgood Mountains Quartzite and phyllite, chert, shale, and limestone of the Preble Formation (Fig. 12). These three formations originally were included in the "transitional assemblage" of Roberts and others (1958). They are all allochthonous pieces of cratonally derived rocks, the original depositional location of which is poorly constrained. The age, paleogeographic, and structural history of the Harmony Formation has long been enigmatic (Rowell and others, 1979; McCollum and McCollum, 1989). The Osgood Mountains terrane has a deformational history that includes extensive east- and west-directed pre-upper Paleozoic folding and faulting (Hotz and Willden, 1964; Erickson and Marsh, 1974c; Stahl and others, 1989), which is distinct from the structural history of the Roberts Mountains terrane. Lithologically, it is similar to Paleozoic cratonal sequences now located farther to the southeast. The Roberts Mountains terrane, the Osgood Mountains terrane, and the Harmony Formation were juxtaposed by the early Pennsylvanian when they were all unconformably overlain by the Antler overlap sequence (Saller, 1980), a Pennsylvanian-Permian sequence of carbonate and clastic rocks.

Lithic subfacies provenance

Clastic sedimentation began at the same time in both the Dry Hills subterrane of the Golconda terrane and in the Diamond Range sequence (Trexler and Nitchman, 1990) in the Antler foreland basin to the east. However, the paucity and angularity of the dark chert fragments in the Dry Hills sandstone is atypical of the chert clasts in Antler-derived sediments. Antler-derived chert clasts commonly are abundant, very well rounded, and well sorted (compare clasts in Fig. 9C & D with Fig. 10). The angularity of Dry Hills chert clasts suggests that they were derived directly from chert beds interlayered with the sandstone, and not from beds of reworked chert of the Roberts Mountains terrane.

A gradation from clasts of quartz arenite to clasts with interlocking polycrystalline quartz grains in the Dry Hills sandstone suggests that the

Key

Golconda terrane and unconformably overlying Triassic and Jurassic volcanic and sedimentary rocks

Antler overlap sequence. Includes Antler Peak, Adam Peak, Edna Mountain, Etchart, Highway, and Battle Mountain formations

Roberts Mountains terrane. Includes Valmy, Vinini, Elder, Slaven and Scott Canyon formations

Harmony Formation

Osgood Mountains terrane. Includes Osgood Mountains Quartzite, Preble, and parts of Comus(?) formations

Contacts with Quaternary alluvium, Tertiary volcanic rocks, Jurassic and Cretaceous intrusive rocks, and Quaternary high angle faults

Depositional contact between Antler Overlap sequence and underlying terranes

Upper Permian to Jurassic faults. Probably includes some younger high angle faults

Hot Springs Range

Osgood Mountains

Winnemucca

117° 30'

Sonoma Range

Edna Mtn.

41° 00'

117° 00'

East Range

40° 30'

Tobin Range

N

Battle Mtn.

Shoshone Range

10 mi

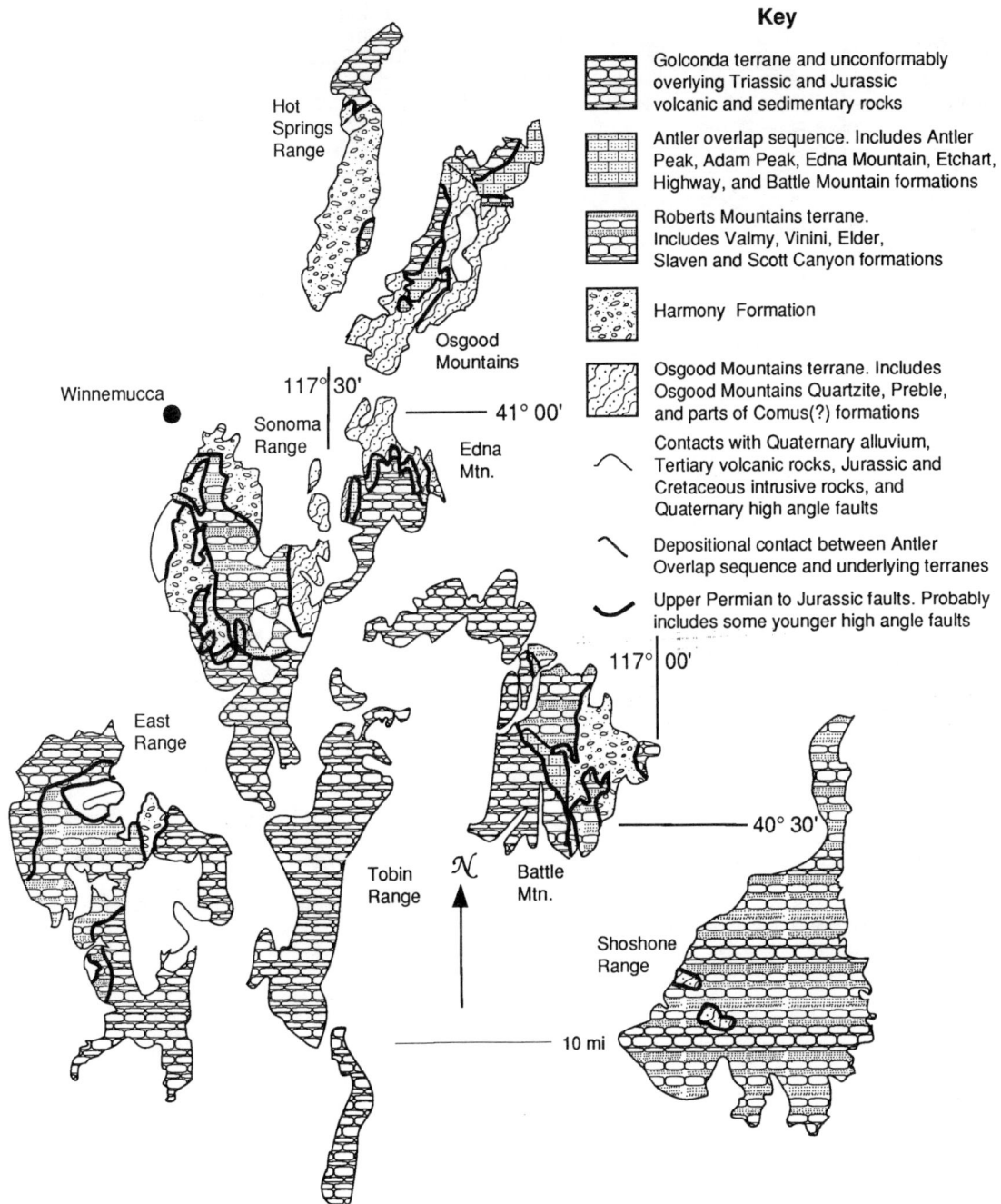

Figure 12. Paleozoic terranes of north-central Nevada.

polycrystalline quartz was derived from a quartz-rich sandstone. The sparse well-rounded, rutilated quartz grains reflect a different source than the angular quartz grains. Extensive reworking is indicated by the roundness of these grains. The Osgood Mountains quartzite, the overlying Battle Mountain Conglomerate of the Antler overlap sequence, and nearby large blocks of the Harmony Formation all contain similar rutilated quartz grains. The rounded quartz grains in all of these units could have had a common origin.

The absence of identifiable volcanic clasts in the Dry Hills sandstone suggests that a) it was not derived from a volcanic

source terrane and b) the feldspar clasts in it were not derived from a primary volcanic source, but more likely were reworked. Hotz and Willden (1964) referred to clasts of "rhyolitic to dacitic pyroclastics" in the Farrel Canyon Formation. These probably were the strongly recrystallized volcanic clasts that are now recognized as part of the Golconda melange. No significant felsic or silicic Paleozoic volcanism is recorded in the Dry Hills subterrane. The presence of abundant clay, detrital potassium feldspar, mica, and rounded pink zircons, also suggests that reworked continentally derived clastic material, not oceanic material was the primary provenance component for the Dry Hills sandstone.

796

Chert-Argillite

Red chert-basalt

Melange blocks

Calcarenite

Lithic

Calcareous-quartzose siliclastic

Carbonate-black chert-basalt

	Permian	Ochoan
		Guadalupian
		Leonardian
		Wolfcampian
	Pennsylvanian	Virgilian
		Missourian
		Desmoinesian
		Atokan
		Morrowan
	Mississippian	Chesterian
		Meramecian
		Osagean
		Kinderhookian
		Famennian

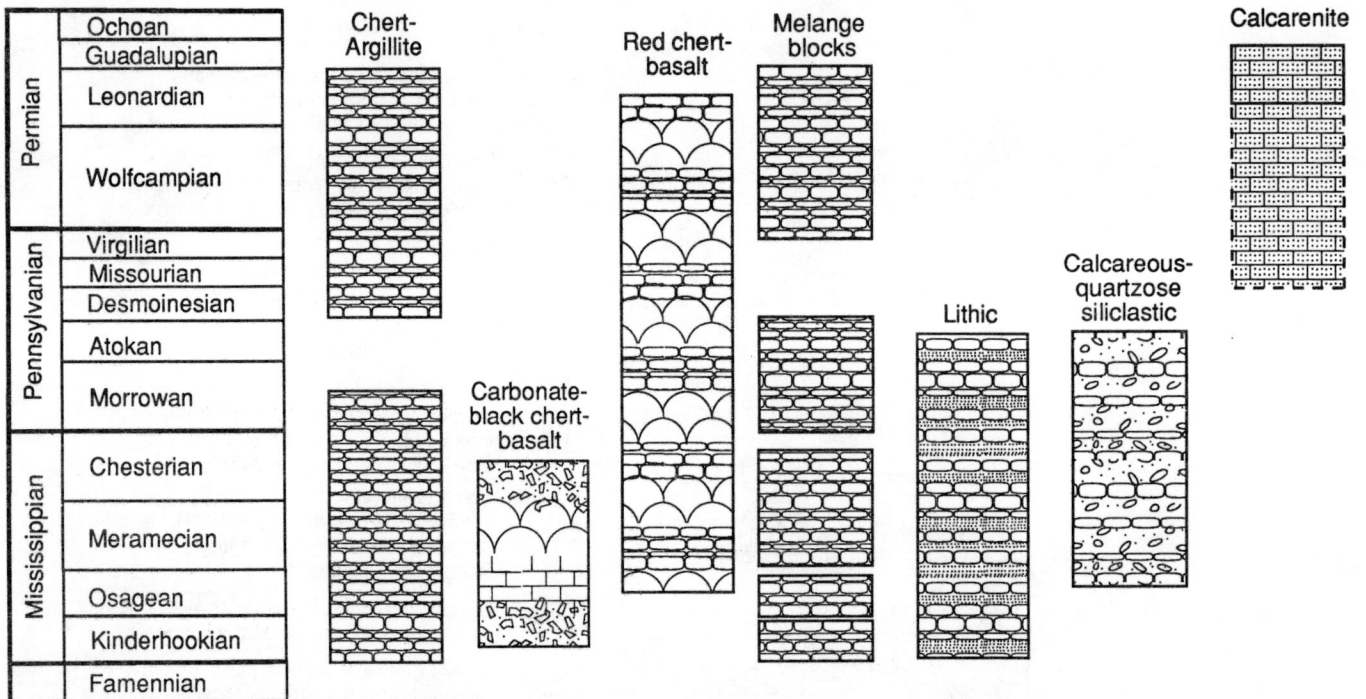

Figure 13. Composite stratigraphic columns for facies of the Golconda terrane. Ages are based on published data (Miller and others, 1982; 1984; Stewart and others, 1986; Murchey, 1989) and author's unpublished data (in prep.) which includes radiolarian, conodont, and macrofossil biostratigraphy. Age of calcarenite subfacies in the Poverty Peak subterrane is presently restricted to the Leonardian (solid lines). Elsewhere, similar lithologic assemblages are as old as middle Pennsylvanian (dashed lines).

The characteristics of the clasts which comprise the Dry Hills sandstone do not support an interpretation that it was derived from the oceanic sedimentary rocks of the Roberts Mountains terrane. The Harmony Formation and the Osgood Mountains terrane, or other similar mixed cratonal sedimentary rocks, represent more suitable source terranes.

Calcareous and quartzose siliclastic subfacies provenance

The well rounded, radiolarian-bearing chert clasts and quartzite pebbles interlayered with bedded radiolarian and spiculitic chert in the calcareous and quartzose siliclastic subfacies reflect a different source region from that of the lithic subfacies in the Dry Hills subterrane. Most of the clasts of this facies were derived from a reworked, clastic oceanic source terrane. Some of the Mississippian chert clasts apparently were derived from the Havallah Formation itself (Murchey, 1989), whereas the Middle and Late Devonian chert clasts and the Silurian conodonts indicate derivation from a lower Paleozoic source such as reworked beds from the Roberts Mountains terrane (Dickinson and others, 1983b; Snyder and Brueckner, 1983; Miller and others, 1984).

Calcarenite subfacies provenance

The differences between the calcarenite subfacies of Poverty Peak Unit (II), and the other externally-derived clastic rocks of the Golconda terrane are: a) the absence of interbedded radiolarian chert layers -Poverty Peak Unit (II) consists of interlayered graded beds of siltstone, clastic limestone and quartz arenite, b) the absence of chert-pebble conglomerate beds, c) the lack of significant amounts of clay, and d) the absence of macrofossils and bioclasts, which are common in the "Jory" siliclastic subfacies at Battle Mountain. It is similar to the Dry Hills sandstone in possessing small amounts of feldspar and mica, indicating a reworked, mature sediment source with a cratonal component.

By the Late Pennsylvanian and Early Permian, the Roberts Mountains terrane, the Harmony Formation, and the Osgood Mountains terrane were largely covered over by clastic and carbonate rocks of the Antler overlap sequence. The Pennsylvanian and Permian calcarenite subfacies of the Golconda terrane has a lithologic composition and age that is compatible with its derivation from reworked Antler overlap rocks or from a similar mature continental margin facies.

Summary

Figure 13 is a diagram showing age and lithologic characteristics for each facies of the Golconda terrane. Thick chert-argillite sequences lacking coarser interbedded rocks suggest deposition in quiet-water, open-marine setting. Intrabasinal clastic sedimentary rocks show evidence for a tectonically active, environmentally diverse basin. Interlayered Mississippian limestone,

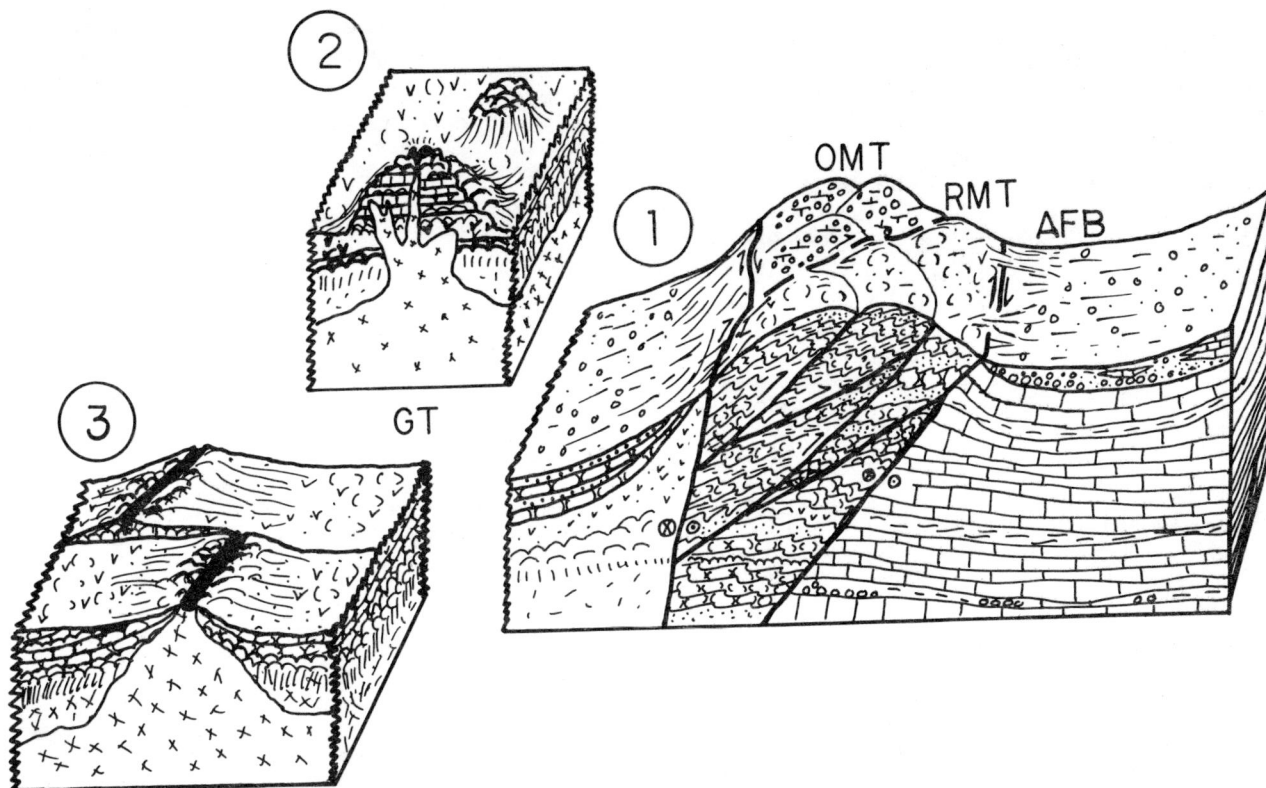

Figure 14. Schematic diagram of paleogeographic settings of the Golconda terrane (GT), the Osgood Mountains terrane (OMT), the Roberts Mountains terrane (RMT) and the Antler foreland basin (AFB). **Block 1** The Antler orogenic belt with the Roberts Mountains terrane (foreground) and the Osgood Mountains terrane (background) shedding debris eastward into the Antler foreland basin, and westward into the Golconda basin into regions that are now the Dry Hills subterrane and the "Jory" siliclastic facies. **Block 2** is a seamount representing the carbonate-black chert-basalt subfacies of the Home Ranch subterrane. **Block 3** is a spreading ridge (or basal seamount) setting representing the red chert-basalt subfacies of Unit (I) of the Poverty Peak subterrane. Terrane boundaries are represented as transpressive, transform faults. The relative paleogeographic relations between these three blocks are unknown. Not to scale.

chert, basalt, volcaniclastic breccia and debris-flow deposits in the carbonate-black chert-basalt subfacies suggest that seamount volcanic activity rose to shallow enough levels to support an environment where abundant invertebrate marine life thrived. Steep gradients on the sides of this intrabasinal volcano created olistostromal debris flows that chaotically mixed volcanic rocks with limestone and spiculitic debris. In the red chert-basalt subfacies, reworked crystal tuffs and basalt pillows and breccias interbedded with radiolarian cherts suggest actively forming oceanic or seamount crust in the Golconda basin throughout most of its history. Juxtaposition of clasts of different facies in a soft-sediment-deformed tuffaceous argillite matrix characterizes the melange subfacies. Soft-sediment and syn-tectonic deformation that created the melange subfacies is probably related to the accretion of the Golconda terrane to North America.

The extrabasinal sedimentary rocks of the Golconda terrane record paleogeographic changes and tectonic evolution in parts of the basin that were near the continental margin (Fig. 14, Block 1). In the early Mississippian and into the early Pennsylvanian deep-ocean crust was forming (Fig. 14, block 3) and a seamount (chain?) was rising to shallow-water depths offshore (Fig. 14, block 2). At the same time, near the continental margin, an influx of clastic debris from a source rich in cratonal material, such as the Harmony Formation or the Osgood Mountains terrane, formed clay-rich lithic sandstone interlayered with radiolarian-rich chert. Elsewhere along the margin, the Golconda basin was receiving reworked oceanic material (bedded cherts, argillites, and quartz arenites) at least some of which was lower Paleozoic in age, and likely derived from the rocks of the Roberts Mountains terrane. Offshore, new ocean or seamount crust was forming in the Golconda basin well into the Early Permian, by which time much of the Antler orogenic belt had been worn down and covered by the clastic and calcareous Antler overlap sequence. As offshore, deep-water portions of the basin floor moved closer to the craton, deep marine, red chert and basalt were covered by mature calcarenites derived from a tectonically stabilized continental margin.

798

CONCLUSION

The close parallel in time between erosion of the Antler orogenic belt, deposition in the Antler foreland basin, and deposition of the Golconda extrabasinal clastic facies, suggests a common tectonic evolution among these three assemblages. Currently debated models of the Antler orogeny summarized by Nilsen and Stewart (1980) involve the closure of a basin and the accretion of an arc in Late Devonian-early Mississippian time to the North American craton. The geologic records in the Paleozoic continental margin, the Roberts Mountains terrane, and the northern and eastern portions of the Golconda terrane do not support these models. The sedimentologic record on the craton and in the adjacent Paleozoic terranes requires the accretion of oceanic and continental sedimentary rocks by Early Mississippian time, but there is no evidence in this record that a volcanic arc was involved in the accretion. The Mississippian clastic sedimentary rocks of the Golconda terrane should have abundant arc-derived volcanic sedimentary rocks mixed with ocean- and craton-derived rocks. If an arc had completely subsided shortly after accreting (Speed and Sleep, 1982), it would still have left a sedimentologic record in the adjacent terranes. Instead, the early and middle Mississippian clastic rocks of the Golconda terrane reflect only reworked continent- and ocean-derived sediment sources, and show no evidence of arc-derived volcanic debris.

By the early Mississippian, the Golconda basin was large enough to support active ocean-floor and seamount volcanism. Because the rocks of the Golconda and Roberts Mountains terranes overlap in age (Murchey, 1989), the basin(s) where they formed could not have evolved by closing and re-opening in the same place as most models propose (Nilsen and Stewart, 1980). The Golconda basin already existed by the time the Roberts Mountains terrane was accreted to the North American margin.

The western margin of North America was tectonically very active from the Late Devonian through the Permian. Clastic sedimentary rocks in part of the Golconda terrane record influx of sediment from an active continental margin during this period, whereas other parts of the terrane record the independent history of part of a basin removed from cratonal influences. A scenario for the Antler orogeny that is more compatible with geologic evidence than existing models is one that involves a long-lived, episodic, transpressive, transform regime (McCollum and McCollum, 1989; Murchey, 1989; Jones, 1990), broadly similar to the tectonic regimes of the late Mesozoic and Cenozoic western North American margin. Discussions of tectonic events of Paleozoic western North America refer to several different, poorly understood orogenies (Ketner, 1977; Snyder and others, 1989; Trexler and others, 1990) that could be explained by a transpressive fault system. This transpressive regime would have been responsible for the accretion of oceanic and continental fragments including the Roberts Mountains terrane, the Golconda terrane, the Osgood Mountains terrane, and the Harmony Formation, and would also have caused episodic deformational events along the continental margin throughout the upper Paleozoic.

ACKNOWLEDGMENTS

The author gratefully acknowledges support for this project from the American Association of University Women, U. C. Berkeley Chancellor's Patent Fund, a Geological Society of America Penrose grant, and a Sigma Xi Grant-in-Aid-of Research. Fossil identification and dating provided by Anita Harris, Bruce Wardlaw, Cal Stevens, and Rex Hanger was critical to this study. Able field assistance was provided by Argyro Adams, Barbara Atkin, Mark Caruso, Chris Garvin, Cameron Jones and Daniel Stein. Gracious accommodations and support during field work was provided by Stan and Janice Klauman and family, Barbara and Steve Atkin, and George and Randy Sanders. The Geology Department faculty and staff at the University of California, Santa Barbara, including James Kennett, Cathy Busby-Spera, and David Pierce were especially generous in providing laboratory equipment and work space. Helpful review by Dave Jones greatly improved the manuscript.

REFERENCES CITED

Brew, D. A., 1971, Mississippian stratigraphy of the Diamond Peak area, Eureka County, Nevada: U.S. Geological Survey Professional Paper 661, 79 p.

Burchfiel, B. C., and Davis, G. A., 1972, Structural framework and evolution of the southern part of the Cordilleran orogen, western United States: American Journal of Science, v. 272, p. 97-118.

-------, and -------, 1975, Nature and controls of Cordilleran orogenesis, western United States: Extensions of an earlier synthesis: American Journal of Science, v. 275A, p. 363-396.

Cowan, D. S., 1985, Structural styles in Mesozoic and Cenozoic melanges in the western Cordilera of North America: Geological Society of America Bulletin, v. 96, n. 4, p. 451-462.

Dickinson, W. R., Beard, L. S., Brakenridge, G. R., Erjavec, J. L., Ferguson, R. C., Inman, K. F., Knepp, R. A., Lindberg, F. A., and Rybert, P. T., 1983a, Provenance of North American Phanerozoic sandstones in relation to tectonic setting: Geological Society of America Bulletin, v. 94, n. 2, p. 222-235.

Dickinson, W. R., Harbaugh, D. W., Saller, A. H., Heller, P. L., and Snyder, W. S., 1983b, Detrital modes of upper Paleozoic sandstones derived from Antler orogen in Nevada: Implications for nature of Antler orogeny: American Journal of Science, v. 283, n. 6, p. 481-509.

Erickson, R. L., and Marsh, S. P., 1974a, Geologic map of the Golconda quadrangle, Humboldt County, Nevada: U.S. Geological

Survey Geologic Quadrangle Map GQ-1174, scale 1:24,000.

———, and ———, 1974b, Geologic map of the Iron Point quadrangle, Humboldt County, Nevada: U.S. Geological Survey Geologic Quadrangle Map GQ-1175, scale 1:24,000.

———, and ———, 1974c, Paleozoic tectonics in the Edna Mountain Quadrangle, Nevada: Journal of Research, U.S. Geological Survey, v. 2, n. 3, p. 331-337.

Fagan, J. J., 1962, Carboniferous cherts, turbidites, and volcanic rocks in northern Independence Range, Nevada: Geological Society of America Bulletin, v. 73, n. 5, p. 595-612.

Harbaugh, D. W., and Dickinson, W. R., 1981, Depositional facies of Mississippian clastics, Antler foreland basin, central Diamond Mountains, Nevada: Journal of Sedimentary Petrology, v. 51, n. 4, p. 1223-1234.

Holdsworth, B. K., and Jones, D. L., 1980a, Preliminary radiolarian zonation for Late Devonian through Permian time: Geology, v. 8, n. 6, p. 281-285.

———, and ———, 1980b, A provisional radiolaria biostratigraphy, Late Devonian through Late Permian: U.S. Geological Survey Open-File Report 80-876, 33 p.

Hotz, P. E., and Willden, R., 1964, Geology and mineral deposits of the Osgood Mountains quadrangle, Humboldt County, Nevada: U. S. Geological Survey Professional Paper 431, 128 p.

Jones, A. E., 1988, The Poverty Peak subterrane of the Golconda allochthon in the northern Hot Springs Range, Humboldt County, Nevada: Geological Society of America Abstracts with Programs, v. 20, n. 3, p. 171.

Jones, D. L., 1990, Synopsis of late Paleozoic and Mesozoic terrane accretion within the Cordillera of western North America: Philosophical Transactions of the Royal Society of London, v. A 331, p. 479-486.

Kennett, J., 1982, Marine Geology: Englewood Cliffs, N. J., Prentice-Hall Inc., 813 p.

Ketner, K. B., 1977, Late Paleozoic orogeny and sedimentation, southern California, Nevada, Idaho, and Montana, in Stewart, J. H., Stevens, C. H., and Fritsche, A. E., eds., Paleozoic Paleogeography of the Western United States: Society of Economic Paleontologists and Mineralogists, Pacific Section, Pacific Coast Paleogeography Symposium 1, p. 363-369.

Laule, S. W., Snyder, W. S., and Ormiston, A. R., 1981, Willow Canyon Formation, Nevada: An extension of the Golconda allochthon: Geological Society of America Abstracts with Programs, v. 13, p. 66.

McCollum, L. B., and McCollum, M. B., 1989, The Antler Orogeny: A transpressional orogen within a transcurrent fault system: Geological Society of America Abstracts with Programs, v. 21, n. 5, p. 114.

Miller, E. L., Kanter, L. R., Larue, D. K., Turner, R. J., Murchey, B., and Jones, D. L., 1982, Structural Fabric of the Paleozoic Golconda Allochthon, Antler Peak Quadrangle, Nevada: Progressive Deformation of an Oceanic Sedimentary Assemblage: Journal of Geophysical Research, v. 87, n. B5, p. 3795-3804.

Miller, E. L., Holdsworth, B. K., Whiteford, W. B., and Rodgers, D., 1984, Stratigraphy and structure of the Schoonover sequence, northeastern Nevada: Implications for Paleozoic plate-margin tectonics: Geological Society of America Bulletin, v. 95, n. 9, p. 1063-1076.

Murchey, B., 1982, Chert facies in the Havallah sequence near Battle Mountain, Nevada: Geological Society of America Abstracts with Programs, v. 14, p. 219.

Murchey, B. L., 1989, Late Paleozoic siliceous basins of the western Cordillera of North America (Nevada, California, Mexico, and Alaska): Three studies using radiolarians and sponge spicules for biostratigraphic, paleobathymetric, and tectonic analyses: Ph.D. thesis, University of California, Santa Cruz, 189 p.

Murchey, B. L., and Harwood, D. S., in press, Age and depositional setting of siliceous sediments in the upper Paleozoic Havallah sequence near Battle Mountain, Nevada: Implications for the paleogeography and structural evolution of the western margin of North America, in Harwood, D. S., and Miller, M. M., eds., Paleozoic and early Mesozoic paleogeographic relations; Sierra Nevada, Klamath Mountains, and related terranes: Boulder, Colorado, Geological Society of America, Special Paper 255

Nilsen, T. H., and Stewart, J. H., 1980, Penrose Conference Report: The Antler orogeny-Mid Paleozoic tectonism in western North America: Geology, v. 8, n. 6, p. 298-302.

Pessagno, E. A., Jr., and Newport, R. L., 1972, A technique for extracting radiolaria from radiolarian cherts: Micropaleontology, v. 18, n. 2, p. 231-234.

Poole, F. G., 1974, Flysch deposits of Antler foreland basin, western United States, in Dickinson, W. R., eds., Tectonics and Sedimentation: Society of Economic Paleontologists and Mineralogists, Special Publication 22, p. 58-82.

Roberts, R. J., 1964, Stratigraphy and structure of the Antler Peak Quadrangle, Humboldt and Lander Counties, Nevada: U.S. Geological Survey Professional Paper 459-A, 93 p.

Roberts, R. J., Hotz, P., E., Gilluly, J., and Ferguson, H. G., 1958, Paleozoic rocks of north-central Nevada: Bulletin of the American Association of Petroleum Geologists, v. 42, n. 12, p. 2813-2857.

Rowell, A. J., Rees, M. N., and Suczek, C. A., 1979, Margin of the North American continent in Nevada during Late Cambrian time: American Journal of Science, v. 279, p. 1-18.

Saller, A. H., 1980, Depositional setting of post-Antler Pennsylvanian strata in

800

north-central Nevada: M.S. thesis, Stanford University, Stanford, California, 118 p.

Sano, H., and Kanmera, K., 1988, Paleogeographic reconstruction of accreted oceanic rocks, Akiyoshi, southwest Japan: Geology, v. 16, n. 7, p. 600-603.

Silberling, N. J., 1975, Age Relationships of the Golconda Thrust Fault, Sonoma Range, North-Central Nevada: Geological Society of America Special Paper 163, 28 p.

Snyder, W. S., and Brueckner, H. K., 1983, Tectonic evolution of the Golconda allochthon, Nevada: Problems and perspectives, in Stevens, C. A., ed., Pre-Jurassic rocks in western North America suspect terranes: Society of Economic Paleontologists and Mineralogists, Pacific Section, v. 32, p. 103-123.

Snyder, W. S., Spinosa, C., and Gallegos, D. M., 1989, Late Pennsylvanian to Permian tectonism along the western United States continental margin: Geological Society of America Abstracts with Programs, v. 21, n. 5, p. 147.

Speed, R. C., 1979, Collided Paleozoic microplate in the western United States: Journal of Geology, v. 87, p. 279-292.

Speed, R. C., and Sleep, N. H., 1982, Antler orogeny and foreland basin: A model: Geological Society of America Bulletin, v. 93, n. 9, p. 815-828.

Stahl, S. D., Huysken, K. T., and Billingham, A., 1989, The upper "Preble" Formation in the Sonoma Range of north-central Nevada: Structural geology, age, and implications: Geological Society of America Abstracts with Programs, v. 21, n. 5, p. 148.

Stewart, J. H., Murchey, B., Jones, D. L., and Wardlaw, B. R., 1986, Paleontologic evidence for complex tectonic interlayering of Mississippian to Permian deep-water rocks of the Golconda allochthon in Tobin Range, north-central Nevada: Geological Society of America Bulletin, v. 97, n. 9, p. 1122-1132.

Trexler, J. H., Jr., and Nitchman, S. P., 1990, Sequence stratigraphy and evolution of the Antler foreland basin, east-central Nevada: Geology, v. 18, n. 5, p. 422-425.

Trexler, J. H., Snyder, W. S., Cashman, P. H., Gallegos, D. M., and Spinosa, C., 1990, An orogenic hierarchy: Examples from western North America: Geological Society of America, Abstracts with Programs, v. 22, n. 7, p. A328.

Watkins, R., and Browne, Q. J., 1989, An Ordovician continental-margin sequence of turbidite and seamount deposits in the Roberts Mountains allochthon, Independence Range, Nevada: Geological Society of America Bulletin, v. 101, n. 5, p. 731-741.

Whiteford, W. B., 1984, Age relations and provenance of upper paleozoic sandstones in the Golconda allochthon: the Schoonover sequence, northern Independence Mountains, Elko Co., Nevada: M.S. thesis, Stanford University, Stanford, Ca, 126 p.

Whiteford, W. B., in press, Paleogeographic setting of the Schoonover sequence, Nevada, and implications for the late Paleozoic margin of western North America, in Harwood, D. S., and Miller, M. M., eds., Paleozoic and early Mesozoic paleogeographic relations; Sierra Nevada, Klamath Mountains, and related terranes: Geological Society of America Special Paper 255

EARLY CAMBRIAN BIOGEOGRAPHY
AND THE PREHISTORY OF EARLY SKELETOGENOUS ANIMALS

Philip W. Signor

Department of Geology, University of California
Davis, California 95616
and White Mountain Research Station
3000 East Line Street, Bishop, California 93514

ABSTRACT

Paleobiogeographic distributions of animals reflect the complex interplay of biological and physical processes acting over geological time. In particular, plate tectonics and the origination and dispersal of new clades play fundamental roles in generating patterns of faunal distribution. Ranges expand through vicariance events or dispersal and contract through local and regional extinctions. Viewed as historical phenomena, paleobiogeographic distributions can be employed to infer prior tectonic and evolutionary events. For example, the distribution of modern terrestrial and marine faunas reflects the interaction of Mesozoic and Cenozoic plate tectonic history and the contemporaneous evolutionary history of the biosphere.

The paleobiogeographic distribution of skeletogenous animals has been strongly provincial since their first appearance in the early Phanerozoic. Classic Early Cambrian examples are the olenellid and redlichiid trilobite provinces, and other examples include echinoderms, archaeocyathans, and "small shelly fossils". The existence of these biogeographic provinces suggests an extended prior history of evolutionary and vicariance events.

The Proterozoic history of skeletogenous organisms and their ancestors is a contentious subject, with many authors arguing that skeletogenous clades have no significant prehistory prior to their appearance in the fossil record. The existence of trilobite provinces dominated by different suborders, for example, suggests that trilobites evolved, dispersed or were separated by plate tectonics, and then evolved independently for an extended period prior to their appearance in the fossil record. Similar arguments can be applied to other clades.

The paleobiogeographic distribution of Early Cambrian organisms also bears upon the timing of the breakup of the putative late Proterozoic supercontinent. Separate provinces would not have developed had the various plates remained united until the end of the Proterozoic.

INTRODUCTION

The sudden appearance of Metazoa in the fossil record, after more than three billion years of life on Earth without animals, has been the focus of a longstanding controversy among paleontologists. The nearly simultaneous appearance of a multitude of clades, including the ancestors of many modern phyla and classes, has been a salient issue in that controversy (Stanley, 1976; Brasier, 1979; Glaessner, 1984; Valentine et al., 1990). Why did three billion years elapse between the appearance of life on Earth and the appearance of animals in the fossil record? Why would so many different kinds of animals appear nearly simultaneously? Discovery of late Proterozoic (Vendian or Ediacaran) large, multicellular organisms in Africa (Gurich, 1930, 1933), Australia (Sprigg, 1947), eastern Europe (Urbanek and Rozanov, 1979), Canada (Narbonne and Hofmann, 1987), and elsewhere has not provided the long-sought resolution. Indeed, considerations of the evolutionary significance of these Proterozoic fossils have evolved into a controversy about their phylogenetic relationships to later faunas (e.g., Seilacher, 1985, 1989; Jenkins, 1985; McMenamin, 1986; Conway Morris, 1985, 1987).

To this day, there are persistent debates about the tempo of metazoan evolution in the late Proterozoic and Early Cambrian. Some paleontologists hold that the sudden appearance of Metazoa in the fossil record is a valid reflection of the actual tempo of evolutionary change in the late Proterozoic and early Phanerozoic (Stanley, 1976; McMenamin and Schulte McMenamin, 1989; Valentine et al., 1990). Certainly, ongoing studies of early faunas have tended to accentuate the apparent suddenness, rather than reveal more gradual trends (Valentine et al., 1990). Others maintain that biases in the fossil record make the record of an inherently gradual phenomenon appear abrupt (e.g., Walcott, 1910; Axelrod, 1958; Durham, 1978). A variation on this latter view is that gradual evolutionary change led to a deceptively spasmodic fossil record (Hutchinson, 1961; Runnegar, 1982a; Boaden, 1989). Finally, Sepkoski (1978) portrayed the metazoan radiation as an exponential process, with a late Proterozoic origin of animals followed by the appearance of increasingly larger numbers of new higher taxa through the Middle Cambrian.

Other types of data and analysis have been brought to bear on the problem. Runnegar (1982b) has attempted to infer the age of the origin of animals using biochemical techniques. He placed the age of the last common ancestor of the annelids,

In Cooper, J.D., and Stevens, C.H., eds., 1991, **Paleozoic Paleogeography of the Western United States-II:** Pacific Section SEPM, Vol, 67, p. 801-810.

mollusks, phoronids, arthropods, and vertebrates at .9-1.0 Ga. However, Erwin (1989) has criticized this conclusion on methodological grounds. The fossil record of traces also has been employed to infer the presence of Metazoa in the Proterozoic. There are no convincing traces older than the late Proterozoic (Byers, 1982; Crimes, 1989), which suggests that the first Metazoa actually evolved at that time.

The thesis of this paper is that, despite evidence from the fossil record for an apparently sudden origin and diversification of animals, there was a significant duration of evolutionary history predating the actual appearance of animals in the fossil record. This conclusion is drawn from three propositions: 1) biogeographic distributions reflect an accumulation of dispersal and vicariance events, together with climatic influences, over geologic time; 2) Early Cambrian faunas are provincial in their distribution; and 3) most Early Cambrian organisms apparently had a limited capacity for dispersal, relative to the paleogeographic distribution of the continents; thus, Early Cambrian biogeography primarily represents the accumulation of vicariance events that occurred over geological, not ecological, time (i.e., on the scale of millions of years).

In the subsequent sections of this paper, the evidence for each proposition will be presented and examined in detail. The arguments presented herein rely extensively upon Early Cambrian faunas from western North America, which constituted a distinct faunal province at the time (Signor, 1991). Subsequently, the timing of the origin of animals will be reconsidered in light of Early Cambrian paleobiogeography.

I. BIOGEOGRAPHY AND EVOLUTION

The role of physical geography in determining the distribution of species and higher taxa is sufficiently well documented that a detailed discussion here is not necessary. The reader who wishes to delve into this field can consult any of a number of recent books on biogeography (e.g., MacArthur, 1972; Valentine, 1973; Cox and Moore, 1980). The significance of biogeographic patterns in evolution is also the frequent subject of research, and only the major points are repeated here. In essence, geographical barriers to dispersal guarantee the independent evolutionary histories of species in different biogeographic regions. Broad oceans prevent the dispersal of either shelf-dwelling marine invertebrates or terrestrial organisms (Simpson, 1943). Likewise, land masses bar the dispersal of marine organisms. The absence of dispersal and concomitant gene flow ensures the evolutionary independence of the different regions (for recent references, see McKerrow and Scotese, 1990).

Biogeographers define faunas on the basis of endemic taxa, but there is no concordance regarding the names of biogeographic units or the numbers of endemic taxa that define such units (see Burrett and Richardson, 1980; Bambach, 1990). In this paper, the term province refers to a geographic region with a significant fraction of endemic taxa. These provinces are probably realms, sensu Bambach (1990), but the distinction is not important for the arguments presented here.

II. PROTEROZOIC AND EARLY CAMBRIAN PALEOGEOGRAPHY AND PALEOBIOGEOGRAPHY

The paleopositions of cratons in the Early Cambrian (Fig. 1) are not well constrained by paleomagnetic data, although reconstructions that are based solely upon paleomagnetic data and other geological evidence are desirable for this analysis (e.g., Smith et al., 1981). Recent attempts (e.g., Ziegler et al., 1979; Parrish et al., 1986; Scotese and McKerrow, 1990) to determine the ancient positions of cratons and terranes rely upon the biogeographic distributions of fossil taxa, in addition to other data (e.g., the distributions of facies, paleomagnetic data, and evidence from tectonic and paleoclimatic studies). Thus, there is an element of circularity involved in employing the distribution of cratons to infer the existence and extent of biogeographic provinces among those land masses. Nevertheless, several important points emerge from a consideration of Early Cambrian paleogeography.

There has been considerable speculation regarding the existence of a supercontinent late in the Proterozoic. Piper (1976, 1983, 1987), Klootwijk (1980), Bond et al. (1983, 1984), Donovan (1987), and McMenamin and Schulte McMenamin (1989) envision most of the continents gathered together into a single land mass, with possibly one or more continents as outliers. McMenamin and Schulte McMenamin (1989) recently have given the name Rodinia to this poorly known, and perhaps apocryphal, supercontinent. Their terminology will be employed here. (The name Palaeopangea has also been applied to this land mass [e.g., Debrenne et al., 1990].) McMenamin and Schulte McMenamin (1989: their Fig. 6.1) suggest Rodinia incorporated Africa, India, South America, Antarctica, and Australia with Baltica, North America, Siberia, and Kazakhstania as outliers.

The history of Rodinia is as uncertain as its composition. Some have argued that the breakup of the continent occurred in the latest Proterozoic (e.g., Bond et al., 1983, 1984; Angevine and Heller, 1986; Lindsay et al., 1987) while others contend the breakup began as early as 900 Ma (Stewart, 1972; Piper, 1983, 1987; also see Van der Voo et al., 1984). The sequence of the breakup is unknown, which is not surprising in light of the disagreement over the composition of Rodinia. Whatever the timing of the breakup, there is considerable agreement that the continents were dispersed by the end of the

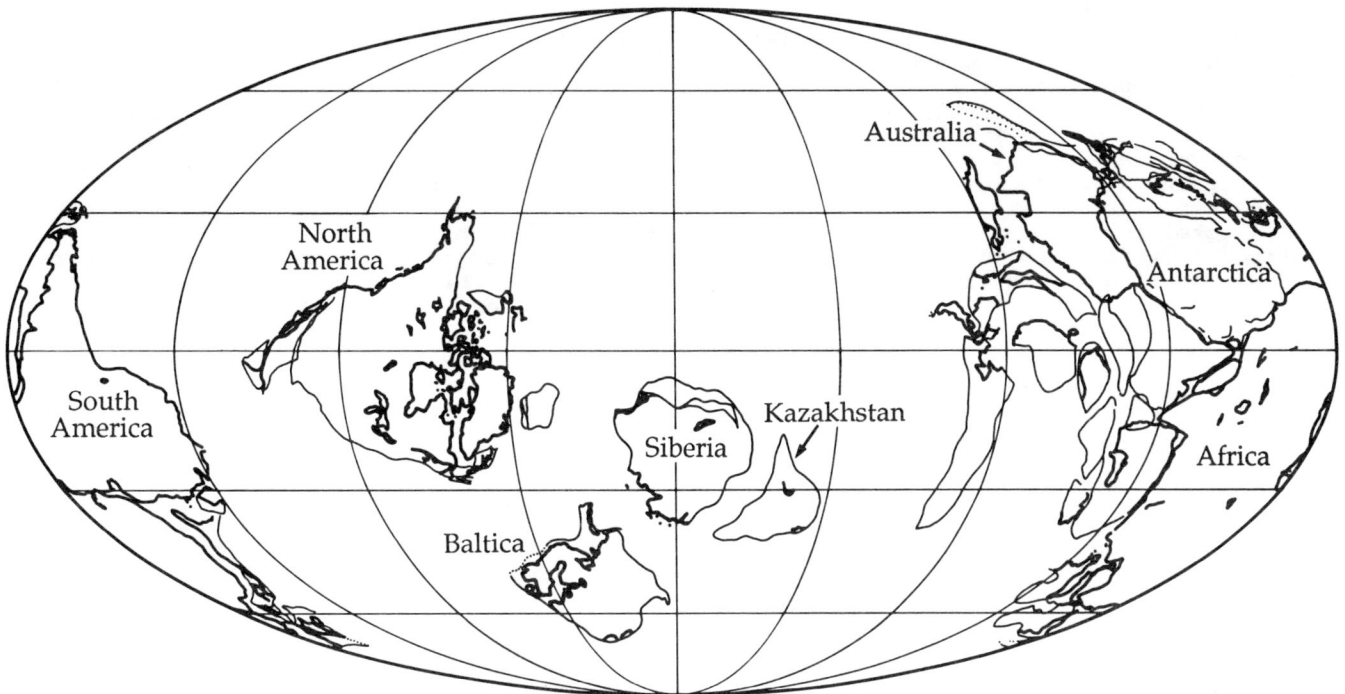

Early Cambrian

Figure 1. Reconstructed positions of continents in the Early Cambrian (after Scotese and McKerrow, 1990). The continents are mostly distributed in temperate to tropical latitudes.

Proterozoic (Ziegler *et al.*, 1979; Smith *et al.*, 1981; Parrish *et al.*, 1986; Scotese and McKerrow, 1990). The time that elapsed between the breakup of Rodinia and the beginning of the Early Cambrian must be on the order of tens of millions of years, inasmuch as rates of plate tectonic movement, while variable in time, have never been rapid.

In the Early Cambrian, the continents were apparently dispersed in low latitudes, with Baltica and South America extending into southern temperate latitudes (Fig. 1: Ziegler *et al.*, 1979; Parrish *et al.*, 1986; Scotese and McKerrow, 1990: see Smith *et al.* 1981 for a contrasting view). This reconstruction is supported by the distribution of algal-archaeocyathan reefs on most continents, with the notable and not unexpected exceptions of Baltica and South America (Zhuravlev, 1986). Early Cambrian archaeocyathan bioherms are found on both coasts of North America and along the entire present-day western margin of the continent, from Sonora, Mexico to northwestern Canada (Signor, 1991). The distribution of cratons in low paleolatitudes suggests that climatically induced faunal variation was minimal in the Early Cambrian.

Late Proterozoic Paleobiogeography

Late Proterozoic paleobiogeography is poorly known, primarily because the floras and faunas of the Proterozoic are relatively rare. Nevertheless, Vidal and Knoll (1983) conclude that late Proterozoic acritarchs tend to be cosmopolitan in their distribution, as might be expected for the fossil remains of planktonic algae. The only skeletogenous organism known from the Proterozoic, *Cloudina*, also has a cosmopolitan distribution (Grant, 1990: Grant considers *Sinotubulites*, possibly a second skeletogenous form, to be a junior synonym of *Cloudina*).

McMenamin (1982) argued that there were two late Proterozoic faunal provinces, one characterized by a benthic, soft-bodied Vendian fauna and the other by a skeletogenous fauna. McMenamin's benthic, soft-bodied fauna includes *Spriggina*, *Dickinsonia*, and other bilaterally symmetrical forms but excludes medusoids and frond-like fossils that are known to be cosmopolitan. The skeletogenous fauna includes *Anabarites*, *Protohertzina* and other early elements associated with late Proterozoic (Nemakit-Daldyn) and earliest Cambrian (Tommotian) faunas, and their temporal equivalents. This idea has not gained general acceptance because the fossil record of acritarchs fails to support the suggestion that the late Proterozoic soft-bodied and Tommotian skeletogenous faunas were coeval.

Precise ages of cosmopolitan Proterozoic fossils might be useful constraints on the earliest development of biogeographic provinciality in the fossil record. However, radiometric dates on late Proterozoic and earliest Cambrian sediments and faunas vary considerably and are not sufficiently accurate for this purpose (see Cowie and Harland, 1989). Furthermore, the evidence from the fossil record does not constrain the dates of the development of biogeographic provinciality among early, small, askeletogenous bilaterians, a group that probably includes the ancestors of clades appearing in the Cambrian radiation (e.g., Valentine, 1989).

Early Cambrian Biogeography

Trilobites

For many years, trilobites have been the best known clade of Early Cambrian metazoans, primarily because of their well-established utility as index fossils. Not surprisingly, the existence of geographically differentiated faunas in the Early Cambrian was first recognized by trilobite workers (Walcott, 1914; Repina, 1972; Cowie, 1971; Palmer, 1973, 1979: also see Theokritoff, 1979, 1985). There were two Early Cambrian trilobite provinces, each recognized by characteristic clades of trilobites. North America, South America, Avalon, and northwestern Europe formed the olenellid province (Cowie, 1971; Palmer, 1973, 1979). Australia, China, and Antarctica constituted the redlichiid province (Palmer, 1979). Some mixing of the two faunas occurred in Siberia, although the degree of mixing varied through time (Cowie, 1971; Repina, 1972; Palmer, 1979).

Paleoecological controls on the distribution of trilobite faunas complicate the biogeographic interpretations (Palmer, 1973, 1979). In western North America, the inner and outer detrital belts and the medial carbonate belt tend to have characteristic faunas, with more cosmopolitan species occupying offshore habitats (Palmer, 1979). Also, there are temporal trends in Early Cambrian trilobite biogeography. The earliest western North American forms (e.g., *Fallotaspis*, *Holmia*) are also found in Europe (Palmer, 1979) while younger genera are less widely distributed. Nevertheless, the olenellid and redlichiid provinces are distinct from the first appearance of the eponymous clades in the fossil record. The patterns established in the Early Cambrian persisted through the Middle Cambrian, although different names have been applied to the younger provinces (Jell, 1975). (Burrett and Richardson [1980] recognized four Early Cambrian realms, but the geography of their realms generally corresponds to the provinces recognized by earlier workers, with subdivision of the olenellid province into two realms and reassignment of the redlichiid province and some intermediate areas into two more realms.)

Trilobite systematists have traditionally treated olenellids and redlichiids as closely related suborders within the order Redlichiida. However, more recent workers have questioned this relationship. Lauterbach (1980: see Fortey, 1990) has even suggested that the olenellids should not be included within the Trilobita! A recent cladistic analysis of the phylogenetic relationships of trilobites indicates that the olenellids are the sister-group to the redlichiids plus all other trilobites (Fortey, 1990). The implication of this conclusion is that the two clades diverged very early in the evolutionary history of the Trilobita.

Archaeocyathans

Early Cambrian archaeocyathan provinces appear to form later than the trilobite provinces (Zhuravlev, 1986; Debrenne et al., 1990), but this pattern is clouded by controversy over the actual age of the Tommotian Stage at its type locality in Siberia. Moczydlowska and Vidal (1988) correlate the upper Tommotian of the Siberian Platform with Atabanian strata of the Baltic Platform. If their correlations prove correct, our present understanding of archaeocyathan patterns and sequence of dispersal will need to be reconsidered.

Occurrences of archaeocyathans are also strongly tied to the presence of carbonate facies, deposited in shallow, warm environments, in the stratigraphic record. Among Early Cambrian clades, archaeocyathans are the most tightly linked to a limited paleoenvironment (Mount and Signor, in press). The sequence of appearances of archaeocyathans on different continents could, therefore, reflect environmental influences and not evolutionary or biogeographic patterns.

On the basis of currently accepted correlations, archaeocyathans appear first in Tommotian strata on the Siberian Platform and begin to appear in other areas of the world in the Atabanian Stage. By the end of the Atabanian, they attain a global distribution and immediately become highly endemic (Zhuravlev, 1986). An endemic fauna dominated by ethmophyllid and syringocnemid archaeocyathans developed in western North America (Debrenne and Rozanov, 1983; Zhuravlev, 1986) and, to a lesser extent, in eastern North America. Together with calcareous algae, these organisms formed bioherms that occur from Sonora, Mexico to northwestern Canada. The taxonomic composition of these bioherms is remarkably homogeneous (e.g., Debrenne et al., 1989), with a number of species occurring throughout the region.

Zhuravlev (1986) also recognizes an Afro-Siberian-Antarctic province that he subdivides into Afro-European (Morocco, Spain, and Sardinia), Austral-Antarctic, and Siberian Provinces. The archaeocyathans of Australia and Antarctica are quite similar at the species level (Debrenne and Kruse, 1986). Zhuravlev (1986) contends that the archaeocyathan faunas of southern China are also similar to the Austral-Antarctic faunas. These patterns of faunal similarity match those previously recognized on the basis of the trilobite faunas (e.g., Cowie, 1971; Palmer, 1973, 1979; Jell, 1975).

Interestingly, Zhuravlev (1986) notes that the later North American archaeocyathan faunas were dominated by genera with aporose septae, unlike other regions of the world. At the same time, clades characterized by pectinate tabulae were common in Siberia and Australia. Presumably, this reflects the characters of the archaeocyathans that originally colonized the region.

Morphological evolution in archaeocyathans is as rapid as observed for any group, as indicated by their extraordinary diversification and equally rapid collapse in the Early Cambrian (Brasier, 1982; Zhuravlev, 1986; Debrenne *et al.*, 1990). Yet the evolution of endemic faunas, distinct at the level of genera, required at least the duration of a stage. While undue weight cannot be placed on comparisons of taxonomic levels between different clades, one might surmise that the morphological differences between redlichiid and olenellid trilobites would require a longer duration of independent evolution.

Small Shelly Fossils

Recent interest in other Early Cambrian taxa has produced a vastly expanded data base on fossils known collectively as "small shelly fossils" (SSFs). These include early mollusks, inarticulate brachiopods, and an impressive array of *incertae sedis* (e.g., Matthews and Missarzhevsky, 1975; Rozanov, 1986). Unfortunately, the taxonomy of SSFs remains in a primitive state; oversplit taxa, regional synonyms, and misinterpreted fossils still abound (see summary in Brasier, 1989a). Biogeographic analyses based upon such data are necessarily suspect. Nevertheless, it is becoming clear that the Early Cambrian distribution of small shelly fossils is strikingly provincial. Brasier (1989b) has recognized three paleogeographic belts stretching from southern Europe to Asia that are characterized by similar faunas: an Inner Paleotethyan Belt, an Outer Paleotethyan Belt, and an Outer Mongolian Belt. The faunas characteristic of each of these belts are distinct from other areas, such as North America, the Siberian Platform, or Australia and Antarctica. Likewise, Evans and Rowell (1990) observe strong faunal similarities among Early Cambrian kennardiids from Antarctica and Australia, similar to the pattern in archaeocyathan distribution observed by Debrenne and Kruse (1986).

Some SSF genera have a global distribution. Small, corrugated, cap-shaped, phosphatic fossils known as *Lapworthella* are cosmopolitan, but individual species are restricted to particular areas. *Lapworthella filigrana* was originally described from northwest Canada (Conway Morris and Fritz, 1984) and is now known to occur in Lower Cambrian sediments from northwestern Canada to Sonora, Mexico. Early Cambrian faunas are quite similar through this region, and apparently represent a single paleobiogeographic region within the olenellid province (Signor, 1991). The extended range of *L. filigrana* indicates a substantial capacity for dispersal, but the species is restricted to western North America.

The areal distribution of Phylum Agmata (*Salterella*, *Volborthella*, and related forms) is a characteristic example of geographically restriction among small shelly fossils. The phylum was the subject of extensive studies by Yochelson, and is known to be restricted to North America, Greenland, and Europe (Yochelson, 1981). *Volborthella* occurs in the Great Basin of western North America and on the Baltic Platform. The distribution of *Salterella* is limited to North America, Greenland, and the British Isles. In western North America, *Salterella* is a common and useful index fossil for the medial portion of the *Bonnia-Olenellus* Zone (Fritz and Yochelson, 1988). Thus, the distribution of Agmata, although somewhat less broad, is congruent with the distribution of olenellid trilobites in the Early Cambrian.

III. VICARIANCE VERSUS DISPERSAL

The relative importance of dispersal, which is the organisms' capacity to reach distant areas as larvae or adults, and vicariance, which is the movement of continental crust and its associated organisms, as agents of geographic isolation is the focus of (yet another) longstanding debate. In this study, vicariance appears to be the more reasonable hypothesis to account for the observed distribution of Early Cambrian fossils. The crucial evidence for the vicariance hypothesis is the development of biogeographic provinces in low latitudes (where climate should play no significant role) and the failure of most clades to colonize in other provinces during the Early Cambrian. The provinces that are established in the Early Cambrian persist throughout the remainder of the Cambrian (e.g., Rowell *et al.*, 1973; Jell, 1975; Palmer, 1979; Taylor and Forester, 1979). Dispersal is therefore an unattractive hypothesis to account for the appearance of most clades around the margins of the various continental land masses early in the Cambrian, although dispersal clearly plays an important role in the evolutionary history of some clades (e.g., archaeocyathans). Also, dispersal cannot account for the evolutionary divergence of trilobites prior to their appearance in the fossil record. It would be useful to know, for example, how long before their appearance in the fossil record the trilobites colonized the olenellid and redlichiid provinces, but the fossil record has yet to yield that information.

Dispersal of Early Cambrian Invertebrates

Archaeocyathans

The capacity of archaeocyathans to disperse is amply demonstrated by their appearance first on the Siberian Platform in the Tommotian stage and their subsequent spread to other continents in the Atabanian and Toyonian stages (see above). Archaeocyathans were sessile organisms that are generally rooted to the substrate (the aberrant, free-lying genus *Retilamina* is one the few exceptions: see Savarese and Signor, 1989). Thus, dispersal must have been accomplished in the larval stage or by attachment to floating debris (e.g., Jokiel, 1990). Regardless of the means of dispersal, the fossil record shows that the capacity to disperse was unequal among clades and regions. Dispersal into North America was limited to relatively few clades, while a larger number reached Australia, Antarctica, and other regions (Debrenne and Rozanov, 1983; Zhuravlev, 1986; Debrenne *et al.*, 1990). The result of the limited dispersal of clades into western North America was the development of a unique regional fauna.

Brachiopods

Early Cambrian brachiopod faunas are still not well known, but preliminary work indicates genera of this age tend to be endemic (e.g., *Hadrotreta, Schizopholis, Spinulothele,* and *Eothele*: Rowell, 1980). Interestingly, Rowell (1980) notes that younger taxa tend to be less endemic (also see Rowell and Henderson, 1978), and speculates that this might reflect a change in their mode of life or in their capacity for dispersal.

Echinodermata

The fossil record of early echinoderms is patchy, and it might be misleading to place too much emphasis upon it. Nevertheless, the earliest representatives of Cambrian echinoderm clades appear in North America (e.g., helicoplacoids, eocrinoids, edrioasteroids), and their appearance on other continents was delayed considerably in most instances (see Sprinkle, 1976). Early echinoderms clearly had the capacity to disperse, but must have done so only infrequently over geologic time.

Mollusca

Chaffee and Lindberg (1986) suggest that the small size of Early Cambrian mollusks strongly implies these organisms lacked a planktonic larval stage in their ontogeny. Among modern mollusks, the capacity to produce eggs is correlated directly with shell size. Species that produce relatively few eggs invariably lack a planktonic stage in their development. Chaffee and Lindberg predicted that high levels of endemism would be found among Early and Middle Cambrian mollusks, reflecting the lack of a planktonic larval stage and resulting incapacity for dispersal. Preliminary work suggests early mollusks are provincial in their distribution (e.g., Brasier, 1989b), but

problems of nomenclature and sampling biases must be addressed before their prediction can be confirmed.

Trilobita

The early appearances of *Fallotaspis* and *Judomia* in Siberia and North America attests to the capacity of trilobites to disperse. Nevertheless, the longstanding taxonomic integrity of the Early Cambrian trilobite provinces, discussed above, demonstrates the limited nature of this capacity. Cowie (1971), Repina (1972) and Palmer (1973, 1979) have chronicled the evolutionary and biogeographic history of Early Cambrian trilobites. Palmer's studies demonstrates that shelf dwelling polymeroid trilobites were highly provincial throughout the Cambrian, while offshore forms tended to be more cosmopolitan. However, the actual mechanisms responsible for the endemic distributions are unknown.

DISCUSSION

Faunal Similarity in Space and Time

The arguments presented here depend, in part, upon the assumption that a geologically significant period of time is necessary to form separate faunal provinces. This assumption has been tested in the Mesozoic, by following the development of separate provinces on each side of the North Atlantic as North America and Europe were rifted apart (Fallaw, 1979). Even after 60 million years (the approximate duration of the Jurassic), there were a significant number of genera in common between the eastern and western Atlantic. This contrasts with the Early Cambrian when, for example, western North America shared few genera with other regions. The greatest degree of faunal similarity was with the Baltic Platform and the least similarity was with Australia (Signor, 1991). While these provinces might not have required sixty million years to develop, there is no evidence suggesting a shorter duration of time.

Timing of the Origin of Animals

The hypothesis that many clades had evolved and differentiated prior to their appearance in the fossil record presupposes that multiple subclades gained the capacity for biomineralization simultaneously, or nearly so. On first consideration, this appears to be an outrageously improbable suggestion. Yet there is ample evidence that exactly this happened among mollusks, arthropods, and many other clades that clearly did begin to biomineralize with calcium carbonate, calcium phosphate, silica, and other minerals at the beginning of the Cambrian. Furthermore, olenellid and redlichiid trilobites simultaneously (or nearly so) developed a well-mineralized cuticle. Agglutinated skeletons also appear at this time. Given this well documented, simultaneous appearance of biomineralization and agglutination in many clades and the simultaneous appearance of well mineralized cuticles in the two major clades of early trilobites, the suggestion of independent development of biomineralization in multiple lineages within clades seems tenable and even likely.

The arguments presented here do not constrain the age of the first occurrence of animals. However, the arguments do suggest that the first evidence of skeletogenous Metazoa preserved in the fossil record is substantially younger than the time when the first members of those clades evolved. The most compelling evidence for a young origin of animals is the late Proterozoic record of trace fossils (Valentine et al., 1990; Signor and Lipps, in press). It now appears that Metazoa must substantially predate these first traces.

The limited capacity of organisms to disperse, as observed in the Early Cambrian, strongly suggests that the movement of continents is a primary control on the development of Early Cambrian paleobiogeography. As noted above, the timing of the breakup of the putative Proterozoic supercontinent is not known. The time required to develop recognizable faunal provinces, discussed above, provides some indication of the time necessary to develop the Cambrian provinces.

Durham (1978) and Glaessner (1983) argued that the morphological complexity of early Metazoa indicated a significant, hidden Proterozoic evolutionary history of animals. However, organisms can evolve extremely rapidly in some circumstances, and it is difficult to determine rigorously how much time the evolutionary development of animals might require. The foregoing analysis depends not upon rates of evolution, but upon demonstrably slower rates of continental movement. The strength of this analysis is the link between Early Cambrian provinciality and these less rapid geological processes.

Timing of the Breakup of Rodinia

There is a wide range of opinion on the timing of the breakup of Rodinia, with estimates running from 900 Ma to 550 Ma. The presence of differentiated paleobiogeographic provinces, within the same climatic belt, in the Early Cambrian is strong evidence for a breakup prior to 550 Ma. The fossil record of Metazoa does not provide a precise control on the time elapsed between the separation of the continents and the beginning of the Early Cambrian. Certainly, rates of evolution observed in the fossil record are sufficiently fast that a few tens of millions of years is adequate to accommodate formation of the Early Cambrian provinces. Rates of continental movement away from the former supercontinent, the other constraint on the formation of provinces, are also sufficiently rapid to allow the formation of provinces within perhaps 30 m.y. Thus, a reasonable minimum age for the breakup of Rodinia is 580 Ma, or 30 m.y. prior to the

beginning of the Cambrian. This estimate conforms with the conclusions of Bond et al. (1984), who envision a late Proterozoic rifting of Rodinia. (Clearly, the logic presented here depends, in part, upon the age of the Early Cambrian. Some recent radiometric dates indicate possible ages as young as 520 Ma for basal Cambrian strata [Conway Morris, 1988; Cowie and Harland, 1989]. Should those dates prove to be correct, the timing of the breakup of Rodinia would be correspondingly younger.) Regardless of the precise timing, the initial breakup of Rodinia must have preceded the beginning of the Cambrian by some tens of millions of years.

SUMMARY

1. Early Cambrian metazoan faunas are provincial in their biogeographic distribution.

2. Provinces are, in part, historical phenomena. The presence of provinces in the Early Cambrian strongly indicates a long prior history of evolution, dispersal, and (or) vicariance that is not reflected in the fossil record.

3. This provinciality most likely reflects the prior rifting apart of the hypothetical supercontinent and the independent evolution of clades upon the resulting rifted continents. The time necessary to accomplish the movement of land masses and the concomitant evolutionary divergence inherent in the formation of biogeographic provinces is on the order of tens of millions of years. Longer periods are not necessary, but cannot be rejected on the basis of the evidence at hand.

4. The Proterozoic supercontinent Rodinia, if it existed, began to separate some tens of millions of years prior to the beginning of the Cambrian period.

ACKNOWLEDGEMENTS

I thank B. R. Signor and G. J. Vermeij for discussions of material presented here. Thanks to J. D. Cooper, J. H. Cooper, and G. J. Vermeij for criticizing earlier drafts of this paper. This research was supported by NSF EAR 88-04798.

REFERENCES

Angevine, C. L. and Heller, P. L., 1986, Global sealevel changes and supercontinent breakup: EOS, v. 67, p. 1210.

Axelrod, D. I., 1958, Early Cambrian marine fauna: Science, v. 128, p. 7-9.

Bambach, R. K., 1990, Late Paleozoic provinciality in the Marine Realm, in W. S. McKerrow and C. R. Scotese, eds., Palaeozoic Palaeogeography and Biogeography: Geolological Society Memoirs, v. 12, p. 307-323.

Boaden, P. J. S., 1989, Meiofauna and the origins of the Metazoa: Zoological

808

Journal of the Linnean Society, v. 96, p. 217-227.

Bond, G. C., Christie-Blick, N., Kominz, M. A. and Devlin, W. J., 1983, Thermal subsidence and eustasy in the lower Paleozoic miogeocline of western North America: Nature, v. 316, p. 742-745.

Bond, G. C., Nickeson, P. A. and Kominz, M. A., 1984, Breakup of a supercontinent between 625 Ma and 555 Ma: New evidence and implications for continental histories: Earth and Planetary Science Letters, v. 70, p. 326-345.

Brasier, M. D., 1979, The Cambrian radiation event. in M. R. House, ed., The Origin of Major Invertebrate Groups: London, Academic Press, p. 103-159.

Brasier, M. D., 1982, Sealevel changes, facies changes, and the late Precambrian-Early Cambrian evolutionary explosion: Precambrian Research, v. 17, p. 105-123.

Brasier, M. D., 1989a, Towards a biostratigraphy of the earliest skeletal biotas, in J. W. Cowie and M. D. Brasier, eds., The Precambrian-Cambrian Boundary: Oxford, Clarendon Press, p 117-165.

Brasier, M. D. 1989b. China and the Palaeotethyan belt (India, Pakistan, Iran, Kazakhstan, and Mongolia), in J. W. Cowie and M. D. Brasier, eds., The Precambrian-Cambrian Boundary: Oxford, Clarendon Press, p 40-74.

Burrett, C. and Richardson, R., 1980, Trilobite biogeography and Cambrian tectonic models: Tectonophysics, v. 63, p. 155-192.

Byers, C. W. 1982. Geological significance of marine biogenic sedimentary structures, in P. L. McCall and M. J. S. Tevesz, eds, Animal-Sediment Relations: New York, Plenum Press, p. 221-256.

Chaffee, C. and Lindberg, D. R., 1986. Larval biology of Early Cambrian molluscs: The implications of small body size: Bulletin of Marine Science, v. 39, p. 536-549.

Conway Morris, S., 1985, The Ediacaran biota and early metazoan evolution: Geological Magazine, v. 122 p. 77-81.

Conway Morris, S., 1987, The search for the Precambrian-Cambrian boundary: American Scientist, v. 75, p. 157-167.

Conway Morris, S., 1988, Radiometric dating of the Precambrian-Cambrian boundary in the Avalon Zone: New York State Museum Bulletin, v. 463, p. 53-58.

Conway Morris, S., and Fritz, W. H., 1984, *Lapworthella filigrana* n.sp. (*incertae sedis*) from the Lower Cambrian of the Cassiar Mountains, northern British Columbia, Canada, with comments on possible levels of competition in the early Cambrian: Palaeontologische Zeitschrift, v. 58, p. 197-209.

Cowie, J. W., 1971, Lower Cambrian faunal provinces, in F. A. Middlemiss et al., eds., Faunal Provinces in Space and Time: Liverpool, Seel House Press, Liverpool, p. 31-46.

Cowie, J. W. and Harland, W. B., 1989, Chronomety. in J. W. Cowie and M. D.

Brasier, eds., The Precambrian-Cambrian Boundary: Oxford, Clarendon Press, p. 186-198.

Cox, C. B. and Moore, P. D., 1980, Biogeography: An Ecological and Evolutionary Approach: Third Ed., Oxford, Blackwell Scientific Publications, 234 p.

Crimes, T. P., 1989, Trace fossils, in J. W. Cowie and M. D. Brasier (eds.) The Precambrian-Cambrian Boundary: Oxford, Clarendon Press, p. 166-185.

Debrenne, F., Gandin, A., and Rowland, S. M., 1989, Lower Cambrian bioconstructions in northwestern Mexico (Sonora). Depositional setting, paleoecology, and systematics of archaeocyaths: Geobios, v. 22, p. 137-195, 12 pl.

Debrenne, F., and Kruse, P. D., 1986, Shackleton limestone archaeocyaths. Alcheringa, v. 10, p. 235-278.

Debrenne, F., and Rozanov, A. Yu., 1983, Paleogeographic and stratigraphic distribution of regular Archaeocyatha (Lower Cambrian fossils): Geobios, v. 16, p. 727-736, 1 pl.

Debrenne, F., Rozanov, A. Yu., and Zhuravlev, A., 1990, Regular archaeocyaths: Cahiers de Paleontologie, 218 p., 32 pl.

Donovan, S. K., 1987, The fit of the continents in the late Precambrian: Nature, v. 327, p. 139-141.

Durham, J. W. 1978. The probable metazoan biota of the Precambrian as indicated by the subsequent record: Annual Reviews of Earth and Planetary Sciences, v. 6, p. 21-42.

Erwin, D. H., 1989, Molecular clocks, molecular phylogenies, and the origin of phyla: Lethaia, v. 22, p. 251-257.

Evans, K. R. and Rowell, A. J. 1990. Small shelly fossils from Antarctica: an Early Cambrian faunal connection with Australia: Journal of Paleontology, v. 64, p. 692-700.

Fallaw, W. C. 1979. Trans-North Atlantic similarity among Mesozoic and Cenozoic invertebrates correlated with widening of the basin. Geology, v. 7, p. 398-400.

Fortey, 1990, Ontogeny, hypostome attachment and trilobite classification: Palaeontolgy, v. 33, p. 529-576.

Fritz, W. H., and Yochelson, E. L., 1988, The status of Salterella as a Lower Cambrian index fossil: Canadian Journal of Earth Sciences, v. 25, p. 403-416.

Glaessner, M. F., 1983, The emergence of Metazoa in the early history of life: Precambrian Research, v. 20, p. 427-441.

Glaessner, M. F., 1984, The Dawn of Animal Life, A Biohistorical Survey: Cambridge, Cambridge University Press, 244 p.

Grant, S. W. F., 1990, Shell structure and distribution of *Cloudina*, a potential index fossil for the terminal Proterozoic: American Journal of Science, v. 290-A, p. 261-294.

Gurich, G., 1930, Die bislang altesten Spuren von Organismen in Sudafrika.

International Geological Conggess, Compte Rendu, v. 15, p. 670-680.

Gurich, G., 1933, Die Kuibis-Fossilien der Nama-Formation von Sudwestafrika: Palaeontologische Zeitshcrift, v. 15, p. 137-154.

Hutchinson, G. E., 1961, The biologist poses some problems. *in* M. Sears (ed.), Oceanography: Washington, D. C., American Association for the Advancement of Science, Washington, D. C., p. 85-94.

Jell, P. A., 1974, Faunal provinces and possible planetary reconstruction of the Middle Cambrian: Journal of Geology, v. 82, p. 319-350.

Jenkins, R..J. F., 1985, The enigmatic Ediacaran (late Precambrian) genus *Rangea* and related forms: Paleobiology, v. 11, p. 336-355.

Jokiel, P. L., 1990, Transport of reef corals into the Great Barrier Reef: Nature, v. 347, p. 665-667.

Klootwijk, C. T., 1980, Early Palaeozoic palaeomagnetism in Australia: Tectonophysics, v. 64, p. 249-332.

Lauterbach, K.-E., 1980, Key events in the evolution of the groundplan of the Arachnata (arthropods): Abhandlungen des Naturwissenschaftlichen Vereins in Hamburg, v. 23, p. 163-327.

Lindsay, J. F., Korsch, R. J., and Wilford, J. R., 1987, Timing of the breakup of a Proterozoic supercontinent: evidence from Australian intracratonic basins: Geology, v. 15, p. 1061-1064.

MacArthur, R. H., 1972, Geographical Ecology: Patterns in the Distribution of Species: New York, Harper and Row.

Matthews, S. C. and Missarzhevsky, V. V., 1975, Small shelly fossils of late Precambrian and early Cambrian age: a review of recent work: Journal of the Geological Society of London, v. 131, p. 289-304.

McKerrow, W. S. and Scotese, C. R., eds., 1990, Palaeozoic Palaeobiogeography and Biogeography: Geological Society Memoirs, v. 12., 435 p.

McMenamin, M.A. S., 1982, A case for two late Proterozoic-earliest Cambrian faunal province loci: Geology, v. 10, p. 290-292.

McMenamin, M. A. S., 1986, The Garden of Ediacara: Palaios, v. 1, p. 178-182.

McMenamin, M. A. S. and Schulte McMenamin, D. L., 1989, The Emergence of Animals: The Cambrian Breakthrough: New York, Columbia Press, 217 pp.

Moczydlowska, M. and Vidal, G., 1988, How old is the Tommotian?: Geology, v. 16 p. 166-168.

Mount, J. F., and Signor, P. W., in press, Faunas and facies, fact and artifact, in: J. H. Lipps and P. W. Signor, eds., Emergence of the Metazoa: Plenum Press, New York.

Narbonne, G. M., and Hofmann, H. J., 1987, Ediacaran biota of the Werneke Mountains, Yukon, Canada: Palaeontology, v. 30, p. 646-767.

Palmer, A. R., 1973, Cambrian trilobites. *in* A. Hallam, ed., Atlas of Palaeobiogeography: New York, Elsevier; pp. 3-11.

Palmer, A. R., 1977, Biostratigraphy of the Cambrian System- a progress report: Annual Review of Earth and Planetary Science, v. 5, p. 13-33.

Palmer, A. R., 1979. Cambrian: Treatise on Invertebrate Paleontology, Lawrence, Kansas, University of Kansas Press and the Geological Society of America, p. A119-A135

Parrish, J. T., Ziegler, A. M., Scotese, C. R., Humphreville, R. G. and Kirshvink, J. R., 1986, Proterozoic and Cambrian phosphorites- specialist studies. Early Cambrian paleogeography, paleooceanography, and phosphorites, *in* P. J. Cook and J. H. Shergold, eds., Phosphate Deposits of the World: Cambridge, Cambridge University Press, p. 280-294.

Piper, J. D. A., 1976, Paleomagnetic evidence for a Proterozoic supercontinent: Philosophical Transactions of the Royal Society of London, v. A280, p. 469-490.

Piper, J. D. A., 1983, Proterozoic paleomagentism and single continent plate tectonics: Geophysical Journal of the Royal Astronomical Society, v. 74, p. 163-197.

Piper, J. D. A., 1987, Paleomagnetism and continental crust: New York, Wiley. 434 p.

Repina, L. N., 1972, Biogeography of Early Cambrian of Siberia according to trilobites: XXIII International Geological Congress, 1968, Proceedings of the International Paleontological Union, p. 289-300.

Rowell, A. J., 1980, Inarticulate brachiopods of the Lower and Middle Cambrian Pioche Shale of the Pioche District, Nevada: University of Kansas Paleontological Contributions, Paper 98, 26 pp., 8 pl.

Rowell, A. J. and Henderson, R. A., 1978, New genera of acrotretids from the Cambrian of Australia and the United States: University of Kansas Paleontological Contributions, Paper 93, 12 pp., 2 pl.

Rowell, A. J., McBride, D. J. and A. R. Palmer, 1973, Quantitative study of Trempealeauian (Latest Cambrian) trilobite distribution in North America: Geological Society of America Bulletin, v. 84, p. 3429-3442.

Rozanov, A. Yu., 1986, Problematica of the Early Cambrian: *in* A. Hoffman and M. H. Nitecki, eds., Problematic Fossil Taxa: Oxford Monographs on Geology and Geophysics, v. 5, p. 87-96.

Runnegar, B., 1982a, The Cambrian explosion: animals or fossils?: Journal of the Geological Society of Australia, v. 29, p. 395-411.

Runnegar, B., 1982b, A molecular-clock date for the origin of the animal phyla: Lethaia, v. 15, p. 199-205.

Savarese, M. and Signor, P. W., 1989, New archaeocyathan occurrences in the Upper Harkless Formation (Lower Cambrian of

810

western Nevada): Journal of Paleontology, v. 63, p. 539-549.

Scotese, C. R. and McKerrow, W. S., 1990, Revised world maps and introduction: Geological Society Memoirs, v. 12, p. 1-21.

Seilacher, A., 1985, Discussion of Precambrian Metazoa: Philosophical Transactions of the Royal Society of London, v. B311, p. 47-48.

Seilacher, A., 1989, Vendozoa: Organismic construction in the Proterozoic biosphere: Lethaia, v. 22, p. 229-239.

Sepkoski, J. J. Jr., 1978, A kinetic model of Phanerozoic diversity. I. Analysis of marine orders: Paleobiology, v. 4, p. 223-251.

Signor, P. W., 1991, Lower Cambrian biogeography of western North America: White Mountain Research Statation Symposium Volume, in press.

Signor, P. W., and Lipps, J. H., in press, The origin and early radiation of the Metazoa, in, J. H. Lipps and P. W. Signor, eds., The Emergence of Metazoa: Plenum Press, New York.

Simpson, G. G., 1943, Mammals and the nature of continents. Science, v. 241, p. 1-31.

Smith, A. G., Hurley, A. M. and Briden, J. C., 1981, Phanerozoic paleocontinental world maps: Cambridge, Cambridge University Press, 102 pp.

Sprinkle, J., 1976, Biostratigraphy and paleoecology of Cambrian echinoderms from the Rocky Mountains: Brigham Young University Geology Studies, v. 23, p. 61-74.

Sprigg, R. C., 1947, Early Cambrian "jellyfishes" of Ediacara, South Australia, and Mount John, Kimberly District, Western Australia: Transactions of the Royal Soceity of South Australia, v. 71, p. 212-224.

Stanley, S. M., 1976, Fossil data and the Precambrian-Cambrian transition: American Journal of Science, v. 276, p. 56-76.

Stewart, J. H., 1972, Initial deposits of the Cordilleran Geosyncline: evidence of a late Precambrian (<850 m.y.) continental separation: Geological Society of America Bulletin, v. 83, p. 1345-1360.

Taylor, M. E. and Forester, R. M., 1979, Distributional model for marine isopod crustaceans and its bearing on early Paleozoic paleozoogeography and continental drift: Geological Society of America Bulletin, v. 90, p. 405-413.

Theokritoff, G., 1979, Early Cambrian provincialism and biogeographic boundaries in the North Atlantic region: Lethaia, v. 12, p. 281-295.

Theokritoff, G., 1985, Early Cambrian biogeography in the North Atlantic region: Lethaia, v. 18, p. 283-294.

Urbanek, A. and Rozanov, A. Yu., eds., 1983, Upper Precambrian and Cambrian paleontology of the East-European Platform: Warsaw, Wydawnictwa Geologiczne, 158 p., 94 pl.

Valentine, J. W., 1973, Evolutionary paleoecology of the marine biosphere, Englewook Cliffs, Prentice Hall, 511 p.

Valentine, J. W., 1989, Bilaterians of the Precambrian-Cambrian transition and the annelid-arthropod transition: Proceedings of the National Academy of Science, v. 86, p. 2272-2275.

Valentine, J. W., Awramik, S. M., Signor, P. W., and Sadler, P. M., 1990, The Biological Explosion at the Precambrian-Cambrian Boundary: Evolutionary Biology, v. 25, p. 279-356.

Van der Voo, R., McCabe, C. and Scotese, C. R., 1984, Was Laurentia part of an Eocambrian supercontinent? in R. Van der Voo, C. R. Scotese, and R. Bonhommet, eds., Plate Reconstruction from Paleozoic Paleomagnetism: American Geophysical Union, Geodynamics Series, v. 12, p. 131-136.

Vidal, G. and Knoll, A. H., 1983, Proterozoic plankton: Geological Society of America Memoirs, v. 161, p. 265-277.

Walcott, C. D., 1910, Cambrian geology and paleontology II. No. 1. Abrupt appearance of the Cambrian fauna of the North American continent: Smithsonian Miscellaneous Collections, v. 57, p. 1-16.

Walcott, C. D., 1914, Cambrian geology and paleontology. 1. The Cambrian faunas of eastern Asia: Smithsonian Miscellaneous Collections, v. 64.

Yochelson, E. L., 1981, A survey of Salterella (Phylum Agmata), in Taylor, M. E., ed., Short Papers for the Second International Symposium on the Cambrian System: United States Geological Survey Open-File Report, no. 81-743 p. 244-248.

Zhuravlev, A., 1986, Evolution of archaeocyaths and palaeobiogeography of the Early Cambrian: Geological Magazine, v. 123, p. 377-385.

Ziegler, A. M., Scotese, C. R., McKerrow, W. S., Johnson, M. E. and Bambach, R. K., 1979, Paleozoic paleogeography: Annual Reviews of Earth and Planetary Science, v. 7, p. 473-502.

CYCLOSTRATIGRAPHY OF LATE CAMBRIAN CARBONATE SEQUENCES: AN INTERBASINAL
COMPARISON OF THE CORDILLERAN AND APPALACHIAN PASSIVE MARGINS

David Osleger
Department of Geological Sciences
University of Southern California
Los Angeles, CA 90089-0740

ABSTRACT

A comparison of the stratigraphic
components, subsidence histories and platform
morphologies of the Late Cambrian Cordilleran
passive margin of central Utah-Nevada and the
Appalachian passive margin suggests that
fundamental differences exist between the two
coeval platforms. The Late Cambrian of the
Utah-Nevada Cordilleran passive margin is
characterized by a spectrum of shallow to deep
subtidal cycles spread out across a distally
steepened ramp. In contrast, the Appalachian
Late Cambrian passive margin, from Tennessee
to Pennsylvania, is characterized by
widespread peritidal cycles across a flat-
topped, aggraded reef-rimmed shelf. The
relationship between cycle type and platform
morphology may be a function of energy regime
combined with differential rates of
subsidence. The simultaneous development of
peritidal and subtidal cycles on separate Late
Cambrian carbonate platforms supports a
eustatic control on the origin of the meter-
scale cyclicity. High-frequency eustatic
oscillations may be controlled by Milankovitch
astronomical rhythms based on spectral
analysis that show strong clustering of
periods around a narrow range of values.

Systematic changes in the stacking
patterns of meter-scale cycles can be used in
conjunction with Fischer plots to define long-
term sea-level cycles. The relative
thicknesses and compositions of the stacked
meter-scale cycles are dependent on the amount
of accommodation space generated by eustasy
and subsidence. Correlated Fischer plots of
cyclic successions from different localities
support a eustatic control on Late Cambrian
sequence development. Combining the sea-level
curves defined by Fischer plots with
paleobathymetric curves of Late Cambrian
cyclic strata suggests that the curves may
approximate the *form* of the eustatic sea-level
curve. A composite "eustatic" sea-level curve
for the Late Cambrian was created by
qualitatively combining the sea-level curves
defined by the different techniques for each
of the four localities.

INTRODUCTION

Various scales of stratigraphic cyclicity
have been recognized throughout the geologic
record (Fischer, 1964; Vail et al., 1977; Chow
and James, 1987; Koerschner and Read, 1989;
Goldhammer et al., 1987; 1990) that appear to
be arranged in a hierarchical framework.
Systematic changes in meter-scale cycles
define stacking patterns that characterize the
internal components of larger scale sequences.
The predictable arrangement of the superimposed scales of cyclicity appears to
manifest the combined effects of several
orders of relative sea-level oscillations. An
interbasinal study of Late Cambrian pericratonic carbonates was performed to evaluate
the composition and stacking patterns of
meter-scale cycles and to determine the degree
of resolution and correlatability of the
larger scale sequences (Vail et al., 1977).
The fundamental goal was to determine whether
the Late Cambrian sequences likely reflect
global eustatic events or relative sea-level
fluctuations intrinsic to individual basins
responding to local or regional tectonism.

The focus of this paper is on the
fundamental differences between the Late
Cambrian passive margin sequences of the
central Utah-Nevada Cordillera and the
Virginia-Tennessee Appalachians. Additional
sections were logged in the Texas cratonic
embayment and the southern Oklahoma aulacogen
to evaluate the degree of synchroneity of the
long-term sequences but will be mentioned only
to supplement the discussion of ultimate
controlling mechanisms. The objectives of
this paper are to: 1) define the
representative peritidal to deep subtidal,
meter-scale cycles that characterize the two
Late Cambrian passive margins; 2) using
spectral analysis, demonstrate probable
Milankovitch control of meter-scale cycle
formation; 3) by using paleobathymetric curves
tied together with graphic correlation,
determine the degree of interbasinal
correlatability of Late Cambrian sequences and
4) illustrate how Fischer plots can be used to
correlate cyclic successions and define rising
and falling portions of relative sea-level
curves.

Stratigraphic and Tectonic Settings

Location of Sections

Complete sections of Late Cambrian strata
were measured and logged bed-for-bed in
Cordilleran passive margin strata of the House
Range of west central Utah and in strata of
the Appalachian passive margin in Tennessee,
Virginia and eastern Pennsylvania (Fig.1). A
total of 2200 meters of section was logged and

In Cooper, J.D., and Stevens, C.H., eds., 1991, **Paleozoic
Paleogeography of the Western United States-II:** Pacific
Section SEPM, Vol. 67, p. 801-828.

811

numerous other sections previously described in the Appalachians (Markello and Read, 1982; Demicco, 1985; Koerschner and Read, 1989) and in the Utah-Nevada Cordillera (Palmer, 1971a; 1984; Kepper, 1972; Brady and Rowell, 1976; Lohmann, 1976; Rees, 1986) were field checked. Hand samples of individual lithofacies were collected for slabbing and thin-section analysis to provide additional detail for paleoenvironmental interpretations. Details regarding the exact locations of sections and drafted logs of stratigraphic intervals can be found in Osleger (1990).

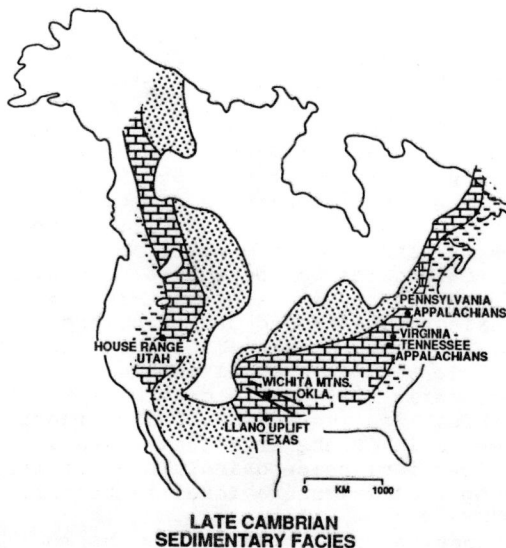

LATE CAMBRIAN
SEDIMENTARY FACIES

Figure 1: Location map of sections measured in the study. Late Cambrian base map modified from Lochman-Balk (1971) and Palmer (1974) to show the inner and outer detrital belts, the middle carbonate belt and the southern Oklahoma aulacogen.

Tectonic Settings and Platform Morphologies

The Cordilleran and Appalachian passive margins originated in response to breakup of a late Proterozoic supercontinent around 625 to 555 Ma (Stewart and Suczek, 1977; Bond et al., 1984). On both the Cordilleran and Appalachian passive margins, wedge-shaped prisms of post-rift subtidal and peritidal carbonates and interlayered siliciclastics (Fig. 2) developed over thick accumulations of syn-rift, Upper Proterozoic/Lower Cambrian, terrigenous siliciclastics. The Cordilleran passive margin extended essentially E-W at about 10° to 15°N latitude, whereas the Appalachian passive margin extended roughly NW-SE at about 15° to 25°S latitude (Scotese and McKerrow, 1990).

The Cordilleran passive margin of the Great Basin accumulated approximately 2 km of post-rift Middle and Late Cambrian subtidal to peritidal carbonates and fine siliciclastics above almost 6 km of late Proterozoic-Early Cambrian syn-rift terrigenous and shallow marine siliciclastics (Stewart and Poole, 1974; Levy and Christie-Blick, 1989). The passive margin hinge line was located along the current Wasatch Range of central Utah where Cambrian through Devonian rocks exhibit

an abrupt thickness increase (Stewart and Poole, 1974). The platform extended almost 400 km (non-palinspastic) from the Wasatch Line to the off-platform facies of the Hot Creek Range of central Nevada, the acknowledged edge of the Cambro-Ordovician platform (Cook and Taylor, 1975).

Lithofacies patterns for the Early to Late Cambrian of Utah-Nevada are distributed along the classic inner and outer detrital belts and middle carbonate belt of Palmer (1971a). An east-west trending reentrant, the House Range embayment, extended across the carbonate belt from east-central Nevada into west-central Utah during Middle to early Late Cambrian time. The embayment may have acted as a bypass zone for the seaward transport of siliciclastic detritus (Lohmann, 1976; Rees, 1986). The Cordilleran platform of the Great Basin is considered to have been a distally steepened ramp during the Late Cambrian-Early Ordovician (Fig. 2) (Cook and Taylor, 1975; Read, 1985). Thermally controlled subsidence continued into Late Devonian time when a change to foreland basin conditions occurred in response to Antler thrust loading (Stewart and Suczek, 1977; Bond et al., 1989).

LATE CAMBRIAN PLATFORM MORPHOLOGIES

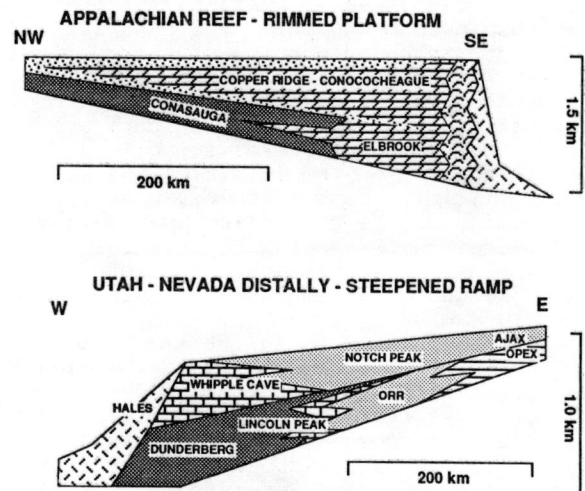

Figure 2: Late Cambrian platform morphologies of the Appalachian and Cordilleran passive margins. Formation and group names superimposed on lithologic symbols.

The Appalachian passive margin contains up to 1.6 km of Middle to Late Cambrian shallow-water carbonates and intrashelf basin shale and siltstone. Passive margin sedimentation was influenced by depocenters in Tennessee and Pennsylvania and arches in central Virginia and New Jersey (Palmer, 1971b; Read, 1989). The morphology of the Appalachian Cambro-Ordovician platform appears to have evolved from an Early Cambrian ramp into a high relief rimmed shelf throughout Early to Late Cambrian time (Read, 1989) (Fig. 2). During the Middle to early Late Cambrian, the shaly intrashelf Conasauga basin developed in southwestern

Virginia/northeastern Tennessee (Markello and Read, 1982) and was surrounded toward the shelf margin and northeastward toward the Virginia Arch by peritidal carbonates. By later Late Cambrian time, the Appalachian passive margin had developed into an aggraded, flat-topped platform of dominantly peritidal facies (Demicco, 1985; Koerschner and Read, 1989). Passive margin conditions ended with Taconic thrust loading during the early Middle Ordovician.

Biostratigraphy and Absolute Ages

Biostratigraphic control for Upper Cambrian strata consists of 10 major trilobite biozones, a few subzones, and conodont zonation in the upper stages of the series (Fig. 3). This relative time control is enhanced by the occurrence of three biomere boundaries, narrow intervals of abrupt changes in trilobite faunas (Palmer, 1965; 1984). Generally agreed to be isochronous time markers, biomere boundaries provide excellent datums for chronostratigraphic correlation. Relative age assignments for the Appalachian and Cordilleran sections were determined from published biostratigraphic data (Palmer, 1965; 1971a; 1971b; 1984; Robison, 1964; Derby, 1965; Rassetti, 1965; Hintze, 1974; Hintze and Palmer, 1976; Hintze et al, 1980; Eby, 1981; Taylor and Miller, 1981; Miller et al, 1982; Orndorff, 1988; Sundberg, 1990). Upper and lower age limits of 525 Ma to 505 Ma were used for the *Cedaria* through *Saukia* interval based on the DNAG time scale (Palmer, 1983).

EARLY ORDO	STAGE	BIOMERE	TRILOBITE ZONE	HOUSE RANGE, UTAH	LLANO UPLIFT, TEXAS		WICHITA MTS. OKLA.	SW VIRGINIA NE TENN.	EASTERN PENN.
			MISSISSIQUOIA						
	TREMPEAL-EAUAN	PTYCHASPID	SAUKIA	LAVA DAM	SAN SABA	NOTCH PEAK	SIGNAL MOUNTAIN	COPPER RIDGE /	
				RED TOPS	POINT PEAK		ROYER	CONOCO-CHEAGUE	
	FRANCONIAN		SARATOGIA	HELLN-MARIA	MORGAN CREEK		FORT SILL		ALLENTOWN DOLOMITE
			TAENICEPHALUS	SNEAK-OVER		WILBERNS FM	HONEY CK		
LATE CAMBRIAN			ELVINIA	CORSET SPRING	WELGE		REAGAN	MAYNARD-VILLE	
		PTERO-CEPHALIID	DUNDERBERGIA	JOHNS WASH		ORR FM		CONASAUGA	
	DRESBACHIAN		APHELASPIS	CANDLAND	LION MT.			NOLICHUCKY SHALE	
			CREPICEPHALUS	BIG HORSE	CAP MT.				
		MARJUMIID	CEDARIA	WEEKS FM.	HICKORY SST	RILEY FM		ELBROOK	
			BOLASPIDELLA						

Figure 3: Biostratigraphic chart of Late Cambrian strata in this study. Note the three biomere boundaries in the Late Cambrian that separate the trilobite zones.

SHALLOWING-UPWARD CYCLES

Meter-scale Cycles

One of the fundamental differences between the Cordilleran and Appalachian Late Cambrian passive margin sections in the study locations is the composition of meter-scale cycles (Figs. 4-6). Repetitive successions of hundreds of fining-upward **peritidal** cycles (Wilson, 1952; Demicco, 1985; Koersch-

ner and Read, 1989) extend across the broad Appalachian reef-rimmed shelf. In contrast, peritidal cycles are restricted to a narrow band along the hinge line of the coeval Cordilleran distally steepened ramp of Utah-Nevada (Palmer, 1971a; Kepper, 1972). Most of the platform in central Utah and Nevada is composed of a seaward gradation of shallow to deep **subtidal** cycles. All of the meter-scale cycles fall within the range of average periods (20 - 400 ky) and thicknesses (0.4 - 15.0 m) expected for shallowing-upward, meter-scale cycles (Algeo and Wilkinson, 1988). Detailed descriptions of lithofacies and paleoenvironmental interpretations can be found in Osleger (1990).

Peritidal Cycles of the Appalachian Passive Margin

Tidal flat-capped cycles (0.4-7.0 m) of the Appalachian Late Cambrian platform are composed of a basal ooid-intraclast grainstone lag deposit overlain by either ribbon carbonates or thrombolite boundstones capped by mudcracked thick laminites and/or cryptalgal laminites (Fig. 4). Quartz arenites or carbonate-clast breccias cap some cycles, particularly during long-term relative sea-level fall. Peritidal cycles fine upward and are asymmetric, with generally abrupt upper and lower boundaries but with gradational internal boundaries between lithofacies (Wilson, 1975; James, 1984; Demicco, 1985; Hardie and Shinn, 1986; Koerschner and Read, 1989). These cycles are recognized within the Elbrook, Copper Ridge, Conococheague and Allentown Formations of the Appalachians. Each cycle records rapid transgression followed by the progradation of tidal flats over a shallow subtidal sandy shelf with patchy thrombolitic bioherms (Koerschner and Read, 1989; Osleger and Read, in review).

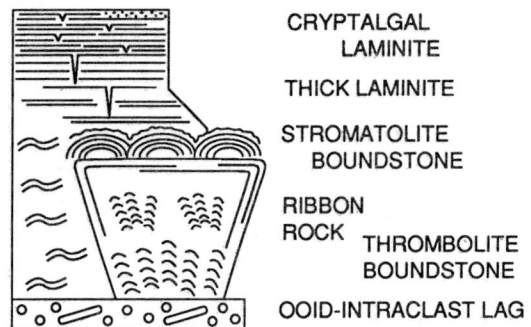

CRYPTALGAL LAMINITE

THICK LAMINITE

STROMATOLITE BOUNDSTONE

RIBBON ROCK

THROMBOLITE BOUNDSTONE

OOID-INTRACLAST LAG

LAMINITE-CAPPED PERITIDAL CYCLE

Figure 4: Vertical arrangement of lithofacies within a typical peritidal cycle of the Late Cambrian Appalachian passive margin.

Subtidal Cycles of the Utah-Nevada Passive Margin

Subtidal cycles of the Cordilleran Late Cambrian platform exhibit an upward increase in grain size, bed thickness, cross-bedding and other high energy sedimentary structures. Subtidal cycles are not capped by intertidal lithofacies nor do the subtidal cycles exhibit exposure features such as micro-karsting or vadose dissolution/cementation. The cycles are gradational across the Utah-Nevada platform and are genetically linked to one another by shared lithofacies (Fig. 5).

Thrombolite-capped cycles (1.5-12.0 m) of the Cordilleran Late Cambrian passive margin consist of a basal dark gray, peloidal packstone overlain by stacked thrombolite-stromatolite bioherms and laterally equivalent light gray, cross-bedded peloidal-oncolitic grainstone (Fig. 5). The basal peloidal lithofacies is characterized by horizontal to low angle cross-lamination, lack of recognizable skeletal material, and dark gray bioturbated textures that suggest restricted, quiet-water (but not necessarily deep) deposition. Shallow subtidal conditions for the thrombolites are supported by the laterally equivalent light gray, cross-bedded peloidal-oncolitic-oolitic-intraclastic grainstones that resemble modern, high energy, non-skeletal grainstones enveloping growing stromatolitic bioherms in tidal channels in the Bahamas (Dill et al., 1986). More than thirty of these cycles are recognized within the upper Hellnmaria Member of the Notch Peak Formation throughout the House Range of west central Utah.

Ooid grainstone-capped cycles (0.5-4.2 m) of the Cordilleran passive margin consist of burrowed wackestone/packstone grading up into oncolite-skeletal packstone/grainstone capped by oolitic grainstone (Fig. 5). Burrowed wackestones/ packstones are subtidal facies deposited below fairweather wave base under normal marine conditions as suggested by pervasive bioturbation, abundant bioclastic debris, and clusters of pellets. Upward-increasing, laterally discontinuous skeletal packstone lenses with erosional bases and burrowed tops are rapidly deposited storm beds that escaped homogenization by burrowers. The upward transition from open marine skeletal packstones to oncolitic-peloidal grainstones indicates increasingly shallow, restricted conditions, perhaps peripheral to active ooid shoals (Hine, 1977). Ooid grainstone-capped cycles are common in the Big Horse Member, Orr Formation of the House Range of Utah (Lohmann, 1976).

Skeletal packstone-capped cycles (1.0-10.0 m) are composed of basal nodular argillaceous wackestone overlain by burrowed, storm-deposited wackestone/packstone coarsening upward into a skeletal packstone/grainstone cap (Fig. 5). These cycles developed on the mid-ramp at intermediate water depths above storm wave base, seaward of ooid grainstone-capped cycles. The basal nodular, argillaceous wackestone is a distal storm facies deposited on the middle ramp between burrowed wackestones and packstones and deeper water siliciclastic muds. The abundant quartz silt may have been transported from the craton across the inner detrital belt and onto the carbonate platform through a west-trending subtidal channel that debouched near the House Range (Lohmann, 1976). Upward within these cycles, scoured surfaces, skeletal lag deposits with abundant shelter porosity and perched mud, hummocky cross-stratification, and rippled laminae attest to active storm sedimentation. The cycle cap of crossbedded skeletal-intraclast packstones manifests the development of storm-reworked skeletal sand sheets and migrating sand shoals. Excellent examples of these cycles are recognized in the lower Big Horse Member of the Orr Formation of the House Range and bear striking similarities to storm-influenced cycles described by Aigner (1985).

Spiculitic wackestone-capped cycles (0.7 - 3.1 m) are composed of basal nodular

GRADATION OF CYCLE TYPES ACROSS A LATE CAMBRIAN SHALLOW TO DEEP RAMP

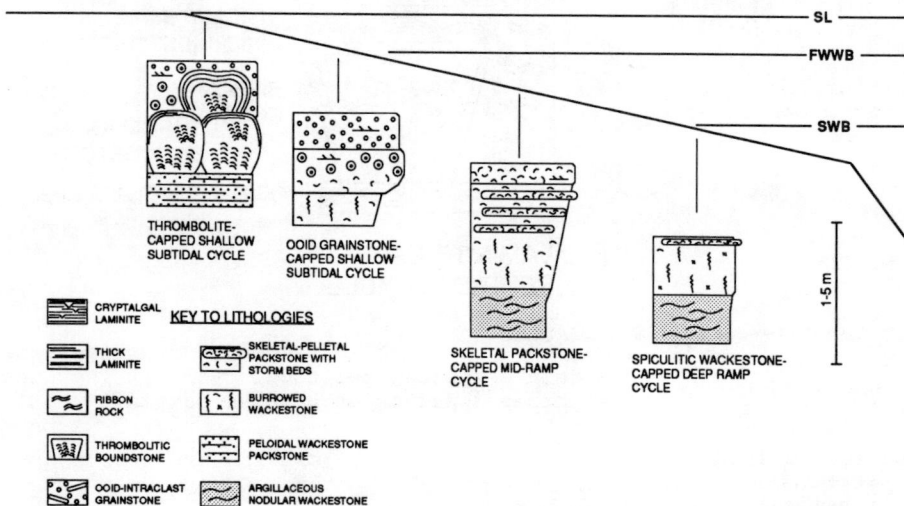

SL
FWWB
SWB

THROMBOLITE-CAPPED SHALLOW SUBTIDAL CYCLE

OOID GRAINSTONE-CAPPED SHALLOW SUBTIDAL CYCLE

SKELETAL PACKSTONE-CAPPED MID-RAMP CYCLE

SPICULITIC WACKESTONE-CAPPED DEEP RAMP CYCLE

1-5 m

KEY TO LITHOLOGIES

CRYPTALGAL LAMINITE
THICK LAMINITE
RIBBON ROCK
THROMBOLITIC BOUNDSTONE
OOID-INTRACLAST GRAINSTONE
SKELETAL-PELLETAL PACKSTONE WITH STORM BEDS
BURROWED WACKESTONE
PELOIDAL WACKESTONE PACKSTONE
ARGILLACEOUS NODULAR WACKESTONE

Figure 5: Arrangement of Late Cambrian shallow to deep subtidal cycle types across a hypothetical platform. Note the location of fairweather and storm wave base and their relation to cycle types.

argillaceous mudstone overlain by burrowed spiculitic wackestone with skeletal packstone lenses that become more abundant upward (Fig. 5). These cycles developed on the deep ramp near maximum storm wave-base, seaward of the skeletal packstone-capped cycles. The similarity in lithofacies to the skeletal packstone-capped cycles suggests that they are incomplete cycles deposited at greater depths on the deep ramp. The abundant bioturbation and trilobite and echinoderm debris within storm beds attest to well oxygenated, normal marine conditions. These cycles occur in the Sneakover Member of the Orr Formation and in the basal Hellnmaria and Lava Dam Members of the Notch Peak Formation of the House Range.

Carbonate-capped Shaly Cycles

Cycles consisting of basal shales abruptly overlain by clear-water carbonates are recognized within intrashelf basin facies of the Appalachian passive margin (Markello and Read, 1982) and in deep ramp facies of the Cordilleran passive margin (Figure 6). **Flat-pebble conglomerate-capped shaly cycles** (0.8-5.5 m) of the Late Cambrian Conasauga intrashelf basin of the Appalachians consist of a basal calcareous green-brown shale sharply grading upward into micro-hummocky cross-laminated peloidal grainstones and quartz siltstones capped by amalgamated flat pebble conglomerate beds (Fig. 6). These cycles appear to record deposition above and below a fluctuating storm wave base in a relatively shallow intrashelf basin. Whereas the laminated shales were deposited below the zone of storm wave reworking, progressive shallowing to above storm wave base is indicated in the peloidal grainstone/quartz siltstone by an increase in grain size and in storm-generated sedimentary structures. The flat-pebble conglomerate caps of the cycles were deposited above storm wave base during severe storms when the underlying semi-lithified peloidal grainstone was eroded and redeposited as tabular to lenticular beds of rounded, elongate clasts. These cycles have been recognized in the Nolichucky Formation of Virginia and Tennessee (Markello and Read, 1982) and in the Point Peak Member of the Wilberns Formation, central Texas, as well as in the Cambrian of Montana (Sepkoski, 1982) and the southern Canadian Rockies (Aitken, 1978).

Carbonate-capped shaly cycles (2.5-15.0 m) of the House Range consist of a thick basal shale abruptly overlain by coarsening-upward skeletal wackestone/packstone (Fig. 6). The thick basal shale units formed below storm wave base during short durations of relative sea-level rise that brought deep water siliciclastic clays up onto the carbonate platform. The fine clastics may have been originally derived from the craton and were transported across the carbonate belt (perhaps through the House Range Embayment trough) and onto the deep ramp as dilute clouds or bottom-hugging nepheloid layers. Siliciclastic clays accumulated in a dysaerobic environment as indicated by the olive green to dark gray color, mildly bioturbated lamination, and sparse trilobite and phosphatic brachiopod fauna. The abrupt transition in paleowater depths between the deep, quiet water shales (probable water depths of >40 to 60 m) and the shallow, clear water carbonates (probable water depths between 5 to 20 m) suggests that these cycles probably did not form by simple aggradation, which would provide a maximum of only 15 meters of shallowing, but rather experienced a relative short-term sea-level rise (shales) followed by relative sea-level fall (carbonates). An alternative and equally viable interpretation of these limestone-shale cycles can be found in Cooper (1989). These cycles occur in the Candland Shale, Corset Spring Shale and Steamboat Pass Members of the Orr Formation of the House Range.

Mechanisms Controlling Meter-scale Cyclicity

Mechanisms proposed for meter-scale carbonate cycle generation include the tidal flat-controlled autocyclic model (Ginsburg, 1971), variations in sedimentation-redistribution (Cloyd et al, 1990), episodic tectonism (Cisne, 1986), and eustatic oscillations of sea-level (Goldhammer et al., 1987; Koerschner and Read, 1989). Each of the proposed mechanisms must explain the apparent upward shallowing of individual cycles, the repetitive stacking of similar cycles throughout a vertical sequence, and the simultaneous development of tidal flat-capped cycles on the Appalachian platform and subtidal cycles across the Cordilleran platform.

SHALE-BASED CYCLES

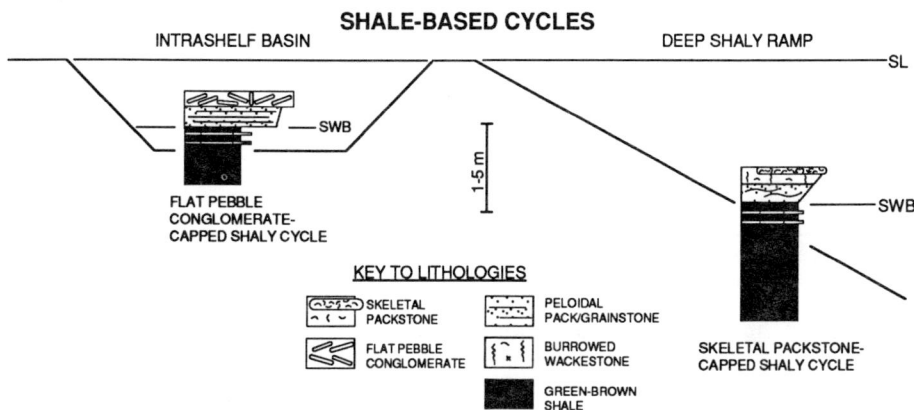

Figure 6: Late Cambrian shaly cycles of the Conasauga intrashelf basin of the Appalachians and of the Cordilleran deep ramp of Utah. Siliciclastic shales are abruptly overlain by "clear-water carbonates" with storm-deposited caps. Note the possible shallower position of storm wave base in the protected intrashelf basin.

The autocyclic model (Ginsburg, 1971) depends upon the periodic progradation of tidal flats over the subtidal carbonate factory to restrict the size of the carbonate source area, effectively shutting down carbonate production until tectonic subsidence provides water depths sufficient to resume production. By definition, subtidal cycles lack a tidal flat cap that might have inhibited productivity during progradation and therefore precludes auto- cyclicity as a potential controlling process on Late Cambrian cycle development. In addition, the auto-cyclic variability of shifting loci of carbonate sedimentation and deposition (Cloyd et al., 1990) cannot explain the persistent vertical and lateral rhythmicity and predict-ability of stacked subtidal cycles. Auto-cyclic mechanisms may only be viable as an explanation for stratigraphic "noise" within individual cycles and do not control the development of repetitive stacks of cycles or the synchronous development of peritidal and subtidal cycles thoughout the Late Cambrian platforms.

Repeated pulses of subsidence have been proposed (Hardie et al., 1986; Cisne, 1986) to generate the accommodation potential for sediment aggradation. If the stress limits between faulting episodes were rhythmic, based on some threshold value, then this model could conceivably explain the coexistence of peri-tidal and subtidal cycles. However, the lateral extent of such events would be extremely limited and could not explain the widespread nature of carbonate cycles across entire platforms. On the Utah-Nevada plat-form, episodic movement along the House Range Embayment southern boundary fault may have contributed "noise" to the subtidal cycle stratigraphy in the Big Horse Member of the Orr Formation (early Late Cambrian), but certainly could not have affected subtidal cycles in later Late Cambrian strata because the House Range Embayment was infilled after the early Late Cambrian (Rees, 1986). Finally, other tectonic mechanisms such as intraplate stress variations (Cloetingh, 1986) are much too slow (0.01 - 0.1 m/ky) and non-periodic to produce high frequency meter-scale cycles.

High-frequency oscillations in eustatic sea-level, probably controlled by fluctuations in glacial ice volume, provide the simplest explanation for the simultaneous development of meter-scale peritidal and subtidal cycles on coeval carbonate platforms (Fischer, 1964; Grotzinger, 1986; Goldhammer et al., 1987; 1990; Koerschner and Read, 1989). However, the forcing mechanism behind high-frequency sea-level oscillations is far from certain. Milankovitch astronomical rhythms have been incontrovertibly proven to have controlled stratigraphic cyclicity during the Plio-Pleistocene (e.g., Hays et al., 1976) and probable Milankovitch influence has been documented for the Cretaceous (Schwarzacher and Fischer, 1982; Herbert and Fischer, 1986), the Triassic (Olsen, 1986; Goldhammer et al., 1987), the Permian (Anderson, 1986), and the Permo-Carboniferous cyclothems (Heckel, 1986).

However, other attempts at showing a Milankovitch influence on ancient cyclic sequences have depended upon the average period- icities of cycles that roughly coincide with one of the Milankovitch periods of 19 - 23 k.y., 41 k.y., 95 - 123 k.y. or 413 k.y. As cautioned by Algeo and Wilkinson (1988), calculations of average cycle period are meaningless in terms of identifying a controlling mechanism on cycle formation.

Spectral Analysis of Late Cambrian subtidal cycles

An objective way of determining cycle periods is by spectral analysis of cyclic successions where dominant periodicities can be extracted and ratios between the periods can be used to establish Milankovitch control (Schwarzacher and Fischer, 1982; Herbert and Fischer, 1986). An inherent weakness of the method is the assumption of constant sediment accumulation throughout the duration of the cyclic succession. This precludes the use of peritidal cycles where much of the cycle period is taken up by non-deposition (Read et al., 1986; Goldhammer et al., 1987; 1990). However, stacks of deep subtidal cycles provide a good data set for time series analysis because the assumption of relatively constant accumulation is more reasonably justified.

Spectral analysis was performed on time series constructed from flat-pebble conglomerate-capped shaly cycles from two localities more than 70 km apart in the Late Cambrian Conasauga intrashelf basin of Virginia and Tennessee (Fig. 7). The time series were created using relative water depth ranks of component lithofacies as described by Olsen (1986). Periods of the peaks on the resultant power spectra were calculated using

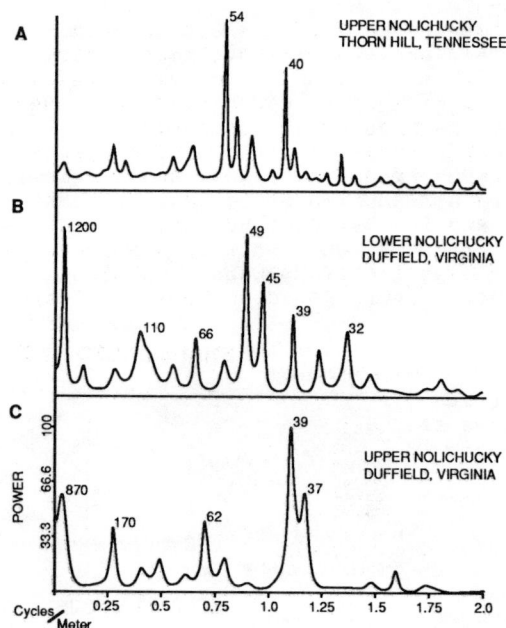

Figure 7: Representative power spectra from two locations of the Nolichucky Formation in Virginia and Tennessee. The Virginia section was divided into lower and upper intervals and analyzed separately.

accumulation rates for that locality. Error ranges of 50%, largely related to the geologic time scale, were used to calculate the probable range of periods. Assuming a 0.044 m/k.y. accumulation rate, all three power spectra record strong peaks near calculated periods of around ±40 k.y.(20 - 60 k.y. range). The two plots for the upper and lower intervals at Duffield (B & C on Fig. 7) record subordinate periods around 62/66 k.y. (30 - 90 k.y. range) and 110/170 k.y. (50 - 300 k.y. ranges) as well as a low frequency signal around 1 m.y. It should be made clear that the range of periods derived from the spectral analysis of these subtidal cycles does not necessarily prove a Milankovitch control on their origin. However, the sharpness of the spectra and the correlatability of peaks between the Tennessee and Virginia localities suggests that sea-level fluctuations controlled by dominant periodicities acted over a broad area, influencing cyclic depositional patterns.

Glacio-eustasy during the
"Non-glacial" Late Cambrian

Estimates of magnitudes of relative sea-level fluctuations that generated the Late Cambrian meter-scale cycles must account for the simultaneous development of peritidal cycles as well as deep ramp subtidal cycles that formed on different parts of the Late Cambrian platforms. One-dimensional and two-dimensional computer modelling (Read et al., 1986; Osleger, 1990; Read et al., in press; Osleger and Read, in review) of cyclic sequences suggests that sea-level oscillations on the order of 15 to 20 (±5) meters could produce both peritidal and subtidal cycles similar to those exhibited in Late Cambrian strata of the Appalachian and Cordilleran passive margins.

A major problem with the scenario of high-frequency sea-level oscillations of 15 to 20 (±5) meters during the Late Cambrian is the lack of a suitable reservoir for the storage and release of moderate volumes of seawater. The Late Cambrian was a time of low-latitude continents and global warmth and no direct evidence of continental glaciation has been recognized. A direct link between changes in solar insolation related to Milankovitch astronomical rhythms and changes in sea-level and sedimentation during globally warm periods of Earth history has yet to be found (Barron et al., 1985). Nevertheless, the evidence provided by the repetitious arrangement of lithofacies into predicable cycles strongly indicates that sea-level had to have fluctuated within the range of 10^4 to 10^5 years. It is acknowledged that the tenuous connection between Milankovitch orbital variations and the presumably ice-free Late Cambrian will remain a major weakness in the model for controlling mechanisms of Late Cambrian cyclicity.

Cycle Type and Platform Morphology

The primary reason why subtidal cycles tend to dominate the Utah-Nevada Cordilleran platform and peritidal cycles tend to dominate the Appalachian platform may be the difference in platform morphologies (Osleger, in review). The Appalachian platform is generally believed to have been a flat-topped reef-rimmed shelf (Demicco, 1985; Read, 1989). Platform slopes are estimated to have been 1 to 2 cm/km (compared to 3 to 4 cm/km for the modern Bahama platform) and, given modern rates of tidal flat progradation, peritidal cycles could easily have prograded over the 300 to 400 km width of the flat slope within a few tens of thousands of years.

In contrast, the Cordilleran passive margin along a transect from central Utah out to the slope facies of the Hot Creek Range of central Nevada (Cook and Taylor, 1977) is believed to have been a distally steepened ramp (Read, 1985). The lack of a paleo-depositional reefal rim, the strike-parallel distribution of lithofacies, and estimated paleowater depths of shaly cycles common on the outer ramp suggest significantly higher depositional slopes than in the Appalachians. Progradation of peritidal cycles from the craton edge near the Wasatch hinge line may have been restricted by the higher slopes and the commensurately greater volumes of accommodation space available to be filled (cf. Hardie and Shinn, 1986).

The connection between platform morphology and cycle type may be related to energy regime. Open, deeply submerged ramps would be subject to higher energy conditions because they are vulnerable to strong wave and current activity generated in response to swells originating in the open ocean. Storm-related swells would travel unimpeded onto the ramp with little loss of energy until they impinged on the ramp bottom (20-200 m for 5-12 second waves) when they would begin to lose energy by frictional dissipation. In this way, fines generated on the shallow ramp would be winnowed and redeposited along narrow tidal flats adjacent to the shoreline or carried out to the deeper ramp where they would settle out of suspension. The higher energies would preclude the nucleation and progradation of tidal flats and thereby maintain subtidal conditions across the ramp. A possible modern analog is the west Florida ramp where the Loop Current and intermittent storms sweep sediment off the platform and redeposit it as thick wedges along the southwestern shelf and upper slope (Mullins et al., 1988).

In contrast, flat-topped, reef-rimmed platforms would be dominated by comparatively low-energy conditions because swells generated in the open oceans would rapidly lose energy during contact with the protective reefal rim. Fine-grained sediments would accumulate within restricted shallow subtidal lagoons where they could be fed onto rapidly prograding tidal flats. The platform would maintain a fully aggraded, low energy, flat-topped profile enhancing the development of widespread peritidal lithofacies.

LATE CAMBRIAN DEPOSITIONAL SEQUENCES

Scales of Cyclicity

Late Cambrian meter-scale cycles (fifth-order scale) are grouped into shallowing-upward successions at the fourth-order scale (0.1 - 1.0 m.y.; tens of meters) as well as at the third-order scale (1.0 - 10.0 m.y.; tens to hundreds of meters). For this discussion, depositional sequences are defined as "a relatively conformable succession of genetically related strata bounded by unconformities and their correlative conformities" (Mitchum, 1977). The majority of sequence boundaries within Late Cambrian strata of the Cordilleran and Appalachian passive margins are Type 2 (Van Wagoner et al., 1987) in that they are not major unconformity surfaces with long-term subaerial exposure (Type 1), but rather are expressed as transitional zones of maximum regressive lithofacies and intermittent exposure features generated by high-frequency sea-level fluctuations. They are the "correlative conformities" equivalent to the more extensive erosional unconformities on the craton and reflect third-order sea-level fall rates that are less than subsidence rates on the pericratonic platforms.

The Big Horse Member of the Orr Formation of the House Range nicely illustrates the various scales of cyclicity (Fig. 8). The interval comprises the upper part of one long-term third-order shallowing-upward sequence whose lower transgressive phase begins in shaly mudstones of the underlying Weeks

Formation. The overlying Big Horse Member is composed of deep ramp cycles toward the base gradually shallowing up into oolite-capped shallow ramp cycles with large thrombolite bioherms marking the probable sequence boundary. Deep ramp shaly cycles of the overlying Candland Shale Member manifest renewed drowning and the beginning of a new depositional sequence.

The Big Horse Member has superimposed within it 11 fourth-order depositional cycles (tens of meters thick; average of 360 k.y.) that are characterized by gradual shallowing from deeper ramp lithofacies to shallow ramp lithofacies followed by rapid drowning and the abrupt appearance of thick, deeper ramp lithofacies. Characteristic meter-scale fifth-order cycles systematically change upward within the third- and fourth-order sequences and can be used to identify long-term relative sea-level cycles. Cycle stacking patterns recognized within the cyclic succession provide the crucial link between the meter-scale cycles and the larger scale sequences.

Late Cambrian Sequences and Grand Cycles

Two major second-order "supersequences" separated by a cratonwide unconformity at the Dresbachian-Franconian boundary have long been recognized within Late Cambrian strata (Lochman-Balk, 1971; Palmer, 1971a,b; 1981b). Subaerial erosion associated with unconformity development has removed the *Dunderbergia* and part of the *Aphelaspis* zones from localities on the North American craton and on the craton margin (Lochman-Balk, 1971; Palmer, 1971a; 1971b).

Shorter-term (1-15 m.y. duration) depositional sequences, including Grand Cycles of the southern Canadian Rockies, also have been defined in Late Cambrian strata (Aitken, 1978; 1981; Palmer, 1981a; Cooper et al., 1982; Chow and James, 1987; Bond et al., 1989; Koerschner and Read, 1989; Read, 1989; McCutcheon and Cooper, 1989). Grand Cycles, consisting of a lower shaly half-cycle and an upper carbonate half-cycle and spanning two or more trilobite zones over thicknesses of 90 to 720 meters (Aitken, 1981), appear to be synonymous with short-term second-order or long-term third-order sequences of Vail and others (1977).

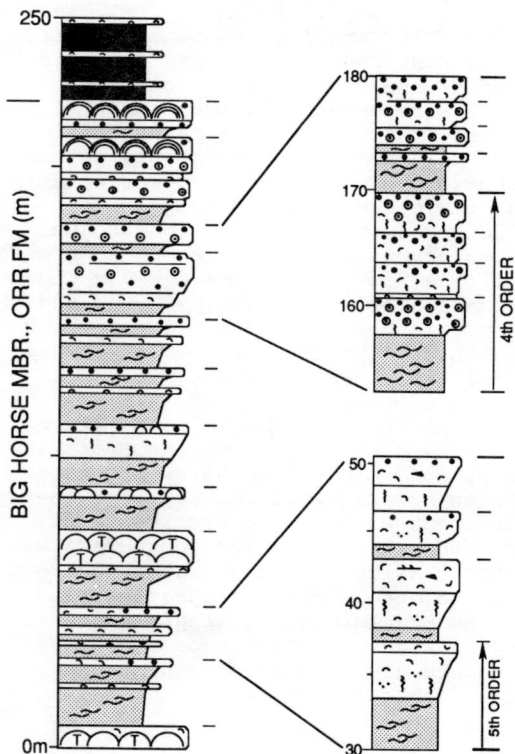

Figure 8: Scales of cyclicity within the Big Horse Member of the Orr Formation of the House Range, Utah. Column on the left shows long-term third-order shallowing evident from the storm-influenced mid-ramp cycles with open marine faunas in the lower Big Horse progressively giving way to shallow subtidal cycles characterized by restricted lithofacies upward in the Big Horse Member. Dashes to the right of the left column denote interpreted

Attempts at interbasinal correlation of Late Cambrian Grand Cycles (Aitken, 1981; Chow and James, 1987) have focused on the isochroneity of Grand Cycle boundaries as datums for correlation (cf. Cooper, 1989). Other attempts at interbasinal correlation of Late Cambrian sequences (Palmer, 1981a) have been hampered by a lack of bio- and lithostratigraphic resolution. However, Aitken (1981) and Bond and others (1989) used detailed lithostratigraphic correlation of biostratigraph- ically well-constrained sections of Late Cambrian strata to recognize numerous third-order shallowing-upward sequences both within Grand Cycles and dominantly carbonate successions elsewhere on the continent. In the following discussion, the degree of time equivalence of Late Cambrian depositional sequences and the relative sea-level fluctuations that generated them are determined using a variety of qualitative and quantitative techniques. These techniques were combined and crosschecked to establish a composite Late Cambrian sea-level curve that defines six major eustatically controlled Late Cambrian sequences. Details of the various methods, correlation diagrams and descriptions of specific equivalent intervals can be found in Osleger (1990).

Cordilleran Chronostratigraphy

To determine whether relative sea-level changes that affected the House Range section also were expressed basinwide, a chronostratigraphic cross-section (Fig. 9) was constructed for Late Cambrian strata from hinge line near the East Tintics of central Utah to the distally steepened ramp edge of the Hot Creek Range of central Nevada (Cook and Taylor, 1977). The cross-section was constructed as a check on the lateral traceability of major large-scale shallowing-upward sequences defined from vertical sections in the House Range. It extends along the axis of the House Range embayment, avoiding all peritidal areas along the southern flank. The House Range Embayment existed from the mid Middle Cambrian through *Crepicephalus* time, and is marked by an abrupt bathymetric transition near the House Range (Palmer, 1971a; Rees, 1986). Data for the plot were derived from detailed logs of the House Range section, field checks of published sections, isopach trends, paleogeographic maps, and pinch-out directions.

The chronostratigraphic cross-section should not be interpreted to imply continuous sedimen- tation across the platform. Certainly episodes of non-deposition occurred, but the dominantly deep to shallow subtidal lithofacies suggest generally submergent conditions throughout the Late Cambrian of the Utah-Nevada Cordilleran passive margin. Throughout the following discussion of the six major sea-level events that can be correlated between the four localities, reference will be made to Figure 9 to illustrate the regional effects of the sea-level events across the Cordilleran platform in Utah and Nevada.

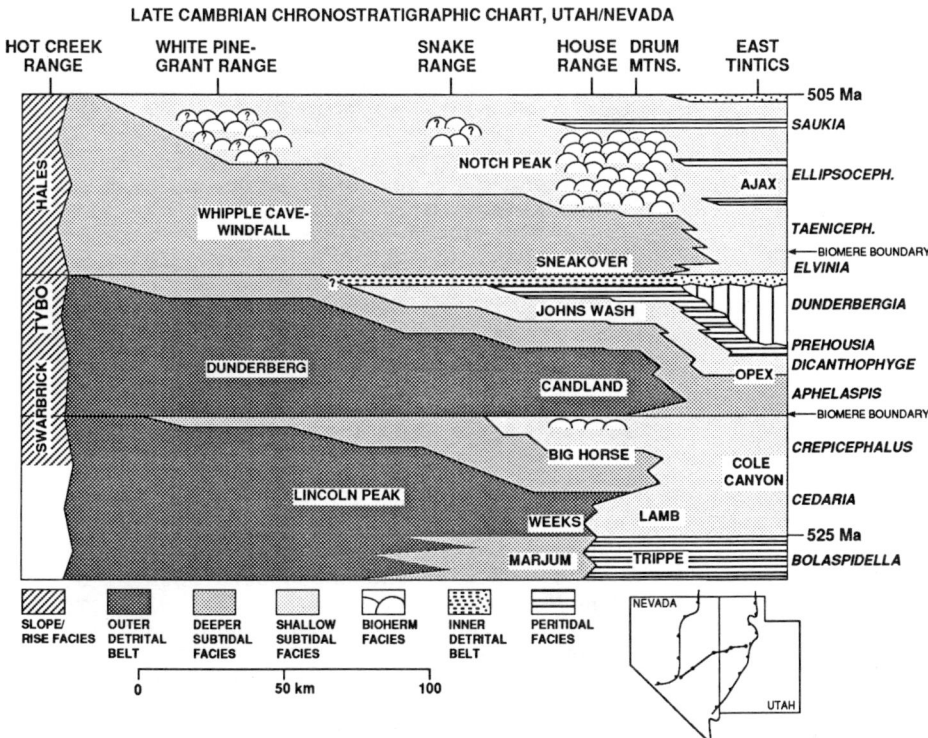

Figure 9: Chronostratigraphic chart for Late Cambrian strata of Utah-Nevada. Trend of cross-section is down the axis of the House Range Embayment from the hinge line in central Utah to the distally-steepened ramp edge in central Nevada. Distances between ranges were estimated using the palinspastic base map of Levy and Christie-Blick (1989). The cross-section defines three major transgressive-regressive sequences in the Late Cambrian of Utah-Nevada.

Graphic Correlation of Depositional Sequences

Paleobathymetric curves were constructed for stratigraphic sections in Utah, Virginia, Texas and Oklahoma to distinguish deepening-upward and shallowing-upward successions of genetically-related stratigraphic intervals. Only the curves for the Utah and Virginia sections are shown in Figure 10. Relative fluctuations in water depth were determined from interpreted changes in depositional environments of lithofacies and associated changes in meter-scale cycle types. Relative water depths of lithofacies were estimated from Holocene analogs, stratigraphic distance of lithofacies below tidal flat caps of peritidal cycles (Grotzinger, 1986; Koerschner and Read, 1989), and by the relative ordering of facies successions in cycles.

Graphic correlation (Shaw, 1964) was used to determine whether the sequences are recognized interbasinally and to establish the relative degree of synchroneity. The Utah section from the House Range was chosen as the standard reference section based on its good biostratigraphic control, its completeness of section with no visible major erosional unconformities and its excellent exposure. It also has the advantage of having three biomere boundaries each defined to within less than one meter which provide excellent isochronous datums useful for establishing a reliable line of correlation. Depositional sequences between different localities were considered to be essentially coeval if their points of intersection fell within reasonable proximity to the line of correlation. Correlations were made to the finest resolution possible within the limits of the biostratigraphic control and equivalent intervals were correlated lithostratigraphically to compare the component cycle types and lithofacies that comprise each section (Osleger, 1990).

The two major second-order "supersequences" (Palmer, 1981b) separated by the Dresbachian/ Franconian transition and six distinct third-order sequences were interbasinally correlated between the four major localities and were named according to the trilobite biozone nearest the top of the sequence. Only the relevant intervals in the Appalachians and in the House Range sections are described below.

Late Cedaria

The earliest correlatable shallowing-upward event occurs toward the end of the Cedaria biozone and is recognized in the Utah, Appalachian, and Texas sections. In Utah, upward shallowing of deep ramp lithofacies in the basal Big Horse Member of the Orr Formation culminates in a thick interval of stacked thrombolite bioherms that occur near the Cedaria - Crepicephalus contact (Eby, 1981). In Conasauga intrashelf basin strata of the Virginia-Tennessee Appalachians, upward shallowing toward the end of Cedaria time (Sundburg, 1990) is suggested by an upward transition from deep intrashelf basin shales into stacked quartzose peloidal packstones and flat pebble conglomerates of the lower limestone member of the Nolichucky Formation.

Mid-Crepicephalus

Deepening followed by distinct shallowing occurs in the Big Horse Member of Utah within a transition from open marine, deeper ramp, skeletal packstone-capped cycles into shallow ramp, oolitic-oncolitic cycles that indicate increasingly restricted conditions. Graphic correlation suggests that this same event in the Appalachians is represented by intrashelf basin shales that grade up into interbedded oolitic grainstones and thrombolitic bioherms of the middle limestone member of the Nolichucky Formation (Fig. 10).

Late Crepicephalus

In Utah, upward shallowing culminates in a transition from shallow subtidal, oolitic grain- stone-capped cycles into large thrombolitic bioherms capped by club-shaped stromatolites (Lohmann, 1976). A trans-mgressive oolitic grain- stone blankets the bioherms and is immediately overlain by deep ramp shaly cycles of the Candland Shale Member. This shallowing event is evident on the Cordilleran chronostratigraphic diagram (Fig. 9) by seaward progradation of shallow subtidal facies that mark the termination of the House Range Embayment at the end of Crepicephalus time (Rees, 1986).

Graphic correlation with Appalachian sections of the intrashelf Nolichucky Formation suggests that this event is manifested by upward shallowing from distal to proximal storm deposits capped by amalgamated flat-pebble conglomerate storm beds, oolitic grainstones, and thrombolitic bioherms. In peritidal sections of the Appalachians, such as the Allentown Formation of eastern Pennsylvania, this event is represented by upward transitions from subtidal-dominated peritidal cycles into tidal flat-dominated cycles with brecciated regolithic caps and quartz sands reworked into laminite facies.

Aphelaspis to earliest Elvinia

Abrupt deepening during early Aphelaspis time in Utah is exhibited by deep ramp shaly cycles of the Candland Shale Member (Fig. 10). These grade upward into burrowed wackestones/packstones of the Johns Wash Member that ultimately shallow up into oolitic grainstones and fenestral lime mudstones of earliest Elvinia age (Hintze and Palmer, 1976). The chronostratigraphy of the Utah-Nevada passive margin (Fig. 9) shows this event as major onlapping of outer detrital belt shales onto the platform, followed by subsequent offlap of shallow subtidal and peritidal lithofacies and development of an unconformity near the Cordilleran hinge line (East Tintic region). The Dresbachian/Franconian transition is

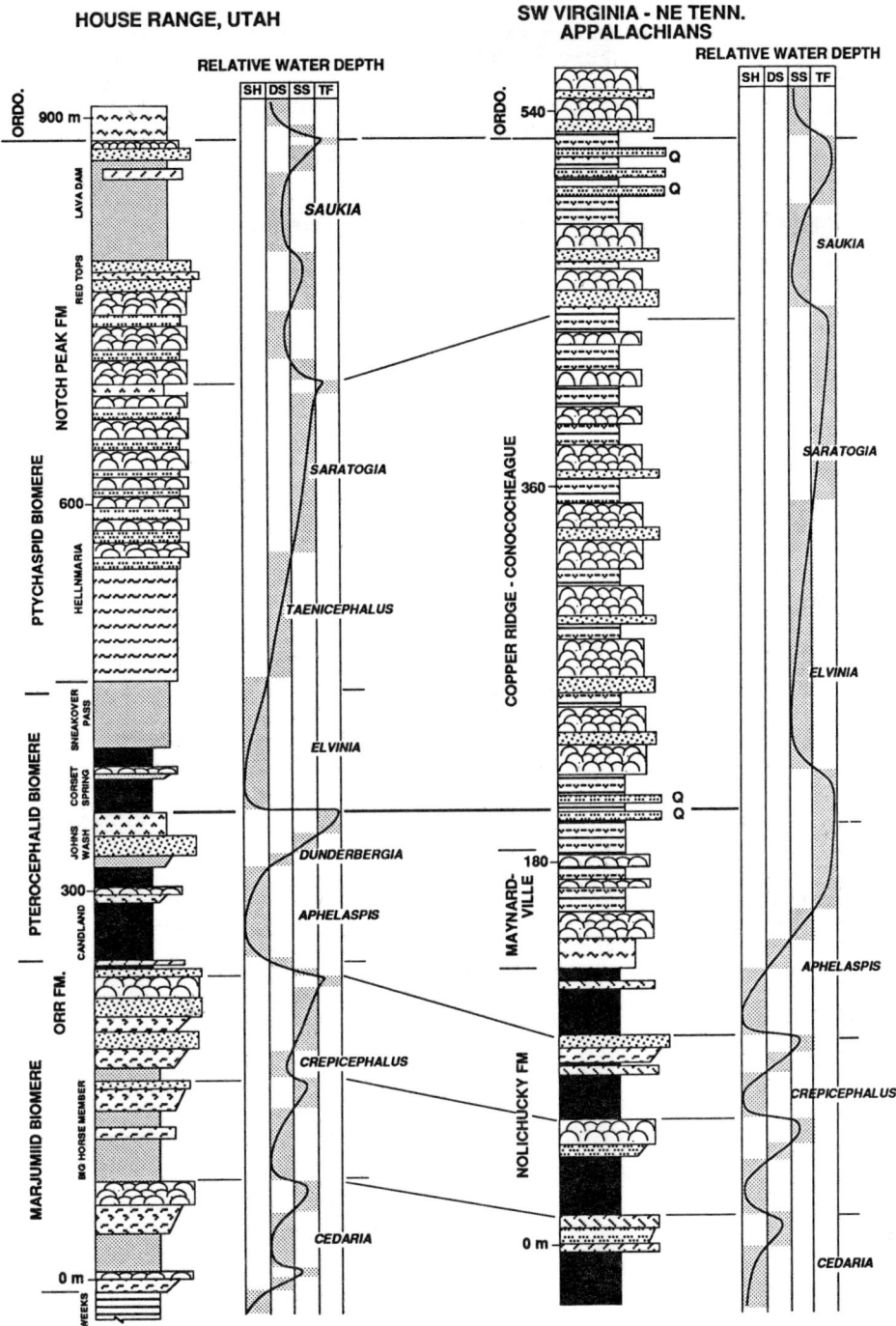

Figure 10: Plots of relative interpreted paleo-water depths of lithofacies for stratigraphic sections in the House Range of Utah and the Virginia-Tennessee Appalachians. Lithofacies on the diagrams are generalized with the shaded rectangles identifying tidal flat-dominated lithofacies (TF) from shallow subtidal (SS), deep subtidal (DS) and shaly deep subtidal-dominated lithofacies (SH). The sections have been expanded for ease of correlation but notice the difference in thickness between each of the sections. The Cambrian-Ordovician boundary establishes the upper datum. Biostratigraphy noted to the right of each relative water depth curve. The two long-term shallowing upward events of the Late Cambrian are separated by the heavy dark horizontal line crossing the middle of the diagram (wavy near unconformities). Other major shallowing-upward events are correlated by thin horizontal tie lines. The Appalachian column is a composite of several sections from the Conasauga intrashelf basin and the peritidal platform.

apparently conformable in the House Range and deeper ramp settings.

Aphelaspis time in the Conasauga basin of the Appalachians is marked by widespread expansion of upper Nolichucky shales onto the surrounding peritidal platform. These intrashelf basin shales grade up into deep ramp ribbon rocks and overlying cryptalgal laminites of the Maynardville Formation. The Dresbachian/Franconian unconformity on the craton (Lochman-Balk, 1971) and in the Texas sections is represented in the Tennessee – Virginia Appalachians as a very thin interval in the basal Copper Ridge Formation (earliest *Elvinia*; Palmer, 1971b; pers. comm.). This interval is character- ized by an influx of quartz sand, thin tidal flat-dominated peritidal cycles, an absence of *Dunderbergia* faunas and subtle erosional truncation of some cycle tops. However, the biostratigraphic control in the Appalachians is insufficient to estimate a duration of exposure, and unequivocal evidence for a major exposure surface is lacking.

Elvinia to early *Saukia*

In Utah, the transgressive base of the long-term Franconian-Trempealeauan second-order event is represented by shaly cycles of the inner detrital belt Corset Spring Member. The shales grade up into argillaceous carbonates of the Sneakover Pass Member "subtidal blanket" of Brady and Rowell (1976), which they recognized to extend across the entire Cordilleran passive margin during middle-late *Elvinia* time. This drowning event shallows within the basal Hellnmaria Member of the Notch Peak Formation before grading into shallow subtidal/peritidal carbonates with extensive stromatolite-thrombolite bioherm development within the upper Hellnmaria Member. This shallowing is recognized on the chronostratigraphic cross-section (Fig. 9) as a widespread expansion of aggraded, shallow subtidal lithofacies (Hales, Whipple Cave and Windfall Formations) with no direct evidence of detrital siliciclastics recognized anywhere on the platform. The top of this sequence in the House Range is marked by dolomitized fenestral lime mudstones that feature scalloped erosion surfaces overlain by coarse oncolitic lag deposits. By graphic correlation with the well-constrained Texas section, this shallowing occurs during earliest *Saukia* time

The predominance of peritidal lithofacies in the Copper Ridge-Conococheague Formations of the Appalachians makes the qualitative recognition of long-term transgressive-regressive sequences difficult, but systematic changes in stacking patterns of peritidal cycles appear to define long-term fluctuations in relative sea-level. Relative deepening above the basal sands of the Copper Ridge-Conococheague is subtle and may be recognized by the appearance of stacks of thick cycles dominated by thrombolitic bioherms. Upward shallowing is suggested by a change in cycle stacking pattern to thin, tidal flat-dominated cycles. Relative shallowing was terminated by the abrupt appearance of thick, slightly deeper water, thrombolitic peritidal cycles in earliest *Saukia* time.

Saukia to Cambrian-Ordovician Boundary

The beginning of the *Saukia* event in the Utah section is marked by subtle deepening into thick stacks of thrombolitic bioherms above the fenestral facies of the previous event. After a brief shoaling into terrigenous tidal flat-capped cycles (Red Tops Member), a return to deeper subtidal conditions is suggested by the argillaceous wackestone cycles of the Lava Dam Member. However, water depths of this lithofacies may not have been very deep because coeval thrombolite bioherms exist in other sections in the House Range and nearby Wah Wah Mountains.

The third-order shallowing evident throughout *Saukia* time culminates at the Cambrian-Ordovician boundary and has been recognized in many sections worldwide (Lange Ranch Eustatic Event of Miller, 1984). Abrupt shallowing from the Lava Dam argillaceous wackestones is manifested by the abrupt appearance of oncolitic packstones and club-shaped stromatolites that are overlain by earliest Ordovician mudstones/wackestones of the upper Lava Dam Member and House Limestone. This lowstand at the end of the Cambrian is recognizable cratonward by a major influx of quartz sand (Wilson, 1952; Palmer, 1971b) and is exhibited toward the Cordilleran hinge line in the East Tintic region by 2 to 6 meters of quartzite (Opahanga Formation; Fig. 9) that are reworked along the Cambrian-Ordovician boundary (Palmer, 1971a).

The sea-level lowstand at the end of the Cambrian in the Copper Ridge-Conococheague Formations of the Appalachians is marked by thin tidal flat-dominated cycles that contain abundant quartz sand derived from the craton (Wilson, 1952; Palmer, 1971; Orndorff, 1988). In the Tennessee section, the Cambrian-Ordovician boundary is marked by quartz-filled tidal channels cut into laminite facies.

Sea-level Cycles Defined from Fischer Plots

Fischer plots provide an excellent technique for identifying relative sea-level events using systematic changes in composition and thickness of meter-scale cycles. Fischer plots (Fig. 11a) are graphic displays of accommodation space, corrected for linear subsidence, through time (Fischer, 1964; Goldhammer et al., 1987; 1990; Read and Goldhammer, 1988). Each fifth-order cycle is assigned an average cycle period by dividing the total estimated duration of the cyclic succession by the number of meter-scale cycles. This average cycle period is merely a device for assigning time per cycle and does not imply that each cycle was deposited over the same duration.

Stacks of thick cycles plot as positive slopes and are presumed to have formed under conditions of increased accommodation space provided by relative sea-level rise. Stacks of thin cycles plot as negative slopes and are presumed to reflect reduced accommodation space during relative sea-level fall. The systematic arrangement of similar cycles on rising and falling limbs of Fischer plots suggests that they record changes in accommodation space generated by third-order relative sea-level oscillations (Osleger, 1990; Osleger and Read, in review).

Correlated Fischer plots (Figs. 11b,c) were constructed from meter-scale cycles defined for each measured stratigraphic section. Relative time lines and correlation lines on Figures 11b and 11c are based on available biostratigraphy, the presence of regional quartz sands, and similarities between patterns of relative sea-level rise and fall on the curves. Data tables and details concerning the construction of the correlated Fischer plot diagrams can be found in Osleger (1990).

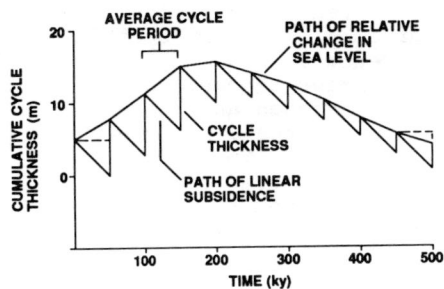

Figure 11a) Explanatory diagram of the Fischer plot technique. For each cycle the amount of accommodation space provided by linear subsidence is plotted over the duration of the average cycle period. Cycle thickness is plotted vertically. The net difference defines the path of relative sea level through time.

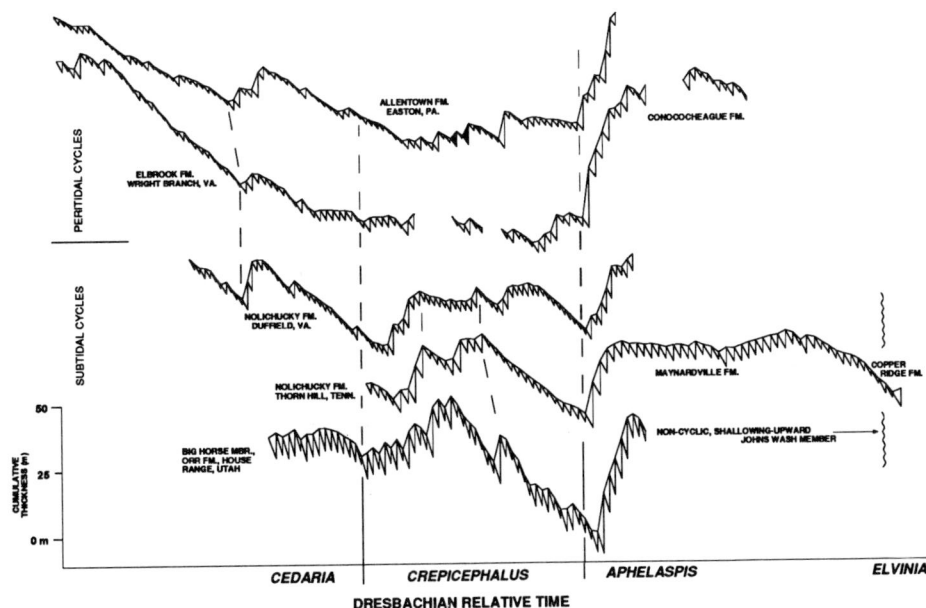

11b) Correlation of Fischer plots for Dresbachian time. Trilobite zonation along the horizontal axis and cumulative thickness in meters, corrected for linear subsidence, along the vertical axis. Upper two Fischer plots (and the latest portion of the Tennessee Fischer plot) are from peritidal cyclic sections whereas the lower three Fischer plots are from shallow to deep subtidal cyclic sections. Quartz sands denoted by black filled triangles.

11c) Correlated Fischer plots for Franconian – Trempealeauan time. Note position of Cambrian-Ordovician boundary and the appearance of quartz sandy cycles during the sea level lowstand at the end of the Late Cambrian. Cycles from the Appalachian sections are peritidal whereas the Utah section is composed of shallow to deep subtidal cycles.

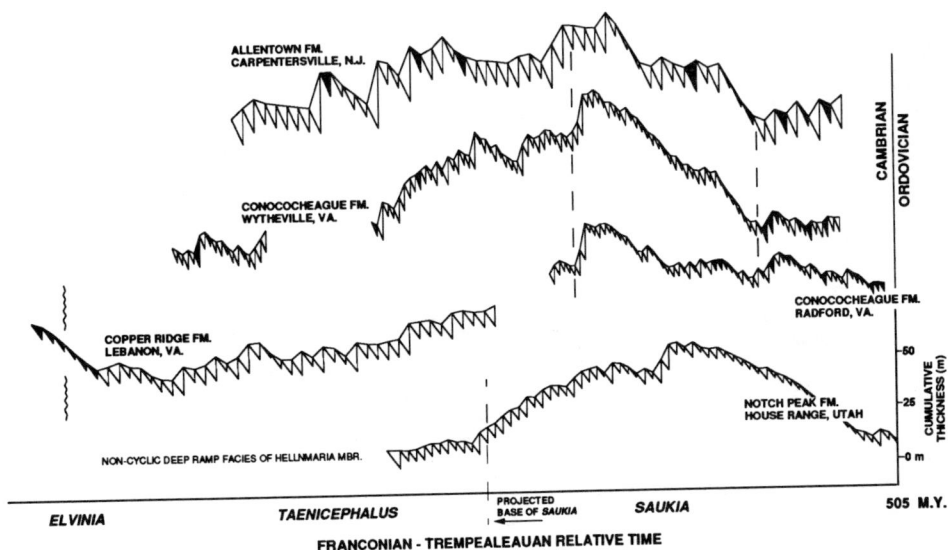

Several sea-level events can be correlated between widely-separated localities using the Fischer plots which strongly supports a eustatic control on their development. The plot for Dresbachian time (Fig. 11b) illustrates generally falling sea-level toward the end of *Cedaria* time in both peritidal and subtidal cyclic successions. This gradually falling trend on the Fischer plots corresponds to the late *Cedaria* regression recognized on the paleobathymetric curves. A well defined rise in relative sea-level at the beginning of *Crepicephalus* time followed by a relative sea-level fall toward the *Crepicephalus/Aphelaspis* biomere boundary is evident on the subtidal cyclic sections of the Nolichucky in Virginia and Tennessee and of the Big Horse Member in Utah. The long-term *Crepicephalus* event is not evident within the peritidal Allentown and Elbrook localities although covered intervals in the Elbrook may mask the event.

All five localities show the rapid sea-level rise at the beginning of *Aphelaspis* time that also was recognized on the correlated paleobathymetric curves. The Fischer plot for the Thorn Hill section of Tennessee shows a complete rise and fall of relative sea-level that culminates in thin, disconformity-capped, restricted tidal flat cycles of the basal Copper Ridge Formation that correspond to the cratonwide unconformity between Dresbachian and Franconian strata.

The plot for Franconian-Trempealeauan time (Fig. 11c) illustrates a long-term relative sea-level rise and fall in all five sections that appears to correspond to the second-order event recognized on the paleobathymetric curves. Quartz sands in the basal part of the Copper Ridge Formation at Lebanon, Virginia connect Figure 11c to the basal Copper Ridge in the Tennessee section on Figure 11b and mark the relative sea-level lowstand at the end of the Dresbachian stage. Generally rising sea-level is evident during *Taenicephalus* time before culminating at a relative highstand during early *Saukia* time. The appearance of quartz sand in peritidal cycles of the Virginia and Pennsylvania Appalachians corresponds with the major sea-level lowstand near the Cambrian-Ordovician boundary. This event may also explain the appearance of shallow water shales and argillaceous carbonates in equivalent strata of the Utah Cordillera.

Composite Late Cambrian Sea-level Curve

The excellent degree of correlatability of Late Cambrian depositional sequences using the qualitative and quantitative techniques described above strongly supports a eustatic control on their origin. Acknowledging the limitations of the different methods used to determine sea-level history, the various sea-level curves for each section were qualitatively combined and a composite "eustatic" sea-level curve was derived for the Late Cambrian (Fig. 12). Magnitudes of sea-level fluc- tuations and rates of sea-level rise and fall on the curves are strictly relative. Determining the magnitude of a "eustatic" sea-level curve from individual stratigraphic sections is not really possible given the complicated interactions of subsidence, sedimentation and eustasy (Burton et al., 1987). Only depositional sequences recognized by a combination of techniques and correlatable with the other three localities are included on the composite curve. Non-correlatable events or fourth-order events only correlatable between two localities are considered regional in origin and are not included. The long-term second-order event, on which all of the other curves are superimposed, was defined from residual eustatic curves generated from subsidence analysis of the entire Cambrian-Ordovician interval (Osleger, 1990).

The composite "eustatic" curve for the Late Cambrian is intended to be used as a template for the correlation of Late Cambrian depositional sequences recognized elsewhere. The curve was generated without the preconceived bias associated with the shale/carbonate alternations of the Grand Cycles of the southern Canadian Rockies. Grand Cycles would provide an ideal test for the validity of the "eustatic" curve defined by Late Cambrian sections around the pericratonic platform of the United States.

Synchroneity of Sequences

A number of factors contribute to the relative timing of the initiation and termination of sequences in different depositional basins. First, the degree of synchroneity of events is difficult to determine due to the inherent errors in biostratigraphy. Biomere boundaries are considered isochronous (Palmer, 1984) making correlation based on biomere boundaries the only true measure of the relative degree of synchroneity between sections. Correlated intervals spanning entire biomeres can be considered chronostratigraphic equivalents. However, sequences defined without the time control of biomere boundaries may be diachronous since the zonal boundaries are subject to the origin, dispersal and extinction of the trilobite populations.

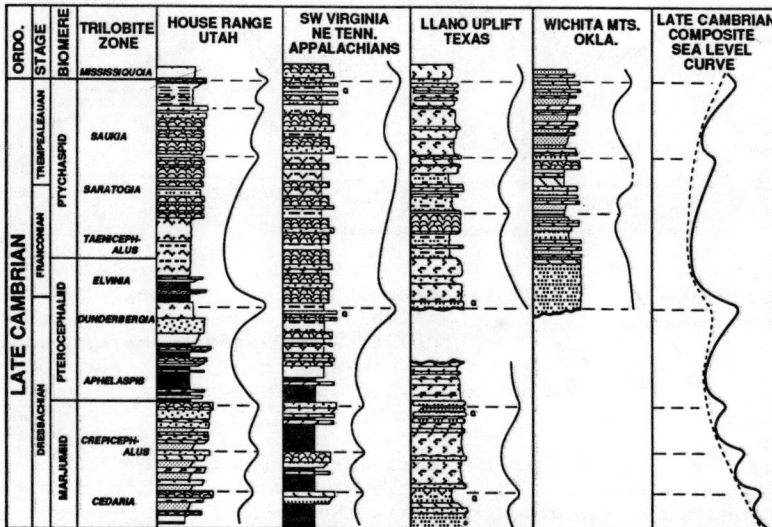

Figure 12: Chronostratigraphic comparison of the "eustatic" sea level curves defined for the Utah Cordillera, the Virginia-Tennessee Appalachians, the Texas craton and the southern Oklahoma aulocogen. Composite Late Cambrian "eustatic" sea level curve was derived from qualitatively combining the different curves from the various locations.

Another major limitation to determining whether correlated sequences are synchronous or not is the difference in sedimentologic response between individual platforms to rising or falling relative sea-level. Pitman and Golovchenko (1988) have maintained that differences in depositional slope and platform width affect the lateral migration of facies; steep and narrow platforms will show a different timing of depositional events than broad, flat platforms. All of the different factors for diachroneity between "equivalent" sequences – limits of biostratigraphy, differential subsidence rates between basins and different platform morphologies – imply that sequences correlated between basins are probably never exactly synchronous.

SUMMARY AND CONCLUSIONS

1) The Late Cambrian of the Utah-Nevada Cordilleran passive margin is characterized by a spectrum of shallow to deep subtidal cycles spread out across a distally steepened ramp. In contrast, the Appalachian Late Cambrian passive margin from Tennessee to Pennsylvania is characterized by widespread peritidal cycles across a flat-topped, aggraded, reef-rimmed shelf. The relationship between cycle type and platform morphology may be a function of energy regime.

2) The simultaneous development of peritidal and subtidal cycles on separate Late Cambrian carbonate platforms supports a eustatic control on the origin of the meter-scale cyclicity. Spectral analysis reveals a strong clustering of periods around a narrow range of values that suggests that high-frequency eustatic oscillations may be controlled by Milankovitch astronomical rhythms.

3) Interbasinal correlation of Late Cambrian cyclic carbonates in the Appalachian and Cordilleran passive margins, the Texas cratonic embayment and the southern Oklahoma aulacogen suggests that second- and third-order depositional sequences are eustatic in origin. Within cyclic successions of Late Cambrian strata (and presumably for other ancient cyclic successions as well), systematic stacking patterns of meter-scale cycles may be used in conjunction with Fischer plots to predict third-order eustatic sea-level events.

ACKNOWLEDGEMENTS

This paper is an outgrowth of dissertation research done at Virginia Polytechnic Institute and conducted under the guidance of Fred Read. His contributions to the ideas expressed in this paper are invaluable. Field assistance and insight into the Cambrian was kindly provided by Jim Miller, Pete Palmer, and Lehi Hintze. The text benefited greatly from critical review by John Cooper and Isabel Montanez. Spectral analysis was done on a VAX computer in the Geophysical Lab at Virginia Polytechnic Institute under the guidance of Dr. C. Coruh. The Cordilleran chronostratigraphic chart was critically reviewed by Pete Palmer but its content reflects only the author's observations and interpretations. Billy Newcomb and Steve van Aken provided able field assistance. Financial assistance was provided by NSF grants EAR 88-16664 and 87-07737 to J.F. Read; by a grant from the American Chemical Society (PRF grant 21282-AC2); by Texaco, Chevron, Marathon and Mobil Oil Companies; and by grants-in-aid from the Geological Society of America, Sigma Xi and the Appalachian Basin Industrial Associates.

REFERENCES CITED

Aigner, T. A., 1985, Storm depositional systems: Dynamic stratigraphy in modern and ancient shallow-marine sequences: Berlin, Springer-Verlag, 174 pp.

Aitken, J. D., 1978, Revised models for depositional Grand Cycles, Cambrian of the southern Rocky Mountains, Canada: Bulletin Canadian Petroleum Geology, v.26, p.515-542.

Aitken, J. D., 1981, Generalizations about Grand Cycles, in Taylor, M. E., ed., Short papers for the Second International Symposium on the Cambrian System: U. S. Geological Survey Open-File Report 81-743, p.8-14.

Algeo, T. J., and Wilkinson, B. H., 1988, Periodicity of mesoscale Phanerozoic sedimentary cycles and the role of Milankovitch orbital modulation: Journal of Geology, v.96, p.313-322.

Anderson, R. Y., 1986, The varve microcosm: propagator of cyclic bedding: Paleoceanography, v.1, p.373-382.

Barron, E. J., Arthur, M. A., and Kauffman, E. G., 1985, Cretaceous rhythmic bedding sequences: a plausible link between orbital variations and climate: Earth & Planetary Science Letters, v.72, p.327-340.

Bond, G. C., and Kominz, M. A., 1984, Construction of tectonic subsidence curves for the early Paleozoic miogeocline, southern Canadian Rocky Mountains: Implications for subsidence mechanisms, age of breakup and crustal thinning: Geological Society of America Bulletin, v.95, p.155-173.

Bond, G. C., Kominz, M. A., Grotzinger, J. P., and Steckler, M. S., 1989, Role of thermal subsidence, flexure and eustasy in the evolution of Early Paleozoic passive margin carbonate platforms, in Crevello, P., Wilson, J. L., Sarg, J. F., and Read, J. F., eds., Controls on Carbonate Platform and Basin Development: Society of Economic Paleontologists and Mineralogists, Special Publication 44, p.39-62.

Bond, G. C., Nickeson, P. A., and Kominz, M. A., 1984, Breakup of a supercontinent between 625 Ma and 555 Ma: new evidence and implications for continental histories: Earth and Planetary Science Letters, v.70, p.325-345.

Brady, M. J., and Rowell, A. J., 1976, Upper Cambrian subtidal blanket carbonate of the

Cordilleran miogeocline, eastern Great Basin: Brigham Young University Geological Studies, v.23, p.153-163.

Burton, R., Kendall, C. G. St.C., and Lerche, I., 1987, Out of our depth: on the impossibility of fathoming eustatic sea-level from the stratigraphic record: Earth Science Reviews, v.24, p.237-277.

Chow, N., and James, N. P., 1987, Cambrian Grand Cycles: a northern Appalachian perspective: Geological Society of America Bulletin, v.98, p.418-429.

Cisne, J. L., 1986, Earthquakes recorded stratigraphically on carbonate platforms: Nature, v.323, p.320-322.

Cloetingh, S., 1986, Intraplate stresses: a new tectonic mechanism for relative fluctuations of sea-level: Geology, v.14, p.617-620.

Cloyd, K.C., Demicco, R.V. and Spencer, R.J., 1990, Tidal channel, levee, and crevasse-splay deposits from a Cambrian tidal channel system: A new mechanism to produce shallowing-upward sequences: Journal of Sedimentary Petrology, v.60, p. 73-83.

Cook H.E., and Taylor, M. E., 1977, Comparison on continental slope and shelf environments in the Upper Cambrian and lowest Ordovician of Nevada, in Cook, H.E., and Enos, P., eds., Deep-Water Carbonate Environments: Society of Economic Paleontologists and Mineralogists, Special Publication 25, p. 51-81.

Cooper, J.D., 1989, Does the Upper Cambrian Dunderberg Shale-Halfpint carbonate couplet in the southern Great Basin qualify as a grand cycle?, in Cooper, J.D., ed., Cavalcade of Carbonates, Society of Economic Paleontologists and Mineralogists, Pacific Section, volume and guidebook, p. 87-100.

Cooper, J.D., Miller, R.H., and Sundberg, F.A., 1982, Environmental stratigraphy of the lower part of the Nopah Formation (Upper Cambrian), southwestern Great Basin, in Cooper, J.D., Troxel, B.W. and Wright, L.A., eds., Geology of Selected Areas in the San Bernardino Mountains, Western Mojave Desert, and Southern Great Basin: Geological Society of America, Cordilleran Section, volume and guidebook, field trip 9, p. 97-115.

Demicco, R. V., 1985, Patterns of platform and off-platform carbonates of the Upper Cambrian of western Maryland: Sedimentology, v.32, p.1-22.

Derby, J. R., 1965, Paleontology and stratigraphy of the Nolichucky Formation in southwest Virginia and northeast Tennessee (Ph.D. dissertation): Blacksburg, Virginia, Virginia Polytechnic Institute and State University, 468 p.

Dill, R. F., Shinn, E.A., Jones, A.T., Kelly, K., and Steinen, R.P., 1986, Giant subtidal stromatolites forming in normal salinity waters: Nature, v.324, No.6, p.55-58.

Eby, R.G., 1981, Early Late Cambrian Trilobite Faunas of the Big Horse Limestone and Correlative Units in Central Utah and Nevada (Ph.D. dissertation): State University of New York, Stony Brook,

613 pp.

Fischer, A. G., 1964, The Lofer cyclothems of the Alpine Triassic, in Merriam, D. F., ed., Symposium of cyclic sedimentation: State Geological Survey of Kansas, Bulletin 169, p. 107-150.

Ginsburg, R. N., 1971, Landward movement of carbonate mud: New model for regressive cycles in carbonates (abstract): American Association Petroleum Geologists Bulletin, v.55, p.340.

Goldhammer, R. K., Dunn, P. A., and Hardie, L. A., 1987, High frequency glacio-eustatic sea-level oscillations with Milankovitch characteristics recorded in Middle Triassic platform carbonates in northern Italy: American Journal Science, v.287, p.853-892.

Goldhammer, R. K., Dunn, P. A., and Hardie, L. A., 1990, Depositional cycles, composite sea-level changes, cycle stacking patterns, and the hierarchy of stratigraphic forcing: Examples from Alpine Triassic platform carbonates: Geological Society of America Bulletin, v.102, p.535-562.

Grotzinger, J. P., 1986, Cyclicity and paleoenvironmental dynamics, Rocknest platform, northwest Canada: Geological Society of America Bulletin, v.97, p.1208-1231.

Hardie, L. A., Bossellini, A., and Goldhammer, R.K., 1986, Repeated subaerial exposure of subtidal carbonate platforms, Triassic, northern Italy: Evidence for high frequency sea-level oscillations on a 10^4 year scale: Paleoceanography, v. 2, p. 447-457.

Hardie, L. A., and Shinn, E. A., 1986, Carbonate depositional environments, modern and ancient; part 3: tidal flats: Colorado School of Mines Quarterly, v.81, p.1-74.

Hays, J. D., Imbrie, J., and Shackleton, J. J., 1976, Variations in the Earth's orbit: Pacemaker of the ice ages: Science, v.194, p.1121-1132.

Heckel, P. H., 1986, Sea-level curve for Pennsylvanian eustatic marine transgressive-regressive depositional cycles along midcontinent outcrop belt, North America: Geology, v. 14, p. 330-334.

Herbert, T. D., and Fischer, A. G., 1986, Milankovithc climatic origin of mid-Cretaceous black shale rhythms in central Italy: Nature, v. 321, p. 739-743.

Hine, A. C., 1977, Lily Bank, Bahamas; history of an active oolite sand shoal: Journal of Sedimentary Petrology, v.47, No.4, p.1554-1581.

Hintze, L. F., 1974, Preliminary geologic map of the Notch Peak quadrangle, Millard County, Utah: U. S. Geological Survey Miscellaneous Field Studies Map MF-636.

Hintze, L. F., and Palmer, A. R., 1976, Upper Cambrian Orr Formation: its subdivisions and correlatives in western Utah: U. S. Geological Survey Bull. 1405-G, p.1-25.

Hintze, L. F., Miller, J. F., and Taylor, M. E., 1980, Upper Cambrian-Lower Ordovician Notch Peak Formation in western Utah: U. S. Geological Survey Open-File Report 80-776, p.67.

James, N. P., 1984, Shallowing-upward sequences in carbonates: *in* Walker, R.G., ed., Facies Models: Geoscience Canada, v.4, No.3, p.126-142.

Kepper, J. C., 1972, Paleoenvironmental patterns in Middle to lower Upper Cambrian interval in eastern Great Basin: American Association of Petroleum Geologists Bulletin, v.56, p.503-527.

Koerschner W.F., and Read, J. F., 1989, Field and modelling studies of Cambrian carbonate cycles, Virginia Appalachians: Journal of Sedimentary Petrology, v. 59, p. 654-687.

Levy, M., and Christie-Blick, N., 1989, Pre-Mesozoic palinspastic reconstruction of the eastern Great Basin (western United States): Science, v. 245, p. 1454-1462.

Lochman-Balk, C., 1971, Cambrian of the craton, *in* Holland, E.R. ed., Cambrian of the New World: London, Wiley-Interscience, p. 79-167.

Lohmann, K. C., 1976, Lower Dresbachian (Upper Cambrian) platform to deep-shelf transition in eastern Nevada and western Utah: An evaluation through lithologic cycle correlation: Brigham Young University Geological Studies, v.23, p.111-122.

Markello, J. R., and Read, J. F., 1982, Upper Cambrian intrashelf basin, Nolichucky Formation, southwest Virginia Appalachians: American Association of Petroleum Geologists Bulletin, v.66, p.860-878.

McCutcheon, K.F., and Cooper, J.D., 1989, Environmental carbonate stratigraphy and cyclic deposition of the Smoky Member of the Nopah Formation (Upper Cambrian), Nopah Range, southern Great Basin: *in* Cooper, J.D., ed., Cavalcade of Carbonates: Society of Economic Paleontologists and Mineralogists, Pacific Section, volume and guidebook, p. 87-100.

Miller, J. F., 1984, Cambrian and earliest Ordovician conodont evolution, biofacies, and provincialism, *in* Clark, D.L., Conodont Biofacies and Provincialism: Geological Society of America Special Paper 196, p. 43-68.

Miller, J. F., Taylor, M. E., Stitt, J. H., Ethington, R. L., Hintze, L. F., and Taylor, J. F., 1982, Potential Cambrian-Ordovician stratotype sections in the western United States, *in* Bassett, M.G., and Dean, W.T., eds., The Cambrian-Ordovician boundary: Sections, fossil distributions, and correlations: National Museum of Wales Geological Series no. 3, p. 155-180.

Mitchum, R.M., 1977, Glossary of terms used in seismic stratigraphy, *in* Payton, C. E., ed., Seismic stratigraphy - applications to hydrocarbon exploration: American Association Petroleum Geologists Memoir 26, p. 205-212.

Mullins, H.T., Gardulski, A.F., Hinchey, E.J., and Hine, A.C., 1988, The modern carbonate ramp slope of central west Florida: Journal of Sedimentary Petrology, v. 58, p. 273-290.

Olsen, P. E., 1986, A 40-million-year lake record of Early Mesozoic orbital climatic forcing: Science, v.234, p.842-848.

Orndorff, R. C., 1988, Latest Cambrian and Earliest Ordovician conodonts from the Conococheague and Stonehenge Limestones of northwestern Virginia: U.S. Geological Survey Bulletin 1837, A1-18.

Osleger, D.A., 1990, Cyclostratigraphy of Late Cambrian cyclic carbonates: an interbasinal field and modelling study, U.S.A. (Ph.D. dissertation): Virginia Polytechnic Institute and State Univ., Blacksburg, Virginia, 303 pp.

Osleger, D.A., (in review), Subtidal cycles: Implications for allocyclic versus autocyclic controls: submitted to Geology.

Osleger, D.A. and Read, J.F., (in review), Relation of eustacy to stacking patterns of Late Cambrian cyclic carbonates, U.S.A.: submitted to Journal of Sedimentary Petrology.

Palmer, A. R., 1965, Trilobites of the Late Cambrian Pterocephaliid biomere in the Great Basin, United States: United States Geological Survey Professional Paper 493, 105 pp.

Palmer, A. R., 1971a, The Cambrian of the Great Basin and adjacent areas, western United States, *in* Holland, E.R. ed., Cambrian of the New World, Wiley-Interscience, p. 1-79.

Palmer, A. R., 1971b, The Cambrian of the Appalachian and eastern New England regions, eastern United States, *in* Holland, E.R. ed., Cambrian of the New World, Wiley-Interscience, p. 289-332.

Palmer, A. R., 1981a, On the correlatibility of Grand Cycle tops, *in* Taylor, M. E., ed., Short Papers for the Second International Symposium on the Cambrian System: U.S. Geological Survey Open File Report 81-743, 156-157.

Palmer, A. R., 1981b, Subdivision of the Sauk sequence, *in* Taylor, M. E., ed., Short Papers for the Second International Symposium on the Cambrian System: U.S. Geological Survey Open File Report 81-743, p. 160-163.

Palmer, A. R., 1983, The Decade of North American Geology 1983 geologic time scale: Geology, v.11, p.503-504.

Palmer, A. R., 1984, The biomere problem: evolution of an idea: Journal of Paleontology, v.58, No.3, p.599-611.

Pitman, W. C., and Golovchenko, X., 1988, Sea-level changes and their effect on the stratigraphy of Atlantic-type margins, *in* Sheridan, R. E., and Grow, J. A., eds., The Atlantic continental margin: The Geology of North America Volume 1-2, Geological Society of America, p. 429-436.

Rassetti, F., 1965, Upper Cambrian trilobite faunas of northeastern Tennessee: Smithsonian Miscellaneous Collections, v.148, p.140.

Read, J. F., 1985, Carbonate platform facies models: American Association of Petroleum Geologists Bulletin, v.69, p.1-21.

Read, J. F., 1989, Controls on evolution of Cambrian-Ordovician passive margin, U.S. Appalachians, *in* Crevello, P., Wilson, J. L., Sarg, J. F., and Read, J. F., eds., Controls on Carbonate Platform and Basin Development: Society of Economic

828

Paleontologists and Mineralogists, Special Publication 44, p. 147-166.

Read, J. F., and Goldhammer, R. K., 1988, Use of Fischer plots to define third order sea level curves in peritidal cyclic carbonates, Early Ordovician, Appalachians: Geology, v.16, p.895-899.

Read, J. F., Grotzinger, J. P., Bova, J. A., and Koerschner, W. F., 1986, Models for generation of carbonate cycles: Geology, v.14, p.107-110.

Read, J.F., Osleger, D.A., and Elrick, M., in press, Two-dimensional computer modelling of cyclic carbonate sequences: Kansas Geological Survey Anniversary publication on computer modelling of cyclic sequences.

Rees, M. N., 1986, A fault-controlled trough through a carbonate platform: the Middle Cambrian House Range embayment: Geological Society of America Bulletin, v.97, p.1054-1069.

Robison, R. A. 1964, Upper Middle Cambrian stratigraphy of western Utah: Geological Society of America Bulletin, v.75, p.995-1010.

Schwarzacher, W., and Fischer, A.G., 1982, Limestone-shale bedding and perturbations of the Earth's orbit, in Einsele, G., and Seilacher, A. eds., Cyclic and Event Stratification, Berlin, Springer-Verlag, p.72-95.

Scotese, C.R., and McKerrow, W.S., 1990, Revised world maps and introduction, in McKerrow, W.S. and Scotese, C.R. eds., Palaeozoic Palaeogeography and Biogeography: Geological Society of London Memoir no. 12, p. 1-24.

Sepkoski, J.J. Jr., 1982, Flat-pebble conglomerates, storm deposits and the Cambrian bottom fauna, in Einsele, G. and Seilacher, A., eds. Cyclic and Event Stratification, Berlin, Springer-Verlag, p. 371-385.

Shaw, A.B., 1964, Time in Stratigraphy, New York, McGraw-Hill, 365 pp,

Stewart, J. H., and Poole, F. G., 1974, Lower Paleozoic and uppermost Precambrian Cordilleran miogeocline, Great Basin, western United States, in Dickenson, W. R., ed., Tectonics and sedimentation: Society of Economic Paleontologists and Mineralogists, Special Publication 22, 28-58.

Stewart, J. H., and Suczek, C. A., 1977, Cambrian and latest Precambrian paleogeography and tectonics in the western United States, in Stewart, J. H., Stevens, C. H., and Fritsche, A. E., eds., Pacific Coast Paleogeography Symposium 1: Paleogeography of the western United States, Society of Economic Paleontologists and Mineralogists, Pacific Section, p.1-18.

Sundberg, F.A., 1990, Morphological Diversification of the Ptychopariid Trilobites in the Marjumiid Biomere (Middle to Upper Cambrian) (Ph.D. dissertation): Virginia Polytechnic Institute, Blacksburg, Virginia, 425 pp.

Taylor, M. E., and Miller, J. F., 1981, Upper Cambrian and lower Ordovician stratigraphy and biostratigraphy, southern House Range, Utah, in Taylor, M.E., and Palmer, A.R., eds., Cambrian stratigraphy and paleontology of the Great Basin and vicinity, western United States: Guidebook for Field Trip 1, 2nd International Symposium on the Cambrian System, p. 73-77.

Vail, P. R., Mitchum, R. M., and Thompson, S. III, 1977, Seismic stratigraphy and global changes of sea-level, Part 4; Global cycles of relative changes of sea-level, in Payton, C. E., ed., Seismic stratigraphy – applications to hydrocarbon exploration: American Association Petroleum Geologists Memoir 26, p. 83-97.

Van Wagoner, J. C., Mitchum, R. M., Posamentier, H. W. and Vail, P.R., 1987, The key definitions of sequence stratigraphy, in Bally, A. W., ed., Atlas of Seismic Stratigraphy (Vol. 1): American Association Petroleum Geologists Studies in Geology 27, p. 11-14.

Wilson, J.L., 1952, Upper Cambrian stratigraphy in the central Appalachians: Geological Society of America Bulletin, v. 63, p. 275-322.

Wilson, J.L., 1975, Carbonate Facies in Geologic History, New York, Springer-Verlag, 471 pp.

HOW DOES THE PALEOGEOGRAPHY OF PALEOZOIC MEXICO RELATE TO WESTERN UNITED STATES?

Matthew C. Taylor, Robert R. Rector, Gina F. Carollo, Baron C. Colchagoff, and R. Gordon Gastil
Department of Geological Sciences, San Diego State University
San Diego, California 92182

ABSTRACT

Cratonal Precambrian basement in Mexico is of two types: Proterozoic volcanic-arc strata intruded by granitic rocks of 1400-1750 Ma in northern Sonora, and 1000-1300 Ma Grenville-like igneous and granulite grade metamorphic rocks exposed in the Zapoteco and Mixteco terranes of southern Mexico, and in several small localities in eastern Mexico.

The Paleozoic rocks of northern Sonora share similarities with rocks in Baja California Norte, southern Arizona, California, and Nevada, but are not clearly related to rocks elsewhere in Mexico. Paleozoic rocks in the northern tier of Mexican states east of Sonora are correlatable to rock across the border in New Mexico and Texas, but there is little continuity of exposure southward, and no convincing correlation with rocks farther south in Mexico.

Most of Mexico is covered by Mesozoic and Cenozoic rocks with little evidence of the type of Paleozoic or older strata hidden beneath. It is permissible that the pre-Mesozoic rocks of Mexico are largely exotic to North America.

INTRODUCTION

From west Texas to the Sierra Nevada, and from the Arizona-Sonora border to Canada, abundant exposures of Paleozoic strata permit definition of a variety of credible tectonic boundaries. In northern Baja California Norte, Sonora, and Chihuahua, Paleozoic rocks are similar to those immediately to the north, but farther south Paleozoic rocks are covered by Jurassic and Cretaceous strata and thick accumulations of Cenozoic volcanic rock. With the exceptions of Baja California Norte, Sonora, Oaxaca, and northern Chihuahua, Paleozoic rocks occur only in isolated exposures that are very difficult to relate to one another.

The objective of this paper is to provide a south-of-the-border perspective on questions such as the location of the hypothetical Mojave-Sonora megashear (Figs. 1-3) and the southern edge of the Paleozoic North American continent. To do this several questions must be addressed. (1) Where is the cratonal southern margin of Paleozoic North America? (2) Was there a strand line that ran southeastward from southern California, south of northern Sonora, and continued northeastward to link up with the Marathon uplift and Ouachita belt (Peiffer-Rangin, 1979; Dickinson, 1981; Poole and Madrid, 1986, 1988; Stewart, 1981, 1988)? (3) Is there a spine of cratonal North America that extends at least as far south as where Oaxaca Grenville-like rocks crop out (Guzmán and de Cserna, 1963; Shurbet and Cebull, 1987)? (4) Are the collisions and truncations resulting from the closing of the proto-Atlantic and the separation of North and South America near the close of the Paleozoic so complex that we have no comprehension of what is hidden beneath the extensive post-Paleozoic cover?

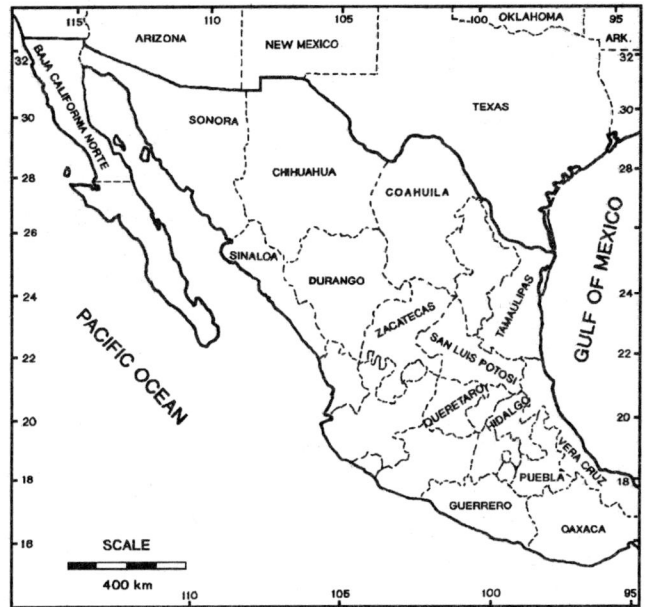

Figure 1. Index map showing those states of Mexico and the United States referenced in the text.

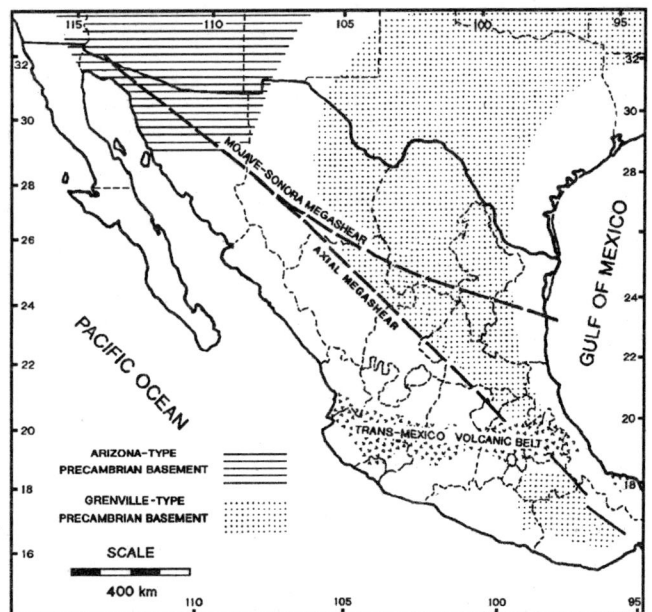

Figure 2. Map of Mexico showing the location of Arizona- and Grenville-type Precambrian basement, the hypothetical Mojave-Sonora megashear, the alternative axial megashear, and the Trans-Mexico volcanic belt.

In Cooper, J.D., and Stevens, C.H., eds., 1991, **Paleozoic Paleogeography of the Western United States-II:** Pacific Section SEPM, Vol. 67, p. 829-838.

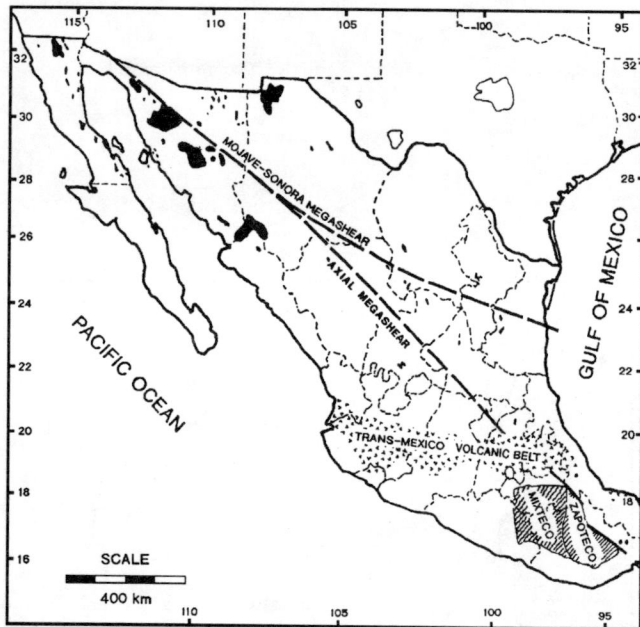

Figure 3. Map of Mexico showing the distribution of exposed Precambrian and Paleozoic rocks in black. The Zapoteco (Oaxaca) and Mixteco (Mixteca) terranes of southern Mexico consist of Grenville-type Precambrian basement overlain by unmetamorphosed Paleozoic rocks (Zapoteco) and metamorphosed pre-Mississippian rocks (Mixteco). The megashear on the northeastern margin of the Zapoteco terrane places it against the Mesozoic rocks of the Cuicateco (Juárez) terrane.

Recent attempts to construct terrane maps for Mexico are largely limited to a portrayal of Mesozoic and younger history (Campa and Coney, 1983; Campa, 1985; Sedlock, 1990). Evidence for lateral offset of hundreds of kilometers along megashears has been presented, but authors have failed to locate the actual shear zones in the field (Silver and Anderson, 1974; Anderson and Schmidt, 1983). In this paper we will make no presumptions concerning megashears.

Here we divide the Paleozoic history of Mexico into early Paleozoic (Cambrian - Silurian), middle Paleozoic (Devonian - Pennsylvanian), and Permian. No attempt is made to discuss every Paleozoic exposure of Mexico, and indeed, there probably are important exposures of which we are unaware. We hope that this summary will provide an adequate starting point for discussion.

EARLY PALEOZOIC STRATA

Three km of uppermost Precambrian and Lower and Middle Cambrian strata in the Caborca area of northern Sonora (Fig. 4, Locs. 6-7) generally are correlated formation for formation with similar rocks in the Death Valley, Mojave, and San Bernadino Mountains (Stewart and others, 1984). The geographic location of these outcrops has been cited both as support for the Mojave-Sonora megashear, and the alternative southeast-bending Paleozoic strand line (Stewart and others, 1990).

Work in Baja California Norte during the past decade has shown the presence of Early Paleozoic strata of both deep- and shallow-water facies. Ordovician strata of deep-water facies (very much like

the Valmy Formation of Nevada) occurs in northwestern Baja California Norte (Fig. 4, Loc. 1; Gastil and others, in press). Essentially coeval miogeoclinal facies marble (Pogonip?) crop out on the top of Carrizo Peak, just north of the international border in Imperial County, California (Fig. 4, Loc. 2; Miller and Dockum, 1983). In northeastern Baja California the miogeoclinal facies are metamorphosed and lack fossils (Fig. 4, Loc. 3-5), but localities 4-5 bear a strong lithologic resemblence to the Cambrian section of northern Sonora (Gastil and others, in press). In central Sonora there are both miogeoclinal quartz arenites resembling the Eureka Quartzite (Fig. 4, Loc. 10 and Sierra Lopez, northwest of Loc. 8; Ketner, 1986; Stewart and others, 1990), and graptolite-bearing chert-argillite sequences (Fig. 4, Locs. 8-9; Stewart and others, 1990).

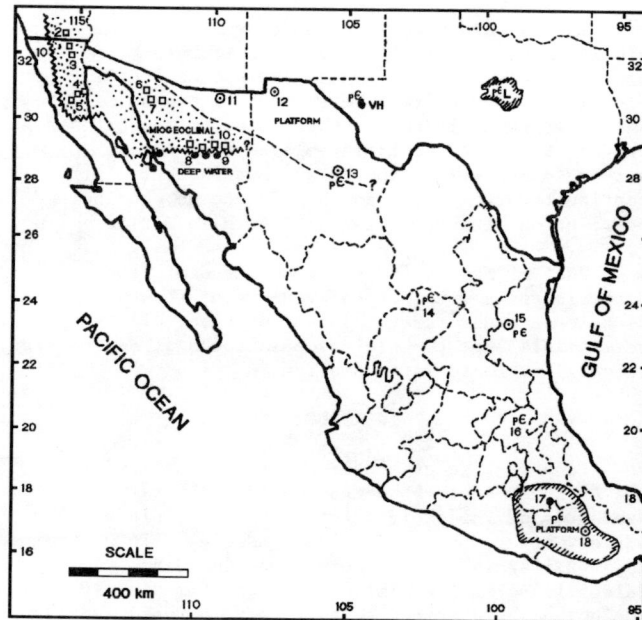

Figure 4. Map of Mexico showing numbered exposures of Cambrian - Silurian strata: solid dot = deep water facies; square = miogeoclinal facies; circled dot = platform facies; L = Llano Uplift; and VH = Van Horn Mountains. The pЄ symbol indicates isolated localities of Grenville-type Precambrian basement. Jagged line marks boundary between miogeoclinal (dotted) and deep-water facies. Dashed line indicates the southern limit of known platform strata. 1, Rancho San Marcos; 2, Carrizo Peak; 3, Sierra Cucapah; 4, San Felipe; 5, southern Sierra San Pedro Martir; 6, Caborca area; 7, Caborca area; 8, Sierra la Flojera; 9, Sierra El Encinal; 10, Sierra Agua Verde; 11, E of Cabullona; 12, Sierra de Palomas; 13, Placer de Guadalupe; 14, Cedros; 15, NW of Ciudad Victoria; 16, N of Molango; 17, Totoltepec; 18, Nochixtlán.

Thus, Gastil and others (in press) have drawn a facies boundary between miogeoclinal and deep-water rocks that extends down the axis of Baja California to 31° latitude then eastward across central Sonora (Fig. 4). Note that the facies boundary in Baja California reflects the 300 km of right-lateral Neogene translation beneath the Gulf of California.

Outside of northwestern Mexico only a few isolated exposures of early Paleozoic strata have been found. At localities 11 and 12 (Fig. 4), strata of platform-facies resemble those immediately north of the international border. In central Chihuahua (Fig. 4, Loc. 13), 550 m of Cambrian and Ordovician cratonal-facies shallow water sandstone and carbonate rock are overlain conformably by 50 m of outer shelf limestone (López-Ramos, 1969). About 15 km northwest of Ciudad Victoria (Fig. 4; Loc. 15), the entire Ordovician to Devonian section is only 218 m, and consists of conglomerate, limestone, and shale (Carrillo-Bravo, 1961, 1965). It is the opinion of some who have studied this area that this section is not in-place shallow-water strata, but olistoliths (J. H. Stewart, written communication, 1990). Within the Zapoteco (Oaxaca) terrane of eastern Puebla and Oaxaca, Cambrian and Ordovician carbonate and shallow-water siliciclastic strata are measured in tens of meters, almost certainly of platform-facies (Fig. 4, Loc. 18; Pantoja-Alor, 1971; Ortega-Gutiérrez, 1981), while within the Mixteco (Mixteca) of southern Puebla and northwestern Oaxaca, the Acatlán Complex is of deep-water facies (Fig. 4, Loc. 17; Ortega-Gutiérrez, 1978).

All of the Cambrian-Ordovician strata in Mexico, except for those in Baja California, are known to either rest on rocks of Precambrian age, or are in close proximity to such rocks. In Sonora, the Cambrian strata rest conformably on latest Precambrian rocks which in turn rest unconformably upon metasedimentary and metavolcanic strata intruded by granitic rocks as old as 1750 Ma (Anderson and Silver, 1981). In contrast, the early Paleozoic strata of central Chihuahua, eastern Mexico, and southern Mexico, rest on so-called "Grenvillian" basement. This basement typically consists of granulitic gneisses, anorthosites, and other igneous and metasedimentary rocks with U-Pb zircon ages in the range 1000 to 1300 Ma (Fig. 4, Locs. 13-20; de Cserna, 1989; Ruiz and others, 1990; K. L. Robinson and M. S. Girty, written communication, 1990). Ruiz and others (1990) have shown that the Grenville-type rocks have Nd model ages of about 1300-1600 Ma, as do similar age rocks near Van Horn (Fig. 4, Loc. VH) and the Llano uplift (Fig. 4, Loc. L) north of the international border.

Silurian exposures apparently are absent in northwestern Mexico. In many sections, a major unconformity separates Ordovician and Devonian strata (Malpica and de la Torre, 1980). This relationship is most evident in the shallow-water miogeoclinal and deep-water sequences of central Sonora (Fig. 4, Locs. 8-10). Thus, in northwestern Mexico, the Silurian system appears to be represented by a regional lacuna that contrasts with the lower Paleozoic stratigraphy of the western Great Basin where such a regional Silurian lacuna is not evident (Poole and others, 1977). The lack of Silurian rocks in northwestern Mexico may be the result of regional nondeposition and/or erosion from Late Silurian through Early Devonian.

In contrast to the sections in northwestern Mexico, the lower and middle Paleozoic sequences in northeastern Sonora (Fig. 4, Loc. 11), east-central Chihuahua (Fig. 4, Loc. 13), and near Ciudad Victoria in Tamaulipas (Fig. 4, Loc. 15), contain sections of conformable Silurian through Devonian rocks, typically part of the same formation. For example, the Solis Limestone, exposed in several places in east-central Chihuahua (Fig. 4, Loc. 13), is a conformable sequence of Silurian-Devonian shallow-water rocks consisting of interbedded clastic, dolomitic, and cherty limestone that is lithologically dissimilar to coeval rocks of northwestern Mexico (López-Ramos, 1969; Malpica and de la Torre, 1980).

MIDDLE PALEOZOIC STRATA

In Baja California Norte, Devonian and/or Mississippian strata that have been dated by fossils are exposed in three areas: Sierra las Pintas (Leier-Engelhardt, 1985); Arroyo Calamaque (Hoobs, 1986); and west of Canal de Ballenas (Crocker, 1989) (Fig. 5, Locs. 23-25, respectively). On the basis of lithologic correlation, similar rocks appear to be distributed from the Sierra Mayor in the north (Fig. 5, Loc. 22) to Bahia de Los Angeles and Isla Angel de la Guarda in the south (Fig. 5, Loc. 25). In each of these areas, "flysch-type" siliciclastic strata contain carbonate turbidites, bedded chert, and pillow basalt. Locally, there are calcarenites, massive quartz arenites, and olistostromal deposits of intraformational blocks up to several meters. The alkaline composition of the basalts and the craton-derived detrital zircons suggest that these slope and basin deposits accumulated close to a continent. It is tempting to compare these rocks with those of the Havallah and Schoonover assemblages of Nevada, and to the section in the El Paso Mountains of southern

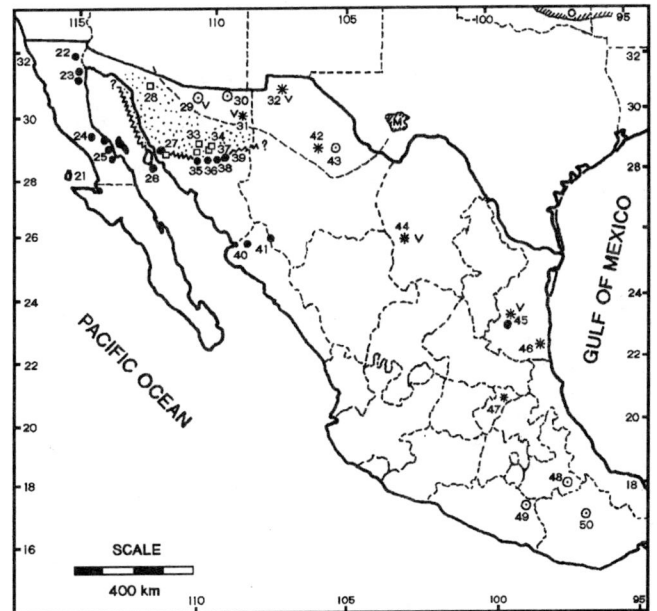

Figure 5. Map of Mexico showing numbered exposures of Devonian - Pennsylvanian strata: solid dot = deep water facies; square = miogeoclinal facies; circled dot = platform facies; asterisk = outer shelf or slope facies; V = presence of volcanic strata; M = Marathon Uplift; and O = Ouachita orogenic belt. Jagged line marks boundary between miogeoclinal (dotted) and deep-water facies. Dashed line indicates the southern extent of known platform strata, including some volcanic strata. 21, Isla Cedros; 22, Sierra Mayor; 23, Sierra las Pintas; 24, Arroyo Calamajue; 25, Puerto Calamajue to Bahia Los Angeles; 26, Isla Turner; 27, S of Punta Onah; 28, Bazini; 29, W of Cananea; 30, Cabullona; 31, Sierra de Teras; 32, Sierra de Palomas; 33, Sierra El Encinal; 34, Cerro Penasco Blanco and Rancho Real Viejo; 35, Cerro Cobachi; 36, Barita de Sonora; 37, Cerro El Aliso; 38, Sierra San Javier; 39, Sierra Agua Verde; 40, El Fuerte; 41, San José de Gracia; 42, W of Via Aldama; 43, SW of Mina Palomas; 44, Las Delicias area; 45, NW of Ciudad Victoria; 46, Poso Gonzalas; 47, Canalli; 48, Tehuacan; 49, Olinalá; 50, Nochixtlán.

California. Bedded chert and associated thin-bedded fine siliciclastic and carbonate rocks of probable middle Paleozoic age crop out at the northwest corner of Isla Tiburón, on the southeast corner of Isla Turner, and south of Punta Onah, opposite Isla Tiburón on the coast of Sonora (Fig. 5, Locs. 26-27), but these rocks have not yet yielded fossils. Siliciclastic and carbonate turbidites and sequences of Devonian bedded chert are exposed southeast of Hermosillo (Fig. 5, Locs. 33-39; Poole and Madrid in Stewart and others, 1990). However, unlike the coeval rocks of Baja California Norte, no pillow basalts have been reported.

From Bazini in northwestern Sonora (Fig. 5, Loc. 28) across northernmost Chihuahua (Fig. 5, Loc. 32), the lower Paleozoic strata are more or less conformably overlain by middle Paleozoic shallow-water platform to miogeoclinal strata that can be correlated with formations north of the international border. The strata are composed of largely carbonate rock, commonly cherty limestone. However, the Upper Mississippian El Tigre Formation, which is exposed 120 km east-southeast of Cananea (Fig. 5, Loc. 30), consists of 1500 m of limestone, mudstone, and volcanic rock (Aponte in Malpica and de la Torre, 1980). Within the Pedregosa basin, 40 km west of Palomas, Chihuahua (Fig. 5, Loc. 32), both the Mississippian Paradise-Escabrosa Formation and the Pennsylvanian Horquilla Formation contain turbiditic sequences and minor volcanic rock, as well as cherty limestone and mudstone (Tovar in Malpica and de la Torre, 1980; Reynolds in Malpica and de la Torre, 1980).

In central Sonora (Fig. 5, Locs. 33-39), coeval Paleozoic shallow- and deep-water facies rock are in tectonic juxtaposition; the deep-water rocks are interpreted to have moved relatively northward over the shallow-water strata (Poole and others, 1986). Two contractional events are recorded: the first occurred in the early Late Mississippian, and the second (stronger) during the Late Permian or Early Triassic (Stewart and others, 1990). This compares with the Late Devonian or Early Mississippian Antler orogeny and the Late Permian or Early Triassic Sonoma orogeny in Nevada, and the Late Pennsylvanian orogeny of west Texas. In central Sonora, however, the boundaries between miogeoclinal and eugeoclinal rocks trend roughly east-west, rather than north-northeast as in Nevada and west Texas.

At Sierra Agua Verde in central Sonora (Fig. 5, Loc. 33), very mature Lower Cambrian quartzite is structurally overlain by a 3,500 m section representing a nearly complete sequence of lower to middle Paleozoic strata. The depositional environment for these rocks is considered to be a subsiding continental shelf that is part of the miogeoclinal southern margin of middle Paleozoic North America and contiguous with the coeval platform deposits to the north (Fig. 5, Locs. 29-32; Stewart and others, 1990). Immediately north of Barita de Sonora in the Sierra Martinez (Fig. 5, Loc. 36) are miogeoclinal rocks ranging in age from Ordovician to Permian (Poole and Madrid in Stewart and others, 1990). Here, the Devonian to Early Pennsylvanian sequence consists of limestone, dolomite, with minor beds of sandstone, siltstone, and mudstone (Poole, 1983). The depositional character is described as shallow-subtidal to supratidal continental shelf (Stewart and others, 1990). The middle Devonian rocks in Cerro Penasco Blanco and nearby Rancho Real Viejo areas, in east-central Sonora (Fig. 5, Loc. 34), represent the eastern-most exposures of miogeoclinal rocks in

Sonora. This Devonian section consists of 2000 m of limestone, dolomite, quartz sandstone, and shale (Almazán-Vázquez, 1990).

Immediately adjacent to and generally south of the miogeoclinal rocks of central Sonora are deep-water facies rocks. At Barita de Sonora (Fig. 5, Loc. 36), a 115 m-thick deformed section of Late Devonian mudstone, siltstone, bedded chert, conglomerate, turbidite sandstone, and exhalitive sedimentary barite deposits is overlain by Lower Mississippian limestone, Upper Mississippian chert and shale, and Pennsylvanian chert, siltstone, and barite (Poole amd others, 1990; Stewart and others, 1990). The characteristic thinning- and fining-upward depositional cycles of the barite deposits and the associated rocks are interpreted as outer-fan deposits on a continental slope-rise to oceanic basin (Poole and others, 1990). This sequence was thrust northward toward the craton (Poole and Madrid in Stewart and others, 1990). The barite deposits of Barita de Sonora have been compared to coeval barite deposits in the Cordilleran eugeoclinal rocks of California and Nevada (Poole, 1981).

At Cerro Cobachi (Fig. 5, Loc. 35), southwest of Barita de Sonora, Upper Devonian and Mississippian quartzose limestone, limy quartz sandstone, bedded barite, and bedded chert are structurally juxtaposed against miogeoclinal Ordovician to Permian rocks (Ketner and Noll, 1987; Stewart and others, 1990). Strong stratigraphic similarities have been noted between these rocks and rocks of similar age and lithology in California and Nevada, whereas dissimilarities were noted between the rocks of Cerro Cobachi and those of west Texas and New Mexico (Ketner, 1990). Southeast of Barita de Sonora, similar sequences of Late Devonian, Mississippian, and Pennsylvanian deep-water rocks occur in Sierra El Aliso, Sierra San Javier, and Sierra El Encinal (Fig. 5, Locs. 37-39, respectively; Bartolini, 1988; Stewart and others, 1990). The rocks at Sierra El Aliso include siltstone, turbiditic limestone, and radiolarian chert (Bartolini, 1988). The latter two locations contain massive quartzite with interbedded chert. All three areas are in part correlative with the deep-water Devonian rocks of Barita de Sonora and Cerro Cobachi but partly of a different facies (Stewart and others, 1990).

The deep-water assemblages of the early to middle Paleozoic Sonoran orogen of central Sonora have been compared to those of both the Antler and Marathon orogens by Poole and Madrid (1988). They conclude that all three orogens share similar paleogeographic settings but note that the stratigraphic sequences, structural character, and facies patterns of the Sonoran package are unique.

In northern Sinaloa, thin-bedded carbonate rock, fine clastic rocks, and bedded chert that occur near both El Fuerte (Fig. 5, Loc. 40; Mullen, 1978) and at San José de Gracia (Fig. 5, Loc. 41) have yielded both Late Mississippian - Early Pennsylvanian (?) foraminifera (Carrillo-Martínez, 1971) and mid-Pennsylvanian conodonts (R. H. Miller, written communication, 1986). The sequence near San José de Gracia includes great thicknesses of bedded chert, but also includes quartz arenite and cobble conglomerate.

Exposed middle Paleozoic strata in northeastern Mexico are largely limited to outcrops near Ciudad Victoria (Fig. 5, Loc. 45) and Poso Gonzalas (Fig. 5, Loc. 46), Tamaulipas (Carrillo-Bravo, 1961, 1965; Gursky and Ramírez-Ramírez, 1986), and Canalli (Fig. 5, Loc. 47), Hidalgo (López-Ramos, 1969). The section near Ciudad Victoria was earlier reported to contain 100 m of novaculite along with shale, sandstone, conglomerate, and sandy limestone, but more recent investigation has shown that the novaculite is actually volcanic rock (Gursky and Ramírez-Ramírez, 1986). The Canalli locality contains massive shale and turbidites. Both localities have been interpreted as deep-water outer shelf or continental slope deposits (López-Ramos, 1969). The oldest strata in the Las Delicias area, Coahuila (Fig. 5, Loc. 44), are Pennsylvanian conglomerate-bearing flysch beds containing limestone clasts, volcaniclastic rocks that accumulated on the submarine flank of volcanic highlands, and detrital limestone that accumulated on the slope of a carbonate bank adjacent to a basin-margin volcanic highlands during a quiescent period (McKee and others, 1988).

The facies boundary in Baja California Norte and Sonora between miogeoclinal and deep-water rocks, as exemplified by Ordovician rocks (Fig. 4), appears to have shifted eastward by middle Paleozoic time, as no miogeoclinal rocks of this age have been found in the peninsula (Fig. 5). Along the Arizona-Texas border region it is possible to draw boundaries between platform and miogeoclinal sections, albeit some of the platform facies sequences include turbidite and volcanic rock. These Mississippian-Pennsylvanian turbidites and volcanic rocks', along with rocks of the localities in Hildago, Tamaulipas, and Coahuila, have been considered by Anderson and Schmidt (1983) to indicate a tectonic belt related to the Ouachita-Marathon orogenic rocks (Fig. 5, Locs. M and O). However, there are no reports of Mexican sequences of middle Paleozoic age east of the state of Sonora that look much like either the miogeoclinal strata of Nevada, California, and central Sonora, nor like the deep-water strata of Arkansas, Oklahoma, and Texas.

Platform facies of Devonian to Mississippian age in the states of Guerrero, Puebla, and Oaxaca are largely conformable with older Paleozoic strata (Fig. 5, Locs. 48-50). These lie unmetamorphosed over the Zapoteco (Oaxaca) terrane but are dynamothermally metamorphosed in the Mixteco (Acatlán Complex) terrane. They are unconformably overlain by unmetamorphosed Pennsylvanian strata of both marine and nonmarine facies. The metamorphic tectonism of the Mixteco terrane of southern Mexico has been compared to that found in the Appalachians, but in a geographic mirror image: the unmetamorphosed lower Paleozoic strata rests on a Grenville-type basement lying to the east, and the metamorphosed strata lie to the west (Figs. 2-3; Coney and others, 1983, 1984).

PERMIAN STRATA

Permian rocks are the most widespread Paleozoic strata in Mexico. On the western part of Isla Cedros (Fig. 6, Loc. 51), a block of fossiliferous shallow-water limestone of Early Permian age is included in a Jurassic melange (Rangin, 1978). At 30° latitude in eastern Baja California Norte, near El Marmol (Fig. 6, Loc. 52), a sequence of Permian, and possibly Pennsylvanian, strata consist of flysch-like rock, pebbly mudstone, olistrostromal breccia, bedded chert, quartz-calcareous sandstone, and a variety of carbonate rocks (DeLattre, 1984). The Leonardian fusulinids, giant crinoids, and other shallow-water fossils contained therein are believed to have been redeposited in a slope or basin environment. At Ejido Serdan in northwestern Sonora, 40 km east of the Colorado River (Fig. 6, Loc. 53), strata of the, Supai Group, Coconino Sandstone, and Kaibab Formation are exposed (Fitts, 1989). This is the only known

occurrence in Mexico of Permian rocks directly correlative to the Grand Canyon section, and hence unquestionably part of the North American craton.

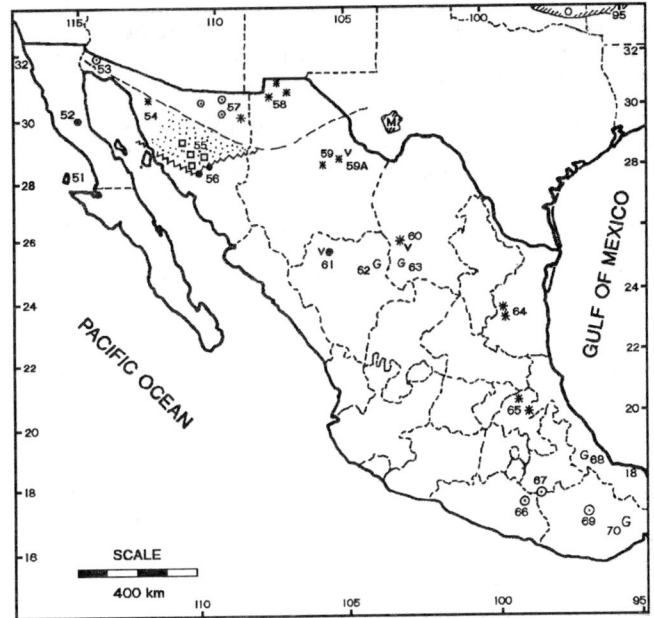

Figure 6. Map of Mexico showing numbered exposures of Permian strata: solid dot = deep water facies; squares = miogeoclinal facies; circled dot = platform facies; asterisk = outer shelf or slope facies; V = presence of volcanic strata; G = presence of granitic rocks; M = Marathon Uplift; and O = Ouachita orogenic belt. Jagged line marks boundary between miogeoclinal (dotted) and deep-water facies. Dashed line indicates the southern extent of known platform strata. 51, Isla Cedros; 52, E of El Marmol; 53, Ejido Serdan; 54, El Antimonio; 55, Sierra Lopez, Las Norias, Sierra Santa Teresa, Sierra Martinez; 56, Sierra la Flojera, Cerro Cobachi, Sierra Encinal; 57, W of Cananea, Cabullona, W of Nacozari, Sierra de Teras; 58, Sierra de Palomas, Sierra de Los Chinos, Sierra de la Salada; 59, Aldama; 59A, Placer de Guadalupe; 60, Las Delicias area; 61, Santa Maria del Oro area; 62, 100 km Nw of Torreón; 63, 100 km NNE of Torreón; 64, NW of Ciudad Victoria; 65, Huayacocotla anticlinorium; 66, Olinalá; 67, join of Puebla, Guerrero, and Oaxaca; 68, NW of Jalapa; 69, Nochixtlán; 70, W of Isthmus of Tehuantepec in Oaxaca.

At El Antimonio, 260 km southeast of Ejido Serdan in northwestern Sonora (Fig. 6, Loc. 54), Late Permian (Guadalupian) strata consist of 500 m of siltstone, limy siltstone, and thin layers of detrital limestone (probably moderate- to deep-water debris-flow deposits) including one limestone bed that contains gigantic fusulinids similar to those of the eastern Klamath Mountains, northern California (Ross and Ross, 1983), suggesting that these rocks are allochthonous to North America (Stewart and others, 1990).

De Cerna (1989) cites Velasco (1966) describing the Paleozoic sedimentary sequence in northeastern Sonora (Fig. 6, Loc. 57; Imlay, 1939) as having clearly accumulated on the border of the craton. Stewart (1988) shows this area as containing Upper Mississippian to Middle Triassic cratonal platform carbonate-siliciclastic rocks. Permian carbonate rocks are extensive in central Sonora. Miogeoclinal sections are present in the Sierra Lopez, Las Norias, Sierra Santa Teresa, and Sierra Martinez (Fig. 6, Loc. 55), and deep-water facies are present at Sierra la Flojera, Cerro Cobachi and Sierra El Encinal (Fig. 6, Loc. 56) (Stewart and others, 1990). Poole and Madrid (1986) consider the deep-water facies to be allochthonous continental rise or marginal ocean-basin deposits that were transported northward and accreted to the southwest margin of North America.

In north-central and northeastern Mexico, the Permian strata are shallow-water and shelf-slope carbonate-siliciclastic rocks characterizing the subsiding Pedregosa basin and adjoining shelf and margin (Fig. 6, Loc. 58-59A; Mellor and Breyer, 1981; Armin, 1987; Dickerson, 1987; Dyer and Reyes, 1987; Espinoza and others, 1987; Montgomery, 1987; Montgomery and Longoria, 1987; Wilson, 1987), including a rhyolite flow at Placer de Guadalupe (Fig. 6, Loc. 59A; Bridges, 1964). Farther south in the Las Delicias area of Coahuila (Fig. 6, Loc. 60), exposed upper Paleozoic rocks consist of a thick sequence of mid(?)-Pennsylvanian through Permian marine strata that mostly accumulated as mass-gravity deposits derived from an active volcanic arc. Principal components of these deposits are andesitic and dacitic debris, fine-grained siliciclastic pelagic sediment, and limestone debris derived from the basin margin (McKee and others, 1988). Due to the presence of the igneous rocks, early work (Rowett and Hawkins, 1975) suggested the existence of a volcanic arc that constituted part of the Ouachita-Marathon orogenic system. However, the more recent work (McKee and others, 1988) interprets the Las Delicias area rocks as probably not part of the Ouachita-Marathon orogen, and possibly allochthonous to North America.

The section exposed within the core of the Huizachal-Peregrina anticlinorium near Ciudad Victoria, Tamaulipas (Fig. 6, Loc. 64), includes Permian flysch that extends through Wolfcampian time. The rocks are composed of pelitic and shallow-water carbonate, low- to high-grade metamorphic, and granitic and volcanic (rhyolitic) detritus (Gursky and Ramírez-Ramírez, 1986; Gursky and Michalzik, 1989). Gursky and Michalzik describe the area as a forearc-basin that was situated west of a carbonate platform-fringed magmatic arc. These rocks have also been described by Campa and Coney (1983) as either the southernmost occurrence of the North American craton or the frontal zone of the Ouachita-Marathon orogenic belt displaced far to the southeast. Within the core of the Huayacocotla anticlinorium, in the states of Hildago and Vera Cruz (Fig. 6, Loc. 65), upper Paleozoic rocks consist of a thick turbidite sequence, the lower part of which is Early Permian and rest on Precambrian gneiss (Martínez-Pérez, 1962). Campa (1985) interprets the rocks of the Santa Maria del Oro area, Durango (Fig. 6, Loc. 61), as part of an allochthonous composite terrane that locally is composed of late Paleozoic island-arc submarine volcanics and melange.

In the Olinalá area of Guerrero (Fig.6, Loc. 66), the Acatlán Complex basement is overlain by a Pennsylvanian to Late Permian sequence of littoral to near-shore marine clastic rocks including a limestone interval containing fossils similar to that of northwestern Mexico (Corona-Esquivel, 1981). Where the states of Guerrero, Puebla, and Oaxaca join (Fig. 6, Loc. 67), Permian (Leonardian) rocks indicate that sedimentation was continental to the east and marine towards the west-southwest (Encisco de la Vega, 1988).

Granitic rocks of late Paleozoic age are reported at localities 62, 63, 68, and 70 (Fig. 6; Damon and others, 1981; Secretaria de Programación y Prespuesto, 1981). Damon and others (1981) characterize the Late Permian batholith complex of the Sierra Madre del Sur of southwestern Chiapas as one of the earliest magmatic events of the southern Cordillera. They cite evidence indicating that the batholith continues on the west side of the Isthmus of Tehuantepec in Oaxaca (Fig. 6, Loc. 70). Also, a late Permian pluton northwest of Jalapa, Vera Cruz, has been dated by K-Ar (Fig. 6, Loc. 68).

In northwest Mexico, there are four very different occurrences of Permian deep-water strata (Fig. 6, Locs. 51-54) that have no clear relationship to one another; certainly, these rocks are candidates for exotic origin. However, the Permian - Lower Triassic section at 30° latitude of eastern Baja California (Fig. 6, Loc. 52) matches coeval formations in the Inyo Mountains area of California (Gastil and others, in press).

In northeastern and eastern Mexico, there are rocks suggestive of local deep water-deposition (Fig. 6, Locs. 59-61 and 64-65), and in a few localities the marine sections include volcanic rocks (Fig. 6, Locs. 59-60), but these rocks have no clear ties to the Ouachita-Marathon orogenic belt. The rocks of southwestern Mexico appear to have been deposited on a relatively stable area with no known ties to rocks exposed farther north.

SUMMARY

Is there a continuum of cratonal (Precambrian) North America beneath most of Mexico? Is there a basis for extending strand or tectonic trends from Baja California and Sonora across north-central Mexico to connect with the deposition and deformation of the Ouachita-Marathon belt? Does the Paleozoic rocks of Mexico place constraints on the existence of the proposed Mojave-Sonoran and other proposed megashears?

Using the available Paleozoic data, it is difficult to reconstruct Paleozoic paleogeography and constrain subsequent tectonic structures. The distribution and nature of Precambrian rocks of Mexico may assist in deciphering the boundaries of the Paleozoic craton. On Figure 2 we have outlined the known extent of two very different Precambrian basements: the Arizona-type Precambrian rocks of Sonora (includes both Precambrian age intervals distinguished by Silver and Anderson, 1974), and the Grenville-type Precambrian rocks of eastern and southern Mexico. Perhaps the apparent mutually exclusive extent of these provinces is simply a function of the cover, or the nature of the cover may be a function of the distribution of Precambrian continental crust. Ruiz and others (1990) suggested that the Grenville-like province of eastern Mexico was actually part of northeastern South America in Paleozoic time. The fact that early Paleozoic rocks are today exposed largely in areas where cratonic Precambrian rocks also occur adds weight to the concept that these areas have inherited a stability from their crustal antiquity.

The gap of exposures between the deep-water Paleozoic rock of southern Sonora and possibly similar rocks in eastern Mexico is large, and the dissimilarities of stratal facies is great. There is little evidence to either support or negate the idea that the Paleozoic strand lines of middle Paleozoic age ever extended from northwestern to northeastern Mexico, or for that matter, that a continuation of the

Ouachita-Marathon deposition-tectonic belt is preserved in northeastern Mexico.

Let us consider the alternatives for the eastward continuation of the central Sonora facies boundary (Stewart and others, 1990). This is a fundamental boundary which places an oceanic terrane against the edge of the North American continent. As mapped, it extends from the west coast of Sonora to eastern Sonora, where it disappears beneath Cenozoic cover. If we could strip away that cover, where would it go? If it continued due east, it would be truncated by the hypothetical Mojave-Sonora megashear (Fig. 2, 3), and would be displaced left-laterally 800 km to the northwest. We can see from the geologic map of North America that this is not correct. If Hagstrom and others (1987) are correct, an eastward continuation would intersect a fundamental megashear in eastern Sonora along which Baja California, Sinaloa, and western Sonora moved northwestward together 14° of latitude during latest Cretaceous and early Cenozoic time. This would not only require that the eastward continuation of the Paleozoic facies boundary (Figs. 4-6) is in southern Mexico, it would imply that the terrane of miogeoclinal Paleozoic rocks resting on Arizona-type Precambrian crust also occurs in southern Mexico. The little that we know about the Paleozoic of southern Mexico contributes nothing to support this palinspastic reconstruction.

If we employ the Mojave-Sonora megashear scenario (Anderson and Schmidt, 1983; Stewart and others, 1990), it is required that the central Sonora facies boundary trends southeast across Mexico just south of the Mojave-Sonora megashear to the state of Taumalipas at which point it is truncated by the Mojave-Sonora megashear and displaced left-laterally to north-central Chihuahua. This requires that the Precambrian crustal edge of North America was similarly offset in the Jurassic. It is difficult for us to believe that the contemporary and subsequent north-south structural and sedimentary trends of eastern Mexico and the western Gulf of Mexico (for example, Leroy, 1990) would trend across this fundamental crustal margin at a high angle and totally obscure it.

We believe that the dominantly north and northwest Jurassic to Cenozoic topographic and structural trends (north of the Trans-Mexico Volcanic Belt) were probably not constructed at a high angle to earlier crustal configuration. Sedloch and others (1990) suggest that the Mojave-Sonora megashear is part of a family of northwest-trending left-lateral faults that trend axially to Mexico (Fig. 3). Perhaps this axial megashear forms the southwestern boundary of the Grenvillian basement. Furthermore, the terrane south of the central Sonora facies boundary and southwest of the axial megashear was exotic, joining North America between Mississippian and Triassic time. Our principal point, however, is that the outcrops are so scattered, key outcrops such as those near Ciudad Victoria so inadequately understood, and the spectrum of possible paleotectonic reconstructions so diverse, that all present proposals must be considered entirely speculative.

ACKNOWLEDGEMENTS

James W. McKee, Calvin H. Stevens, and John H. Stewart each read an earlier version of this manuscript and made valuable suggestions and contributions for its improvement.

REFERENCES CITED

Almazán-Vázquez, E., 1989, El Cámbrico-Ordovícico de Arivechi, en la región centrooriental del Estado de Sonora, Universidad Nacional Autónoma de México, Instituto de Geología, Revista, v. 8, n. 1, p. 58-66.

_____, 1990, Pre-Mesozoic sequences of the Sonora mountainous region and their tectonic significance: Geological Society of America Abstracts with Programs, v. 22, p. 1.

Anderson, T. H., and Schmidt, V. A., 1983, The evolution of Middle America and the Gulf of Mexico - Caribbean Sea region during Mesozoic time: Geological Society of America Bulletin, v. 94, p. 941-966.

Anderson, T. H., and Silver. L. T., 1978, The nature and extent of Precambrian rocks in Sonora, Mexico [abs.], in Nájera-Garza, J., and Ortlieb, L., eds., Primer Simposio sobre Geología y Potencial Minero del Estado de Sonora, Resúmenes: Universidad Nacional Autónoma de México, Instituto de Geología, p. 9-10.

_____, 1981, An overview of Precambrian rocks in Sonora, Universidad Nacional Autónoma de México, Instituto de Geología, Revista, v. 5, n. 2, p. 131-139.

Armin, R. A., 1987, Sedimentology and tectonic significance of Wolfcampian (Lower Permian) conglomerates in the Pedregosa basin: southeastern Arizona, southwestern New Mexico, and northern Mexico: Geological Society of America Bulletin, v. 99, p. 42-65.

Bartolini, C., 1988, Regional structure and stratigraphy of Sierra EI Aliso, central Sonora, Mexico [M.S. thesis]: Tucson, University of Arizona, 189 p.

Boles, J. R., and Landis, C. A., 1981, Coloradito and Eugenia formations, an Upper Jurassic slope channel facies, Cedros Island, Baja California: Geological Society of America Abstracts with Programs v. 13, p. 46.

Bridges, L. W., 1964, Stratigraphy of Mina Plomosas-Placer de Guadalupe area, in Geology of Mina Plomosas-Placer de Guadalupe area, Chihuahua, Mexico: West Texas Geological Society Publication 64-59, p. 50-64.

Campa, M. F., 1985, The Mexican thrust belt, in Howell, D. G., ed., Tectono-stratigraphic terranes of the circum-Pacific region: Circum-Pacific Council for Energy and Mineral Resources Earth Science Series, n. 1, p. 199-313.

Campa, M. F., and Coney, P. J., 1983, Tectonostratigraphic terranes and mineral resource distributions in Mexico: Canadian Journal of Earth Science, v. 20, p. 1040-1051.

Carrillo-Bravo, J., 1961, Geología del Anticlinorio Huizachal-Peregrina al N-W de Ciudad Victoria, Tamaulipas: Boletín de la Asociación Mexicana de Geólogos petroleros, v. 13, p. 1-98.

_____, 1965, Estudio geológico de una parte del Anticlinorio de Huayacoctla: Boletín de la Asociación Mexicana de Geólogos Petroleros, v. 17, p. 73-96.

Carillo-Martínez, M., 1971, Geología de la Hoja San José de Gracia, Sinaloa [tesis profesional]: México, D. F., Facultad de Ingeniería, Universidad Nacional Autónoma de México, 56 p.

Coney, P. J., and Campa, M. F., 1984, Lithotectonic terrane map of Mexico: United States Geological Survey Open-File Report 84-523, p. D1-D14, scale 1:2,000,000.

Corona-Esquival, R. J. J., 1981 (1984), Estratigráfica de la región de Olinalá Teconcoyunca, noreste del estado de Guerrero: Instituto de Geología,

Universidad Nacional Autónoma de México, Revista, v. 5, p. 17-24.

Crocker, J., 1987, Stratigraphy and structure of Paleozoic metasediments south of Bahia Calamajue, Baja California, Mexico [M.S. thesis]: San Diego State University, 129 p.

Damon, P. E., and Clark, K. F., 1981, Age trends of igneous activity in relation to metallogenesis in the southern Cordillera, in Dickinson, W. R., and Payne, W. D., eds., Relations of tectonics to ore deposits in the southern Cordillera: Arizona Geological Society Digest, v. 14, p. 137-154.

de Cserna, Z., 1989, An outline of the geology of Mexico, in Bally, A. W., and Palmer, A. R., eds., The geology of North America; an overview: Geological Society of America, p. 233-264.

DeLattre, M. P., 1984, Lower Permian metasedimentary rocks of Arroya Zamora northeastern Baja California, Mexico, in Frizzel, V. A., Jr., ed., Geology of the Baja California Peninsula: Los Angeles, Pacific Section, Society of Economic Paleontologists and Mineralogists, p. 23-29.

Dickerson, P. W., 1987, Structural and depositional setting of SW U.S. and northern Mexico along a Paleozoic transform plate margin, in Paleozoico de Chihuahua: Sociedad Geológica Mexicana A. C., Facultad de Ingeniería, Unversidad Autónoma de Chihuahua, Gaceta Geológica, v. 1, n. 1, Excursion Geológica n. 2, p. 129-159.

Dickinson, W. R., 1977, Paleozoic plate tectonics and the evolution of the Cordilleran continental margin, in Stewart, J. H., Stevens, C. H., and Fritsch, A. E., eds., Paleozoic paleogeography of the western United States: Los Angeles, Pacific Section, Society of Economic Paleontologists and Mineralogists, Pacific Coast Paleogeography Symposium 1, p. 137-155.

_____, 1981, Plate tectonic evolution of the southern Cordillera, in Dickinson, W. R., and Payne, W. D., eds., Relations of tectonics to ore deposits in the southern Cordillera: Arizona Geological Society Digest, v. 14, p. 113-135.

Dyer, R., and Reyes, I. A., 1987, The geology of Cerro El Carrizalillo, Chihuahua, Mexico: preliminary findings, in Paleozoico de Chihuahua: Sociedad Geológica Mexicana A. C., Facultad de Ingeniería, Universidad Autónoma de Chihuahua, Gaceta Geológica, v. 1, n. 1, Excursión Geológica n. 2, p. 108-128.

Encisco de la Vega, S., 1988, Una nueva localidad Pérmica con fusulínidos en Puebla: Instituto de Geología, Universidad Nacional Autónoma de México, Revista, v. 7, p. 28-34.

Espinoza, J. A. T., Hinojosa, C. R. S., and Flores, A. L., 1987, Estratigrafía preliminar del Paleozoico en las areas de La Vinata y Sierra Azcarate, Noroeste de Chihuahua, México, in Paleozoico de Chihuahua: Sociedad Geológica Mexicana A. C., Facultad de Ingeniería, Universidad Autónoma de Chihuahua, Gaceta Geológica, v. 1, n. 1, Excursión Geológica n. 2, p. 203-217.

Fitts, D. E., 1989, Metasedimentary strata of Ejido Serdan (Mexico) [M.S. thesis]: San Diego State University, 119 p.

Fries, C., Jr., Rincón-Orta, C., Solorio-Munguía, J., Schmitter-Villada, E., and de Cserna, Z., 1970, Una edad radiometrica Ordovícica de Totoltepec, Estado de Puebla: Sociedad Geológica Mexicana, Sobretiro del Libro-guía de la excursion México-Oaxaca, p. 164-166.

Garrison, J. R., Jr., Ramírez-Ramírez, C., and Long, L. E., 1981, Rb-Sr isotopic study of the ages and provenance of Precambrian granulite and Paleozoic greenschist near Ciudad Victoria, Mexico, in Pilger, R. H., ed., The origin of the Gulf of

Mexico and the early opening of the central north Atlantic Ocean: Houston Geological Society Continuing Education Series Symposium, p. 37-49.

Gastil, R. G., 1986, Terranes of peninsular California and adjacent Sonora, in Howell, D. G., ed., Tectono-stratigraphic terranes of the circum-Pacific region: Circum Pacific Council for Energy and Mineral Resources, Earth Sciences Series, n. 1, p. 273-283

Gastil, R. G., and Miller, R. H., 1983, Pre-batholithic terranes of southern and Peninsular California, U.S.A. and Mexico: Status Report, in Stevens, C. H., Pre-Jurassic rocks in western North American suspect terranes: Los Angeles, Pacific Section, Society of Economic Paleontologists and Mineralogists, p. 49-61.

_____, 1984, Pre-batholithic paleogeography of Peninsular California and adjacent Mexico, in Frizzel, V. A., Jr., ed., Geology of the Baja California Peninsula: Los Angeles, Pacific Section, Society of Economic Paleontologists and Mineralogists, v. 39, p. 9-16.

Gastil, R. G., Miller, R., Leier-Engelhardt, P. J., Anderson, P., Crocker, J., Campbell, M., Lothringer, C., DeLattre, M. P., and Buch, P., 1991, The relation between the Paleozoic strata on opposite sides of the Gulf of California, in Amaya-Martínez, R., ed., Geology of Sonora: Geological Society of America Memoir (in press).

Gursky, H.-J., and Michalzik, D., 1989, Lower Permian turbidites in the northern Sierra Madre Oriental, Mexico: Stuttgart, Zbl. Geol. Paläont. Teil I, 1989 (5/6), p. 821-838.

Gursky, H.-J., and Ramírez-Ramírez, C., 1986, Notas preliminares sobre el disobriniento de volcanitas acidas en El Cañon de Caballeros (Núcleo del Anticlinorio Huizachal-Peregrina, Tamaulipas, México): Actas de la Facultad de Ciencias de la Tierra, Universidad Autónoma de Nuevo León, Linares, v. 1, p. 11-22.

Guzmán, E. J., and de Cserna, Z., 1963, Tectonic history of Mexico, in Childs, O. E., and Beebe, B. W., eds., Backbone of the Americas, a symposium: American Association of Petroleum Geologists, Memoir 2, p. 113-129.

Hagstrom, J. T., Sawlan, M. G., Hauback, B. P., Smith, J. G., and Gromme, C. S., 1987, Miocene paleomagnetism and tectonic settings of the Baja California Peninsula, Mexico: Journal of Geophysical Research, v. 92, n. B3, p. 2627-2639.

Hoobs, J. H., 1986, Carboniferous island-arc and associated rocks from the Mission Calamujue area, Baja California, Mexico [M.S. thesis]: San Diego State University, 122 p.

Imlay, R. W., 1939, Paleogeographic studies in northeastern Sonora: Geological Society of America Bulletin, v. 50, p. 1723-1744.

Ketner, K. B., 1968, Origin of Ordovician quartzite in the Cordilleran miogeosyncline, in Geological Survey Research 1968: United States Geological Survey Professional Paper 600-B, p. B169-B177.

_____, 1980, Stratigraphic and tectonic parallels between Paleozoic geosynclinal siliceous sequences in northern Nevada and those of the Marathon uplift Texas and Ouachita Mountains, Arkansas and Oklahoma, in Fouch T. D., and Magathan, E. R., eds., Paleozoic paleogeography of west-central United States: Denver, Rocky Mountain Section, Society of Economic Paleontologists and Mineralogists, West-Central United States Symposium 1, p. 363-369.

_____, 1986, Eureka Quartzite in Mexico? - Tectonic implications: Geology, v. 14, p. 1027-1030.

Ketner, K. B., and Noll, J. H., Jr., 1987, Preliminary geologic map of the Cerro Cobachi area, Sonora,

Mexico: United States Geological Survey Miscellaneous Field Studies Map MF-1980, scale 1:20,000.

Leier-Engelhardt, P. J., 1985, Middle Paleozoic strata of the Sierra Las Pintas, northeastern Baja California Norte, Mexico [M.S. thesis]: San Diego State University, 169 p.

Leroy, S. D., 1990, Crustal structure of the Gulf of Mexico: New Prospectives on structural trends and salt geometry from gravity: Geological Society of America Abstracts with Programs, v. 22, p. A186.

López-Ramos, E., 1969, Marine Paleozoic rocks of Mexico: American Association of Petroleum Geologists Bulletin, v. 53, n. 12, p. 2399-2417.

Lund, S. P., and Bottjer, D. J., in press, Paleomagnetic evidence for microplate tectonic development of southern and Baja California, in The Gulf and Peninsular Province of the Californias: American Association of Petroleum Geologists, Memoir n. 47.

Malpica C., R., and de la Torre L., G., 1980, La integration estratigráfica del Paleozoica de México, Partes 1-3 y Anexos 1-48: Instituto Mexicano del Petróleo, Subdirección de Tecnología de Exploración, Proyecto C-1079.

Martínez-Pérez, J., 1962, Estudio geológica de una porción de la Sierra Madre Oriental al oriente de Zacualtipán y Tianguistengo, Hidalgo [tesis profesional]: Mexico, D. F., Instituto Politécnico Nacional, Escuela Superior de Ingeniería y Arquitectura, 48 p.

McKee, J. W., Jones, N. W., and Anderson, T. H., 1988, Las Delicias basin: A record of late Paleozoic arc volcanism in northeastern Mexico: Geology, v. 16, p. 37-40.

Mellor, E. I., and Breyer, J. A., 1981, Petrology of late Paleozoic basin-fill sandstones, north-central Mexico: Geological Society of America Bulletin, Part I, v. 92, p. 367-373.

Miller, R. H., and Dockum, M. S., 1983, Ordovician conodonts from metamorphosed carbonates of the Salton Trough, California: Geology, v. 11, p. 410-412.

Montgomery, H., 1987, Microfacies and paleogeographic significance of the Permian patch reefs at Sierra Plomosa, Chihuahua, in Paleozoico de Chihuahua: Sociedad Geológica Mexicana A. C., Facultad de Ingeniería, Universidad Autónoma de Chihuahua, Gaceta Geológica, v. 1, n. 1, Excursión Geológica n. 2, p. 70-81.

Montgomery, H., and Longoria, J. F., 1987, Paleozoic tectonics of Chihuahua - a paleogeographic perspective, in Paleozoico de Chihuahua: Sociedad Geológica Mexicana A. C., Facultad de Ingeniería, Universidad Autónoma de Chihuahua, Gaceta Geológica, v. 1, n. 1, Excursión Geológica n. 2, p. 54-69.

Mullan, H. S., 1978, Evolution of part of the Nevada orogen in northwestern Mexico: Geological Society of American Bulletin, v. 89, p. 1175-1188.

Ortega-Gutiérrez, F., 1978, Estratigrafía del Complejo Acatlán en la Mixteca Baja, Estadoes de Puebla y Oaxaca: Instituto de Geología, Universidad Autónoma de México, Revista, v. 2, n. 2, p. 112-131.

_____, 1981, Metamorphic belts of southern Mexico and their tectonic significance: México, D. F., Geofísica Internacional, v. 20-3, p. 177-202.

Pantoja-Alor, J., 1970, Rocas sedimentarios Paleozoicas de la región centro-septentrional de Oaxaca, in Segura, L. R., and Rodríguez-Torres, R., eds., Libro-guía de la excursión México-Oaxaca: México, D. F., Sociedad Geológica

Mexicana, p. 67-84.

Peiffer-Rangin, F., 1979, Les zones isopiques du Paleozoique inferieur du nord-ouest Mexicain, te moins du relais les Appalaches en la Cordillere ouest-americaine: Paris, Academia des Seances Comptes Rendus, v. 288, n. 20, ser. D, p. 1517-1519.

Poole, F. G., 1981, Silurian and Devonian rocks in southwestern United States and northwestern Mexico: Geological Society of America Abstracts with Programs, v. 13, p. 101.

Poole, F. G., and Madrid, R. J., 1986, Paleozoic rocks in Sonora (Mexico) and their relation to the southwestern continental margin of North America: Geological Society of America Abstracts with Programs, v. 18, p. 720-721.

Poole, F. G., Madrid, R. J., and Morales-Ramirez, J. M., 1990, Sonoran orogeny in the Barita de Sonora mine area, central Sonora, Mexico: Geological Society of America Abstracts with Programs, p. 76.

Poole, F. G., Murchey, B. L., and Stewart, J. H., 1983, Bedded barite of middle and late Paleozoic age in central Sonora, Mexico: Geological Society of America Abstracts with Programs, v. 15, p. 299.

Poole, F. G., Sandberg, C. A., and Boucot, A. J., 1977, Silurian and Devonian paleogeography of the western United States, in Stewart, J. H., Stevens, C. H., and Fritsch, A. E., eds., Paleozoic paleogeography of the western United States: Los Angeles, Pacific Section, Society of Economic Paleontologists and Mineralogists, Pacific Coast Paleogeography Symposium 1, p. 39-65.

Rangin, C., 1978, Speculative model of Mesozoic geodynamics, central Baja California to northeastern Sonora (Mexico), in Howell, D. G., and McDougall, K. A., eds., Mesozoic paleogeography of the western United States: Los Angeles, Pacific Section, Society of Economic Paleontologists and Mineralogists, Pacific Coast Paleogeography Symposium 2, p. 85-106.

Rogers, R. W., 1990, Tectonics and stratigraphy of the Sierra Catorce Uplift - north-central Mexico: Geological Society of America Abstracts with Programs, v. 22, p. A113.

Ross, C. A., and Ross, J. R. P., 1983, Late Paleozoic accreted terranes of western North America, in Stevens, C. H., ed., Pre-Jurassic rocks in western North America suspect terranes: Los Angeles, Pacific Section, Society of Economic Paleontologists and Mineralogists, p. 7-22.

Rowett, C. L., and Hawkins, C. M., 1975, Late Paleozoic volcanic island arc-trench system in northern Mexico: Geological Society of America Abstracts with Programs, v. 7, p 230-231.

Ruiz, J., Patchett, J. P., and Ortega-Gutiérrez, F., 1987, Proterozoic and Phanerozoic basement terranes of Mexico from Nd isotope studies: Geological Society of America Bulletin, v. 100, p. 274-281.

_____, 1990, Proterozoic and Phanerozoic terranes of Mexico based on Nd, Sr, and Pb isotopes: Geological Society of America Abstracts with Programs, v. 22, n. 7, p. A113.

Schumacher, D., 1978, Devonian stratigraphy and correlation in southeastern Arizona and southwestern New Mexico: New Mexico Geological Society Guidebook, 29th Field Conference, Land of Cochise, p. 175-181.

Secretaria de Programación y Presupuesto, 1981, Cartografía Geológico de México, scale 1:1,000,000.

Sedlock, R. L., and Ortega-Gutiérrez, F., 1990,

838

Sinistral displacement in eastern Mexico: Geological Society of America Abstracts with Programs, v. 22, p. A113.

Shurbet, D. H., and Cebull, S. E., 1987, Tectonic interpretation of the westernmost part of the Ouachita-Marathon (Hercynian) orogenic belt, west Texas-Mexico: Geology, v. 15, p. 458-461.

Silver, L. T., and Anderson, T. H., 1974, Possible left-lateral early to middle Mesozoic disruption of the southwestern North American craton margin: Geological Society of America Abstracts with Programs, v. 6, p. 955-956.

Stewart, J. H., 1981, Early and Middle Paleozoic margin of the North American continent in the southwestern United States and northern Mexico, in Howard, K. A., Carr, M. D., and Miller, M. D., eds., Tectonic framework of the Mojave and Sonoran Deserts, California and Arizona: United States Geological Survey Open-file Report 81-274, p. 101-103.

_____, 1988, Latest Proterozoic and Paleozoic southern margin of North America and the accretion of Mexico: Geology, v. 16, p. 186-189.

_____, 1990, Latest Proterozoic and Paleozoic history of the southern margin of North America in Mexico and the United States: Geological Society of America Abstracts with Programs, v. 22, p. A113.

Stewart, J. H., McMenamin, M. A. S., and Morales-Ramirez, J. M., 1984, Upper Proterozoic and Cambrian rocks in the Caborca region, Sonora, Mexico - physical stratigraphy, biostratigraphy, paleo-current studies, and regional relations: United States Geological Survey Professional Paper 1309, 36 p.

Stewart, J. H., Poole, F. G., Ketner, K. B., Madrid, R. J., Roldan-Quintana, J., and Amaya-Martínez, R., 1990, Tectonics and stratigraphy of the Paleozoic and Triassic southern margin of North America, Sonora, Mexico, in Gehrels, G. E. and Spencer, J. E., eds., Geologic excursions through the Sonoran desert region, Arizona and Sonora: Arizona Geological Survey Special Paper 7, p. 183-202.

Tardy, M., Carfantan, J., and Rangin, C., 1986, Essai de synthèse sur la structure du Mexique: Bulletin Societe Géologique du France, v. 11, n. 6, p. 1025-1031.

Velasco, J. R., 1966, Geology of the Cananea district, in Titley, S. R., and Hicks, C. L., eds., Geology of the porphyry copper deposits southwestern North America: Tucson, University of Arizona Press, p. 245-249.

Wilson, J. L., 1987, The late Paleozoic geologic history of southern New Mexico and Chihuahua, in Paleozoico de Chihuahua: Sociedad Geológica Mexicana A. C., Facultad de Ingeniería, Universidad Autónoma de Chihuahua, Gaceta Geológica, v. 1, n. 1, Excursión Geológica n. 2, p. 36-53.

PALEOECOLOGIC IMPLICATIONS OF HIGH LATITUDE AND MIDDLE LATITUDE AFFINITIES OF THE AMMONOID URALOCERAS

Claude Spinosa
Department of Geology and
  Geophysics
Boise State University
Boise, Idaho 83725

Walter W. Nassichuk
Institute of Sedimentary
  and Petroleum Geology
Geological Survey of Canada
3303 33rd Street NW Calgary
Alberta Canada, T2L 2A7

Walter S. Snyder
Department of Geology and
  Geophysics
Boise State University
Boise, Idaho 83725

Dora M. Gallegos
Department of Geology and
  Geophysics
Boise State University
Boise, Idaho 83725

ABSTRACT

The ammonoid genus Uraloceras has been recognized as characterizing Lower Permian boreal paleogeography. New Lower Permian occurrences in Alaska, dominated by boreal Uraloceras species, are compatible with and reinforce the boreal distribution of some species of the genus.

A new Uraloceras species from Nevada, also recognized from the Yukon, and Uraloceras from Nei Monggol, China, inhabited regions of lower paleolatitudes. The fauna from Nevada, which was geographically and faunally transitional between boreal and equatorial settings, includes Uraloceras in mixed association with cosmopolitan ammonoid taxa and endemic (equatorial) perrinitid ammonoids. This fauna is referred to as a "transitional association."

A lower stratigraphic assemblage of Wrangellia in eastern Alaska, (Slana Spur and Eagle Creek Formations), contains abundant Uraloceras and Paragastrioceras of boreal affinities. These are overlain by younger rock including the Triassic Nikolai Greenstone, for which, paleomagnetic data indicate an equatorial origin with paleolatitudes of 10 to 17° N or S. The Nikolai Greenstone appears to have a far-traveled history contradictory with ammonoid evidence that suggests cool water settings on Pangaea.

INTRODUCTION

The formation of the supercontinent Pangaea II (Greater Pangaea) during Late Paleozoic to Triassic time generated new patterns of oceanic currents and accentuated the development of latitudinal climatic zones. These zones are reflected in the distributions of various Late Paleozoic faunas. Cooler oceanic waters from high paleolatitudes penetrated seas on the northern and northwestern Pangaean shelf disrupting relatively warmer and equable patterns of oceanic currents that had been established during Carboniferous time. These northern oceanic currents caused the development of cooler water paleobiogeographic provinces with faunas that contrast markedly with warmer water faunas to the south.

Faunas with varying degrees of endemism have received various designations in the literature, including "Boreal realm" and "Tethyan realm." The term "boreal" may be employed to describe some occurrences of the ammonoid genus Uraloceras (Fig. 1, localities 4, 5, 6 and 7) because these are restricted to high latitude, though not arctic, regions in the Northern Hemisphere. The term Tethyan, however, has a longitudinal as well as a latitudinal connotation and its use should be restricted to occurrences with paleogeographic connection to the Alpine-Himalayan orogenic belt. The terms "realm" and "province", amongst others, have been used loosely and somewhat interchangeably in the literature to denote various levels of faunal endemism. Kauffman (1973), however, established rather precise, arbitrary guidelines regarding the use of certain paleobiogeographic terms. He suggested the term endemic center for 5-10% endemism; subprovince 10-25%; province 25-50%; region 50-75%; and realm >75%. We will not use of the terms "realm" or "province" in Kauffman's sense, rather we will employ the term "association" to designate distinct faunas, whose level of endemism or distinction has not been subjected to the numerical assessment required by Kauffman's scheme.

Early Permian paleobiogeographic patterns have been recognized in North America for various forms including brachiopods, fusulinaceans, corals, bryozoans and ammonoids. Brachiopod faunas composed of boreal-restricted taxa and associated cosmopolitan genera are dominated by spiriferaceans and a few large productaceans (Stehli, 1973). Ross (1967) recognized a Tethyan verbeekinid fusulinacean association that contrasted with non-verbeekinid faunas common to North America. Gobbett, 1973, indicated that Fusiella and Fusulinella, "relict from the Upper Carboniferous, ... were typical of the Boreal realm" during lowermost Permian on the northern edge of Eurasia and North America. Ross and Ross (1990) suggested that an Early Permian northward migration of Pangaea along with a closure of the Uralian-Tethyan connection resulted in the isolation and cooling of shelf bryozoan faunas on the northern Pangaean coast. Stevens and Rycerski (1983) divided North American Permian rugose colonial corals into several autochthonous faunal

In Cooper, J.D., and Stevens, C.H., eds., 1991, **Paleozoic Paleogeography of the Western United States-II:** Pacific Section SEPM, Vol. 67, p. 839-846.

839

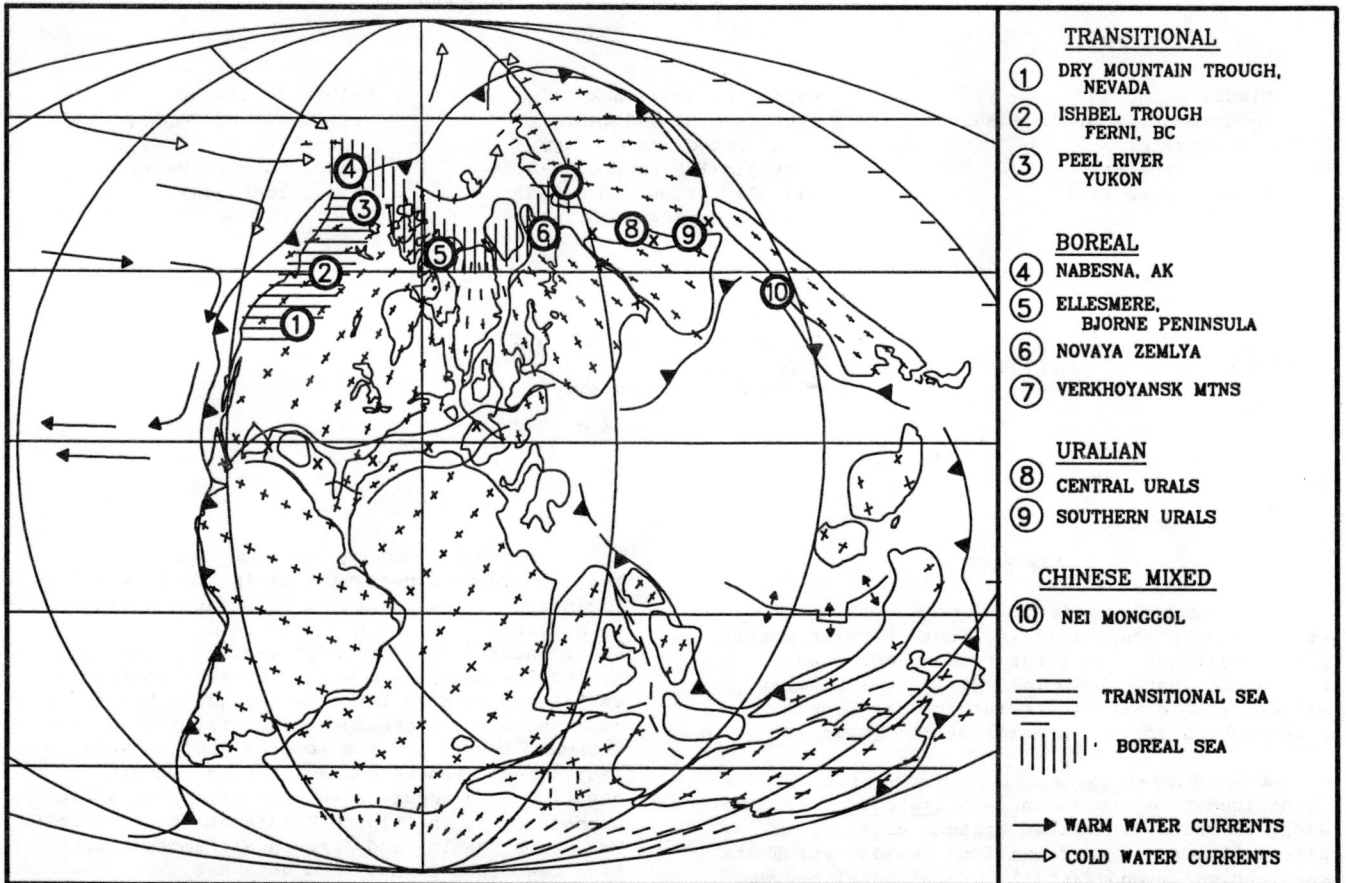

Figure 1. Lower Permian (Sakmarian and Artinskian) distribution of major <u>Uraloceras</u> occurrences plotted on a reconstruction of Greater Pangaea. Localities 4, 5, 6, and 7 represent a boreal association. Localities 1, 2 and 3 represent a transitional association. Localities 8 and 9 represent the central and southern Urals. Locality 10 represents mixed occurrences of <u>Uraloceras</u> and perrinitid ammonoids in Nei Monggol, China. Base map modified from Scotese, 1990.

groups including Arctic and Great Basin "subprovinces"; these, in a general way, correspond in part to the geographic distribution of the boreal <u>Uraloceras</u> assemblage and the transitional <u>Uraloceras</u> assemblage.

Ammonoid faunas dominated by <u>Uraloceras</u> and <u>Paragastrioceras</u> have been associated with a "Boreal Realm" (Nassichuk and others, 1965), with "no direct implication of north-polar position". Presently we recognize a Lower Permian boreal assemblage characterized by <u>Uraloceras</u> (Fig.1, localities 4, 5, 6 and 7). Near its eastern extent, this boreal assemblage was connected with, but faunally distinct from the "Uralian" assemblage (Fig. 1, localities 8 and 9); at its western extent, the boreal assemblage was connected with but faunally distinct from a "transitional" assemblage (Fig. 1, localities 1, 2 and 3). This North American transitional <u>Uraloceras</u> assemblage was located south of the boreal assemblage and north of the equatorial ammonoid faunas of Mexico and West Texas.

A new <u>Uraloceras</u> fauna from Wrangellia in eastern Alaska suggests deposition within a cool to cold water setting of approximately 45-50° N paleolatitude, in close geographic and paleoecologic proximity to other boreal assemblages from Ellesmere Island, Novaya Zemlya,

Siberia and the Urals (Fig. 1, localities 5, 6 and 7). This boreal fauna seems to contradict an equatorial setting for Wrangellia as suggested by various paleomagnetic interpretations. Explanations for this apparent inconsistency between paleomagnetic and paleontologic data include: (1), rapid tectonic migration of Wrangellia; (2), tectonic discontinuities within Wrangellia; (3), the possibility that the units containing the boreal fauna was scraped off the craton at latitudes higher than equatorial and carried with Wrangellia to its present location; and (4), that the portion of Wrangellia in question is not of equatorial origin. Interpretations for the origin and emplacement of Wrangellia that are based on paleomagnetic signatures, should be reexamined within the context of new paleontologic data.

DISTRIBUTION AND TAXONOMIC NOTES

<u>Uraloceras</u> is particularly abundant and taxonomically diverse in the southern Ural Mountains (Fig. 1, localities 8 and 9) from where it was originally described by Ruzhencev (1936) and from where Ruzhencev (1951, 1956) recognized the following species: <u>Uraloceras</u> <u>simense</u>, <u>U</u>. <u>limatulum</u>, (Tastubian and Sterlitamakian); <u>U</u>. <u>involutum</u> and <u>U</u>. <u>burtiense</u>, (Tastubian through Baigendzhinian); <u>U</u>. <u>gracilentum</u>, (Aktastinian);

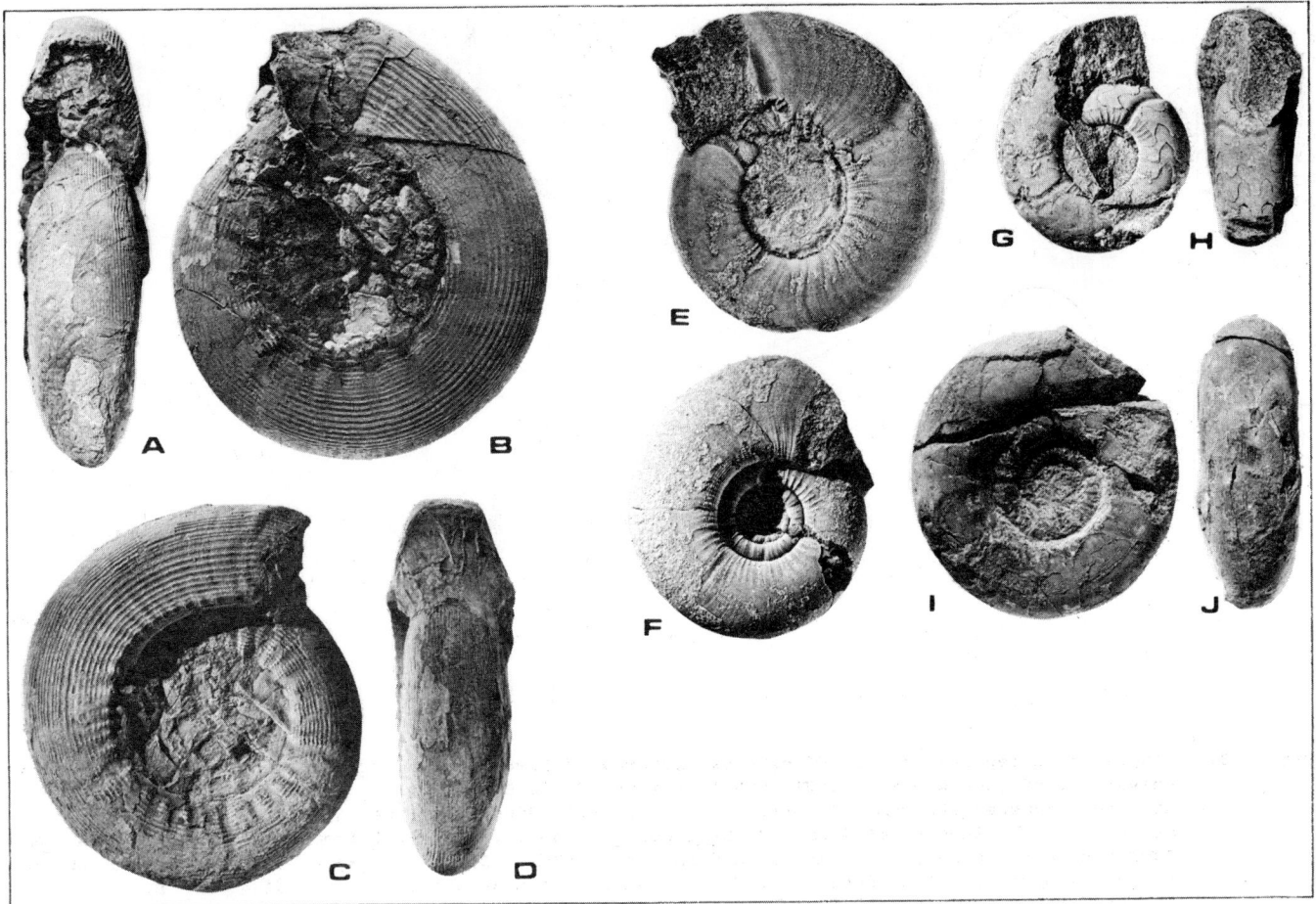

Figure 2. A - D, Uraloceras involutum from the Eagle Creek Formation (Lower Permian) Nabesna, Alaska. A, B, apertural and lateral views, conch diameter of 75mm, (.75X). C, D, lateral and apertural views, conch diameter of 70mm, (.75X). E - J, a new species of Uraloceras from strata of the Lower Permian, transitional association, Dry Mountain trough, northeastern Nevada. E, lateral view, conch diameter of 30mm, (1.50X). G, H, lateral and apertural views, conch diameter of 22mm, (1.43X). F, Lateral view, conch diameter of 25mm, (1.50X). I, H, Lateral and end views, University of Iowa specimen (SUI 58200), conch diameter of 65mm, (.60X).

U. complanatum, (Aktastinian and Baigendzhinian); U. fedorowi, U. belgushkense, U. extenuatum, U. vietum, U. suessi, (Baigendzhinian). Uraloceras involutum is the most abundant species, with 1650 specimens reported from Akstastinian strata (Ruzhencev, 1956, p. 26); this species and U. burtiense are geographically widespread, occurring also in North America. Similarly, both U. involutum and U. fedorowi occur in China (Nei Monggol). Some proposed endemic Chinese species from northwest Gansu (Liang 1981) and from the Mt. Qomolangma (Mt. Everest) area in Tibet (Chin and others, 1977), probably should be assigned to other genera. Bogoslovskaya (1962) recognized U. suessi, U. involutum, U. fedorowi, U. complanatum, U. extenuatum from the central Urals. Representatives of Uraloceras fedorowi also occur in Novaya Zemlya (Yermolaev, 1937 [fide Miller and Furnish, 1940]), in the Verkhoyansk Mountains (Andrianov, 1968), in Guangxi (Zhou and Glenister pers. com.) and on Ellesmere Island (Nassichuk, Furnish and others, 1965). We recognize a new, as yet undescribed species of Uraloceras that occurs in the northern Yukon, south-central British Columbia, and Nevada. This species, described in the Yukon as Uraloceras cf. U. irwinense by Nassichuk (1971) and in British Columbia as Uraloceras sp. is important to the conclusions reached in this report.

The original concept of Uraloceras accommodated a morphologic spectrum wider than that proposed in a recent revision (Glenister and others, 1990). Glenister and others reassigned closely related paragastrioceratin forms, characterized by conchs that are more compressed (narrower) and of smaller mature diameter, to Svetlanoceras Ruzhencev (1974). The present concept of Uraloceras, following Glenister and others, is simplified and morphologically restricted. It includes relatively large paragastrioceratins whose mature conchs (Fig. 2) approximate 100 mm diameter and possess a large umbilicus (ratio of umbilical diameter to conch diameter approximates 40 percent) and a relatively wide conch (the ratio of width to conch diameter approximates 35 percent). Conch ornament characteristically consists of conspicuous longitudinal lirae, well developed constrictions and ventrolateral extensions of the mature

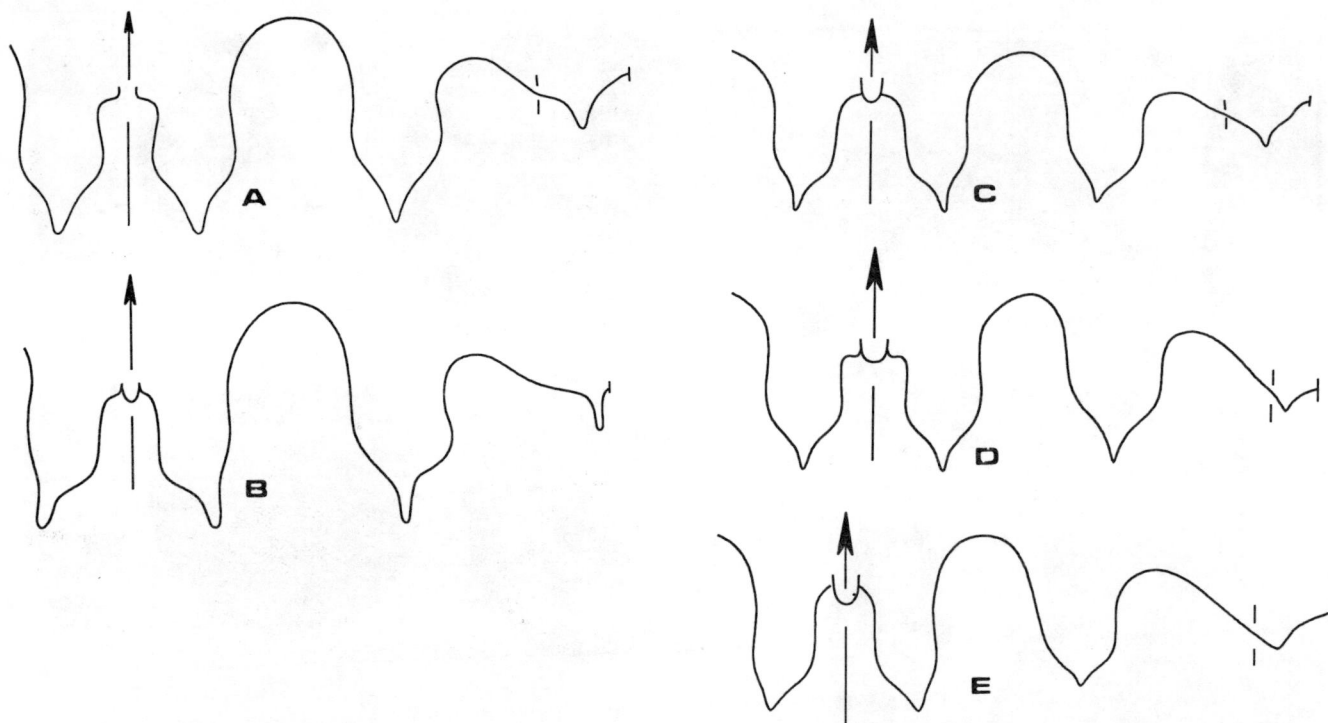

Figure 3.   Diagrammatic representations of external sutures of Lower Permian *Uraloceras*. A, *Uraloceras* sp. University of Iowa specimen (SUI 58001) from the Eagle Creek Formation, Nabesna, Alaska. Drawn at conch diameter of 70mm. B, *Uraloceras* sp., (SUI 59004) from the same formation and locality, drawn at conch diameter of 48mm. C, *Uraloceras* n. sp., (SUI 58200) from Lower Permian strata of Buck Mountain, Nevada, drawn at conch diameter of 57mm. D, *Uraloceras* n. sp., Geological Survey of Canada hypotype (GSC 25663) from Fernie, B.C., drawn at conch diameter of 45mm. E, *Uraloceras* n. sp., (GSC 25518) from the Lower Permian Jungle Creek Formation, of northern Yukon, drawn at conch diameter of 25mm. D and E from Nassichuk, 1971.

aperture (Fig. 2). The sutures (Fig. 3) of this genus are typically paragastrioceratin and are characterized by equal width of each prong of the ventral lobe (V₁) and the first lateral lobe (L).

PALEOBIOGEOGRAPHY

The distribution of Lower Permian (Sakmarian and Artinskian) ammonoids in Greater Pangaea can be differentiated into three distinct faunal associations: A relatively high latitude, or boreal association, an equatorial association and a third association that is geographically and faunally transitional between these two. The boreal association, comprising localities in Alaska, Ellesmere Island, Novaya Zemlya and the Verkhoyansk Mountains, (Fig. 1, localities 4, 5, 6 and 7) is characterized by *Paragastrioceras* and *Uraloceras* with close affinities to species occurring in the Urals. The assemblage from the Urals (Fig. 1, localities 8 and 9) however, shows markedly greater diversity and abundance; furthermore, this association contains a greater number of cosmopolitan taxa. Some of the taxa of the Uralian association, especially certain species of *Uraloceras* and *Paragastriocers*, are closely related to taxa of the boreal association. The boreal association and the Uralian association, however, are conspicuously different in general faunal composition and can be regarded as separate faunas that inhabited different paleoecologic settings. A third North American faunal association, the equatorial association (e.g., West Texas and

Coahuila, Mexico), is characterized by abundant and diverse ammonoid faunas that include cosmopolitan genera as well as more endemic genera such as *Perrinites*. *Uraloceras* does not occur in equatorial faunas. *Perrinites* does not occur in the boreal association nor in the Uralian association. An abundant new species of *Uraloceras*, restricted to North America and associated with common perrinitid ammonoids, characterizes the transitional association. The transitional association is uniquely distinguished by mixing three distinct ammonoid groups: (a), *Uraloceras* species with boreal affinities; (b), perrinitid species with equatorial affinities; and (c), cosmopolitan taxa such as adrianitids and medlicottiids.

Boreal Association

The major North American *Uraloceras* occurrences are shown in Figure 4. A boreal *Uraloceras* and *Paragastrioceras* fauna has been recovered from the Eagle Creek Formation (Richter and Dutro, 1975) and correlative strata of the Wrangellia terrane in Nabesna and the Alaska Range, eastern Alaska. The Eagle Creek Formation (Lower Permian) and the underlying Slana Spur Formation (Middle Pennsylvanian to Lower Permian), compose a lower stratigraphic assemblage of Wrangellia that is disconformably overlain by an upper assemblage comprised of Triassic chert, limestone, the Nikolai Greenstone and younger rock (Fig. 4).

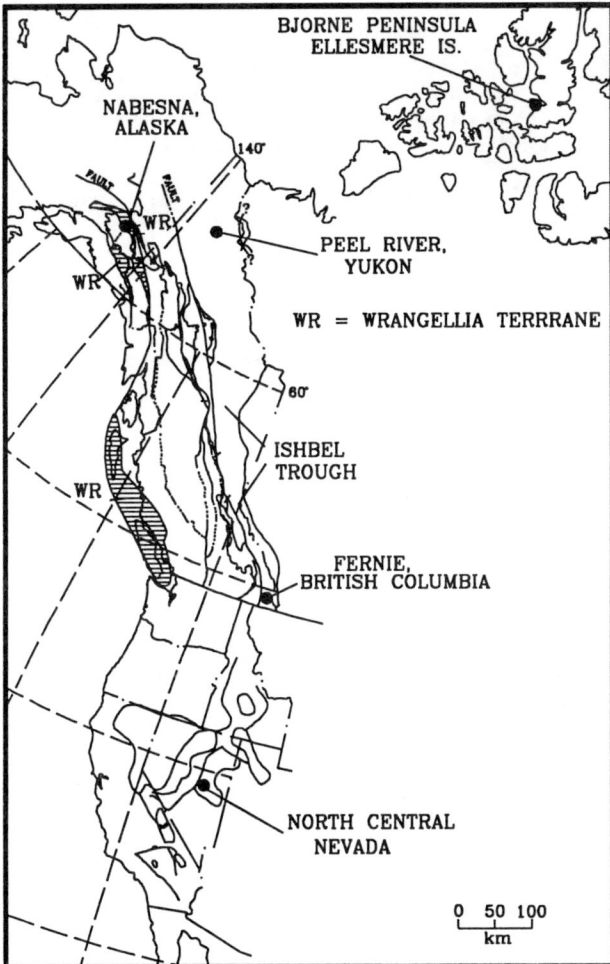

Figure 4. Major occurrences of Lower Permian *Uraloceras* in Western North America and Ellesmere Island.

The boreal association from the Nabesna area consists of more than fifty specimens of *Uraloceras* and *Paragastrioceras* in association with only two other ammonoid representatives: a single adrianitid and a fragmentary medlicottiid. Representatives of *U. involutum* and *U. burtiense* constitute over seventy-five percent of this fauna, and are closely related to species from other boreal faunas of the Verkhoyansk Mountains, Ellesmere Island and Novaya Zemlya (Fig. 1, localities 4, 5, 6, 7).

Transitional Association

The transitional ammonoid association from Nevada and the Yukon (Fig. 1, localities 1 and 2) contrasts with boreal and equatorial associations because of the unique mixed occurrence of *Uraloceras* (with boreal affinities) and perrinitids (with equatorial affinities). The transitional association, additionally, demonstrates greater taxonomic diversity than the boreal association of Alaska but much lower diversity than equatorial associations. The transitional *Uraloceras* representatives are consistently smaller; mean conch diameter of *Uraloceras* from Nevada approximates 20 mm whereas average conch diameter for boreal *U. burtiense* from Alaska is 73 mm. The new *Uraloceras* species from southeastern Nevada appear to be conspecific with ones from British Columbia;

detailed taxonomic and stratigraphic relationships of the latter occurrence, however, are uncertain (Nassichuk, 1971). In Nevada, *Uraloceras* occurs abundantly at one locality. In excess of 500 representatives of a new *Uraloceras* species have been recovered from the Buck Mountain locality within the open marine strata (central basin facies) of the Dry Mountain trough (Fig. 6)(see Gallegos and others, this volume). A single representative of *U. burtiense*, of boreal affinities, has been recovered associated with the abundant new *Uraloceras* species. Only one additional *Uraloceras* representative has been recovered from twelve other Lower Permian ammonoid localities in northeastern Nevada. Such locally abundant occurrences and nearly complete absence from nearby localities apparently is characteristic of the genus in the Urals (Ruzhencev, 1956) and elsewhere.

The faunas of the transitional association occupied discrete basins along the continental margin of western Greater Pangaea. Offshore land areas, perhaps analogous to modern-day Vancouver Island, isolated these basins from the open ocean. These land areas include the Antler highlands in the western U.S., the Cariboo terrane in Canada (Henderson, 1989), and offshore island arcs such as the McCloud terrane (Sonomia). The shelf seas on the western coast of Greater Pangaea were moderated by temperate Panthalassic ocean currents. Some of these discrete basins, such as the Dry Mountain trough of Nevada and the Ishbel trough of British Columbia (Henderson, 1989; Trexler and others, this volume), were situated on the western edge of the continental shelf adjacent to the highlands. Large scale endemism, i.e., the major faunal differences between boreal and transitional associations, can be attributed in large measure, to the effects of latitude and of oceanic current patterns. The low diversity demonstrated by the boreal association, for example, is characteristic of rigorous environments as might have existed on a high latitude, northern shelf impacted by cold or cold-temperate waters of a "paleo-arctic" ocean. This rigorous, high latitude (boreal) environment, contrasts with more optimal environments of the transitional association to the south and with the higher diversity characteristic of the transitional setting.

The Dry Mountain trough and similar discrete basins on western Pangaea are characterized by rapidly changing facies (Gallegos and others, this volume). This reflects the effects of the deeper water environment of these basins in close proximity to shallower water environments to the east and to tectonic uplifts to the west. Local level endemism, for example, the distribution of *Uraloceras* occurrences in Nevada, may be explained by the isolation of these basins and sub-basins from similar basins and sub-basins and by the local paleoecologic differences caused by these rapidly changing environments. The extent of large scale endemism, that is, the differences between boreal and transitional associations, may be limited by two factors. First, *Uraloceras* from the transitional association is closely related to *Uraloceras* of the boreal association; this suggests that some genetic linkage and geographic interconnection must have existed between the two regions. Similarly, the occurrence of perrinitids in the transitional and the equatorial associations mitigates the level of endemism between these two provinces. The occurrence of cosmopolitan forms in the three associations further indicates geographic connection.

					ALASKA RANGE	NABESNA AND ALASKA RANGE

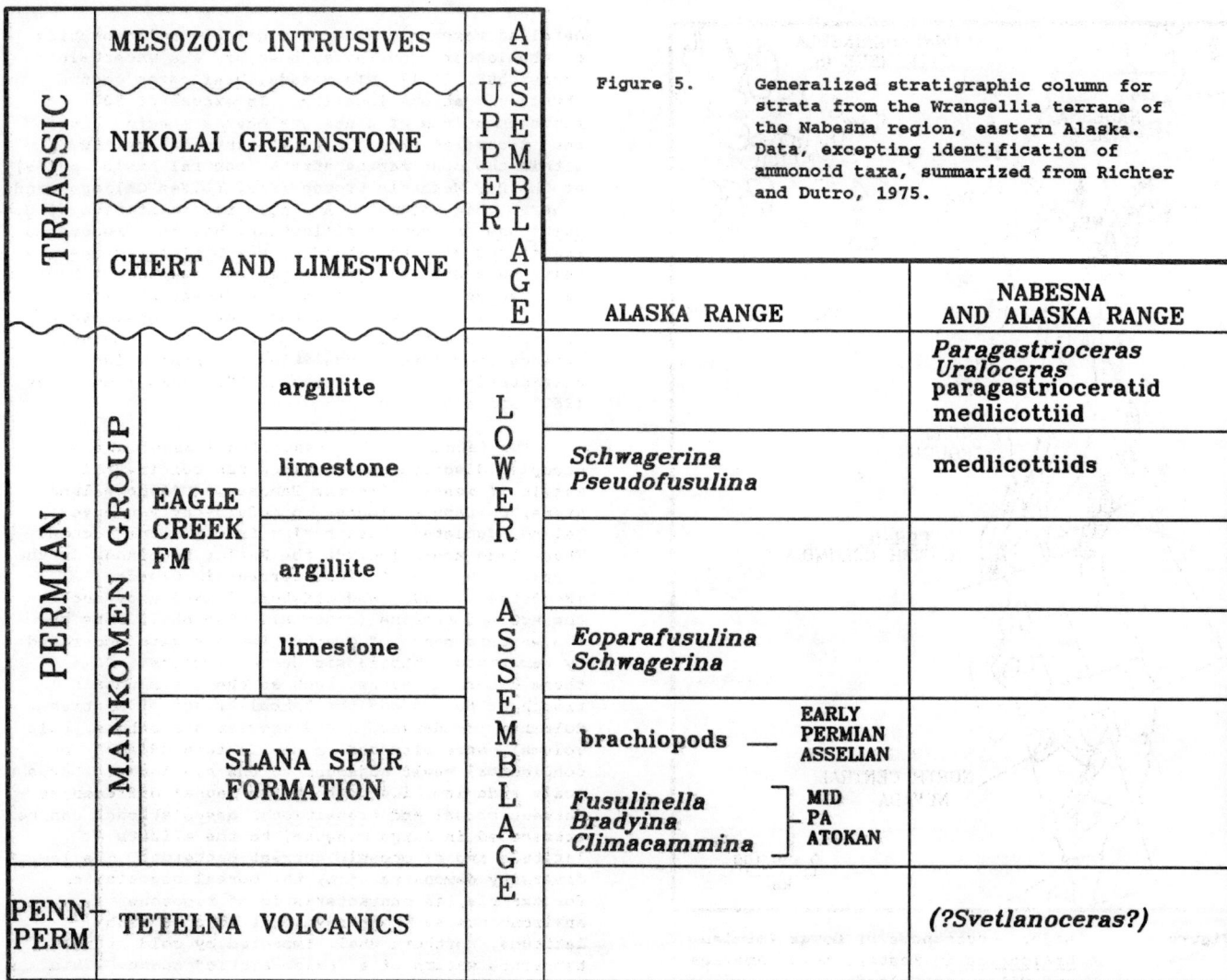

Figure 5. Generalized stratigraphic column for strata from the Wrangellia terrane of the Nabesna region, eastern Alaska. Data, excepting identification of ammonoid taxa, summarized from Richter and Dutro, 1975.

TRIASSIC

UPPER ASSEMBLAGE:
- MESOZOIC INTRUSIVES
- NIKOLAI GREENSTONE
- CHERT AND LIMESTONE

PERMIAN — MANKOMEN GROUP

LOWER ASSEMBLAGE:

EAGLE CREEK FM:
- argillite — *Paragastrioceras Uraloceras* paragastrioceratid medlicottiid
- limestone — *Schwagerina Pseudofusulina* — medlicottiids
- argillite
- limestone — *Eoparafusulina Schwagerina*

SLANA SPUR FORMATION:
- brachiopods — EARLY PERMIAN ASSELIAN
- *Fusulinella Bradyina Climacammina* — MID PA ATOKAN

PENN-PERM
- TETELNA VOLCANICS — (?*Svetlanoceras*?)

Wrangellia of Eastern Alaska

Published paleomagnetic data indicate that the Triassic Nikolai Greenstone had an equatorial origin (Hillhouse and Gromme, 1984), with paleolatitudes of 10 - 17° N or S. Paleomagnetic data for rocks correlative of the Uraloceras-bearing, lower stratigraphic assemblage of Wrangellia in Nabesna, are taken to "provide strong evidence that...Wrangellia was located at lat. ~15° N, during late Pennsylvanian (?) to Triassic time" (Panuska and Stone, 1981, p.562). The Nikolai Greenstone and the lower stratigraphic assemblage of Wrangellia seem to have a far-traveled history including an equatorial origin that seems incompatible with ammonoid evidence indicating origin in cool to cold water setting on the northern continental shelf of Greater Pangaea. Whereas we have not attempted to evaluate all available paleontologic, geologic and paleomagnetic data for this part of Wrangellia, ammonoids indicate that this portion of Wrangellia was not equatorial during Early Permian time. Furthermore, ammonoid data suggest that it may never have moved far from its present relative location on Greater Pangaea.

ACKNOWLEDGEMENTS

The National Science Foundation (grants EAR-8618450, EAR-8746085 and EAR-90004909) and the Idaho State Board of Higher Education (SBOE 91-090) provided financial support. Brian F. Glenister and W.M. Furnish, University of Iowa provided comparable material from Nevada and from the Urals. Zhou Zuren, Nanjing Institute of Geology and Palaeontology provided field guidance and logistical support in China as well as access to fossil collections housed in the Nanjing Institute. David L. Schwarz and D. Crowther provided assistance during field work in Alaska. J.T. Dutro, U.S. Geological Survey, kindly provided comparable material from Nabesna and the Alaska Range. Warren J. Nokleberg, Branch of Alaska Geology, U.S. Geological Survey provided criticisms and reprints of pertinent articles. Dechin Wang, Greg Korn and Dixon Van Hofwegen, BSU, provided laboratory assistance. C.H. Stevens provided suggestions regarding the manuscript. Assistance of these individuals and support from the NSF and the SBOE are sincerely appreciated.

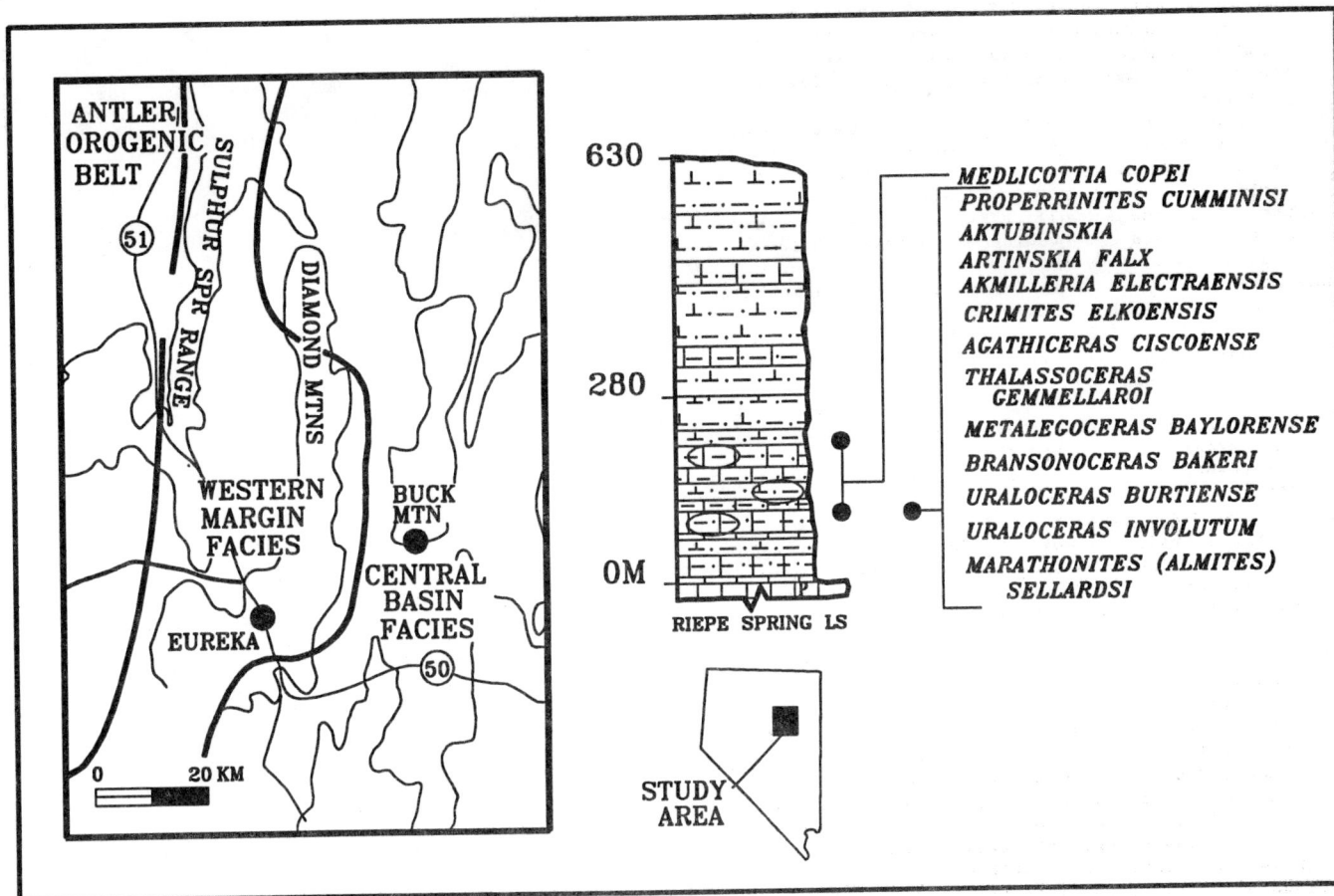

Figure 6.   Generalized location map of the Dry Mountain trough of Nevada, showing position of central basin facies, western margin facies and the Buck Mountain Uraloceras locality.

SELECTED REFERENCES

Amateis, Larry Joe, 1981, The Geology of the Permian Garden Valley Formation, Eureka County, Nevada; unpublished MS Thesis, University Nevada, Reno.

Andrianov, V.N., 1968, The goniatite genus Daubichites from the West Verkhoyansk region and its stratigraphic importance; [Goniatit Daubichites iz Zapadnogo Verkhoyan'ya i yego stratigraphicheskoye znacheniye] Akademiia Nauk SSSR, Doklady, v.178, p.1145-1148.

Bissell, H.J., 1964, Ely, Arcturus, and Park City Groups (Pennsylvanian-Permian) in eastern Nevada and western Utah; Bulletin of American Association of Petroleum Geologists, v.48, p.565-627.

_____, 1970, Realms of Permian Tectonism and Sedimentation in western Utah and eastern Nevada; American Association of Petroleum Geologists Bulletin, v.543, p.285-312.

Glenister, B.F., Baker, C., Furnish, W.M. and Thomas, G.A., 1990, Additional Early Permian ammonoid cephalopods from Western Australia; Journal of Paleontology, v.64, p.392-399.

Henderson, C. M., 1989, Upper Carboniferous and Permian western Canada Sedimentary Basin, in Rickets, B. D., ed., western Canada Sedimentary Basin, A Case History: Canadian Society of Petroleum Geology, p. 203-217.

Kauffman. E.G., 1973, Cretaceous Bivalvia, in Hallam, A., (ed.), Atlas of Palaeobiogeography. Elsevier, New York.

Lowe, P.C., Richter, D.H., Smith, R.L., and Schmoll, H.R., 1982, geologic map of the Nabesna B-5 Quadrangle, Alaska; GQ 1566, US Geological Survey.

Nassichuk, W.W., 1971, Permian ammonoids and nautiloids, southeastern Eagle Plain, Yukon Territory; Journal of Paleontology, v.45, p.1001-1021.

_____, Furnish, W.M. and Glenister, B.F., 1965, The Permian Ammonoids of Arctic Canada; Geological Survey Canada, Bulletin 131, p.56.

Nokleberg, W.J., Albert, N.R.D., Bond, G.C., Herzon, P.L., Miyaoka, R.T., Nelson, W.H., Richter, D.H., Smith, T.E., Stout, J.H., Yeend, W. and Zehner, R.E, 1982, Geologic map of the southern part of the Mount Hayes Quadrangle, Alaska; Department of the Interior, United States Geological Survey, O.F.R. 82-52, p.1-26.

Nokleberg, W.J., Foster, H.L. and Aleinikoff, J.N., 1989, Geology of the northern Copper River Basin, eastern Alaska Range, and southern Yukon-Tanana Upland; In: Nokleberg, W.J. and Fisher, M.A. (eds) Alaskan geological and geophysical transect (Valdez to Coldfoot, Alaska, 1989), field trip guidebook T104, p.34-63.

Nokleberg, W.J., Jones, D.L., and Silberling, N.J., 1985, Origin and tectonic evolution of the Maclaren and Wrangellia terranes, eastern Alaska Range, Alaska; Geological Society of America Bulletin, v.96, p.1251-1270.

Richter, D.H., 1976, Geologic Map of the Nabesna Quadrangle, Alaska; Map I 932, U.S. Geologic Survey.

846

Richter, D.H., and Dutro, J.T., 1975, Revision of the type Mankomen Formation (Pennsylvanian and Permian), Eagle Creek Area, eastern Alaska Range, Alaska; U.S. Geological Survey Bulletin 1395-B, p.B1-B25.

Richter, D.H., Matson, N.A., Jr., and Schmoll, H.R., 1976, Geologic map of the Nabesna C-4 Quadrangle, Alaska; Department of the Interior, United States Geological Survey, MAP GQ-1303.

Richter, D.H. and Schmoll, H.R., 1973, Geologic Map of the Nabesna C-5 Quadrangle, Alaska; GQ 1062, U.S. Geological Survey.

Ruzhencev, V.E., 1936, Paleontologicheskie zametki o kamennougol'nykh i permskikh ammoneyakh (Paleontological Notes on Carboniferous and Permian ammonoids); Problemy Sovetskoi Geologii, n.12, p.1072-1088.

Ruzhencev, V.E., 1938, Ammonei Sakmarskogo yarusa i ikh stratigraphicheskoe znachenie (Ammonoids of the Sakmarian Stage and their stratigraphic importance); Problemy Paleontology, v.4, p.187-285.

Ruzhencev, V.E., 1951, Nizhneperinskie ammonity Yuzhnogo Urala. I. Ammonity Sakmarskogo yarusa (Lower Permian ammonoids of the southern Urals. I. Ammonoids of the Sakmarian Stage); Akademiia Nauk SSSR, Paleontologicheskogo Instituta Trudy, v.33, 188p.

Ruzhencev, V.E., 1952, Biostratigrafiya sakmarskogo yarusa y Aktyubinskoi oblasti Kazakhskoi SSR (Biostratigraphy of the Sakmarian Stage in the Aktyubinsk Region, Kazakh SSR); Trudy Paleontologicheskogo Instituta, v.42, p.1-90.

Ruzhencev, V.E., 1956, Nizhnepermskie ammonity Yuzhnogo Urala. II; Ammonity artinskogo yarusa (Lower Permian Ammonoids of the southern Urals. II Ammonoids of the Artinskian Stage); Trudy Paleontologicheskogo Instituta, v.60, p.1-274.

Ruzhencev, V.E., 1974, The families Paragastrioceratidae and Spirolegoceratidae; Akademiia Nauk SSSR, Paleontologigeskii Zhurnal, no.1, p.19-29.

Scotese, C.R., 1990, Atlas of Phanerozoic plate tectonic reconstructions International Lithospherre Program (IUGG-IUGS), Paleomap Project Technical Report 10-90-1.

Spinosa, Claude, and Nassichuk, W.W., 1985, The Permian Ammonoid Uraloceras in North America and its global significance; Geological Society of America Annual Meeting, Abstracts with Programs, p.724.

Stevens, C.H., 1979, Lower Permian of the central cordilleran miogeosyncline; Geological Society of America Bulletin, v.90, n.II, p.381-455.

Wheeler, J.O., Brookfield, A.J., Gabrielse, H., Monger, J.W.H., Tipper, H.W. and Woodsworth, G.J. (Compilers), 1984, Terrane map of the Canadian Cordillera; Department of Energy, Mines and Resources, Geological Survey of Canada., O.F.

Yermolaev, M.M., 1937, Stratigraphy of Paleozoic deposits of Novaya Zemlya; XVII International geologic Congress, The Novaya Zemlya Excursion, v.I, p.91-134.

COMPARISON OF TWO EARLY PALEOZOIC CARBONATE SUBMARINE FANS--WESTERN UNITED STATES AND SOUTHERN KAZAKHSTAN, SOVIET UNION

Harry E. Cook
U.S. Geological Survey
Menlo Park, California, U.S.A. 94025

Michael E. Taylor
U.S. Geological Survey
Denver, Colorado, U.S.A. 80225

Slava V. Zhemchuzhnikov
Ministry of Geology of the Kazakh S.S.R.
Alma-Ata, Kazakhstan, U.S.S.R. 480077

Mikhail K. Apollonov
Gappar Kh. Ergaliev
Zhannur S. Sargaskaev
Institute of Geological Sciences
Academy of Sciences of the Kazakh S.S.R.
Alma-Ata, U.S.S.R. 480100

Svetlana V. Dubinina
Geological Institute
Academy of Sciences of the U.S.S.R.
Moscow, U.S.S.R. 109017

Ludmila Mel'nikova
Palaeontological Institute
Academy of Sciences of the U.S.S.R.
Moscow, U.S.S.R. 117321.

ABSTRACT

Early Paleozoic passive-margin carbonate platforms of the western United States and southern Kazakhstan, U.S.S.R., developed at low paleolatitudes on rifted Precambrian continental crust adjacent to the proto-Pacific Ocean. In the western United States, early Paleozoic carbonate submarine fans and slides formed on continental slopes in central Nevada. Coeval shoal-water carbonate sediments in eastern Nevada and Utah are interbedded with siliciclastic sediments and onlap the North American craton. In contrast, coeval early Paleozoic carbonate sediments in the Malyi Karatau region, southern Kazakhstan, were deposited on isolated microcontinental blocks that developed during Late Proterozoic rifting of the continental crust. Shoal-water carbonate sediment accumulated on one isolated block which was the site of the 40-km-wide Aisha Bibi seamount. The seamount was flanked by deeper water carbonate submarine fans and slides.

Comparison of stratigraphic sections in the Hot Creek Range of Nevada and the Malyi Karatau of southern Kazakhstan indicates a similar upward-shallowing and seaward-prograding development. The Hot Creek Range section is comprised of the Upper Cambrian Swarbrick Formation and herein-reinstated Tybo Shale, Upper Cambrian and Lower Ordovician Hales Limestone, and the lowermost part of the Lower Ordovician Goodwin Limestone. These depositional facies include: basin plain (Swarbrick and Tybo, totaling about 1000 m thick); lower slope and basin plain carbonate submarine fan (uppermost part of the Tybo and lower part of the Hales, about 205 m thick); upper slope (upper part of the Hales, about 315 m thick); and a relatively deep-water, non-rimmed platform margin (lowermost part of the Goodwin, 25 m thick). No rocks older than the Upper Cambrian Tybo Shale crop out in the Hot Creek Range.

The Kyrshabakty and Batyrbay sections in the Malyi Karatau are comprised of Cambrian and Lower Ordovician rocks of the Shabakty Suite. Middle and Upper Cambrian and Lower Ordovician depositional facies include: basin plain (about 100 m thick); lower slope and basin plain carbonate submarine fan (about 465 m thick); upper slope (about 125 m thick); and shoal-water carbonate seamount margin and seamount interior (about 375+ m thick).

INTRODUCTION

Previous research in the western United States emphasized the synergistic value of integrated sedimentologic and paleontologic studies for recognizing and interpreting passive-margin carbonate platforms (for example, Cook and Taylor, 1975, 1977; Taylor, 1977; Taylor and Cook, 1976). The ocean-margin part of these carbonate platforms (i.e., platform margin/slope/basin plain) comprises the record of interaction between continents and ocean basins. The stratigraphic record of these deposits contains important clues for deciphering the character and origin of early Paleozoic paleo-oceans and their adjacent paleocontinents. Ocean-margin sequences are comprised both of autochthonous oceanic sediments and faunas, as well as allochthonous mass-transport sediments and faunas which commonly contain clasts, sediments, and faunas derived from the continental shelf or terrestrial realm. Information from ocean-margin deposits can provide constraints on plate tectonics, ocean-basin types, characteristics of the water column, oceanic circulation patterns, eustatic events, environmental controls on biofacies distribution, and the nature of contiguous continents, microcontinents, or seamounts (Cook and Taylor, 1975, 1977, 1990a, 1990b; Cook and Egbert, 1981; Taylor, 1976, 1977; Taylor and Cook, 1976, 1990; Hine and Mullins, 1983; James and Mountjoy, 1983; Sarg, 1988; Cook and others, 1989a, 1989b; Crevello and others, 1989).

Numerous early Paleozoic carbonate platforms throughout the world developed on

In Cooper, J.D., and Stevens, C.H., eds, 1991, **Paleozoic Paleogeography of the Western United States-II:** Pacific Section SEPM, Vol. 67, p. 847-872.

847

passive continental margins that formed after a major episode of continental rifting in the latest Proterozoic (Read, 1982, 1985; Bond and others, 1983, 1984, 1989; Llyin, 1990). Following Read (1982) and Bond and others (1989) we refer to these types of carbonate platforms as "passive-margin carbonate platforms".

We have studied early Paleozoic passive-margin carbonate platforms on the widely separated paleotectonic plates of western North America and southern Kazakhstan, Soviet Union (Fig. 1). This summary report compares development of deeper water facies of these partially coeval passive-margin platforms in reference to their (1) paleotectonic settings, (2) sedimentologic facies, and (3) paleogeographic settings.

Low Paleolatitude
Passive-Margin Carbonate Platforms

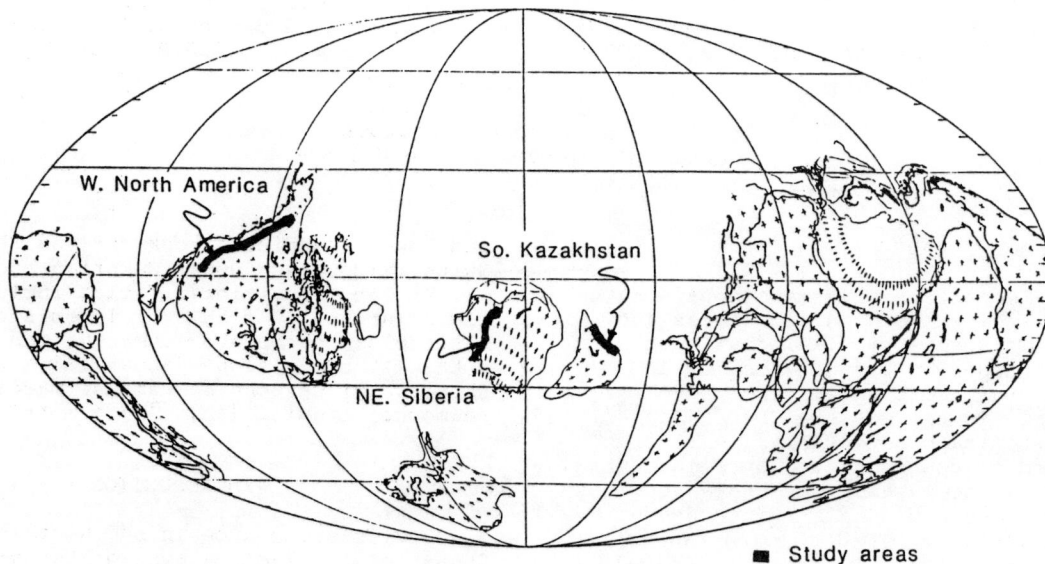

Figure 1. Paleogeographic index map showing location of early Paleozoic low-paleolatitude passive-margin carbonate platforms in western North America, eastern Siberia, and the Malyi Karatau region of southern Kazakhstan, U.S.S.R. Base map is Late Cambrian map from Scotese and McKerrow (1990, p.7).

PALEOTECTONIC SETTINGS

Western North America

The Precambrian passive continental margin of western North America formed when a rift and spreading center developed within Laurentia and fragments drifted apart forming a new ocean--the proto-Pacific (Figs. 1, 2, 3) (Stewart, 1972, 1976, 1980; Stewart and Suczek, 1977; Burchfiel and Davis, 1972; Gabrielse, 1972; Monger and others, 1972). Thus, rifting and spreading defined the currently north-trending passive continental margin of western North America (Fig. 4). Occurrence of diamictite and volcanigenic rocks, stratigraphic backstripping, and the shape of thermally generated subsidence curves show that the breakup and onset of spreading probably was initiated in the Late Proterozoic (650 Ma) and may have extended into the earliest Cambrian (555 Ma) (Sleep, 1971; Dickinson, 1977; Stewart and Suczek, 1977; Armin and Mayer, 1983; Bond and others, 1983, 1985; Bond and Kominz, 1984; Bally, 1987; Hamilton, 1987; Christie-Blick and Levy, 1989; Levy and Christie-Blick, 1989; Stewart, this

volume). The edge of the continent formed during the passive-margin rifting phase is inferred to coincide approximately with the contour of initial $87SR/86SR = 0.706$ at about 117° W. Longitude in central Nevada (Stewart and Poole, 1974; Kistler, 1974; Speed, 1982; Farmer and DePaolo, 1983).

The earliest direct paleontologic and sedimentologic evidence for locating the early Paleozoic passive continental margin of the western U.S. was presented by Cook and Taylor (1975; also see Taylor and Cook, 1976; Cook and Taylor, 1977; Taylor, 1977; Kepper, 1981). The carbonate platform that developed on the Late Cambrian and Early Ordovician continental margin had the profile of a non-rimmed distally steepened carbonate ramp (Fig. 5, Stage 1) (Cook, 1983; Cook and Taylor, 1987; Cook and others, 1989a). This carbonate platform classification follows Read (1985, Fig. 2F).

In the Hot Creek Range of central Nevada, the oceanward margin of the early Paleozoic ramp is characterized by depositional environments below the photic zone and below

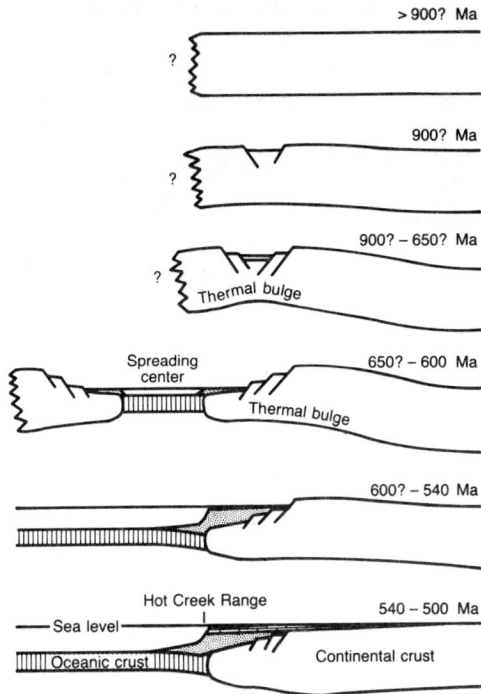

Figure 2. Model of the late Precambrian and Cambrian development of the western United States with the interpreted position of the Hot Creek Range, central Nevada. From Cook (1988, 1989) as modified after Stewart and Suczek (1977).

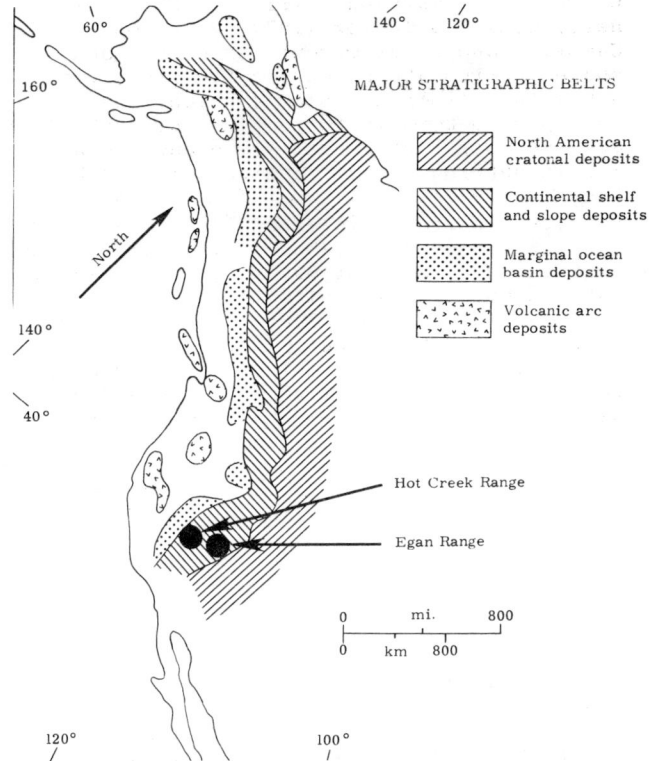

Figure 4. Major regional stratigraphic belts during the Paleozoic in western North America. Hot Creek Range in central Nevada contains Upper Cambrian and Lower Ordovician continental slope and submarine fan facies. Egan Range in eastern Nevada contains coeval shoal water carbonates. From Cook and Taylor (1977).

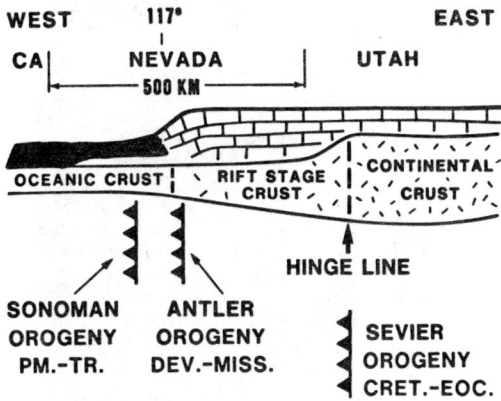

Figure 3. Model of western United States passive continental margin showing interpreted positions of oceanic crust, rift-stage crust, and continental crust during rifting 650-555 Ma. Seaward edge of passive-margin carbonate platform at 117° served as a buttress during the Antler orogeny and Sonoma orogeny; continental crust-to-rift stage crust transition (hinge line) limited the eastward thrust motions during the Sevier Orogeny. Based on data and interpretations in Stewart (1972), Cook and Taylor (1975, 1987), Allmendinger and others (1987), and Cook (1988, 1989).

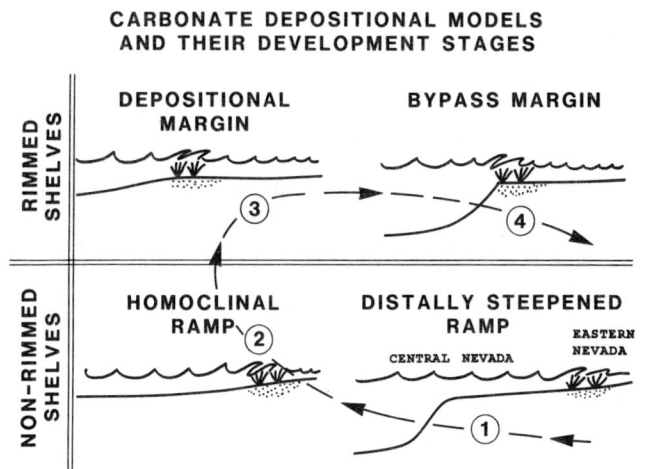

Figure 5. Diagrammatic carbonate depositional model showing the four stages of evolution for the Great Basin passive-margin carbonate platform: (1) Late Cambrian through Early Ordovician; (2) Early Ordovician through Early Silurian; (3) Late Silurian through Early Devonian; (4) Middle Devonian through Late Devonian. From Cook and Taylor (1991a).

normal storm wave-base (Fig. 5, Stage 1). Here, Cook and Taylor (1977) described Upper Cambrian and Lower Ordovician cherts and shales, carbonate submarine-fan facies, and submarine slides and slumps which they interpreted to have formed in a proto-Pacific, deep-marine, continental slope and ocean-margin setting. These sediments in the Hot Creek Range occur at about 116° 30' W. Longitude near the inferred $^{87}SR/^{86}SR = 0.706$ contour (Figs. 3, 4) (Kistler, 1974). In the central Egan Range of eastern Nevada, shoal water depositional environments occur on the ramp landward of the platform margin at depths within the photic zone and above storm wave base (Fig. 5, Stage 1). Thus, by Middle to Late Cambrian time, rifting and spreading were well established, thermal subsidence provided accommodation space for the sedimentation of a thick sequence of ocean-margin carbonates, and open ocean faunal distribution paths between Asia and western North America were established (Taylor and Cook, 1976; Cook and Taylor, 1977; Taylor, 1976, 1977; Taylor and Forester, 1979; Taylor and others, 1989).

Southern Kazakhstan

Kazakhstan is composed of several amalgamated microcontinental blocks, called "Kazakhstania" (Zieglar and others, 1977, 1979), that exhibit a complex development, which began with Late Proterozoic (Riphean) rifting of continental crust (Fig. 6a) (Kompaneitsev, 1986; Abdulin and others, 1986; Azerbaev, 1988; Sargaskaev, 1988, 1990; Sargaskaev and Ergaliev, 1988; Abdulin and others, 1990, Fig. 3; Cook and others, 1989b; Cook and Taylor, 1990b; Llyin, 1990). During rifting, a passive continental margin developed in the region of the Tien Shan foldbelt (Terman, 1974) in southern Kazakhstan and northwestern China. In contrast to the western North American platform which developed as a relatively stable continental margin attached to the craton (Figs. 2, 3), the central Asian continental margin was broken into several large isolated blocks during rifting (Fig. 6a) (Cook and others, 1989b; Abdulin and others, 1990, Fig. 3; Sargaskaev, 1990). In the Malyi Karatau region Late Proterozoic siliciclastic submarine fans were deposited in structural lows around the blocks (Figs. 6b, 7) (Sargaskaev, 1988, 1990). With continued spreading and post-rift thermal subsidence, the surfaces of the isolated structural blocks became loci for shallow-marine carbonate sedimentation during the Late Proterozoic (Vendian), Cambrian, and Early Ordovician (Arenig) (Fig. 6c). The deeper water margins of the blocks became the location of slope sedimentation and development of carbonate submarine fans and slides (Fig. 6c) (Zhemchuzhnikov, 1986; Apollonov and Zhemchuzhnikov, 1988; Cook and others, 1989b; Abdulin and others, 1990, Fig. 3). These microcontinental blocks were probably separated by considerable widths of oceanic crust in the early Paleozoic. Sometime in the latest Carboniferous or at least by the Early Permian, these microcontinents and adjacent oceanic areas had amalgamated and collided with Siberia (Burrett, 1974).

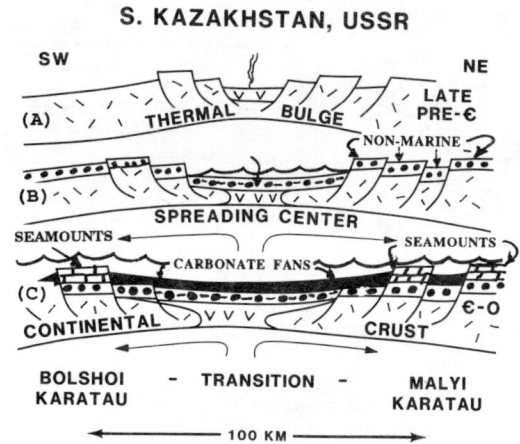

Figure 6. Development of southern Kazakhstan passive continental margin. (A) Initial thermal bulge, rifting of continental crust, and volcanic activity (v-pattern) develops during the Riphean and Vendian (Late Proterozoic) in the region of the Malyi Karatau-Bolshoi Karatau structural belt. (B) Late Proterozoic siliciclastic submarine fans deposited in structural lows in the spreading center (black dots and dashes pattern) over volcanics; nonmarine siliciclastics (black dots pattern) deposited on surfaces of isolated structural blocks. (C) With continued spreading and post-rift thermal subsidence, isolated blocks became the locations for carbonate seamounts in the Vendian to Ordovician; carbonate submarine fans (thick black-band pattern) formed at the margins of the seamounts.

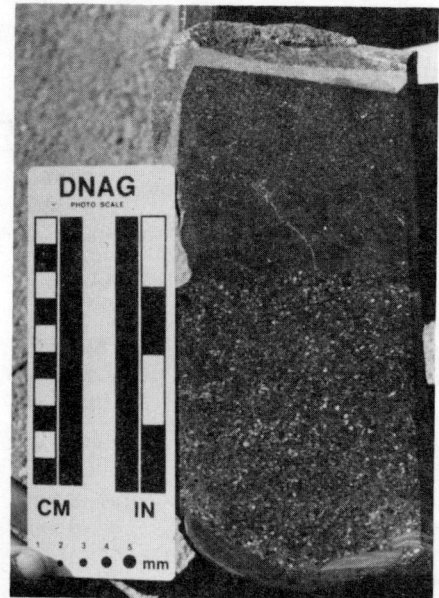

Figure 7. Late Proterozoic siliciclastic submarine fan turbidite from Malyi Karatau area; showing Bouma Ta division with rip-up shale clasts. Turbidite depicted in figure 8b. Photo by Sargaskaev, Zh. S.

CARBONATE PLATFORM MARGINS

Nevada, Western United States

The Swarbrick Formation, Tybo Shale, and
Hales Limestone in the Hot Creek Range,
central Nevada, represent deeper water
deposition on the continental slope and basin-
plain edge of the distally steepened ramp
(Fig. 5, Stage 1, and Fig. 8). In contrast,
140 km to the east, the coeval Whipple Cave
Formation and lower part of the House
Limestone in the central Egan Range, eastern
Nevada, are examples of shoal-water deposition
characterizing the shallow-water part of the
ramp (Fig. 5, Stage 1, and Fig. 9) (Cook and
Taylor, 1975, 1977; Taylor and Cook, 1976).
Both the deep-water and shoal-water localities
exhibit a similar upward-shoaling and seaward-
prograding development.

Tybo Canyon, Hot Creek Range

The Tybo Canyon section of the Hot Creek
Range in central Nevada represents deep-water
sedimentation on the edge of the early
Paleozoic continental margin. In this region,
the Cambrian and Lower Ordovician
stratigraphic succession in the Hot Creek
Range, originally described by Ferguson
(1933), consists of the Upper Cambrian
Swarbrick Formation, the Upper Cambrian Tybo
Shale and the Upper Cambrian and Lower
Ordovician Hales Limestone (Fig. 10). We here
reinstate the name "Tybo Shale" in the sense
of Ferguson (1933) for rocks mapped by
Quinlivan and Rogers (1974) as the Dunderberg
Shale in the Hot Creek Range because the Tybo
Shale represents a substantially deeper water
facies than the Dunderberg Shale and because
the Tybo Shale is bounded by different
formations than the Dunderberg.

The Tybo Shale and Hales Limestone of the
Hot Creek Range contain vastly different rocks
and faunas from those in the coeval Whipple
Cave Formation and lower part of the House
Limestone in the central Egan Range, 140 km to
the east (Cook and others, 1989a; Taylor and
others, 1989).

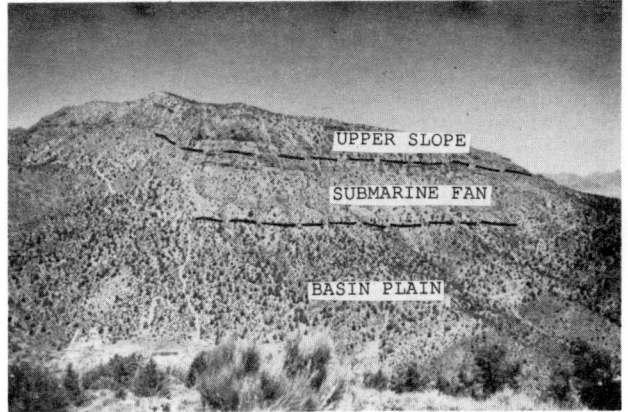

Figure 8. Hot Creek Range, Tybo Canyon
section of the Swarbrick Formation, Tybo
Shale, and Hales Limestone. The approximate
boundary between the basin plain, carbonate
submarine fan, and upper slope facies is
indicated by the dashed lines. Near Warm
Springs, central Nevada. View toward north.

Figure 9. Central Egan Range section of
theWhipple Cave Formation, near Lund, eastern
Nevada. View toward east. From Cook and Taylor
(1977).

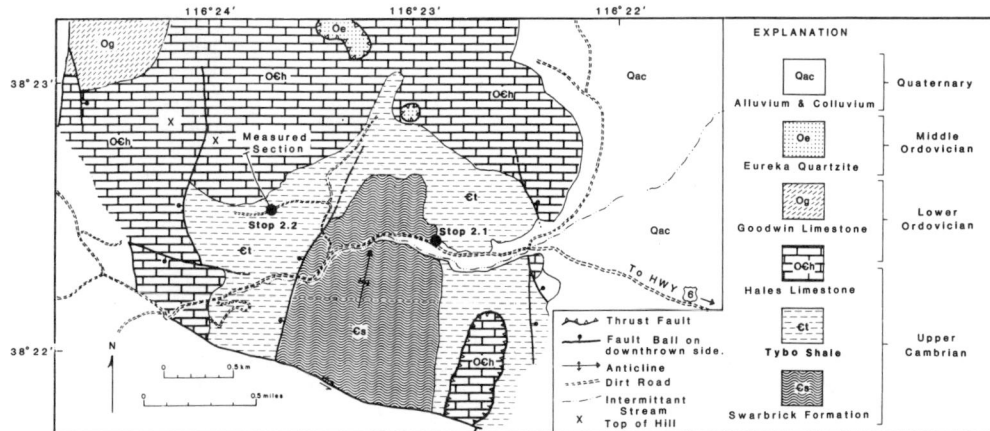

Figure 10. Geologic map of Tybo Canyon and vicinity, Hot Creek
Range, central Nevada. Map from Cook and others (1989a) as modified
after Quinlivan and Rogers (1974).

Basin-Plain Facies (about 1000 m):

The Swarbrick Formation (about 500 m thick) and the overlying Tybo Shale (about 500 m thick) comprise the basin-plain facies (Fig. 8). Two lithologies that characterize the Swarbrick Formation are gray, thinly laminated lime mudstone and black laminated chert. Origin of the laminated lime mudstone is uncertain. It may represent deposition from dilute turbidity currents that formed only Bouma Tde divisions or it may have originated as pelagic, hemipelagic and periplatform ooze. Both types of lime mudstones are present. However, the chert contains remnants of abundant sponge spicules, which probably originated in a relatively deep-water setting on the basin plain far removed from allochthonous siliciclastic sedimentation.

The overlying Tybo Shale is characterized by olive-green calcareous shale and thin (1-5 cm thick) interbedded calcisiltite turbidites. The turbidites consist of Bouma divisions Tbc, Tc, and Tcde. In addition, rare directional sole markings (flute and tool casts) and ichnofossils are preserved at the base of some beds. These thinly bedded calcisiltite turbidites become more abundant upsection.

Lower Slope and Basin-Plain Facies (205 m):

The lower slope sequence of the uppermost part of the Tybo Shale and the overlying Hales Limestone consists of about 50 to 70 percent of mass-transport deposits and 30 to 50 percent of interbedded in-situ slope sediment. These mass-transport deposits form the carbonate submarine fan described below.

Sediment gravity-flow deposition along the deep-water flanks of carbonate platforms typically does not produce submarine fans (Mullins and Cook, 1986; Cook and Mullins, 1989). Instead, wedge-shaped carbonate aprons develop parallel to the adjacent shelf-slope break (Schlager and Chermak, 1979; Crevello and Schlager, 1980; Cook, 1983; Mullins, 1983b, 1986; Mullins and Cook, 1986; Sarg, 1988; Cook and Mullins, 1989). The major difference between submarine fans and carbonate aprons is a point source with channelized sedimentation on fans, versus a line source with sheet-flow sedimentation on aprons. Mullins and Cook (1986) discuss the similarities and differences between carbonate aprons and siliciclastic submarine fans.

The carbonate submarine fan in the Hot Creek Range is an unusual type of fan for two main reasons. First, most of the fan debris did not originate in shoal-water platform margin sites. More than three-fourths of the fan debris is derived from numerous slope-generated submarine slides that remolded into debris flows and turbidity-current flows (Figs. 11, 12) (Cook, 1979; Cook and Mullins, 1983). Thus, the lithology of the inner-, middle-, and outer-fan facies is mainly derived from the deep-water slope environment with a lesser percentage transported from sediment-gravity-flows that originated in shoal-water sites to the east. Second, the large inner-fan feeder channels do not extend up the slope onto the platform margin. Rather,

these feeder channels are restricted to the slope because they originated on the slope by the erosive action of massive volumes of slide-generated megabreccia debris flows that moved across semilithified sediment (Fig. 11).

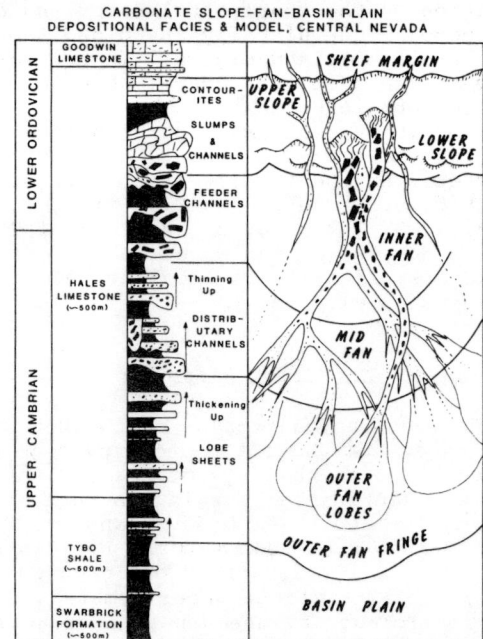

Figure 11. Columnar section and interpreted depositional environments in Upper Cambrian and Lower Ordovician rocks of Tybo Canyon, Hot Creek Range, central Nevada. Interpreted environments in planimetric view., No relative scales are implied. From Cook and others (1981, 1989a)

Outer-Fan Fringe Facies (25 m):

The uppermost 25 m of the Tybo Shale consists of thinly bedded calcarenite and olive-green shale. The thin beds of calcarenite range from 2 to 15 cm thick and are generally thicker than the turbidite beds in the basal part of the Tybo Shale (Cook and Mullins, 1983, Fig. 132). Several thicker beds (10-15 cm) are turbidites composed predominantly of trilobite exoskeletal debris (Taylor and Cook, 1976, Fig. 23).

Outer-Fan Lobe Facies (33 m):

This interval was deposited as outer-fan lobe sheets and does not contain significant evidence of channeling. The facies is further characterized by thinly bedded calcarenite turbidites that exhibit mostly Bouma divisions Ta, Tab, and Tabc. Several upward-thickening sequences of turbidites occur within this stratigraphic interval (Fig. 13) (Cook and Mullins, 1983, Figs. 129-131). Many of the turbidites are composed of Nuia grains (a probable calcareous green alga) derived from shoal-water areas to the east. Flute and tool marks are occasionally found on the soles of some beds.

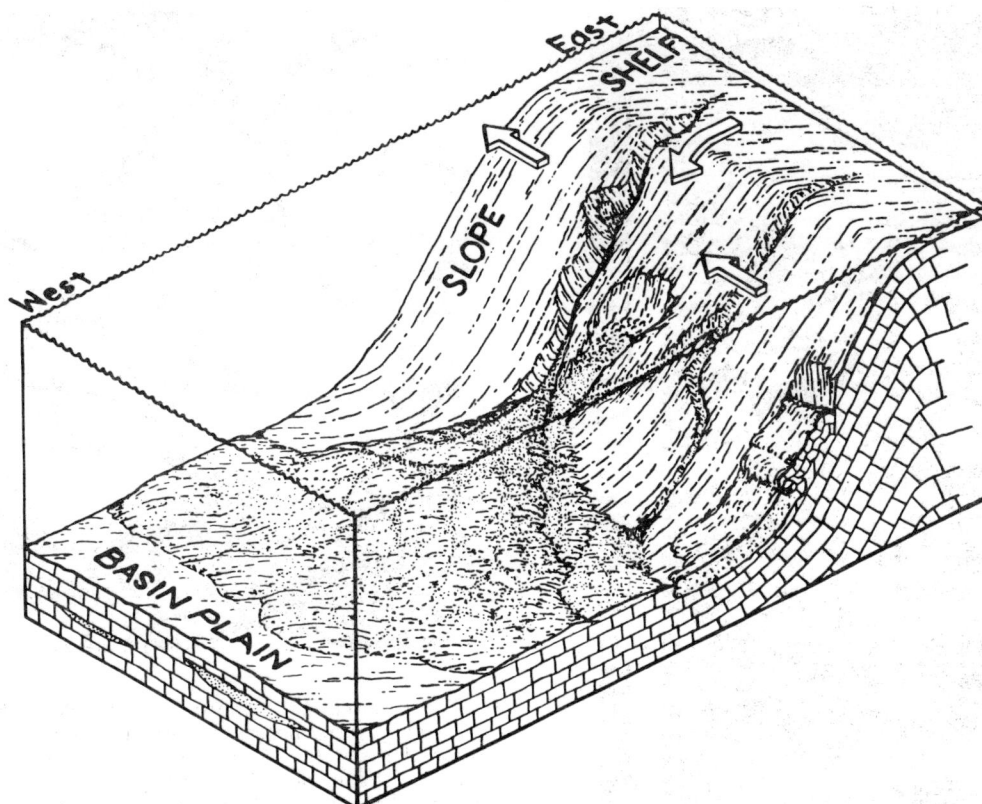

Figure 12. Model of interpreted seaward edge of the passive-margin carbonate platform in the Late Cambrian and Early Ordovician of central Nevada. This shows slope incised by numerous gullies but no major canyons; carbonate submarine fan develops at base of slope; fan sediment is a mixture of shoal-water carbonate platform carbonates, deeper water slide-generated debris, and fine-grained siliciclastics swept across the carbonate platform during sea-level lowstands; contour currents flow northerly along upper slope. From Cook and Mullins, 1983.

Figure 13. Outer-fan lobe facies. Upward-thickening nonchannelized turbidite sheets. Note flat bases and tops of turbidite beds. Lower part of the Hales Limestone, Upper Cambrian, Hot Creek Range, central Nevada. From Cook (1983) and Cook and Mullins (1983).

Mid-Fan Distributary Channel Facies (20 m):

The mid-fan facies consists of channelized debris-flow conglomerates and thinly bedded turbidites. Some upward-thinning cycles are present. The channels are 1-5 m deep, 20-100 m wide, and rapidly coalesce laterally and vertically. The debris-flow deposits are composed of imbricated tabular limestone clasts, dipping to the east in an upslope direction (Fig. 14). These clasts generally grade upward into fine-grained calcarenite that forms a ripple-laminated cap to the debris flow (Bouma Tac division) (Cook and Mullins, 1983, Figs. 125, 126). The debris beds thin and fine laterally into sand-sized textures exhibiting ripple laminations and climbing ripples (Cook and Mullins, 1983, Fig. 127).

The channelized debris flows are interpreted to represent distributary channel

Figure 14. Mid-fan distributary channel facies. Upper part of 1.5 m-thick channel deposit. Clasts are normally graded and imbricated in an upslope direction at the top of the channel. Flow westerly downslope from right to left. Rippled calcarenites cap the bed. Clasts and grains are slope- and shoal-water derived. Scale in inches. Lower part of the Hales Limestone, Upper Cambrian, Hot Creek Range, central Nevada. From Cook and Taylor (1977).

Figure 15. Inner-fan feeder channel facies, showing megabreccia debris-flow deposit about 10 m thick and 500 m wide. Dashed line outlines slope-derived clast 3 x 15 m. Solid black line is along base of channel. Top of channel not shown. Middle part of the Hales Limestone, Lower Ordovician, Hot Creek Range, central Nevada. From Cook (1979, 1983) and Cook and Mullins (1983).

deposits in a mid-fan depositional setting (Cook and Mullins, 1983, Figs. 24, 71, 122, 123, 125, 126, 128). Interbedded with the debris-flow deposits are thin wavy-bedded turbidites that are laterally discontinuous and composed dominantly of ripple laminations. These units probably represent overbank-levee and interdistributary-channel deposits.

Inner-Fan Feeder-Channel and Submarine Slide Facies (130 m):

Inner-fan feeder channels 10 m deep and 500 m wide are filled with disorganized boulder-bearing conglomerate and megabreccia (Cook and Mullins, 1983, Figs. 58, 59, 78). Clasts in these channelized megabreccia debris-flow deposits are up to 3 by 15 m across (Fig. 15) (Cook and Mullins, 1983, Fig. 58). The clasts are mainly slope-derived black lime mudstone with lesser amounts of shoal-water derived material. Channel thalwegs are

up to 15 m deep with channel walls having erosional gradients of 25 degrees (Cook, 1979, Fig. 19). A detailed description of these feeder channels and the mechanics of sediment movement in them is discussed and illustrated by Cook (1979) and Cook and Mullins (1983).

Several translational slides about 10 m thick and 500 m wide occur within pelagic and hemipelagic lower-slope and inner-fan deposits (Fig. 16). These slides exhibit progressive deformation of overfolds to virtually complete remolding into conglomeratic debris flows (Cook, 1979; Cook and Mullins, 1983, Figs. 26-37). These remolded slides provided much of the debris that forms the submarine fan. Slip-line orientations in the slides, clast imbrications and cross laminations in the sediment-gravity flows all indicate transport directions from east to west (Cook and Taylor, 1977, Fig. 38; Cook and Mullins, 1983, Fig. 80).

In-situ facies in the lower slope are different from upper slope in-situ facies in that lower slope sediments are finer grained and darker colored. Lower slope in-situ sediments consist of black lime mudstone (Figs. 17, 18) (Cook and Taylor, 1977, Figs. 25, 26; Cook and Mullins, 1983, Figs. 9, 13). The mudstone is thin-bedded (beds are normally less than 2 cm thick), contains millimeter-thick laminae, and is virtually devoid of burrows.

Figure 16. Slope-derived submarine slide on lower part of slope; about 10 m thick and 500 m wide. Large open overfolds developed in semilithified slope lime mudstone during downslope movement. Base and top of slide not shown. Pick for scale. Middle part of the Hales Limestone, Lower Ordovician, Hot Creek Range, central Nevada. From Cook (1979, 1983) and Cook and Mullins (1983).

Upper Slope Facies (about 315 m):

Upper-slope facies of the upper Hales Limestone consist of about 90 percent of light gray hemipelagic and periplatform ooze derived from the shelf. Transport of ooze material from the shelf onto the slope was probably by tidal and seasonal currents and storms. Upper slope sediments are characteristically medium-bedded (2 to 8 cm thick) and contain a considerably higher degree of bioturbation than occurs in lower slope sediments.

Sediment-gravity-flow deposits comprise about 10 percent of the upper slope sequence. Thin-bedded turbidites, with constituents that were mainly derived from the shelf, are relatively common in upper slope facies;

Figure 17. In-situ slope deposits of black argillaceous lime mudstone and wackestone. This limestone has paper-thin laminations and is fissile when weathered. Lower part of the Hales Limestone, Upper Cambrian, Hot Creek Range, central Nevada. From Cook and Taylor (1977).

Figure 18. Oblique view of polished slab from Hot Creek Range section, showing (left of bar scale) articulated specimen of *Hedinaspis regalis* (Troedsson, 1937) and fabric of entombing deeper water sediment. Note thinly laminated parallel bedding and pervasive sponge spicules (small light-colored blebs) in black mud matrix. Larger white-appearing spherules are authigenic pyrite. Bar scale equals 1 cm. Hales Limestone (lower part), Upper Cambrian, Hot Creek Range, central Nevada. From Taylor and Cook (1976) and Cook and Taylor (1977).

debris-flow deposits are less abundant. Many of the redeposited sediments accumulated in a small perched basin on the upper slope (Cook 1983, Figs. 5-54 - 5-56).

An important subfacies, apparently restricted to a 10-m-thick interval of the upper slope facies, is well sorted, rippled contourite grainstones whose transport direction was to the northwest parallel to the paleoslope trend (Cook and Taylor, 1977, Figs. 42-44; Cook, 1983, Fig. 5-53; Cook and Mullins, 1983, Fig. 96). These contourites are composed of well-sorted, silt- to fine sand-sized, shoal-water derived algal (*Nuia*) grains.

Paleontology:

Cambrian and Lower Ordovician biostratigraphic correlations of the western United States are based mainly on trilobites and conodonts (see Taylor, 1989, Fig. 1-2, for a recent compilation of biostratigraphic zonal schemes). A summary of biostratigraphic correlations between the western North America and the Malyi Karatau are given by Ergaliev (1980, 1981).

In Tybo Canyon, Hot Creek Range, basin plain sediments of the upper 3 m of the Swarbrick Formation contain a trilobite assemblage consisting of *Aphelaspis subditus* Palmer, *Glyptagnostus reticulatus reticulatus* (Angelin), *Liostroa toxoura* Palmer, and *Pseudagnostus* sp. (Palmer, 1962, p. F43). The fauna is assigned to the *Aphelaspis* Zone of the Upper Cambrian Dresbachian Stage (Palmer, 1965, p. 14).

From the middle part of the Tybo Shale, Palmer (1965) reported *Simulolenus wilsoni* (Henningsmoen), *Simulolenus granulatus* (Palmer), *Morosa brevispina* Palmer, *Litocephalus granulomarginatus* Palmer, *Elviniella laevis* Palmer, *Elburgia quinnensis* (Resser), and *Erixanium multisegmentus* Palmer. The faunal assemblage is assigned to the *Dunderbergia* Zone of the upper Dresbachian Stage and represents an open ocean assemblage preserved in basin-plain sediments.

Trilobite occurrences in the Hales Limestone are of two types (Taylor, 1976, 1977; Taylor and Cook, 1976): (1) disarticulated and broken exoskeletal material that occurs as redeposited sediment in debris-flow and turbidite deposits; and (2) in-situ articulated dorsal exoskeletons and unbroken exuvae associated with fine-grained, dark-gray lime mudstone. Allochthonous faunas have paleozoogeographic affinities with the North America, whereas autochthonous faunas, characterized by *Hedinaspis* and *Charchaqia*, are characteristic of Late Cambrian ocean-margin and basin sites in Alaska, South Korea, eastern and southeastern China, northwestern China, and southern Kazakhstan (Lu, 1974; Taylor, 1976; Ergaliev, 1980, 1983; Apollonov and others, 1984). In Nevada, the *Hedinaspis-Charchaqia* fauna occurs most commonly in pelagic and hemipelagic sediments associated with the distal parts of the submarine-fan complex in the Hales Limestone (Fig. 18).

Known stratigraphic ranges of polymeroid trilobites and conodonts from the Hales Limestone in Tybo Canyon are summarized by in Cook and others (1989a).

Conodonts from the Swarbrick Formation, the Tybo Shale, and Hales Limestone are under study by J. F. Miller. Recovered conodonts are rare to common but occur in sufficient numbers to provide a workable biostratigraphic framework for correlation (for summary, see Cook and others, 1989a, Fig. 5-3). The oldest biostratigraphically useful conodonts occur 13 m above the base of the Hales Limestone and are assigned to the *Proconodontus muelleri* Zone of the lower-middle part of the Trempealeauan Stage. All of the younger Cambrian and basal Ordovician conodont zones and subzones have been recognized through the *Clavohamulus hintzei* Subzone of the *Cordylodus intermedius* Zone and the lower part of the overlying *Cordylodus lindstromi* Zone. Younger strata have not been sampled for conodonts. Conodonts of this age are geographically widespread, hence these strata can be correlated precisely to other parts of North America and to Central and East Asia, Australia, and Scandinavia (Miller, 1988; Dubinina, 1991).

Central Egan Range, Eastern Nevada

The Upper Cambrian and Lower Ordovician Whipple Cave Formation and the lowermost part of the overlying Lower Ordovician House Limestone comprise about 630 m of shoaling-upward marine carbonate rocks that evolved on the shallow-water part of the ramp (Fig. 5, Stage 1, and Fig. 9) (Cook and Taylor, 1977; Taylor and others, 1989). These rocks are composed of argillaceous lime wackestone, microbial stromatolite and thrombolite boundstone (bioherms and biostromes), algal (*Nuia*) grainstone, fenestral limestone, flat-pebble limestone conglomerate and breccia, and dolomitized lime mudstone (Fig. 19) (Cook and Taylor, 1977, Figs. 4-21). Environments ranged from low energy, shallow-subtidal, open-shelf settings to higher energy beaches and tidal-flats.

The conodont and trilobite biostratigraphy of the shoal-water carbonate rocks of the Whipple Cave Formation and House Limestone of the central Egan Range is summarized in Taylor and others (1989).

Malyi Karatau, Southern Kazakhstan

Joint field work was conducted in the Malyi Karatau region, southern Kazakhstan, in 1987 and 1990. The Malyi Karatau ("Little Black Mountains") is located near the edge of the lithospheric plate called Kazakhstania, which has been most-recently interpreted by Scotese and McKerrow (1990) as a microcontinent located between the paleocontinents of Siberia and Gondwana during the early Paleozoic (Fig. 1).

The lower Paleozoic stratigraphy and paleontology of the Malyi Karatau, and nearby Bolshoi Karatau ("Big Black Mountains"), have been described by Ergaliev (1980, 1981, 1983, 1989), in a multiauthored monograph edited by Apollonov and others (1983), Apollonov and others (1984), Dubinina (1991), Chugaeva and Apollonov (1982), Ergaliev and Kasymov (1984), Zaitsev and others, (1984), Zhemchuzhnikov

Figure 19. Algal buildups in middle part of Egan Range section. Massively-bedded bioherms in upper part of photograph are composed of coalescing, hemispherical microbial stromatolite and thrombolite boundstone. Thinly bedded lime grainstone layers lap over and interfinger with bioherms. Lime grainstone is composed of peloids and skeletal grains of trilobites, gastropods, echinoderms, and the possible alga *Nuia*. Upper Cambrian part of the Whipple Cave Formation, central Egan Range, eastern Nevada. From Taylor and Cook (1976) and Cook and Taylor (1977).

(1986), Apollonov and Zhemchuzhnikov (1988), and Sargaskaev (1988, 1990).

Integrated sedimentologic, paleontologic, and structural data suggest that the early Paleozoic carbonate platform of the Malyi Karatau formed on isolated microcontinental blocks which developed during Late Proterozoic (Riphean) rifting of continental crust (Cook and others, 1989b) (Fig. 6). V. G. Zhemchuzhnikov (personal communication, 1987) first suggested that the Malyi Karatau carbonate platform may be an isolated platform. Our joint field studies in 1987 confirmed Zhemchuzhnikov's earlier proposal that the carbonate sediment accumulated on one isolated block which was the site of the 40 km-wide Aisha Bibi seamount (Cook and others, 1989b) (Fig. 20-22). Virtually no siliciclastic sediments accumulated with the carbonate sediments of the seamount, which supports an interpretation that the carbonate platform was isolated from a siliciclastic source, rather than attached to a craton as suggested by Sovetov (1990). The shoal-water seamount platform was flanked by deeper-water submarine fans, slides, and slumps. The Aisha Bibi seamount, which is thought to be the oldest Paleozoic seamount reported, may be part of an early Paleozoic archipelago extending from the Karatau region of southern Kazakhstan southeastward into the Tien Shan foldbelt of northwestern China (M. K. Apollonov, personal communication, 1987).

Kyrshabakty and Batyrbay Sections, Malyi Karatau

Our interpretations are based on the study of eight coeval and partly coeval stratigraphic sections over a broad area of the Malyi Karatau and mapping of key horizons between the sections (Fig. 21). The following summary is largely based on the Kyrshabakty and Batyrbay sections which are excellent examples of well exposed, upward-shallowing and seaward-prograding sequences at the seamount margin (Figs. 22, 23). From Middle Cambrian through the Lower Ordovician both sections shoal upward from basin-plain facies to seamount-interior facies. The rocks comprise the Cambrian and Early Ordovician Shabakty Suite. With the exception of the Dongulek section which also shoals upward from basin-plain to seamount-interior facies the other five stratigraphic sections consist entirely of slope or seamount-interior facies (Fig. 22).

The Aisha Bibi seamount margin at the Kyrshabakty and Batyrbay sections trends northeast (present-day coordinates). Determination of this trend is based on (1) paleocurrent data in the fan facies at Kyrshabakty and Batyrbay which indicate transport direction to the northwest and, (2) paleontological data which show that the seamount margin and submarine fan prograded seaward from Kyrshabakty to Batyrbay, a distance of about 5 km to the northwest (Figs. 21, 22); the seamount margin and submarine fan is younger at Batyrbay than at Kyrshabakty (Fig. 22). Paleocurrent directions in the submarine fan facies at Dongulek indicate a southeast transport direction suggesting that the Dongulek fan was forming on the opposite southeastern slopes of the seamount. All of the coeval stratigraphic sections between the seamount-margin sections of Dongulek, Batyrbay, and Kyrshabakty are seamount-interior sections that consist entirely of shallow-subtidal and tidal-flat facies (Figs 21, 22; stratigraphic sections 2-5.).

Sovetov (1990) interprets the Dongulek section as late Precambrian in age rather than Early Ordovician. Although paleontologic data are few, chitinozoans reported by Zhemchuzhnikov and others (1989) support an Early Ordovician age for the Dongulek section.

Basin-Plain Facies (about 100 m):

Basin-plain facies are characterized by 0.5 to 5 cm-thick black lime mudstone, lime wackestone, and argillaceous lime mudstone. The rocks have millimeter-thick laminations and a virtual absence of bioturbation (Fig. 24). Locally these facies contain abundant sponge spicules and an occasional in-situ complete sponge. A minor lithology in the basin-plain facies is black fissile shale with millimeter-thick laminations and an absence of burrowing. Thin-bedded (0.5-10 cm thick) lime wackestone turbidites also occur in the basin-plain sequence. Turbidites comprise less than 10 percent of the basin-plain facies and are randomly distributed stratigraphically. They contain shoal-water-derived ooid and algal

858

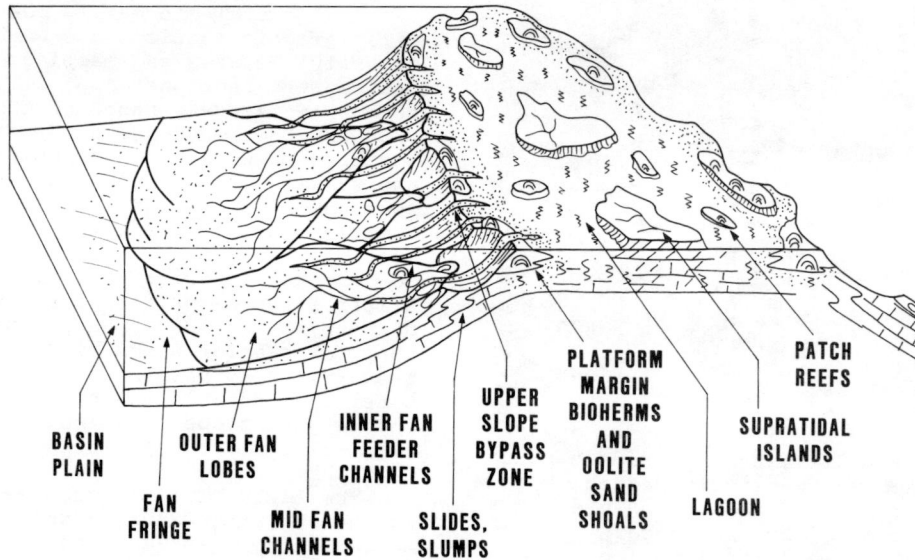

АЙША БИБИ   SEAMOUNT - KAZAKHSTAN, USSR
AISHA BIBI

BASIN PLAIN · FAN FRINGE · OUTER FAN LOBES · MID FAN CHANNELS · INNER FAN FEEDER CHANNELS · SLIDES, SLUMPS · UPPER SLOPE BYPASS ZONE · PLATFORM MARGIN BIOHERMS AND OOLITE SAND SHOALS · LAGOON · SUPRATIDAL ISLANDS · PATCH REEFS

Figure 20. Model of Aisha Bibi seamount, showing its depositional environments and associated lithofacies. Seamount interior is about 40 km-wide. Carbonate submarine fans extend out into the basin at least 5 km. Submarine fan facies derived from episodic seamount-margin collapse during sea-level lowstands and from numerous small line-source channels at the seamount margin. Scalloped seamount-margin morphology results from seamount margin collapse events. Model is based on stratigraphic sections shown in figures 21 and 22. Cambrian and Early Ordovician, Shabakty Suite, Malyi Karatau, southern Kazakhstan, U.S.S.R. From Cook and others, 1989b.

grains, and exhibit Bouma Tad and Tde divisions.

Lower Slope and Basin-Plain Facies (465 m):

The carbonate submarine fan in Malyi Karatau is strikingly different in its lithofacies types in comparison to coeval carbonate submarine fan facies in the Hot Creek Range of Nevada. This is a result of a different paleotectonic origin of the passive continental margin in Nevada versus that of southern Kazakhstan. During the early Paleozoic in the western United States the continental margin had the profile of a distally steepened ramp and the fan facies were mainly derived from the remolding of slope-generated submarine slides. Thus, the fan facies consists mainly of black and gray lime mudstone clasts with lesser amounts of shoal-water derived debris. In contrast, the paleotectonic origin of southern Kazakhstan was such that the central Asian continental margin was broken into a series of large isolated blocks during rifting and the surfaces of the blocks became the loci for shallow-marine carbonate sedimentation. The

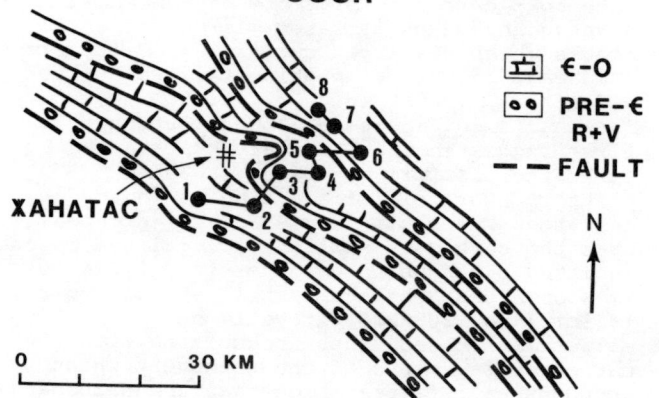

MALYI KARATAU KAZAKHSTAN USSR

ХАНАТАС

€-O
PRE-€ R+V
FAULT
N

0   30 KM

Figure 21. Generalized geologic map of the Malyi Karatau region, southern Kazakhstan, U.S.S.R.. Beds dip steeply to the northeast. Numbers 1 through 8 are the locations of the measured stratigraphic sections across the Aisha Bibi Seamount as shown in figure 22. These sections are: (1) Dongulek, (2) Berkuty, (3) Artugi, (4) Ausokan-Lower, (5) Ausokan-Upper, (6) Kryshabakty, (7) Batyrbay, and (8) Batbaluk. R, Riphean; V, Vendian.

Figure 22. Generalized stratigraphic cross-section through the Aisha Bibi seamount, Malyi Karatau, southern Kazakhstan, U.S.S.R.. The map locations and names of the eight sections are shown in figure 21. Late Proterozoic non-marine siliciclastics (solid black circles) overlain by Lower Cambrian tidal-flat dolomites (horizontal wavy pattern). As the isolated block subsided, the seamount facies developed. Sections 2-5 are tidal flat and lagoon facies of the seamount interior; sections 1, 6 and 7 are seamount-margin facies that shoal-upward from basin-plain facies to submarine-fan facies to seamount-margin facies; section 8 is an upper-slope facies. The seaward progradation of the seamount margin and submarine fan are well documented by sections 6 and 7. R, Riphean; V, Vendian.

Figure 23. Kryshabakty section (section #6 on figure 22). Northeast margin of the Aisha Bibi seamount, Malyi Karatau, southern Kazakhstan, U.S.S.R. View toward southeast. Beds dip 80-90 degrees to the left, upsection, showing the section shoaling up from a basin-plain facies (Middle Cambrian), to a carbonate submarine fan (Middle Cambrian and Upper Cambrian), to a seamount-margin facies (uppermost Cambrian and Lower Ordovician) over a 650-m-thick interval. The approximate boundary between the facies is indicated the dashed lines. Each facies can be traced laterally along strike for several kilometers. Lower Cambrian tidal-flat dolomites form a prominent ridge (upper right corner of photograph).

Figure 24. Basin-plain facies. Black, laminated, argillaceous lime mudstones virtually devoid of bioturbation. Kryshabakty section (section #6 on figure 22). Northeast margin of the Aisha Bibi seamount, Malyi Karatau, southern Kazakhstan, U.S.S.R.

Aisha Bibi submarine fan facies is mainly derived from material that originated on the shoal-water seamount-margin with lesser amounts of debris from slope settings.

The origin of the submarine fan in the Malyi Karatau is similar to that of the Hot Creek Range in one very intrinsic aspect. Allochthonous debris in both fans is the result of episodic mass-failure events during the Late Cambrian and Early Ordovician. Repeated episodes of platform-margin failure in the Malyi Karatau resulted in downslope transport of massive amounts of debris. Likewise, in the Hot Creek Range, coeval deeper-water slope failures contributed the majority of debris in the fan facies.

Outer-Fan Fringe Facies (50 m):

The outer-fan fringe marks that interval in the section where carbonate turbidites become abundant. This part of the section consists of about equal amounts of thinly bedded (5-15 cm thick) calcarenite turbidites with Bouma Ta divisions and in-situ argillaceous lime mudstones similar to those in the underlying basin-plain facies. Like the basin-plain turbidites, outer-fan turbidites contain shoal-water-derived ooid and algal grains.

Outer-Fan Lobe Facies (175 m):

The outer-fan lobe facies contains about 25 upward-thickening turbidite cycles (Fig. 25). These cycles range from 1 to 5 m thick with the basal beds 2 to 5 cm thick and the upper beds ranging from 10 to 40 cm thick. Individual turbidites show Bouma Ta, Tab, or Tbc divisions without significant evidence of

Figure 25. Outer-fan lobe facies, showing one of about 25 upward-thickening turbidite cycles. Stratigraphic up to right. Cycle is about 2 m thick. Pick for scale. Kryshabakty section (section #6 on figure 22). Northeast margin of the Aisha Bibi seamount, Malyi Karatau, southern Kazakhstan, U.S.S.R. View toward northwest.

In this interval, the underlying sediment-gravity-flow deposits of the outer fan lobe facies comprise 50 to 90 percent of the section whereas the remainder of the facies is in-situ lime mudstone and wackestone. The first evidence of burrowing in the in-situ beds occurs midway through the mid-fan distributary channel and interchannel facies.

The channelized conglomeratic debris flows are interpreted to represent meandering distributary channel deposits. The thinner-bedded calcarenite turbidites at the tops of the cycles probably represent lateral margins of the channels and interdistributary-channel turbidites.

At the transition between the Sackian and Malykaratauian Stages of Ergaliev (1981, Fig. 2), several separate submarine slides 5 m thick and up to 2 km wide can be traced along strike (Fig. 26). The slides exhibit virtually no internal deformation and were well lithified prior to their transport downslope. The submarine slides contain oolite packstone and grainstone, which show that large segments of the shoal-water seamount margin collapsed and were transported downslope into deeper water environments. Field relations suggest these slides were transported at least 5 km downslope into basin-plain settings.

channeling at the base of the beds. Turbidites in this interval comprise 50 to 90 percent of the section with the remainder of the sediment being in-situ, unburrowed, black, argillaceous lime mudstone and wackestone.

The lower parts of the cycles are calcarenite packstone turbidites with sand-sized grains of ooids, algae, and peloids. The uppermost thick beds in the cycles are commonly imbricated flat-pebble breccia that may represent transport by viscous turbidity-currents or debris flows. The breccia is a combination of slope-derived lime mudstone clasts and seamount-margin-derived lime packstone and grainstone. The clasts are up to 10 cm across and immersed in a matrix of oolite and dark gray lime mudstone.

Mid-Fan Distributary Channel Facies (105 m):

The mid-fan distributary channel facies contains about 10 upward-thinning turbidite cycles. The cycles range from 5 to 10 m thick, with the basal beds being up to 4 m thick. The basal beds are channelized and have widths from 25 to 100 m. In the thalwegs of the channels, the deposits contain clasts of *Epiphyton* cyanobacterial boundstones, oolite grainstone, and basinal lime mudstone up to 0.2 by 1.0 m. Clasts are in a pervasive dark lime mudstone and ooid-rich matrix. This conglomerate and breccia represents debris flows or viscous turbidity flows. The thin calcarenite turbidite beds at the tops of the cycles show Bouma Ta and Tab divisions. Small-scale slumping at the margins of the channels is locally evident.

Figure 26. Late Cambrian submarine slide 5 m-thick and about 1.5 km wide within the mid-fan facies. Stratigraphic top to the left. Slide represents well-lithified segment of seamount margin grainstone and packstone facies that collapsed and moved down slope. Slide occurs at the transition between the Sackian and Malykaratauian Stages of Ergaliev (1981, fig. 2), which correlates with the *Elvinia-Taenicephalus* Zone transition in North America. Slide is one of several at this age horizon. Kryshabakty section (section #6 on figure 22). Northeast margin of the Aisha Bibi seamount, Malyi Karatau, southern Kazakhstan, U.S.S.R. View toward southeast.

Figure 27. Inner-fan feeder-channel facies. Next to cross mark (+) in upper left of photograph are two men standing on a 10 m-thick, massively bedded, megabreccia debris-flow deposit. This channelized deposit is bracketed by arrows (>> <<). Inner-fan feeder channel can be traced laterally along strike for about 500 m at which point it depositionally pinches out. Stratigraphic up to left. Kryshabakty section (section #6 on figure 22). Northeast margin of the Aisha Bibi seamount, Malyi Karatau, southern Kazakhstan, U.S.S.R. View toward southeast.

Figure 28. Inner-fan feeder-channel facies. Man at bottom of photograph is standing in front of an allochthonous *Epiphyton* boundstone clast which is within a 15 m-thick debris flow channel deposit. This boulder-sized clast is 10 x 12 m (outlined by black dashed line). Clast is part of a *Epiphyton* bioherm displaced from the seamount margin during an interpreted sea-level lowstand. Stratigraphic up to right. Batyrbay section (section #7 on figure 22). Northeast margin of the Aisha Bibi seamount, Malyi Karatau, southern Kazakhstan, U.S.S.R. View toward northwest.

Inner-Fan Feeder-Channel and Submarine Slide Facies (135 m):

This 135-m-thick interval contains about 10 inner-fan feeder channels. The channels, up to 15 m deep and 500 m wide, are filled with disorganized boulder-bearing conglomerate and megabreccia (Fig. 27). Clasts are up to 10 by 12 m. The boulder-sized clasts are oolite grainstone and *Epiphyton* cyanobacterial boundstone, which in turn are immersed in a matrix of lime mudstone and shoal-water-derived grains (Figs. 28-31). The immense volume of shoal-water debris in the channelized facies suggests the debris was generated during times of catastrophic failure of the seamount margin.

In contrast to the Nevada submarine fan, very few slope-derived translational slides are still recognizable as intact slides in the Malyi Karatau section. Virtually all of the slope-derived slides completely remolded into individual debris flows or were incorporated into debris flows that originated at the seamount margin.

The inner-fan interval consists of at least 80 to 90 percent sediment-gravity flow deposits. The remaining interbedded, in-situ sediments are bioturbated lime mudstone and wackestone which represent periplatform ooze derived from the seamount margin and interior.

Upper Slope Facies (about 60 m):

The upper slope sequence consists of about equal amounts of in-situ light gray periplatform ooze derived from the seamount margin and interior and thin-bedded carbonate turbidites and debris flows. It is further characterized by a considerably higher degree of bioturbation than observed in lower slope sediments.

Seamount Margin and Interior (375+ m):

The oceanward margins of the seamount consist of *Epiphyton* bioherms up to 20 m thick and 50 m wide (Figs. 32, 33). Areally extensive mobile ooid-sand shoals occur between and directly behind bioherms (Figs. 34, 35). Seamount interior environments were characterized by shallow-subtidal lagoons with scattered, small *Epiphyton* mounds and broad tidal flats with shallow subtidal to supratidal environments (Figs. 36-39).

Paleontology:

The early Paleozoic paleontology of the Malyi Karatau and nearby Bolshoi Karatau have been described by Ergaliev (1980, 1981), in a multiauthored monograph edited by Apollonov and others (1983), Apollonov and others (1984), Chugaeva and Apollonov (1982),

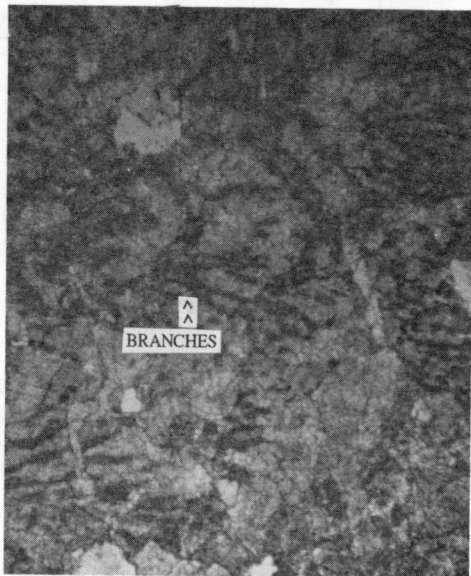

Figure 29. Inner-fan feeder-channel facies. Photomicrograph from a 4 x 12 m *Epiphyton* boundstone clast within a megabreccia channel deposit. Stratigraphic top of the channel is toward the top of the photo. Growth orientation of the *Epiphyton* branches shows that they are not in their original growth positions; *Epiphyton* branches were growing from the left side toward the right side of the photograph. Thus the biohermal clast is lying on its side in the channel deposit. *Epiphyton* branches are about 0.04-0.08 mm in width. Kryshabakty section (section #6 on figure 22). Northeast margin of the Aisha Bibi seamount, Malyi Karatau, southern Kazakhstan, U.S.S.R.

Figure 30. Inner-fan feeder-channel facies, showing part of a 15 m-thick megabreccia debris-flow deposit and displaying the variety of clasts and textural appearance of the debris. Large ovoid-shaped clasts are oolite grainstones that originated in shoal waters on the seamount margin; smaller tabular-shaped clasts are laminated lime mudstones that formed in deeper waters on the slope. Not in photo but in this bed is a oolite grainstone boulder 9 x 10 m. As the debris flow moved downslope from the seamount margin it eroded the partially lithified slope sediment and incorporated a variety of clasts in the flow. Batyrbay section (section #7 on figure 22). Northeast margin of the Aisha Bibi seamount, Malyi Karatau, southern Kazakhstan, U.S.S.R.

Ergaliev and Kasymov (1984), Zhemchuzhnikov (1986), Apollonov and Zhemchuzhnikov (1988), Ergaliev (1989), and Dubinina (1991). Biostratigraphic correlations between the Malyi Karatau and western North America are presented by Ergaliev (1980, 1981; Apollonov and others, 1984).

Upbuilding of the seamount from basin plain to carbonate platform interior took place from the Middle Cambrian to the Early Ordovician (Arenig). By the Arenig, the seamount was a major topographic feature, standing about 1,000 m above the adjacent basin plain of the paleo-Asian ocean.

The shallowing upward, 1000-m-thick Batyrbay section records: (1) a Middle Cambrian anaerobic to dysaerobic basin plain; (2) a Middle and Late Cambrian carbonate submarine fan developed in dysaerobic waters of the base-of-slope and lower slope; (3) a Late Cambrian and Early Ordovician aerobic upper slope; and (4) an Early Ordovician (Arenig) well-oxygenated shoal-water seamount platform.

Fossils of the dysaerobic basin plain consist mainly of sponge spicules, pelagic (Robison, 1972) agnostoid trilobites, and rare benthic ostracodes (L. M. Mel'nikova, unpublished data). Ichnofossils are generally absent, suggesting anaerobic conditions existed below the sediment-water interface. The few ichnofossils that occur are associated with thin-bedded distal turbidites and probably reflect minor influx of oxygenated turbidity currents.

Trilobites, inarticulate brachiopods, and conodonts are increasingly more common up section in the fan-lobe and interchannel facies. Fossils occur as both in-situ and redeposited exoskeletal debris in thin- to medium-bedded turbidites. The inner fan distributary channels and slide deposits contain redeposited clasts of *Epiphyton* and unidentified shelly fossil debris. Redeposited ostracodes occur rarely in the interchannel facies.

Conodonts occur in the Upper Cambrian part of the submarine carbonate fan facies and Lower Ordovician (Arenig) seamount margin and seamount interior facies. Assemblages include mixtures of warm and cool biofacies, suggesting the Aisha Bibi seamount had unrestricted access to the open ocean and was exposed to both warm- and cool-water oceanic currents (Dubinina, 1991).

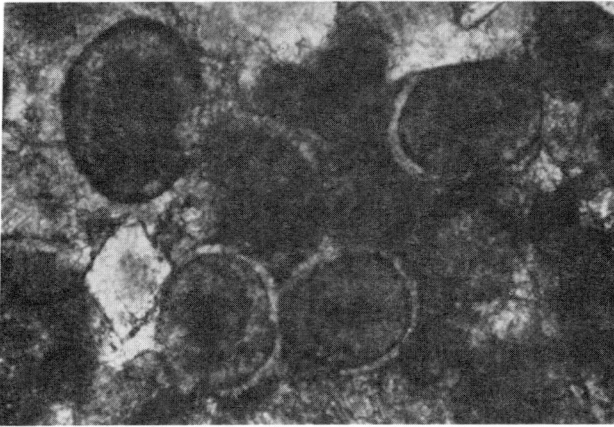

Figure 31. Inner-fan feeder-channel facies. Photomicrograph of an oolite grainstone clast in a 12 m-thick megabreccia debris-flow deposit. Ooids are about 0.5 mm in diameter. Kryshabakty section (section #6 on figure 22). Northeast margin of the Aisha Bibi seamount, Malyi Karatau, southern Kazakhstan, U.S.S.R.

Figure 33. Seamount margin. Photomicrograph of in-situ *Epiphyton* cyanobacterial seamount-margin bioherm shown in figure 32. Stratigraphic top of the section is toward the top of the photo. Note the upward growth pattern of these *Epiphyton* branches as compared to the orientation of the *Epiphyton* branches in the allochthonous bioherm block in figure 29. *Epiphyton* branches are about 0.04-0.08 mm in width. Batyrbay section (section #7 on figure 22). Northeast margin of the Aisha Bibi seamount, Malyi Karatau, southern Kazakhstan, U.S.S.R.

Figure 32. Seamount margin. In-situ *Epiphyton* cyanobacterial bioherm about 10 m-thick and 50 m wide. Upper 1 m of bioherms at the seamount margin exhibit extensive karsting. Solution voids in the karsted bioherm are filled with partially dolomitized limestone breccia (light-colored patches in the upper half of the photograph). Stratigraphic top of bioherm (not shown in photograph) is onlapped by laminated supratidal dolomite. Early Ordovician, Batyrbay section (section #7 on figure 22). Northeast margin of the Aisha Bibi seamount, Malyi Karatau, southern Kazakhstan, U.S.S.R.

Figure 34. Seamount margin, showing beds dipping about 90 degrees. Stratigraphic top to right. White scale bar 15 cm long. In center of photograph are calcarenite sand waves with heights of 0.8 m and lengths of 9.7 m. Three prominent foreset beds shown in the center of the photograph are dipping 25-30° towards the northwest (into photograph). Oolite sand waves moving in a N 45 W direction (into photograph) which is the approximate direction that the seamount margin is prograding. Kryshabakty section (section #6 on figure 22). Northeast margin of the Aisha Bibi seamount, Malyi Karatau, southern Kazakhstan, U.S.S.R.

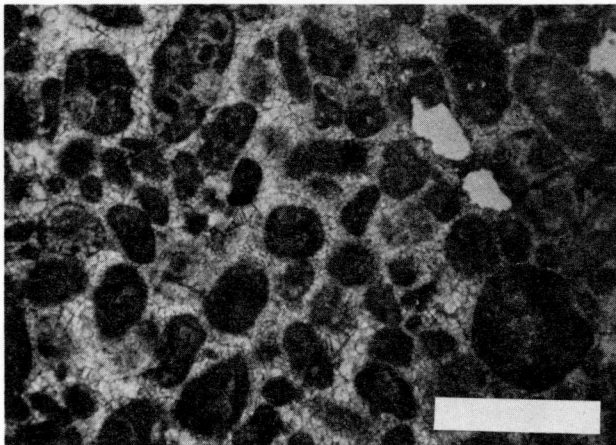

Figure 35. Seamount margin.
Photomicrograph of calcarenite grainstone in
figure 34. Constituents include ooids,
peloids, grapestones, *Epiphyton* clasts, and
lime mudstone lithoclasts. Bar scale 1.0 mm.
Kryshabakty section (section #6 on figure 22).
Northeast margin of the Aisha Bibi seamount,
Malyi Karatau, southern Kazakhstan, U.S.S.R.

Figure 37. seamount interior, showing
close-up view of one of the cycles in figure
36. Basal part of cycle is a white, laminated,
supratidal dolomitized lime mudstone. This is
gradationally overlain by a dark-brown,
burrowed, dolomitized, shallow subtidal
wackestone. The upper surface of the burrowed
facies often exhibits karsting and solution
breccia. Ausokan-Lower section (section #4 on
figure 22). Interior part of the Aisha Bibi
seamount, Malyi Karatau, southern Kazakhstan,
U.S.S.R.

Figure 36. Seamount interior, showing
shallow-subtidal lagoon and tidal-flat facies
with numerous 0.5 to 1.5 m thick cycles. Many
cycles consist of a basal gray to white,
laminated supratidal dolomite or dolomitic
flat-pebble breccia overlain by dark-brown,
burrowed, shallow-subtidal limestone. This
limestone is commonly dolomitized. Ausokan-
Lower section (section #4 on figure 22).
Interior part of the Aisha Bibi seamount,
Malyi Karatau, southern Kazakhstan, U.S.S.R.

Figure 38. Seamount interior, showing
dolomitized flat-pebble breccias that formed
in tidal flats. Ausokan-Lower section (section
#4 on figure 22). Interior part of the Aisha
Bibi seamount, Malyi Karatau, southern
Kazakhstan, U.S.S.R.

Figure 39. Seamount interior. Solution breccia in the upper part of one of the cycles in figure 36. Rock is a dark-brown, burrowed, bioclastic wackestone that is dolomitized. Ausokan-Lower section (section #4 on figure 22). Interior part of the Aisha Bibi seamount, Malyi Karatau, southern Kazakhstan, U.S.S.R.

DISCUSSION AND CONCLUSIONS

In summary, preliminary comparison of the early Paleozoic passive-margin carbonate platforms in the western United States and southern Kazakhstan have led to the following conclusions:

(1) The four geologic controls primarily responsible for the development and paleogeography of the Nevada and Kazakhstan passive-margin carbonate platforms include: (a) low paleolatitude locations; (b) the style of tectonic rifting and spreading; (c) post-rift thermal subsidence; and (d) sea-level changes.

(2) The tectonic style of rifting and spreading strongly influenced the type of passive continental margins that were initiated. Both the Nevada and southern Kazakhstan carbonate platforms developed at low paleolatitudes on rifted Precambrian continental crust adjacent to the proto-Pacific Ocean (Fig. 1). However, the style of passive-margin rifting between the two areas was different. During rifting and spreading of Laurentia, the western margin of North America developed as a relatively stable, passive continental margin hundreds of kilometers wide and attached to the craton (Figs. 2, 3). By the Late Cambrian, the continental margin had the profile of a non-rimmed distally steepened carbonate ramp (Fig. 5, Stage 1). In central Nevada, the platform-margin facies developed in relatively deep water below the photic zone and beneath normal storm wave-base. Coeval shoal-water facies were restricted to the interior parts of the distally-steepened ramp hundreds of kilometers to the east. By Late

Cambrian time a sinuous but overall north-northeast-trending passive-margin carbonate platform existed in central Nevada (Fig. 4).

In contrast, during rifting and spreading in southern Kazakhstan, the passive margin did not develop as a broad continental margin attached to the craton. Rather, the Precambrian crust was broken into isolated blocks (Fig. 6). In southern Kazakhstan, one shoal-water carbonate platform formed on an isolated block. A variety of narrow shallow-subtidal and tidal-flat facies belts developed inboard from the shoal-water platform-margin. The feature, named the Aisha Bibi seamount (Fig. 20), may be part of an early Paleozoic archipelago that extended from southern Kazakhstan southeastward into China along an early Paleozoic plate suture preserved in the Tien Shan foldbelt.

(3) Thermally controlled subsidence during rifting and spreading is considered to have been the dominant geologic control that provided accommodation space for each passive-margin carbonate platform to develop. In Nevada, sediment accumulation on the Late Cambrian and earliest Ordovician carbonate platform and platform margin kept pace with thermal subsidence by building a prograded sediment wedge at least 1,000 m thick.

During thermal subsidence in the Malyi Karatau, the oceanward margins of the seamount became the sites of shoal-water carbonate deposition that kept pace with post-rift subsidence, resulting in the seamount accumulating at least 1,000 m of carbonate rocks during the Middle and Late Cambrian and Early Ordovician. Seamount margins prograded at least 5 kilometers oceanward during upbuilding of the seamount interior (Fig. 22). By Early Ordovician (Arenig) time, the seamount was a major topographic feature standing as much as 1,000 meters above the adjacent basin plain of the paleo-Asian ocean. Post-rift subsidence was terminated in the Malyi Karatau during the Middle Ordovician when the region underwent tectonic compression and uplift (Abdulin and others, 1990). In western North America subsidence continued over 200 m.y. (Cambrian through Devonian) until it was terminated by uplift during the Late Devonian to Early Mississippian Antler orogeny (Poole, 1974; Armin and Mayer, 1983).

(4) In the last several years geologists have debated the effects of sea-level highstands versus lowstands on the frequency of turbidity currents generated at carbonate platform margins (for example, Mullins, 1983a; Shanmugam and Moiola, 1982, 1983; Droxler and Schlager, 1985). There is a growing consensus that most small-scale carbonate turbidites are generated at platform margins during highstands of sea level. The carbonate platforms in the Bahamas shed considerably more sediment into the surrounding basins during the early phase of a sea-level high than during sea-level lows (Droxler and Schlager, 1985; Wilber and others, 1990). Much of this carbonate sediment is transported down the Bahama platform slopes by turbidity currents (Droxler and Schlager, 1985) where it forms carbonate aprons (Mullins and Cook, 1986).

In contrast to small-scale carbonate turbidites, occasionally large-scale collapse of carbonate platform margins and slopes occurs and immense volumes of material move downslope as submarine slides and megabreccia debris flows (for example, Cook and others, 1972, 1987; Davies, 1977; Crevello and Schlager, 1980; Johns and others, 1981; Cook and Mullins, 1983; Mullins and others, 1986; Bosellini, 1989; McNeill, 1990; Yose and Hardie, 1990; Cook and Taylor, 1991b). Such catastrophic events can form sequences tens to hundreds of meters thick and cover hundreds of square kilometers. These spectacular deposits can be triggered by earthquakes or tsunamis (Cook and others, 1972; Johns and others, 1981). However, when such deposits are regionally widespread or on separate lithospheric plates, and at times of sea-level lowering, the trigger mechanism is most likely eustatic sea-level fluctuations. Crevello and Schlager (1980) and McNeill (1990) propose that the mass-transport events they have studied in the Bahamas occurred during a sea-level lowering. Mullins and others (1986) suggest that minor sea-level falls during the overall middle Miocene highstand could have triggered collapse of the west Florida carbonate platform margin. Bosellini (1989) in his lucid paper on Cretaceous carbonate platforms around the Mediterranean makes a strong case for platform margin collapse during sea-level lowstands. During these lowstands large parts of the platform margins collapsed and were transported downslope as megabreccias. Platform-interior sites exhibit general emersion surfaces with associated bauxitic horizons (Bosellini, 1989).

Yose and Hardie (1990) caution the need to correctly differentiate between (1) platform-margin collapse associated with sea-level falls, and (2) platform-margin collapse related to overproduction and slope instabilities generated during sea-level rises. Evidence of faunal hiatuses and karsting in shoal-water platform-interior sites as well as the influx of siliciclastics (where applicable) across the platform can provide additional criteria to establish the connection between sea-level lowering and platform-margin collapse.

We propose that sea-level lowstands were the triggering mechanism for large-scale collapse of carbonate platform margins and slopes in our study. Stratigraphic sections in both the western United States and Malyi Karatau, southern Kazakhstan, record episodes of sea-level lowstands. The lowstands, which we interpret to be eustatic in origin, are recognized by seaward collapse of large segments of platform margins and slopes, and solution breccias and faunal discontinuities in shoal-water sites of platform interiors. Such interpretations can be tested by recognition of coeval evidence for low relative sea-level positions on platforms in different paleotectonic settings. For example, one episode of massive-scale submarine sliding in the Malyi Karatau, discussed above, reflects a seamount-margin collapse near the transition between the Sackian and Malykaratauian Stages of Ergaliev (1981, Fig. 2). This horizon correlates with the *Elvinia-*

Taenicephalus Zone transition in North America. The evidence for coeval sea-level lowstand in North American at this time is recorded by an increase in frequency and coarseness of carbonate debris flows in the lower part of the deep water Hales Limestone, excursion of siliciclastic sediment onto the carbonate platform in central and western Utah (Worm Creek Quartzite Member of the St. Charles Formation and Corset Spring Shale Member of the Orr Formation), and occurrences of limestone breccia in deep water shale in Arkansas (Hart and others, 1987), which we interpret to represent erosion or collapse of a carbonate platform margin.

We concur with Crevello and Schlager (1980) and Shanmugam and Moiola (1983) that during the *initial* stage of sea-level lowering, slope and/or platform-margin collapse can occur. It is at this period when lowered wave base destabilizes the upper slopes and platform margins and sediments are most susceptible to collapse. Thus, collapse during a sea-level lowering may result from gravitational instability of partially cemented carbonate sediments (Cook and others, 1987; Cook and Taylor, 1991b).

ACKNOWLEDGEMENTS

We thank John D. Cooper and Brian D. Edwards who kindly reviewed our manuscript and offered many positive suggestions for its improvement. Arcady V. Zakharov was very helpful with technical aspects of the study. Rarely have we had the opportunity to work with such a patient and encouraging editor as John Cooper. His support and gentle prodding are greatly appreciated. This report is an outgrowth of a cooperative research project sponsored by the U.S. Geological Survey and the National Academy of Sciences/Academy of Sciences of the U.S.S.R. Soviet Exchange Program. H. E. Cook and M. E. Taylor conducted field work in Kazakhstan in 1987 and 1990 in collaboration with the Soviet coauthors. The authors thank Academician A. A. Abdulin, Director, K. I. Satpaev Institute of Geological Sciences, Academy of Sciences of the Kazakh S.S.R., Alma-Ata, Kazakhstan, U.S.S.R.; and Dr. A. Yu. Rozanov, Vice-Director, Palaeontological Institute, Academy of Sciences of the U.S.S.R., Moscow, for their help in facilitating the research project.

REFERENCES CITED

Abdulin, A. A., Apollonov, M. K., and Ergaliev, G. Kh., eds., 1990, Malyi Karatau, Kazakh SSR Excursion: Field Trip Guidebook, Excursion 2, 3rd International Symposium on the Cambrian System: International Subcommission on the Cambrian System and International Commission on the Stratigraphy of the International Union of Geological Sciences, Academy of Sciences of the Kazakh SSR, Alma-Ata, 61 p.

Abdulin, A. A., Chimbulatov, M. A., Azerbaev, N. A., Ergaliev, G. Kh., Kasymov, M. A., and Tsirelson, B. S., 1986, Geologiya i metallogeniya Karatau, v. 1, Alma-Ata, 239 p. (In Russian).

Allmendinger, R. W., Haugue, T. A., Hauser, E. C., Potter, C. J., Klemper, S. L., Nelson, K. D., Knuepfer, P., and Oliver, J., 1987, Overview of the COCORP 40° N Transect, western United States: The fabric of an orogenic belt: Geological society of America Bulletin, v. 98, p. 308-319.

Apollonov, M. K., and Zhemchuzhnikov, V. G., 1988, Lithostratigraphy of the Batyrbay Cambrian-Ordovician boundary section, Malyi Karatau: Izvestiya Academii Nauk KazakhSSR, Seriya Geol., n. 1, p. 22-36 (In Russian).

Apollonov, M. K., Bandaletov, W. M., and Ivshin, N. K., eds., 1983, The lower Paleozoic stratigraphy and palaeontology of Kazakhstan: Academy of Sciences of the Kasakh SSR, Alma-Ata, 176 p., 34 pls. (A collection of separately authored reports in Russian on stratigraphy, trilobites, graptolites, conodonts, problematica, brachiopods, and corals).

Apollonov, M. K., Chugaeva, M. N., and Dubinina, S. V., 1984, Trilobites and conodonts from the Batyrbay section (uppermost Cambrian-Lower Ordovician) in Malyi Karatau Range (atlas of the palaeontological plates): Academy of Sciences of the Kazakh SSR, "Nauka" Kazakh SSR Publishing House (Alma-Ata), 32 pls. (In English).

Armin, R. A., and Mayer, L., 1983, Subsidence analysis of the Cordilleran miogeocline: Implications for timing of late Proterozoic rifting and amount of extension: Geology, v. 11, p. 702-705.

Azerbaev, N. A., 1988, The structural-facies zonation of Bolshoi Karatau: Vestnik Academii Nauk KazakhSSR (Alma-Ata), n. 6, p. 37-42 (In Russian).

Bally, A. W., 1987, Phanerozoic basin evolution in North America: Episodes, v. 10, p. 248-253.

Bond, G. C., Kominz, M. A., and Devlin, W. J., 1983, Thermal subsidence and eustasy in the lower Paleozoic miogeocline of western North America: Nature, v. 306, p. 775-779.

Bond, G. C., and Kominz, M. A., 1984, Construction of tectonic subsidence curves for the early Paleozoic miogeocline: Implications for subsidence mechanisms, age of breakup, and crustal thinning: Geological Society of America Bulletin, v. 95, p. 155-173.

Bond, G. C., Christie-Blick, N., Kominz, M. A., and Devlin, W. J., 1985, An Early Cambrian rift to post-rift transition in the Cordillera of western North America: Nature, v. 315, p. 742-746.

Bond, G. C., Kominz, M. A., Steckler, M. S., and Grotzinger, J. P., 1989, Role of thermal subsidence, flexure, and eustasy in the evolution of early Paleozoic passive-margin carbonate platforms, in Crevello, P. D., Wilson, J. L., Sarg, J. F., and Read, J. F., eds., Controls on carbonate platform and basin development: Society of Economic Paleontologists and Mineralogists Special Publication 44, p. 39-61.

Bosellini, A., 1989, Dynamics of Tethyan carbonate platforms, in Crevello, P. D., Wilson, J. L., Sarg, J. F., and Read, J. F., eds., Controls on carbonate platform and basin development: Society of Economic Paleontologists and Mineralogists Special Publication 44, p. 3-13.

Burchfiel, B. C., and Davis, G. A., 1972, Structural framework and evolution of the southern part of the Cordilleran orogen, western United States: Am. Jour. Sci., v. 272, p. 97-118.

Burrett, C. F., 1974, Plate tectonics and the fusion of Asia: Earth and Planetary Science Letters, v. 21, p. 181-189.

Christie-Blick, N. and Levy, M., 1989, Stratigraphy and tectonic framework of Upper Proterozoic and Cambrian rocks in the western United States, in Christie-Blick, N., Mount, J. F., Levy, M., Signor, P. W., and Link, P. K., eds., Late Proterozoic and Cambrian tectonics, sedimentation, and record of metazoan radiation in the western United States. Field Trip Guidebook T331, 28th International Geological Congress: American Geophysical Union, Washington, D.C., p. 7-22.

Chugaeva, M. N., and Apollonov, M. K., 1982, The Cambrian-Ordovician boundary in the Batyrbaisai section, Malyi Karatau Range, Kazakhstan, USSR, in Bassett, M. G., and Dean, W. T., eds., The Cambrian-Ordovician boundary-sections, fossil distributions, and correlations: National Museum of Wales (Cardiff), Geological Series n. 3, p. 77-85.

Cook, H. E., 1979, Ancient continental slope sequences and their value in understanding modern slope development, in, Doyle, L. J., and Pilkey, O. H., eds., Geology of continental slopes: Society of Economic Paleontologists and Mineralogists Special Publication 27, p. 287-305.

Cook, H. E., 1983, Ancient carbonate platform margins, slopes, and basins, in Cook, H. E., Hine, A. C., and Mullins, H. T., Platform margin and deep water carbonates: Society of Economic Paleontologists and Mineralogists, Short Course No. 12, p. 5-1 - 5-189.

Cook, H. E., 1988, Overview: Geologic history and carbonate petroleum reservoirs of the Basin and Range Province, western United States, in Goolsby, S. M, and Longman, M. W., eds., Occurrence and petrophysical properties of carbonate reservoirs in the Rocky Mountain region: Rocky Mountain Association of Geologists, p. 213-227.

Cook, H. E., 1989, Geology of the Basin and Range Province: an Overview, in Taylor, M. E., ed., Cambrian and Early Ordovician stratigraphy and paleontology of the Basin and Range Province, western United States. Field Trip Guidebook T125, 28th International Geological Congress: American Geophysical Union, Washington, D.C., p. 6-13.

Cook, H. E., and Egbert, R. M., 1981, Late Cambrian-Early Ordovician continental margin sedimentation, in Taylor, M. E., ed., Short Papers for the Second International Symposium on the Cambrian System: U.S. Geological Survey Open-File Report 81-743, p. 50-56.

Cook, H. E., and Mullins, H. T., 1983, Basin margin environment, in Scholle, P. A., Bebout, D. G., and Moore, C. H., eds., Carbonate depositional environments: American Association of Petroleum Geologists Memoir 33, p. 539-617.

Cook, H. E., and Mullins, H. T., 1989, Carbonate aprons--their petroleum reservoir potential, in Magoon, L. B., ed., The petroleum system--status of research and methods, 1990: U. S. Geological Survey Bulletin 1912, p. 13-19.

Cook, H. E., and Taylor, M. E., 1975, Early Paleozoic continental margin sedimentation, trilobite biofacies, and the thermocline, western United States: Geology, v. 3, p. 559-562.

Cook, H. E., and Taylor, M. E., 1977, Comparison of continental slope and shelf environments in the Upper Cambrian and lowest Ordovician of Nevada, in Cook, H. E., and Enos, Paul, eds., Deep-water carbonate environments: Society of Economic Paleontologists and Mineralogists Special Publication 25, p. 51-81.

Cook, H. E., and Taylor, M. E., 1987, Stages in the evolution of Paleozoic carbonate platform and basin margin types: western United States passive continental margin: American Association of Petroleum Geologists Bulletin, v. 71, p. 542-543.

Cook, H. E., and Taylor, M. E., 1990a, Comparative evolution of western North America and southern Kazakhstan, U.S.S.R.--two early Paleozoic carbonate passive margins, in Carter, L. M. H., ed., U. S. Geological Survey Research on Energy Resources--1990, U. S. Geological Survey Circular 1060, p. 17-18.

Cook, H. E., and Taylor, M. E., 1990b, Comparative Late Cambrian development of the western United States and southern Kazakhstan, U.S.S.R.--sedimentology: Abstracts, Third International Symposium on the Cambrian System, Novosibirsk, U.S.S.R., p. 82.

Cook, H. E., and Taylor, M. E., 1991a, Paleozoic carbonate passive-margin evolution and resulting petroleum reservoirs--Great Basin, western United States: Geological Society of Nevada and U.S. Geological Survey Great Basin Symposium, Proceedings (in press).

Cook, H. E. and Taylor, M. E., 1991b, Carbonate slope failures as indicators of sea-level lowerings: American Association of Petroleum Geologists Bulletin, v. 75 (in press).

Cook, H. E., McDaniel, P. N., Mountjoy, E. W., and Pray, L. C., 1972, Allochthonous carbonate debris flows at Devonian bank ("reef") margins, Alberta, Canada: Canadian Petroleum Geology Bulletin, v. 20, n. 3, p. 439-497.

Cook, H. E., Taylor, M. E., and Egbert, R. M., 1981, Upper Cambrian and Lower Ordovician biostratigraphy and depositional environments. Tybo Canyon, Hot Creek Range, Nevada, in Taylor, M. E., and Palmer, A. R., eds., Cambrian stratigraphy and paleontology of the Great Basin and vicinity, western United States: Second International Symposium on the Cambrian System (Denver), Guidebook for Field Trip, 1, p. 51-67.

Cook, H. E., Taylor, M. E., and Magoon, L. B., 1987, The role of major carbonate debris-flow events in recognizing and dating eustatic lowering of sea level--perspectives from Pre-Panthalassa ocean-margin terranes: Carbonate Gravity Deposits. Society of Economic Paleontologist and Mineralogists Research Conference, Ainhoa, France, Abstracts, p. 7-8.

Cook, H. E., Taylor, M. E., and Miller, H. E., 1989a, Late Cambrian and Early Ordovician stratigraphy, biostratigraphy, and depositional environments, Hot Creek Range, Nevada, in Taylor, M. E., ed., Cambrian and Early Ordovician stratigraphy and paleontology of the Basin and Range Province, western United States; Field Trip Guidebook T125, 28th International Geological Congress: American Geophysical Union, Washington, D.C., p. 28-36.

Cook, H. E., Taylor, M. E., Zhemchuzhnikov, V. G., Apollonov, M. K., Ergaliev, G. Kh., Dubinina, S. V., and Mel'nikova, L. M., Sargaskaev, Z. S., and Zakharov, A. V., 1989b, Evolution of an early Paleozoic carbonate seamount, Malyi Karatau Range, southern Kazakhstan, U.S.S.R.: new evidence for early history of Kazakhstania: Abstracts, 28th International Geological Congress, Washington, D.C., v. 1, p. 1-322--1-323. Authorship corrected, in Hanshaw, B. B., 1990, General Proceedings, 28th International Geological Congress, Washington, D.C., p. 81.

Crevello, P. D., and Schlager, W., 1980, Carbonate debris sheets and turbidites, Exuma Sound, Bahamas: Journal of Sedimentary Petrology, v. 50, p. 1121-1147.

Crevello, P. D., Wilson, J. L., Sarg, J. F., and Read, J. F., eds., 1989, Controls on carbonate platform and basin development: Society of Economic Paleontologists and Mineralogists Special Publication 44, 405 p.

Davies, G. R., 1977, Turbidites, debris sheets and truncation structures in upper Paleozoic deep-water carbonates of the Sverdrup Basin, Arctic Archipelago, in Cook, H. E., and Enos, P., eds., Deep-water carbonate environments: Society of Economic Paleontologists and Mineralogists Special Publication 25, p. 221-247.

Dickinson, W. R., 1977, Paleozoic plate tectonics and the evolution of the Cordilleran continental margin, in Stewart, J. H., Stevens, C. H., and Fritsche, A. E., eds., Paleozoic paleogeography the western United States: Society of Economic Paleontologists and Mineralogists, Pacific Section, Pacific Coast Paleogeography Symposium 1, p. 137-155.

Droxler, A. W., and Schlager, W., 1985, Glacial versus interglacial sedimentation rates and turbidity frequency in the Bahamas: Geology, v. 13, p. 799-802.

Dubinina, S. V., 1991, Late Cambrian and Early Ordovician conodont associations from open-ocean paleoenvironments, illustrated by Batyrbay and Sarykum sections in Kazakhstan, in Barnes, C. R., and Williams, S. H., eds., Advances in Ordovician Geology. Geological Survey of Canada Paper 91- . (in press).

Ergaliev, G. Kh., 1980, Trilobiti srednego i berkhnego Kembrija Malogo Karatau: Akademija Nauk Kazakh SSR (Alma-Ata), 211 p., 20 pls.

Ergaliev, G. Kh., 1981, Upper Cambrian biostratigraphy of the Kyrshabakty section, Maly Karatau, southern Kazakhstan, in Taylor, M. E., ed., Short Papers for the Second International Symposium on the Cambrian System: U.S. Geological Survey Open-File Report 81-743, p. 82-88.

Ergaliev, G. Kh., 1983, in Apollonov. M. K., Bandaletov, S. M., and Ivshin, N. K., eds., The lower Paleozoic stratigraphy and palaeontology of Kazakhstan: Academy of Sciences of the Kazakh SSR, Nauk, Kazakh SSR Publishing House, Alma-Ata, p. 35-66 (in Russian).

Ergaliev, G. Kh., 1989, The stratigraphy of the Shabakty series in Malyi Karatau (south Kazakhstan): Izvestiya Akademii Nauk Kazakh SSR, Serya Geol., n. 6, p. 23-26 (in Russian).

Ergaliev, G. Kh., and Kasymov, M. A., (excursion leaders), 1984, Geology and phosphorite deposits in Malyi Karatau Ridge: 27th International Geological Congress (Moscow), Guidebook for Excursion 045A, p. 21-56.

Farmer, G. L., and DePaolo, D. J., 1983, Origin of Mesozoic and Tertiary granite in the western United States and implications for pre-Mesozoic crustal structure, 1. Nd and Sr isotopic studies in the geocline of the northern Great Basin: Journal of Geophysical Research, v. 88, p. 33793401.

Ferguson, H. G., 1933, Geology of the Tybo District, Nevada: University of Nevada Bulletin, v. 27, n. 3, 61 p.

Gabrielse, H., 1972, Younger Precambrian of the Canadian Cordillera: American Journal of Science, v. 272, p. 521-536.

Hamilton, W., 1987, Plate-tectonic evolution of the western U.S.A.: Episodes, v. 10, p. 271-277.

Hart, W. D., Stitt, J. W., Hohensee, S. R., and Ethington, R. L., 1987, Geological implications of Late Cambrian trilobites from the Collier Shale, Jessieville area, Arkansas: Geology, v. 15, p. 447-450.

Hine, A. C., and Mullins, H. T., 1983, Modern carbonate shelf-slope breaks, in Stanley, D. J., and Moore, G. T., eds., The shelfbreak: critical interface on continental margins: Society of Economic Paleontologists and Mineralogists Special Publication 33, p. 169-188.

James, N. P., and Mountjoy, E. W., 1983, Shelf-slope break in fossil carbonate platforms: an overview, in Stanley, D. J., and Moore, G. T., eds., The shelfbreak: critical interface on continental margins: Society of Economic Paleontologists and Mineralogists Special Publication 33, p. 189-206.

Johns, D. R., Mutti, E., Rosell, J., and Seguret, M., 1981, Origin of a thick, redeposited carbonate bed in Eocene turbidites of the Hecho Group, southcentral Pyrenees, Spain: Geology, v. 9, p. 161-164.

Kepper, J. C., 1981, Sedimentology of an outer shelf margin with evidence for synsedimentary faulting, eastern California and western Nevada: Journal of Sedimentary Petrology, v. 51, p. 807-821.

Kistler, R. W., 1974, Phanerozoic batholiths in western North America--A summary of some recent work on variations in time space, chemistry, and isotopic composition: Annual Review, Earth and Planetary Science, v. 2, p. 403-418.

Kompaneitsev, V. P., 1986, The Bolshoi Karatau structural-formational zone--the Late Baikalian-Caledonian magmatic complexes: Geology and Metallogeny of Karatau, v. 1, Alma-Ata, p. 148-154 (In Russian).

Levy, M., and Christie-Blick, N., 1989, Pre-Mesozoic palinspastic reconstruction of the eastern Great Basin (western United States): Science, v. 245 (29 Sept.), p. 1454-1462.

Llyin, A. V., 1990, Proterozoic supercontinent, its latest Precambrian rifting, breakup, dispersal into smaller continents, and subsidence of their margins: evidence from Asia: Geology, v. 18, p. 1231-1234.

Lu, Yen-Hao, 1974, Bio-environmental control hypothesis and its application to Cambrian biostratigraphy and palaeozoogeography: Memoirs of the Nanking Institute of Geology and Palaeontology, Academy Sinica, v. 5, n. 2, p. 27-110 (In Chinese with Latin nomenclature).

McNeill, D. F., 1990, Contribution of sea level-induced erosional events in platform/periplatform deposits to lowstand sedimentation in the Bahamas: American Association of Petroleum Geologists Bulletin, v. 74, p. 717.

Miller, J. F., 1988, Conodonts as biostratigraphic tools for redefinition and correlation of the Cambrian-Ordovician boundary: Geological Magazine, v. 7, p. 349-362.

Monger, J. W. H., Southes, J. G., and Gabrielse, H., 1972, Evolution of the Canadian Cordillera--a plate-tectonic model: American Journal of Science, v. 272, p. 577-602.

Mullins, H. T., 1983a, Comment on "Eustatic control of turbidites and winnowed turbidites": Geology, v. 11, p. 57-58.

Mullins, H. T., 1983b, Modern carbonate slopes and basins of the Bahamas, in Cook, H. E., Hine, A. C., and Mullins, H. T., eds., Platform margin and deep water carbonates: Society of Economic Paleontologists and Mineralogists, Short Course No. 12, p. 4-1 - 4-138.

Mullins, H. T., 1986, Periplatform carbonates: Colorado School of Mines Quarterly, v. 48, n. 1/2, p. 37-79.

Mullins, H. T., and Cook, H. E., 1986, Carbonate apron models: alternatives to the submarine fan model for paleoenvironmental analysis and hydrocarbon exploration: Sedimentary Geology, v. 48, p. 37-80.

Mullins, H. T., Gardulski, A. F., and Hine, A. C., 1986, Catastrophic collapse of the west Florida carbonate platform margin: Geology, v. 14, p. 167-170.

Palmer, A. R., 1962, *Glyptagnostus* and associated trilobites in the United States: U.S. Geological Survey Professional Paper, 374-F, 49 p.

Palmer, A. R., 1965, Trilobites of the Late Cambrian Pterocephaliid Biomere in the Great Basin, United States: U.S. Geological Survey Professional Paper 493, 105 p., 20 pls.

Poole, F. G., 1974, Flysch deposits of Antler foreland basin, western United States, in Dickinson, W. R., ed., Tectonics and sedimentation: Society of Economic Paleontologists and Mineralogists Special Publication 22, p. 58-82.

Quinlivan, W. D., and Rogers, C. L., 1974 (1975), Geologic map of the Tybo Quadrangle, Nye County, Nevada: U.S. Geological Survey Miscellaneous Investigation Series Map I-821 (map scale 1:48,000).

Read, J. F., 1982, Carbonate platforms of passive (extensional) continental margins--types, characteristics and evolution: Tectonophysics, v. 81, p. 195-212.

Read, 1985, Carbonate platform facies models: American Association of Petroleum Geologists Bulletin, v. 69, p. 1-21.

Robison, R. A., 1972, Mode of life of agnostid trilobites: 24th International Geological Congress, Montreal, Section 7, p. 33-40.

Sarg, J. F., 1988, Carbonate sequence stratigraphy, in Wilgus, C. K., Hastings, B. S., Kendall, C. G. St. C., Posamentier, H. W., Ross, C. A., and Van Wagoner, J., eds., Sea-level changes: an integrated approach: Society of Economic Paleontologists and Mineralogists Special Publication 42, p. 155-181.

Sargaskaev, Zh., S., 1988, The environments of the formation of the Kokzhotian series in Karatau (Kazakhstan): Litologiya i poleznyie iskopaemyie, n. 1, p. 70-81 (In Russian).

Sargaskaev, Zh., S., 1990, The stratigraphy and evolution of the Kokzhotian basin in the Vendian-Early Paleozoic: Geologiya prognozirovanie i otsenka mestorozdeniy poleznykh iskopaemykh Kazakhstana, Alma-Ata, p. 49-60 (In Russian).

Sargaskaev, Zh., S., and Ergaliev, G. Kh., 1988, The paleocurrents in the Vendian-Early Paleozoic Kokzhotian paleobasin: Izvestiya Academii Nauk Kazakhskoi SSR Seriya geologicheskaya, n. 6, p. 70-78 (In Russian).

Schlager, W., and Chermak, A., 1979, Sediment facies of platform-basin transition, Tongue of the Ocean, Bahamas, in Pilkey, O. H., and Doyle, L. S., eds., Geology of continental slopes: Society of Economic Paleontologists and Mineralogists Special Publication 27, p. 193-208.

Scotese, C. R., and McKerrow, W. S., eds., 1990, Paleozoic paleogeography and biogeography: The Geological Society (London), Memoir 12, 435 p.

Scotese, C. R., Bambach, R. K., Barton, C., Van Der Voo, R., and Ziegler, A. M., 1979, Paleozoic base maps: Journal of Geology, v. 87, n. 3, p. 217-277.

Shanmugam, G., and Moiola, R. J., 1982, Eustatic control of turbidites and winnowed turbidites: Geology, v. 10, p. 231-235.

Shanmugam, G., and Moiola, R. J., 1983, Reply to Comment on "Eustatic control of turbidites and winnowed turbidites": Geology, v. 11, p. 58-60.

Sleep, N. H., 1971, Thermal effects of the formation of Atlantic continental margins by continental breakup: Geophysical Journal of the Royal Astronomical Society, v. 24, p. 325-350.

Sovetov, J. F., 1990, The Cambrian-Precambrian boundary and the Precambrian sedimentary assemblages in the Malyi Karatau: stratigraphic, sedimentologic and paleotectonic aspects: Institute of Geology and Geophysics of the Siberian Branch of the Academy of Sciences of the USSR, n. 14, 36 p.

Speed, R. C., 1982, Evolution of the sialic margin in the Central and Western United States, in, Watkins, J. S., and Drake, C. L., eds., Studies in Continental Margin Geology: American Association of Petroleum Geologists Memoir 34, p. 457-468.

Stewart, J. H., 1972, Initial deposits on the Cordilleran Geosyncline: Evidence of a Late Precambrian (850 m.y.) continental separations: Geological Society of America Bulletin, v. 83, p. 1345-1360.

Stewart, J. H., 1976, Late Precambrian evolution of North America: Plate tectonics implication: Geology, v. 4, p. 11-15.

Stewart, J. H., 1980, Geology of Nevada: Nevada Bureau of Mines and Geology Special Publication, v. 4, 136 p.

Stewart, J. H., and Poole, F. G., 1974, Lower Paleozoic and uppermost Precambrian Cordilleran miogeocline, Great Basin, western United States, in Dickinson, W. R., ed., Tectonics and sedimentation: Society of Economic Paleontologists and Mineralogists Special Publication 22, p. 28-57.

Stewart, J. H., and Suczek, C. A., 1977, Cambrian and latest Precambrian paleogeography and tectonics in the western United States, in Stewart, J. H., Stevens, C. H., and Fritsche, A. E., eds., Paleozoic Paleogeography of the Western United States: Pacific Coast Paleogeography Symposium 1, Pacific Section, Society of Economic Paleontologists and Mineralogists, p. 1-18.

Taylor, M. E., 1976, Indigenous and redeposited trilobites from Late Cambrian basinal environments of central Nevada: Journal of Paleontology, v. 50, p. 668-700.

Taylor, M. E., 1977, Late Cambrian of western North America: Trilobite biofacies, environmental significance, and biostratigraphic implications, in Kauffman, E. G., and Hazel, J. E., eds., Concepts and methods of biostratigraphy: Dowden, Hutchinson and Ross, Inc., Stroudsburg, Pennsylvania, p. 397-425.

Taylor, M. E., ed., 1989, Cambrian and Early Ordovician stratigraphy and paleontology of the Basin and Range Province, western United States: 28th International Geological Congress, Washington, D.C., Field Trip Guidebook T125, 86 p.

Taylor, M. E., and Cook, H. E., 1976, Continental shelf and slope facies in the Upper Cambrian and lowest Ordovician of Nevada, in Robison, R. A., and Rowell, A. J., eds., Paleontology and depositional environments: Cambrian of western North America, A Symposium: Brigham Young University, Geology Studies, v. 23, pt. 2, p.181-214.

Taylor, M. E., and Cook, H. E., 1990, Comparative Late Cambrian development of the western United States and southern Kazakhstan, USSR--palaeontology: Abstracts, Third International Symposium on the Cambrian System, Novosibirsk, U.S.S.R., p. 160.

Taylor, M. E., and Forester, R. M., 1979, Distributional model for marine isopod crustaceans and its bearing on early Paleozoic paleozoogeography and continental drift: Geological Society of America Bulletin, v. 90, p. 405-413.

Taylor, M. E., Cook, H. E., and Miller, J. F., 1989, Late Cambrian and Early Ordovician biostratigraphy and depositional environments of the Whipple Cave Formation and House Limestone, central Egan Range, Nevada, in Taylor, M. E., ed., 1989, Cambrian and Early Ordovician stratigraphy and paleontology of the Basin and Range Province, western United States. Field Trip Guidebook T125, 28th International Geological Congress: American Geophysical Union, Washington, D.C., p. 37-44.

Terman, M. C., compiler., 1974, Tectonic map of China and Mongolia: Geological Society of America, scale 1:5,000,000, 2 sheets.

Troedsson G. T., 1937, On the Cambro-Ordovician faunas of western Quroq Tagh, eastern T'ien Shan: Paleontology Sinica, New Series B, n. 2, Whole Series n. 106, 74 p.

Wilber, R. J., Milliman, J. D., and Halley, R. B., 1990, Quaternary stratigraphy of western Great Bahama Bank: eustatic signatures in progradational slope units along a leeward margin: American Association of Petroleum Geologists Bulletin, v. 74, p. 790.

872

Yose, L. A., and Hardie, L. A., 1990, The significance of carbonate megabreccia in sequence stratigraphy: examples from the Triassic of the Dolomites, northern Italy: American Association of Petroleum Geologists Bulletin, v. 74, p. 795.

Zaitsev, Yu. A., Ergaliev, G. Kh., and Shlygin, A. E., eds., 1984, Geology and phosphorite deposits in Maly Karatau Ridge: Guidebook for Excursion 045A, Kazakh SSR. 27th International Geological Congress, Moscow. (In English; bound together with Guidebook for Excursion 101A).

Zhemchuzhnikov, V. G., 1986, Carbonate breccias in the section of upper Cambrian and lower Ordovician deposits in the Batyrbay ravine of the Lesser Karatau area: Lithology and Mineral Resources, n. 6, p. 76-87 (In Russian).

Zhemchuzhnikov, V. G., Kraev, O. M., Zaslavskaya, N. M., and Uzhkenov, B. S., 1989, The situation of the "Dzhanytas" series in the Malyi Karatau section relative to the chitinozoan finds: Izvestiya Academii Nauk Kazakhskoi SSR, Seriya geol., n. 1, p. 70-80 (In Russian).

Ziegler, A. M., Hansen, K. S., Johnson, M. E., Kelly, M. A., Scotese, C. R., and Van Der Voo, R., 1977, Silurian continental distributions, paleogeography, climatology, and biogeography: Tectonophysics, v. 40, nos. 1 and 2, p. 13-51.

Ziegler, A. M., Scotese, C. R., McKerrow, W. S., Johnson, M. E., and Bambach, R. K., 1979, Paleozoic paleogeography: Annual Review of Earth and Planetary Sciences, v. 7, p. 473-502.